Primitive meteorites are full of tiny igneous spherules called chondrules. These have excited and challenged scientists since they were first described nearly 200 years ago. Chondrules were made by some pervasive process in the early solar system that formed melted silicate droplets. This is the first comprehensive review of chondrules and their origins since a consensus developed that they were made in the disk of gas and solids that formed the Sun and planets 4.5 billion years ago.

Fifty scientists from assorted disciplines have collaborated to review how chondrules could have formed in the protoplanetary disk. When and where in the disk did they form? What were they made from and how fast were they heated and cooled? What provided the energy to melt chondrules – nebular shock waves, lightning discharges, protostellar jets?

Following an exciting international conference in Albuquerque, New Mexico, the latest answers to these questions are presented in thirty-four articles.

Chondrules and the Protoplanetary Disk

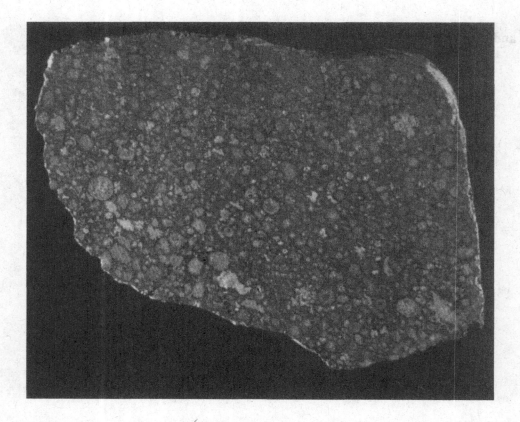

Abundant chondrules and a small fraction of irregular white Ca-, Al-rich inclusions (CAI) are visible in this 7 × 5 cm slab of the Axtell CV3 chondrite. Photo from Steven Simon, sample from Field Museum of Natural History, Chicago (S.B. Simon *et al.*, Meteoritics 1995, vol. 30, pages 42–46).

Chondrules and the Protoplanetary Disk

R. H. HEWINS,
R. H. JONES
and E. R. D. SCOTT

Editors

CAMBRIDGE
UNIVERSITY PRESS

CAMBRIDGE UNIVERSITY PRESS
Cambridge, New York, Melbourne, Madrid, Cape Town, Singapore,
São Paulo, Delhi, Dubai, Tokyo, Mexico City

Cambridge University Press
The Edinburgh Building, Cambridge CB2 8RU, UK

Published in the United States of America by Cambridge University Press, New York

www.cambridge.org
Information on this title: www.cambridge.org/9780521174893

First published 1996
First paperback edition 2011

A catalogue record for this publication is available from the British Library

Library of Congress Cataloguing in Publication data

Chondrules and the protoplanetary disk / R. H. Hewins, R. H. Jones,
  E. R. D. Scott, editors.
      p.   cm.
    ISBN 0 521 55288 5 (hc: alk. paper)
    1. Chondrules – Congresses.   2. Chondrites (Meteorites) –
Congresses.   3. Planets – Origin – Congresses.   I. Hewins, Roger H.
II. Jones, R. H. (Rhian H.)   III. Scott, E. R. D.
    QB758.5C46C458   1996
    523.5'1–dc20   95-52605 CIP

ISBN 978-0-521-55288-2 Hardback
ISBN 978-0-521-17489-3 Paperback

Contents

Authors

Associate editors

Preface

The oldest meteorites are stuffed full of little igneous particles that have puzzled scientists for almost 200 years. The particles, called chondrules, seem to have formed in the disk of dust and gas from which the Sun and the planets accreted.

To learn more about the origin of chondrules, an international conference entitled "Chondrules and the Protoplanetary Disk" was convened by Roger Hewins on October 13–15, 1995 in Albuquerque at the University of New Mexico. The conference was organized by the Associate Editors of this book with the expert assistance of LeBecca Simmons at the Lunar and Planetary Institute in Houston and smoothly run by Rhian Jones and her colleagues at the University of New Mexico. The meeting was held in Albuquerque to help celebrate the 50th anniversary of the Institute of Meteoritics and was cosponsored by the Lunar and Planetary Institute and NASA's Origins of Solar Systems Program.

Nearly 100 meteorite researchers, astrophysicists and astronomers from 9 countries attended the meeting. Their goal was to review and assess the numerous constraints imposed by laboratory analyses and experiments on the formation and thermal history of chondrules and to develop plausible models for the formation of chondrules and the equally enigmatic Ca- Al-rich inclusions that accompany them in primitive meteorites.

The organization of papers in this book is broadly similar to the structure of the conference. There are five parts: an overview by Hewins (part I); astronomical, astrophysical and meteoritical reviews of protoplanetary disks and constraints on disk processes (part II); the nature of chondrule precursor materials (part III); constraints on thermal histories of chondrules deduced from experiments and chemical data (part IV); and specific models of chondrule formation and reviews of models (part V). Some papers discuss several topics. For example, the three papers by Sears *et al.*, Alexander, and Grossman, which are focussed on chemical aspects of chondrule formation, constrain thermal histories and chondrule precursors. All papers were thoroughly and enthusiastically reviewed. We warmly thank the reviewers and the Associate Editors for their help in assisting authors and evaluating papers.

I. Introduction

1: Chondrules and the Protoplanetary Disk: An Overview

ROGER H. HEWINS
Geological Sciences, Rutgers University, Piscataway, NJ 08855–1179, U.S.A.

ABSTRACT

Chondrules and calcium-aluminum-rich inclusions (CAI), which are the major components of chondritic meteorites, provide an important record of heating events in the protoplanetary disk. There is currently strong support for a flash heating origin for chondrules, with the leading mechanism being nebular shock waves. Competing chondrule formation mechanisms need to fit constraints provided by petrological and chemical data. Some of the questions important to modellers which can be addressed by meteoriticists include the relative timing of chondrule and CAI formation, whether they were formed in small or large events, whether they were formed from nebular condensates agglomerated at different temperatures, whether their compositions were modified by extensive evaporation and/or reduction during melting, how rapidly they were heated and cooled, and to what extent the heating process was repeated and chondrules remelted.

INTRODUCTION

Chondrules are small particles of silicate material that experienced melting before incorporation into chondritic meteorite parent bodies. The abundance of chondrules in chondrites and hence in many asteroids implies that melting of small particles was a common phenomenon in the early solar system. Understanding chondrule formation therefore has a potential astrophysical and/or planetary significance.

The first conference on Chondrules and their Origins was organized by E. King and the Lunar and Planetary Institute in 1982. The questions debated then included what is a chondrule, as well as how did chondrules form, in what setting, and from what precursor materials. Both planetary (e.g. impact) and nebular (e.g. lightning) models for chondrule origins were considered and similarly the term chondrule was applied to all kinds of melt spherules. The 1982 conference led to a very influential paper (Taylor *et al.*, 1983) which summarized the arguments that chondrules formed in the solar nebula, rather than in a planetary/asteroidal environment. A consensus has since developed on what a chondrule is (e.g., Grossman *et al.*, 1988): the term is now usually restricted to the majority of chondrite ferromagnesian silicate particles that are assumed to have formed by a common process; it includes fragmented as well as complete melted spherules, and similar objects

which lack a droplet form because of incomplete melting, but excludes particles of other compositions, e.g. calcium-aluminum-rich or refractory inclusions (CAI) and basaltic fragments, as well as melt spherules found in impact and volcanic deposits.

The conference on Chondrules and the Protoplanetary Disk held in Albuquerque in 1994 was focussed on what chondrules can tell us about nebular processes and whether astrophysical models can explain chondrule origins, although as this book shows, there is still not unanimity on a strictly nebular origin for chondrules. The nature of the precursor materials of chondrules (relevant to prior nebular events), and the timing and conditions of their melting were among the topics discussed in Albuquerque. The range of opinions expressed by meteoriticists over chondrules and the melting process was considerable, but much less wide than the range of heating mechanisms proposed by astrophysicists and modellers. Perhaps the most significant progress achieved in conference discussion was the realization of the difficulties experienced in explaining some aspect of chondrule origins by all but a few of the diverse mechanisms considered, as summarized in a chapter by Boss (Concise Guide). The most favored mechanism appears to be some form of shock wave which heated solid particles by friction (e.g., Boss and Graham, 1993; Hood and Horanyi, 1993) but lightning discharges (e.g., Morfill *et al.*, 1993) also attracted much attention. The present introduction will concentrate on petrologic-cosmochemical results presented in this book, and attempt to show where there is consensus and where there is controversy on the details of chondrule formation, and whether enough is known to constrain heating processes in the disk.

CHONDRULE DIVERSITY AND CLASSIFICATION

A clearer understanding of the nature of chondrules has developed over the last decade, the result of much work stimulated to a considerable extent by the 1982 conference. There are now many more data on chondrules (bulk chemical, trace element, isotopic, petrographic, mineral chemical, and physical), as well as experimental simulations of chondrules. There are several recent critical reviews of chondrule data (Grossman *et al.*, 1988; Grossman, 1988; Hewins, 1989; Wasson, 1993) and of the nebular significance of chondrules and CAI (Wood, 1988; Palme and Boynton, 1993). An essential part of interpreting chondrule data involves the relationship between groups of chondrules with different properties. Therefore, although it is not our intent to go over the enormous chondrule literature here, the diversity of chondrule properties and classification of chondrules is reviewed below.

Early approaches to chondrule classification depended mainly on either bulk composition or texture (see Fig. 1). McSween (1977) recognized two main groups of chondrules, one FeO-poor (type I) and the other FeO-rich (type II), while Gooding and Keil (1980) employed a system involving texture and dominant mineral, e.g. PO, PP, BO, RP, etc., for porphyritic olivine, porphyritic pyroxene, barred olivine, (excentro)radial, etc. chondrules. BO chondrules (Fig. 1a) contain one to a few olivine crystals with parallel plate morphology indicating rapid growth from the silicate melt droplet.

A combination of classification criteria has been used in a series of detailed studies of chondrules of specific types (Scott and Taylor, 1983; Jones and Scott, 1989; Jones, 1990, 1992, 1994). Thus for unequilibrated chondrites, chondrules with porphyritic texture can be classified as type I (FeO-poor) or II (FeO-rich) each subdivided into A and B (SiO_2-poor or SiO_2-rich; i.e., olivine-rich and pyroxene-rich respectively). Type IAB and IB have a poikilitic texture in which small olivine grains are enclosed by larger pyroxene crystals (Fig. 1b). Type IIA PO chondrules tend to contain bigger olivine crystals than type IA PO (Fig. 1c,d). Chondrules with other textures can be described using composition in a similar way: e.g., radial pyroxene chondrules in carbonaceous chondrites, called type III by McSween (1977), have very Si-rich type IB compositions and not surprisingly were totally melted. Very fine-grained chondrules, described as granular or dark zoned from optical observation, are seen to be porphyritic from SEM observation (Fig. 1e,f).

Chondrules of each type occur in most kinds of chondrites, with similar textures and mineral compositions (e.g., Scott and Taylor, 1983) but subtle differences in bulk compositions occur from group to group (e.g., Grossman *et al.*, 1988) and distinct differences in O isotopic composition exclude a common source (Clayton *et al.*, 1991).

This classification has the advantage that most chondrules can be still be classified by texture and dominant mineral, even when the bulk composition is modified, e.g. by Fe-Mg diffusive exchange in equilibrated chondrites. Compositional changes during metamorphism have been documented for given chondrule types (Scott and Jones, 1990; McCoy *et al.*, 1991). A very different approach to chondrule classification (Sears *et al.*, 1992) uses chemical and physical properties of dominant chondrule phases to define two main groups, A and B. Because these properties change during metamorphism, classification of most chondrule textural types changes in the Sears scheme with the petrologic type of the chondrite.

CHONDRULES, CAI AND PROTOPLANETARY DISKS

Many authors (cf. Grossman, 1988) argue that chondrules were melted by some form of flash heating of solid material in the nebula: the midplane is the setting generally preferred. Cuzzi *et al.* argue that turbulence would prevent settling of dust aggregates to the midplane: they suggest that chondrules were made throughout the nebula and then concentrated between eddies to make clumps that settled to the midplane. Most authors favor melting of dust aggregates and then remelting of dust-covered chondrules, but Wood doubts that there was ever sufficient time to accrete dust into aggregates without first forming microdroplets.

Hartmann discusses the evolution of the protoplanetary disk, with emphasis on the FU Orionis outbursts when disk material is accreted by the growing sun, and the subsequent more quiescent T Tauri phase which lasted several million years. Wood suggests that it is more plausible that chondrules and CAI formed in the early phase when more energy is released than in the longer more quiescent stage, as commonly assumed. If this is correct, it would require mechanical dissipation of energy at large distances from the sun, perhaps as shock waves that melted chondrules, because the main heating caused by FU Orionis events is very close to the star (Cassen, Hartmann). Though such shocks appear possible, later shocks due to clump infall appear stronger candidates for chondrule generation (Hood and Kring, Hartmann). Given the uncertainty, about the stage in disk evolution when chondrules were formed, any direct information on the timescale of formation of CAI and chondrules becomes extremely important.

Detailed information on the heating and cooling conditions of chondrules is needed to discriminate between various possible heating mechanisms. Wasson therefore compares the energy requirements of very small scale events generating a very small number of chondrules, e.g. by lightning, and megameter events. The amount of energy needed depends on the temperature interval over which the chondrules are raised and the latent heat of fusion, but also on the amount of dust evaporated, and most critically whether hydrogen is also heated and dissociated (see also Scott *et al.*). The energy delivered must be only marginally capable of totally melting chondrules, and heating of brief duration is favored, to minimize volatile loss.

Davis and MacPherson review data on calcium-aluminum-rich or

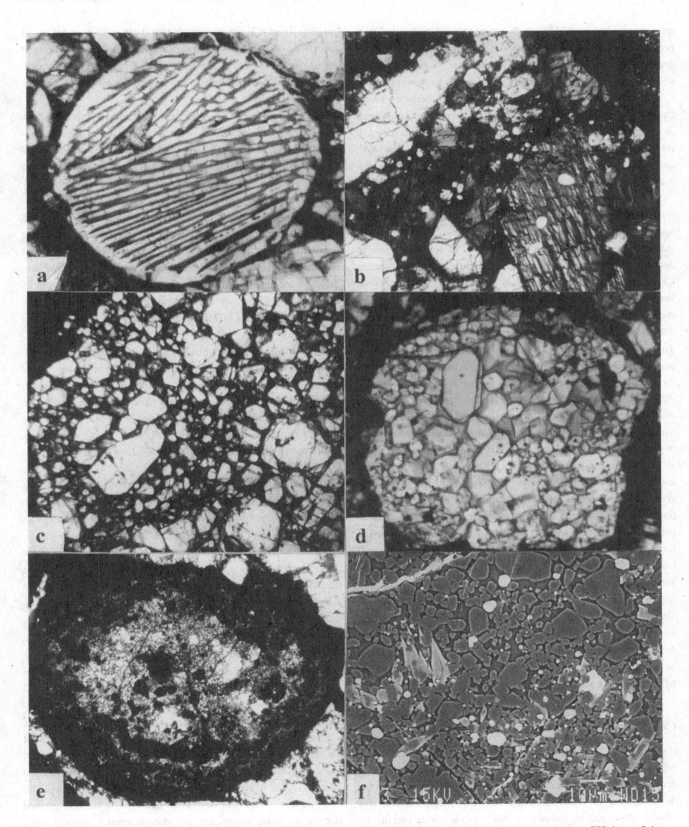

Fig. 1. (a) Barred olivine chondrule in Sharps H3.4 chondrite; white olivine, grey glass; plane polarized light (PPL) image, 0.4 mm long. (b) Porphyritic pyroxene chondrule (FeO-poor, type IB) in Semarkona LL3.0 chondrite; small olivine grains, mostly white, enclosed in large inverted protopyroxene crystals; cross polarized light image, 0.8 mm long. (c) Porphyritic olivine chondrule (FeO-rich, type IIA) in Semarkona; olivine crystals, white, up to 400 μm long; PPL image, 1.6 mm long. (d) Microporphyritic olivine chondrule (FeO-poor, type IA) in Sharps; olivine crystals, white, up to 100 μm long, in light grey glass; PPL image 0.4 mm long. (e) Dark-zoned chondrule in Semarkona, too fine grained for igneous texture to be apparent; white silicates, black metal and sulfide; PPL image 0.8 mm long. (f) Same chondrule as (e); back-scattered electron image 0.12 mm long (by B. Zanda and M. Bourot-Denise) reveals cryptoporphyritic texture; grey olivine crystals mostly 1–10 μm long, dark grey glass, light grey pyroxene, white metal and sulfide.

refractory inclusions (CAI), which appear to require slightly lower peak temperatures and distinctly lower cooling rates than chondrules, suggestive of differences in origin. Some chondrules contain small inclusions of CAI material (Misawa and Nakamura) and some CAI have been modified by a second (flash heating) event. Conceivably then CAI were formed before chondrules but at least in some cases reheated in a chondrule-forming event (Scott *et al.*). Hutchison points out that several chondrites contain igneous particles, which might mean that melting was occurring on planetesimals before the accretion of chondrules to their parent asteroids.

Several isotopic systems with short halflives potentially can resolve the small time differences which might exist between CAI and chondrules, as summarized by Swindle *et al.* The I-Xe system measures primarily alteration events over a 50 m.y. time span, so does not help to answer this question. The Mn-Cr system suggests Chainpur chondrule formation 7 m.y. after Allende CAI. Mg-Al isotopic data suggest a similar time gap of at least 2–3 m.y., as there is evidence of ^{26}Al in many CAI but excess ^{26}Mg has not been detected in chondrules (Hutcheon and Jones, 1995; Swindle *et al.*). This time is essentially the same as some estimates of the accretion period based on a model in which ^{26}Al is the heat source for asteroid metamorphism and melting (Grimm and McSween, 1993). Although some authors doubt that ^{26}Al was uniformly distributed and question the validity of the inferred Al-Mg ages (e.g. Wood; Clayton and Jin, 1995)), a growing number have accepted them (e.g., Cameron, 1995). However, it is difficult to detect the daughter product in Mg-rich chondrules, and more data for chondrules and CAI from the same meteorite are required before we can be certain that chondrules formed at a later stage than CAI. It is also possible that chondrule formation occurred over a considerable part of disk history, with chondrules with primitive precursors forming soon after CAI and recycling of chondrules continuing long after. If CAI really formed much earlier than chondrules, some process must have prevented them from drifting into the sun before chondrules and chondrites are formed (Skinner, 1994; see also Scott *et al.*, Wood). Clayton and Lin (1995) have developed a model in which the bulk compositions and ^{26}Al contents of CAI are developed simultaneously by shocks and irradiation on the surface of the disk, and the objects sink into the disk to be melted, probably by the chondrule heating process.

CHONDRULE PRECURSORS AND MULTIPLE MELTING

If we can identify the materials from which chondrules were made, we have information on nebular events and processes prior to chondrule melting. Some models of chondrule formation, e.g. frictional heating of material falling into the protoplanetary disk (Wood, 1984) require interstellar grains as chondrule precursors. Despite the occurrence of small numbers of undoubted interstellar grains in chondrites, unprocessed interstellar material is an unpopular choice for chondrule precursors, because of high temperatures in the central regions of protoplanetary disks and transient heating events elsewhere. The innermost part of the disk achieved temperatures high enough to evaporate solids (Boss, Large Scale

Processes) and it is not impossible that these conditions existed at about 3AU (Cassen). Fractionations observed in chondrule compositions are considered evidence of recondensation after evaporation, and thus nebular condensates remain plausible as chondrule precursor materials.

Heating due to clumpy disk accretion and passage of shock waves into the midplane has been considered as a possible chondrule-forming mechanism, but could also have caused evaporation-condensation in low pressure regions (cf. Hewins, 1989). The biggest question about condensates as precursors is whether they were fractionated or unfractionated, i.e. whether chondrule formation continued throughout the condensation epoch (e.g. Wasson) or whether it occurred afterwards (e.g. Sears). Another interesting possibility is that, as the flash heating process forming chondrules probably was repeated, the immediate precursors of many chondrules could have been earlier chondrule material (Alexander). Planetary precursors are rarely invoked for chondrules, because they would leave geochemical or isotopic signatures which are not observed.

Many CAI and some chondrules show evidence of ^{16}O-rich components. This is often interpreted as evidence of the survival or limited processing of a presolar component in the precursors (Wood, 1981; Clayton *et al.*, 1991). However, Thiemens argues that the mass independent fractionations observed in chondrule (and CAI) oxygen isotopes result solely from chemical processes in the nebula that accompanied the formation of chondrules and CAI.

The case for nebular condensates is strengthened by the observations of Misawa and Nakamura, who find fractionated rare earth element abundance patterns in carbonaceous chondrite chondrules similar to those in CAI. Grossman considers that at least six independent fractionations of groups of elements occurred before chondrule formation. Sears, on the other hand, thinks that much of the diversity in chondrule compositions could arise in processing from a single precursor, due to reduction and volatile loss during chondrule formation. However, the wide ranges of many elements among chondrules, e.g. refractories, are hard to understand without at least some initial variation in composition (e.g., Grossman).

In addition to the question of chemical components present in chondrule precursors, there is also great interest in their physical nature, and whether any unmelted chondrule precursors can be recognized in chondrites. Experimental reproduction of chondrule textures (Connolly and Hewins) has shown that the grain size of the starting material has a big influence on the melting behavior, nucleation density and resulting texture. In particular, fine grained textures typical of FeO-poor chondrules require a very high nucleation density which has only been achieved using fine grained starting material, though the sporadic occurrence of large relict grains in these type I chondrules suggests a mix of occasional large grains amongst finer material. Objects answering this description are described by Weisberg and Prinz under the general term agglomeratic chondrules. These include fine grained unmelted olivine aggregates, particles with interstitial glass between the fine grained olivine crystals, and large olivine crystals with fine grained rims. Wood suggests that the latter fine grained layers could have been formed by accretion of finely dispersed partly molten microdroplets rather than microcrystals. A possible candidate for fine grained precursor material is chondrite matrix, but Brearley shows that this is a complex mixture of condensates, chondrule debris and presolar grains, which probably evolved separately from chondrules. Chondrules evidently spent time in dusty regions and transported much dust into chondrites as chondrule rims (Metzler and Bischoff). A close relationship between chondrules and coarse-grained (igneous) rims is suggested by the fact that rims are slightly depleted in ^{16}O and their silicates usually have slightly higher Fe/Mg than chondrules (Weisberg and Prinz, Rubin and Krot).

If the chondrule heating mechanism was capable of recurring, then it is possible that some chondrules were heated more than once. Unambiguous evidence of remelting of chondrules then becomes a significant piece of evidence of the nature of the heating process. Large grains not crystallized from chondrule melts have long been interpreted as relicts of precursors (e.g. Nagahara, 1981) and opinion is divided over what fraction of these were derived from chondrules (Grossman *et al.*, 1988; Jones). Jones considers that at least 15% of chondrules contain material that experienced two chondrule-forming events and the difference in composition of many relict grains and their host chondrules indicates that FeO-rich and FeO-poor chondrules formed in close proximity. Additional chondrules are enveloped by igneous rims or embedded in other chondrules and there is some evidence of remelting of the primary object. Rubin and Krot consider that these chondrules also experienced a second heating event, after the acquisition of a dust mantle on the primary object. Wood, however, argues in the case of some multiple layers that the material accreted was already partially molten. It is possible that some 25% of chondrules experienced a second heating of low enough intensity to preserve the evidence (Rubin and Krot), and that some other chondrules experienced a second very thorough melting which destroyed the evidence of the primary chondrule. Thus an unknown fraction of chondrules might have chondrule precursors. Wasson takes the extreme view that chondrule textures were significantly coarsened by reheating, but experiments to date have been unable to verify this suggestion (Yu *et al.*).

HEATING, COOLING AND VOLATILES

Chondrules have clearly been heated approximately to their liquidus temperatures, which vary enormously with their bulk compositions. For porphyritic chondrules, it has generally been considered that temperatures were a little below liquidus temperatures, so as to permit survival of nuclei which generated crystalline textures on cooling. Hewins and Connolly show that the most easily melted chondrules (the most FeO- and SiO_2-rich) were probably superheated and crystals subsequently nucleated by collisions with dust grains. For more refractory compositions, especially with relatively coarse grained precursors, their experiments show that melting can be incomplete for short duration heating even with

superliquidus temperatures. Hewins and Connolly suggest a range of 1550–1900°C for chondrule peak temperatures. Greenwood and Hess use a model for the kinetics of congruent melting to show that relict forsteritic olivine grains in chondrules were not heated above 1900°C for more than a few seconds. Their temperature limit for chondrules with enstatitic pyroxene is 1557°C.

Cooling rates for chondrules determined in the laboratory are much slower than radiative cooling of isolated spherules but much faster than global nebular cooling. It is therefore widely accepted that chondrules were generated in some large quantity in relatively opaque nebular domains (e.g., Wood), though Wasson favors virtually quenching chondrules and Eisenhour *et al.* (1995) propose cooling by the fading of a UV heating mechanism. Definition of very precise cooling rates is difficult. Experiments suggest (linear) cooling rates of 100–2000°C/hr (Hewins, 1988) or close to 100°C/hr (Lofgren). Yu *et al.* define conditions for Na retention and S loss, as well as suitable olivine morphology and zonation, using rapid heating, and curved cooling paths rather than a constant cooling rate. They conclude that many chondrules were heated for a minute or less at temperatures of 1550°C or more, with initial cooling rate about 5000°C/hr declining to about 500°C/hr near the solidus. Wasson and Wood have disagreed about the relative importance of the volatile element and olivine zonation constraints on chondrule cooling rates. It appears that both constraints are reconciled with rapid heating and a curved cooling path, as well as a local nebular gas of non-canonical composition. The occurrence of chondrules containing unusual minerals (Wood) suggests very small scale events for chondrule formation. On the other hand, if cooling is primarily radiative (as opposed to by gas expansion and mixing, or by fading of the heat source) relatively large volumes (a few hundred kilometers across) of hot chondrules are required (Sahagian and Hewins, 1992).

One of the most intensely debated questions at the Albuquerque meeting was whether melted chondrules acted essentially as closed systems, so that their bulk compositions are very close to those of their precursors, or whether they were open to loss of volatile elements so that the range of chondrule compositions was largely derived by processing of a single precursor composition. Sears proposes that chondrule precursors were FeO-rich similar to type II (or group B) chondrules. Heating in an environment with high oxygen fugacity caused little volatile loss or reduction, giving type II chondrules; with lower oxygen fugacites Na was extensively lost and FeO reduced to Fe metal, giving type IAB chondrule compositions, and in the extreme, Si was also lost by evaporation giving type IA compositions. Alexander also uses volatile loss to explain some chondrule composition variations, but to explain details which require chemical heterogeneity in the precursors, he appeals to recycling of chondrules. He shows that chondrules had varied initial ratios of Fe metal to oxidized Fe, and hence were not all like type II chondrules.

The general trends expected for open systems correspond to the average compositions of some of the main chondrule types, but Grossman argues that chondrule compositions were essentially fixed by prior fractionations and but little modified during melting.

He tests one simple model for volatile loss from chondrules and finds that the compositions of chondrites cannot be easily explained by adding and subtracting observed chondrule compositions or a calculated lost volatile component. If processes other than evaporation had a large effect on chondrule compositions, it is difficult to determine the extent of any volatile losses.

The absence of K isotopic fractionation suggests that chondrule melts suffered no evaporative losses (Humayun and Clayton, 1995). If so, volatile loss was minimized either by rapid heating and cooling (Wasson, Yu *et al.*) or by ambient gas conditions. Wood argues that though chondrule melts are unstable relative to gas at nominal nebular pressures, they become stabilized by concentration of dust in the gas, so that small amounts of evaporation during heating give liquid-gas equilibrium. In that case, flash heating is less needed to explain retention of moderately volatile elements though it may still be required to prevent large relict grains from dissolving in the melt.

SUMMARY

A synthesis of the ideas discussed in Albuquerque is presented below but, caveat emptor, a thorough reading of this book will show the extent to which individual contributors differ in their interpretations.

Most participants at Albuquerque appeared to believe that CAI formed earlier, perhaps in the infall stage, and chondrules later, perhaps in the relatively quiescent stage of the disk. Both CAI and chondrules cooled in localized domains within the disk, rather than by global nebular cooling, though the exact position of these domains is not certain. The heating process was therefore localized, though the scale of the events is not well established. Of all the heating mechanisms considered, shock waves are the most promising (Boss, Concise Guide).

A number of geochemical fractionations, related to early high temperature processing in the disk, preceded chondrule formation and some chondrule precursors contained CAI-like material. At least 15% of chondrules show evidence of two chondrule formation events, e.g. in the form of relict grains. Loss of moderately volatile elements such as S, Na and Fe from molten chondrules would be expected to happen to some extent, but there is no conclusive proof that it actually occurred.

The temperatures experienced by chondrules are between about 1500 and 1900°C. The peak temperatures were attained for very short lengths of time (minutes or less), or relict grains would have been dissolved in the melt. Cooling rates required to grow zoned olivine crystals were in the range 10–1000°/hr, but initially rapid cooling (5000°/hr) decreasing to produce appropriate zoned olivine crystals aids retention of moderately volatile elements. Heating in a dusty environment might produce high concentrations of O, S, Na in the gas by evaporation, thus tending to preserve the original chondrule compositions against volatile loss and relax the need for a heating pulse measured in seconds.

What do chondrules tell us about the protoplanetary disk? The lack of agreement about the detailed interpretation of chondrule petrology and geochemistry is perhaps the reason that, despite the consensus in favor of nebular origins, "planetary" models of chondrule formation are still being considered (Hutchison, Kitamura and Tsuchiyama, Sanders). One major constraint now widely accepted is that many chondrules were heated more than once, and therefore a shock heating mechanism which could be repeated, e.g. clump infall (Hood and Kring; Boss, Concise Guide), is preferable to one which acts essentially once, e.g. grain infall (Ruzmaikina and Ip). A second constraint is that the chondrule cooling rates indicated by experimentalists (Lofgren, Yu *et al.*) require either relatively large volumes heated or very long discharges, and it is not clear whether lightning can supply these conditions (Horanyi and Robertson; Boss, Concise Guide).

ACKNOWLEDGEMENTS

I am indebted to P. Pellas and colleagues at the Muséum National D'Histoire Naturelle for hospitality and to R.H. Jones and E.R.D. Scott for critical review of the manuscript.

REFERENCES

Boss A. P. and Graham J. A. (1993) Clumpy disk accretion and chondrule formation. *Icarus* **106**, 168–178.

Cameron A.G.W. (1995) The first ten million years in the solar nebula. *Meteoritics* **30**, 133–161.

Clayton D.D. and Jin L. (1995) Origin of CAIs and of their high ^{26}Al concentrations (abstract). *Meteoritics* **30**, 499–500.

Clayton R.N.,Mayeda T.K., Goswami J.N. and Olsen E.J. (1991) Oxygen isotope studies of ordinary chondrites. *Geochim. Cosmochim. Acta* **55**, 2317–2337.

Eisenhour D.D. and Buseck P.R. (1995) Chondrule formation by radiative heating: a numerical model. *Icarus*, in press.

Gooding J.L. and Keil K. (1980) Relative abundances of chondrule primary textural types in ordinary chondrites and their bearing on conditions of chondrule formation. *Meteoritics* **16**, 17–43.

Grimm R.E. and McSween H.Y. Jr. (1993) Heliocentric zoning of the asteroid belt by aluminum–26 heating. *Science* **259**, 653–655.

Grossman J.N. (1988) Formation of chondrules. In *Meteorites and the Early Solar System*, 680–696, ed. J.F. Kerridge and M.S. Matthews, Univ. of Arizona.

Grossman J.G., Rubin A.E., Nagahara, N. and King, E.A. (1988) Properties of chondrules. In *Meteorites and the Early Solar System*, 619–659, ed. J.F. Kerridge and M.S. Matthews, Univ. of Arizona.

Hewins R.H. (1988) Experimental studies of chondrules. In *Meteorites and the Early Solar System*, 660–679, ed. J.F. Kerridge and M.S. Matthews, Univ. of Arizona.

Hewins R.H. (1989) The evolution of chondrules. *Proc. NIPR Symp. Antarct. Meteorites* **2**, 202–222.

Hood L. L. and Horanyi M. (1993). The nebular shock wave model for chondrule formation. *Icarus* **106**, 179–189.

Humayun M. and Clayton R.N. (1995) Potassium isotope cosmochemistry: genetic implications of volatile element depletions. *Geochim. Cosmochim. Acta* **59**, 2131–2151.

Hutcheon I.D. and Jones R.H. (1995) The ^{26}Al–^{26}Mg record of chondrules: Clues to nebular chronology (abstract). *Lunar Planet. Sci.* **26**, 647–648.

Jones R.H. (1990) Petrology and mineralogy of type II, FeO-rich chondrules in Semarkona (LL3.0): Origin by closed-system fractional crystallization, with evidence for supercooling. *Geochim. Cosmochim. Acta* **54**, 1785–1802.

Jones R.H. (1992) Classification of porphyritic pyroxene-rich chondrules in the Semarkona ordinary chondrite (abstract). *Lunar Planet. Sci.* **23**, 629–630.

Jones R.H. (1994) Petrology of FeO-poor, porphyritic pyroxene chondrules in the Semarkona chondrite. *Geochim. Cosmochim. Acta* **58**, 5325–5340.

Jones R.H. and Scott E.R.D. (1989) Petrology and thermal history of type IA chondrules in the Semarkona (LL3.0) chondrite. In *Proc. 19th Lunar Planet. Sci. Conf.*, 523–536, Lunar and Planetary Institute, Houston,.

McCoy T.J., Scott E.R.D., Jones R.H., Keil K. and Taylor G.J. (1991) Composition of chondrule silicates in LL3–5 chondrites and implications for their nebular history and parent body metamorphism. *Geochim. Cosmochim. Acta* **55**, 601–619.

McSween H. Y. (1977) Chemical and petrographic constraints on the origin of chondrules and inclusions in carbonaceous chondrites. *Geochim. Cosmochim. Acta* **41**, 1843–1860.

Morfill G., Spruit H. and Levy E.H. (1993) Physical processes and conditions associated with the formation of protoplanetary disks. In *Protostars and Planets III* (eds. E.H. Levy and J. Lunine), 939–978.

Nagahara H. (1981) Evidence for secondary origin of chondrules. *Nature* **292**, 135–136.

Palme H. and Boynton W. V. (1993) Meteoritic constraints on conditions in the solar nebula. In *Protostars and Planets III* (eds. E. Levy and J. I. Lunine), pp. 979–1004. University of Arizona Press.

Sahagian D. L. and Hewins R. H. (1992) The size of chondrule-forming events (abstract). *Lunar Planet. Sci.* **23**, 1197–1198.

Scott E.R.D. and Jones R.H. (1990) Disentangling nebular and asteroidal features of CO3 carbonaceous chondrite meteorites. *Geochim. Cosmochim. Acta* **54**, 2485–2502.

Scott E. R. D. and Taylor G. J. (1983) Chondrules and other components in C, O, and E chondrites: Similarities in their properties and origins. *Proc. Lunar Planet. Sci. Conf.* **14**, B275–B286.

Sears D.W.G., Lu J., Benoit P.H., DeHart J.M. and Lofgren G.E. (1992) A compositional classification scheme for meteoritic chondrules. *Nature* **357**, 207–210.

Skinner W. R. (1994) Pre-Allende planetesimals with refractory compositions: The CAI connection (abstract). *Lunar Planet. Sci.* **25**, 1283–1284.

Taylor G. J., Scott E. R. D. and Keil K. (1983) Cosmic setting for chondrule formation. In *Chondrules and their Origins* (ed. E. A. King), 262–278. Lunar and Planetary Institute, Houston.

Wasson J. T. (1993) Constraints on chondrule origins. *Meteoritics* **28**, 13–28.

Wood J.A. (1981) The interstellar dust as a precursor of Ca,Al-rich inclusions in carbonaceous chondrites. *Earth Planet. Sci. Lett.* **70**, 11–26.

Wood J. A. (1984) On the formation of meteoritic chondrules by aerodynamic drag heating in the solar nebula. *Earth Planet. Sci. Lett.* **70**, 11–26.

Wood J. A.(1988) Chondritic meteorites and the solar nebula. *Ann. Rev. Earth Planet. Sci.* **16**, 53–72.

II. Chondrules, Ca–Al–rich Inclusions and Protoplanetary Disks

2: Astronomical Observations of Phenomena in Protostellar Disks

L. HARTMANN

Havard-Smithsonian Center for Astrophysics, 60 Garden St. MS-15, Cambridge, MA 02138, U.S.A.

ABSTRACT

Astronomical constraints on phenomena in disks of pre-main sequence stars are reviewed. At least half of all young stars seems to have substantial circumstellar disks at an age of about 10^6 yr. The best current estimates suggest that disks with substantial amounts of small dust particles generally "disappear" on a timescale of $\sim 10^7$ yr, with substantial uncertainty. The masses of these disks are thought to range in order of magnitude from values close to the minimum mass solar nebula $\sim 10^{-2}\, M_\odot$ up to appreciable fractions of a solar mass, with sizes of the order of 100 AU. During early evolutionary phases (probably the first 10^5 yr), the disk dumps its material into the central star in FU Orionis outbursts, with accretion rates reaching $10^{-4}\, M_\odot\, yr^{-1}$. FU Orionis outbursts may only occur in the inner disk (inside of 1 A.U.), but this is not well-constrained by observations. As infall to the disk ceases, disk accretion slows down, with typical rates of $10^{-7}\, M_\odot\, yr^{-1}$ during the T Tauri phase, which lasts for a few million years.

INTRODUCTION

In this article I discuss general astronomical constraints on the properties of pre-main sequence disk – masses, temperatures, and energetic phenomena. This review is offered in a somewhat apologetic spirit, because the best-studied astronomical phenomena may not be particularly relevant to the problems of chondrule formation. Nevertheless, a summary of current astronomical knowledge about the disks of young stellar objects may at least help set the context for ideas of chondrule formation. It also seems worthwhile to try to point out some areas of real uncertainty that may not be apparent to the casual reader of the astronomical literature.

As discussed by John Wood in his review (see Fig. 1), the current paradigm of star formation invokes the collapse of a cold molecular cloud core star plus disk system (Shu *et al.* 1987). The cloud collapse phase is thought to last about a hundred thousand years or so for a solar-mass star, during which time the star acquires most of its mass and the disk tends to have high accretion rates. The disk subsequently evolves over several to ten million years to the point where it is difficult or impossible to detect astronomically, and may have already formed planets.

A more detailed view of accretion processes in the life of a "typical" pre-main sequence disk is summarized in Fig. 1 (Fig. 5 of Wood's review). Rotating cloud collapse initially results in mas-

sive infall to the disk. The accretion of mass through the disk onto the central star does not appear to be steady, but instead is punctuated by outbursts of rapid disk accretion, the so-called "FU Orionis" events. During FU Ori outbursts disk accretion rates rise by two to three orders of magnitude, up to $10^{-4}\, M_\odot\, yr^{-1}$. Rapid disk accretion seems to occur mostly in the earliest evolutionary stages, when infall may still be occurring to the disk. After proto-

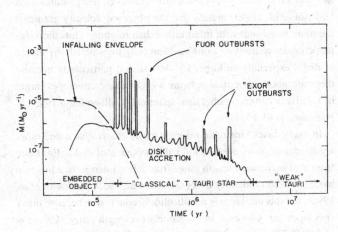

Fig. 1. Schematic overview of accretion events in the life of a typical low-mass star, as described in the text. The starting time is taken to be the formation of the initial protostellar core.

stellar cloud collapse ceases, the central object becomes visible as a pre-main sequence "T Tauri star". Over timescales of roughly three to ten million years detectable disk accretion ceases ($M \lesssim 10^{-8}$ $M_\odot$ yr^{-1}), and the infrared excess characteristic of optically-thick disks is no longer observed, because material has been accreted, or ejected, or perhaps simply because the dust has agglomerated into larger particles.

It must be emphasized that the observations suggest wide variations in disk properties among individual stars. Many young stars of a million years of age exhibit no astronomically-detectable disks, suggesting that the disk evolutionary timescale is much shorter than that indicated in Fig. 1 for perhaps half of all solar type stars. Rapid disk evolution can be driven by tidal forces from a companion star, but it is not clear that this accounts for all cases of rapidly-disappearing disks. From the point of view of solar nebula models, an even more important consideration is that the astronomical observations cannot demonstrate that the solar system disk must have experienced FU Ori eruptions; the most we can say is that a non-negligible fraction of low-mass stars must undergo multiple FU Ori outbursts. There is very little information on energetic events in later disk evolution.

DISKS VS. ENVELOPES

In the last few years astronomers have begun to appreciate the need to distinguish disks from the infalling, rotating, dusty envelopes from which the disks formed. Because infalling envelopes are bigger than the disks they produce (by definition), it is much easier to detect envelopes by imaging at radio, optical, or near-infrared wavelengths. Although the disk dust emission from main sequence stars like β Pic cannot be confused with envelopes, since these stars are so old that infall ceased long ago, the situation is not so clear for the much younger pre-main sequence stars in molecular clouds.

For example, Sargent and Beckwith (1987, 1991) discovered a 2000 AU flattened structure of gas around the young star HL Tau from radio-wavelength interferometry in ^{13}CO, and suggested that this was a rotationally-supported disk. However, using similar data Hayashi *et al.* (1993) argued that the observed velocity gradients are more consistent with infall rather than rotation. Thus the kinematic evidence must be examined carefully to establish the nature of disks, especially on large (10^3 AU) scales, particularly because there are good theoretical reasons why infalling envelopes might be relatively flattened rather than spherical (Galli and Shu 1993a,b; Hartmann *et al.* 1994).

In many cases kinematic information is not available and other arguments must be used to infer the presence of disks. Emission from dust in the wavelength range from a few microns to a few mm has been used to identify disks (Strom *et al.* 1989; Beckwith *et al.* 1990), but this emission is not absolutely conclusive because dusty envelopes can also emit in the same wavelength range (Calvet *et al.* 1994). The argument for disks is strengthened if it can be shown from observation that masses on small scales are relatively large. Assuming radial free-fall and typical infall parameters (e.g.,

Adams, Lada, and Shu 1987; Kenyon *et al.* 1993a,b), the mass of a collapsing protostellar envelope contained within a given radius τ is

$$M_r = \int_o^r \dot{M} / (2GM_* / r)^{1/2} = 10^{-3}\ M_\odot\ M_{-5} (M_* / 0.5 M_\odot)^{-1/2}\ r_{100}^{3/2}, \tag{1}$$

where M–5 is the mass infall rate in units of 10–5 solar masses per year, M^* is the central mass, and t100 is the outer radius in units of 100 AU. Using techniques discussed below, determinations of dust masses $\lesssim 0.01\ M_\odot$ (a minimum mass solar nebula) within $\lesssim 100$ AU of the central star are plausible indications of a circumstellar disk, since large masses are difficult to explain with an infalling envelope. For example, Lay *et al.* (1994) have apparently resolved a ~ 60 AU structure in the sub-mm (dust) continuum emission of HL Tau; the amount of mass needed to produce this emission probably ensures that this structure is the actual rotationally-supported disk (Sargent and Beckwith 1987, 1991; Beckwith *et al.* 1990).

The most likely situations where envelopes are confused with disks are those in which the central star is heavily extincted by dust and/or surrounded by an extensive reflection nebula. Many pre-main sequence T Tauri stars can be identified which do not have these characteristics, but still exhibit evidence for circumstellar dust emission; for these objects, the disk interpretation of the observed long-wavelength radiation is probably safe (Beckwith *et al.* 1990).

DISK FREQUENCIES AND LIFETIMES

Disks around pre-main sequence stars have been most systematically identified from spatially-unresolved observations of long-wavelength emission from circumstellar dust. The most extensive surveys have been undertaken at near-infrared wavelengths (e.g., Strom *et al.*, 1989), which measure the optically-thick radiation from the innermost disk, and mm wavelength observations (Beckwith *et al.* 1990; Osterloh and Beckwith 1995), which detect optically-thin emission from dust. Both types of surveys suggest that, during the first million years or so of evolution, about half of all T Tauri stars have optically-thick disks which emit strongly in the wavelength range between roughly 2 μm and about 100 μm. A few stars with long wavelength emission do not have near-infrared emission excesses and vice versa. Apparently some stars may have inner disks, which produce the near-infrared excesses (temperatures of ~ 1000 K) without having large enough or massive enough disks to be detected at longer wavelengths. Conversely, some stars with large-scale outer disks may exist without near-infrared excesses, particularly in the case of a close binary, for which the only stable configuration is a large-scale, circumbinary disk.

The usual statement that disks "disappear" on timescales of three to ten million years (Skrutskie *et al.* 1990) (actually, that small dust particles which dominate the disk opacity disappear) rests on an uncertain observational basis. The best statistics are derived from the Taurus molecular cloud complex, in which there are hardly any stars with ages ~ 3×10^6 – 10^7 yr (revised ages result in fewer old stars than indicated in Fig. 8 of Podosek and Cassen [1994]).Still older stars in the region show no evidence for disk emission; but the positions of these stars in the HR diagram are so close to the main sequence that contraction ages are unreliable (see, e.g., Fig. 5 of Strom *et al.* 1989), and these stars could be as old as 10^8 yr. The absence of infrared excess emission in young open clusters with ages $\gtrsim 5 \times 10^7$ yr (Stauffer, personal communication) probably provides a more certain upper limit on disk evolutionary timescales. Thus, as Podosek and Cassen (1994) argued, disk lifetimes could be 10^7 yr or longer.

GLOBAL DISK PROPERTIES

Pre-main sequence disks are probably optically thick at short wavelengths and optically thin at long wavelengths, based on observational arguments (Beckwith *et al.* 1990). Thus, at infrared wavelengths $\lesssim 100$ μm, the disk spectrum only depends upon the temperature distribution (e.g., Lynden-Bell and Pringle 1974), while the mm and sub-mm fluxes from optically-thin dust will depend upon the dust opacity and total mass present in addition to the temperature distribution. In principle, temperatures can be derived from the optically-thick region of the spectrum, and used as an input to estimate disk masses from sub-mm and mm wavelength observations.

In the case where the disk temperature distribution can be represented as a power law over a large range in radii R as $T \propto R^{-q}$, the disk emission spectrum has a power law form in the infrared. Expressed in the typical form used in the astronomical literature, the observed infrared flux S has the form

$$\nu S_\nu = \lambda S_\lambda \propto \lambda^{(2/q-4)} \propto \lambda^s \quad (2)$$

Simple theory for geometrically-thin, steady accretion disks makes the robust prediction $q = 3/4$ (see Fig. 2), so that the infrared spectral index should be $s = -4/3$. This is true whether the disk heating is dominated by local accretion energy dissipation (Lynden-Bell and Pringle 1974) or by absorption of light from the central star (Adams and Shu 1986). Unfortunately, most infrared spectra of T Tauri stars suggest flatter disk temperature distributions than this. At near- to mid-infrared wavelengths, the typical spectral index is $s \sim -2/3$ (Kenyon and Hartmann 1987; Adams, Lada and Shu 1988), implying $q \sim 3/5$, while at far-infrared wavelengths, $s \approx 0$ (Beckwith *et al.* 1990; Osterloh and Beckwith 1995), implying $q \sim 1/2$ (see Fig. 2).

It is important to understand disk temperature distributions for two reasons. First, in many cases the identification of "disks" rests only on the detection of the infrared excess spectrum; it is therefore somewhat unsettling for the identification of the disk if the spectrum differs from what is expected. Second, some knowledge of the temperature distribution is needed to determine masses from long-wavelength observations, let alone infer physical conditions relevant to planetary formation.

The source of the discrepancy between simple disk theory and observed spectra is not clear at present. One possibility is that some sort of non-standard accretion process is involved. For example,

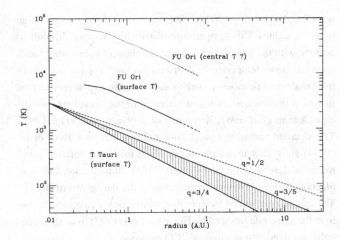

Fig. 2. Inferred and guessed disk temperature distributions for characteristic objects. The dotted region bounded by dark solid lines shows the likely range of surface disk temperatures for typical T Tauri stars; $q = 3/4$ corresponds to an infrared spectral index $s = -4/3$, while $q = 3/5$ corresponds to $s = -2/3$ (see text). The dashed line immediately above indicates the $q = 1/2$ temperature distribution for a "flat spectrum" ($s = 0$) T Tauri star; this line probably is an upper limit, since flat spectrum emission probably should be attributed to a dusty circumstellar (and probably infalling) envelope, rather than the disk (see text). The uppermost dark solid line indicates the surface temperature distribution of FU Ori, as inferred from its spectral energy distribution (see Fig. 3); the region beyond 0.5 AU is dashed, because it is not clear whether the high accretion rates of FU Ori outbursts extend to disk radii much larger than this (see Fig. 3). The upper dotted line represents a guess as to the central temperature of a massive FU Ori disk, estimated from α viscosity thermal instability models (Bell and Lin 1994; Bell *et al.* 1995).

spiral waves might transfer the necessary accretion energy to outer disk regions (Adams *et al.* 1988; Adams *et al.* 1989; Shu *et al.* 1990); but this mechanism may require an unrealistically sharp outer edge to the disk, and detailed predictions of the disk heating have not yet been made. This kind of picture also probably implies non-steady accretion in a way which raises other difficulties (Kenyon and Hartmann, 1987).

Other mechanisms have been suggested which enhance the absorption of light by the central star. If the disk is sufficiently thick and its surface is curved upward, it may absorb substantially more radiation from the central star than would a geometrically thin, flat disk (Kenyon and Hartmann 1987). However, it is not clear that dust can be suspended to the necessary ~ 3 scale heights above the disk midplane in the presence of dust settling (e.g., Weidenschilling and Cuzzi 1993). Alternatively, a tenuous and roughly spherical envelope of dust could effectively scatter additional short-wavelength starlight *into* the disk (Natta, 1993). However, the physical basis for the tenuous dusty envelope is not clear. The envelope could be the infalling material left over from the collapse of the protostellar cloud; but the calculations of Natta (1993) suggest that the expected spectrum is not quite appropriate. Natta suggests that the dust might arise in a disk wind, but it is not clear that such dusty winds actually exist (e.g., Hartmann, 1995), or whether they have the necessary density distribution.

Some T Tauri stars have "flat" spectra, $s \approx 0$, from near-infrared wavelengths out to ~ 100 μm, implying $q = 1/2$ over a very large range in radius. This temperature distribution is very difficult to achieve with the disk theories discussed above. On the other hand, such shallow temperature distributions are typical of dusty infalling envelope models, such as those used to explain protostars, in which the infrared emission arises from the envelope, *not* the disk (Adams *et al.* 1987; Kenyon *et al.* 1993a; Calvet *et al.* 1994). The detailed radiative equilibrium calculations of Calvet *et al.* (1994) show that it is easy to get $s \approx 0$ using the rotating infall model of Terebey, Shu, and Cassen (1984). In particular, Calvet *et al.* show that the far-infrared emission of the flat spectrum star HL Tau can be reproduced by emission from a dusty envelope falling in at the same rate inferred by Hayashi *et al.* (1993) from their spatially-resolved observations of ^{13}CO emission.

I suspect that the infrared spectra of the flat-spectrum T Tauri stars are dominated by emission dusty infalling envelopes, an interpretation supported by the observation that most of these objects have dusty envelopes clearly discernible in scattered light. (The number of flat spectrum sources, constituting about 5–10% of the total T Tauri population, suggests lifetimes comparable to if not shorter than typical infall times of $\lesssim 10^5$ yr.) The further question is: how does infall decay with time? Is there an abrupt termination of infalling matter, or is there a gradual decrease? If the latter, it is conceivable that *thermal* emission from relatively tenuous infalling envelopes could dominate the far infrared emission for typical T Tauri stars, making the determination of outer disk temperatures from far-infrared emission problematic.

Considering these uncertainties, Fig. 2 is offered as a rough guide to typical disk surface temperatures for T Tauri stars. Pre-main sequence stars will vary in luminosity as a function of mass and age, changing the heating of the disk, and one may expect variations in the accretional heating), so the T Tauri temperature distributions shown in Fig. 2 are only approximate. In addition, central disk temperatures may be somewhat larger in earlier evolutionary phases, where the disk accretion rate is relatively high and disk optical depths are sufficient to trap the energy derived from viscous heating.

In principle, disk masses can be determined from measurements of long-wavelength radiation. The most comprehensive surveys are those of Beckwith *et al* (1990), Andre and Montmerle (1994), and Osterloh and Beckwith (1995). These measurements are based on the reasonable assumption that sub mm to mm wavelength radiation on small scales comes from optically-thin dust particles in the circumstellar disk. Although the analysis of disk masses is relatively simple, there are still a number of complicating factors in proceeding from observed long-wavelength fluxes to disk masses. In addition to the complications of the temperature distribution discussed immediately above, there are also uncertainties due to the unknown distribution of mass with radius.

Probably the most significant uncertainty in estimating disk masses arises from poor knowledge of the dust opacity. In theory, the dependence of the flux on frequency determines the index β of the dust grain opacity $\kappa_\nu \propto \nu^\beta$. Observations in the diffuse interstellar medium, as well as theory for wavelengths much longer than

the particle size, suggest $\beta \sim 1.5 - 2$ (Pollack *et al.* 1994). Observations of young stars suggest a range of β between –0.5 and 2, with a peak near 1.0 (Beckwith and Sargent 1991). These results are interesting, because smaller values of β suggest that grain processing may be occurring in the circumstellar disks of young stars, with a buildup of larger grains. However, the possible change in grain properties makes the conversion between dust opacity to (gas) mass more uncertain. Pollack *et al.* (1994) suggest that the magnitude of the uncertainty of the opacity near 1 mm is about a factor of four (on an uncertain basis). To summarize, the long-wavelength continuum fluxes provide our best constraints on pre-main sequence disk masses. With our current understanding of disk structure and grain opacities, the derived masses should be regarded as order of magnitude estimates.

The recent studies of Andre *et al.* (1994) and Osterloh and Beckwith (1995) suggest typical T Tauri disk masses of 10^{-3} to 10^{-2} $M_\odot$ at ages ~ 10^6 yr. It appears that (in Taurus at least) only about half of the stars have detectable disks in mm continuum emission ($\gtrsim 10^{-2} M_\odot$) at this age. This result is similar to one originally derived by Strom *et al.* (1989) on the basis of near-infrared disk emission. (At such wavelengths, the disks are thought to be very optically-thick, so mass estimates are not possible.) The presence of detectable mm disk emission depends on the presence of a companion star; Osterloh and Beckwith argue that stars with companions closer than about 100 AU are less frequently detected in mm emission. This result supports theoretical ideas that tidal forces in binary or multiple systems can play an important role in truncating disks or otherwise driving disk evolution (Mathieu 1994).

ACCRETION AND ACCRETION EVENTS

T Tauri stars with strong emission in excess of stellar photospheric radiation are thought to be accreting from their circumstellar disks. In principle, accretion rates can be determined from infrared disk emission (e.g., Lynden-Bell and Pringle 1974); however, the departure of the "disk" spectrum from the standard steady accretion disk predictions suggests problems with this procedure, and may well indicate that the disk infrared emission is mostly the result of heating by the central star, not local release of accretion energy. Therefore, the most reliable estimates of accretion rates for T Tauri stars come from observations which constrain the amount of disk material landing on the central star. It is believed that the optical-ultraviolet excess continuum emission observed in T Tauri stars with inner disks arises from the accretion of disk material, probably through a magnetosphere (e.g. Königl 1991). Mass accretion rates then can be inferred from estimates of the optical-uv accretion luminosity.

Attempts to measure accretion luminosities in T Tauri stars are complicated by large ultraviolet interstellar extinction and the need to distinguish between excess accretion energy and the normal stellar photospheric emission. Despite these problems, several different analyses (Bertout *et al.* 1988; Hartmann and Kenyon 1990; Hartigan *et al.* 1991) all suggest typical accretion rates ~ 10^{-7} $M_\odot$ yr^{-1} for T Tauri stars with inner disks as indicated by near-infrared excess emission. (T Tauri stars without evidence for inner disk near-infrared emission show no signs of accretion energy release in optical-uv spectra). If we adopt a typical lifetime of $\sim 10^6$ yr for the active T Tauri phase, the accretion rates imply that the disk masses must originally be at least $\sim 10^{-1}\,M_\odot$. This is somewhat larger than typical T Tauri disk masses estimated from mm observations (see above), which are more typically $10^{-2}\,M_\odot$; it is not known whether the discrepancy is dominated by errors in the assumed dust opacity or the difficulties in estimating accretion rates. My own guess is that disk masses need to be revised upward somewhat from the mm estimate.

Essentially al pre-main sequence stars vary in brightness, and much (though not all) of this variability is now attributed to variations in the mass accretion rate through the circumstellar disk. However, we have relatively little understanding of this variability, since most observations of these phenomena are at optical wavelengths, which mostly constrains accretion directly onto the central star. Monitoring of variability out to the near-infrared regions (e.g., Gahn *et al.* 1989; Kenyon *et al.* 1994), constrains at best the innermost disk regions, at distances of only a few stellar radii. My guess is that extreme variations in optical emission (called "EXOR" outbursts by Herbig [1977]; Fig. 1) correspond to variations in accretion rate of about an order of magnitude from the base T Tauri disk accretion rate.

The most spectacular variations in light are observed in the classical FU Orionis outbursts (Herbig 1977, 1989), during which the objects brighten by a factor of one hundred or more at optical wavelengths. These outbursts indicate dramatic increases in disk accretion of up to three orders of magnitude on a time scale of a year or so, lasting for many decades; we have not followed any FU Ori object long enough to know when the outbursts will cease.

The evidence that FU Ori outbursts are the result of rapid increases in the mass accretion rate through a disk around a T Tauri star is summarized by Hartmann *et al.* (1993). One basic piece of the picture is the spectral energy distribution. Fig. 3 shows the spectrum of FU Ori, which is clearly broader than that of a single-temperature blackbody (or star). The infrared excess can be quite well matched by a simple *steady* accretion disk model of the type discussed in the previous section. (Although the disk of FU Ori cannot be completely steady because of the outburst, it is plausible that the system has settled down to a quasi-steady state several decades after the initial optical outburst.) There is evidence for a modest excess in the emission over that predicted by the simple disk model at wavelengths between 10 and 100 μm (Fig. 3). In other FU Ori systems, this excess is even more pronounced, and has been interpreted in terms of emission from a remnant dusty envelope exterior to the disk (Adams *et al.* 1987; Kenyon and Hartmann 1991), and will be discussed further below. The observed mm and sub-mm emission falls below the optically thick disk model, strongly suggesting that, as in the case of T Tauri stars in general, the disk is optically thin at such wavelengths.

The nature of the central stars of FU Ori objects is uncertain, since in outbursts the disks are much brighter than the stars. Only one pre-outburst spectrum of an FU Ori object is known (Herbig 1977), and this spectrum is typical of a moderately strong-emission T Tauri star. Measurements of rotational line broadening can be used to constrain the central stellar masses. There are uncertainties in this procedure because the system inclinations are unknown, but the results generally suggest that the central objects are low-mass stars.

Because FU Ori disks are much brighter than the central star, one does not need to worry about irradiation affecting the infrared spectrum; moreover, the good fit of the spectral energy distribution to simple, steady accretion disk models indicates that the infrared disk emission can be used to infer the accretion rate directly (Fig. 3), unlike the situation for T Tauri stars. The results of infrared spectral analysis (cf. Kenyon *et al.* 1988) indicate accretion rates of $\sim 10^{-4}\,M_\odot\,\mathrm{yr}^{-1}$ at maximum light. With decay times of decades to centuries, the observations indicate that masses of 10^{-3} to $10^{-2}\,M_\odot$ are accreted in an outburst; i.e., FU Ori itself has accreted something like a minimum mass solar nebula into its central star during the ~ 60 years that it has been observed.

Herbig (1977) originally suggested that FU Ori outbursts were repetitive, based on statistics rather than direct observation, since outbursts last so long. The best estimate comes from adopting a mean star formation rate within a Kpc or so of the Sun, and assuming that we have detected all of the FU Ori outbursts in this volume. Currently, it appears that eleven currently-accreting FU Oris are known in this region. Outbursts have been directly observed in FU Ori, V1057 Cyg, V1515 Cyg, and Elias 1-12; the brightening of V346 Nor and RNO 1B/C are difficult to distinguish from possible changes in line-of-sight dust extinction. With 4-5 outbursts over the last fifty years, the repetition rate implied is about ~ 10 per low-mass star (Hartmann and Kenyon 1985). This is probably a lower limit to the frequency, since we have probably missed some objects. There is no certainty that *all* low-mass stars like the Sun must have had such eruptions, especially since the driving mechanism for the outbursts is not really understood (see below).

One clue to the outburst mechanism comes from the frequent observation of far-infrared excess emission above that predicted by the disk (e.g., Fig. 3), and the universal result that all FU Ori objects have reflection nebulae (Goodrich 1987). The implication is that FU Oris are relatively younger than T Tauri stars, which generally do not have such dusty envelopes/reflection nebulae. Comparison with statistics of T Tauri stars in the Taurus molecular cloud suggests that the FU Ori phenomenon is concentrated to the first 10^5 years of the evolution of a low-mass star. Thus, perhaps Fig. 1, which is a revision of the Protostar and Planets III figure of Hartmann *et al.* (1993), should be further revised to concentrate the outbursts to even earlier stages.

The observed excess dust emission at wavelengths $\gtrsim 10$ μm suggests the presence of a dusty envelope not more than $\sim 10^1$ AU from the central star. This result, coupled with the general youth implied for FU Oris, led Kenyon and Hartmann (1991) to suggest that many, if not all, objects have infalling envelopes landing on the outer disk. The infalling envelope could then provide a mass source, replenishing the disk material accreted during outburst, leading to repetitive eruptions.

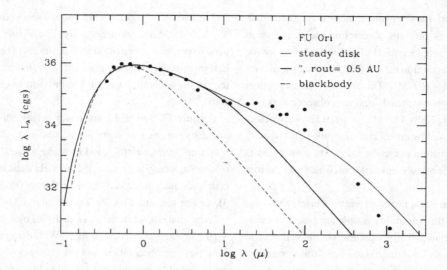

Fig. 3. Spectral energy distribution of FU Ori. The spectrum is clearly broader than that of a single temperature blackbody (dashed line). (Because the disk luminosity is so large in outburst, the contribution of the stellar photosphere is negligible). The overall spectrum can be matched quite well with an optically-thick, steady-accretion disk model (light solid line) for wavelengths $\lesssim$ 10 μm. Between 10 and 100 μm, the observed fluxes are larger than the simple disk model, but this may be due to emission from an outer dusty envelope. Because of the possible contribution of envelope emission longward of 10 μm, it is not clear whether the disk is in outburst (high accretion rate state) at radii larger than that predicted by thermal instability models; this is demonstrated by the heavy solid line, which shows that a disk with an outer radius of 0.5 AU can adequately explain the spectrum at wavelengths shortward of 10 μm (see text). The observed steep spectral index near $\lambda \sim 1$ mm is probably due to the disk becoming optically thin.

This basic picture was elaborated on by Bell and Lin (1994) and Bell *et al.* (1995) in the context of thermal instability models for the accretion outbursts. In the Bell/Lin picture, the inner disk has a "natural" accretion rate that is lower than the $\sim 10^{-5} M_\odot$ yr^{-1} infall rate onto the outer disk. Thus, material piles up in the disk. As the disk gains mass, it becomes hotter because of radiative trapping of the heat generated by accretional energy release. Eventually, the disk becomes hot enough to partially ionize hydrogen, at which point due to complicated reasons the disk can no longer be in thermal equilibrium until it becomes much hotter- this leads to a quasi-steady state at a very high accretion rate. The outburst drains material out of the disk because the high-state accretion rate is larger than the rate at which mass is added to the outer disk by infall.

Bell and collaborators show that many features of the observations can be explained by this thermal instability model. However, it is not clear that the thermal instability picture is the correct one for FU Ori outbursts. The infall-driven model described above can account for the slow rise time (decades) of the outburst of V1515 Cyg quite easily. The rapid optical (one year rise times) outbursts of FU Ori and V1057 Cyg are another matter; the disk may need some sort of perturbation to make such abrupt increases in accretion rate. Moreover, since these models are characterized by the "α" viscosity formulation, with an arbitrary variable α, the physical basis of the calculation remains somewhat suspect.

Bonnell and Bastien (1992) suggested that binary system formation might produce FU Ori outbursts quite naturally; passage of the companion star on an eccentric orbit can kick the disk into an outburst, and this may even be repeated a few times. But the applica-bility of the binary hypothesis remains unclear, since no measurable radial velocity shift (greater than a couple of km/s) has been observed in an FU Ori, and it is not clear that there are enough binary companions on appropriate orbits to account for the estimated frequency of events (every 10^3 to 10^4 yr).

Modelling of the spectral energy distributions of FU Ori objects, such as that displayed in Fig. 3, result in calculated surface temperature distribution like that shown for FU Ori in Fig. 2. It is clear that, during such an event, the disk surface must be quite remarkably hot, reaching 10^3K or higher at radial distances out to ~ 0.5 AU. Whether the hot, high-accretion rate part of the disk extends to larger radii is not clear, because of the confusing effects of the dusty envelope emission. In Fig. 3 the heavy solid line shows the spectrum of a rapidly accreting disk with an outer limiting radius of $\tau = 0.5$ AU. It is possible that the emission longward of 10 μm comes from the envelope, not the disk; this is suggested by the apparent change in spectral index, and by comparison with other FU Ori objects for which the evidence for envelope emission at 10 μm is quite strong (Kenyon and Hartmann 1991). On the other hand, there could be substantial disk emission and just a weak envelope component. At the present, we simply cannot distinguish these possibilities. This uncertainty is unfortunate, because the thermal instability model of Bell and Lin (1994) predicts that only the inner 30 $R_\odot \sim 0.2$ AU region of the disk goes into the high accretion state; current observations do not provide a definitive test.

We do not know what the central temperature of an FU Ori disk is, since we would need to know its surface density to estimate the trapping of heat. The models of Bell *et al.* (1995), which have a low disk viscosity and therefore a high disk surface density, suggesting that the internal temperatures might be an order of magnitude larger than the surface temperatures. A crude estimate of what the central temperature might be is plotted in Fig. 2. The surface and central temperature distributions are plotted as dotted lines outside of 0.5 AU because of the difficulty of separating disk and envelope emission discussed in the previous paragraph.

SPECULATIONS

What, if anything, do the astronomical observations imply for chondrule formation? Some admittedly speculative thoughts, possibly subject to my poor understanding of meteoritic constraints:

In his review, John Wood suggests that chondrules might have been formed early in the solar nebula's evolution. He argues that FU Ori events provide an indication of the energetic processes that would be required to make chondrules, although FU Ori outbursts themselves do not cool on the appropriate timescales- and this seems quite reasonable to me. I would add that, if chondrules formed early on, it would be necessary to assume that FU Ori outbursts, with their long cooling timescales, did not vaporize everything (because the cooling of this global heating event would take a long time). One way out is to assume that chondrule formation happens after the last FU Ori event, but this may not be necessary, since the disk might not get hot enough at a few AU to melt grains. As pointed out above, the observational constraints on temperatures in this region are poor, and the disk thermal instability models do not extend out to 1 AU. Disk temperatures at 2.5 AU may or may not be high enough to eliminate grains, depending upon the (unknown) accretion rate in FU Ori disks.

Wood suggests that shocks, such as those considered by Hood and Horanyi (1993), might be responsible for chondrule heating, and that such shocks might be the result of gravitational instabilities in massive disks. This seems quite possible to me; the FU Ori disk models of Bell *et al.* (1995) would be gravitationally unstable if extrapolated out to a few AU from the central star. The shock velocities required by Hood and Horanyi, of order 5–6 km s^{-1}, seem rather large to me in this context (~ 1/3 the circular velocity), but I am not aware of explicit calculations for gravitational instabilities which would either support or contradict this suggestion. In this vein, it seems to me that the accretion shock of infalling material to the disk may deserve further consideration. This is the one place where we can be certain that shocks of the required magnitude actually occurred. If the initial rotation of the protosolar cloud was not too large – such as to produce a disk initially of no larger than 30 AU – then essentially all of the mass would be processed through a shock of several km s^{-1}. Most of this mass ended up in the Sun, but it can't all go in by conservation of angular momentum. Whether or not chondrule chemistry, "aerodynamic sorting", and radionuclide timing rule out this early formation scenario (e.g., chapters by Huss and MacPherson and Davis) is beyond my competence to judge.

Boss and Graham (1993) have discussed an accretion shock picture in which the infall is clumpy, and in the (later) T Tauri phase of evolution. I have no difficulty in believing that infall is not quite steady, but I don't see why this is necessary. I think that the reason why Boss and Graham emphasize clumps may be that they want to have shocks driven into the disk, which has the proto-chondrule grains, and that in a steady situation the shocks might not propagate far enough into the disk. But it is unlikely that the infall shock can be steady even for steady infall, because of the mismatch in angular momentum between underlying disk material in Keplerian rotation and the infalling material at that radius, which I would think must drive turbulent motions. Furthermore, if this process is occurring *while the disk itself is being built up by the infall*, much of the outer regions of the disk may be affected by either the initial accretion shock, or possibly subsidiary shocks driven by the turbulent mixing, and in any event would be time-dependent. The model of Galli and Shu (1993a,b), in which a substantial fraction of the envelope collapses into a non-rotationally supported disk because of the magnetic field, and then accretes in the disk midplane, eventually crashing into the rotationally-supported disk, might have different implications than the usual Terebey *et al.* (1984) picture. In any event, given the apparent difficulties of many (most?) chondrule formation pictures, perhaps it is worth taking a closer look at early formation scenarios. This would be true also for mechanisms using bipolar outflows to eject chondrules from inner disk regions, because such outflows are likely to be much more powerful and energetic in the earliest phases of evolution (cf. Hartmann 1995).

ACKNOWLEDGEMENTS

I am grateful to John Wood for showing me his article before publication, to Pat Cassen for informative discussions, and to Gary Huss for a very helpful referee report. This work was supported in part by NASA grant NAGW-2919.

REFERENCES

Adams F. C., Lada C., and Shu F. H. (1987) Spectral Evolution of Young Stellar Objects. *Astrophys. J.* **312**, 788–806.

Adams F. C., Lada C. J., and Shu F. H. (19888) The Disks of T Tauri Stars with Flat Infrared Spectra. *Astrophys. J.* **326**, 865–883.

Adams F. C., Ruden S. P., and Shu F. H. (1989) Eccentric Gravitational Instabilities in Nearly Keplerian Disks. *Astrophys. J.* **347**, 959–975.

Adams F. C. and Shu F. H. (1986) Infrared Spectra of Rotating Protostars. *Astrophys. J.* **308**, 836–853.

Andre P., and Montmerle T. (1994) From T tauri Stars to Protostars: Circumstellar Material and Young Stellar Objects in the ρ Ophiuchi Cloud. *Astrophys. J.* **420**, 837–862.

Beckwith S., Sargent A. Chini R., and Güsten R. (1990) A Survey for Circumstellar Disks around Young Stellar Objects. *Astron. J.* **99**, 924–945.

Beckwith S., and Sargent A. (1991) Particle Emissivities in Circumstellar Disks. *Astrophys. J.* **381**, 250–258.

Bell K. R. and Lin D. N. C. (1994) Using FU Orionis Outbursts to Constrain Self-Regulated Protostellar Disk Models. *Astrophys. J.* **427**, 987–1004.

Bell K. R., Lin D. N. C., Hartmann L., and Kenyon S. (1995) The FU Orionis Outburst as a Thermal Accretion Event: Observational Constraints for Protostellar Disk Models. *Astrophys. J.*, in press.

Bertout C., Basri G., and Bouvier J. (1988) Accretion Disks around T Tauri Stars. *Astrophys. J.* **330**, 350–373.

Bonnell I., and Bastian P. (1992) A Binary Origin for FU Orionis Stars. *Astrophys. J.* **401**, L31–34.

Boss A. P. and Graham J. A. (1993) Clumpy Disk Accretion and Chondrule Formation. *Icarus* **106**, 168–178.

Calvet N., Hartmann L., Kenyon S. J., and Whitney B. A. (1994) Flat Spectrum T Tauri Stars: The Case for Infall. *Astrophys. J.* **434**, 330–340.

Goodrich R. W. (1987) The Ring-Shaped Nebulae Around FU Orionis Stars. *Pub. Astro. Soc. Pacific* **99**, 116–125.

Gahm G. F., Fischerström C., Liseau R., and Lindroos K. P. (1989) Long- and Short-term Variability of the T Tauri Star RY Lupi. *Astron. Astrophys.* **211**, 114–130.

Galli D., and Shu F. H. (1993a) Collapse of Magnetized Molecular Cloud Cores. I. Semianalytical Solution. *Astrophys. J.* **417**, 220–242.

Galli, D., and Shu F. H. (1993b) Collapse of Magnetized Molecular Cloud Cores. II. Numerical Results. *Astrophys. J.* **417**, 243–258.

Hartigan P., Kenyon S. J., Hartmann L., Strom S. E., Edwards S., Welty A. D., and Stauffer J. (1991) Optical Excess Emission in T Tauri Stars. *Astrophys. J.* **382**, 617.

Hartmann L. (1995) Observational Constraints on Disk Winds. Invited review at IA-UNAM meeting on Circumstellar Disks, Outflows, and Star Formation. To appear in *Revista Mexicana de Astronomia y Astrofisica*.

Hartmann L., Boss A. P., Calvet N., and Whitney B. (1994) Protostellar Collapse in a Self-Gravitating Sheet. *Astrophys. J.* **430**, L49–52.

Hartmann L., and Kenyon s. J. (1985) On the Nature of the FU Orionis Objects. *Astrophys. J.* **299**, 462–478.

Hartmann L., and Kenyon S. J. (1988) Accretion Disks Around Young Stars. In *Formation and Evolution of Low-Mass Stars* (eds. A. K. Dupee and M. T. V. T. Lago), pp. 163–179. Reidel.

Hartmann L., and Kenyon S. J. (1990) Optical Veiling, Disk Accretion, and the Evolution of T Tauri Stars. *Astrophys. J.* **349**, 190–196.

Hartmann L., Kenyon S., and Hartigan P. (1993) Young Stars: Episodic Phenomena, Activity, and Variability. In *Protostars and Planets* **III** (eds. E. H. Levy and J. Lunine), pp. 497–518. University of Arizona Press.

Hayashi M., Ohashi N., and Miyama S. (1993) A Dynamically Accreting Gas Disk Around HL Tauri. *Astrophys. J. Letters* **418**, L71–L74.

Herbig G. H. (1977) Eruptive Phenomena in Early Stellar Evolution. *Astrophys. J.*, **217**, 693–715.

Herbig G. H. (1989) FU Orionis Eruptions. In *ESO Workshop on Low-Mass Star Formation and Pre-Main Sequence Objects* (ed. B. Reipurth), pp. 233–236. ESO.

Hood L. L., and Horanyi, M. (1993) The Nebular Shock Wave Model for Chondrite Formation: One-Dimensional Calculations. *Icarus* **106**, 179–189.

Kenyon S. J., Calvet N., and Hartmann L. (1993a) The Embedded Young Stars in the Taurus-Auriga Molecular Cloud. I. Models for Spectral Energy Distributions. *Astrophys.. J.* **414**, 676–694.

Kenyon S. J., and Hartmann L. (1987) Spectral Energy Distributions of T Tauri Stars: Disk Flaring and Limits on Accretion. *Astrophys. J.* **323**, 714–733.

Kenyon S. J., and Hartmann L. (1991) The Dusty Envelopes of FU Orionis Variables. *Astrophys. J.* **383**, 664–673.

Kenyon S. J., Hartmann L., and Hewett, R. (1988) Accretion Disk Models for FU Orionis and V1057 Cygni: Detailed Comparisons Between Observations and Theory. *Astrophhys. J.* **325**, 231–251.

Kenyon S. J., Hartmann L., Hewett R., Carrasco L., Cruz-Gonzalez I., Recillas E., Salas L., Serrano A., Strom K. M., Strom S. E., and Newton, G. (1994) The Hot Spot in DR Tau. *Astron. J.* **107**, 2153–2163.

Kenyon S. J., Hartmann, L. W., Strom, K. M. and Strom, S. E. (1990) An IRAS Survey of the Taurus-Auriga Molecular Cloud. *Astron. J.* **99**, 869–887.

Kenyon S. J., Whitney B., Gomez M., and Hartmann L. (1993b) The Embedded Young Stars in the Taurus-Auriga Molecular Cloud. II. Models for Scattered Light. *Astrophys. J.* **414**, 773–792.

Königl A. (1991) Disk Accretion onto Magnetic T Tauri Stars. *Astrophys. J.* **370**, L39–42.

Lay O. P., Carlstrom J. E. Hills R. E. and Phillips T. G. (1994) Protostellar Accretion Disks Resolved with the JCMT-CSO Interferometer. *Astrophys. J.* **434**, L75–78.

Lynden-Bell D., and Pringle J. E. (1974) The evolution of Viscous Disks and the Origin of the Nebular Variables. *MNRAS* **168**, 603–637.

Mathieu R. D. (1994) Pre-Main Sequence Binary Stars. *Annual Review Astronomy and Astrophysics* **32**, 465–530.

Natta A. (1993) The Temperature Profile of T Tauri Disks. *Astrophys. J.* **412**, 761–770.

Osterloh M., and Beckwith S. V. W. (1995) Millimeter-Wave Continuum Measurements of Young Stars. *Astrophys. J.*, in press.

Podosek F. A., and Cassen P. (1994) Theoretical, Observational, and Isotopic Estimates of the Lifetime of the Solar Nebula. *Meteoritics* **29**, 6–25.

Pollack J. B., Hollenbach D., Beckwith S., Simonelli D. P., Roush T., and Fong W. (1994) Composition and Radiative Properties of Grains in Molecular Clouds and Accretion Disks. *Astrophys. J.* **421**, 615–639.

Pringle J. E. (1981) Accretion Disks in Astrophysics. *Annual Review Astronomy and Astrophysics* **19**, 137–162.

Sargent A. I., and Beckwith S. (1987) Kinematics of the Circumstellar Gas of HL Tauri and R Monocerotis. *Astrophys. J.* **323**, 294–305.

Sargent A. I., and Beckwith S. (1991) The Molecular Structure around HL Tauri. *Astrophys. J.* **382**, L31–35.

Shu F. H., Tremaine S., Adams F. C., and Ruden S. P. (1990) SLING Amplification and Eccentric Gravitational Instabilities in Gaseous Disks. *Astrophys. J.* **358**, 495–514.

Shu F. H., Adams F. C. and Lizano S. (1987) Star Formation in Molecular Clouds: Observation and Theory. *Ann. Rev. Astronomy and Astrophysics* **25**, 23–81.

Skrutskie M. F., Dutkevitch D., Strom S. E., Edwards S., Strom K. M., and Shure M. A. (1990) *Astron. J.* **99**, 1187–1195.

Terebey S., Shu F. H., and Cassen P. (1984) The Collapse of the Cores of Slowly Rotating Isothermal Clouds. *Astrophys. J.* **286**, 529–551.

Strom K. M., Strom S. E., S. Cabrit S. and Skrutskie M. F. (1989) Circumstellar Material Associated with Solar-Type Pre-Main Sequence Stars: A Possible Constraint on the Timescale for Planet Building. *Astron. J.* **97**, 1451–1470.

Weidenschilling S. J. and Cuzzi J. N. (1993) Formation of Planetesimals in the Solar Nebula. In *Protostars and Planets* **III** (eds. E. H. Levy and J. Lunine), pp. 1031–1060. University of Arizona Press.

3: Overview of Models of the Solar Nebula: Potential Chondrule-Forming Environments

PATRICK CASSEN

NASA - Ames Research Center, 245–3, Moffett Field, CA 94035–1000, U.S.A.

ABSTRACT

The characterization of the masses, lifetimes, luminosities, and temporal variations of circumstellar disks around T Tauri stars provides an observational basis for the construction of solar nebula models. In particular, the modeling of thermal structure identifies conditions under which an early, hot phase existed throughout the terrestrial planet region. If the product of the mass accretion rate through the nebula (in solar masses/year) times the nebula optical depth exceeded about 0.1, the evaporation of silicates to a distance of about 3 AU is predicted. Such conditions are plausible in the early nebula, but not guaranteed. Condensation from a hot state produces chemical and elemental fractionations like those retained in chondritic meteorites. Although base state densities, pressures and temperatures are defined by nebula models, it cannot be claimed that they provide definitive constraints on the chondrule formation mechanism or site, or a basis for the rejection of hypotheses that require extreme values of these variables.

INTRODUCTION

It is now possible to discuss the formation of the meteorites in the context of solar nebula models which have more than a merely theoretical basis. Astronomical observations of disks around young, solar-like stars provide useful information even in the absence of a comprehensive, quantitative theory, and in spite of inevitable uncertainties in the interpretation of the data. Of course, a compelling theory of the formation of the Solar System must ultimately rely on both the astronomical evidence and that provided by meteorites and other planetary objects, if only because such unpredictable factors as the angular momentum of the protosolar cloud, the random effects of collisions during accumulation, and possible interactions with the extra-solar environment must be deduced from the remaining record. Chondrules loom large in that record, by virtue of their abundance and (near) ubiquity in primitive meteorites. But the degree to which chondrules reflect nebula evolution, or even nebula properties, is still an open question; the "wealth of information [contained in chondrules] about conditions in the nebula" (Taylor *et al.*, 1983) has yet to be revealed in a coherent picture. This state of affairs stems largely from (1) the conclusion

that chondrule production occurred in brief, localized events, and (2) the failure thus far to establish the physical nature of those events, much less their relation to other nebula processes.

Nevertheless, nebula models do have implications for the thermal and chemical history of Solar System solids in general, and for chondrule precursor material in particular. They also provide the context for the discussion of parameter regimes, the interpretation of meteoritically derived timescales, and the identification of possibly relevant interfaces (such as the accretion shock, condensation fronts, and wind-disk boundaries). Finally, they can be used to examine the consequences for Solar System material of astronomically observed episodic behavior. These matters are the subject of this Chapter.

NEBULA EVOLUTION

Thermal structure

Multi-wavelength observations of T Tauri stars have confirmed that many of them are associated with circumstellar material in the form of a disk of gas and dust with global properties not unlike those of the putative solar nebula (e.g., Beckwith and Sargent

1993, Strom *et al.* 1993). Information regarding the disk masses, temperatures, and duration can be derived from these observations. Although all such information is subject to both observational and interpretive uncertainties, the view presented here is intended to be noncontroversial; we adopt only the broadest implications, and attempt to avoid fragile conclusions, in order to outline the most general aspects of nebula evolution.

Disk masses are usually estimated from millimeter wavelength observations of (optically thin) thermal emission from fine dust. Lower limits are obtained if observations indicate that the source is optically thick at infrared wavelengths, as is frequently the case. Difficulties in quantitative interpretations are caused by surrounding (non-disk) material, uncertain optical properties associated with grain composition and growth, the unresolved spatial distribution of optically thick dust, and possible deviations from the cosmic dust/gas ratio (which must be considered to determine total disk mass). For our purposes, the most important conclusion is that disks around T Tauri stars commonly have masses sufficient to make planets like those in the Solar System, but seldom exceed that of the star itself (Beckwith *et al.*, 1990). The latter limit has been inferred even for the embedded stage (Terebey *et al.*, 1993), during which time the disk may be fed material from the collapsing cloud. Thus there is an indication that during star formation material is fed through the disk to the star efficiently enough so that the disk does not acquire most of the mass, as it would if mass transfer were inefficient. Such a conclusion has important implications for thermal structure, described next.

Models of the internal temperature structure of a protostellar disk can be calculated starting with the effective (i.e., photospheric) temperature distribution, $T_e(r)$. Then, if the disk radiation is due primarily to the local release of gravitational energy associated with accretion, the internal temperatures can be calculated for various assumptions regarding the vertical distributions of opacity and dissipation within the disk. Standard radiative transfer theory can be used because, in most situations of interest, radiation dominates the vertical heat transport, even in the presence of convection (Cassen, 1993). The answers are sensitive to the opacity distribution, but insensitive to the vertical distribution of dissipation unless it is extremely localized.

How is $T_e(r)$ specified? One might rely on accretion disk theory (Lynden-Bell and Pringle 1974), which predicts that $T_e \propto r^{-3/4}$ for uniform accretion through the disk. The constant of proportionality is determined in terms of the disk luminosity by

$$L_d = \int_{r_*}^{\infty} 4\pi r \sigma T_e^4 \, dr \qquad (1)$$

(where r_* is the stellar radius and σ is the Stefan-Boltzmann constant), and in terms of the mass accretion rate through the disk by the relation (derived from conservation of energy)

$$L_d = \frac{GM_*}{2r_*}\frac{dM_*}{dt} \qquad (2)$$

Here, M_* is the stellar mass, dM_*/dt is the rate of mass transfer from the disk to the star, and G is the gravitational constant. The result is

$$T_e(r) = \left(\frac{GM_*}{8\pi\sigma r^3}\frac{dM_*}{dt}\right)^{1/4} \qquad (3)$$

On the other hand, $T_e(r)$ can be inferred from the spectral energy distribution of emitted radiation. For an optically thick disk with $T_e \propto r^{-q}$, the power per unit wavelength F_λ is given by $\lambda F_\lambda \propto \lambda^{-n}$, where λ is wavelength and $n = 4 - 2/q$. As noted, the prediction of accretion disk theory is that $q = 3/4$; coincidentally, this is the same value produced by the absorption and re-emission of stellar radiation by an optically thick (but spatially thin) disk (Safronov, 1969). But several factors may modify the value of q, including unsteady accretion, flared disks, substantial radial transport of energy within the disk and the presence of residual or infalling circumstellar material. The observed values of n (in the infrared part of the spectrum, where disk emission occurs) for many T Tauri stars are between 4/3 and 0 (Beckwith *et al.*, 1990), which corresponds to $3/4 \geq q \geq 1/2$. In the following, we adopt $q = 3/4$, which yields a lower limit to the temperatures in the terrestrial planet region.

It remains to determine the internal temperature distribution, and for this the opacity (or optical depth) must be specified. Cassen (1994) has given approximate formulas for calculating internal temperatures (from T_e) for situations in which dust coagulation has not yet occurred (and under the assumption that dissipation is proportional to the local mass density), using opacities given by Pollack *et al.* (1994). In general, one may represent the nebular midplane temperature T_m by the expression

$$T_m = T_e(\eta\tau)^{1/4}, \qquad (4)$$

where τ is the effective optical depth to the midplane and η is a number that depends on the vertical distributions of opacity and dissipation, but is of order unity in normal situations. From relations (3) and (4), one may show that T_m is expected to exceed 1350 K, the temperature above which most silicates would evaporate, whenever

$$\tau\frac{dM_*}{dt} \geq 5 \times 10^{-3} \, r_{au}^3 \,(\text{solar masses / year}) \qquad (5)$$

(for a solar mass star). The condition is conservative in the sense that other energy sources, such as stellar illumination and compressional heating during collapse (Boss 1993), which have not been accounted for in (5), would make the disk hotter (as would $q < 3/4$).

Estimates of the accretion rate dM_*/dt have been made for T Tauri stars, based on either their infrared or ultraviolet excess radiation, or both. For instance, Basri and Bertout (1989) derived values between 10^{-6} and 10^{-9} solar masses/year for several T Tauri stars, based primarily on the excess ultraviolet radiation produced near the star as mass is transferred from the disk. At an earlier stage, during protostellar collapse, dM_*/dt could be even higher. From counts of embedded (and therefore presumably still collapsing) protostars compared to their revealed counterparts (the T Tauri stars), it is deduced that a solar mass is accumulated in, typically, a few times 10^5 years (Kenyon *et al.*, 1994), which agrees with theoretical arguments (Shu, 1977). Recalling that this mass is apparently transferred efficiently through the disk to the star, at least when averaged over time, one deduces values of several times 10^{-6} solar masses/year for dM_*/dt during the collapse stage. To estimate

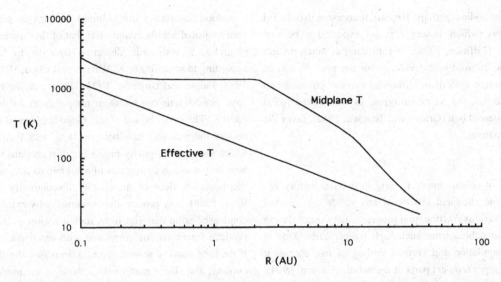

Fig. 1. The photospheric and calculated midplane temperature for a nebula model at an early, hot stage. Parameter values are: $dM_*/dt = 1.8 \times 10^{-6}$ solar masses/year; total mass = 1 solar mass; nebula mass = 0.2 solar masses; total angular momentum = 1.2×10^{53} gm–cm^2/sec. Surface density is assumed to be $\propto r^{-3/2}$.

the optical depth, consider that, at 1 AU in a minimum mass solar nebula with condensed, uncoagulated dust, the appropriate τ (at infrared wavelengths) is several times $10^3 – 10^4$. Thus, it is likely that condition (5) would be satisfied at 1 AU for any such nebula during the collapse stage. Nebula masses as high as 0.1 solar mass might be required to vaporize silicates in the asteroid belt.

Fig. 1 shows the nebula thermal profiles (T_e and T_m) after a total of 1 solar mass has accumulated, 0.2 being retained in the disk, calculated for q = 3/4, and $dM_*/dt = 1.8 \times 10^{-6}$ solar masses/year. Subsequent to the collapse stage, both dM_*/dt and τ must diminish, and the disk consequently cools. Thus, one is led to consider a solar nebula model in which silicates and other rock-forming elements are initially vaporized at the midplane out to a few AU, but exist as fine dust at higher altitudes. Water ice initially exists at the midplane beyond several AU, but also occurs closer to the Sun at high altitudes. As dust coagulates and tends to settle, thereby decreasing the local optical depth, and as the accretion rate through the disk diminishes, isotherms move inward. The precise location and evolution of condensation fronts cannot be predicted, because they depend on such unknown factors as the initial angular momentum of the solar system (on which the local surface density ultimately depends), and the particular history of accretion experienced by the Sun, but the qualitative behavior of global thermal structure is sufficiently constrained to permit more detailed modeling of local nebula processes, such as coagulation, advection and migration. These processes have direct consequences for the nature of chondrule precursor material, and are discussed in the next section.

Another observed aspect of circumstellar disks that is likely to be important for models of the solar nebula is episodic evolution. Ever since Herbig (1977) first pointed out the possibility, it has often been suggested that the episodic energetic events known as the FU Orionis phenomenon (Hartmann, this volume) might have been associated with the formation of chondrules. But, because FU

Orionis outbursts are believed to be associated with rapid disk accretion, it is also acknowledged that preserving any solid record of the event is problematical. Here, we wish to emphasize two points: (1) the effects of outbursts on material in the terrestrial planet region are possibly quite indirect, and (2) the perceived difficulty of preserving solids from the earliest history of the nebula is severe, and exists independently of FU Orionis associated behavior.

With regard to (1), it is difficult to diagnose the state of material in disks at distances of 1 AU and greater from outbursting stars, because the circumstellar environment is generally confused by the existence of obscuring material (Hartmann, this volume). However, models that attempt to account for the observations by disk instabilities (Bell and Lin, 1994) suggest that the disk beyond about 0.5 AU does not participate in the outburst: neither the accretion rate nor the internal temperatures, at those distances, are substantially increased by the dynamical response of the disk. Whether or not those models are correct in detail, they do demonstrate that current observations do not require that the disk be substantially heated beyond the innermost regions (Bell *et al.* 1995). Indirect effects, such as that produced by the increased external radiation field, have yet to be evaluated.

Preservation of solids as discrete, chondrule-sized objects for a million years or more is not generally predicted, even by models of a quiescent nebula. For millimeter-sized particles in a typical model, Weidenschilling (1977a) calculates characteristic lifetimes of order 10^5 years, set by drag-induced, inward radial drift. It might be possible to prolong that time by adjusting nebula properties, but it seems difficult to avoid substantial radial migration for such particles under any circumstances, particularly when one considers that radial gas motions associated with accretion should be added to the inward drift velocity. Preservation for million year intervals in the presence of the nebula seems to require either prompt incorporation into larger bodies (perhaps 10^4 cm), immersion in outward flowing nebula gas (which, however, is not expected to be sustained) or a wind (Liffman, 1992), or formation at much greater distances than the asteroid belt. Evidence for the preservation of apparently primordial, systematic differences in the physical and chemical characteristics of meteorite types, as well as the radial structure of the asteroid belt (Gradie and Tedesco, 1982), favor the first of these alternatives.

Fractionation

The vaporization of silicate material early in nebular history has implications for the chemical history of any solids that survive from that epoch. Various fractionation processes may occur during the cooling of the nebula from such high temperatures. One is caused by the coagulation and vertical settling of fine dust as it condenses in the upper (cooler) parts of the nebula (Cassen, 1994), and another is associated with the volatility-selective accumulation of material near the midplane. Both are potentially important for the composition of chondrule precursor material (Wasson and Chou, 1974; Cassen 1995).

Consider a time, perhaps shortly after nebula formation was complete, when rock-forming elements in the terrestrial planet region were vaporized at the midplane but condensed at the cooler altitudes near the surfaces of the nebula, as shown in Fig. 1. The temperature at the surface, or photosphere, of the nebula is that required to radiate the dissipated energy; the temperature at the bottom of the dusty layer is the vaporization temperature of the opacity-producing dust. Coagulation of the condensed dust would tend to lower the optical depth of the upper layer and promote cooling. But, if internal dissipation remains constant (or changes more slowly than particle growth), the bounding temperatures of the layer of coagulating dust remain fixed during coagulation. Because these two temperatures are related by the optical depth of the layer, cooling must proceed by the lowering of the condensation front in a manner that preserves the optical depth, and therefore the column density of fine dust, of the dusty layer. If the *total* column density was also preserved (negligible net influx due to radial transport), then the column density of evaporated dust in the lower layer would also remain constant, thereby producing an enrichment of heavy elements in the vapor phase as that layer thins with the descent of the condensation front to the midplane (Cassen, 1994). Note that the preferential accumulation of refractory material is favored in the lower layer.

The significance of this fractionation process for chondrule precursor material has yet to be quantitatively evaluated. As described above, it does not necessarily result in elemental fractionations in meteorites, but affects the chemical conditions under which grains condense and heterogeneous reactions take place, and possibly the production of CAI-like objects (Skinner, 1994). Predictive models must include the effects of both radial mass transport and the gradual diminishment of dissipated energy, as well as the cosmochemical consequences of non-solar gas phase compositions.

Chondritic meteorites are characterized by several elemental fractionation patterns, apparently established prior to chondrule formation (Grossman, this volume). Perhaps the pattern most relevant to global nebula evolution is that of the "moderately volatile" elements; a systematic depletion (relative to CI abundances) according to volatility (e.g., Wasson and Chou, 1974; Palme *et al.*, 1988; Palme and Boynton, 1993). These are the elements which have equilibrium condensation temperatures in the range 650 – 1350 K. The upper bound of this range has special significance for nebula thermal structure because it is near that temperature at which most of the opacity-producing dust condenses, and therefore near that at which a large part of a hot nebula may reside due to the thermostatic effect of an opacity discontinuity (Morfill, 1985; Boss, 1988). Any process that systematically extracts sequentially condensed solid material from such a region as it cools, tends to produce fractionations correlated with condensation temperature. If the total mass of source material decreases simultaneously with cooling, abundance patterns like those of the moderately volatile elements (in which all elements are present, but with modest systematic depletions) are produced. Quantitative aspects of the depletions depend on the detailed histories of cooling and mass transport, not only locally, but throughout the nebula (because the depletion history of a particular parcel of nebula determines the amount of condensibles locally available). Preliminary calculations that attempt to address these issues (Cassen 1995) suggest that the patterns observed in carbonaceous meteorites can be accounted for by models of nebula evolution that are consistent with the distribution of mass in the present solar system and the properties inferred for solar nebula-like disks around T Tauri stars. The parameters that govern the models include the characteristic timescale for coagulation, the efficiency of accumulation (what fraction of condensible material winds up in planetary bodies and meteorites), the rate of accretion from the nebula to the Sun, and the total angular momentum of the primordial Solar System. By calculating the abundance patterns predicted for the various meteorite classes, it may be possible to simultaneously constrain these nebula evolution parameters and account for the bulk elemental compositions of chondrule precursor material.

LOCAL PROCESSES

General quantitative constraints

Global nebula evolution models have little predictive power regarding the chondrule formation process itself. Practically speaking, they cannot resolve events that last less than an orbital period (3.16×10^7 seconds at 1 AU), a time disconcertingly longer than the accepted range of times for chondrule crystallization ($10^2 – 10^4$ seconds). It should also be realized that their predictions regarding the environmental parameters (e.g., pressure and density) of the chondrule-forming epoch can usually be derived directly from the underlying assumptions made about the total mass and angular momentum of the primordial solar system. For instance, if one assumes that the terrestrial planets formed from rocky material that was distributed smoothly throughout their present locations, one derives an estimate for the total surface density at 1 AU of about 3×10^3 gm/cm^2, and perhaps 1×10^3 gm/cm^2 in the asteroid zone

(Weidenschilling, 1977b). The volume density $\rho = \Sigma/2h$, where the scale height h depends on the square root of the temperature. If the latter was 650 K, (the maximum consistent with the presence of sulfur in precursor material) the scale height at (3 AU) was about 0.3 AU, and $\rho = 1 \times 10^{-10}$ gm/cm^3. Presumably this value decreased with the decay of the nebula. On the other hand, neither theory nor observation excludes nebula masses ten or twenty times the minimum required to form the planets, or temperatures at 3 AU as low as 200 K, so ρ might be as high as 4×10^{-9} gm/cm^3. The corresponding range of pressures is 2×10^{-6} to 3×10^{-5} bars. These rough estimates of the range of early nebula base states in the asteroid region are probably no worse than any derived from current theoretical models of nebula evolution, since the details of the models are still subject to the same kinds of assumptions regarding the mass and extent of the nebula, as well as uncertainties associated with unconfirmed theories of angular momentum transport. Estimates for other parts of the nebula can be obtained by similar considerations, but depend on such factors as the compositions of the gas giant planets.

Physical models of local, chondrule-forming processes are apparently constrained by the following quantitative conditions: (1) solids at a base state of 650 K or less were heated to temperatures in the range 1850 K to 2050 K and melted, which required an energy of about 1×10^{10} ergs/gm; (2) cooling through crystallization occurred at rates between 5 and 3,000 K/hour, with values for the best constrained cases (porphyritic olivines) between 100 and 1,000 K/hour (Lofgren, this volume); (3) the vast majority of surviving drops are restricted to the diameter range 0.2 to 1.0 mm. (Note that chondrules are somewhat smaller than the gas mean free paths inferred from the range of densities given above: 1.6 – 60 cm.) Implications of some of these conditions are briefly discussed, in the context of specific nebular environments, below.

Shocks

Condition (1), above, has immediate consequences for all hypotheses that rely on the kinetic energy of relative motion as a heat source (these include gasdynamic acceleration of solids, collisions among solid bodies and ablation): the pertinent relative velocitiy must be at least $V_{min} = 1.4 \times 10^5$ cm/sec, comparable to the nebular sound speed [$= 1.2 \times 10^5 \times (T/300)^{1/2}$ cm/sec]. The most plausible way for discrete, chondrule-sized objects to achieve high velocities relative to gas is by encountering shock waves.

What might produce shocks? Candidates include: (1) the accretion shock (Wood, 1984; Ruzmaikina and Ip, this volume) and (2) other shocks generated by "clumpy" infalling material (Boss and Graham, 1993), (3) shocked spiral density waves, (4) lightning, (5) explosive collisions, and (6) bow shocks of planetesimals on eccentric orbits. Candidates (1–3) are attractive because they would have been pervasive; (2–4) would have been repetitive. Only (1) is inherently predicted in current nebula models. All are either subject to some criticism as chondrule producers, or face serious unresolved issues. The accretion shock requires presolar precursors; "clumpy accretion" demands a continuous production of density discontinuities in infalling gas, strong enough to deliver

shocks into the dense nebula; it has not been shown that self-excited density waves shock, and driven density waves in the solar system are not likely to be strong, except perhaps very close to Jupiter; lightning requires charge separation and the development of a large breakdown potential in spite of the presence of mobile charges; and powerful collisions and strong bow shocks require the maintenance of high planetesimal eccentricities in spite of gas drag. The viability of most of these sources has yet to be critically evaluated by a complete model. Nevertheless, the physics of shocks is understood well enough to evaluate their potential without reference to a specific source (e.g., Hood and Horanyi, 1993), and the following estimates illustrate the conditions under which they might have produced chondrules.

For the purpose of examining the effects of fundamental parameters, first consider a shock in an optically thin medium. The flux of kinetic energy intercepted by a particle must at least balance the radiative losses at the melting temperature:

$$\frac{1}{2}\rho_s V^3 \pi a^2 > 4\pi a^2 \sigma T_{melt}{}^4.$$

Here, ρ_s is the volume density of shocked gas, V is the component of shock velocity normal to the shock front, a is the particle radius and T_{melt} is the chondrule temperature. For a fast, isothermal shock, the density ρ_s is related to the pre-shock density ρ_o (approximately) by

$$\rho_s = \rho_o \frac{2\gamma}{\gamma+1} M_o{}^2,$$

where $M_o = V/C_s$ is the shock Mach number, C_s is the sound speed in the pre-shock gas, and γ is the ratio of specific heats. Thus

$$\rho_o V^5 > 4\frac{\gamma+1}{\gamma} C_s^{-2} \sigma T_{melt}{}^4,$$

along with $V > V_{min}$, define a permitted region of ρ_o, V space, shown in Fig. 2. Cooling in an optically thin medium would be rapid enough so that crystalization was regulated by the duration of the heating event itself, in which case the crysrtalization time was approximately equal to the stopping time (the time for a particle to encounter its own mass of gas):

$$t = \frac{4a\rho_p}{3\rho_s V},$$

where ρ_p is the particle density. Thus lines of constant t are given by

$$\rho_o V^3 = \frac{4a\rho_p}{3t} \frac{\gamma+1}{2\gamma} C_s^2,$$

and these are also shown in Fig. 2, for $t = 10^2$ and 10^4 sec. One may conclude from such a plot that high shock velocities (> 10 km/sec) and low densities are necessary if chondrules were made by shock heating in optically thin regions, in agreement with the more rigorous calculations of Ruzmaikina and Ip (1994; this volume).

Lower velocities would suffice if precursors were not free to radiate to a cold environment while they were being melted (Hood and Horanyi, 1993). The heating takes place downstream of the shock, mainly within a layer of thickness equal to the stopping distance Vt, and therefore within Vt of the cold, upstream gas. Thus, a necessary condition for the retardation of radiative losses is that the optical

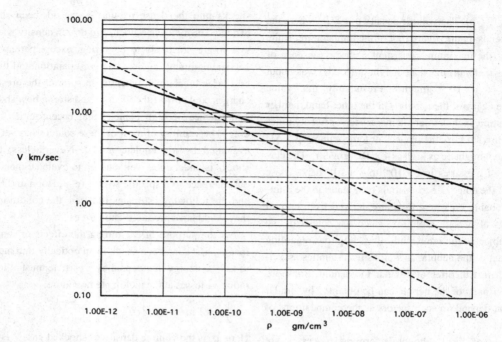

Fig. 2. Conditions on shock velocity and nebula density for melting chondrules in optically thin environments from a base temperature of 650 K. Shock velocities must be above the heavy solid line; the upper and lower dashed lines give the velocity and density corresponding to heating durations of 10^4 and 10^2 sec, respectively. The horizontal dotted line is the sound speed, 1.7 km/sec.

depth of this layer be unity or greater. If chondrules themselves supply the opacity, the optical depth is $3\Sigma_{ch}/4a\rho_p = \rho_{ch}/\rho_s$, where Σ_{ch} is the surface density of chondrules within (Vt) of the shock and ρ_{ch} is the volume density of chondrules. For an adiabatic shock, $\rho_s = \rho_o(\gamma+1)/(\gamma-1) = 6\rho_o$ (for $\gamma = 7/5$), so for optical depth unity:

$$\rho_{ch} = 6\rho_o.$$

That is, the mass density of solids must exceed that of the gas. (See the more detailed treatment of Hood and Horanyi, 1993.) A small amount of fine dust can greatly increase the opacity and therefore reduce the required mass of solids, but fine particles would be rapidly accelerated to velocities that would tend to destroy chondrule precursors.

It appears that shock production of chondrules therefore requires either high velocities and rather low densities, or regions heavily loaded with chondrule precursors. Challenges remain for the shock heating model. The principal ones are rigorous treatments of the dust component and cooling process, and, of course, identification of the shock source.

Condensation fronts

The condensation of major components (silicates and ice) takes place over vast regions of the inner solar system, and possibly over long periods of time. However, global models of condensation zones reveal no obvious conditions for chondrule production, the main problems being accommodation of the short timescale for chondrule cooling and the apparent base temperature of 650 K or less. The fronts are expected to be locations where the kind of free energy associated with weather is available: latent heat, the separation and release of electrical charge (Morfill *et al.*, 1993) and small

scale changes in the radiative environment (clouds) all might contribute to generally stormy conditions. But it is not obvious that impulsive energetic events, strong temperature discontinuities or supersonic transport would actually occur.

Dusty midplane

Theoretical analyses of dust coagulation and vertical settling in the nebula indicate that the timescale for these processes might be as little as 10^4 years in the terrestrial planet region (Weidenschilling and Cuzzi, 1993, and references therein). Therefore it has been supposed that a solid-particle-enriched layer, with a dusty component maintained by collisions and erosion, formed at the nebular midplane. Such a layer is frequently regarded as a favorable place for chondrule formation. A heavy dust (and ice) loading could produce non-solar composition oxidation states during energetic events that vaporized dust, prolonged cooling of molten droplets in the optically thick environment, and the opportunity for rapid accumulation of chondrules into meteorites. It has been argued that such an environment could support powerful lightning discharges, provided the dust-laden medium possessed low electrical conductivity, and a correspondingly large breakdown potential (Wasson and Rasmussen, 1994).

But gravity near the midplane is extremely weak. Cuzzi *et al.* (this volume; see also Weidenschilling, 1980) have argued that even the slightest turbulent motions would prevent a strong midplane concentration of fine dust and/or particles as small as chondrules. In fact, Cuzzi *et al.* propose that chondrule formation, followed by a size-selective process that collected chondrules into larger, loosely bound chondritic clusters, *preceded* settling and the formation of a solids-enriched midplane. It is argued, and supported by numerical simulations, that chondrule-sized objects were preferentially concentrated into clumps with sizes comparable to the smallest turbulent eddies. In this scheme, the apparent abundance of chondrules is due to this selection process; dust that was not turned into chondrules was collected only secondarily by chondrules themselves (in the form of rims), and by bodies that had already been built of chondrules. The dependences of this conclusion on nebula parameters remains to be explored.

Extremely low density environments: above the nebula

Solar-generated energetic particles and radiation penetrate very little matter, so mechanisms that invoke these energy sources, and still assume the presence of the nebula, are restricted to reside outside it. Ultraviolet radiation is effectively absorbed by 10^{-2} gm/cm^2 of gas containing a cosmic abundance of dust in 1 micron grains; solar flare protons penetrate no more than about 1 gm/cm^2 (Woolum and Hohenberg, 1993), or less than 10^{-5} to 10^{-3} of the nebular column densities estimated earlier. Energetic particles might be generated in flares associated with the annihilation of nebular magnetic fields, but, for plausible field strengths, flare energy can be released at a rate sufficient to melt chondrules only where the gas density is very low, less than 10^{-18} gm/cm^3 (Morfill *et al.*, 1993). The transport of solids to such regions in sufficient quantity to make chondrules is problematical, but once produced, the chondrules would rapidly settle into the nebula.

Protosolar wind

Liffman (1992) has suggested that chondrules might be ablated droplets stripped from solid bodies that find themselves immersed in a nebular wind close to the Sun, and other suggestions that implicate a protosolar wind, or wind-disk interface, in chondrule formation have been offered (e.g., Skinner, 1990; Cameron, 1995). Some of these mechanisms would involve the transport of the products over great distances before they were incorporated into meteorites, a proposition contrary to interpretations that favor the prompt assembly of chondrites (e.g., Wood, 1985), but perhaps not impossible, in view of the coexistence of apparently disparate components (e.g., CAIs and chondrules) in meteorites. Evaluation of these hypotheses in the context of nebula models will require progress in two areas: (1) Models of the source region of the wind, which observations indicate is the innermost part of an active accretion disk; (Edwards *et al.* 1993; Konigl and Ruden, 1993; see also Shu *et al.*, 1994, for a new model). (2) Analysis of the nature of the wind-nebula interaction, how it changes as both components evolve, and the role of the wind in removing the nebula.

CONCLUSIONS

The theory of nebula evolution, in spite of being vastly incomplete, places constraints on the range of physical states that might have been experienced by surviving Solar System material. Thermal history, in particular, is usefully constrained by theoretical considerations, combined with inferences from observations of T Tauri stars. Models support the idea that chondrule precursor material was

fractionated by various processes now reflected in the compositions of chondritic meteorites. The fractionations occurred early, during a hot epoch of nebular history, when accretion rates and optical depths were high. Although nebula models are useful for delineating the ranges and evolutions of ambient physical states, they have yet to offer definitive boundary conditions regarding the many proposals for chondrule formation. Because physical conditions in the nebula varied dramatically (as a function of altitude, for example, even at the same radius and time), and because the formation time of chondrules is only loosely constrained on nebula evolution timescales, even hypotheses that seem to require extreme conditions cannot necessarily be rejected. Modeling of local phenomena, such as that being pursued with regard to shocks (Hood and Horanyi, 1993; Ruzmaikina and Ip, 1994), lightning (Morfill *et al.*, 1993; Horanyi and Robertson, this volume) and gas-solid turbulent interaction (Cuzzi *et al.*, 1993) is needed if a compelling theory of chondrule formation is to emerge. It is worth keeping in mind that some of the most salient observational characteristics of chondrules – their once molten state, retention of volatiles, evidence of non-solar composition oxidation conditions, sedimentary-like physical aspects, accumulation of dusty rims, and possibly rapid agglomeration – evoke conditions that one would not immediately associate with the tenuous nebula. Consideration of other much denser, perhaps transient, "atmospheres", planetary or collisional, should not be rejected without further analysis.

ACKNOWLEDGEMENTS

Thanks to Jeff Cuzzi, Lee Hartmann and Tamara Ruzmaikina for constructive reviews. This work was supported by NASA's Origins of Solar Systems Program and the Astrophysics Theory Program, which supports the Center for Star Formation Studies at NASA-Ames Research Center, University of California at Berkeley and University of California at Santa Cruz.

REFERENCES

Basri G. and Bertout C. (1989) Accretion disks around T Tauri stars. II. Balmer emission. *Astrophys. J.* **341**, 340–358.

Beckwith S. V. W. and Sargent A. I. (1993) The occurrence and properties of disks around young stars. In *Protostars and Planets III* (eds. E. H. Levy and J. I. Lunine), pp. 521–541. University of Arizona Press.

Beckwith S., Sargent A. I., Chini R. S. and Güsten R. (1990) A survey for circumstellar disks around young stellar objects. *Astron. J.* **99**, 924–945.

Bell K. R. and Lin D. N. C. (1994) Using FU Orionis outbursts to constrain self-regulated protostellar disks. *Astrophys. J.* **427**, 987–1004.

Bell K. R., Lin D. N. C. Hartmann L. and Kenyon S. J. (1995) The FU Orionis outburst as a thermal accretion event: observational constraints from protostellar disk models. *Astrophys. J.* (in press).

Boss A. P. (1988) High temperatures in the early solar nebula. *Science* **241**, 565–567.

Boss A. P. (1993) Evolution of the solar nebula. II. Thermal structure during nebula formation. *Astrophys. J.* **417**, 351–367.

Boss A. P. and Graham J. A. (1993) Clumpy disk accretion and chondrule formation. *Icarus* **106**, 168–178.

Cameron A. G. W. (1995) Leonard Award Address – The first ten million years in the solar nebula. *Meteoritics* **30**, 133–161.

Cassen P. (1993) Why convective heat transport in the solar nebula was inefficient. *LPSC XXIV* 261–262.

Cassen P. (1994) Utilitarian models of the solar nebula. *Icarus* **112**, 405–430.

Cassen, P. (1995) On the abundances of moderately volatile elements in meteorites. *LPSC XXVI* (in press).

Cuzzi J. N., Dobrovolskis A. R. and Champney J. M. (1993) Particle-gas dynamics in the midplane of a protoplanetary nebula. *Icarus* **106**, 102–134.

Edwards S., Ray T. and R. Mundt (1993) Energetic mass outflows from young stars. In *Protostars and Planets III* (eds. E. H. Levy and J. Lunine), pp. 567–602. University of Arizona Press.

Gradie J. and Tedesco E. F. (1982) Compositional structure of the asteroid belt. *Science*, **216**, 1405–1407.

Herbig G. (1977) Eruptive phenomena in early stellar evolution. *Astrophys. J.* **217**, 693–715.

Hood L. L. and Horanyi M. (1993) The nebular shock wave model for chondrule formation: one-dimensional calculations. *Icarus* **106**, 179–189.

Kenyon S. J., Gomez M., Marzke R. O. and Hartmann L. (1994) New pre-main sequence stars in the Taurus-Auriga molecular cloud. *Astron. J.* **108**, 251–261.

Königl A. and Ruden S. P. (1993) Origin of outflows and winds. In *Protostars and Planets III* (eds. E. H. Levy and J. Lunine), pp. 641–687. University of Arizona Press.

Liffman K. (1992) The formation of chondrules via ablation. *Icarus*, **100**, 608–620.

Lynden-Bell D. and Pringle J. E. (1974) The evolution of viscous discs and the origin of the nebular variables. *MNRAS* **168**, 603–637.

Morfill G. E. (1985) Physics and chemistry in the primitive solar nebula. In *Birth and Infancy of Stars*, (R. Lucas and A. Omont, eds.), pp. 693–794. North-Holland.

Morfill G., Spruit H. and Levy E. H. (1993) Physical processes and conditions associated with the formation of protoplanetary disks. In *Protostars and Planets III* (eds. E. H. Levy and J. Lunine), pp. 939–978. University of Arizona Press.

Palme H. and Boynton W. V. (1993) Meteoritic constraints on conditions in the solar nebula. In *Protostars and Planets III* (eds. E. H. Levy and J. Lunine), pp. 979–1004. University of Arizona Press.

Palme H., Larimer J. S., and Lipschutz M. E. (1988) Moderately volatile elements. In *Meteorites and the Early Solar System* (eds. J. F. Kerridge and M. S. Matthews), pp. 436–458. University of Arizona Press.

Pollack J. B., Hollenbach D., Simonelli D., Beckwith S., Roush T., and Fong W. (1994) Optical properties of grains in molecular clouds and accretion disks. *Astrophys. J.* **421**, 615–639.

Ruzmaikina T. V. and Ip W. H. (1994) Chondrule formation in radiative shock. *Icarus* **112**, 430–447.

Safronov V. S. (1969) Evolution of the Protoplanetary Cloud and Formation of the Planets. Translated (1972) by the Israel Program for Scientific Translations as NASA TT–F–677 (pp 36–41).

Shu F. H., Najita J., Ostriker E., Wilkin F., Ruden S. and Lizano S. (1994) Magnetocentrifugally driven flows from young stars and disks. I. A generalized model. *Astrophys. J.* **429**, 781–796.

Shu, F. H. (1977) Self-similar collapse of isothermal spheres and star formation. *Astrophys. J.* **214**, 448–497.

Skinner W. R. (1990) Bipolar outflows and a new model for the early Solar System. Part II: the origins of chondrules, isotopic anomalies, and chemical fractionations. *LPSC XXI*, 1168–1169.

Skinner W. R. (1994) Pre-Allende planetesimals with refractory compositions: the CAI connection. *LPSC XXV* 1283–1284.

Strom S. E., Edwards, S. and Skrutskie M. F. (1993) Evolutionary timescales for circumstellar disks associated with intermediate- and solar-type stars. In *Protostars and Planets III* (eds. E. H. Levy and J. I. Lunine), pp. 837–866. University of Arizona Press.

Taylor G. J., Scott, E. R. D. and Keil, K. (1983) Cosmic setting for chondrule formation. In *Chondrules and Their Origins* (ed. E. A. King), pp. 262–278. Lunar and Planetary Institute.

Terebey S., Chandler C. J. and André P. (1993) The contribution of disks and envelopes to the millimeter continuum emission from very young low-mass stars. *Astrophys. J.* **414**, 759–772.

Wasson J. T. and Rasmussen K. L. (1994) The fine nebula dust component: a key to chondrule formation by lightning. *Papers presented to Chondrules and the Protoplanetary Disk*, LPI Contribution No. 844, p. 43, Lunar and Planetary Institute.

Wasson J. Y. and Chou C.-L. (1974) Fractionation of moderately volatile elements in ordinary chondrites. *Meteoritics* **9**, 69–84.

Weidenschilling S. J. (1977a) Aerodynamics of solid bodies in the solar nebula. *MNRAS* **180**, 57–70.

Weidenschilling S. J. (1977b) The distribution of mass in the planetary system and solar nebula. *Astrophys. and Space Sci.* **51**, 153–158.

Weidenschilling S. J. (1980) Dust to planetesimals: settling and coagulation in the solar nebula. *Icarus* **44**, 172–189.

Weidenschilling S. J. and Cuzzi J. N. (1993) Formation of planetesimals in the solar nebula. In *Protostars and Planets III* (eds. E. H. Levy and J. Lunine), pp. 1031–1088. University of Arizona Press.

Wood J. A. (1984) On the formation of meteoritic chondrules by aerodynamic drag heating in the solar nebula. *Earth Planet. Sci. Lett.* **70**, 11–26.

Wood J. A. (1985) Meteoritic constraints on processes in the solar nebula. In *Protostars and Planets II* (eds. D. C. Black and M. S. Matthews), pp. 687–702. University of Arizona Press.

Woolum D. S. and Hohenberg C. (1993) Energetic particle environment in the early Solar System: extremely long pre-compaction ages or an enhanced early particle flux. In *Protostars and Planets III* (eds. E. H. Levy and J. Lunine), pp. 903–919. University of Arizona Press.

4: Large Scale Processes in the Solar Nebula

ALAN P. BOSS

DTM, Carnegie Institution of Washington, 5241 Broad Branch Road, N.W. Washington, DC 20015–1305, U.S.A.

ABSTRACT

Most proposed chondrule formation mechanisms involve processes occurring inside the solar nebula, so the large scale (roughly 1 to 10 AU) structure of the nebula is of general interest for any chondrule-forming mechanism. Chondrules and Ca,Al-rich inclusions (CAIs) might also have been formed as a direct result of the large scale structure of the nebula, such as passage of material through high temperature regions. While recent nebula models do predict the existence of relatively hot regions, the maximum temperatures in the inner planet region may not be high enough to account for chondrule or CAI thermal processing, unless the disk mass is considerably greater than the minimum mass necessary to restore the planets to solar composition. Furthermore, it does not seem to be possible to achieve both rapid heating and rapid cooling of grain assemblages in such a large scale furnace. However, if the accretion flow onto the nebula surface is clumpy, as suggested by observations of variability in young stars, then clump-disk impacts might be energetic enough to launch shock waves which could propagate through the nebula to the midplane, thermally processing any grain aggregates they encounter, and leaving behind a trail of chondrules.

INTRODUCTION

Primitive meteorite components such as chondrules and calcium, aluminum-rich inclusions (CAIs) offer the tantalizing prospect of yielding unique insights about the physical and chemical processes occurring in the solar nebula during the earliest phases of planetary formation. In particular, chondrules and CAIs provide strong constraints on the temperatures required for their formation; on the frequency, duration, and sequencing of the heating events; and on the oxidation state and degree of closed system chemistry of the chondrule-forming process.

Chondrules and CAIs have spheroidal shapes indicative of having experienced a molten phase (Wood, 1988). Melting temperatures for chondrule components require heating the precursor aggregates to temperatures in the range of 1500 K to 2100 K, and to as high as 2400 K for one CAI (Simon *et al.*, 1994). However, this heating phase could not have lasted for more than about an hour, because chondrule compositions are not in equilibrium with nebular gas at such high temperatures (Wood, 1988), and because partial melting (or collisions with unmolten seed grains) may be necessary to explain most chondrule textures (Connolly and Hewins, this volume). The granular and porphyritic textures of chondrules also require cooling rates of about 1000 K per hour

(Hewins, 1988). Cooling rates for CAIs with spheroidal shapes and certain compositions (Type B) have been shown to be somewhat slower, about 0.1 K to 10 K per hour (Stolper and Paque, 1986). Compared to global nebula evolutionary time scales of 10^5 to 10^6 years, these timescales for heating and cooling of chondrules and CAIs are so short as to suggest that a localized heating event must have been responsible. Transient heating of chondrules is also required by the absence of isotopic fractionation of potassium in chondritic meteorites (Humayun and Clayton, 1995); any significant evaporative loss of potassium during a prolonged heating event would have produced measurable isotopic fractionation, because of preferential evaporation of the lighter isotope.

While CAIs are not uncommon, chondrules are quite common in most primitive meteorites (Grossman *et al.*, 1988), implying that chondrule and CAI formation processes could not have been rare events. The high temperature rims found on many chondrules (Kring, 1991) imply that chondrule formation must have been episodic as well. CAIs and chondrules are usually found separately in carbonaceous chondrites, implying independent origins, but the discovery of a relict CAI inside a chondrule implies that at least some CAIs were formed prior to chondrules (Misawa and Fujita, 1994).

Chondrules are chemically complementary (in Fe and Cr) to

dark inclusions and matrix, implying formation in a closed system (Palme *et al.*, 1992). There is also a need for a range of oxygen partial pressures in order to explain the amount of FeO contained in different types of chondrules (Nagahara *et al.*, 1994), with the low FeO (type IA) chondrules requiring an oxygen partial pressure at least an order of magnitude lower than the high FeO (type II) chondrules. The presence of reduced carbon in the precursor material can also affect the final oxidation state of the chondrule (Connolly *et al.*, 1994). It is unclear if this range of oxygen fugacities requires processing at a variety of heights above the dust-rich nebula midplane, or whether small-scale (<< 1 AU) processes can achieve the desired effects.

Theoretical models of the structure of the solar nebula should be able to provide the physical context in which to help evaluate the efficacy of any mechanism proposed for the formation of chondrules or CAIs. These models often use the equations of radiative hydrodynamics to calculate the large scale structure of the solar nebula (i.e., the temperature, density, pressure, and velocity fields) throughout the planet-forming region. Prior to the realization of the need for both rapid heating and rapid cooling of chondrules, it was suggested that the global structure of the nebula was directly responsible for the heating experienced by chondrules and CAIs. This paper first discusses the status of suggestions that chondrules were thermally processed in a hot inner nebula (Cameron and Fegley, 1982; Morfill, 1983) or in a strongly nonaxisymmetric disk (Boss, 1988). Considering the constraints on heating and cooling rates, it will be shown that a more promising mechanism is nebula shock wave processing (Hood and Horanyi, 1991; 1993), with the shock waves being driven by clumps of gas and dust that have been hypothesized to impact the nebula at quite high velocities (Boss and Graham, 1993). Heating of chondrule precursors by passage through the accretion shock separating the disk from the infalling molecular cloud core is discussed by Ruzmaikina and Ip (this volume).

GLOBALLY HOT NEBULA

Cameron and Fegley (1982) studied condensation in a nebula model with a midplane temperature of 1600 K at 1 AU and a vertical variation in the temperature. They noted that hot gas could condense onto grains by moving either outward by about 0.2 AU, or upward in the nebula by a similar distance, and suggested that chondrules and refractory inclusions like the CAIs could thereby be formed and diffusively transported throughout the nebula. Morfill (1983) suggested that chondrules and CAIs formed in the high temperature regions close to the protosun, and described the diffusive transport process whereby samples of the hot inner nebula might reach the region of the asteroid belt. Assuming the freshly condensed chondrules are carried along by the turbulent motions of the gas, some of the chondrules would be expected to undergo a random walk outward. This random walk must be accomplished against a headwind though, because a nebula containing substantial turbulence is presumably being driven by the turbulent viscous stresses in a such way that in the inner disk, gas and dust flow inward toward the protostar (e.g., Morfill, 1983).

The degree to which dust grains diffuse upstream depends upon the ratio of the eddy diffusivity (D) to the turbulent viscosity (v) (Morfill and Volk, 1984). Stevenson (1990) used a viscous accretion disk model to calculate the fractional "contamination" upstream of a source of "pollution" (here, chondrules), and found that the degree of contamination was much less for D = v than for D = 3v far upstream. The former case is expected when the turbulence is driven by thermal convection, implying that little contamination should occur. Prinn (1990) simultaneously argued in favor of extensive contamination, pointing out that some turbulent eddies will actually subtract from the viscous shear that drives the overall mass influx, thereby reducing the effective viscosity, while still promoting diffusion; in this case D would be much greater than v. Prinn (1990) points out that if the nebula turbulence is driven in part by infalling matter, then the nonaxisymmetric accretion of clumps will create these "negative viscosity" eddies whenever the angular velocity of the infalling clumps exceeds that of the underlying disk. For the initial cloud considered by Cassen and Moosman (1981), for example, the infalling matter arrives at the disk with an angular velocity less than Keplerian. The degree of upstream mixing thus appears to depend on a knowledge of the types of turbulent eddies that occurred in the solar nebula and on the detailed angular velocity profile of infalling matter. Definitive knowledge about either of these phenomena is unlikely to be forthcoming soon, so the degree of upstream mixing retains a large degree of uncertainty. There are even recent calculations implying that convective viscous accretion disks do not transport angular momentum outward as expected (Kley *et al.*, 1993; Ryu and Goodman, 1992).

Outward transport of hot condensates can also occur during the planetary accumulation phase. Wetherill (1994) has shown that the source regions for the terrestrial planets extend from 0.5 AU to at least 2.5 AU, so at least some asteroidal material may have originated within 1 AU. However, these orbital evolution calculations refer to massive planetesimals, not chondrule-sized objects, and the outward flux is not sufficient to explain the ubiquity of chondrules in primitive meteorites. These calculations also probably apply to the phase when nearly all of the nebula has been removed.

If upstream mixing is not significant, then the logical possibility remains of attaining high nebula temperatures closer to the region where the chondrules and CAIs presumably formed. Midplane temperatures on the order of 1500 K inside about 1 AU region have been calculated for viscous accretion disk models of the solar nebula (Morfill, 1988), given certain assumptions (e.g., a turbulent viscosity alpha parameter of about 0.01 and a mass accretion rate of 10^{-5} solar masses per year). Temperatures at 2 AU to 3 AU are on the order of 700 K in such a model. Such temperatures are about as high as a viscous accretion disk model can be expected to produce without making more radical assumptions, yet these temperatures fall substantially below those required for chondrule and CAI thermal processing.

One might wonder whether turbulent viscous dissipation is the only likely source of nebula heating, especially since the underlying turbulence is poorly understood, and its very driving mechanism is in some doubt. In comparison, heating caused by gas compression during cloud collapse and contraction is certain to have occurred, is straightforward to calculate, and is routinely included in radiative hydrodynamical models of the formation and evolution of protostellar and protoplanetary disks. The state-of-the-art for radiative hydrodynamical models of the formation of the solar nebula is given by the two dimensional calculations of Yorke *et al.* (1993). However, these models did not solve for the region inside 5 AU, and could not be advanced past the phase where the mass of the disk was still comparable to the mass of the central protostar; they depict a phase much earlier than that where solids accumulated and avoided being swallowed by the general mass influx toward the protostar.

The two dimensional radiative hydrodynamical models of Boss (1993) sidestep the problem of getting most of the disk mass into the central protostar by the artifice of skipping over the earlier phases and assuming the existence of a solar-mass protostar. The calculation consists of starting from an approximate, quasi-equilibrium disk configuration in Keplerian rotation about the protostar, subject to continued infall from the placental cloud envelope, and then searching for the quasi-steady state temperature distribution within the disk. A minimum mass disk (0.02 solar masses inside 10 AU) was studied, appropriate for the planet-forming epoch, while mass accretion rates on the order of 10^{-6} to 10^{-5} solar masses per year were employed, meaning that the models correspond to a putative phase where significant accretion is still occurring, but most of the mass that is going to accrete has already reached the central protostar. Compressional energy and the stellar luminosity are the primary sources of heating in these models as well as those of Yorke *et al.* (1993). The temperature distribution calculated in the Boss (1993) nebula model is consistent with the effective temperatures inferred observationally at 1 AU for disks around T Tauri stars (Beckwith *et al.*, 1990) and is consistent with an analytical calculation of the midplane temperature expected in an accretion disk (Cassen, 1995). The temperature distribution also produces an infrared spectral energy distribution quite similar to that of T Tauri itself (Boss and Yorke, 1993).

Fig. 1 shows the midplane temperature profiles calculated for two different disk masses (Boss, 1995) using the same techniques as Boss (1993). It can be seen that midplane temperatures in the asteroidal region are likely to be only on the order of 1000 K when the nebula mass is about 0.02 solar masses (Boss, 1993), not high enough to melt the chondrule and CAI precursors. However, when the nebula mass is increased to 0.13 solar masses, maximum temperatures of 1700 K extend outward to almost 2 AU. Such a nebula mass is about a factor of two greater than that usually ascribed to

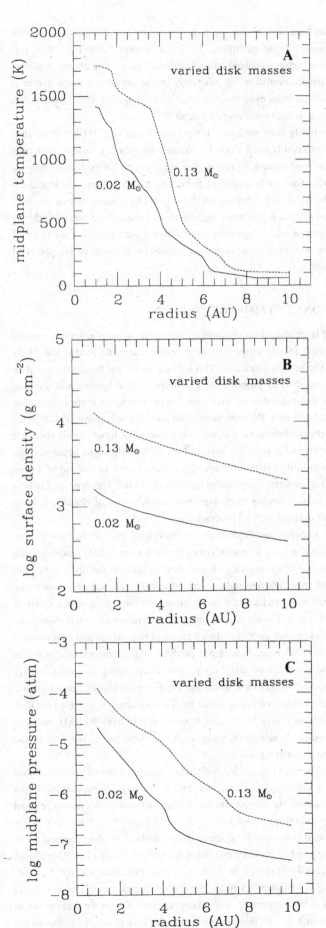

Fig. 1. Midplane temperatures (a) as a function of distance from a central solar-mass protostar for protoplanetary disks with different masses, heated by accretion from the cloud envelope (Boss, 1995). Maximum temperatures in the inner planet region are on the order of 1000 K to 1500 K in a minimum mass (0.02 solar mass) disk, rising to 1700 K only in a relatively massive (0.13 solar mass) disk. Total surface densities (b) and gas pressures (c) are also shown as a function of heliocentric radius.

the "maximum" for the minimum mass nebula (about 0.07 solar masses; Weidenschilling, 1977a). However, Wetherill (1994, private communication) has pointed out that the late phases of planetary accumulation are inherently inefficient in that much matter is ejected from even the terrestrial planet zone, in which case such a large nebula mass might be needed.

Fairly high midplane temperatures (about 1700 K) could thus conceivably exist around 2 AU in a roughly 0.1 solar mass nebula still undergoing accretion. The requirement for both rapid heating and rapid cooling appears to be fatal to any attempt to form chondrules in such a hot region, however. The relative motion required between the precursor aggregate and the nebula gas in order to enter or exit a high temperature region with a size on the order of an AU within one hour is a significant fraction of the speed of light!

NONAXISYMMETRIC NEBULA

If large scale nonaxisymmetry occurs in the nebula, such as a bar or spiral density wave, then the nebula temperature field is also likely to be nonaxisymmetric. This is because the gas temperature will be controlled by processes (e.g., compressional heating and trapping of radiation) that are maximized in the regions of highest density of gas and dust. Because large solid particles will move on Keplerian orbits, whereas the gas will orbit somewhat more slowly (because of its radial pressure support), relative motion will occur and the possibility exists for grain aggregates to orbit in and out of a nonaxisymmetric temperature field (Boss, 1988). The degree of nebula nonaxisymmetry to be expected thus becomes of possible interest for cyclical thermal processing.

Laughlin and Bodenheimer (1994) have studied the three dimensional growth of nonaxisymmetry in a solar nebula model formed during axisymmetric collapse, finding that the nebula is subject to the growth of one and two-armed spiral instabilities that can transport mass and angular momentum on time scales on the order of 10^5 years. The spiral structure results in rather modest density variations in azimuth, less than a factor of two, so the nonaxisymmetry that arises is not expected to produce major azimuthal variations in the temperature distribution. The calculations assume a locally isothermal gas, and so do not predict what the nonaxisymmetric temperature variation would be. The calculations modeled the disk only on a very large scale, from 11 AU to over 200 AU, so these models do not directly constrain the behavior of the disk in the inner planet region.

Tomley *et al.* (1994) studied the two dimensional growth of nonaxisymmetry in a thin disk orbiting around a central star, including in a heuristic manner the interaction between disk cooling (caused by radiative losses) and gravitational instability of the disk. They found that initially unstable disks (without cooling) heated up to the point of gravitational stability. A finite degree of cooling allows the disks to remain unstable, and even to fragment when the cooling is sufficiently strong. However, the disk models of Tomley *et al.* (1994) represent disks extending from the stellar surface out to about 1 AU, so that the entire disk lies well inside the radius of the

Fig. 2. Density contours exhibiting a central bar and trailing spiral arms in a three dimensional model of a protoplanetary disk in the process of formation (Boss, 1989). At this time the central protostar has a mass of 0.2 solar masses and the disk has a mass of 0.9 solar masses. The radius of the region shown is 10 AU; contours correspond to changes in density by factors of two.

present-day asteroid belt, though similar behavior might be expected throughout the disk.

The three dimensional models of solar nebula formation by Boss (1988, 1989) covered the region from 0.5 AU to 40 AU, and by this criterion are better matched to studies of chondrule formation than the models by Laughlin and Bodenheimer (1994) or Tomley *et al.* (1994). However, the Boss (1989) models were not started from appropriate quasi-equilibrium initial states, like those of Laughlin and Bodenheimer (1994) and Tomley *et al.* (1994); instead, the Boss (1989) models depict the formation of the nebula from the collapse of a cloud with an unrealistically high density (chosen to shorten greatly the calculation time). It would be far more preferable to use the axisymmetric disk models of Boss (1993), which extend from 0.5 AU to 10 AU, as quasi-equilibrium initial conditions for three dimensional calculations of the growth of nonaxisymmetry in the planetary region — indeed, such calculations are currently underway by the author. In the interim, we have only the results of Boss (1989) as a guide (Fig. 2) in the 1 AU to 3 AU region. While Boss (1988) found an extreme degree of nonaxisymmetry (i.e., formation of a transient binary system) to occur for certain models, in general the degree of nonaxisymmetry was much milder, especially in models where the disk mass was much less than that of the central protostar, as would be most appropriate for the planet-forming phase of nebula evolution. The models suggest that azimuthal density variations would be on the order of factors of two (Fig. 2) or so (similar to Laughlin and Bodenheimer, 1994), significant for long term angular momentum transport, but again not likely to lead to large azimuthal temperature variations.

Even if future nonaxisymmetric nebula models should reveal the presence of large scale azimuthal temperature gradients, there is still the problem of achieving rapid heating and rapid cooling, which ruled out the globally hot nebula mechanism for chondrule formation. The maximum relative orbital motion between the gas and solid particles is about 10^4 cm/sec, and this relative motion occurs only for fairly large (roughly meter-size) bodies; the relative motion falls to very low values for mm-sized aggregates (Weidenschilling, 1977b). Unless the high temperature region was exceedingly sharply defined in azimuth (e.g., a shock front), chondrule and CAI-sized aggregates would enter and exit any high temperature region much too slowly to experience rapid heating and cooling cycles.

CLUMPY DISK ACCRETION

A hint as to another possible mechanism for heating chondrule precursors has arisen from recent observations of young variable stars (T Tauri stars) undergoing phases of evolution probably experienced long ago by the early sun. Three different types of astronomical observations suggest the presence of clumps of optically thick matter close to young stars (Graham and Phillips, 1987; Graham, 1992). First, irregular variations occur in the brightness of young stars over short time periods (days) without the accompanying spectral changes that would indicate that the variability is caused by phenomena on the stellar surface, such as magnetic flares. Second, changes in the illumination of extended reflection nebulae are also seen, and are apparently caused by shadowing of the central star by opaque material between the star and the reflection nebula. Finally, night-to-night changes are observed in spectral lines that are excited in low density circumstellar regions, indicating rapid changes in this material. These observations are consistent with the presence of circumstellar clumps of gas and dust moving with speeds up to 250 km/sec. Boss and Graham (1993) proposed that the impact of such clumps with a protoplanetary disk could drive shock waves into the nebula.

While direct proof of the existence of such clumps may be extremely difficult to obtain, subsequent observations have clarified many aspects of the causes of T Tauri variability and lent further support to the basic idea of opaque clumps moving close to young stars. Gahm *et al.* (1995) continuously monitored the luminosity and certain spectral lines for a number of young variable stars. They found evidence for rapid changes (over minutes) in ultraviolet luminosity that appear to be caused by flares on the stellar surface, as well as for slower changes (over hours) thought to be caused by inhomogeneous accretion of disk matter onto the stellar surface, and by changes in the dust obscuration toward the star. The latter two suggestions could be consistent with infalling clumps of dusty gas.

Herbst *et al.* (1994) monitored the luminosity of T Tauri stars over periods of days to months, finding evidence for three types of variations. The first is caused by cool (dark) star spots that rotate in and out of view for thousands of stellar rotation periods. The second is caused by hot (bright) star spots attributed to the accretion of matter onto the star; these appear and disappear on much shorter time scales. The third type involves large changes in luminosity without accompanying spectral changes, presumably caused by variable dust obscuration, i.e., clumps. Herbst *et al.* (1994) argue that these dust clumps cannot be confined to a disk, because then such variations would be seen much less often than is observed, and point out that a more likely option is the continuing infall of dust clumps.

These observations of young stars seem to be consistent with the presence of opaque clouds with speeds up to 250 km/sec and masses greater than 10^{22} g. The clump lifetime is likely to be very short (years), implying that a mechanism for their continual generation would have to exist, such as the return of disk or infalling matter previously entrained by the stellar outflow. If such clumps exist, and if their high velocity orbits are inclined to the disk midplane, the clumps will eventually impact and drive shock waves into the nebula (Fig. 3). Shock speeds greater than or equal to about 5 km/sec are required to thermally process chondrule precursor aggregates (Hood and Horanyi, 1991), and this constraint can be used to derive the fraction of the nebula that would be shocked to this speed (Boss and Graham, 1993). Assuming that clump-disk impacts occur on a weekly basis over a time period of a million years, essentially every dust grain aggregate in the nebula could be thermally processed once. However, clump masses on the order of 10^{26} g may be needed in order to process aggregates residing in a dust sub-disk at the nebula midplane.

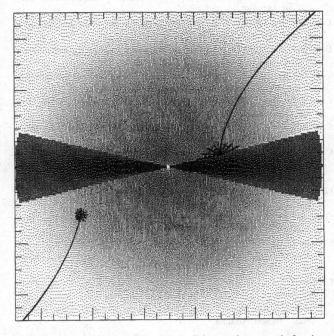

Fig. 3. Schematic diagram of the clumpy disk accretion scenario for chondrule formation (Boss and Graham, 1993). Opaque clumps of accreting gas and dust impact the disk at high velocities, dissipating kinetic energy and driving shock waves into the disk. The shock waves thermally process precursor grain aggregates as they propagate toward the midplane (Hood and Horanyi, 1991; 1993). The region shown is 10 AU in radius; the grey scale represents the cross sectional density in the models of Boss (1995).

CONCLUSIONS

Even if a region exists around 2.5 AU in the solar nebula which is hot enough to melt chondrule and CAI precursor aggregates, the near-impossibility of moving the solids into and out of the large scale hot region on time scales of hours to days seems to rule out globally hot regions as furnaces for thermally processing chondrules and Type B CAIs. The requirement for rapid heating and cooling cycles points instead toward phenomena such as nebula shock waves. Given the occurrence of clumpy accretion of gas and dust onto the protoplanetary disk at the times when chondrules and CAIs were formed, nebular shock waves launched by clump-disk impacts appear to be quite capable of thermally processing precursor aggregates into chondrules.

ACKNOWLEDGEMENTS

This work was partially supported by the NASA Planetary Geology and Geophysics Program under grant NAGW-1410.

REFERENCES

Beckwith S. V. W., Sargent A. I., Chini R. S. and Güsten R. (1990) A survey for circumstellar disks around young stellar objects. *Astron. J.* **99**, 924–945.

Boss A. P. (1988) High temperatures in the early solar nebula. *Science* **241**, 565–567.

Boss A. P. (1989) Evolution of the solar nebula I. Nonaxisymmetric structure during nebula formation. *Astrophys. J.* **345**, 554–571.

Boss A. P. (1993) Evolution of the solar nebula. II. Thermal structure during nebula formation. *Astrophys. J.* **417**, 351–367.

Boss A. P. (1995) Thermal profiles in protoplanetary disks. *Lunar Planet. Sci.* **26**, 151–152.

Boss A. P. and Graham J. A. (1993) Clumpy disk accretion and chondrule formation. *Icarus* **106**, 168–178.

Boss A. P. and Yorke H. W. (1993) An alternative to unseen companions to T Tauri stars. *Astrophys. J.* **411**, L99–L102.

Cameron A. G. W. and Fegley M. B. (1982) Nucleation and condensation in the primitive solar nebula. *Icarus* **52**, 1–13.

Cassen P. (1994) Utilitarian models of the solar nebula. *Icarus* **112**, 405–429.

Cassen P. and Moosman A. (1981) On the formation of protostellar disks. *Icarus* **48**, 353–376.

Connolly H. C., Hewins R. H., Ash R. D., Zanda B., Lofgren G. E. and Bourot-Denise M. (1994) Carbon and the formation of reduced chondrules. *Nature* **371**, 136–139.

Gahm G. F., Lodén K., Gullbring E. and Hartstein D. (1995) Activity on young stars. *Astron. Astrophys.*, in press.

Graham J. A. (1992) Clumpy accretion onto pre-main-sequence stars. *Pub. Astron. Soc. Pacific* **104**, 479–488.

Graham J. A. and Phillips A. C. (1987) Rapid variability of the R Coronae Australis reflection nebula NGC 6729. *Pub. Astron. Soc. Pacific* **99**, 91–98.

Grossman J. N., Rubin A. E., Nagahara H. and King E. A. (1988) Properties of chondrules. In *Meteorites and the Early Solar System* (eds. J. F. Kerridge and M. S. Matthews), pp. 619–659. Univ. Arizona Press, Tucson, Arizona.

Herbst W., Herbst D. K., Grossman E. J. and Weinstein D. (1994) Catalogue of UBVRI photometry of T Tauri stars and analysis of the causes of their variability. *Astron. J.* **108**, 1906–1923.

Hewins R. H. (1988) Experimental studies of chondrules. In *Meteorites and the Early Solar System* (eds. J. F. Kerridge and M. S. Matthews), pp. 660–679. Univ. Arizona Press, Tucson, Arizona.

Hood L. L. and Horanyi M. (1991) Gas dynamic heating of chondrule precursor grains in the solar nebula. *Icarus* **93**, 259–269.

Hood L. L. and Horanyi M. (1993) The nebular shock wave model for chondrule formation: one-dimensional calculations. *Icarus* **106**, 179–189.

Humayun M. and Clayton R. N. (1995) Potassium isotope cosmochemistry: genetic implications of volatile element depletion. *Geochim. Cosmochim. Acta*, in press.

Kley W., Papaloizou, J. C. B. and Lin D. N. C. (1993) On the angular momentum transport associated with convective eddies in accretion disks. *Astrophys. J.* **416**, 679–688.

Kring D. A. (1991) High temperature rims around chondrules in primitive chondrites: evidence for fluctuating conditions in the solar nebula. *Earth Planet. Sci. Lett.* **105**, 65–80.

Laughlin G. and Bodenheimer P. (1994) Nonaxisymmetric evolution in protostellar disks. *Astrophys J.* **436**, 335–354.

Misawa K. and Fujita T. (1994) A relict refractory inclusion in a ferromagnesian chondrule from the Allendre meteorite. *Nature* **368**, 723–726.

Morfill G. (1983) Some cosmochemical consequences of a turbulent protoplanetary cloud. *Icarus* **53**, 41–54.

Morfill G. (1988) Protoplanetary accretion disks with coagulation and evaporation. *Icarus* **75**, 371–379.

Morfill G. E. and Völk H. J. (1984) Transport of dust and vapor and chemical fractionation in the early protosolar cloud. *Astrophys. J.* **287**, 371–395.

Nagahara H., Kushiro I. and Mysen B. O. (1994) Evaporation of olivine: low pressure phase relations of the olivine system and its implications for the origin of chondritic components in the solar nebula. *Geochim. Cosmochim. Acta*, **58**, 1951–1963.

Palme H., Spettel B., Kurat G. and Zinner E. (1992) Origin of Allende chondrules. *Lunar Planet. Sci.* **23**, 1021–1022.

Prinn R. G. (1990) On neglect of nonlinear momentum terms in solar nebula accretion disk models. *Astrophys. J.* **348**, 725–729.

Ryu, D. and Goodman J. (1992) Convective instability in differentially rotating disks. *Astrophys. J.* **388**, 438–450.

Simon S. B., Yoneda S., Grossman L. and Davis A. M. (1994) A $CaAl_4O_7$-bearing refractory spherule from Murchison: Evidence for very high-temperature melting in the solar nebula. *Geochim. Cosmochim. Acta* **58**, 1937–1949.

Stevenson D. J. (1990) Chemical heterogeneity and imperfect mixing in the solar nebula. *Astrophys. J.* **348**, 730–737.

Stolper E. and Paque J. M. (1986) Crystallization sequences of Ca-Al-rich inclusions from Allende: the effects of cooling rate and maximum temperature. *Geochim. Cosmochim. Acta* **50**, 1785–1806.

Tomley L., Steiman-Cameron T. Y. and Cassen P. (1994) Further studies of gravitationally unstable protostellar disks. *Astrophys. J.* **422**, 850–861.

Weidenschilling S. J. (1977a) The distribution of mass in the planetary system and solar nebula. *Astrophys. Space Sci.* **51**, 153–158.

Weidenschilling S. J. (1977b) Aerodynamics of solid bodies in the solar nebula. *Icarus* **180**, 57–70.

Wetherill G. W. (1994) Provenance of the terrestrial planets. *Geochim. Cosmochim. Acta*, **58**, 4513–4520.

Wood J. A. (1988) Chondritic meteorites and the solar nebula. *Ann. Rev. Earth Planet. Sci.* **16**, 53–72.

Yorke H. W., Bodenheimer P. and Laughlin G. (1993) The formation of protostellar disks. I. 1 Solar Mass. *Astrophys. J.* **411**, 274–284.

5: Turbulence, Chondrules, and Planetesimals

JEFFREY N. CUZZI[1], ANTHONY R. DOBROVOLSKIS[2], AND ROBERT C. HOGAN[3]

[1]245-3, Space Science Division, NASA–Ames Research Center, Moffett Field, CA 94035, U.S.A.; [2]Lick Observatory, University of California, Santa Cruz, CA 95064, U.S.A.; [3]SYMTECH, 245-3, Space Science Division, NASA–Ames Research Center, Moffett Field, CA 94035, U.S.A.

ABSTRACT

This paper presents a descriptive, qualitative, "review and preview" discussion of the epoch and environment of chondrule formation and aggregation into meteorite parent bodies. Several new perspectives on the accretion of planetesimals, at least in the very earliest stages, are presented. This chapter focuses primarily on insights we have gained from recent studies of nebula turbulence and coupled particle-gas dynamics. It does not address the chondrule-forming process per se, but does suggest that it was a key event or series of events which enabled the formation of chondrite parent bodies as we know them.

INTRODUCTION AND OVERVIEW

The evolution of the protoplanetary nebula is often broken up into a series of temporal stages. We will focus on the history of chondritic material as opposed to highly refractory or isotopically anomalous material. We find it useful to distinguish the following stages, some of which are probably overlapping or contemporaneous (see chapters by Cassen and by Wood in this volume for different sequences):

Stage 1: Interstellar material collapses into an accretion disk. Temperatures in the terrestrial planet region are sufficiently high to vaporize most incoming solid material. Very strong turbulence may be driven by ongoing infall and/or high temperatures. Solids recondense into monomineralic grains of tens of microns or less in size. The disk spreads radially to roughly its present extent by one or more of a number of candidate processes.

Stage 2: Infall and turbulence diminish; the nebula cools. Diffusion-limited growth of fractal aggregates from nebula grains follows. Sticking coefficients are uncertain but not much different from unity at low gas-dominated relative velocities. Aggregates are "fluffy" and firmly trapped to the gas, showing no tendency to settle to the midplane and diffusing widely in nebula radius.

Stage 3: At least (and maybe only) in the asteroid-chondrite parent body region (say 2–3 AU) the mystery chondrule formation (CF) process begins. Nebula temperatures at 2 AU are below 680 K (FeS is present), but the CF process is sufficiently intense (2000° K, 10^{10} erg/g) to melt rocky fragments in repeated events, each of short duration. The melting of previously "fluffy" aggre-

gates dramatically changes their aerodynamic stopping time and their response to turbulence. Fragmentation and remelting of previously solidified chondrules are common.

Stage 4: Nebula-wide turbulence concentrates chondrules and similar size fragments by large factors into many small, dense, transient zones throughout the vertical extent of the nebula. The concentration process is highly size-selective according to the aerodynamic stopping time of the particles. Assuming nominal, currently accepted properties of weak turbulence in protoplanetary nebulae, we demonstrate that particles of radius 0.1–1 mm are concentrated by five orders of magnitude or so relative to all other particle sizes in this way. During this stage, chondrule-size particles possess sufficient mean and fluctuating velocities relative to the gas to accumulate "rims" of fine dust in relatively short times.

Stage 5: These dense particle accumulations have no strength, and are still far from solid, with densities of 10^{-6} to 10^{-5} g cm^{-3} and mass ~ 10^{3} kg (see "Dense Agglomerates", below). Under the vertical component of solar gravity, some of these accumulations settle towards the nebula midplane. While so doing, they compress and merge with similar accumulations, or possibly fragment into bits to be recycled through stages 3 and 4. All accumulations produced at a given location and epoch will have constituents that are similar in composition and size distribution. Stages 4 and 5 are responsible for "primary accretionary texture" (Metzler *et al.* 1992).

Stage 6: Surviving, somewhat compacted, aggregates and their fragments remain in the midplane and accrete other, similar objects which drift radially inward under gas drag. Depending on conditions, radial drift can be slower than often assumed. Nevertheless, planetesimal sized objects (10–100 km radius) can result in only several 10^4 years (see "Midplane Growth", below), depending on the (unknown) efficiencies of sticking versus destruction.

Stage 7: Gas disappears and/or becomes dynamically less relevant; gravitational perturbations stir planetismals up until high-velocity collisions result. Meteorite parent bodies suffer thermal and aqueous alteration, partial and/or total melting, and collisional abrasion/brecciation/fragmentation, obliterating much of earlier primary accretionary texture. Planets are built from planetesimals.

NATURE OF TURBULENCE

Two different turbulent regimes are likely to be of importance. In the early stages of a warm, dusty nebula, turbulence probably extends over the entire nebula scale height, driven either by thermal convection or by differential rotation. The turnover time of the largest eddies is probably comparable to the orbital period. At a later stage, after most of the solids have grown to macroscopic sizes and settled to the midplane, turbulence can be driven within a $10^4 - 10^5$ km thick boundary layer by the differentiating rotating, densely settled particle layer. This turbulence has smaller length and time scales.

Turbulence is known to possess structure on many scales. The outer, or integral, scale L is the scale on which energy is deposited (by stirring, instability, or whatever process produces the turbulence). It is usually assumed to be the largest relevant dimension of the system and could be on the order of a vertical gas scale height $H \approx 0.1R$ where R is the distance from the sun (it is probably smaller, though, as noted below). Turbulent kinetic energy cascades towards smaller scales; for fully developed, homogeneous turbulence, the turbulent kinetic energy E peaks at the integral scale L and is distributed within its "inertial range" according to the Kolmogorov energy spectrum $dE(l) \propto (l/L)^{-1/3}dl$. The smallest scale in a turbulent regime is the Kolmogorov microscale η, sometimes referred to as the inner scale. On this scale, molecular viscosity dissipates turbulent kinetic energy. Another important length scale is known as the Taylor microscale λ. The vorticity, or characteristic velocity gradient, is dominated by this scale.

Turbulent flows are charcterized by their Reynolds numbers $Re \equiv UL/\nu$, where U is a typical velocity scale and ν is molecular viscosity. The Reynolds number describes the ability of the flow to transport anything by actual fluid motions (UL) relative to molecular diffusion (ν). (There are subtle distinctions between transport of scalars, such as heat, particles, *etc.*, as opposed to vector properties such as momentum, which we will not discuss here.) Generally, fluid transport processes are related to the turbulent (eddy) viscosity $\nu_T = UL$, where U and L are the velocity and length scales of the turbulence, as opposed to those of the system (e.g., Cuzzi *et al.* 1993). A dimensionless parameter α is often introduced such that $\nu_T = UL = \alpha cH$, where c is the speed of sound. Then $Re = \nu_T/\nu = \alpha cH/\nu$, where only α is poorly constrained.

In the nebula as a whole, rotation can influence the larger length and velocity scales of the turbulence. This makes their values (and

hence Re or α) problematic. Two independent groups have recently attempted to constrain α in moderately realistic nebula environments. Cabot *et al.* (1987) studied a nebula in which thermal convection is both the cause and the effect of turbulent transport and of dissipation of the basic orbital shear motion. They found that rotation reduces the most important (radial) length and velocity scales significantly from those of the global system, so that $\alpha = 10^{-4}$ to 10^{-2}. The applicability of this result depends on the presence of abundant, micron-sized dust to provide substantial thermal opacity; thus, it is questionable during the chondrule stage when significant particle growth has already occurred and the thermal opacity has decreased.

Dubrulle (1992, 1993) has suggested that the orbital shear motion itself is tapped as an energy source for the turbulence. This apparently obvious claim is actually far from straightforward to support; rotating shear flow (such as the gaseous nebula) is formally stable to the infinitesimal perturbations (*i.e.*, noise) which are usually relied upon to induce turbulence. Instead, Dubrulle (1992, 1993) points out that *finite* perturbations from any cause *do* lead to turbulent instabilities in rotating shear flow. Turbulent instability of rotating shear flow to small but finite perturbations with fractional amplitude as small as 10^{-6} has been verified in laboratory experiments (S. Davis, personal communication, 1994). Based on a turbulence model that is calibrated on the Earth's rotating shear flow, which is in a comparable flow regime, Dubrulle (1992) finds that $\alpha = 2 \times 10^{-3}$ for the nebula given the presence of such perturbations, and Dubrulle (1993) discusses several possible sources of the finite perturbations required; we propose yet another one below. (See, however, note added in proof.)

For nominal nebula models at around 2 AU, having gas density around 2×10^{-10} g cm^{-3} and temperature of ~ 200 K, the molecular viscosity is $\sim 6 \times 10^5$ cm^2/sec. Assuming α in the range $10^{-4} - 10^{-2}$, then $Re \approx 4 \times 10^7 - 4 \times 10^9$. These are very large Reynolds numbers – well beyond the normal transition to turbulence at $Re \approx 10^3$. Higher Re flows are more energetic and must drive turbulence to smaller Kolmogorov scales before molecular viscosity can provide the needed dissipation. To within a factor of order unity, the characteristic eddy overturn velocities u on different eddy length scales l may be obtained from the turbulent kinetic energy $E(l)$ characterizing that scale size: $u(l) = \sqrt{2lE(l)} \propto (l/L)^{1/3}$. Thus the turnover time for an eddy of size l is $t_e(l) \approx l/u(l) = t_e(L)(l/L)^{2/3}$. The turnover time $t_e(L)$ for the largest eddy is probably the nebula orbital period at that location. This scale-dependent turnover time is one of the key parameters of the turbulent size-selective concentration process. Using the above expressions, we find that the Kolmogorov scale $\eta = LRe^{-3/4}$ is on the order of 1 km (with turnover time $t_e(\eta) = t_e(L)Re^{-1/2} \approx 1 - 10$ hr), while the Taylor scale $\lambda = 4LRe^{-1/2} = 4\eta Re^{1/4}$ is on the order of 10^3 km (with turnover time $t_e(\lambda) \approx 12-60$ hr). These spatial and temporal scales are far smaller than the global scales of the system, which are likely to be affected by the rotation of the nebula. Thus, turbulence on these scales is very likely to be isotropic and homogeneous. (*cf.* Tennekes and Lumley 1972, p. 65).

PARTICLE STOPPING TIME AND TURBULENT DIFFUSION

An oft-heard presumption is that chondrules, or their precursors, "settle to the midplane" and reside all or part of the time in a layer which has a relatively high mass density compared to the gas. This has been viewed as a way to concentrate solids relative to the gas in order to facilitate the relatively high oxidation states of many chondritic materials (Wood, this volume). It is also a fundamental aspect of the original "particle layer (or Goldreich-Ward) instability", in which centimeter-sized objects were envisioned to settle into a layer near the midplane which was so dense that it would fragment under its own self-gravity and collapse directly to solid planetesimals of around one km in size. First challenged by Weidenschilling (1980) using scale arguments, this gravitational fragmentation has also been shown unlikely by Cuzzi *et al.* (1993, 1994) in detailed numerical models of the particle mass concentration, which include self-consistent diffusion of the particles by turbulence. The particle diffusion coefficient D is related to the turbulent viscosity ν_T by

$$D = \frac{\nu_T}{1 + St}$$

(Cuzzi *et al.* 1993). Here St is called the *Stokes number*, defined as

$$St \equiv t_s / t_e(L),$$

where t_s is the particle aerodynamic stopping time due to gas drag and $t_e(L) \approx 1/\Omega$ is the overturn time of the large scale eddies. For the case of chondrule-sized objects, the stopping time is simply written as

$$t_s = r_p \rho_s / c \rho_g$$

(Weidenschilling 1977), where r_p is the radius of a particle, ρ_s is its density, and ρ_g is the density of the gas (10^{-9} to 10^{-10} g cm^{-3} in a "minimum mass" nebula at 1–3 AU). As before, c is the gas sound speed.

Diffusion of particles is important in both the vertical and radial directions. Being well trapped to the gas ($D \approx \nu_T$), the particles will spread radially by $\Delta R = \sqrt{\nu_T t} \approx 1$ AU in 10^5 yr if $\alpha = 10^{-3}$ ($\nu_T \approx 10^{14}$ cm^2/sec). In contrast, appreciable vertical settling requires very low intensities of turbulence. Recently, Dubrulle *et al.* (1995) have developed an analytical approximation to the vertical diffusion equations which is valid in the limit of small particles such as chondrules and their precursors. Their family of solutions of $\mathcal{J}$ is shown in Fig. 1; the density enhancement due to settling is approximately $\sqrt{\mathcal{J}} = \sqrt{St/\alpha}$. In the 1-3 AU region, unit density particles of radius 0.1 cm have stopping times $t_s \approx 0.3 - 3$ hr and Stokes numbers $St \approx$ a few times 10^{-4} referred to the orbital frequency Ω. Sufficient vertical settling to provide a midplane enhancement of solid/gas density ratio by more than a factor of several requires $\alpha < 10^{-5}$, much less than the current expected values in the solar nebula. Thus, no tendency for chondrules or their precursors to settle towards the midplane can be justified unless current α values are seriously overestimated.

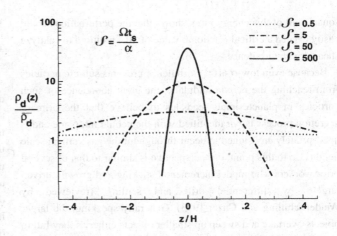

Fig. 1. Vertical profiles of steady state particle mass density $\rho_d(z)$ normalized to average mass density $\bar{\rho}_d$, for several different values of the parameter $S = St/\alpha$. Ω is the orbital frequency, t_s is the particle stopping time, and α is the turbulence parameter. Equilibrium results from the balance of settling under the vertical component of solar gravity with upward diffusion in nebula turbulence. Significant "settling" towards the midplane against diffusion requires a Stokes number $St = \Omega t_s$ which is hundreds or more times larger than α. For current expectations of $\alpha \approx 10^{-3}$, the solid curve would refer to particles of approximately 10 cm radius at 1 AU. This plot is adapted from Dubrulle *et al.* (1995).

THE CHONDRULE FORMATION (CF) PROCESS

As mentioned above, the Cabot *et al.* (1987) formulation requires strong thermal convection and a preponderance of micron-sized particles; various workers have suggested that "episodes" of quiescence can occur as accretion renders the nebula transparent to thermal radiation. In comparison, the Dubrulle (1992) mechanism is far more robust and relies on one of the most fundamental and pervasive aspects of the protoplanetary nebula – its orbital shear motion. We note that the mystery CF process itself might provide precisely the finite perturbations required to maintain turbulence at the level whch we assume here.

Although other interpretations may be placed on the CF process (several in other chapters in this volume), "flash heating" is a popular one (see "Concise guide" chapter by Boss, this volume). The size s of an individual heating and cooling event inferred from cooling rates (Sahagian and Hewins 1992; also Hewins, personal communication, 1994) is on the order of $s = 100$ km (the specific value of s is dependent on the particle opacity assumed). Roughly speaking, the energy per unit area on a sphere of this radius, multiplied by the area and divided by the mass of the chondrule precursor, must exceed the 10^{10} erg/gm needed to melt silicates. We estimate that the burst energy must thus exceed $4.8 \times 10^{24}(\rho_s/3\text{gcm}^{-3})(r/0.1 \text{ cm})(s/100 \text{ km})^2$ ergs, or hundreds of Megatons of TNT (1 MT $= 4 \times 10^{22}$ ergs) for $\rho_s = 3$, $r = 0.1$ cm, and $s = 100$ kM! These are not gentle events, and there must have been many of them, and over an extended period, to process all the chondrules. Even merely comparing the overall energy (per gram of nebula) required to melt all solids with the orbital energy per unit mass, and assuming *velocity* perturbations scale with the square root of this energy ratio, shows that the perturbations could easily exceed a critical fractional value of 10^{-3} required to catalyze shear-driven turbulence.

Because even low-α (10^{-3}) turbulence prevents sub-cm particles from reaching the nebula midplane, the great abundance of such particles in planetesimals leads us to believe that the primary accretion process is *not* identified with the nebula midplane and is appropriately envisioned to occur throughout the gas vertical scale height. Up to this point, our perspective is similar to that of several prior workers who model incremental sticking and growth, driven largely by differential drift and settling (reviewed by Weidenschilling and Cuzzi 1993). Generally speaking, the larger objects overtake and sweep up smaller objects either as they fall to the midplane or within the midplane. However, in prior studies there is no special role for size selectivity; the fragment size distribution in chondrites must be regarded as an artifact of the initial conditions or of the CF process itself.

Some difficulties for a "special production" hypothesis are posed by several meteoritic observations: (a) Both "microchondrules" and apparent fragments of much larger rounded objects are found in meteorites. Thus there is little to favor initial conditions as a reason for the chondrule size distributions. (b) Spheroidal chondrules, clastic fragments of larger chondrules, lithic fragments of other types, and other objects are all sorted according to the product of their radius and density (Skinner and Leenhouts 1993). This correlation is easily explained with gas drag (since aerodynamic stopping time $t_s \propto r_p\rho_s$), but only in an *ad hoc* fashion without it.

The CF process itself may have triggered a dramatic change in the evolutionary history of early aggregates. Fractal aggregates with a dimension of 2 are of special interest; they are commonly seen to result from diffusion limited grain growth, and their aggregate density decreases as r_p^{-1} (Weidenschilling and Cuzzi 1993). Thus their stopping time t_s remains constant as they grow, and they are as firmly tied to the gas as are their individual grain elements. Settling is practically impossible and they diffuse well throughout the nebula under even minimal turbulence. The process of forming chondrules from such aggregates would have changed their aerodynamical properties dramatically. We take r_c, r_a, and r_g respectively as the radii of a chondrule, of its precursor aggregate, and of an individual grain. Then, some algebra leads to $t_s(r_c)/t_s(r_a) \approx (r_c/r_g)$ (the aggregate radius r_a cancels out). If $r_c \approx 300 - 1000\ \mu$, and $r_g \approx 10 - 30\ \mu$, the stopping time increases by a factor of $10 - 100$ upon melting of the aggregate.

The specifics are clearly uncertain, but the point remains that the heating and melting process can cause significant changes in the aerodynamic behavior of milligram chunks of nebula silicate flotsam. In our perspective, it is not necessary for the CF process, or for the precursor population, to have any particular or special size dependence or preference. Rather, as described below, we believe the preferred size seen in chondrules within a given meteorite or class is caused by aerodynamic sorting subsequent to their formation, which concentrates a specific type of particle, independent of lithology, by orders of magnitude relative to all other types.

TURBULENT CONCENTRATION

The process of interest is that of turbulent concentration of particles, which has been found to be highly size-selective. This is of special interest because the narrow size distribution of chondrite constitutents is one of their defining characteristics (Grossman *et al.* 1989), and the size sorting is believed to be aerodynamic in nature (Skinner and Leenhouts 1993). It is important to realize that there are few readily accessible analogs to this regime in terrestrial natural or laboratory environments, partly due to the influence of terrestrial gravity and partly due to the different regimes of gas density and Reynolds number accessible to us compared to the nebula. Thus, most studies of this process to date, including our own, have been conducted numerically. Below, we first describe the phenomenology of the process, covering particle and fluid properties of interrest. We then present some numerical results and a feasible scaling or extrapolation to nebula conditions. Finally we discuss some meteoritical implications, including chondrule size and density relationships, chondrule rims, and possible timescales of interest.

To our knowledge, the first demonstration of turbulent concentration of particles with preferred aerodynamic stopping times was in a series of papers by Squires and Eaton (1990, 1991). In fully 3D numerical simulations of homogeneous, isotropic turbulence (as well as of decaying turbulence), they pointed out that particles with Stokes numbers in a certain narrow range (a factor of two or so) were concentrated into clumps in which the particle density was up to 30 times larger than the average density. These early calculations were for $Re \approx 100$, too small for the development of a well-defined inertial range in the turbulence spectrum. Particle trajectories were not followed and little was known about clump evolution with time. Since then, additional work (recently reviewed by Eaton and Fessler 1994) has shown that the most highly concentrated particles have $St = 1$ when referred to the turnover time of the Kolmogorov scale, or smallest eddies; that is, $t_s = t_e(\eta)$. The fundamental scaling persists over a range of Reynolds number and particle size, unlike the scaling by $t_e(L)$. This key result is under closer study for validation by us and other groups; its physical significance is poorly understood. The preferred size may be that of the largest particle which is sufficiently small to be stopped by gas drag within a Kolmogorov-size zone. Another result of our numerical modeling to date is that the typical size of the density enhancements is on the order of η (Dobrovolskis *et al.* 1993).

PREFERRED PARTICLE SIZE

Because larger Reynolds number flows drive turbulence to smaller scales, which have shorter eddy turnover times, the stopping time and thus the physical size of the particles which are preferentially concentrated decreases as the flow Reynolds number increases. To estimate the particle sizes and clump scales which would be favored by this process in the nebula, we have made use of the fundamental scaling laws which hold within an extensive inertial range of length and time scales. Such an inertial range is expected for the high Reynolds numbers implied by current nebula models (e.g., Cabot et al. 1987; Dubrulle 1992, 1993).

The maximally-concentrated particles are those with stopping times $t_s = r_p \rho_s / c \rho_g$ comparable to the Kolmogorov timescale $t_e(\eta) = \Omega^{-1} Re^{-1/2}$. Equating these times gives

$$r_p = Re^{-1/2} H \rho_g / \rho_s,$$

where $H = c/\Omega$ is the scale height of the nebula. Note also that

$$Re = \alpha c H / \nu \approx 2\alpha H / \ell,$$

where the molecular viscosity $\nu \approx c\ell/2$. Here ℓ is the mean free path of a gas molecule, given by

$$\ell = \frac{m}{\sigma \rho_g},$$

where $m \approx 3.2 \times 10^{-24}$ g and $\sigma \approx 5.7 \times 10^{-16}$ cm^{-2} are respectively the mass and collisional cross-section of a hydrogen molecule. Combining the above relations then gives

$$r_p \approx (2\alpha H \rho_g \sigma / m)^{-1/2} H \frac{\rho_g}{\rho_s}$$

$$= 0.15 \text{mm} \left(\frac{10^{-3}}{\alpha}\right)^{\frac{1}{2}} \left(\frac{2.5 \text{gcm}^{-3}}{\rho_s}\right) \left(\frac{\Sigma}{1000 \text{gcm}^{-2}}\right)^{\frac{1}{2}},$$

where $\Sigma = 2H\rho_g$ is the surface density of the gaseous nebula.

In the expression above, the coefficient is a combination of fundamental constants and is known exactly. For nominal nebula parameters ($\alpha \approx 10^{-3}$, $\rho_s \approx 2.5$ g cm^{-3}, and $\Sigma \approx 1000$ g cm^{-2}), the expression above yields the remarkable result that the most strongly concentrated particles are about 0.3 mm in diameter (see Fig. 2), in excellent agreement with the typical sizes of chondrules and fragments in primitive meteorites (e.g. Grossman et al. 1989,

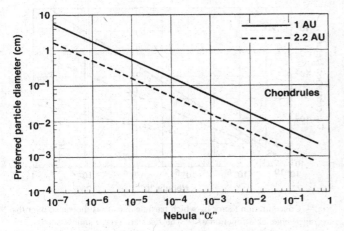

Fig. 2. Diameter of particles that are most strongly concentrated in turbulence, for two nebula locations, as a function of α. Note that for the values of α that are currently anticipated to characterize the protoplanetary nebula (10^{-4} to 10^{-2}), the preferred size for strong concentration is in excellent agreement with the chondrule size range. No additional parameters were adjusted to obtain this result; standard nebula properties were assumed.

Table 9.1.1, page 627). Chondrules of one millimeter diameter would be selectively concentrated by turbulence at a level of $\alpha = 10^{-4}$, well within the range now deemed reasonable. If the properties of the nebula are anything like the currently accepted values chosen above, chondrule diameters in the 0.1–1 millimeter range are easily understood from the perspective of this hypothesis. The most uncertain parameter in the above expression is clearly α; if this hypothesis could be verified by future studies, it might even serve to constrain the value of α.

CONCENTRATION FACTOR

Confident extrapolation of the factor by which preferred size particles may be concentrated requires a deeper understanding of this key process. The flow field may be divided into different zones according to the local fluid pressure, vorticity vector (rotation of the velocity), and strain tensor (Hunt et al. 1988, Wray and Hunt 1989, Dobrovolskis et al., in preparation). Two flow zones are of particular interest: eddies and stagnant zones. An "eddy" is a region of low pressure, low strain and high vorticity. A "stagnant zone" is a region of high pressure, high strain and low vorticity surrounding a stagnation point in the flow. Of course, the classification of a volume element changes with time as the flow evolves and eddies form, shift, blend, split, and disperse.

Numerical simulations (Squires and Eaton 1990, 1991, Dobrovolskis et al. 1993) demonstrate that particles which are initially distributed uniformly across the volume tend to vacate eddies and linger in stagnant zones, where their local density is considerably enhanced (Fig. 3). This eddy avoidance implicates centrifugal force as the driving mechanism. The clear indication of the importance of rotation in this process has led us to focus on the Taylor microscale λ associated with the dominant vorticity rather than on the large or integral scale L (see, e.g., Vincent and Meneguzzi 1991; Tennekes and Lumley 1972).

Any differential motion which separates particles from gas must be driven by some acceleration. Particles in eddies suffer a constant centrifugal acceleration of order $\lambda / te^2(\lambda)$. This may be compared with the more normally cited vertical component of the solar gravity $(z/R)(GM/R^2) = z \Omega^2$, where z is the distance from the midplane, R is the distance from the sun, G is the gravitational constant, M is the sun's mass, and Ω is the local orbital frequency. From scaling laws given above, it may be shown that the centrifugal acceleration due to eddies exceeds the vertical component of the sun's gravity by more than an order of magnitude everywhere in the nebula; turbulence is indeed "stronger than gravity" for particles in this size range. Of course, much larger particles are not swirled around in Taylor-scale eddies and thus vertical gravity is their primary source of acceleration (e.g. Cuzzi et al. 1993).

Extending the above "centrifugation" analogy somewhat, we give one simple estimate for the particle concentration factor C. If particles are spun out of volumes on the Taylor scale λ and concentrated into fluid volumes on the Kolmogorov scale η, C could be on the order of $(\lambda/\eta)^3 \propto Re^{3/4}$. An improved scaling would include a correction factor for the relative volumes in the flow occupied by

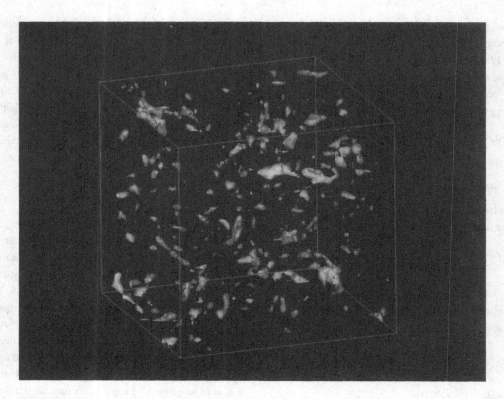

Fig. 3. Zones where particle concentration is 15 or more times larger than the average, as obtained from our three dimensional fluid dynamical models of homogeneous, isotropic turbulence. Each zone is localized at or near a stagnation point in the flow, avoiding "eddy" zones of strong vorticity or rotation. The concentrations appear to be on the order of a Kolmogorov scale in size, and move around with the flow. As the Reynolds number increases, concentration factors increase and zones become smaller and more numerous.

fluid structures on these respective scales, and might, for example, alter the value of the exponent. Given expected values of $Re \approx 10^7 - 10^9$, it is not inconceivable that $C \approx 10^5 - 10^7$ in the nebula. Some results from our numerical experiments to date are shown in Fig. 4. The values of C plotted are averages over multiple clumps. Also shown are a variety of alternate Re-dependent scaling laws that we have come up with, including the simple one mentioned above. Note that averages over more numerous clumps (shown by circles) produce (a) lower C and (b) weaker dependence of C on Re than averages over fewer, denser clumps (shown by triangles). This implies that the actual volume distribution of particles is complicated, and it may be that only the densest clumps show a strong dependence on Re such as implied by the above simple reasoning.

We are currently working to push the simulations to larger values of Re and to find better ways to describe and classify density enhancements. Meanwhile, we merely denote the concentration factor as some general probability $p(C, re)$ that any unit volume of flow with Reynolds number Re has concentration factor C.

TIME EVOLUTION OF PARTICLES AND CLUMPS

Dense particle concentrations arise in fluid stagnant zones, on spatial scales comparable to the Kolmogorov scale and smaller (Fig. 3). In the current simulations, concentrations typically persist for several times longer than it takes a particular particle to cross the zone. It is a little harder to distinguish from our simulations so far

whether the clumps form over the large eddy timescale $t_e(L)$ or on the Taylor timescale $t_e(\lambda)$. In our models, particles clustered at one time disperse again shortly afterwards, but the morphology of the particle "family" is not symmetric in time. These particles are closely followed by successors with similar trajectories (Fig. 5).

Our numerical results show that a particle of the optimally con-

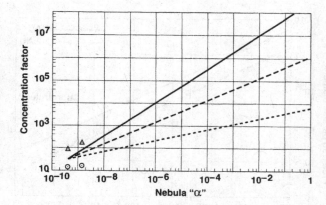

Fig. 4. Concentration factor by which particle density is increased over the average value, as a function of α. The lines are (very) simple scaling-law models, and must be constrained better through the use of numerical model results. Several of our current results to date are shown by the triangles and circles. The circles denote concentration factors averaged over several hundred of the densest clumps; the triangles are averaged over the dozen or so densest clumps. The fewer, densest clumps seem to show a steeper dependence on α. Modeling higher α is very computationally intensive; we are currently working on a value of α twice the largest one shown.

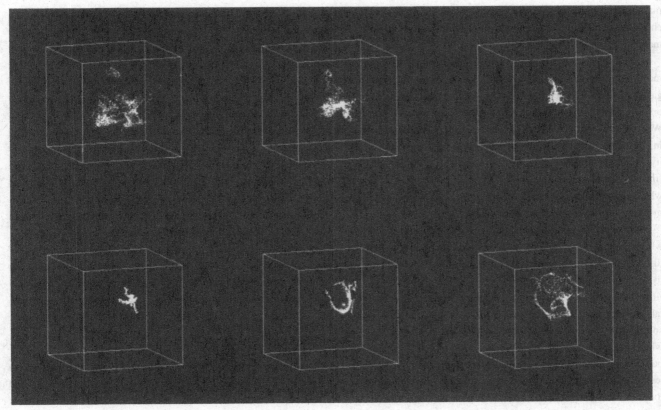

Fig. 5. Snapshots of a family of particles selected by their presence in a dense clump in the fourth timestep (lower left), and then traced over time. Their initial locations are widely and diffusely separated; they converge into the dense zone but continue through it and disperse in characteristic strands. Subsequently, these particles ultimately find their ways into other, different clumps and the process repeats.

centrated radius, while tossed and thrown around in turbulence, spends much of its time in stagnant zones, where the particle density tends to be high. The time history of the particle densities it passes through will be a function of $p(C, Re)$. Most of the clumps it encounters will have a density far larger than the nebula average value, but only occasionally will it encounter a clump with the peak density represented by the estimate $C \approx (\lambda/\eta)^3$.

Our simulations to date (with concentration factors up to 10^2 or 10^3) do not contain or require interparticle collisions or damping/modification of the turbulent flow *by* the particles. These latter feedback effects will become increasingly important as the concentrated particle density becomes much larger than the gas density, as we have suggested. Then interparticle collisions and/or turbulence damping in sufficiently dense clumps might trap particles in the clump and prevent, or retard, their dispersal. At this point, we expect that the particles end their independent lives to become permanent members of a particular accumulation.

As a ballpark example, a clump with $C \approx 10^5 - 10^7$ has a mass density of $10^{-6} - 10^{-4}$ g cm^{-3}. For 0.3 mm radius chondrules, the volume density is $3 \times 10^{-6} - 3 \times 10$ cm^{-3}, and the random velocity is $\eta/t_e(\eta)$ m $\approx \eta/t_s$; thus there is on the order of one collision per particle per clump transit. Collisions might be fairly inelastic for dust-coated chondrules. The dispersal of clumped particles might be further impeded by damping of fluid motions in and around these clumps, which have mass density 10^3 or more times that of

the gas. Given $p(C, Re)$ and the various timescales and relative velocities, in this way we may estimate the expected lifetime of an independent particle prior to accumulation. Subsequent evolution of the accumulation is speculated on below.

DUST RIMS

Another aspect of meteorite constituents which may provide an independent constraint on the evolutions of chondrules and chondrites is the rims of fine-grained dust which envelop many chondrules and lithic fragments. In the best-characterized cases of 'primary accretionary texture" in CM meteorites (Metzler *et al.* 1992), *all* objects are enveloped in dusty rims to the extent that these rims effectively account for the entire content of micronsized, non-chondritic particles in the rock. The accumulation of such dust rims on chondrules and similarly sized lithic fragments is an obvious and straightforward outcome of the process described above.

Recall that porous, "fluffy" aggregates with fractal dimension approximately 2 will have aerodynamic properties nearly identical with those of their constituent grains. Thus relative velocities will be limited by Brownian motion, and will be fairly small (of order 1 cm/sec or less). There will be no distinction bertween "rim" and "core" for such objects, and their growth will be a diffusion-limited process (Weidenschilling and Cuzzi 1993). Once milligram or so mass particles are melted and coalesce into solid objects such as chondrules, their stopping time increases by more than an order of magnitude.

Simplifying previous arguments even further, we can approximate the history of such "chondrules" as one of constant centrifugal acceleration $\lambda/t_e^2(\lambda)$; their velocity relative to the gas is damped on the stopping time t_s so they acquire a terminal velocity $\lambda t_s/t_e^2(\lambda)$. We find a relative velocity of this order between particles and gas in our numerical runs. For the parameters of the nebula, this velocity is meters or tens of meters per second. Assuming the fine (micron-sized) grains are firmly trapped to the gas, the rate at which rims accumulate depends on this relative velocity and the mass density of the grains per unit volume.

Ultimately, we believe that a consistent picture will emerge, in which the time to entrapment (a function of the clump probability distribution) is in agreement with the duration of exposure required to account for the observed thickness and density of fine-grained rims. Rough order-of-magnitude estimates suggest that it would take several years to produce 60μ rims on 300μ cores if all the solids are in fine dust. A more reasonable guess for this stage, in which much or most of the solids are already in chondrule-sized particles or larger, might be a dust mass fraction of a percent (or less) and rimming times of hundreds of years (or proportionately longer). Clearly, this subject requires a more detailed treatment; it is the subject of ongoing work by us and also by other groups (K. Metzler, R. Durisen, personal communications).

Up to this point, then, we have hypothesized a process which aerodynamically selects particles with a specific stopping time for significant (orders of magnitude) concentration in a gas-dominated nebula. We have shown that the actual sizes observed for chondrules (0.1–1 mm radius) follow directly from turbulent intensities (Reynolds numbers) which have been inferred recently by several different groups, for different reasons and under different assumptions (Fig. 2). Intrinsic to this concept is the idea that particles *not* in this preferred size-density range will not be concentrated. Thus, if this process dominates "primary accretion", other size ranges, while possibly present in the nebula, will be relatively under-represented in meteorite physical constituents unless they grow or fragment into the appropriate size range. A natural aspect of this process is the accumulation of rims of fine dust on such fragments; the rim thickness will be functionally dependent on the core radius through the relative velocity. Following this stage, however, these relatively dense groupings of particles must be compacted into actual rocks – another several orders of magnitude of compression. Below we speculate on possible elements of this subsequent stage.

DENSE AGGLOMERATES

The evolution of dense particle accumulations resulting from the turbulent concentration effect has not been studied; the following is only a speculative scenario, presented for completeness and to stimulate discussion and further investigations. Such a clump may behave as an aerodynamic unit, due to its density and damping of turbulence. The stopping time of a "bound" clump with radius 1 km and density 10^{-1} g cm^{-3} (using Stokes rather than Epstein drag coefficients) is comparable to that of a solid particle of unit density 10 cm in radius (although its mass is much greater). Unit density particles of 10 cm radius were studied closely by Cuzzi *et al.* (1993) and found to settle into a midplane layer several tens of thousands of km (about 10^{-3} AU) thick over a time of approximately one year.

We speculate that such clumps, which are still far from solid objects, provide the vehicle for bringing rimmed, size-sorted chondrules to the nebula midplane *en masse*. Clearly, much needs to be studied concerning this stage. Strengthless, compressible, essentially fluid objects sinking through a medium of much lower density could suffer a variety of effects, ranging from compression and compaction to shear-driven fragmentation and destruction. They could also collide with other clumps, resulting in merger or breakup. One expects, though, that with a mass density thousands or more times larger than the surrounding average gas, they may have the potential to retain some identity on a settling timescale.

MIDPLANE GROWTH

We envision many similar agglomerates falling to the midplane at any epoch, carrying their burden of size-sorted chondrules and fragments. They will be similar because the chondrule precursors are well mixed throughout the nebula by diffusion, and local heating events and turbulence will be creating and concentrating objects with similar aerodynamic properties on a scale of H or so. Thus, there will be little to distinguish neighbouring agglomerates of 10^3 kg mass from each other. We envision that objects accreting from such clumps will then also be fairly similar at any epoch, although to some extent material could be swept up onto previously accreted objects in the midplane. As objects are compressed, collide, and merge and/or fragment with differing degrees of intensity, another degree of inefficiency is introduced. Because of this (unknown) inefficiency, the midplane growth stage could last considerably longer than the time it takes any particular clump to form and settle.

As discussed by Weidenschilling (1988) and Cuzzi *et al.* (1993), midplane accretion is aided by gas drag; growing planetismals are continually "fed" by smaller material drifting inward. This depends on a variety of poorly known effects; Cuzzi *et al.* (1993) studied the process carefully in the limit that the turbulence was all generated by the particle layer (no global turbulence) and found that 100 km planetesimals could grow in such an environment in 10^3 (10^4) years at 1 (10) AU, with negligible radial drift. Weidenschilling (1988) found slower growth and greater drift in an environment more stirred by global turbulence, in which the particle density was comparable to the gas density near the midplane. These planetesimals could themselves be meteorite parent bodies; cosmic ray exposure ages lead to a current belief that there is of order one such 100-km radius parent body for each chondrite class (see, e.g. Wood, this volume). The scenario described above is consistent, in our opinion, with the need to create such homogeneous objects composed primarily of "chondrule"-sized bits.

ALTERATION OF PLANETESIMALS

Subsequent complications affect much of the meteorite record, and few (if any) samples are genuinely representative of the stages discussed so far. 100-km objects are large enough to produce thermal and aqueous alteration; also, at some point, 100-km objects must cross orbits and merge into planet-sized objects. This implies very energetic, destructive, collisions. Most meteorite samples are strongly evolved by these processes. Abrasion, alteration, melting, fragmentation, brecciation, and so on, are common; increased abundance of solar-type rare gases in brecciated meteorites may indicate that the nebula gas had dispersed by this stage. Interestingly, if the "primary accretionary texture" identified by Metzler *et al.* (1992) indeed shows less evidence for such rare gases, it may be more reliably placed in the earlier, gaseous-nebula stage in which we propose that aerodynamic size-sorting occurred. Stated yet another way, we would expect aerodynamic size-sorting signatures to be most prominent in those meteorites which show the least evidence for abrasion/brecciation/other mechanical evolution (collisional, etc.) in a nebula-gas-free environment. Conversely, one should not be surprised if aerodynamic signatures are lost or obscured in material which has been heavily evolved subsequent to the primary accretion stage itself.

SUMMARY

This has been an overview of recent work (primarily our own) dealing with the environment in which chondrules come together to form planetesimals: "primary accretion". In other work (Cuzzi *et al.* 1993, 1994; *cf.* Weidenschilling 1980) we have noted that gravitational instability is an unlikely mechanism for primary accretion. Instead, we sketch here a sequence beginning with turbulent concentration, continuing through "dense (but not solid) aggregates", and terminating in midplane growth, compaction, and modification which we believe shows promise for comprehending primary accretion and planetesimal formation. We argue that the still-widespread prejudice for a "dusty nebula midplane" as the site of the chondrule epoch is in glaring disagreement with most or all current theoretical expectations about the properties of the nebula and its particulates. We described the physics of a new process: turbulent concentration of aerodynamically size-sorted particles. Our best estimate of the aerodynamically preferred particle size is in the chondrule size range. Particles of this size might be concentrated by factors of 10^5 or greater, while accumulating rims of fine dust prior to being aggregated together into a planetesimal (chondrite parent body). We identified certain aspects of the meteorite data which are worthy of further study and which might provide direct evidence of the process of primary accretion; these include size-density correlations, and dust rim thickness and density.

ACKNOWLEDGEMENTS

We are very grateful to our colleagues Ted Bunch, Pat Cassen, John Eaton, Julie Paque, Ed Scott, William Skinner, Kyle Squires, John Wasson, and John Wood for many helpful conversations and discussions. This work was supported by the Planetary Geology and Geophysics Program and the Origins of Solar Systems Program. The numerical work would have been impossible without the support of the National Aerodynamical Simulator (NAS) Facility at NASA Ames Research Center.

REFERENCES

Cabot, W., V. M. Canuto, O. Hubickyj, and J. B. Pollack (1987) The role of turbulent convection in the primitive solar nebula; II: Results. *Icarus* **69**, 423–457.

Cuzzi, J. N., A. R. Dobrovolskis, and J. M. Champney (1993) Particle-gas dynamics in the midplane of a protoplanetary nebula. *Icarus* **106**, 102–134.

Cuzzi, J. N., A. R. Dobrovolskis, and R. C. Hogan (1994) What initiated planetesimal formation? *LPSC Abstracts* **XXV**, 307–308

Dobrovolskis, A. R., J. N. Cuzzi, and R. C. Hogan (1993) Particle sorting and segregation in a preplanetary nebulla. *Bull. Amer. Astron. Soc.* **25**, 1122.

Dubrulle, B. (1992) A turbulent closure model for thin accretion disks. *Astron. Astrophys.* **266**, 592–604.

Dubrulle, B. (1993) Differential rotation as a source of angular momentum transfer in the solar nebula. *Icarus* **106**, 59–76.

Dubrulle, B., G., Morfill, and M. Sterzik (1995) The dust sub-disk in the protoplanetary nebula. *Icarus* **114**, 237–246.

Eaton, J. K. and J. R. Fessler (1994) Preferential concentration of particles by turbulence. *Int. J. Multiphase Flow* **20**, Suppl., 169–209.

Grossman, J. N., A. E. Rubin, H. Nagahara, and E. A. King (1989) Properties of Chondrules. In *Meteorites and the Early Solar System*; Kerridge and Matthews, eds; Univ. of Arizona Press.

Hunt, J. R., A. A. Wray, and P. Moin (1988) Eddies, streams, and convergence zones in turbulent flows. *Proc. Summer Program*, *Center for Turbulence Research*, Ames Research Center.

Metzler, K., A. Bischoff, and D. Stöffler (1992) Accretionary dust mantles in CM chondrites: evidence for solar nebula processes. *Geochim. Cosmochim. Acta* **56**, 2873–2897.

Sahagian, D. L., and R. Hewins (1992) The size of chondrule-forming events. *LPSC Abstracts* **XXIII**, 1197–1198.

Skinner, W. R., and W. Leenhouts (1993) Size distribution and aerodynamic equivalence of metal chondrules and silicate chondrules in ACFER 059 (abstract). *LPSC* **XXIV**, 1315–1316.

Squires, K. D. and J. A. Eaton (1990) Particle response and turbulence modification in isotropic turbulence. *Phys. Fluids* **A2**, 1191–1203.

Squires, K. D. and J. A. Eaton (1991) Preferential concentration of particles by turbulence. *Phys. Fluids* **A3** 1169–1178.

Tennekes, H., and J. L. Lumley (1972) *A first course in turbulence*. MIT Press, Cambridge, Mass.

Vincent, A., and M. Meneguzzi (1991) The spatial structure and statistical properties of homogeneous turbulence. *J. Fluid Mech.* **225**, 1–20.

Weidenschilling, S. and J. N. Cuzzi (1993) Planetesimal formation in the protoplanetary nebula. In *Protostars and Planets*, Univ. of Arizona Press.

Weidenschilling, S. J. (1977) Aerodynamics of solid bodies in the solar nebula. *Mon. not. Roy. Astr. Soc.* **180**, 57–70.

Weidenschilling, S. J. (1980) Dust to planetesimals; settling and coagulation in the solar nebula. *Icarus* **44**, 172–189.

Wray, A. A., and J. C. R. Hunt (1989) Algorithms for classification of turbulent structures. *Proc. IUTAM Conf on Topology of Fluid Mechanics*; 95–104; Cambridge.

Note added in proof:

Balbus, Hawley, and Stone (1996, *Astrophys. J.*, submitted) have shown that Keplerian disks are stable to gross instabilities induced by finite perturbations. However, the turbulent kinetic energies they find seem to level out at small but finite values. Thus, the issue of just how much turbulence may be maintained by shear appears worthy of further study.

6: Chondrule Formation: Energetics and Length Scales

JOHN T. WASSON

Institute of Geophysics and Planetary Physics* University of California, Los Angeles, CA 90095–1567, U.S.A.

* Also Department of Earth and Space Sciences and Department of Chemistry and Biochemistry

ABSTRACT

The energy necessary to heat chondrule precursors from 500 to 1900 K is estimated to be ~1630 J/g; the additional heat required for complete fusion is 480 J/g yielding a total of 2110 J/g. Heat was also required to vaporize portions of chondrule surfaces and nearby dust grains. In a centimeter-scale (one that formed 1–1000 chondrules) process this might have required an additional 800 J for each g of chondrules. If 1 g of H_2 gas was also heated for each g of chondrule, this consumed an amount of energy far in excess of that required to melt the chondrule precursor. Chondrule formation in the dust-rich midplane requires about an order of magnitude less heat per g of chondrule than formation in megameter-scale (10^3 km-scale) regions that extend into the gaseous part of the nebula. Formation in megameter-scale regions also tends to cook chondrules for periods of the order of 30 min at maximum temperatures, above or near 1900 K. Such prolonged, high temperature cooking would have depleted chondrules in volatiles and relict grains, and is thus contrary to observations.

Megameter-scale models have been produced to account for the low inferred cooling rates ($0.03–1$ K s^{-1}) resulting from the incorporation of textural information produced by furnace experiments into a chondrule formation model calling for a single heat pulse that produced the melting and the mineral growth. There is now abundant evidence that most chondrule surfaces are melted 2 or more times. Multiple heating events allow chondrule textures to be coarsened in repetitive centimeter-scale heating events, which offer the high cooling rates needed to preserve volatiles, especially FeS.

INTRODUCTION: ENERGY AND TEMPERATURE DIFFERENCES AMONG CHONDRULE MODELS

Chondrule formation models currently discussed within the community range greatly in scale. Some, which I designate centimeter-scale models, call for formation of most chondrules as individual objects (e. g., Rasmussen and Wasson, 1982; 1995). Others, which I designate megameter-scale models call for the simultaneous formation of ~10^{20} chondrules in regions having linear dimensions of ~10^3 km (Sahagian and Hewins, 1992; Morfill *et al.*, 1993; Horanyi *et al.*, 1995). The chief pieces of evidence cited in support of these models are the inferred thermal histories of the chondrules. Centimeter-scale models are well suited to explain high chondrule volatile contents, which are difficult to preserve at high temperatures (1900 K) for times longer than minutes. Megameter-scale

models were created to explain the low chondrule cooling rates inferred from furnace simulations of chondrule textures. I present arguments indicating that the preservation of volatiles in chondrule interiors and the very low cooling rates commonly inferred from furnace experiments are mutually inconsistent. The problem seems to be with the model used to interpret the furnace experiments, i.e., texture formation during monotonic cooling following a single heating event. The alternative is that grain growth resulted from a multiplicity of minor heating events.

A large fraction of the chondrules formed by melting dust grains in a nebula having an ambient temperature low enough (<680 K) to permit the existence of solid FeS (some chondrules having very low contents of volatiles may have formed at higher nebular temperatures). It has long been accepted that the minimum energy needed to melt chondrules must allow for heating precursor solids from low nebular temperatures to temperatures near the liquidus as well as the latent heat of melting. I reexamine these quantities but also make estimates of the average amount of energy expended in vaporizing solids and in heating the adjacent gas. Proposed megameter-scale energy-release processes would occur in regions having higher gas/dust ratios than in regions proposed for centimeter-scale processes, and would thus require more energy per g of chondrule.

I will largely limit the discussion to observations on ordinary chondrite (OC) chondrules. There are several reasons why this is appropriate: (1) the three ordinary chondrite groups are the most common chondrite groups and the chondrites with the largest fraction (75–80%) of chondrules; (2) with rare exceptions that retain a comparable record the most unequilibrated ordinary chondrites preserve the nebular record better than the members of other groups; and (3) there is less ambiguity as to what constitutes a chondrule in the ordinary chondrites as compared to those in the carbonaceous chondrites that are almost as primitive. The latter ambiguity relates to the fact that the there is no sharp compositional hiatus between the igneous refractory inclusions and the chondrules in carbonaceous chondrites.

A NEBULAR MODEL

The discussion is couched in terms of the following cosmochemical model of the solar nebula. I assume that, at the time most chondrules present in chondrites formed, the nebula had evolved through the accretion-disk phase and that temperatures had fallen below 1000 K such that condensation of all but the most volatile elements had been completed. Solids in the chondrule formation region (nominally 2 AU) had largely settled into a midplane layer having a thickness above the midplane of $<3\times10^4$ km ($<2\times10^{-4}$ AU), about 0.001 the canonical thickness of the gas. The density of solids was greater than that of the gas in the central layer. Because most of the particles were still small (<1 mm) this layer was opaque. I assume that throughout the period during which solids condensed and were melted to produce chondrules, the ambient oxygen fugacity was largely buffered by the canonical pH_2/pH_2O ratio of 1500.

EQUILIBRIUM MELTING OF CHONDRULES

There is a general consensus that the "droplet" chondrules with spherical shapes (i.e., the radial-pyroxene (RP), cryptocrystalline (C) and many barred-olivine (BO) chondrules), were fully melted during their formation. In contrast, the porphyritic chondrules (see list in Table 1) that have large mineral crystals (some of which are demonstrably relict) and more irregular shapes were generally incompletely melted. Although kinetics played a major role in the formation of chondrules, it is nonetheless useful to examine the fraction of melt formed under equilibrium conditions as a function of temperature. Similar calculations using another algorithm and with different emphasis were published by Hewins and Radomsky (1990).

Table 1. *Sources of chondrule compositional data and notes regarding chondrule types. Abbreviations: porph = porphyritic, px = pyroxene, ol = olivine*

Source	Chondrule	Type	Notes
Jones & Scott (1989)	low-FeO	PO (IA)	porph ol; ol>>px; LL3.0 Semarkona
Jones (1994)	low-FeO	POP (IAB)	porph ol-px; modal ol 20–80%; LL3.0 Semarkona
Jones (1994)	low-FeO	PP (IB)	porph. px; modal <20%; LL3.0 Semarkona
Jones (1990)	high-FeO	PO (II)	porph ol; ol>px; LL3.0 Semarkona
Jones unpub., 1994	high-FeO	PP (IIB)	porph px; LL3.0 Semarkona
Lux *et al.* (1981)	high-FeO	RP	radial px; OC types3.4 Semarkona, Bishunpur, Krymka, Chainpur

As chondrule compositions I used data on LL3.0 Semarkona chondrules published by Jones (1990, 1994) and Jones and Scott (1989) as well as unpublished data of Jones. A description of the chondrule sets is listed in Table 1. Because Jones studied few droplet chondrules, I included compositions of four radial-pyroxene chondrules in type 3.4 OC chondrites published by Lux *et al.* (1981). For these chondrule compositions G.W. Kallemeyn calculated the equilibrium melt fraction as a function of temperature using the MELTS program of Ghiorso and Sack (1995).

In Fig. 1 I plot equilibrium melt fractions of the individual chondrules against temperature; Fig. 1a shows melt fractions for low-FeO chondrules, and Fig. 1b those for high-FeO chondrules. Different symbols indicate the different textural-chemical classes. The mean FeO/(FeO+MgO) ratios for low-FeO and high-FeO chondrules are 3 and 20 mol%, respectively. Perhaps reflecting a desire to characterize end members, Jones included only one low-FeO chondrule with FeO/(FeO+MgO) > 8% mol and only two high-FeO chondrules with ratios <17 mol% whereas randomly chosen Semarkona chondrules tend toward a flat distribution without a pronounced mode (e.g., Scott and Taylor, 1983).

In the low FeO chondrules (Fig. 1a) the total range in liquidus temperatures is about 200 K; this range gradually expands as the melt fraction decreases. At melt fractions of 70% and above, with a single exception, low-FeO PO chondrules melt at higher temperatures than the POP and PP chondrules. In high-FeO chondrules (Fig. 1b) the story is similar; PO chondrules and the one POP chondrule (its normative mineral composition yields olivine >> pyroxene), tend to be much more refractory than PP and RP chondrules.

Summarized in Fig. 2 are medians and quartiles for the data from Fig. 1. The median liquidus temperatures of low-FeO and high-FeO chondrules differ by only 50 K; thus the effect of FeO on the liquidus temperature is relatively minor. A more important influence is the fraction of olivine in the normative composition; note that Fig. 1a shows that the median liquidus temperature of

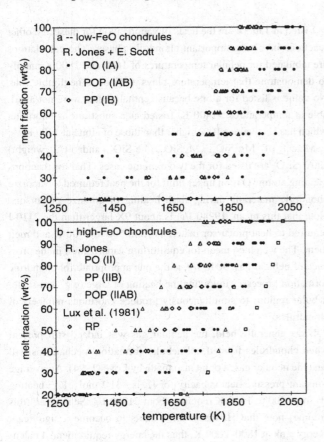

Fig. 1. Melt fractions as a function of temperature calculated for chondrules in highly unequilibrated (type 3.4) ordinary chondrites using the MELTS program of Ghiorso and Sack (1995). Chondrule compositional data sources are listed in Table 1. Porphyritic-olivine (PO) chondrules are more refractory than other textural types. These data suggest that successful chondrule models need to produce melts having temperatures of ~1900 K.

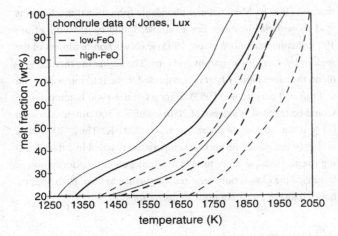

Fig. 2. Medians (heavy curves) and quartiles (light curves) summarizing individual results plotted in Fig. 1. Median liquidus values for low-FeO chondrules are only ~50 K higher than those for high FeO chondrules, but the difference in calculated temperatures increases to 100 K for melt fractions < 60%.

low-FeO PO chondrules is about 70 K higher than the medians for low-FeO POP and PP chondrules. At melt fractions of 60% and below the effect of FeO is more important.

I suggest that all objects that are properly classified chondrules had initial melt fractions >60% and that a large majority had melt fractions in the range 70–95%. The 90% melt medians are 1927 and 1874 K; thus, 1900 K is a reasonable melt temperature to ask model calculations to produce. The igneous or coarse-grained rims surrounding ~10% of chondrules in OC (Rubin, 1984) had melt fractions that covered a larger range, perhaps 20–80% with a best-guess mode about the middle of this range (Krot and Wasson, 1995). Thus, typical temperatures involved in the formation of igneous rims might be 1600–1800 K.

HEATING, MELTING AND VAPORIZATION OF NEBULAR SOLIDS: ENERGY REQUIREMENTS

Although some of the chondrules that survived until planetesimal accretion fixed chondrite compositions have low volatile contents indicating formation when the nebular temperatures were above 1000 K, and in some cases, as high as 1200 K (at 1300 K the major silicates would evaporate in a canonical nebula with $pH_2 \sim 10^{-5}$ atm (e.g., Wasson, 1978)), the chondrules with appreciable amounts of interior S and high FeO/(FeO+MgO) ratios seem to have formed when ambient nebular temperatures were 500 K. As discussed in the previous section, a typical temperature required to achieve ca.–90% melting is 1900 K; thus, chondrule formational models must supply enough heat to increase the temperature of the precursor solids from 500 to 1900 K.

In Table 2 are heat capacities calculated at four temperatures within this range based on algorithms listed in Robie et al. (1978) and Kubaschewski and Alcock (1979). Although chondrule heating events undoubtedly also melted metal, most recognizable chondrules consist mainly of the first four silicates listed in Table 2. Because these have similar heat capacities, the listed average is probably within 5 % of the actual value for all normal chondrules.

Table 2. Heat capacities in $J\ g^{-1}\ K^{-1}$ of the major minerals of chondrules. Values at 1800 K outside stability range of fayalite and albite. Data for Fe from Kubaschewski and Alcock (1979), remaining data from Robie et al. (1978)

Temp. (K)	600	1000	1400	1800
Main silicate minerals				
$MgSiO_3$	1.04	1.20	1.26	1.28
Mg_2SiO_4	1.11	1.25	1.32	1.37
Fe_2SiO_4	0.83	0.92	1.03	1.17
$NaAlSi_3O_8$	1.06	1.19	1.25	1.29
mean	**1.01**	**1.14**	**1.21**	**1.28**
Other minerals				
Fe(γ)	0.53	0.59	0.65	0.71
SiO_2(cr)	1.03	1.16	1.21	1.23

Integrating the product of these values from 500 to 1900 K yields the result that 1630 J/g are required to heat chondrule precursor solids to 1900 K. The amount would be about 550 J lower if the nebular temperature were 1000 K when the chondrule formed.

Heats of fusion of several chondritic minerals are listed in Table 3. These show more variation among minerals than do the heat capacities, and some are reported to have uncertainties >25%. I again averaged the four main minerals, and to allow for the fact that plagioclase normally comprises only 10–15% of chondrule silicates, I assigned albite 1/2 weight in the mean. The tabulated mean is probably within 10–12% of the correct value for most chondrules. The sum of the internal heat, 1630 J, and the heat of fusion, 480 J, yields the nominal amount of heat required to melt 1 g of chondrule precursor materials: 2110 J.

It is rarely treated in models of chondrule formation, but it seems probable that the amount of heat that goes into vaporization of solids was comparable to or greater than that required for chondrule melting. A lucid discussion of the vaporization of nebular solids is given by Hashimoto (1991). He notes that for oxides, the actual evaporation reaction controlled by kinetics may not be the one predicted by thermodynamics. In particular, the evaporation reaction may produce atomic O or O_2, whereas in the presence of nebular gas the liberated O would react with H_2 to produce H_2O.

Consider $MgSiO_3$; the actual evaporation reaction at the atomic level is probably:

$$MgSiO_3(s)=Mg(g)+SiO(g)+2O(g) \qquad (1)$$

This reaction yields a relatively large heat of vaporization. It is possible that the O is able to form molecular O_2 already at the moment of evaporation according to the reaction:

$$MgSiO_3(s)=Mg(g)+SiO(g)+O_2(g) \qquad (2)$$

This yields a much smaller heat of vaporization. If the nebular gas was still present, the net equilibrium reaction is:

$$MgSiO_3(s)+2H_2(g)=Mg(g)+SiO(g)+2H_2O(g) \qquad (3)$$

which yields a still smaller heat of vaporization.

Table 3. *Heat of fusion of common chondrite minerals. Data from Robie et al. (1978) except forsterite and Fe from JANAF (1985)*

Mineral	Formula	Form. wt. (g/mol)	T fusion (K)	Heat of fusion (kJ/mol)	(J/g)
fayalite	Fe_2SiO_4	203.8	1490	92.2	452
forsterite	Mg_2SiO_4	140.7	2163	71.1	505
(clino)enstatite	$MgSiO_3$	100.4	1830	61.5	613
albite	$NaAlSi_3O_8$	262.2	1391	59.3	226
weighted mean of fayalite, forsterite, enstatite and albite					**480**
anorthite	$CaAl_2Si_2O_8$	278.2	1830	81.0	291
cristobalite	SiO_2	60.08	1996	8.16	136
diopside	$MgCaSi_2O_6$	216.6	1664	77.4	357
Fe metal	Fe	55.85	1809	13.8	247

Listed in Table 4 are the heats of vaporization for these and other reactions involving important chondrule phases. The calculations are tabulated for nebular temperatures of 1600 and 1800 K mainly to demonstrate that temperature plays a minor to negligible role. No value is listed for albite because enthalpy data were not available at temperatures >1400 K. Listed as a substitute is anorthite, which has H_{vap} values 10% higher than those of albite at 1400 K.

Means of Mg_2SiO_4, $MgSiO_3$, Fe_2SiO_4 and (1/2 weight) $CaAl_2Si_2O_8$ are listed for the two extreme cases. That for reactions yielding atomic O is an upper limit for the heat required to vaporize chondrule materials during flash heating of cm-size precursors. Note that this mean, 19240 J/g, is about 9X larger than the 2110 J required to heat precursor materials from 500 to 1900 K and melt them. The weighted mean for equilibrium vaporization in the presence of nebular gas, 9040 J/g, is the one appropriate for chondrule formation processes involving the heating of large (e.g., km-scale) nebular regions to simultaneously produce enormous numbers of chondrules.

If, as generally held, the nebular gas was indeed still present when chondrules formed, all nebular formation mechanisms will heat the nebular gas, a point also made by Love (1994). An average constant-pressure heat capacity for H_2 is ~31 J mol^{-1} K^{-1}; heating H_2 from 500 to 1900 K requires about 21700 J/g. Scott *et al.* (this volume) note that H_2 dissociation starts to become a significant energy sink at 1800–2000 K, thus the energy requirements I calculate for heating the gas are conservatively low.

HEAT TRANSPORT INTO THE INTERIORS OF PROTOCHONDRULES: SURFACE TEMPERATURES DURING MELTING EVENTS

In order to transport heat into chondrule interiors by conduction the surface temperatures need to be distinctly above the maximum temperatures reached in the interiors. The calculations summarized above show that the interior temperatures of most chondrules reached 1900 K. Most nebular chondrule formational mechanisms lead to surficial heating. For example, lightning (Morfill *et al.*, 1993; Rasmussen and Wasson, 1995) involves bombardment of the surface by energetic electrons and ions. The higher the thermal gradient, the more rapidly heat is transported to the interior. A diagram of Fujii and Miyamoto (1983) suggests that a 1900 K temperature would be reached at a depth of 20 µm within a 100 µm grain within 0.1 s if the surface temperature was ~2000 K. Thus, 2000 K is probably the minimum surface temperature reached in a flash heating event. Because more time is available, chondrule formation in large opaque clouds only requires the surface to have been slightly hotter than the final chondrule temperature.

EVAPORATION OF HOT CHONDRULES

As discussed by Langmuir (1913), Knudsen-effusion equations can be used to calculate the evaporation rate of metallic phases. I calculated halflives for four sizes of spherical metallic grains or droplets. The first two are assumed to be free-floating in space and thus los-

Table 4. *Heat of vaporization (H_{vap}) per gram of major chondrule phases at temperatures of 1600 and 1800 K. Oxide vaporization reactions calculated three ways; (a) assuming O is evaporated as O(g); and (b) assuming that O reacts with $H_2(g)$ to form H_2O (g); and (c) assuming that O forms O_2 (g). See text for details*

No.	Mineral	Vaporization reaction	H_{vap} (J/g) 1600 K	1800 K
1a	forsterite	$Mg_2SiO_4(s) = 2Mg(g)+SiO(g)+3O(g)$	21840	21770
1b	forsterite	$Mg_2SiO_4(s)+3H_2(g) = 2Mg(g)+SiO(g)+3H_2O(g)$	11070	10980
1c	forsterite	$Mg_2SiO_4(s) = 2Mg(g)+SiO(g)+3/2O_2(g)$	16410	16330
2a	enstatite	$MgSiO_3(s) = Mg(g)+SiO(g)+2O(g)$	20620	20580
2b	enstatite	$MgSiO_3(s)+2H_2(g) = Mg(g)+SiO(g)+2H_2O(g)$	10560	10500
2c	enstatite	$MgSiO_3(s) = Mg(g)+SiO(g)+O_2(g)$	11100	11060
3a	fayalite	$Fe_2SiO_4(s) = 2Fe(g)+SiO(g)+3H_2O(g)$	13790	13700
3b	fayalite	$Fe_2SiO_4(s)+3H_2(g) = 2Fe(g)+SiO(g)+3H_2O(g)$	4410	4330
4a	anorthite	$CaAl_2Si_2O_8(s) = Ca(g)+2Al(g)+2SiO(g)+6O(g)$	22660	22600
4b	anorthite	$CaAl_2Si_2O_8(s)+6H_2(g) = Ca(g)+2Al(g)+2SiO(g)+6H_2O(g)$	11770	11690
5	Fe metal	$Fe(s) = Fe(g)$	7100	7020
weighted mean H_{vap} of reactions 1a, 2a, 3a, and 4a at 1800 K				**19240**
weighted mean H_{vap} of reactions 1b, 2b, 3b, and 4b at 1800 K				**9040**

ing mass through 4 π steradians; a 1–mm radius representing a chondrule and a 5-μm-radius sphere representing the maximum size of matrix grains. The other two examples are 100 μm and 10 μm spheres assumed to be embedded in the molten surface of a chondrule and thus having one hemisphere exposed to space (2 steradians). For each of these I calculated halflives, i.e., the periods necessary for the radii to shrink to 79.4% of the initial value. These are tabulated in Table 5. At T 1800 K the data are for solid Fe, at T 1900 K they are for molten Fe. At 1900 K the halflife of the metallic "chondrule" with 1 mm radius is only 118 s, and those for the smaller sizes are much smaller. The calculations are for vacuum conditions, but the correction for ambient gas is negligible for those in which the vapor pressure is greater than the ambient (H_2) pressure of ~10^{-5} atm, i.e., all Fe values and Mg_2SiO_4 values 2100 K. Evaporation rates will be lower and halflives longer for Mg_2SiO_4 at lower temperatures. If the ambient pressure of O-bearing species has been enhanced by evaporation that has already occurred this would also depress the Mg_2SiO_4 evaporation rate.

Hashimoto (1991) used furnace experiments to determine evaporation rates of heated SiO_2, MgO and Mg_2SiO_4. He showed that evaporation is inhibited relative to the rates calculated from thermodynamics data. His data form a linear array on a plot of log evaporation rate (dm/dt) vs. 1/T; I used this relationship, log dm/dt = 7.34–30000/T to calculate halflives for forsterite, Mg_2SiO_4. These are listed in Table 5 for the same sizes of spheres and conditions as used for metallic Fe. Forsterite evaporates at a much lower rate than Fe.

Evaporation rates for glassy silicates having typical compositions must be much greater than that of Mg_2SiO_4 because of lower bond strengths and because these include relatively volatile plagioclase and ferroan silicate components. I suggest that typical chondrule halflives are 10× less than those listed for solid Mg_2SiO_4.

Table 5. *Evaporation halflives (in s) of spheres consisting of Fe and Mg_2SiO_4 as a function of temperature and size. The first two sizes are free floating and thus losing mass isotropically (4π); their radii (1 mm and 5 μm) are chosen to represent large chondrules and very large matrix grains. The latter two sizes (radii of 100 and 10 μm) represent phenocrysts embedded in the surface melt of chondrules that are losing mass only from their outer hemisphere (2)*

Temp. (K)	Radius of Fe 1 mm	5 μm	100 μm	10 μm	Radius of Mg_2SiO_4 1 mm	5 μm	100 μm	10 μm
1800	690	3.5	140	14	$9.9 \cdot 10^5$	5000	$2.0 \cdot 10^5$	$2.0 \cdot 10^4$
1900	120	0.94	38	3.8	$1.3 \cdot 10^5$	660	$2.6 \cdot 10^4$	2600
2000	59	0.30	12	1.2	$2.1 \cdot 10^4$	110	4300	430
2100	21	0.10	4	0.4	$4.1 \cdot 10^3$	21	830	83

CHEMICAL EVIDENCE FOR OR AGAINST VOLATILE LOSS

If chondrules remain at or near the melting temperatures for appreciable periods in a nebular environment volatile loss will occur (e.g., Yu *et al.*, this volume). As summarized by Grossman (1988), there are several kinds of volatile-element indicators that seem to require that chondrules cooled rapidly after the melting event. For example, the absence of appreciable interelement fractionations among the alkali elements is inconsistent with appreciable loss of volatiles (Misawa and Nakamura, 1988), as is the absence of $^{40}K/^{41}K$ fractionations in bulk chondrites (Humayun and Clayton, 1995). The best volatile indicator is FeS, because it is (1) highly volatile; (2) abundant enough to allow its siting to be determined microscopically; and (3) present as a pure phase that can dissociate directly to a gaseous species, whereas trace volatiles such as Zn or

In would have to diffuse to grain surfaces in order to volatilize.

Some authors, most notably J.A. Wood (priv. comm.), have argued that the FeS present in grain interiors was transported there during metamorphism and thus its presence does not place time constraints on the duration of chondrule heating. Wood's assertion should be testable by detailed petrograpic observations (e.g., FeS should always form corrosion films such as those produced by Imae, 1994), but he has not published such supporting evidence. Grossman (1988) "strongly disagreed" with a metamorphic origin of chondrule FeS and cited observations of FeS enclosed inside glass or clear crystals.

In low-type-3 OC the bulk of the FeS in chondrites is not inside silicate chondrules, but associated with extrachondrular fine and coarse metal. The scenario that would lead to FeS being incorporated into chondrules seems to consist of the following assumptions and steps: (1) Immediately after the accretion of a highly unequilibrated chondrite such as Semarkona, some S was in (undefined) highly labile sites that were unstable with respect to FeS. (2) During even the mildest metamorphism the labile S diffused through the rock (presumably as $S^=$ or S_2) until it found a metallic Fe surface on which it could nucleate FeS; some of these metallic sites were inside chondrules. (3) This "new" FeS grew to sizes large enough to become microscopically visible.

There are several problematical elements in this scenario. (a) What was the form of this labile S? It obviously could not have been present as FeS, since the main metamorphic reaction would then have been grain growth rather than nucleation of new FeS grains (the latter has a kinetic barrier, but the former does not). (b) The flux of S must have been greatest near the source, i.e., in the fine-grained matrix. In chondrites of petrographic type ≤ 3.4 minimal diffusion has occurred; thus, most of the labile S must have reacted near the source. Because surface/volume ratios increase with decreasing grain size, the reaction rate would have been highest with fine metal. Since a large fraction of the fine metal was in the matrix, the bulk of the new FeS should form in the matrix. (c) If the sulfur diffused as $S^=$ or S_2 we can be sure that the diffusion was confined to grain boundaries; body diffusion would have been much too slow to allow transport through crystals or glass. Metal that was trapped inside grains or glass would have been unaffected and FeS that is sealed inside grains or glass could not have formed by the hypothetical process. (d) Formation of FeS from Fe involves a volume increase of a factor of 2 (and thus a radius increase of a factor of 1.26); this is possible in porous materials such as matrix but difficult in low-porosity materials such as chondrules.

Thus each of the assumptions and steps in this scenario is dubious. There is no reason to suppose that appreciable amounts of labile S were present. If all S were initially present as FeS, there was no chemical potential gradient to cause it to diffuse to metal rather than to FeS (the latter would reduce surface energy). In type 3.4 chondrites the labile S would not have diffused farther than necessary to find fine metal, and could not have diffused through crystalline or glass barriers. Because of the volume increase FeS grains inside chondrules could not grow inside nonporous solids. In summary, it seems highly likely that most of the FeS in the interiors of chondrules in type 3.0–3.4 chondrites was incorporated during the melting event or events that formed the chondrules.

MINIMUM SIZES OF CHONDRULE FORMING EVENTS

The fewest number of chondrules that can form in a single event is one, but all chondrules cannot have formed alone. Sibling compound chondrules constitute 1.4% of OC chondrules (Wasson *et al.*, 1995). A model that could explain this observation is that 10% of chondrules formed as large clouds of chondrules having the same composition, with a 14% probability of colliding near the point of formation before both objects were able to solidify.

Thus, although many of nebular precursors may have been just large enough to produce a single mg-size chondrule, a sizable fraction was larger, with order-of magnitude precursor masses of ca. 1 g sufficient to form 10^3–10^4 chondrules.

COOLING RATES AND THE DIMENSIONS OF REGIONS HEATED DURING CHONDRULE FORMATION

Models for the formation of chondrules offer a wide range in sizes of the nebular regions heating during the flash heating events. At the one extreme are nebular models calling for bolt diameters of cm to m (Rasmussen and Wasson, 1995) and at the other extreme, nebular shock fronts (Hood and Horanyi, 1991) and lightning (Morfill *et al.*, 1993) encompassing volumes >10^6 km^3. I will designate the small-volume events "centimeter-scale" models and the large-volume events "megameter-scale" models. In the former case the hot chondrule cools quickly by radiating its heat into cooler surroundings; Fig. 3 shows cooling curves calculated by Rasmussen and Wasson (1995) for this case. At a chondrule temperature that is >200 K above the background temperature, cooling rates are the same

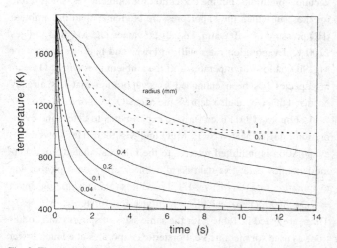

Fig. 3. The cooling rates of chondrules depend on their radii (in mm) and, to a much smaller degree, on the background temperature (solid curves 400 K, dashed curves 1000 K). In this calculation we assumed that 60% of the latent heat of fusion was released in equal amounts for each temperature interval between 1880 and 1680 K, and that only internal heat was released at lower temperatures. Because the temperature enters to the fourth power, the cooling rate of a chondrule at 1600 K radiating into a background sink at 1000 K is 90% of that of a chondrule radiating into a sink at 400 K.

within a factor of a few, independent of the background value. In contrast, heating events in megameter-scale opaque (dusty) regions produce cooling rates determined by the size of the region; Sahagian and Hewins (1992) showed that the inner 80% of a midplane slab 2000 km thick would cool at about 0.6 K s^{-1} (2000 K hr^{-1}) if chondrules were simultaneously formed throughout the layer and fine particles were evaporated. Although a similar cooling rate was obtained in a 200 km midplane layer if fine dust was added to increase the opacity, the authors noted that fine dust should evaporate during the heating event.

DIFFICULTIES ASSOCIATED WITH LARGE-SCALE CHONDRULE-FORMING EVENTS

Several chondrule-formation models call for melting to occur in megameter-scale regions. The dimensions in most of these models were chosen to meet the cooling rates inferred from furnace experiments (e.g., Hewins, 1988; Lofgren 1989) and, to a lesser extent, to allow chondrules to collide while molten, as seemingly indicated by the Gooding-Keil (1981) studies of compound chondrules (Wasson *et al.*, 1995, noted that this latter constraint is not valid if compound chondrules are formed either in two heating events or in single events producing many (10^2–10^4) chondrules from a single precursor lump).

There are, however, serious problems associated with chondrule formation in large opaque regions. There are two categories of problems, those associated with the source of energy and those associated with the effects on the chondrules of being immersed in a hot reservoir for periods of minutes to hours.

The energy source problem has never been examined in detail. The energy must be generated, stored and then rapidly discharged (within minutes?) into an opaque region containing the chondrule precursors. As discussed above, the surfaces of all the chondrules in these Mm regions were simultaneously at temperatures ≥ 1900 K; after a few seconds the gas (which probably had a surface density similar to that of the solids) temperature was the same as that of the solids.

As discussed above, 2110 J are needed to heat and melt 1 g of chondrule precursor solids. This amount must be augmented somewhat to allow for the volatilization of the surfaces of precursor grains during the heating event. For example, if 4 % of these grains are evaporated at an energy investment of 19240 J/g (in a centimeter-scale environment), an additional 770 J must be input for each g of resulting chondrule, a total of 2880 J/g. It seems inevitable that some gas will also be heated during these events; Rasmussen and Wasson (1995) proposed that chondrules formed at the edge of a dusty midplane layer in which the mass density of the dust exceeded that of the gas by a factor of 100, but I will assume that 0.1 g of H$_2$ is heated 1450 K for each g of chondrule, requiring an input of 2250 J. Thus, for these conditions the energy source must be able to supply 5130 J per g of melted chondrules.

In a megameter-scale opaque cloud the energy requirements are far greater. Although evaporation of solids requires only half as much energy per g because equilibrium vaporization (equation 3)

can occur, the savings are far outweighed by the much larger amounts of solids that would evaporate for each g of chondrule that is melted. Because the chondrules are held at temperatures near or above 1900 K for a much longer period than in centimeter-scale heating events, a larger fraction of the precursor grains will evaporate, and all of the ambient free-floating fine material smaller than a few μm will evaporate. Table 5 shows that at 1900 K, 5 μm-radius grains of metal and Mg$_2$SiO$_4$ evaporate half their mass in 0.94 and 660 s, respectively. The halflife of a chondrite melt or FeO-rich matrix would probably be an order of magnitude shorter than that of Mg$_2$SiO$_4$; since matrix grains are <5 μm, it seems certain that all matrix evaporated in megameter-scale chondrule forming events.

And, in most of these models the gas/dust mass ratios would have been unity or higher. No heat source localized in the dusty midplane has been proposed; for example, megameter-scale lighting (Morfill *et al.*, 1993) or clumpy accretion (Boss and Graham, 1993) occur well above the dusty midplane. Thus, in megameter-scale models a large amount of ambient gas must be heated to the same temperature as the chondrule surfaces. A conservatively low assumption regarding the required energy is that, for every g of chondrule melted, 1 g of fine materials and grain surfaces is evaporated, and 1 g of H$_2$ is heated from 500 to 1950 K. The evaporation of solid grains requires 1700 J/g to heat the precursors to 1950 K and 9000 J/g to evaporate, a total of 10700 J/g. The heat required to heat 1 g of H$_2$ 1450 K at constant pressure is 22500 J/g (as noted above I am neglecting H$_2$ dissociation). Thus, for each g of chondrule that is produced a total of 2100+10700+22500 = 35300 J/g is required, 7× higher than the 5130 J/g required to melt chondrules and vaporize 0.1 g of solids in centimeter-scale events.

One can tinker with these values, but it seems inescapable that the heat required to form 1 g of chondrule in megameter-scale settings is far greater than the amount required in centimeter-scale settings. For convenience, I will refer to this difference as an order-of-magnitude difference in the required energy input. Not only do megameter-scale events require an order of magnitude more energy per g of chondrule, they require that this energy be discharged within a short period of time into nebular masses ≥ 10^{17} g (calculated by multiplying an assumed surface density of 10 g cm^{-2} times a midplane area of 10^6 km^2, i.e., (10^3 km on a side). This energy is >10^{16} times larger than required if chondrules are formed in sets of one to 1000.

Rasmussen and Wasson (1982; 1995) suggested that the power source for small-scale events is turbulence generated by the difference in orbital velocity between the dust-rich midplane and the overlying gas-rich layers. It does not stretch the imagination too much to think that this "steady-state" long-time-scale turbulence can store up enough (~5 kJ) energy to form a set of 1000 large chondrules (from ~10 g of precursor). In contrast, it seems clear that the enormously greater energy requirements of the megameter-scale chondrule formation models can only be met by the nebula-wide, catastrophic events. The accretion of large parcels of materials to the inner solar nebula envisioned by Boss and Graham (1993) or similar events required to power FU-Orionis-type luminosity flashes (Hartmann *et al.*, 1993) are indicated. And this process must be repeated one or more times to account for evidence of repetitive heating of most chondrules.

The other problem with megameter-scale models is that they cook the chondrules at high temperatures for long (hour-scale) periods which would lead to substantial depletions of volatiles and to the dissolution of relict grains that would be inconsistent with observations. Before discussing these I note an important aspect of the cooling of megameter-scale systems. The subregions that have cooling rates that best fit values inferred from furnace experiments are the hot central regions ("cores"), but because the surrounding surficial subregions are initially equally hot, the core remains at the maximum temperature for a long period. The curves published by Sahagian and Hewins (1992) allow one to infer that, if it takes 60 min to cool the core from 1900 to 1500 K, during the first 30 min its temperature will have remained above ~1850 K. I will therefore assume that this temperature ranged from just above to just below 1900 K for 30 min in assessing the alteration produced in chondrules.

The halflives tabulated in Table 5 show that all grains of Fe-Ni that became exposed on chondrule surfaces would have evaporated under these conditions, whereas Mg_2SiO_4 grains having radii $\geq 10\,\mu m$ would have survived. It is clear that a sizable layer of melt having a chondritic composition would have evaporated; if, as I suggested above, evaporation rates are 10X higher than those of Mg_2SiO_4, a layer roughly $20\,\mu m$ thick would have evaporated. Assuming convective mixing of melt throughout the chondrule, volatile components (e.g., K, Rb, Zn and particularly S) that were not sequestered in relict grains would have had their bulk concentrations reduced by large factors (>10), and their surficial regions depleted by still larger factors (>100).

Grains immersed in silicate melts cannot survive very long. Experiments of Radomsky and Hewins (1990) showed that olivine dissolved at a rate of $0.2\,\mu m/s$ at 60 K above the calculated liquidus, and theoretical studies by Greenwood and Hess (this volume) yielded rates an order of magnitude faster. Lofgren (1989) showed that "finely ground" minerals dissolve into melts held 2 hours at 1870 K, 20 K above the liquidus temperature of his charge. Since it was important to his experiments that no relict olivine nuclei survived, it is certain that these heating periods were several times longer than the minimum necessary to dissolve all grains. It seems clear that grains $\leq 10\,\mu m$ in radius would totally dissolve within 30 min under these conditions.

Although not explicitly discussed in the megameter-scale models, it seems that they share the view that all kinds of the common chondrules, i.e., all textures and compositions, would have been produced in each such event. Fig. 1 shows that, if the megameter-scale core temperature was 1900 K, about 28% of the porphyritic chondrules would have been at temperatures 50–150 K above their liquidus. It seems unlikely that these could have retained the nuclei that furnace experiments show to be necessary to form porphyritic textures (e.g., Lofgren, 1989). Although megameter-scale events may be repeatable, the power source problem is then amplified and there is the additional complication that chondrules having low liquidus temperatures will completely remelt during each such event.

A final problem faced by megameter-scale models is how to form the igneous rims. As documented by Krot and Wasson (1995), in ordinary chondrites these generally (1) contain large amounts of glass and zoned crystals requiring moderately high (50%) degrees of melting and (2) tend to be rich in FeS showing rounded shapes indicating rapid solidification as a low-temperature liquid. These were clearly formed in heating events that occurred later than those that formed the host chondrule. And the abundant, near-surface FeS shows that temperature-time products must have been very low (i.e., high temperatures could only have been present for short times). It is hard to conceive that these rims were formed in megameter-scale events. But, if they formed in a different class of event, why did this class not also produce chondrules?

I conclude that, because of the implausibly large amounts of energy that would need to be discharged within an hour or less, and because the volatile, relict-grain and igneous-rim evidence are inconsistent with long cooking times, it is highly unlikely that chondrules were mainly formed in megameter-scale events.

MULTIPLE HEATING EVENTS, AN ALTERNATIVE TO MONOTONIC COOLING MODELS

Chondrule formation models involving large megameter-scale opaque regions have been devised to account for the low cooling rates inferred on the basis of furnace experiments but the discussion in the previous section shows that these require very large and undefined heat sources and are inconsistent with several key chondrule properties. It therefore seems that the lesser evil is to abandon the simplistic idea that textures must be understood in terms of a single heating event followed by monotonic cooling. Wasson (1993a) suggested that one should consider the alternative that chondrule textures reflect multiple heating events.

Because a large fraction of ordinary-chondrite chondrules are independently compound (i.e., attached to another chondrule formed in an independent heating event), because a large fraction of chondrules is surrounded by rims that experienced extensive (20–80%) melting in a separate heating event, and because relict grains (Nagahara, 1981; Rambaldi, 1981) are probably survivors from previous chondrule generations, Wasson (1993b), Wasson *et al.* (1995) and Krot and Wasson (1995) concluded that most chondrules experienced two or more flash-heating events.

The hypothesis of Wasson (1993a) is that, for every chondrule formed by melting in a particular heating event there will have been many more that were heated to moderate degrees. For example, although lightning may mainly heat chondrule precursors by the impact of ions and electrons, the discharges also produce large amounts of light that impinge on preexisting chondrules that happen to be nearby. Eisenhour and Buseck (1995) report detailed model calculations for heating and even melting by visible and infrared radiation. Chondrule surfaces will also be heated by the impact of ions and electrons that may produce diffusive transport and grain growth in interiors even though, because of low surface-volume ratios, large-scale melting was rarely produced.

Hewins and Radomsky (1990), in summarizing their and earlier furnace experiments stated that "it is easy to make glass chondrules in the lab but relatively few natural chondrules are glass." Although some textures can be produced by collisions of super-cooled molten droplets with dust grains (Connolly *et al.*, this volume), some textures have not been produced and some chondrules must have chilled to glasses in dust-free environments. The usual interpretation is that the laboratory experiments failed to reproduce the temperature-time conditions of the initial chondrule-forming events. I suggest that, because of the difficulties associated with forming chondrules in megameter-scale regions, a more plausible interpretation is that many chondrule melting events did form glass-rich chondrules, and that the textures we see today are mainly the result of later processing.

Possible scenarios for this later thermal processing include (1) the nebula microenvironment (e.g., radiant heat from nearby lightning strokes), (2) larger scale nebular events (e.g., shock fronts of the sort envisioned by Wood (1984) and Hood and Horanyi (1991) but with smaller dimensions and lower maximum temperatures (perhaps 1200–1300 K) and (3) heating in a parent-body setting (e.g., impact compaction of km to 10 km-size bodies may have raised temperatures to ~1000 K for appreciable periods, perhaps ka).

I find centimeter-scale heating events to be the most probable, and offer the following example scenario: (a) after flash heating and initial cooling, the chondrule consisted of pure glass or of a mixture of relict crystals and glass; (b) structural defects in the glass were present as residual nuclei or were formed by impacts, cosmic rays, etc.; (c) additional flash heating events heated the chondrule, particularly the glassy mesostasis (because this is darker than mafic silicates, the heating occurred primarily in the mesostasis); and (d) because high-FeO materials are darker than those with low FeO contents, high-FeO chondrules experienced more heating and crystal growth than low-FeO chondrules. And one can easily show that the number of heating events increases exponentially as the energy fluence per event decreases.

SUMMARY

There are major difficulties associated with using the cooling rates inferred from furnace experiments as constraints for chondrule formation. By adopting the simplistic view that chondrules were heated only once, and that their textures must be understood in terms of monotonic cooling following this singular event, one has been forced to models that require chondrules to form in large, megameter-scale regions of the solar nebula.

It is most difficult to find an energy source that can be rapidly discharged to provide the enormous amounts of heat required by these models (ca. 36 kJ for each g of chondrule in regions that produced $\geq 10^{17}$ g of chondrules); the only plausible source seems to be gravitational energy release associated with the infall into the inner nebula of large parcels of interstellar matter. The problem is compounded by the need to melt the surfaces of most chondrules two or more times.

More tangible are the anticipated effects on the chondrules. The cores of these parental clouds would have remained at ca.1900 K

for periods of about 30 min, with the result that chondrule surfaces would have evaporated to a depth of 20 μm and, if chondrule melt was convectively mixed, bulk volatiles would have been depleted by large factors and relict grains would have dissolved.

For these reasons it behooves us to seriously consider that the interpretation that has been applied to the furnace experiments is incorrect. Rather than trying to explain textures by monotonic cooling following a singular event we should examine the possibility that each chondrule experienced many events that heated its glass enough to produce an incremental amount of diffusion and grain growth, and in many cases melting. It is now recognized that most chondrules experienced at least two events that heated their surfaces to 1800–1900 K or above. For each such event the chondrule may have experienced tens or hundreds of events that heated the glassy portions to 1500 K, sufficient to allow some growth onto mineral surfaces.

Such a multiple-heating scenario for chondrule evolution would bring the temperature-time record inferred from textures into agreement with that inferred from volatile-retention models. Removal of the constraint to heat megameter-scale nebular regions allows modelists to focus on cm-to-m scale physical processes in their quest for the elusive solution to the formation of chondrules.

ACKNOWLEDGEMENTS

I am indebted to R.H. Jones for discussions and for unpublished data on Semarkona chondrules and to G.W. Kallemeyn for the calculation of melt fractions. I thank J.N. Grossman, A.N. Krot, K.L. Rasmussen and A.E. Rubin for numerous discussions, R.H. Hewins, J.F. Kerridge, E.R.D. Scott and J.A. Wood for careful reviews, and B.G. Choi, J.J. Hong and R. Hua for technical assistance. This research mainly supported by NASA grant NAGW–3535.

REFERENCES

Boss A. P. and Graham J. A. (1993) Clumpy disk accretion and chondrule formation. *Icarus* **106**, 168–178.

Eisenhour D. D. and Buseck P. R. (1995) Chondrule formation by radiative heating: a numerical model. *Icarus*, in press.

Fujii N. and Miyamoto M. (1983) Constraints on the heating and cooling process of chondrule formation. In *Chondrules and their Origins* (ed. E. A. King), pp. 53–60. Lunar Planet. Inst.

Ghiorso M. S. and Sack R. O. (1995) Chemical mass transfer in magmatic processes IV. A revised and internally consistent thermodynamic model for the interpolation and extrapolation of liquid-solid equilibria in magmatic systems at elevated temperatures and pressures. *Contrib. Mineral. Petrol.* **119**, 197–212.

Gooding J. L. and Keil K. (1981) Relative abundances of chondrule primary textural types in ordinary chondrites and their bearing on conditions of chondrule formation. *Meteoritics* **16**, 17–43.

Grossman J. N. (1988) Formation of chondrules. In *Meteorites and the Early Solar System* (eds. J. F. Kerridge and M. S. Matthews), pp. 680–696. Univ. Arizona Press.

Hartmann L., Kenyon S. and Hartigan P. (1993) Young stars: Episodic phenomena, activity, and variability. In *Protostars and Planets III* (eds. E. Levy and J. S. Lunine), pp. 497–518. Univ. Arizona Press.

Hashimoto A. (1991) Evaporation kinetics of forsterite and implications for the early solar nebula. *Nature* **347**, 53–55.

Hewins R. H. (1988) Experimental studies of chondrules. In *Meteorites and* the Early Solar System (eds. J. F. Kerridge and M. S. Matthews), pp. 660–679. Univ. of Arizona Press.

Hewins R. H. and Radomsky P. M. (1990) Temperature conditions for chondrule formation. Meteoritics 25, 309–318.

Hood L. L. and Horanyi M. (1991) Gas dynamic heating of chondrule precursor grains in the solar nebula. Icarus 93, 259–269.

Horanyi M., Morfill G., Goertz C. K. and Levy E. H. (1995) Chondrule formation in lightning discharges. Icarus 114, 174–185.

Humayun M. and Clayton R. N. (1995) Potassium isotope cosmochemistry: Genetic implications of volatile element depletion. Geochim. Cosmochim. Acta 59, 2131–2148.

Imae N. (1994) Direct evidence of sulfidation of metallic grain in chondrites. Proc. Japan. Acad. 70B, 133–137.

JANAF (1985) JANAF Thermochemical Tables, 3rd ed., Amer. Chem. Soc., 1856p.

Jones R. H. (1990) Petrology and mineralogy of type II, FeO-rich chondrules in Semarkona (LL3.0): Origin by closed-system fractional crystallization, with evidence for supercooling. Geochim. Cosmochim. Acta 54, 1785–1802.

Jones R. H. (1994) Petrology of FeO-poor, porphyritic pyroxene chondrules in the Semarkona chondrite. Geochim. Cosmochim. Acta 58, 5325–5340.

Jones R. H. and Scott E. R. D. (1989) Petrology and thermal history of type IA chondrules in the Semarkona (LL3.0) chondrite. Proc. Lunar Planet. Sci. Conf. 19, 523–536.

Krot A. N. and Wasson J. T. (1995) Igneous rims on low-FeO and high-FeO chondrules in ordinary chondrites. Geochim. Cosmochim. Acta, 59, 4951–4966.

Kubaschewski O. and Alcock C. B. (1979) Metallurgical Thermochemistry, 5th ed., Pergamon, 449p.

Langmuir I. (1913) The vapor pressure of metallic tungsten. Phys. Rev. 2, 329–342.

Lofgren G. E. (1989) Dynamic crystallization of chondrule melts of porphyritic olivine composition: Textures experimental and natural. Geochim. Cosmochim. Acta 53, 461–470.

Love G. S. (1994) Implications of a phase-transition thermostat for chondrule melting (abstract). In Papers presented to the Conference on Chondrules and the Protoplanetary Disk, LPI Contribution No. 844, Lunar and Planetary Institute.

Lux G., Keil K. and Taylor G. J. (1981) Chondrules in H3 chondrites: textures, compositions and origins. Geochim. Cosmochim. Acta 45, 675–685.

Misawa K. and Nakamura N. (1988) Demonstration of fractionation among individual chondrules from the Allende (CV3) chondrite. Geochim. Cosmochim. Acta 52, 1699–1710.

Morfill G., Spruit H. and Levy E. H. (1993) Physical processes and conditions associated with the formation of protoplanetary disks. In Protostars and Planets III (eds. E. H. Levy and J. I. Lunine), pp. 939–978. Univ. Arizona Press.

Nagahara H. (1981) Evidence for secondary origin of chondrules. Nature 292, 135–136.

Radomsky and Hewins R. H. (1990) Formation conditions of pyroxene-olivine and magnesian olivine chondrules. Geochim. Cosmochim. Acta 54, 3537–3558.

Rambaldi E. R. (1981) Relict grains in chondrules. Nature 293, 558–561.

Rasmussen K. L. and Wasson J. T. (1995) A lightning model of chondrule melting. Unpublished manuscript.

Rasmussen L. and Wasson J. T. (1982) A new lightning model for chondrule formation (abstract). Conf. on Chondrules and their Origins.

Robie R. A., Hemingway B. S. and Fisher J. R. (1978) Thermodynamic Properties of Minerals and Related Substances at 298.15 K and 1 Bar (10^5 Pascals) Pressure and at Higher Temperatures. U. S. Geol. Survey Bull. 1452, U. S. Govt. Printing Office, 456p.

Rubin A. E. (1984) Coarse-grained chondrule rims in type 3 chondrites. Geochim. Cosmochim. Acta 48, 1779–1789.

Sahagian D. L. and Hewins R. H. (1992) The size of chondrule-forming events (abstract). Lunar Planet. Sci. 23, 1197–1198.

Scott E. R. D. and Taylor G. J. (1983) Chondrules and other components in C, O, and E chondrites: Similarities in their properties and origins. Proc. Lunar Planet. Sci. Conf. 14, B275–B286.

Wasson J. T. (1978) Maximum temperatures during the formation of the solar nebula. In Protostars and Planets (ed. T. Gehrels), pp. 488–501. Univ. Arizona Press.

Wasson J. T. (1993a) Multiplicity of chondrule heating events and the coarsening of chondrule textures (abstract). Lunar Planet. Sci. 24, 1489–1490.

Wasson J. T. (1993b) Constraints on chondrule origins. Meteoritics 28, 13–28.

Wasson J. T., Krot A. N., Lee M. S. and Rubin A. E. (1995) Compound chondrules. Geochim. Cosmochim. Acta 59, 1847–1869.

Wood J. A. (1984) On the formation of meteoritic chondrules by aerodynamic drag heating in the solar nebula. Earth Planet. Sci. Lett. 70, 11–26.

7: Unresolved Issues in the Formation of Chondrules and Chondrites

JOHN A. WOOD

Harvard-Smithsonian Center for Astrophysics, 60 Garden Street, Cambridge MA 02138, U.S.A.

ABSTRACT

Fifteen debating-points or issues crucial to an understanding of chondrites are identi-fied and briefly discussed. Particular importance is attached to Issue No. 3: "If chon-drules and CAIs were formed in the nebula, was it during the ~0.5 Ma stage of infall and rapid accretion, or the subsequent ~10 Ma stage of relative quiescence pictured by astrophysicists?" About 99.99% of the mechanical energy associated with interstellar cloud collapse and accretion of the sun is dissipated as heat during the brief, early stage named. Only ~10^{-4} of it is left to be spent in the subsequent residual minimum-mass nebula, which is where chondrite formation is customarily assumed to have occurred. Since copious amounts of energy were required to chemically fractionate the substance of chondrites and form chondrules and CAIs, this paper argues in an inter-pretive section that chondrite formation is most likely to have taken place during the early stage of infall and rapid accretion. However, the isotopic evidence speaks against this concept if Al was isotopically uniform in the solar nebula; to make the components of chondrites in <~0.5 Ma requires that $^{26}Al/^{27}Al$ must have been nonuniform.

INTRODUCTION

The first section of this paper identifies and discusses sixteen important issues that must be resolved before the origin of chon-dritic meteorites can be said to be understood. Some of them are familiar riddles that workers in the field have debated for years. Others, however, may surprise the reader: they are in the nature of boundary conditions that we have allowed to constrain our think-ing without, in my opinion, having established their validity. Some of these started as simple assumptions that needed to be made in order to test models, but they have evolved with time into axioms. If one of these self-imposed constraints is wrong, blind acceptance of it could thwart forever an understanding of the formation of chondrules, Ca, Al-rich inclusions (CAIs), and chondrites. I have demoted these axioms to the status of open issues. This section attempts to be objective.

A shorter second section abandons objectivity and outlines a model of chondrite formation, in the context of the issues discussed in the first section.

THE UNRESOLVED ISSUES

1. Were chondrules formed in the solar nebula?

Alternatives are that they formed in planetary environments, or in interstellar space near the nebula. The planetary environment usu-ally pictured is the surface of a planet during an epoch of heavy bombardment (Urey, 1967; Dodd, 1971, 1978; Fredriksson, 1963; Fredriksson *et al.*, 1973; Hutchison and Bevan, 1983; Hutchison *et al.*, 1988; King *et al.*, 1972; Sears, 1995). In this picture, chon-drules are solidified droplets of melted impact debris. The compo-sitional variability of chondrules may derive from heterogeneity in the target rock caused by earlier igneous fractionation. The strength of this model is that it relies on relatively well-understood processes (igneous fractionation, impact mechanics, devolatiliza-tion). However, it has trouble explaining the great abundance of chondrules in chondrites (chondrules are rare in the lunar regolith), the compositions of some chondrules (*e.g.*, silica-normative chon-drules; Brigham *et al.*, 1986), and the substantial accretional rims on some chondrules (*e.g.*, Rubin, 1984; Metzler *et al.*, 1992), which have no counterparts in the lunar regolith. Moreover, the chemical variability among chondrules does not correspond to planetary igneous fractionation trends (Grossman, 1988). The impact model for chondrule formation is criticized in detail by Taylor *et al.* (1983) and Grossman (1988).

Or, chondrules (Kurat *et al.*, 1984) and/or CAIs (Podolak *et al.*, 1990) might have been formed in dense early planetary atmospheres, by the drag heating, melting, and vapor fractionation of matter as it fell into them at high velocities. The atmospheric composition (a free parameter) might account for aspects of chondrule/CAI chemistry that seem inconsistent with a nebular environment, and drag heating is a calculable source of heat for melting and partial evaporation. Problems are that very large planetesimals must be invoked in the asteroid belt, where now are none, in order to hold atmospheres; the amount of chondritic material that could be made by this mechanism would be small compared to the volumes of the hypothetical planetesimals; and while the nebula was present its gas would damp out eccentric and inclined components of the motion of interplanetesimal protochondrules, which would prevent them from encountering planetary atmospheres at high enough velocities for drag to melt them.

The other alternative named is that an energetic astrophysical process operating in the presence of the nebula, but effectively outside of it, processed chondrules and CAIs. Skinner (1990) has proposed that these objects were formed in bipolar outflows associated with the protosolar disk, and Cameron (1994a) suggests that CAIs were. A disadvantage of this model is that it limits the options for chemical processing. In the near-vacuum of interstellar space melting can produce evaporation residues, but recondensation of the lost volatile component would be into ultrafine highly-dispersed dust which could not be brought together again as a discrete component. Thus condensation as a fractionating tool is not available in this environment, yet it is needed to explain the compositions of some CAIs (Boynton, 1978) and some chondrules (Brigham *et al.*, 1986; Wood and Holmberg, 1994).

Most meteoriticists accept the premise that chondrules and CAIs were processed and accreted in the solar nebula. It is frustrating that the required nebular heating mechanism (Issue 8) has eluded discovery for so long; but the benign properties of the nebula, which permit chemical fractionation *via* condensation as well as evaporation, cooling on an appropriate time scale (Issue 6), and the accretion of dust rims, are considered by most workers to outweigh this absence of understanding.

2. What is the relationship between CAIs and chondrules?

Were they formed by variants of the same basic process, or by different processes altogether? This question is seldom addressed. Each type of object has its own literature; CAIs and chondrules are very rarely discussed in the same paper. An exception is McSween (1977), who found a compositional continuum between amoeboid olivine aggregates (at the Mg-rich end of the compositional trend of refractory inclusions, which tends to follow the condensation sequence) and Type I chondrules (at the Mg-rich end of the compositional trend of chondrules), and suggested all formed by conden-

sation over a range of conditions which, low in the condensation sequence, dipped into the field of liquid stability (see Issue 5). Bischoff and Keil (1984) have also argued for a continuum of compositions and formative processes between CAIs and chondrules, but their two populations merge in a composition range that is Al-rich rather than Mg-rich, and the objects are created by melting of precursor dustballs rather than condensation.

Formation by different processes altogether, at times and/or places far removed from one another, have also been proposed (*e.g.*, Wark, 1979; Cameron, 1994a). More generally, formation by different mechanisms is implicit in the widely-held belief, based on radiometric studies, that chondrules were formed much later in early solar system history than CAIs (see Issue 4).

3. If chondrules and CAIs were formed in the nebula, was it during the ~0.5 Ma stage of infall and rapid accretion, or the subsequent ~10 Ma stage of relative quiescence pictured by astrophysicists?

The scenario of star and protostellar disk formation alluded to is summarized by Shu *et al.* (1987). Fig. 1 is from their paper. Stage *a* shown represents the formation of slowly rotating cloud cores, dense concentrations of matter in a parent molecular cloud. In Stage *b* a cloud core has become large and dense enough to be gravitationally unstable, causing it to collapse upon itself nonhomologously (*i.e.*, from the inside out). Most of the infalling cloud material has angular momentum that prevents it from joining the protostar directly; instead it falls onto a disk that forms around the protostar. The system is surrounded and hidden by an envelope of infalling gas and dust.

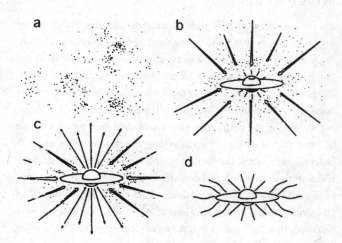

Fig. 1. Four stages of star formation described by Shu *et al.* (1987). *a*, Formation of dense cores in the parent molecular cloud. *b*, Self-gravitational collapse of a cloud core, with most material falling into a protostellar disk. *c*, A bipolar flow is created when the stellar wind is able to break through infalling matter at the poles of the protostar. *d*, After infall and disk accretion onto the central star cease, a gravitationally stable disk remains associated with the star for several Ma. Fig. from Shu *et al.* (1987), *Ann. Rev. Astron. Astrophys.*, **25**, p. 72, Fig. 7.

With time, the initiation of deuterium-burning in the protostar creates a strong stellar wind. Also with time infalling matter arrives from farther and farther out in the cloud source volume, so it has a higher angular momentum content and less of it falls directly onto the protosun. This leaves "holes" in the pattern of ram pressure created by infalling matter, at the poles of the protostar, through which the stellar wind can escape (Stage c), which Shu et al. suggest accounts for the bipolar outflows observed in association with young stellar objects.

Eventually the source of infalling matter is exhausted to large distances, and a strong stellar wind terminates the dwindling infall, exposing a T Tauri star surrounded by a remnant disk (Stage d). Disk evolution during Stages b and c has carried most of the matter that fell onto the disk inward, to be accreted by the central star; what is left in Stage d is a residual minimum-mass nebula. The nebula persists for a time, then is dissipated by poorly-understood processes.

Cameron (1994b) has stressed the importance of knowing the approximate duration of these stages of nebular history for stars in general and the solar system in particular. Podosek and Cassen (1994) have reviewed the astronomical evidence bearing on this question. Surveys of T Tauri stars at IR and mm-radio wavelengths reveal, in the majority of cases, excess luminosity (relative to the luminosity at visible wavelengths) at those wavelengths that can be accounted for only by models in which the stars are surrounded by dust disks which absorb stellar energy and reradiate it at longer wavelengths. The evidence for disks is associated with T Tauri stars up to 10 Ma in age; few stars >10 Ma old display excess long-wavelength radiation. The ages attached to T Tauri stars come from their positions in the Herzsprung-Russell diagram, relative to tracks in the diagram that are calculated from theoretical studies of the evolution of stellar structure.

It appears, then, that the total lifetime of the solar nebula (embracing Stages b, c, and d of Shu et al.) was ~10 Ma. During how much of this time was matter actively infalling to the disk (Stages b and c)? Podosek and Cassen review the evidence for active disks around T Tauri stars, which consists of IR luminosities so excessive that they cannot be entirely due to energy reradiation by dust disks; the extra IR is believed to be a component of radiation generated in the inner disk near the star, where mechanical energy is being dissipated as a large flux of disk material moves inward to the star. This is a process that would be important during Stages b and c (as matter collected by the disk moved into the protostar) but not during Stage d, when the flux of matter through the nebula was small. The evidence for active disks is associated with very young T Tauri stars, 0.1–1 Ma. This is in substantial agreement with the time (~1 Ma) it would take for interstellar material to fall to the sun from the limits of a plausible source volume (radius ~10^4 parsecs) at the free-fall velocity.

Thus in this paradigm two distinctly different phases of nebular history can be distinguished: a brief, relatively energetic period of <1 Ma during which the substance of the solar system fell into the nebula and most of it accreted onto the sun (Stages b and c of Shu et al., 1987); and a longer (~10 Ma), more quiescent period when

the sun had a residual minimum-mass nebula (Stage d of Shu et al.). Conditions were very different in these two periods, and it is unlikely that both were suitable for the formation of chondrites.

Most if not all published discussions of chondrite formation in the context of the solar nebula have assumed conditions appropriate to the quiescent minimum-mass phase of nebular history. Clearly radiometric data bearing on the ages of chondrite components (Issue 4) are crucial to an understanding of the roles the infall and quiescent nebula phases played in the processing of chondritic material.

4. Assuming chondrules and CAIs were created in the nebula, over what span of time did nebular processing form the chondrules and CAIs that ultimately became aggregated into any given small volume of chondritic material?

There are two sub-issues here: How much time separated the formation of CAIs from the formation of the chondrules and matrix (excluding presolar grains); and over what span of time did neighboring chondrules and matrix, which make up most of the volume of chondrites, form? I amplify on the last question first.

A consensus exists as to the chain of events in the solar nebula that produced chondrules, though it is not often plainly stated in the literature. This is as follows (e.g., Wasson, 1985; Grossman, 1988): (a) thermal processing of older (presolar?) dust in the nebula produced batches of dust of specialized composition by partial or total volatilization and recondensation; (b) this dust, remixed in various proportions, together with mineral grains from various sources, coagulated into porous dustballs; (c) unspecified transient high-energy events (Issue 8) melted the dust-balls; (d) they cooled promptly (Issue 6), forming chondrules; (e) dust also condensed, or was mixed from a different reservoir with the chondrules; (f) in some cases the chondrules accumulated rims of dust or perhaps partly-molten material; (g) the chondrules and remaining dust accreted into chondritic planetesimals.

Little thought has been given to the duration of this orderly chain of events, but it is necessarily somewhat protracted; in particular, the steps involving accumulation, (b) and (g), are bottlenecks. The coagulation of a chondrule-size dustball under nonturbulent conditions is estimated to take the order of a thousand years, though it can be more rapid in the presence of vigorous turbulence (see Issue 9).

On the other hand, I have argued that the formation and accretion of chondrules and dust had to be relatively rapid to prevent the characteristic populations of chondrules in the various chondrite types from being lost by mixing (Wood, 1985). The need for rapid formation and accretion of chondrules is especially compelling for the very rare and exotic merrihueite-bearing chondrules which occur concentrated in the Mezö-Madaras chondrite (Wood and Holmberg, 1994). Wood (1985) also pointed out that chondrules and matrix, separately, in CM2 chondrites contain non-cosmic Fe/Si, but in these chondrites the two components occur mixed in complementary proportions, so that Fe/Si in the bulk chondrites is approximately cosmic. This suggests that related, complementary processes formed the two components, which were then accreted rapidly and quantitatively, before they could be physically fractionated from one another or mixed with unrelated major components. Palme *et al.* (1992) have demonstrated that other elements are distributed in complementary proportions between chondrules and matrices of CV3 chondrites.

It is hard to sharply define this particular issue, since actual time scales of formation have not been put forward either for the traditional model of chondrite formation or for my reading of the time constraints. There is only a qualitative choice between the conventional many-stage, time-consuming process and a faster process with fewer stages, of the type proposed by Wood and Holmberg (1994).

The $^{129}I-^{129}Xe$ isotopic system has been used in an attempt to directly determine age differences of chondrules, *i.e.*, the span of time over which they formed. Apparent age differences as great as 10 Ma are routinely found between chondrules that occur in close proximity to one another. However there is reason to question the reality of these large age differences; for example, it seems impossible that the distinctive compositions, chondrule populations, and O isotope compositions (Clayton, 1993) of the various chondrite classes could have been preserved if their components had ~10 Ma to mix in the nebula before accretion. In some cases the age differences have been ascribed to resetting connected with postaccretional alteration (Swindle *et al.*, 1991a,b).

The other sub-issue defined above is the question of how much time separated the formation of CAIs from the formation of the chondrules and matrix. The radiometric data, taken at face value, indicate that CAIs formed early, and chondrules formed several Ma later. The primary evidence for this is the fact that most CAIs contained live ^{26}Al when they formed ($^{26}Al/^{27}Al \sim 5 \times 10^{-5}$), while those chondrules that have been studied appear to have contained much less or none (*e.g.*, Hutcheon *et al.*, 1994). However, the chondrules that have been studied are few and of specialized composition; they must contain modal plagioclase to provide high-$^{27}Al/^{24}Mg$ data points. Initial $^{26}Al/^{27}Al$ has not been determined for garden-variety ferromagnesian chondrules.

For CAIs and chondrules to have formed more closely together in time, it is necessary to place a less straightforward interpretation than this on the Mg isotopic data. Two assumptions are attached to the above interpretation: (1) Al was isotopically homogeneous in the solar nebula, and (2) Mg isotopic anomalies produced by ^{26}Al decay have not been subsequently disturbed by thermal metamorphism or hydrothermal alteration. In has been shown that these are not always safe assumptions, but could the breakdown of either of them plausibly account for the observed differences in Mg isotopic anomalies between CAIs and chondrules?

The evidence that ^{26}Al might have been heterogeneously distributed in the nebula is ably reviewed by Podosek and Cassen (1994). FUN inclusions C–1 and HAL contain little or no anomalous ^{26}Mg attributable to the decay of ^{26}Al, yet they display stable isotope anomalies that seem to militate against formation late in nebular history (surely mixing of precursor material in the nebula for millions of years after infall had ceased would have homogenized its isotopic composition). Fahey *et al.* (1987) describe a rimmed Type A CAI in which the rim contained higher $^{26}Al/^{27}Al$ than the core,

an arrangement that cannot be ascribed to ^{26}Al decay between the times of core and rim formation (however, Goswami *et al.* [1994] challenge this observation). Distinctive populations of hibonite (Ireland, 1988) and corundum (Anders *et al.*, 1991) grains that contained little or no live ^{26}Al have been found in the Murchison CM2 chondrite.

Thus the isotopic difference between (most) CAIs and (all studied) chondrules could point to their formation from precursor materials that, respectively, did and did not contain the canonical level of ^{26}Al, instead of their formation at times differing by millions of years. It may be significant that the CAIs and chondrules under discussion do not occur intimately mixed; practically all of the CAIs that contained ^{26}Al are in carbonaceous chondrites, while the plagioclase-bearing chondrules that have been studied are in ordinary chondrites.

It is somewhat harder to rationalize the apparent age differences on the basis of disturbed isotopic systems. There is abundant evidence for ^{26}Al-Mg systems that have been disturbed during or after ^{26}Al decay (*e.g.*, Podosek *et al.*, 1991). For this to explain the apparent differences in initial $^{26}Al/^{27}Al$ between CAIs and chondrules, however, the record in chondrule plagioclases must always be disturbed, while the record in CAI anorthites usually is not. The difference cannot somehow be related to plagioclase An content, because the unsuccessful search for ^{26}Al in chondrules of Hutcheon *et al.* (1994) measured plagioclases with a wide range of An contents. The only differences between the two sets of objects that might be invoked to rationalize their different Mg isotopic compositions, other than disparate ages, are (1) the presence of the chondrules in ordinary chondrites (more metamorphosed, even Semarkona, than C chondrites?), and (2) the proximity of plagioclase in these chondrules but not in the CAIs to a large reservoir of Mg of normal isotopic composition, in a mineral in which Mg diffusion is relatively rapid (Mg olivine).

Why strain for a way of making CAI and chondrule formation coeval? Why not accept the apparent age differences? Because it is very awkward to have to store CAIs, made early, in the solar nebula for 5–10 Ma before accreting them in late-formed chondrite parent bodies. In a quiescent nominal minimum-mass nebula, drag-induced radial drift would cause chondrule-size objects to migrate from anywhere in the disk to the sun in <10 Ma (Weidenschilling, 1977). Centimeter-size objects (*e.g.*, some CAIs) would make the trip much more rapidly. Further, CAIs were very unevenly distributed among the various chondrite groups. The types, sizes, and abundances of CAIs differ among CV3, CO3, CM2, ordinary, and enstatite chondrites (MacPherson *et al.*, 1988). A way would need to be found not only of preventing the CAIs from drifting in to the sun, but of allocating them selectively to various sites of chondrite formation. It is arguably more straightforward to form characteristic populations of CAIs and characteristic populations of chondrules together, then accrete them promptly.

Niemeyer (1988a,b) has shown that some CV3 chondrules display ^{50}Ti (stable) isotope anomalies. As in the case of CAIs, this seems to require early formation.

5. Did chondrule/CAI formation always occur in a well-mixed system of cosmic composition, or was the bulk chemistry sometimes changed by physical fractionation of components that condense at different temperatures (involatile minerals, organic compounds, ices, uncondensable gases)?

The simplest and most aesthetic assumption to make is that the nebula remained well-mixed. However, a review of natural systems, both geological and astrophysical, reminds us that homogeneity is rare; at some metaphysical level natural systems seem powerfully impelled to unmix. The nebular mixture of dissimilar materials—solid particles and very thin gas—has a legitimate basis for unmixing. Turbulence can keep the nebular system mixed on a large scale, but high-velocity turbulence can also promote gas/dust fractionation on a local scale (Cuzzi et al., 1995 and this volume).

The creation of dust-rich zones by gas/dust fractionation would profoundly affect processes of chondrule, CAI, and chondrite formation. (1) The mean free times between collision of solid particles or molten droplets are shortened. (2) The local IR opacity is increased, and the radiative cooling rates of hot objects are slowed. (3) The bulk chemistry of local nebular material is changed, essentially by decreasing the ratio of H and He to everything else. Metal vapor pressures are increased, perhaps to the point where silicate melts are stable; the Fe^{2+} content of minerals and melts in equilibrium with the gas phase is increased (Wood, 1967); and Na loss from melted silicates is slowed (Tsuchiyama et al., 1981) or perhaps prevented (Lewis et al., 1993; Yu and Hewins, 1994). In general, these effects would make chondrite formation easier to understand. For the most part, only the effects of silicate dust concentration have been contemplated, but condensed organic compounds (e.g., Connolly et al., 1994) and even ices may also have been concentrated in the nebula. These could have interesting effects on local chemistry.

6. What were the time scales of chondrule and CAI formation?

Beginning about 1980 a series of controlled laboratory simulations of chondrule formation (reviewed by Hewins, 1988) have constrained the mean cooling rates of most chondrules (the barred olivine and porphyritic varieties) to the range 100°–2500°C/hr; i.e., the time scale for cooling to the solidus was the order of an hour. This is a very important astrophysical constraint, because it is too slow to allow cooling of the chondrules by radiation to free space but much too fast to correspond to global cooling of the embedding nebula, or any considerable portion of it. Such a (radiative) cooling rate can be obtained in a nebular environment of suitable optical opacity (Sahagian and Hewins, 1992).

Early experiments showed that chondrules melted for the ~1 hr indicated by cooling rate determinations, in systems having the cosmic composition, should become substantially devolatilized of Na (e.g., Tsuchiyama et al., 1981). Wasson (1993 and this volume), noting that volatile elements like Na and S are present in chondrules in approximately their cosmic proportions, concludes that devolatilization did not occur and therefore the chondrules were not above 1000 K for longer than ~100 sec, which would correspond to a mean cooling rate much faster than the one found by

experimental petrologists, >30,000°C/hr. Wasson considers that this constraint based on volatility is more compelling than the laboratory simulations, and suggests that the textures and microchemical zoning patterns of silicate minerals in chondrules could have been formed by a series of very brief heating events as well as by the more protracted cooling histories employed by experimental petrologists.

However, more recently Yu and Hewins (1994) have shown that if experimental charges are flash heated in a relatively oxidizing atmosphere and cooled approximately exponentially instead of linearly (arguably realistic conditions for the nebular setting), Na loss is relatively minor. Moreover, Zanda et al. (1996) found that chondrules did in fact lose most of their S when they were melted, but during cooling much of the S rejoined the chondrules as troilite rims, apparently via reaction of S-rich nebular gas with metal droplets on the chondrule surfaces. A portion of the troilite in chondrules did survive the melting event, but Yu et al. (1994) showed that under the experimental conditions which preserved most Na in their chondrule analogues, some troilite is also retained.

Type B CAIs, which lend themselves to cooling rate determinations via experimental studies of molten analogue systems, appear to have cooled more slowly than chondrules, <~1–50°C/hr (Stolper and Paque, 1986).

7. Did chondrules melt as closed systems or open (to the nebular environment) systems?

"Open" implies both losses to and gains from the nebula. Three cases can be considered. (1) The molten droplets were largely closed to their gaseous environment; they were not hot long enough to react with the nebula. (2) While molten, the droplets were actively volatilizing, but the outward flux of vapors shielded them from reaction with the ambient gas; thus they were open to losses, but not to reaction with the nebula. (3) The droplets were open in both directions to their environment.

Most workers (e.g., Grossman and Wasson, 1985; Grossman, 1988; Jones, 1990) have concluded that chondrules formed as (kinetically) closed systems (case 1). This interpretation has the virtue of simplicity, and it explains why the molten droplets did not evaporate (as the equilibrium phase diagram for a system of cosmic composition says they should, at temperatures high enough to melt).

Droplets held at high temperatures (~1800°C) for the order of an hour lose substantial amounts of Fe, Si, and Mg by vaporization (Hashimoto, 1983). This would fulfill part of the definition of case 2, though it is not clear that the effusion rate of metal oxides is great enough to prevent the composition of the gaseous environment (especially its redox state) from affecting the residues. Some CAIs, or components of CAIs, may have been formed in this way, but chondrules do not have bulk compositions that correspond to evaporation residues.

Experiments show that at the lower temperatures which dominated the thermal processing of chondrules, the evaporation rate of major moderately-volatile elements is not great enough to fulfill the definition of case 3 (a system completely open to both losses to and gains from the environment). However if chondrules were formed by the melting and aggregation of dispersed dust grains instead of preaccreted dustballs (Issue 9), case 3 would apply to the hot dust grains and microdroplets prior to aggregation because of their large surface/volume ratios.

8. What were the pervasive impulsive heating events that created chondrules (and CAIs)?

This is the perennial unanswered question of meteoritics. Several proposed non-nebular heating mechanisms were discussed under Issue 1. Among nebular heat sources, the most frequently cited (*i.e.*, by >1 person) possibilities are lightning (*e.g.*, Whipple, 1966; Morfill *et al.*, 1993; Wasson, 1993; Eisenhour *et al.*, 1994; Horanyi, this volume) and shock heating (Hood and Horanyi, 1993; Boss and Graham, 1993; Ruzmaikina and Ip, 1994 and this volume). In the case of shock, the principal source of heat is not shock compression of the enclosing gas, but acceleration of the gas past solid particles, which heats the particles by friction (drag).

Several other nebular mechanisms for chondrule production have been advanced recently. The chondrules may have been produced near a contact discontinuity between the nebula and an inner annulus cleared by early pre-main sequence solar eruptions, by reaction between the hot shocked wind and clumps of presolar dust embedded in the nebula (Huss, 1988). Levy and Araki (1989) have suggested the dustballs were lofted close enough to the disk surface to have been melted by radiant energy emitted by magnetic reconnection flares outside the nebula. Chondrules may have been ablated from surfaces of planetesimals by hot, dense, high-velocity gas at the disk/protosun interface, entrained in bipolar outflows emanating from that region, and projected ballistically to the disk region where chondrites were formed (Liffman 1992 and this volume). Cameron (1994a) proposes that chondrules were melted in the same setting envisaged by Huss (above), but by radiant energy from nearby magnetic reconnection flares (as in the Levy and Araki model) rather than by hot shocked gas.

9. Did chondrules form by the melting of cold, preformed dustballs, or by accumulation of hot droplets and dust?

Cold, preformed dustballs are universally assumed. Only Skinner (1991) has been willing to critically evaluate the assumption. Workers concerned with the origin of chondrules have been happy to accept the concept as an initial condition, so they can focus their attention on the question of an energy source. The melting of dustballs would provide sizeable droplets of molten material from the outset, and this has seemed to be a requirement, to kinetically protect the melted material from evaporation (Issue 7); dispersed dust, if heated and melted, would vaporize very quickly because of its high surface/volume ratio. However, this concern is rooted in the assumption that the system had an unfractionated, cosmic composition (Issue 5). Fractionated systems can have a high enough abundance of cationic elements to saturate the vapor after only part of the dust has evaporated, and thereafter residual molten microdroplets could exist as stable open systems. Wood and Holmberg (1994) proposed that aggregation and melting of chondrules

occurred in one step, not two, in an effort to simplify and accelerate chondrule formation (Issue 4).

A difficulty with the dustball concept is that melting of these objects by a transient, local heat source, whatever its nature, inevitably should leave some of the dustballs (those that were on the periphery of the event) only incipiently melted. Objects that might represent this condition are rarely observed in the chondrites. The presence of relict mineral grains in chondrules is always interpreted as resulting from their earlier presence in precursor dustballs, and this has come to seem like evidence for the existence of the dustballs; but in fact a dense system of hot, aggregating material could easily include mineral grains from an earlier cycle of nebular activity as well as microdroplets of melt.

Moreover, it is not straightforward to create a population of dustballs in the first place. This problem is commonly ignored. A dustball needs to be aggregated rapidly in order to preserve any special composition that had been established by a previous fractionating process; dustballs that accreted gradually (and the chondrules created by their melting) should be nearly uniform in composition. A number of authors (*e.g.*, Wieneke and Clayton, 1983; Nakagawa *et al.*, 1986) have shown that it would take the order of a thousand years to accumulate dustballs in an unfractionated nebula, a long time. The time scale decreases in inverse proportion to the degree of dust/gas fractionation, but the settling and concentration of unclumped dust is also extremely slow (Weidenschilling, 1988). In the presence of strong turbulence coalescence rates can be increased, but then weak dustballs are vulnerable to disruption by high-velocity collisions (Weidenschilling, 1988). Coalescence becomes much more plausible if sticky molten microdroplets, instead of dry solid particles, are encountering one another.

Some chondrules have rims of coarse-grained material (see especially Rubin, 1984). Rubin interprets these as having originated by the accretion of fine particulate material onto the chondrule cores, after which heating events sintered or partially melted the rims to their present configurations. However, they also could have formed by the accretion of finely-dispersed, partly-molten material onto chondrules formed in earlier cycles of activity. A few chondrules consist *entirely* of coarse rim material (Fig. 2). These could be seen as examples of the aggregation of hot dispersed material, in several stages, at temperatures too low to allow the material to blend into homogeneous igneous systems.

An obvious problem with the chondrules-from-microdroplets concept is that one would expect to find abundant preserved microchondrules in the matrices of primitive chondrites. Microchondrules do exist (*e.g.*, Mueller, 1968; Rubin *et al.*, 1982), but like incipiently-melted dustballs, they are rare.

10. Why does each chondrite class have such a sharply defined size distribution of chondrules?

Size distributions have been measured by, *e.g.*, Martin and Mills (1976) and Hughes (1978). Most workers have given little thought to the meaning of the narrow range of chondrule sizes, yet this observation is probably capable of sharply limiting the number of possible chondrule-forming mechanisms. Dodd (1976a) attributes

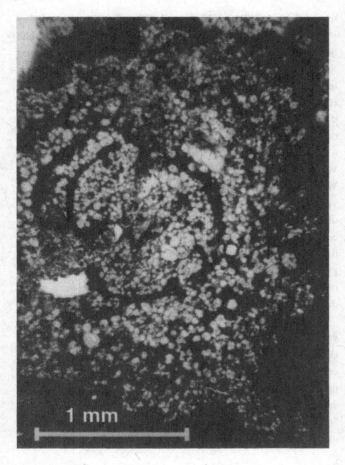

Fig. 2. Chondrule in Allende that consists entirely of concentric layers, *i.e.*, rims, which may have been accreted from a spray of hot microdroplets and mineral grains.

the size distribution to aerodynamic sorting, but Martin and Mills (1976) object that then a way must be found to dispose of the unwanted size fractions. Rubin and Grossman (1987) raise the possibility that size distributions were affected by the collisional breakup of droplets having different strengths (a function of viscosity, hence of composition and temperature). Or surface tension may have been the issue; at a velocity of ~1 m/sec, the kinetic energy of a 1–mm diam. melt droplet equals its surface energy (assuming 350 ergs/cm² for the latter). If chondrules were melted in a crowded, turbulent zone, mutual collisions may have been too frequent to allow the survival of small droplets, yet when droplets grew too large and weak they became vulnerable to dispersal by additional collisions. Liffman (1992) discusses the control that gas dynamical forces and surface tension would exert on droplet size. Cuzzi *et al.* (this volume) show that the process of turbulent concentration of solids would inevitably sort them on the basis of size and mass density.

11. Did the various chondrite classes form at different radial distances in the nebula, or at different times in the history of the nebula, or both; or were they differentiated in some other way altogether?

The idea that the chondrite classes formed at different radial distances in a quiescent minimum-mass (Stage *d*) nebula, essentially in the same time interval, has been examined in the most detail and championed most consistently by Wasson (*e.g.*, 1977). In Wasson's model, nebular conditions produced meteorites that were more oxidized and contained a larger component of refractory elements, farther out in the disk. The enstatite chondrites formed relatively near the sun, the ordinary chondrites in the inner asteroid belt, and the carbonaceous chondrites at >5 AU. The "stratification" of compositional types (from reflectance spectra) observed in the present asteroid belt (Gradie and Tedesco, 1982) supports this concept.

Clearly it is also possible that as the nebula evolved, ongoing chondrite-forming processes and their products also evolved, and a range of different chondritic materials might have been produced at the same radial distance. Again, time and radial distance might both have been factors in producing chondritic diversity.

12. Do all representatives of a given chondrite class (*e.g.*, L chondrites) come from a single parent body?

The literature often makes reference to "the L-chondrite parent body," "the H-chondrite parent body," *etc.*, as though it has been proven that there was only one of each. This is not the case. It is highly probable that the chondrites which form peaks in the distributions of cosmic ray exposure ages, such as the prominent peak in ²¹Ne ages of H-chondrites at 8 Ma (Schulz *et al.*, 1991), were ejected by a single impact from a single parent body. A similar argument can be made for the L-chondrites that form an ~0.5 Gy peak in the histogram of concordant U,Th-He and K-Ar ages (Taylor and Heymann, 1969). However, chondrites of the same chemical class that form other peaks, or plot away from peaks, can have come from other parent bodies. Wasson (1974, p. 185) notes that the bimodal distribution of olivine compositions in L-chondrites observed by Keil and Fredriksson (1964) (see also Mason, 1963) is suggestive of more than one parent body.

If, as is traditionally supposed, each chemical class of chondrites sampled material at a particular radial distance in the nebula during some limited interval of time, then this material originally occupied an annular volume of vast circumference; it is very unlikely that only one planetesimal would nucleate and accrete in such a huge space. (Multiple early planetesimals may have tended to merge, but given the finite eccentricities and inclinations of these bodies there is little reason to expect that the mergers would have involved planetesimals of identical chemical class, since the composition of accretable material is thought to have varied with radial distance [Issue 11]).

Within each chemical class, the material which is most metamorphosed and therefore presumably accreted earliest is almost isochemical with material that accreted later, for all but the most volatile elements. This implies one of two things: (1) Each class had a very large annular reservoir of raw material to draw on, only a small fraction of which was accreted into planetesimals. If each reservoir was so small that the last of it was used up in accreting the Type 3 layers of planetesimals, it would be very difficult to maintain compositional homogeneity throughout the stratigraphic column. (2) Or the accretion mechanism was rapid and indiscriminate, providing no opportunity for physical fractionation mechanisms to bias the compositions of materials that would occupy deep and shallow positions in the planetesimals. The Goldreich and Ward (1973) accretion mechanism might have accomplished this, but the concept has been discredited (Weidenschilling, 1988; Cuzzi *et al.*, 1993 and this volume).

Dodd (1976b), from an examination of 63 wet-chemical analyses of ordinary chondrites, reported Fe is depleted in late-accreted (low metamorphic grade) H, L, and LL chondrites, which would be an exception to the above generalization about isochemistry across petrographic types. However Kallemeyn *et al.* (1989), who compared INAA of 66 ordinary chondrites, were not able to reproduce this effect.

13. Were the volatile-element depletions displayed by chondrites caused by the thermal processing that created chondrules, or were they already present in the bulk material that chondrule formation operated on?

Almost all chondrites have lost volatile elements, or incompletely acquired them when they were formed. Anders and co-workers (*e.g.*, Keays *et al.*, 1971; Laul *et al.*, 1973), drawing upon a suggestion of the author (Wood, 1963), argued that the presence in differing proportions of two chemical components, chondrules that had been devolatilized when they were melted and matrix that retained its cosmic complement of volatiles, largely accounts for the differing patterns of volatile elements in the various chondrite classes. (Largely but not wholly. Anders and co-workers recognized from the outset that the chondrule-forming event seemed to have depleted some elements more than others, so theirs was not strictly a two-component model.) This interpretation was challenged by Wasson and co-workers (*e.g.*, Wasson and Chou, 1974; Wai and Wasson, 1977), who explained volatile element depletions in a very different way, by the systematic withdrawal of gas containing uncondensed volatile elements as the protochondritic system cooled (or by settling of the solids to the midplane on the same time scale as nebular cooling, which comes to the same thing; it gradually increases the solid/gas ratio). Chondrules were then made from this depleted material in a later melting event, which did not involve volatile loss.

A series of papers in the 1970s disputed whether the pattern of volatile element depletions in fact declines monotonically (Wasson) or traces a sigmoid pattern (Anders) when plotted against condensation temperature. More recent analytical data for one particular chondrite class (CV3) are shown in Fig. 3. (Nominal condensation temperatures in Fig. 3 are from Wasson [1985]. I say "nominal" because condensation temperatures are extremely model-dependent and difficult to estimate. The wonder is that such

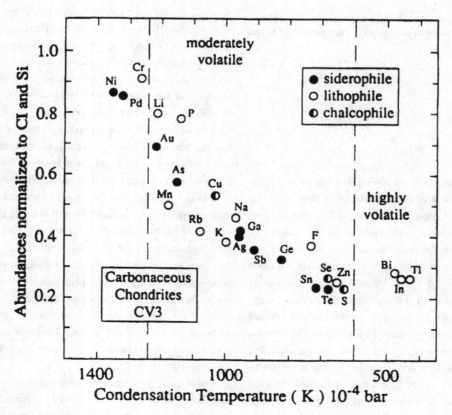

Fig. 3. Abundances of moderately volatile elements in CV3 chondrites, normalized to abundances in CI chondrites and Si=1.0; plotted against the temperature at which 50% of each element would be condensed in an unfractionated nebula at a pressure of 10⁻⁴ b. Fig. from Palme and Boynton (1993), p. 987, Fig. 4, in *Protostars and Planets III* (eds. E. H. Levy and J. I Lunine), University of Arizona Press.

a good correlation as that in Fig. 3 can be produced.) The depletion pattern is not consistent with a strict two-component model, for which it would be a step-function. However, it is significant that the trend in Fig. 3 is asymptotic to an abundance level of ~0.25×CI, not zero, suggesting the importance of a component at the ~25% level (*i.e.*, matrix) that contains volatile elements in the cosmic proportions. This is not consistent with the Wasson concept, which predicts no plateau at low temperatures.

The issue of whether volatile element depletions occurred prior to or during chondrule formation remains unresolved. As a practical matter, the unlikelihood of globally-high nebular temperatures at 3 AU, the very slow settling rates of dust in the nebula (Issue 9), and the ~1-hr cooling time of chondrules (Issue 6) make Wasson's model, sketched above, difficult to implement. The observation that S is concentrated as troilite on the surfaces of chondrules of primitive chondrites, apparently having rejoined the chondrules as they cooled (Zanda *et al.*, 1996), shows that chondrule formation was not unrelated to volatile element depletions.

The models of Anders and Wasson are not the only possible alternatives. Both of these models assume equilibrium condensation (modulated by the kinetics of solid-state diffusion). If chondrite accretion followed rapidly on the heels of chondrule formation (Issue 4), the kinetics of condensation and reaction may have been important. Volatile elements outgassed during chondrule formation may have had time to recondense only partly before accretion isolated a chondritic system from the gas phase. In a monotonically cooling system, the volatile elements with highest condensation temperatures would begin condensing soonest and have the longest time to condense before accretion, so they would be least depleted in the chondritic material formed; elements with lower condensation temperatures would be progressively more depleted, which is the trend shown in Fig. 3 down to the plateau. The presence of a plateau seems to require that unfractionated matrix material was mixed into the system before or during its aggregation.

A related possibility is that hot chondrules and dust were concentrated by turbulence (Cuzzi *et al.*, this volume) and aggre-gated while they were still cooling from the chondrule-forming event(s) and before the condensation of volatiles was complete. This can be considered a variant of Wasson's model in which gas is withdrawn from solids while volatile elements are condensing; differences are that fractionation occurs because of turbulent concentration instead of gravitational settling to the nebula midplane; it happens after chondrule formation instead of before; and the time scale is short instead of long.

14. What produced the differences in Si, Mg, and Fe contents between the chondrite classes?

The proportions of Si, Mg, and Fe vary substantially among the chemical classes of chondrites (Fig. 4). Since these three elements and associated O account for ~90% of the mass of chondrites (except hydrated carbonaceous chondrites), the processes that fractionated them were fundamental to chondrite formation.

The variability of Fe/Si has been attributed to fractionation in the nebula between metal and silicates (*e.g.*, Larimer and Anders, 1970; Wasson, 1972). The evidence for this is very strong: Ni/Si and Co/Si variations among chondrite groups correlate with Fe/Si. A strictly physical fractionation could have been involved, caused by differences in (*e.g.*) the densities or magnetic properties of metal and silicate grains; or the fractionation could have been keyed to temperature in the condensation sequence. Especially if nucleation barriers depressed the condensation temperature of metal (Blander and Katz, 1967), differing values of Fe/Si could be ascribed to variable proportions of high- and low-temperature condensates.

Magnesium and Si do not form separate minerals, as siderophile and lithophile elements do, so there is no basis for a strictly physical mechanism of fractionation. There seems no alternative to fractionation during condensation (or evaporation) to produce the observed variations in Mg/Si. The value of this ratio in equilibrium condensate changes drastically through the temperature interval of ~200 K in which Si condensation reacts forsterite to enstatite; if aggregation favored early or late condensates in this temperature

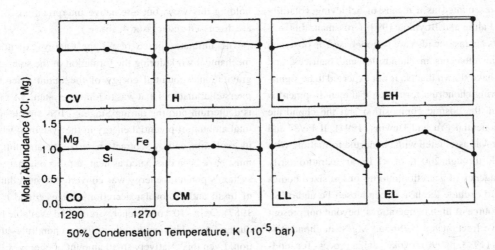

Fig. 4. Relative abundances of the major elements Mg, Si, and Fe in eight chondrite classes, normalized to abundances in CI chondrites and Mg=1.0; plotted against condensation temperature.

range, Mg/Si in the matter aggregated would be strongly affected. Unfortunately Fig. 4 shows no particular tendency for chondrite groups that are enriched or depleted in high-temperature condensate according to the Fe/Si criterion (above) to be biased in the same direction according to the Mg/Si criterion.

The real issue here is how to fractionate high- from low-temperature condensates. Two things are required: a high-temperature event, and a means of separating solid particles (enriched in relatively involatile elements) from coexisting vapor (enriched in more volatile elements). The separation must occur on a time scale comparable to the time in which the system cools by 200 K. If it takes much longer than that, subsequent condensation of the more volatile elements removes the opportunity for fractionation. The traditional, and most efficient, way of separating solids from gas is by aggregation, so only the outermost grains in a clump are left in reactive contact with the gas. It is not enough just to aggregate chondrule precursors, however, because the volatile elements left behind can still condense into matrix grains and end up contributing to the bulk composition of the chondrite. (Actually it seems something like that *did* happen, accounting for the complementary distribution of Fe and other elements among chondrite components discussed under Issue 4; but aggregation still had to be rapid enough relative to cooling to leave the differences in Fe/Si and Mg/Si that define the chondrite groups.) What is needed to preserve compositional differences between the groups is another cycle of rapid aggregation, on a scale much larger than that of chondrule precursors—aggregation into m- or km-scale bodies.

What were the high-temperature events that achieved these element fractionations? Among possibilities it is obligatory to list (1) a nebula that was globally hot, of the sort projected in Cameron's (1962) classic paper. Cameron's nebula cooled through the 200 K of Mg/Si fractionation in ~2 yr. However, most nebular models of more recent vintage have not been hot enough at ~3 AU to vaporize silicates (*e.g.*, Boss, this volume, Large Scale Processes), though Cassen (1992) has argued for a hot inner nebula during the infall stage of nebular evolution. (2) Abundant small, transient thermal events melted the chondrules. The chondrule-forming events are widely discounted as a means of achieving volatility fractionation (*e.g.*, Palme and Boynton 1993; Grossman, this volume); but arguments against the idea are largely based on observed abundances of alkali elements in chondrules and matrices, and recent experiments have shown that Na is not expected to be significantly depleted during chondrule formation if the environment is more oxidizing than the cosmic composition and the chondrule cooling rate is exponential (Yu and Hewins, 1994). If Fe/Si and Mg/Si fractionation was associated with chondrule formation, with a cooling time scale through 200 K of <1 hr, an extraordinarily rapid and efficient agency of accretion into m- or km-sized bodies would be required to exclude the last, uncondensed Fe and/or Si. (3) Or some other source of high temperatures, beyond our present imagining, may have fractionated Fe/Si and Mg/Si in chondrites. Palme and Boynton (1993, p. 990) invoke such a process for moderately volatile elements ("...the duration of the heating event may have been quite short as indicated in Figs. 5 and 6 where significant losses of volatiles are shown to occur within a few days [the time scale for incomplete condensation would depend on kinetics and is thus difficult to predict]").

15. What happened to the relatively volatile elements, both major (Issue 14) and minor or trace (Issue 13) that fractionation excluded from chondrite formation?

This question is rarely asked. Three possible answers come to mind. (1) Chondrite-formation may have been a very wasteful process: only a small fraction of the potential planetary material in the nebula was thermally processed and gathered into chondritic planetesimals. Leftover volatile elements mixed back into the nebula at large and had no appreciable affect on its composition (Larimer and Anders, 1967). (2) Volatile elements were excluded from chondrites because of the promptness of aggregation after high-temperature processing, as suggested under Issue 13. This effect operated no matter how high the concentrations of residual volatile elements became in the nebula. The volatile elements were eventually lost upon dissipation of the nebular gases, by whatever means. (3) The volatile elements eventually condensed, and the condensate aggregated into volatile-enriched volumes of material that were to some extent the chemical complements of normal chondrites (*e.g.*, the "mysterite" of Higuchi *et al.*, 1977 and Grossman *et al.*, 1980).

A PARTISAN DISCUSSION OF CHONDRITE FORMATION

In the writer's opinion, the third issue laid out in the preceding section is by far the most important one. The nebular environment available for chondrite formation during the ~0.5 Ma period of infall and rapid protosolar accretion differed radically from circumstances during the subsequent ~10 Ma Stage *d* of relative quiescence. The latter protracted stage is the one traditionally employed by meteoriticists, partly just because it is a simpler, more familiar setting. (The isotopic evidence for CAI and chondrule formation over a long span of time is a more defensible reason for holding this view, but alternative interpretations of this evidence are discussed under Issue 4, above.)

The ultimate source of all the energy expended as heat or mechanical work during the formation of the solar system was the gravitational potential energy of the initial extended volume of interstellar material that was to join the system. (A minor exception is deuterium-burning during Stage *c*.) It is simple to calculate the total amount of potential energy in the system at various stages of its evolution, and by subtraction derive the amounts of energy that must have been dissipated as heat. It turns out that ~99.99% of the system's potential energy was converted to heat during the period of infall and protosolar accretion (Stages *b* and *c* of Shu *et al.*, 1987). Only ~10^{-4} of the energy remained available for expenditure during Stage *d* in (among other things) chondrule- and CAI-formation. Even this relatively small amount of energy was not all spent; some of it still remains in the planets as orbital and potential energy. Since the application of energy in generous amounts seems essential to formation of the ingredients of chondrites, the author submits that it is much more likely the latter were formed early in the nebula, in Stages *b* and *c*, than in Stage *d*. This is the principal point the present paper attempts to make.

The nebular environment during Stages *b* and *c*

During Stages *b* and *c* the angular momentum content of most infalling matter constrained it to fall onto the disk, not into the protosun directly. There being no (known) orderly process that would promptly remove the angular momentum from this material and allow it to flow into the sun, matter accumulated in the disk, driving the disk/sun mass ratio to high values. When M_{disk}/M_{sun} exceeded $1/3 - 1/4$ the disk became gravitationally unstable (*e.g.*, Cassen *et al.*, 1981; Shu *et al.*, 1990) and probably separated into a pattern of spiral density waves (*e.g.*, Cameron, 1994a). Tidal interactions in this complex gravitational field had the effect of transferring angular momentum outward in the system, and mass inward to the sun. Thus the pileup of infalling mass in the solar nebula was a self-correcting process. If the correction described was efficient and rapid, gravitational stability could be reestablished and some time might elapse before continued infall made the system unstable again. In this case accretion of disk material onto the protosun would have been episodic.

Some young stellar objects display episodic increases in brightness (FU Orionis eruptions), which may be due to increased energy dissipation during periods of rapid mass accretion from their disks onto the central protostars (*e.g.*, Hartmann *et al.*, 1993; Fig. 5). In solar-mass protostars, brightness increases by ~100× on a time scale of months or years and decays on a somewhat longer time scale. The number of FU Orionis eruptions observed in the solar neighborhood in the last ~50 yr is ~10× the rate at which stars are being formed in the same volume (Hartmann and Kenyon, 1985),

so each star experiences roughly ten such eruptions in its youth if this phenomenon accompanies the birth of all stars. (However, since the eruptions would be spaced ~10^5 yr apart, it is impossible to confirm from observations that the phenomenon is universal among young stellar objects.)

This paper entertains the possibility that FU Orionis eruptions are in fact manifestations of rapid energy dissipation that occur when disks become gravitationally unstable, and tidally reconfigure their excessive mass so most of it is rapidly dumped into the sun; and that the high-energy processing of protoplanetary material which is recorded in the chondrites occurred during comparable dissipation episodes that affected our own solar nebula.

The formation of chondrules and CAIs

The copious dissipation of energy that causes the observed luminosity increase during an FU Orionis event is thought to occur where mass is accreting onto a protostar, at its interface with the surrounding disk. The writer considers this an unsuitable setting for chondrule and CAI formation (because of the difficulty of cooling these objects there on the appropriate time scale, and problems associated with transporting them to what is now the asteroid belt). However, lesser amounts of energy dissipation would also be occurring at all radial distances in a disk during the process of tidal reconfiguration pictured. The question of the nature of the mechanical processes that dissipate energy at radial distances which correlate with the present asteroid belt has not even been asked, so it is impossible to examine the process of chondrule and CAI formation in this context. In the most general (and vague) spirit, it seems likely that powerful shocks and turbulence (perhaps supersonic turbulence) would be important, and that interactions between these shocks and concentrations of protoplanetary dust may have created the chondrules and CAIs. Shock as a means of forming chondrules

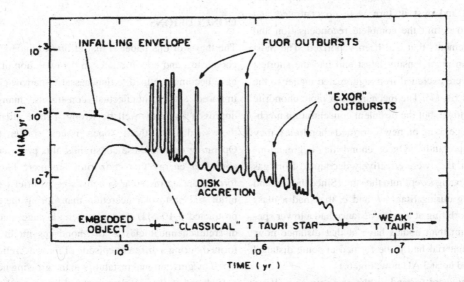

Fig. 5. Schematic illustration of the variation of rates of mass transfer (ordinate) with time (abscissa) during the evolution of a young star/disk system. The dashed curve shows the (declining) rate of infall into the system; the solid curve represents the rate of flow of disk material into the protostar. FU Orionis outbursts are thought to represent transient increases (spikes) in this accretion rate. Fig. from Hartmann *et al.* (1993), p. 499, Fig. 1, in *Protostars and Planets III* (eds. E. H. Levy and J. I Lunine), University of Arizona Press. L. Hartmann (*pers. comm.*) notes that abundant FU Orionis eruptions probably also occur to the left of the disk accretion curve's intersection with the infalling envelope curve, but along with the protostar they are hidden by dense surrounding cloud material.

has been explored by Hood and Horanyi (1993) and Ruzmaikina and Ip (1994 and this volume). As already noted, the main heating mechanism in a shock is not compression of the gas, but drag on solid particles produced when the shocked gas races past them.

Other aspects of chondrite formation

In this section I enlarge upon a model for the formation of chondrites during episodes of disk mass accretion. However, the value of the broad concept of chondrite formation in this setting, and during Stages *b* and *c* more generally, transcends the ultimate correctness of the details to follow.

Chondrules and CAIs were formed by the heating of dispersed but concentrated dust, not dustballs (Issue 9). The degree of turbulence during rapid disk accretion would make the formation of dustballs very difficult, but it might concentrate dust in zones between turbulent eddies, where subsequent (or concurrent) heating could make it into chondrules. Partial melting, partial evaporation, partial recondensation, and aggregation into droplets occurred in the same high energy event. The studies cited above of shock as a means of melting chondrules assumed that drag acted on chondrule-size precursors, but in principle it would work as well on dust particles; length and time scales would of course be much smaller. Dust was concentrated not by leisurely settling in a non-turbulent medium, but was spun by turbulent eddies into dust-rich stagnation zones (Cuzzi *et al.* 1993 and this volume). These authors assume relatively low-velocity turbulence, which is effective in concentrating only chondrule-size solids; but in principle higher-speed (*i.e.*, supersonic) turbulence would be capable of concentrating dust particles (J. N. Cuzzi, *pers. comm.*).

The mixtures of hot particles, microdroplets, and gas pictured were open systems (Issue 7). Volatile elements were selectively vaporized, and only partially recondensed before the particles and microdroplets coalesced into chondrules. As temperature decreased, chondrules and dust in turn were concentrated and aggregated too soon to permit the complete recondensation and incorporation of all elements that had been vaporized. The most volatile elements began recondensing latest and had the shortest time to condense, so were excluded from aggregated matter to the greatest degree (Issues 13, 14). The mechanism of rapid chondrite accretion remains unknown, but the turbulent concentration mechanism of Cuzzi *et al.*, operating on newly-formed chondrules, may have contributed importantly. Once chondritic aggregations achieved sizes >~1 km, they were effectively decoupled from gas motions and safe from being swept into the sun. (Since the sun and disk were still growing during Stages *b* and *c*, they had a mass <1 M$_\odot$, thus an object orbiting at a given distance had a lower specific angular momentum than bodies have at that distance now. Therefore chondritic material had to be formed at some distance >3 AU in the disk to end up at 3 AU now; Fig. 6.)

Refractory inclusions were formed by the same process (Issue 2), operating in a more optically-opaque setting where the coagulation rate of hot dust and microdroplets could outstrip cooling, resulting in the aggregation of refractory minerals and liquids at high temperatures, in the absence of more volatile phases. This

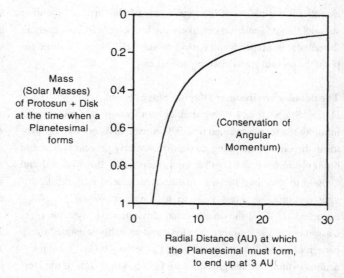

Fig. 6. Relationship between solar system mass at the time when a solid body forms in the disk, and the degree by which its orbit will shrink when the solar system mass grows to its present value.

longer time scale of CAI formation is consistent with the observed slow cooling rates of CAIs compared with those of chondrules (Issue 6).

An interesting but highly speculative idea is that each of the finite number of episodes of disk mass accretion (FU Orionis events) which accompanied solar system formation produced a characteristic type of chondritic material (more realistically, a limited range of chondritic materials). Thus during (*e.g.*) the L-chondrite accretion episode a substantial amount of that type of material might have been produced, which accreted into many L-chondrite planetesimals (Issues 11, 12).

CONCLUSIONS

The thesis of this paper is that efforts to understand the origin of chondrites, and how it relates to the evolution of the solar nebula, have become confined to unnecessarily narrow channels by a self-imposed set of intellectual constraints, many of which are unproven and may well be wrong. I have attempted to identify these and label them "open issues" rather than constraints. Operating outside these constraints, this paper argues from availability of energy that chondrites are more likely to have been formed during the solar systems's early, brief (~0.5 Ma) stage of infall and protosolar accretion than during the subsequent more protracted (~10 Ma) quiescent nebula stage; and it suggests that discrete chemical classes of chondrites might each have been formed during a different episode of mass accretion to the sun triggered by gravitational instability in the growing nebula.

Reduced to the simplest terms, this can be thought of as a *fast* model of chondrite formation as opposed to more traditional interpretations, which are *slow* by comparison. It operates rapidly and chaotically in a very-high-energy environment, and requires a relatively small number of steps to transform interstellar dust into chondrites; whereas most published discussions contemplate a lengthier, more orderly series of separate operations which individually and collectively require a long time to complete. Though the model proffered is simpler in outline, the individual chondrule-forming events called for are extremely complex, so complex that if the model is broadly correct it may never be possible (or worthwhile) to understand all the detailed properties of chondrites that have been published.

ACKNOWLEDGEMENTS

Discussions with P. Cassen, J. N. Cuzzi, and L. Hartmann contributed to the point of view expressed in this paper. Cassen, E. Anders, and D. Kring offered helpful comments on the manuscript, and H. Palme a thoughtful and constructive review. A review by J. N. Grossman was also very helpful, if testy. The research was supported in part by NASA Grant NAGW–3451.

REFERENCES

Anders E., Virag A., Zinner E. and Lewis R. S. (1991) ^{26}Al and ^{16}O in the early solar system: Clues from meteoritic Al_2O_3. *Astrophys. J.* **373**, L77–L80.

Bischoff A. and Keil K. (1984) Al-rich objects in ordinary chondrites: Related origin of carbonaceous and ordinary chondrites and their constituents. *Geochim. Cosmochim. Acta* **48**, 693–709.

Blander M. and Katz J. L. (1967) Condensation of primordial dust. *Geochim. Cosmochim. Acta* **31**, 1025–1034.

Boss A. P. and Graham J. A. (1993) Clumpy disk accretion and chondrule formation. *Icarus* **106**, 168–178.

Boynton W. V. (1978) The chaotic solar nebula: Evidence for episodic condensation in several distinct zones. In *Protostars and Planets* (ed. T. Gehrels), pp. 427–438. University of Arizona Press.

Brigham C. A., Yabuki H., Ouyang Z., Murrel M. T., El Goresy A. and Burnett D. S. (1986) Silica-bearing chondrules and clasts in ordinary chondrites. *Geochim. Cosmochim. Acta* **50**, 1655–1666.

Cameron A. G. W. (1962) The formation of the sun and planets. *Icarus* **1**, 13–69.

Cameron A. G. W. (1994a) The first ten million years in the solar nebula. *Meteoritics* **30**, 133–161.

Cameron A. G. W. (1994b) How long did the solar nebula last? *Meteoritics* **29**, 5.

Cassen P. M. (1992) Thermal models of the primitive solar nebula. *Lunar Planet. Sci.* **23**, 207–208.

Cassen P. M., Smith B. F., Miller R. H., and Reynolds R. T. (1981) Numerical experiments on the stability of preplanetary disks. *Icarus* **48**, 377–392.

Clayton R. N. (1993) Oxygen isotopes in meteorites. *Ann. Rev. Earth Planet. Sci.* **21**, 115–149.

Connolly H. C. Jr., Hewins R. H., Ash R. D., Zanda B., Lofgren G. E. and Bourot-Denise M. (1944) Carbon and the formation of reduced chondrules. *Nature* **371**, 136–139.

Cuzzi J. N., Dobrovolskis A. R. and Champney J. M. (1993) Particle-gas dynamics in the midplane of a protoplanetary nebula. *Icarus* **106**, 102–134.

Dodd R. T. (1971) The petrology of chondrules in the Sharps meteorite. *Contr. Mineral. and Petrol.* **31**, 201–227.

Dodd R. T. (1976a) Accretion of the ordinary chondrites. *Earth Planet. Sci. Lett.* **30**, 281–291.

Dodd R. T. (1976b) Iron-silicate fractionation within ordinary chondrite groups. *Earth Planet. Sci. Lett.* **28**, 479–484.

Dodd R. T. (1978) Compositions of droplet chondrules in the Manych (L3) chondrite and the origin of chondrules. *Earth Planet. Sci. Lett.* **40**, 71–82.

Eisenhour D. D., Daulton T. L. and Buseck P. R. (1994) Electromagnetic heating in the early solar nebula and the formation of chondrules. *Science* **265**, 1067–1070.

Fahey A. J., Zinner E. K., Crozaz G. and Kornacki A. (1987) Microdistributions of Mg isotopes and REE abundances in a Type A calcium-aluminum-rich inclusion from Efremovka. *Geochim. Cosmochim. Acta* **51**, 3215–3229.

Fredriksson K. (1963) Chondrules and the meteorite parent bodies. *Trans. N. Y. Acad. Sci.* **25**, 756–769.

Fredriksson K., Noonan A. and Nelen J. (1973) Meteoritic, lunar and Lonar impact chondrules. *The Moon* **7**, 475–482.

Goldreich P. and Ward W. R. (1973) The formation of planetesimals. *Astrophys. J.* **183**, 1051–1061.

Goswami J. N., Srinivasan G. and Ulyanov A. A. (1994) Ion microprobe studies of Efremovka CAIs: I. Magnesium isotope composition. *Geochim. Cosmochim. Acta* **58**, 431–447.

Gradie J. and Tedesco E. (1982) Compositional structure of the asteroid belt. *Science* **216**, 1405–1407.

Grossman J. N. (1988) Formation of chondrules. In *Meteorites and the Early Solar System* (eds. J. F. Kerridge and M. S. Matthews), pp. 680–696. University of Arizona Press.

Grossman J. N. and Wasson J. T. (1985) The origin and history of the metal and sulfide components of chondrules. *Geochim. Cosmochim. Acta* **49**, 925–940.

Grossman L., Allen J. M. and MacPherson G. J. (1980) Electron microprobe study of a 'mysterite'-bearing inclusion from the Krymka LL-chondrite. *Geochim. Cosmochim. Acta* **44**, 211–216.

Hartmann L. and Kenyon S. (1985) On the nature of the FU Orionis objects. *Astrophys. J.* **299**, 462–478.

Hartmann L., Kenyon S. and Hartigan P. (1993) Young stars: Episodic phenomena, activity and variability. In *Protostars and Planets III* (eds. E. H. Levy and J. I. Lunine), pp. 497–518. University of Arizona Press.

Hashimoto A. (1983) Evaporation metamorphism in the early solar nebula— evaporation experiments on the melt $FeO-MgO-SiO_2-CaO-Al_2O_3$ and chemical fractionations of primitive materials. *Geochem. J.* **17**, 111–145.

Hewins R. H. (1988) Experimental studies of chondrules. In *Meteorites and the Early Solar System* (eds. J. F. Kerridge and M. S. Matthews), pp. 660–679. University of Arizona Press.

Higuchi H., Ganapathy R., Morgan J. W. and Anders E. (1977) "Mysterite": a late condensate from the solar nebula. *Geochim. Cosmochim. Acta* **41**, 843–852.

Hood L. L. and Horanyi M. (1993) The nebular shock wave model for chondrule formation: One-dimensional calculations. *Icarus* **106**, 179–189.

Hughes D. W. (1978) A disaggregation and thin section analysis of the size and mass distribution of the chondrules in the Bjürbole and Chainpur meteorites. *Earth Planet. Sci. Lett.* **38**, 391–400.

Huss G. R. (1988) The role of interstellar dust in the formation of the solar system. *Earth, Moon and Planets* **40**, 165–211.

Hutcheon I. D., Huss G. R. and Wasserburg G. J. (1994) A search for ^{26}Al in chondrites: Chondrule formation time scales. *Lunar Planet. Sci.* **25**, 587–588.

Hutchison R. and Bevan A. W. R. (1983) Conditions and time of chondrule accretion. In *Chondrules and their Origins* (ed. E. A. King), pp. 162–179. Lunar and Planetary Institute, Houston.

Hutchison R., Alexander C. M. O. and Barber D. J. (1988) Chondrules: chemical, mineralogical and isotopic constraints on theories of their origin. *Phil. Trans. R. Soc. Lond.* A **325**, 445–458.

Ireland T. R. (1988) Correlated morphological, chemical, and isotopic characteristics of hibonites from the Murchison carbonaceous chondrite. *Geochim. Cosmochim. Acta* **52**, 2827–2839.

Jones R. H. (1990) Petrology and mineralogy of Type II, FeO-rich chondrules in Semarkona (LL3.0): Origin by closed-system fractional crystallization, with evidence for supercooling. *Geochim. Cosmochim. Acta* **54**, 1785–1802.

Kallemeyn G. W., Rubin A. E., Wang D. and Wasson J. T. (1989) Ordinary chondrites: Bulk compositions, classification, lithophile-element fractionations, and composition-petrographic type relationships. *Geochim. Cosmochim. Acta* **53**, 2747–2767.

Keays R. R., Ganapathy R. and Anders E. (1971) Chemical fractionations in meteorites—IV Abundances of fourteen trace elements in L-chondrites; implications for cosmothermometry. *Geochim. Cosmochim. Acta* **35**, 337–363.

Keil K. and Fredriksson K. (1964) The iron, magnesium, and calcium distribution in coexisting olivines and rhombic pyroxenes of chondrites. *J. Geophys. Res.* **69**, 3487–3515.

King E. A. Jr., Carman M. F. and Butler J. C. (1972) Chondrules in Apollo 14 Samples: Implications for the origin of chondritic meteorites. *Science* **175**, 59–60.

Kurat G., Pernicka E. and Herrwerth I. (1984) Chondrules from Chainpur (LL–3): Reduced parent rocks and vapor fractionation. *Earth Planet. Sci. Lett.* **68**, 43–56.

Larimer J. W. and Anders E. (1967) Chemical fractionations in meteorites—II. Abundance patterns and their interpretation. *Geochim. Cosmochim. Acta* **31**, 1239–1270.

Larimer J. W. and Anders E. (1970) Chemical fractionations in meteorites—III. Major element fractionations in chondrites. *Geochim. Cosmochim. Acta* **34**, 367–387.

Laul J. C., Ganapathy R., Anders E. and Morgan J. W. (1973) Chemical fractionations in meteorites—VI. Accretion temperatures of H-, LL-, and E-chondrites, from abundance of volatile trace elements. *Geochim. Cosmochim. Acta* **37**, 329–357.

Lewis R. D., Lofgren G. E., Franzen H. F. and Windom K. E. (1993) The effect of Na vapor on the Na content of chondrules. *Meteoritics* **28**, 622–628.

Levy E. H. and Araki S. (1989) Magnetic reconnection flares in the protoplanetary nebula and the possible origin of meteorite chondrules. *Icarus* **81**, 74–91.

Liffman K. (1992) The formation of chondrules via ablation. *Icarus* **100**, 608–620.

MacPherson G. J., Wark D. A. and Armstrong J. T. (1988) Primitive material surviving in chondrites: Refractory inclusions. In *Meteorites and the Early Solar System* (eds. J. F. Kerridge and M. S. Matthews), pp. 746–807. University of Arizona Press.

Martin P. M. and Mills A. A. (1976) Size and shape of chondrules in the Bjürbole and Chainpur meteorites. *Earth Planet. Sci. Lett.* **33**, 239–248.

Mason B. (1963) Olivine Composition in meteorites. *Geochim. Cosmochim. Acta* **27**, 1011–1023.

McSween H. Y. Jr. (1977) Chemical and petrographic constraints on the origin of chondrules and inclusions in carbonaceous chondrites. *Geochim. Cosmochim. Acta* **41**, 1843–1860.

Metzler K., Bischoff A. and Stöffler D. (1992) Accretionary dust mantles in CM chondrites: Evidence for solar nebula processes. *Geochim. Cosmochim. Acta* **56**, 2873–2897.

Morfill G., Spruit H. and Levy E. H. (1993) Physical processes and conditions associated with the formation of protoplanetary disks. In *Protostars and Planets III* (eds. E. H. Levy and J. I. Lunine), pp. 939–978. University of Arizona Press.

Mueller, G. (1968) Significance of microchondrules of olivine in Type I carbonaceous chondrites. *Nature* **218**, 1239–1240.

Nakagawa Y., Sekiya M. and Hayashi C. (1986) Settling and growth of dust particles in a laminar phase of a low-mass solar nebula. *Icarus* **67**, 375–390.

Niemeyer S. (1988a) Titanium isotopic anomalies in chondrules from carbonaceous chondrites. *Geochim. Cosmochim. Acta* **52**, 309–318.

Niemeyer S. (1988b) Isotopic diversity in nebular dust: The distribution of Ti isotopic anomalies in carbonaceous chondrites. *Geochim. Cosmochim. Acta* **52**, 2941–2954.

Palme H. and Boynton W. V. (1993) Meteoritic constraints on conditions in the solar nebula. In *Protostars and Planets III* (eds. E. H. Levy and J. I. Lunine), pp. 979–1004. University Arizona Press.

Palme, H., Spettel, B., Kurat, G. and Zinner, E. (1992) Origin of Allende Chondrules. *Lunar Planet Sci.* **23**, 1021–1022.

Podolak M., Bunch T. E., Cassen P., Reynolds R. T. and Chang S. (1990) Processing of refractory meteorite inclusions (CAIs) in parent-body atmospheres. *Icarus* **84**, 254–260.

Podosek F. A. and Cassen P. (1994) Theoretical, observational, and isotopic estimates of the lifetime of the solar nebula. *Meteoritics* **29**, 6–25.

Podosek F. A., Zinner E. K., MacPherson G. J., Lundberg L. L., Brannon J. C. and Fahey A. J. (1991) Correlated study of initial $^{87}Sr/^{86}Sr$ and Al-Mg isotopic systematics and petrologic properties in a suite of refractory inclusions from the Allende meteorite. *Geochim. Cosmochim. Acta* **55**, 1083–1110.

Rubin A. E. (1984) Coarse-grained chondrule rims in type 3 chondrites. *Geochim. Cosmochim. Acta* **48**, 1779–1789.

Rubin A. E. and Grossman J. N. (1987) Size-frequency-distributions of EH3 chondrules. *Meteoritics* **22**, 237–251.

Rubin A. E., Scott E. R. D., and Keil, K. (1982) Microchondrule-bearing clast in the Piancaldoli LL3 meteorite: A new kind of type 3 chondrite and its relevance to the history of chondrites. *Geochim. Cosmochim. Acta* **46**, 1763–1776.

Ruzmaikina T. V. and Ip W. (1994) Chondrule formation in radiative shock. *Icarus* **112**, 430–447.

Sahagian D. L. and Hewins R. H. (1992) The size of the chondrule-forming events. *Lunar Planet. Sci.* **23**, 1197–1198.

Sears, D. W. G. (1995) The formation of chondrules. *Lunar Planet. Sci.* **26**, 1263–1264.

Schultz L., Weber H. W. and Begemann F. (1991) Noble gases in H-chondrites and potential differences between Antarctic and non-Antarctic meteorites. *Geochim. Cosmochim. Acta* **55**, 59–66.

Shu F. H., Adams F. C. and Lizano S. (1987) Star formation in molecular clouds: Observation and theory. *Ann. Rev. Astron. Astrophys.* **25**, 23–81.

Shu F. H., Tremaine S., Adams F. C. and Ruden S. P. (1990) Sling amplification and eccentric gravitational instabilities in gaseous disks. *Astrophys. J.* **358**, 495–514.

Skinner W. R. (1990) Bipolar outflows and a new model of the early solar system. Part I: Overview and implications of the model; Part II: The origins of chondrules, isotopic anomalies, and chemical fractionations. *Lunar Planet. Sci.* **21**, 1166–1169.

Skinner W. R. (1991) Origin of chondrules by droplet coalescence: An alternative to the dust-ball hypothesis. *Lunar Planet. Sci.* **22**, 1269–1270.

Stolper E. and Paque J. (1986) Crystallization sequences of Ca-,Al-rich inclusions from Allende: The effects of cooling rate and maximum temperature. *Geochim. Cosmochim. Acta* **50**, 1785–1806.

Swindle T. D., Grossman J. N., Olinger C. T. and Garrison D. H. (1991a) Iodine-xenon, chemical, and petrographic studies of Semarkona chondrules: Evidence for the timing of aqueous alteration. *Geochim. Cosmochim. Acta* **55**, 3723–3734.

Swindle T. D., Caffee M. W., Hohenberg C. M., Lindstrom M. M. and Taylor G. J. (1991b) Iodine-xenon studies of petrographically and chemically characterized Chainpur chondrules. *Geochim. Cosmochim. Acta* **55**, 861–880.

Taylor G. J. and Heymann D. (1969) Shock, reheating and the gas retention ages of chondrites. *Earth Planet. Sci. Lett.* **7**, 151–161.

Taylor G. J., Scott E. R. D. and Keil K. (1983) Cosmic setting for chondrule formation. In *Chondrules and their Origins* (ed. E. A. King), 262–278. Lunar and Planetary Institute, Houston.

Tsuchiyama A., Nagahara H. and Kushiro I. (1981) Volatilization of sodium from silicate melt spheres and its application to the formation of chondrules. *Geochim. Cosmochim. Acta* **45**, 1357–1367.

Urey H. C. (1967) Parent bodies of the meteorites and the origin of chondrules. *Icarus* **7**, 350–359.

Wai C. M. and Wasson J. T. (1977) Nebular condensation of moderately volatile elements and their abundances in ordinary chondrites. *Earth Planet. Sci. Lett.* **36**, 1–13.

Wark D. A. (1979) Birth of the presolar nebula: The sequence of condensation revealed in the Allende meteorite. *Astrophy. Space Sci.* **65**, 275–295.

Wasson J. T. (1972) Formation of ordinary chondrites. *Rev. Geophys. Space Phys.* **10**, 711–759.

Wasson J. T. (1974) *Meteorites: Classification and Properties*. Springer-Verlag, Berlin. 316 pp.

Wasson J. T. (1977) Relationship between the composition of solar system solid matter and the distance from the sun. In *Comets, Asteroids, Meteorites* (ed. A. H. Delsemme), pp. 551–559. University of Toledo Press.

Wasson J. T. (1985) *Meteorites—Their Record of Early Solar-System History*. W. H. Freeman and Co., New York. 267 pp.

Wasson J. T. (1993) Constraints on chondrule origin. *Meteoritics* **28**, 14–28.

Wasson J. T. and Chou C.-L. (1974) Fractionation of moderately volatile elements in ordinary chondrites. *Meteoritics* **9**, 69–84.

Weidenschilling S. J. (1977) Aerodynamics of solid bodies in the solar nebula. *Mon. Not. Roy. Astron. Soc.* **180**, 57–70.

Weidenschilling S. J. (1988) Formation processes and time scales for meteorite parent bodies. In *Meteorites and the Early Solar System* (eds. J. F. Kerridge and M. S. Matthews), pp. 348–371. University of Arizona Press.

Whipple F. L. (1966) Chondrules: suggestion concerning the origin. *Science* **153**, 54–56.

Wieneke B. and Clayton D. D. (1983) Aggregation of grains in a turbulent pre-solar disk. In *Chondrules and their Origins* (ed. E. A. King), pp. 284–295. Lunar and Planetary Institute, Houston.

Wood J. A. (1963) On the origin of chondrules and chondrites. *Icarus* **2**, 152–180.

Wood J. A. (1967) Olivine and pyroxene compositions in Type II carbonaceous chondrites. *Geochim. Cosmochim. Acta* **31**, 2095–2108.

Wood J. A. (1985) Meteoritic constraints on processes in the solar nebula. In *Protostars and Planets II* (eds. D. C. Black and M. S. Matthews), pp. 687–702. University of Arizona Press.

Wood J. A. and Holmberg B. B. (1994) Constraints placed on the chondrule-forming process by merrihueite in the Mezö-Madaras chondrite. *Icarus* **108**, 309–324.

Yu, Y. and Hewins, R. H. (1994) Retention of sodium under transient heating conditions—experiments and their implications for the chondrule forming environment. *Lunar Planet. Sci.* **25**, 1535–1536.

Yu Y., Hewins R. H., Zanda B. and Connolly H. C. (1994) Can sulfide minerals survive the chondrule-forming transient heating event? *Lunar Planet. Sci.* **25**, 1537–1538.

Zanda B., Yu Y., Bourot-Denise, M., Hewins, R. H. and Connolly, H. (1996) Sulfur behavior in chondrule formation and metamorphism. *Geochim. Cosmochim. Acta*, in press.

8: Thermal Processing in the Solar Nebular: Constraints from Refractory Inclusions

ANDREW M. DAVIS[1] and GLENN J. MACPHERSON[2]

[1]Enrico Fermi Institute, University of Chicago, Chicago, IL 60637, U.S.A. [2]Department of Mineral Sciences, National Museum of Natural History, Smithsonian Institution, Washington, DC 20560, U.S.A.

ABSTRACT

Like true chondrules, many CAIs passed through one or more molten stages during their lifetimes. The absence of significant mass-dependent isotopic fractionation effects shows that such once-molten CAIs formed by melting of solid precursors of appropriate bulk compositions, not by evaporation of less refractory melts. Measurements of isotopic mass-dependent fractionation in these CAIs do not in most cases constrain the duration of the heating events that induced melting. However, the same isotopic data rule out the possibility that the CAI melts originated by fractional evaporation of chondritic or other nonrefractory compositions. The subsequent cooling rates of those CAIs—markedly slower than inferred for chondrules—were governed neither by radiative cooling into a vacuum nor by the cooling of a very large body of enclosing gas. Rather, the cooling was intermediate in rate and may have been controlled by the cooling rates of small pockets of gas within the much larger nebula. Apparently, the conditions that prevailed during the formation of CAIs and chondrules were different, a perhaps not-unexpected conclusion if, as isotopic evidence suggests, the chondrules formed up to several million years later than CAIs.

INTRODUCTION

Calcium-, aluminum-rich inclusions (CAIs) are a diverse population of objects in chondritic meteorites whose chemical and isotopic properties demonstrate that they are among the most primitive objects formed in the solar system. The textures and crystal chemical properties of those CAIs that passed through one or more molten stages can be used, in conjunction with experimental data, to constrain their thermal histories in the earliest solar nebula, including not just cooling rates but maximum temperatures and the nature of the heating events as well.

BACKGROUND: ONCE-MOLTEN CAIS

The petrologic, chemical and isotopic characteristics of CAIs in chondrites have been reviewed in considerable detail by MacPherson *et al.* (1988). A great diversity of CAI types is known, and these represent an equally diverse range of formation histories. From the standpoint of the present review, however, the most interesting inclusions are those that solidified from melts. The reason is

that igneous crystallization is especially amenable to laboratory study. Analogue melts of some appropriate inclusion bulk compositions have been studied experimentally by both equilibrium and dynamic crystallization methods. The results of such experiments, used in conjunction with petrographic and mineral chemical observations of natural inclusions, allow quantitative constraints to be placed on such parameters as cooling rates. In order for the reader to be on comfortable footing in the sections that follow, what follows in this section is a brief synopsis of the major inclusion types that probably formed through melt solidification.

Of all the "igneous" CAIs, the best studied by far are the so-called Type B inclusions found exclusively in CV3 carbonaceous chondrites. These commonly-spheroidal objects are among the largest of refractory inclusions, typically 0.5 cm in diameter or more, with primary phases that are sufficiently coarse-grained (up to 1 mm) for very detailed optical and quantitative chemical study. An example of a Type B CAI is shown in Fig. 1. Such objects were among the very first CAIs studied after the recovery and distribution of material from the newly-fallen Allende meteorite in 1969, and it was in Type B inclusions that the former presence of the

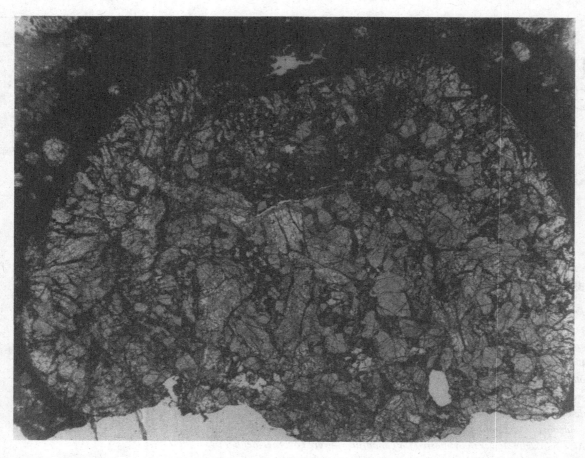

Fig. 1. Photomicrograph of a Type B1 inclusion, USNM 3529–Z, from the Allende CV3 meteorite. The original inclusion is spheroidal in shape (the thin section truncates the bottom edge) and approximately ~1.9 cm in diameter. The depression at the top of the inclusion is a graben, presumably caused by impact and is bounded by microfaults. The primary phases are melilite, fassaite, anorthite, and spinel. Transmitted, plane polarized light.

short-lived radionuclide ^{26}Al was first clearly established (Lee *et al.*, 1976).

The primary phase assemblage of all Type B inclusions is rather simple, consisting of melilite (a solid solution of gehlenite, $Ca_2Al_2SiO_7$, and åkermanite, $Ca_2MgSi_2O_7$), anorthite ($CaAl_2Si_2O_8$), a Ti-Al-rich calcic pyroxene referred to as fassaite (a solid solution of $CaMgSi_2O_6$, $CaAl_2SiO_6$, $CaTi^{4+}Al_2O_6$, and $CaTi^{3+}AlSiO_6$), and spinel (*sensu stricto*: $MgAl_2O_4$). Type B CAIs are subdivided (Wark and Lovering, 1977) into Types B1 and B2. Type B2 CAIs have a more or less homogeneous distribution of spinel + melilite + anorthite + pyroxene throughout, whereas B1s have a nearly monomineralic melilite mantle surrounding a spinel + melilite + anorthite + pyroxene core. Oxidized iron and alkalis are present in the primary phases in trace abundances only, but are abundant in an assemblage of fine-grained secondary minerals that preferentially replace melilite and anorthite. Such secondary minerals are more prevalent in CAIs from oxidized subgroup CV3 meteorites (*e.g.*, Allende, Grosnaja) than in CAIs from reduced subgroup CV3s (*e.g.*, Leoville, Vigarano, Efremovka). The equilibrium crystallization sequence in melts of Type B composition is spinel, spinel + melilite, spinel + melilite + anorthite, and finally spinel + melilite + anorthite + pyroxene (Stolper, 1982). The tex-

tural difference between Types B1 and B2 CAIs is a manifestation of their differing bulk compositions, the B1s having a greater temperature interval over which melilite + spinel cocrystallize to the exclusion of the other phases.

A rarer variant of the Type B inclusions contains olivine (forsterite) as a major phase (see Wark *et al.*, 1987), and are labeled forsterite-bearing Type Bs (FoBs).

The melilite-rich Type A inclusions have been subdivided into two subtypes: irregularly-shaped "fluffy" Type As, which have almost certainly never been melted (MacPherson and Grossman, 1984); and more rounded "compact" Type As, some of which may have been melted. Compact Type A CAIs are common in CV and CO chondrites, but have not been studied in the same detail as Type B CAIs. A recent study of major and trace element zoning patterns within several Allende compact Type As suggests that most did not form by simple melt crystallization (Simon *et al.*, 1995).

Yet another variant are the Type C CAIs, consisting mostly of anorthite + pyroxene + spinel with little or no melilite. These are commonly characterized by ophitic or subophitic textures that are taken to be igneous in origin.

Very different from these silicate-rich, once-molten inclusions in CV3 chondrites are aluminum-rich and silicate-poor spherules found in CM2 meteorites. These objects are rarely larger than 100–200 μm in diameter, making electron microscopy a necessary tool for detailed petrographic characterization. Their dominant phases are spinel and/or hibonite ($CaAl_{12}O_{19}$), with perovskite ($CaTiO_3$) as a common accessory. Melilite-bearing inclusions are rare in CM chondrites. Although many CM inclusions have textures that appear inconsistent with melt crystallization, the spinel-hibonite spherules do appear to have formed in that manner. It should be noted, however, that analogue crystallization experiments on melts of CM spherule composition have not yet been made.

MAXIMUM TEMPERATURES

Type B bulk compositions have liquidus temperatures of 1500–1550°C (Stolper, 1982; Beckett, 1986). However, comparison of melilite mineral chemistry and textures produced in crystallization experiments on Type B bulk compositions with those observed in natural Type B CAIs indicates that the natural inclusions probably never reached liquidus temperatures. Stolper and Paque (1986) showed that melting significantly in excess of ~1400°C (the equilibrium crystallization temperature of melilite) destroys melilite crystallization nuclei and leaves only solid spinel grains, leading on cooling to dendritic melilite textures not observed in natural CAIs.

The smaller, more refractory spherules from CM chondrites show evidence for substantially higher temperature melting events. MacPherson *et al.* (1984a) showed that a corundum-hibonite inclusion from Murchison contained two generations of hibonite. Their explanation for the formation of this inclusion, which is consistent with the $CaO-Al_2O_3$ phase diagram, is that hibonite partially incongruently melted to corundum plus liquid, some of the calcium in the liquid evaporated and the second generation of hibonite formed on cooling. The temperature of incongruent melting of hibonite is ~1850°C. It should be noted, however, that the distribution of trace elements within this inclusion requires that the two generations of hibonite formed from unrelated reservoirs (Hinton *et al.*, 1988). From the spherical shape and the distribution of phases within a Murchison CAI consisting of spinel, hibonite, grossite [$CaAl_4O_7$] and perovskite, Simon *et al.* (1994) inferred that the spherule crystallized from a melt. They used the $CaO-MgO-Al_2O_3$ phase diagram to show that the bulk composition of this CAI had a liquidus temperature of 2110°C and a minimum melting temperature of 2080°C (if some of the spinel was relict). Simon *et al.* also found excess Al_2O_3 in spinel, which they attributed to equilibration of spinel with Al_2O_3-rich melt at temperatures substantially higher than those experienced by spinel in Type B CAIs.

TIME DURATION ABOVE THE SOLIDUS

Until recently it was thought that total nebular pressures of 0.1 atm or more were required for melts to condense from the solar nebula, but recent calculations incorporating a nonideal solution model for silicate melts shows that melts can be in equilibrium with the solar

nebula at pressures as low as 5×10^{-2} atm and perhaps as low as 1×10^{-2} atm (Yoneda and Grossman, 1995). Although melting of CAI (or chondrule) precursors might have occurred at total nebular pressures of 10^{-2} atm or less, the resulting melts will be unstable against evaporation, because the condensation temperatures of these compositions are below the temperatures required to partially melt such compositions. Evaporation from a melt will enrich the residual melt in the heavy isotopes of elements that evaporate to a significant degree, because mass fractionation due to the evaporation reaction at the surface of the melt is controlled by the kinetic isotope effect. The relationship between fraction of an element lost and the degree of mass fractionation for that element can in principle, be described by the Rayleigh equation, $R/R_o = f^{\alpha-1}$, where R is the isotope ratio of the residue (*e.g.*, $^{25}Mg/^{24}Mg$), R_o is the initial isotope ratio of the melt, f is the fraction remaining and α is the gas/residue isotopic fractionation factor (*e.g.*, $(^{25}Mg/^{24}Mg)_{gas}/(^{25}Mg/^{24}Mg)_{residue}$). α is given by the square root of the mass ratio of the evaporating species (*e.g.*, for Mg atoms, $\alpha = \sqrt{24/25}$). Davis *et al.* (1990) demonstrated experimentally that evaporation of melts of forsterite composition leads to substantial enrichments in the heavy isotopes of magnesium, silicon and oxygen in the residues, and that the relationship between degree of evaporation and mass fractionation is in fact very well described by the Rayleigh equation; the most abundant gas phase species in these experiments were Mg, O_2 and SiO_2. The Rayleigh equation applies only when the residue remains isotopically homogeneous (well mixed) during evaporation. This is easily achieved in silicate liquids, where diffusion rates are relatively rapid. However, the experiments of Davis *et al.* (1990) showed that evaporation of *solid* forsterite did not lead to significant bulk isotopic fractionation effects, because diffusion rates in solids are too slow for the residues to homogenize internally. Evaporation experiments have also been performed on melts of other bulk compositions for the purpose of ascertaining isotopic fractionations. For example, Wang *et al.* (1994a) showed that evaporation of melts having chondritic bulk compositions also leads to residues enriched in the heavy isotopes of magnesium, silicon and oxygen. Wang *et al.* (1994b) showed that evaporation of wüstite leads to residues enriched in the heavy isotopes of iron and oxygen. This implies that evaporation of chondritic composition melts will lead to residues enriched in the heavy isotopes of iron, because iron is the most volatile of the major elements for chondritic compositions (Hashimoto, 1983; Wang *et al.*, 1993).

Of the elements for which isotopic compositions are available for CAIs, magnesium is the most sensitive to evaporation effects, because it is among the most volatile of major elements in CAIs and also because its major evaporating species, Mg atoms, have lower mass than the major evaporating species of silicon and oxygen, O_2, SiO and SiO_2. Magnesium also has the advantage that it is easily accessible to analysis by ion microprobe. Silicon is also useful, because the mass fractionation of silicon can be measured with high precision by gas-source isotope ratio mass spectrometry (Molini-Velsko *et al.*, 1983). A small subset of the CAI population called FUN inclusions—in reference to their containing mass

Fractionated and Unknown Nuclear isotopic effects—show mass fractionation effects as large as 37 ‰/amu for magnesium, 16 ‰/amu for silicon and 27 ‰/amu for oxygen. Most CAIs, however, do not have large mass fractionation effects. The largest fractionation effects seen in non-FUN CAIs are $\Delta^{25}Mg = 10‰$ and $\Delta^{29}Si = 2.5‰$; for most, $\Delta^{25}Mg < 5‰$ and $\Delta^{29}Si < 1.5‰$. If we assume that the relationship between magnesium or silicon isotopic fractionation and fraction of magnesium or silicon evaporated is the same as for liquids of forsterite or initially chondritic composition (Davis *et al.*, 1990; Wang *et al.*, 1994a), we calculate that no more than 50% of the magnesium or 25% of the silicon in the most heavily fractionated non-FUN CAIs has evaporated; no more than 25% of the magnesium or 16% of the silicon has evaporated from most non-FUN CAIs. This is much less than the 97% of magnesium and 94% of silicon that would have had to evaporate if Type B CAIs formed by evaporation from an initial composition with CI chondritic proportions of major elements. Hashimoto (1983) and Wang (1995) measured vacuum evaporation rates at 1700–2000°C for melts that had initial solar proportions of MgO, Al_2O_3, SiO_2, CaO and FeO. Extrapolation of this data to the temperature range for the presence of melt during crystallization of Type B CAIs, 1400–1200°C (see MAXIMUM TEMPERATURES, above), and to the bulk composition of Type B CAIs indicates that negligible fractions of magnesium and silicon evaporate at the inferred cooling rates (see COOLING RATES, below) of 2—50°/hr. An initially chondritic melt held at 1800°C for one hour evaporates 24% of its magnesium and 42% of its silicon; evaporation of 30% of the magnesium requires ~100 days at 1400°C. There are two reasons to believe that these evaporation rates are rather uncertain: (1) Hashimoto's experiments were done on melts of solar rather than Type B CAI composition, so the extrapolation from a CaO, Al_2O_3-poor melt to a CaO, Al_2O_3-rich one is rather long; and (2) Nagahara and Ozawa (1994) have shown that the evaporation rate of forsterite can be increased up to a factor of 10 above the vacuum evaporation rate in the presence of hydrogen. Although not very useful in constraining cooling rates or length of time above the solidus, the limited magnesium and silicon fractionation effects in most CAIs do show that evaporation was not a significant process in generating the refractory character of the CAI melts. Extrapolation of measured evaporation rates to petrologically-inferred thermal histories of once-molten Type B CAIs suggests that little magnesium and silicon evaporation of these CAIs should have occurred while they were partially molten in the solar nebula. If the melts resulted from the fusion of solid precursors, those precursors already possessed refractory compositions.

Isotopic mass fractionation effects in chondrules may be able to place more stringent limits on the importance of evaporation, for two reasons. (1) Substantially higher liquidus temperatures have been inferred for some types of chondrules. Hewins and Connolly (1994) inferred maximum temperatures of 1600–1800°C for Type 1 barred olivine chondrules. (2) Iron is considerably more volatile than magnesium (Hashimoto, 1983; Wang *et al.*, 1993, 1994a), so that isotopic mass fractionation effects in iron may be larger than those in magnesium. For ferromagnesian chondrules, there are few

reliable data on magnesium isotopic mass fractionation effects and no data available for iron isotopic mass fractionation. Among Allende chondrules there are no significant differences in silicon isotopic fractionation between barred olivine and porphyritic chondrules (Clayton *et al.*, 1985), although barred olivine chondrules may have been heated to higher temperatures than porphyritic chondrules (Hewins and Connolly, 1994). Dhajala chondrules in the 100–250 μm size range are enriched in heavy silicon isotopes by 0.5 ‰/amu compared to larger chondrules (Clayton *et al.*, 1985). If this fractionation effect is due to evaporation, application of the gas/solid silicon isotopic fractionation factor from evaporation experiments on chondritic compositions (Wang *et al.*, 1994a) implies that small chondrules have lost 6% of their silicon relative to larger chondrules. This degree of silicon loss takes about 10 minutes at 1800°C (Wang, 1995).

COOLING RATES

Most of our knowledge of cooling rates of CAIs comes from comparison of melilite zoning patterns in Type B CAIs with those in dynamic cooling experiments on melts of similar composition. Stolper (1982) noted that although the *equilibrium* crystallization sequence for Type B bulk compositions is spinel, melilite, anorthite, fassaite, kinetic factors such as cooling rate could cause a change in the order of appearance of fassaite relative to anorthite. A combination of experimental and observational evidence now indicates that this reversal in crystallization order did in fact occur in many natural Type B inclusions. Type B CAIs commonly contain melilite that is complexly zoned (MacPherson *et al.*, 1984b). During cooling of aluminous melts in the gehlenite-åkermanite binary system, the early-formed melilite is gehlenite-rich (high Al_2O_3) and late melilite is åkermanite-rich (high MgO). In natural Type Bs, the cores of melilite crystals conform to this pattern with the Al_2O_3/MgO ratio decreasing with progressive crystal growth. However, near the outer boundaries of melilite crystals in some CAIs the trend reverses and the Al_2O_3 content rises again briefly, then finally falls yet again at the outermost crystal rims. Crystallization experiments have shown that this reverse zoning is due to changes in liquid composition caused by fassaite crystallizing before anorthite (MacPherson *et al.*, 1984b; Stolper and Paque, 1986). In these experiments, the reversal of order of crystallization of fassaite and anorthite occurs when the cooling rate is faster than 0.5°C/hr. In order to grow euhedral melilite like that seen in natural Type B CAIs, CAIs must have been only partially melted (in order to preserve nucleation sites for melilite crystallization) and must cool at a rate slower than 50°C/hr (Stolper and Paque, 1986).

Although reversely-zoned melilite can form at cooling rates of 0.5–50°C/hr, trace element evidence suggests cooling rates at the high end of this range. Trace element partition coefficients have been measured for melilite grown at 2°C/hr from Type B bulk composition (Beckett *et al.*, 1990). Measurement of trace element concentrations in Type B CAI melilite gives incompatible trace element concentrations higher than those calculated from the Beckett *et al.* partition coefficients. These excesses are probably due to formation of incompatible-element-rich boundary layers during melilite growth and suggest that the natural Type B CAIs cooled at rates significantly faster than 2°C/hr (Davis *et al.*, 1992).

Although only a few Type B CAIs have been subjected to careful studies of internal trace element distribution, most CAIs contain reversely-zoned melilite. The inferred narrow range of cooling rates of natural Type Bs, >2°–50°C/hr, are substantially slower than those expected if Type B CAIs had cooled by radiation in a nebular gas, yet much faster than the cooling rate of a putative globally-warm solar nebula. A possible model is that the cooling rates of the inclusions were externally controlled by the cooling rates of small pockets of nebular gas that were hotter than the general nebular gas in the near vicinity. Most chondrules appear to have cooled somewhat faster than normal Type B CAIs: Lofgren (1994) suggested cooling rates of 5–3000°C/hr for radiating pyroxene chondrules, 500–2500°C/hr for barred olivine chondrules and 100–1000°C/hr for porphyritic chondrules. Various kinds of isotopic evidence suggest that chondrules may have formed as much as 2–3 million years later than CAIs (*e.g.*, see Swindle *et al.*, this volume; MacPherson *et al.*, 1995), so the implied differences in cooling environments for the two groups of objects is perhaps not surprising.

The oxygen isotopic composition of individual minerals in CAIs are thought to be determined by isotopic exchange between solids that initially had $\delta^{17}O = \delta^{18}O = -40‰$ with a gaseous reservoir with $\delta^{17}O = \delta^{18}O = \sim0‰$. Spinel in Type B CAIs nearly always has $\delta^{18}O = -40‰$, regardless of whether it is surrounded by pyroxene ($\delta^{18}O = -30$ to $-40‰$) or melilite ($\delta^{18}O = \sim0‰$); anorthite usually has the same isotopic composition as melilite (Clayton *et al.*, 1977). Ryerson and McKeegan (1994) have measured the rates of oxygen self-diffusion in melilite, diopside, anorthite and spinel and have applied these data to the CAI oxygen record to infer thermal histories. They assumed that the observed grain sizes are representative of effective diffusion dimensions and investigated three types of thermal history: (1) gas-solid exchange at fixed temperature; (2) gas-solid exchange during cooling; (3) gas-solid exchange during partial melting and recrystallization. They were unable to match the observed oxygen isotopic distribution within CAIs with any of the three thermal histories and suggested that oxygen isotopic exchange occurred during multistage alteration and recrystallization events of the sort proposed by MacPherson and Davis (1993).

In contrast to normal CAIs, the grossite-bearing spherule discussed above was heated to such extreme temperatures that the magnesium isotopic composition strongly constrains possible cooling rates. The temperature range from the liquidus to the solidus for this CAI is 250–600°C above the maximum temperature stability field of any major phase in a gas of solar composition at 10^{-3} atm. Evaporation of magnesium (leading to enrichment of the residue in the heavy isotopes of magnesium) at these temperatures would have been extremely rapid (Hashimoto, 1990; Davis *et al.*, 1990). The lack of magnesium isotopic mass fractionation in this CAI suggests that it cooled radiatively, taking only 0.25 s to cool from the liquidus (~2100°C) to the solidus (1725°C) and a further

0.6 s to cool to the maximum possible temperature of surrounding gas (~1500°C) (Simon *et al.*, 1994).

Transient heating events

CAIs sometimes preserve evidence of intense, short-lived (at most a few seconds long) heating events that produced local partial melting or evaporation effects. A forsterite-bearing CAI from Vigarano shows evidence for local melting and evaporation that produced a 300-μm mantle that is in petrologic and isotopic disequilibrium with the core (Davis *et al.*, 1991). This heating event must have been of short duration, as the mantle appears to have been completely melted and the core completely unmelted. Melilite at the exterior of many CAIs shows a polygonal-granular texture and finer grain size than interior melilite, suggestive of local solid-state recrystallization (*e.g.*, MacPherson and Davis, 1993). Flash-heating has also been proposed as a cause of some of the layers of the Wark-Lovering rims (Wark and Lovering, 1977) that usually surround CAIs (Murrell and Burnett, 1987). The latter authors also discuss the relation between intensity and duration of heating. Multiple episodes of secondary alteration and partial melting were invoked to explain chemical and isotopic features of a Vigarano Type B CAI (MacPherson and Davis, 1993). Processing of this sort may be relatively common among CAIs in CV chondrites and could explain the evolution in bulk composition from Types A to B to C.

CONCLUSIONS

Like true chondrules, many CAIs passed through one or more molten stages during their lifetimes. The once-molten CAIs have some advantages over ferromagnesian chondrules in inferring thermal histories, because they contain several minerals with extensive solid solution effects that reveal subtle details of the melt crystallization process. Comparison of CAI textures with those produced in analogue experiments indicates that Type B CAIs were partially melted and that maximum temperatures probably did not exceed 1400°C. Ferromagnesian chondrules appear to have been heated to 1600–1800°C. The cooling rates of Type B CAIs, 2–50°C/hr, are markedly slower than those inferred for chondrules. Although CAIs cooled more slowly than chondrules, the cooling rates of both types of object were governed neither by radiative cooling into a vacuum nor by the cooling of a very large body of enclosing gas. Rather, the cooling was intermediate in rate and may have been controlled by the cooling rates of small pockets of gas within the much larger nebula. Since isotopic evidence suggests that chondrules formed up to several million years later than CAIs, differences in maximum temperature and cooling rate between CAIs and ferromagnesian chondrules are to be expected.

ACKNOWLEDGEMENTS

We thank G. R. Huss and J. M. Paque for their helpful reviews. This work was supported by the National Aeronautics and Space Administration through the following grants: NAGW–3384 (to

AMD), NAGW–3553 (to GJM) and NAGW–3345 (R. N. Clayton, Univ. of Chicago).

REFERENCES

Beckett J. R. (1986) The origin of calcium-, aluminum-rich inclusions from carbonaceous chondrites: an experimental study. Ph.D. Thesis, Univ. of Chicago.

Beckett J. R., Spivack A. J., Hutcheon I. D., Wasserburg G. J. and Stolper E. M. (1990) Crystal chemical effects on the partitioning of trace elements between mineral and melt: an experimental study of melilite with applications to refractory inclusions from carbonaceous chondrites. *Geochim. Cosmochim. Acta* **54**, 1755–1774.

Clayton R. N., Onuma N., Grossman L. and Mayeda T. K. (1977) Distribution of the pre-solar component in Allende and other carbonaceous chondrites. *Earth Planet. Sci. Lett.* **34**, 209–224.

Clayton R. N., Mayeda T. K. and Molini-Velsko C. A. (1985) Isotopic variations in solar system material: evaporation and condensation of silicates. In *Protostars and Planets II* (ed. D. C. Black and M. S. Matthews), pp. 755–771. University of Arizona Press, Tucson.

Davis A. M., Hashimoto A., Clayton R. N. and Mayeda T. K. (1990) Isotopic mass fractionation during evaporation of Mg_2SiO_4. *Nature* **247**, 655–658.

Davis A. M., MacPherson G. J., Clayton R. N., Mayeda T. K., Sylvester P. J., Grossman L., Hinton R. W. and Laughlin J. R. (1991) Melt solidification and late-stage evaporation in the evolution of a FUN inclusion from the Vigarano C3V chondrite. *Geochim. Cosmochim. Acta* **55**, 621–637.

Davis A. M., Simon S. B. and Grossman L. (1992) Melilite composition trends during crystallization of Allende Type B1 refractory inclusion melts (abstract). *Lunar Planet. Sci.* **23**, 281–282.

Hashimoto A. (1983) Evaporation metamorphism in the early solar nebula—evaporation experiments on the melt $FeO-MgO-SiO_2-CaO-Al_2O_3$ and chemical fractionation of primitive materials. *Geochem. Jour.* **17**, 111–145.

Hashimoto A. (1990) Evaporation kinetics of forsterite and implications for the early solar nebula. *Nature* **247**, 53–55.

Hewins R. H. and Connolly H. C. Jr. (1994) Experimental constraints on models for origin of chondrules: peak temperatures (abstract). In *Papers Presented to Chondrules and the Protoplanetary Disk*, LPI Contribution No. 844, pp. 11–12.

Hinton R. W., Davis A. M., Scatena-Wachel D. E., Grossman L. and Draus R. J. (1988) A chemical and isotopic study of hibonite-rich refractory inclusions in primitive meteorites. *Geochim. Cosmochim. Acta* **52**, 2573–2598.

Lee T., Papanastassiou D. A. and Wasserburg G. J. (1976) Demonstration of ^{26}Mg excess in Allende and evidence for ^{26}Al. *Geophys. Res. Lett.* **3**, 109–112.

Lofgren G. E. (1994) Experimental constraints on models for the origin of chondrules: cooling rates (abstract). In *Papers Presented to Chondrules and the Protoplanetary Disk*, LPI Contribution No. 844, pp. 19–20.

MacPherson G. J. and Davis A. M. (1993) A petrologic and ion microprobe study of a Vigarano Type B refractory inclusion: evolution by multiple stages of alteration and melting. *Geochim. Cosmochim. Acta* **57**, 231–243.

MacPherson G. J. and Grossman L. (1984) "Fluffy" Type A Ca-, Al-rich inclusions in the Allende meteorite. *Geochim. Cosmochim. Acta* **48**, 29–46.

MacPherson G. J., Davis A. M. and Zinner E. K. (1995) The distribution of aluminium-26 in the early solar system—a reappraisal. *Meteoritics* **30**, 365–381.

MacPherson G. J., Grossman L., Hashimoto A., Bar-Matthews M. and Tanaka T. (1984a) Petrographic studies of refractory inclusions from the Murchison meteorite. *Proc. 15th Lunar Planet. Sci. Conf., J. Geophys. Res. Suppl.* **89**, C299–C312.

MacPherson G. J., Paque J. M., Stolper E. and Grossman L. (1984b) The origin and significance of reverse zoning in melilite from Allende Type B inclusions. *J. Geol.* **92**, 289–305.

MacPherson G. J., Wark D. A. and Armstrong J. T. (1988) Primitive material surviving in chondrites: refractory inclusions. In *Meteorites and the Early Solar System* (eds. J. F. Kerridge and M. S. Matthews), pp. 746–807. University of Arizona Press.

Molini-Velsko C. A., Mayeda T. K. and Clayton R. N. (1983) Silicon isotopes in components of the Allende meteorite (abstract). *Lunar Planet Sci.* **14**, 509–510.

Murrell M. T. and Burnett D. S. (1987) Actinide chemistry in Allende Ca-Al-rich inclusions. *Geochim. Cosmochim. Acta* **51**, 989–999.

Nagahara H. and Ozawa K. (1994) Vaporization rate of forsterite in hydrogen gas (abstract). *Meteoritics* **29**, 508–509.

Ryerson F. J. and McKeegan K. D. (1994) Determination of oxygen self-diffusion in åkermanite, anorthite, diopside, and spinel: implications for oxygen isotopic anomalies and the thermal histories of Ca-Al-rich inclusions. *Geochim. Cosmochim. Acta* **58**, 3713–3734.

Simon S. B., Davis A. M. and Grossman L. (1995) Crystallization of compact Type A refractory inclusions: implications from crystal zoning and trace element distribution (abstract). *Lunar Planet. Sci.* **26**, 1303–1304.

Simon S. B., Yoneda S., Grossman L. and Davis A. M. (1994) A $CaAl_4O_7$-bearing refractory spherule from Murchison: evidence for very high-temperature melting in the solar nebula. *Geochim. Cosmochim. Acta* **58**, 1937–1949.

Stolper E. (1982) Crystallization sequences of Ca-Al-rich inclusions from Allende: an experimental study. *Geochim. Cosmochim. Acta* **46**, 2159–2180.

Stolper E. and Paque J. M. (1986) Crystallization sequences of Ca-Al-rich inclusions from Allende: the effects of cooling rate and maximum temperature. *Geochim. Cosmochim. Acta* **50**, 1785–1806.

Wang J. (1995) Chemical and isotopic fractionation during the evaporation of synthetic forsterite and material of solar composition. Ph. D. dissertation, University of Chicago.

Wang J., Davis A. M. and Clayton R. N. (1993) Rare earth element fractionation during evaporation of chondritic material (abstract). *Meteoritics* **28**, 454–455.

Wang J., Davis A. M., Clayton R. N. and Mayeda T. K. (1994a) Chemical and isotopic fractionation during the evaporation of the $FeO-MgO-SiO_2-CaO-Al_2O_3-TiO_2$-REE melt system (abstract). *Lunar Planet. Sci.* **25**, 1457–1458.

Wang J., Davis A. M., Clayton R. N. and Mayeda T. K. (1994b) Kinetic isotopic fractionation during the evaporation of the iron oxide from liquid state (abstract). *Lunar Planet. Sci.* **25**, 1459–1460.

Wark D. A., Boynton W. V., Keays R. R. and Palme H. (1987) Trace element and petrologic clues to the formation of forsterite-bearing Ca-Al-rich inclusions in the Allende meteorite. *Geochim. Cosmochim. Acta* **51**, 607–622.

Wark D. A. and Lovering J. F. (1977) Marker events in the early solar system: evidence from rims on Ca-Al-rich inclusions in carbonaceous chondrites. *Proc. Lunar Sci. Conf. 8th*, 95–112.

Yoneda S. and Grossman L. (1995) Condensation of $CaO-MgO-Al_2O_3-SiO_2$ liquids from cosmic gases. *Geochim. Cosmochim. Acta*, **59**, 3413–3444.

9: Formation Times of Chondrules and Ca-Al-rich Inclusions: Constraints from Short-Lived Radionuclides

TIMOTHY D. SWINDLE[1], ANDREW M. DAVIS[2], CHARLES M. HOHENBERG[3], GLENN J. MACPHERSON[4] and LAURENCE E. NYQUIST[5]

[1]Lunar and Planetary Laboratory, University of Arizona, Tucson AZ 85721, U.S.A., [2]Enrico Fermi Institute, University of Chicago, 5640 S. Ellis Ave., Chicago IL 60637, U.S.A., [3]McDonnell Center for Space Sciences, Washington University, St. Louis MO 63130, U.S.A., [4]Department of Mineral Sciences, National Museum of Natural History, Smithsonian Institution, Washington DC 20560, U.S.A., [5]Mail Code SN4, NASA Johnson Space Center, Houston TX 77058, U.S.A.

ABSTRACT

Relative chronometers based on the decays of the extinct radionuclides ^{26}Al, ^{53}Mn and ^{129}I have been applied to individual chondrules from a variety of chondrites. Results from all three systems suggest that chondrule formation occurred several million years after the formation of calcium-, aluminum-rich inclusions, which are believed to be the first solids formed in the solar nebula.

INTRODUCTION

Astronomical observations of other stars, radiometric dating of meteorites and theoretical considerations place some constraints on the history of the solar nebula (Podosek and Cassen, 1994), suggesting that the nebula itself lasted a few million years (Ma), but that the accretion stage of the nebula probably lasted less than 1 Ma (Cameron, 1995). Measurement of the time of formation of chondrules relative to formation of other primitive objects such as calcium-, aluminum-rich inclusions (CAIs) is key to understanding how both types of objects formed and constrains the lifetime of the solar nebula. For example, if chondrule formation requires the presence of nebular gas or dust, the solar nebula cannot have dissipated before chondrule formation.

In principle, we could determine ages of chondrules based on the same radiometric dating techniques that have established the time scale of the histories of the Earth and Moon. However, for events occurring more than 4500 Ma ago, and differing from each other by no more than a few Ma, most dating schemes based on long-lived radionuclides lack the precision needed to resolve these differences. The one possible exception is the U-Pb system, but it has not been extensively applied to individual chondrules.

A promising approach is to use extinct radionuclides, i.e., those whose half-lives are short enough to have experienced significant decay in a 1 Ma interval, but long enough to have survived from the time of their synthesis until they were incorporated into material we can now analyze. Although several extinct radionuclides have been identified in meteorites (see Swindle, 1993, for a recent review), only three, those based on the decays of ^{26}Al, ^{53}Mn and ^{129}I, have been fruitfully applied to individual chondrules. We will concentrate on these three systems, summarized in Table 1. For all three systems the major issue that governs their utility as chronometers is whether they were uniformly distributed in the solar nebula, or at least among the reservoirs from which the various objects being dated formed. The mechanisms of nucleosynthesis of ^{26}Al, ^{53}Mn and ^{129}I differ (see Swindle, 1993 for a brief review), so consistency between dating schemes based on these elements argues in favor of a chronological interpretation of isotopic data.

Table 1

Parent isotope	Daughter isotope	$T_{1/2}$ (Ma)
^{26}Al	^{26}Mg	0.75
^{53}Mn	^{53}Cr	3.7
^{129}I	^{129}Xe	15.7

Data from Walker et al. (1989).

As it turns out, data from all three systems have been interpreted as implying that chondrule formation occurred several Ma after the formation of CAIs. We will summarize the data and the arguments leading to these interpretations. For the Al-Mg system, the initial $^{26}Al/^{27}Al$ ratio for CAIs is well-determined. For the I-Xe and Mn-Cr systems, establishing the initial $^{129}I/^{127}I$ and $^{53}Mn/^{55}Mn$ ratios of the CAIs is more difficult, and will be discussed in some detail. Because there are other samples with more precisely reproducible initial isotopic compositions (the angrite LEW86010 for Mn-Cr; the L4 chondrite Bjurböle for I-Xe), we will follow historical usage and plot apparent ages relative to these meteorites, with the understanding that it is the time difference between CAI formation and chondrule formation that is our focus. We will not address the question of the absolute ages implied by these results. Several recent or ongoing studies have addressed the calibration of ages based on various extinct radionuclides with absolute Pb-Pb ages and other chronometers (*e.g.*, Lugmair and Galer, 1992; Nichols *et al.*, 1994; Brazzle *et al.*, 1995), but there is not yet a consensus on the issue.

$^{26}Al–^{26}Mg$

Observations

The short-lived radionuclide ^{26}Al decays to ^{26}Mg via positron emission or electron capture, with a half life of about 0.73 Ma (Table 1). Although its presence in the early solar system had been predicted as early as 1955 (Urey, 1955), nearly 20 years elapsed before analytical techniques were sufficiently refined to detect magnesium isotope anomalies in CAIs (Gray and Compston, 1974; Lee and Papanastassiou, 1974) and confirm that excesses of ^{26}Mg were due to *in situ* decay of ^{26}Al in meteoritic material (Lee *et al.*, 1976).

The primary evidence for the existence of live ^{26}Al in the early solar system is found in CAIs in carbonaceous chondrites. CAIs are ideal places to look for ^{26}Al, because they contain several phases that are aluminum-rich and magnesium-poor. MacPherson *et al.* (1995) reviewed all of the available data, over 1500 analyses collected over the past 25 years, and showed that ^{26}Al was widespread throughout the region where CAIs formed, at an initial abundance level of $(^{26}Al/^{27}Al)_I \sim 4.5 \times 10^{-5}$.

Unlike CAIs, however, most chondrules consist exclusively of low-Al/Mg phases in which any radiogenic ^{26}Mg that might be present is swamped by the nonradiogenic magnesium. There exists a subset of chondrules that contain primary plagioclase, however, and these are amenable to analysis of ^{26}Mg through use of an ion microprobe. While comparatively rare with respect to the more common ferromagnesian chondrules, plagioclase-rich chondrules have been found in a number of different meteorites and meteorite types. Such objects therefore offer the chance to compare the Al-Mg isotopic systematics of chondrules and CAIs, in principle meaning that the relative ages of the two types of objects can be established. Doing so requires making two major assumptions. First, it must be taken more or less on faith that plagioclase-rich chondrules and normal chondrules really are contemporaneous. At present the validity of this assumption is unknown but perhaps it is sufficient to note here, and as we will show below, that there is a

significant isotopic difference between CAIs and the plagioclase-rich chondrules. Interpreted in chronologic terms, the data imply a time difference of at least 2–3 Ma between one type of chondrule and the oldest CAIs. The second assumption is that differences in Al-Mg isotopic systematics reflect age differences and not nebular heterogeneity of ^{26}Al or selective reprocessing of chondrules. We will return to the second assumption below.

Al-Mg isotopic data are generally represented graphically as the fractionation-corrected excess, $\delta^{26}Mg$ (deviation from terrestrial $^{26}Mg/^{24}Mg$ in parts in 10^3), plotted as a function of the bulk Al/Mg ratio ($^{27}Al/^{24}Mg$) of the sample, for example Fig. 1. For an inclusion or chondrule that plausibly formed as an equilibrium system, a positive correlation between $\delta^{26}Mg$ and the $^{27}Al/^{24}Mg$ ratios of the host phases is indicative of the *in situ* decay of ^{26}Al, and the slope of such a data array on this "isochron diagram" gives the $^{26}Al/^{27}Al$ ratio at the time of isotopic closure, defined as $(^{26}Al/^{27}Al)_I$.

Fig. 1 compares the Al-Mg isotopic data (see MacPherson *et al.*, 1995) for CAIs and chondrules plus chondrule-related objects. Several filters were applied to the data shown in Fig. 1. First, for clarity, only data with $^{27}Al/^{24}Mg < 1200$ are shown; CAI data with much higher ratios exist, but their addition would add nothing to the discussion and the chondrule data would be reduced to an illegible clump near the origin. Second, only data for what might be called normal CAIs are plotted on Fig. 1; a small subset of CAIs have unusual nuclear isotope anomalies and at the same time have no excesses of ^{26}Mg. Such objects indicate real heterogeneity in the solar nebula, but the scale of such heterogeneity is believed to be small (MacPherson *et al.*, 1995). Finally, some ion probe data for plagioclase grains extracted from ordinary chondrites (Zinner and Göpel, 1992) have not been included because the provenance of such grains is unknown. They may be from chondrules or they may not (in fact some of the data do show excess ^{26}Mg indicative of *in*

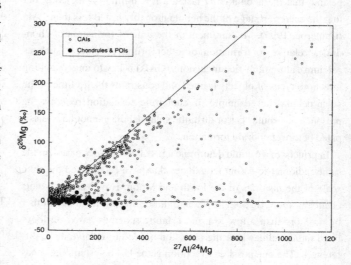

Fig. 1. $\delta^{26}Mg$ (the difference in the $^{26}Mg/^{24}Mg$ ratio from the terrestrial value, in parts per thousand) *vs.* $^{27}Al/^{24}Mg$ for CAIs from carbonaceous chondrites and chondrules and POIs (plagioclase-olivine inclusions, or plagioclase-rich chondrules) from carbonaceous, ordinary and enstatite chondrites (after MacPherson *et al.*, 1995).

situ decay of ^{26}Al), but we have restricted the discussion here to measurements of known chondrules and chondrule-like objects.

There are far fewer chondrule analyses than CAI analyses, but the data given in Fig. 1 clearly show no evidence for radiogenic ^{26}Mg in any of the measured chondrules. The data set includes analyses of chondrules from ordinary and enstatite chondrites as well as carbonaceous chondrites, so the lack of a magnesium isotopic signature is apparently consistent across chondrite types. The chondrule data are in stark contrast to the voluminous CAI data that have well-resolved ^{26}Mg excesses consistent with a $(^{26}Al/^{27}Al)_I$ ratio of $\sim 4.5 \times 10^{-5}$. The latter limit is well represented in all types of carbonaceous chondrites. However, there is but one analysis of a CAI from an ordinary chondrite, and that object appears to be of the isotopically anomalous variety noted above; it gives a $(^{26}Al/^{27}Al)_I$ ratio of $\sim 8.4 \times 10^{-6}$ (Hinton and Bischoff, 1984; Ireland *et al.*, 1992).

A different way of representing the same data is to plot histograms of the calculated $(^{26}Al/^{27}Al)_I$ ratio for each analysis, where this ratio is calculated by assuming a uniform initial $\delta^{26}Mg_I$ value for all samples of 0 (Fig. 2). The data for CAIs are bimodal, with peaks at $(^{26}Al/^{27}Al)_I \sim 4.5 \times 10^{-5}$ and ~ 0. This bimodality is interpreted to mean, first, that initial CAI formation was restricted in time, and second, that CAIs experienced significant late-stage reprocessing that partially reset the isotope systems. The $(^{26}Al/^{27}Al)_I \sim 4.5 \times 10^{-5}$ peak consists largely of data points from CAI minerals that are unambiguously primary, whereas many of the analysis spots with lower $(^{26}Al/^{27}Al)_I$ ratios are from regions that show petrologic evidence for reheating or secondary alteration (MacPherson *et al.*, 1995). The time difference represented by the two peaks, assuming that a chronologic interpretation is valid, is at least 2–3 Ma. However, it is important to understand the significance of the second "peak" at $(^{26}Al/^{27}Al)_I \sim 0$. Any resetting event that occurred after decay of most or all of the ^{26}Al, including events that occurred literally yesterday, would still give the same effective isotopic signature. In other words, the time difference implied by the two peaks is a minimum; it could just as easily be a difference of 3 billion years as 3 million years. Moreover, the "peak" cannot be interpreted to mean that the resetting events were necessarily restricted in time; they could very well have been spread out over a very long time period. In contrast to the CAI data, the chondrule data show a single well-defined peak at $(^{26}Al/^{27}Al)_I \sim 0$, meaning, in purely chronologic terms, that the chondrules have "ages" at least 2–3 Ma younger than the oldest normal CAIs.

Discussion

The difference in Al-Mg isotopic signatures between chondrules and CAIs can be interpreted in several ways. One possibility is that chondrules and CAIs formed out of different nebular reservoirs having different initial ^{26}Al abundances. Alternatively, chondrules and CAIs formed out of the same isotopic reservoir but the chondrules have been selectively reprocessed to eradicate any trace of the original ^{26}Mg excesses that resulted from decay of ^{26}Al. Finally, the data can be taken at face value to mean that chondrules formed at least several million years after initial formation of CAIs.

The strongest evidence for nebular ^{26}Al heterogeneity comes from a small subset of the CAI population, briefly alluded to above and known collectively as FUN objects, because the isotopes of several elements exhibit mass Fractionation and Unknown Nuclear effects. FUN objects contain no excess ^{26}Mg but they do exhibit nuclear anomalies in other isotopes that preserve signatures of the interstellar dust from which they must ultimately have been formed (Lee *et al.*, 1976; Wasserburg *et al.*, 1977; Loss *et al.*, 1994). A chronologic interpretation of the Al-Mg isotopic data would require that these objects form several million years after normal CAIs and yet retain isotope anomalies that the latter do not, so the reasonable conclusion is that the isotopically-unusual objects did indeed form out of a distinct isotopic reservoir. In other words, the nebula must have been heterogeneous to some degree. A far different kind of evidence that might be consistent with, but does not require, nebular heterogeneity is the general nonisochronism exhibited by the Al-Mg isotopic data in most normal CAIs. Hutcheon (1982; and many other workers since) showed that ^{26}Mg excesses and $^{27}Al/^{24}Mg$ ratios within individual inclusions rarely define a single correlation line; in fact, within some inclusions, different crystals each define different correlation lines. Possible interpretations noted by Hutcheon include initial isotopic heterogeneity, isotopic variation induced by later alteration or other disturbance, or differences in age.

MacPherson *et al.* (1995) summarized all of the available data and made two points. First, although the nebular heterogeneity implied by isotopically-anomalous inclusions cannot be denied, it may have been a minor effect. Considering just the CV3 data for example, only a small number of inclusions out of the many that have now been analyzed show large isotope anomalies coupled with the absence of ^{26}Mg excesses. In the rest of the CAIs, there is ample evidence for the former presence of ^{26}Al at an abundance level relative to ^{27}Al of $\sim 4.5 \times 10^{-5}$. On the scale of whole meteorites or their parent bodies, ^{26}Al was reasonably homogeneous

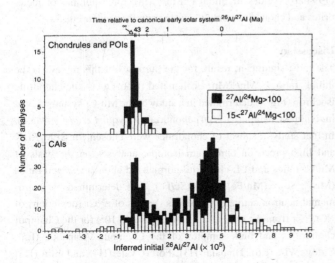

Fig. 2. Stacked histogram of inferred $(^{26}Al/^{27}Al)_I$. For CAIs, the distribution appears to be bimodal, while for the chondrules and POIs, only the lower $(^{26}Al/^{27}Al)_I$ group is represented.

throughout the region of the solar nebula where CAIs formed and their carbonaceous chondrite hosts were ultimately assembled (although it may have largely decayed by the time those carbonaceous chondrite parent bodies accreted). The second point made by MacPherson *et al.* (1995) is that there is now mounting evidence that the nonisochronous behavior of Al-Mg isotopic systematics in most CAIs can be explained as being due to later reprocessing. In a number of inclusions studied recently, different isotopic signatures can be identified with specific petrologic components and those components with apparent $(^{26}Al/^{27}Al)_I$ ratios of ~ 0 are clearly later in a petrologic sense than those with $(^{26}Al/^{27}Al)_I$ ratios of ~ 4.5 × 10^{-5} (*e.g.*, Podosek *et al.*, 1991; MacPherson and Davis, 1993; Caillet *et al.*, 1993). The resetting is highly localized within a given inclusion and is not caused by the secondary mineralization/alteration so prevalent in, for example, Allende inclusions. Rather, the resetting is caused by local solid state recrystallization and even partial melting that may occur multiple times. The cause is unknown but may be shock heating or some other transient event.

Returning then to the isotopic differences between the CAIs and chondrules, are the chondrules simply more processed than CAIs? This interpretation requires that the plagioclase in the chondrules has selectively lost all of its radiogenic ^{26}Mg to secondary processes whereas the plagioclase in CAIs has not. The chief argument against such an interpretation is that, if one looks just at the subset of data for all kinds of materials analyzed from a single meteorite (Allende is the only one for which sufficient data exist), there are systematic differences in the Al-Mg isotopic data between different kinds of material (see discussion by MacPherson *et al.*, 1995). Normal (Types A, B) CAIs have consistent ^{26}Mg excesses, but other kinds of inclusions (olivine-bearing varieties, for example, and so-called Type C inclusions) rarely show any such excesses. Most importantly from the standpoint of the present review, the plagioclase-rich chondrules (the "POIs" of Sheng *et al.*, 1991) show very small or no ^{26}Mg excesses. The fact that different kinds of objects within a single meteorite parent body show different isotopic signatures argues against nebular heterogeneity or selective reprocessing as explanations for the isotopic differences. In terms of a nebular heterogeneity model, for example, the Allende data would require that different types of objects were made exclusively in different and isotopically-distinct regions of the nebula and later brought together. A more extreme interpretation is that Types A and B CAI did not form in the solar system at all, but this interpretation runs up against the observation that no CAI is known to have an absolute age significantly greater than 4560 Ma (*e.g.*, MacPherson *et al.*, 1988). In our opinion the more reasonable interpretation is the chronologic one: chondrules (plagioclase-rich ones, anyway) and other non-refractory inclusions formed at least 2–3 Ma later than did normal CAIs.

^{53}Mn–^{53}Cr

Observations

^{53}Mn decays to ^{53}Cr with a half-life of 3.7 Ma (Table 1). Following several unsuccessful searches for radiogenic ^{53}Cr in primitive

meteorites (*cf.*, Lee and Tera, 1986), Birck and Allègre (1985) reported variations in $^{53}Cr/^{52}Cr$ in Allende refractory inclusions which were linearly correlated with the $^{55}Mn/^{52}Cr$ ratios of the samples.

Observed correlations for bulk inclusions, and for mineral separates of one coarse-grained Type B inclusion (BR1), correspond to initial $^{53}Mn/^{55}Mn$ ratios of $(6.7±2.2) × 10^{-5}$ and $(3.7±1.2) × 10^{-5}$, respectively. Birck and Allègre (1985) adopted the mean value, $(4.4±1.1) × 10^{-5}$, as the best value for the initial $(^{53}Mn/^{55}Mn)_o$ at the time of CAI formation. Concerned with the ubiquitous presence of isotopic anomalies in Allende inclusions, Nyquist *et al.* (1994a,b) prefer the value determined for mineral separates of BR1, on the assumption that isotopic equilibration is more likely to have occurred within this single coarse-grained Type B inclusion than among the assemblage of fine- and coarse-grained inclusions which define the isochron for the bulk inclusions. For consistency, we will use the BR1 mineral separate value. Although BR1 might have suffered late-stage alteration, as argued for many inclusions on the basis of Al-Mg systematics, the use of Birck and Allègre's adopted value, or even the bulk inclusion value, would not significantly change the conclusions about the chronological relationship of chondrules to CAIs.

Chromium isotopic studies were extended to other meteorite types by Birck and Allègre (1988), and in several later publications by them and their colleagues at the University of Paris. The chromium isotopic data obtained by the Paris group by thermal ionization mass spectrometry for carbonaceous chondrites were reviewed by Rotaru *et al.* (1992). *In situ* ion microprobe investigations of manganese-enriched phases in iron meteorites by the Caltech group (*cf.*, Hutcheon *et al.*, 1992) and in enstatite chondrites by the Washington University group (El Goresy *et al.*, 1992) have further confirmed the existence of fossil radiogenic ^{53}Cr in early solar system materials. The existence of ^{53}Cr anomalies in angrites has been demonstrated by Lugmair *et al.* (1992) and Nyquist *et al.* (1994c) and in two eucrites by Lugmair *et al.* (1994a,b). Values of initial $(^{53}Mn/^{55}Mn)_I$ for a number of meteorites and chondritic components are summarized in Fig. 3.

Discussion

The most significant result, for the purposes of this review, is the initial ratio $(^{53}Mn/^{55}Mn)_I$ calculated from a whole-chondrule isochron (Fig. 4) determined in a study of the Mn-Cr systematics of individual Chainpur (LL3.4) chondrules (Nyquist *et al.*, 1994a,b). In that study, a subset of chondrules were chosen for SEM/EDX and high precision chromium isotopic analyses on the basis of Mn/Cr ratios and LL-chondrite-normalized abundances and ratios $(Mn_{LL}, Sc_{LL}, (Mn/Fe)_{LL}$ and $(Sc/Fe)_{LL})$ as determined by instrumental neutron activation analyses. Values of $\varepsilon^{53}Cr$ (deviations of $^{53}Cr/^{52}Cr$ from the terrestrial value in parts in 10^4) for the Chainpur chondrules and bulk samples of the Chainpur, Arapahoe (L5), Colby, Wis. (L6), Dhajala (H3), Forest Vale (H4) and Jilin (H5) chondrites are plotted *vs.* $^{55}Mn/^{52}Cr$ in Fig. 4. The Mn/Cr ratios of the chondrules vary from half to twice the ratios in the bulk chondrites. This variation may be caused by the difference in volatility

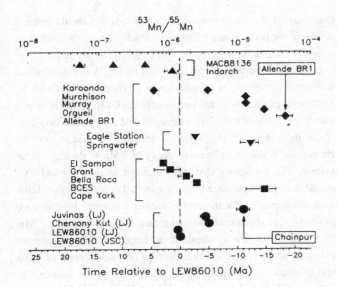

Fig. 3. Summary of $(^{53}Mn/^{55}Mn)_I$ for meteorites, chondrules and CAIs. The Chainpur datum is for the whole chondrule isochron (Fig. 4). Also shown are data for the LEW86010 angrite (Lugmair *et al.*, 1992 (LJ); Nyquist *et al.*, 1994 (JSC)); eucrites Chervony Kut and Juvinas (Lugmair *et al.*, 1994a,b); iron meteorites El Sampal, Grant, Bella Roca, and Cape York (Hutcheon and Olsen, 1991); a "compound isochron" for the BCES irons Bear Creek, Chupaderos, Mt. Edith, and Sandtown (Hutcheon *et al.*, 1992); pallasites Eagle Station (Birck and Allègre, 1988) and Springwater (Hutcheon and Olsen, 1991); carbonaceous chondrites Karoonda, Murchison, and Murray (Birck *et al.*, 1990), and Orgueil (Birck, 1991); Type B Allende inclusion BR1 (Birck and Allègre, 1985); and enstatite chondrites Indarch (Birck and Allègre, 1988) and MAC88138 (El Goresy *et al.*, 1992). Abbreviations: LJ = La Jolla (Scripps Institution of Oceanography, University of California, San Diego), JSC = Johnson Space Center. The data for the Springwater and Cape York meteorites appear to imply that their parent asteroids accreted and melted very early, essentially contemporaneously with Chainpur chondrule formation. A complete discussion of this and more general aspects of the complete data set is beyond the scope of this paper.

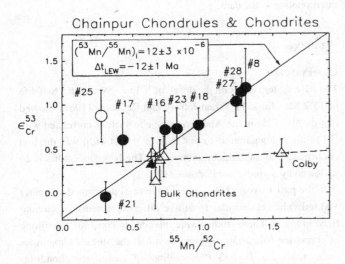

Fig. 4. $\varepsilon^{53}Cr$ (the difference in the $^{53}Cr/^{52}Cr$ ratio from the terrestrial value, in parts per 10,000) *vs.* $^{55}Mn/^{52}Cr$ for Chainpur chondrules (circles), bulk Chainpur (filled triangle) and other bulk chondrites (open triangles). The isochron corresponds to chondrule formation at 12 Ma before that of the angrite LEW 86010, 6 Ma after Allende CAI BR1.

of manganese and chromium, or it may be due to variations in the mixing ratio of precursor materials. The Mn-Cr isotopic data for most of the chondrules lie along a regression line (Williamson, 1968) fitted to the data, including the data for the bulk chondrites. As with all diagrams of this type, the slope of the regression line, interpreted as an isochron, determines the initial $(^{53}Mn/^{55}Mn)_I$ ratio in the chondrules, assuming they had a common initial $(^{53}Cr/^{52}Cr)_I$ ratio at a common time of formation. [Here $(^{53}Cr/^{52}Cr)_I$ is expressed as $\varepsilon^{53}Cr_I$].

The isochron shown in Fig. 4 corresponds to $(^{53}Mn/^{55}Mn)_I = (12\pm3) \times 10^{-6}$ and $\varepsilon^{53}Cr_I = -0.28\pm0.16$. If the variations in Mn/Cr ratio are caused by differences in volatility of manganese and chromium during chondrule formation, and if all the chondrules had the same initial $^{53}Cr/^{52}Cr$ ratio, this value of $(^{53}Mn/^{55}Mn)_I$ determines the formation of the Chainpur chondrules to have occurred at $\sim\Delta t_{LEW} = 12\pm1$ Ma prior to igneous crystallization of the LEW86010 angrite. More importantly for this discussion, this is 6 Ma after crystallization of Allende inclusion BR1. This formation interval relative to CAI is consistent with the >2–3 Ma suggested by Al-Mg systematics. We consider it unlikely that the Mn-Cr isochron dates parent body metamorphism rather than formation of Chainpur chondrules. Chainpur is an unequilibrated chondrite of low metamorphic grade (3.4); it is unlikely that chromium isotopic equilibration would have been achieved over the distances of several mm separating individual chondrules during thermal metamorphism of its parent body. It is perhaps more likely that the Mn-Cr systems of individual chondrules remained closed during their formation from precursor nebular dust. In that case, the Mn-Cr "formation interval" would represent formation of the dust, if the $^{53}Cr/^{52}Cr$ ratio was initially everywhere the same in the nebula during dust formation. In the former (unlikely) case, the Mn-Cr isochron would date an event following chondrule formation; in the latter case, events preceding chondrule formation.

Fig. 4 is an example of a "whole rock" isochron of the type often used in geological studies. In such studies it is assumed that the analyzed samples represent cogenetic systems if the data lie within error limits of a best fit line, and that such a line is indeed an isochron. Conversely, since the chondrule data are not all colinear in the isochron diagram, they cannot all represent cogenetic systems. A high scandium abundance and Sc/Fe ratio of Ch-25 suggests that the departure of the Mn-Cr data of this chondrule from the isochron could be due to late, preferential vaporization of manganese after ^{53}Mn had decayed, so Ch-25 is not included in the fit. There is no indication that this is also true for Ch-17, which also appears to be slightly displaced from the best fit line. The required preferential loss of manganese is less for Ch-17, and may simply not be evident in the scandium and iron data. Alternatively, $(^{53}Cr/^{52}Cr)_I$ for this chondrule may have differed from that of the others. Determining the degree to which initial chromium isotopic homogeneity can be assumed among a set of chondrules is a goal for future investigations. We note that the data for the bulk chondrites establish homogeneity among H, L, and LL chondrites at the current level of precision. Interestingly, the $^{55}Mn/^{52}Cr$ ratio of a bulk sample of the Colby L6 chondrite was unusually high, but the

^{53}Cr/^{52}Cr ratio was similar to that for the other bulk samples, consistent with late Cr-isotopic homogenization in this highly equilibrated chondrite. Values of ε^{54}Cr (^{54}Cr/^{52}Cr) are plotted *vs.* ε^{53}Cr in Fig. 5. The fitted trend line shown in the figure has the equation ε^{54}Cr = $(-0.4\pm1.3)\times\varepsilon^{53}$Cr and a correlation coefficient r = 0.76, suggesting that the ^{53}Cr and ^{54}Cr effects are correlated. chromium isotopic heterogeneities might be expected from the observation of comparatively large positive values of both ε^{53}Cr and ε^{54}Cr for Allende CAI EK1–4–1 (Papanastassiou, 1986) and of both positive and negative values of ε^{54}Cr in stepwise dissolutions of carbonaceous chondrites (Rotaru *et al.*, 1992; Ott *et al.*, 1994). Oxygen isotopic heterogeneities among Chainpur chondrules relative to a bulk sample were reported by Clayton *et al.* (1991). Apparent ^{54}Cr heterogeneities among the chondrules are small (~±1 ε-unit), and the slope of the apparent trend line less than expected from the ε^{54}Cr/ε^{53}Cr ratio of ~3 found for EK1–4–1. Shih *et al.* (1995) have examined the chromium isotopic data of carbonaceous chondrites as summarized by Rotaru *et al.* (1992). They note that whereas large variations in ε^{54}Cr for CI chondrites are not accompanied by variations in ε^{53}Cr, the reverse is true for C3 and C4 chondrites. For the latter, significant variations in ε^{53}Cr are accompanied by smaller variations in ε^{54}Cr, resulting in a somewhat flatter trend than shown in Fig. 5 for the Chainpur chondrules. Thus, there is a possibility that the apparent correlation observed for the Chainpur chondrules could be due to mixing of precursor dust components carrying different chromium isotopic signatures. (In principle, apparently correlated ^{53}Cr and ^{54}Cr anomalies also could be produced by small interferences at masses 50 or 52, but background scans of the mass spectrum in the chromium region give no evidence of such interference. Indeed, such scans show the mass region between mass-50 and mass-54 to be unusually free of interference. We believe that uncompensated mass fractionation effects are also unlikely because the values of ε^{53}Cr and ε^{54}Cr for samples are calculated relative to standards run under very similar conditions, so that chromium evaporation and ionization should proceed similarly for both. A more complete description of the analytical details of the chromium analyses will be given elsewhere.)

Because the apparent trend exhibited by Fig. 5 is incompatible with trends observed for components of CI chondrites and with the FUN trend of EK1–4–1, and because the C3/C4 trend could be primarily the result of radioactive decay of ^{53}Mn, we tentatively conclude that most of the variation in ^{53}Cr/^{52}Cr ratios for the chondrules is due to radioactive decay of ^{53}Mn in the early solar system. The possibility that the variations in the ^{53}Cr and ^{54}Cr abundances are of presolar origin needs further investigation, however, since models for the NSE (Nuclear Statistical Equilibrium) production of chromium and manganese suggest that ^{53}Cr, ^{53}Mn, ^{54}Cr, and ^{55}Mn are synthesized at large and similar neutron excesses, whereas ^{50}Cr and ^{52}Cr, used for normalization in the mass spectrometric chromium analysis, are synthesized at smaller neutron excesses (*cf.*, Clayton, 1981). It should be noted that if the trend line in Fig. 4 indeed is produced in part by mixing precursor materials with positively correlated ^{53}Cr, ^{54}Cr, and ^{55}Mn anomalies, the initial $(^{53}$Mn/^{55}Mn$)_I$ of the chondrules would have to be less and the formation interval relative to CAI longer than calculated here. That is, the value presented here for the time between the formation of Allende CAI BR1 and that of Chainpur chondrules is a lower limit for the absolute value of the formation interval.

Finally, we note that the values of $(^{53}$Mn/^{55}Mn$)_I$ and ε^{53}Cr$_I$ obtained from the chondrule isochron are compatible within error limits with radiogenic growth in a solar nebula with the solar photospheric value of $(Mn/Cr)_{solar}$ = 0.55 by weight (Anders and Grevesse, 1989) from initial values $(^{53}$Mn/^{55}Mn$)_o$ = 3.7 × 10^{-5} and ε^{53}Cr$_o$ = -1.5, respectively, for the Allende BR1 inclusion (Birck and Allègre, 1985). The $(^{53}$Mn/^{55}Mn$)_I$ and ε^{53}Cr$_I$ values of the LEW86010 angrite (Nyquist *et al.*, 1994c) lie on the same radiogenic growth curve. Thus, a chronological interpretation of the correlation of ^{53}Cr/^{52}Cr with ^{55}Mn/^{52}Cr is the most straightforward interpretation of the data.

^{129}I–^{129}Xe

Observations

The I-Xe system relies on the decay of ^{129}I to ^{129}Xe, with a half-life of 15.7 Ma (Table 1). When Jeffery and Reynolds (1961) showed that excess ^{129}Xe in the Abee enstatite chondrite is correlated with ^{127}I (with data analogous to that in Figs. 1 and 4), it was the first confirmation of an extinct radionuclide. The I-Xe system has been reviewed by Swindle and Podosek (1988).

In the past 14 years I-Xe studies have been performed on nearly 100 individual chondrules from five different meteorites, summarized in Fig. 6. These studies were intended to constrain the timing of chondrule formation, so more than half the studied chondrules come from the Type 3 ("unequilibrated") ordinary chondrites Chainpur, Semarkona and Tieschitz. However, "unequilibrated" does not necessarily mean "unaltered" (Hutchison *et al.*, 1987; Alexander *et al.*, 1989; Ruzicka, 1990)—it is now generally believed that most of these chondrites experienced some postfor-

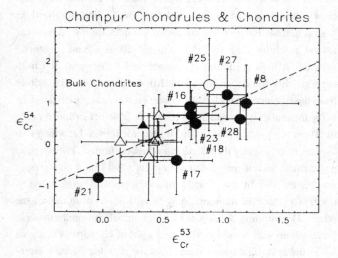

Fig. 5. ε^{54}Cr *vs.* ε^{53}Cr for Chainpur chondrules (circles) and bulk chondrites (filled triangles). The apparent trend line may result from mixing different nebular components in chondrule precursors.

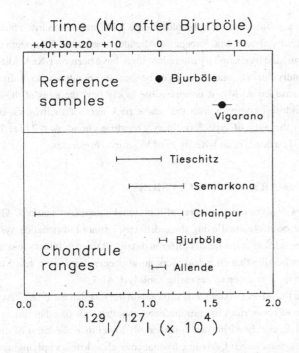

Fig. 6. Range in $(^{129}I/^{127}I)_I$, or apparent ages, of individual chondrules from various meteorites (Caffee *et al.*, 1982; Swindle *et al.*, 1983, 1991a, 1991b; Nichols *et al.*, 1990, 1991; Gilmour and Turner, 1994). For Vigarano, the horizontal lines are 1σ error bars (the error bars for Bjurböle are smaller than the symbol). For the chondrules, the horizontal lines represent the range in ages measured from different chondrules.

mation processing (*e.g.*, aqueous alteration, very low-grade metamorphism, shock), albeit slight. Most of the I-Xe ages of chondrules, even from Type 3 ordinary chondrites, have ultimately been interpreted in terms of postformation alteration.

Studies of petrographically characterized individual chondrules from the LL3 chondrite Chainpur (Swindle *et al.*, 1991a), the LL3 chondrite Semarkona (Swindle *et al.*, 1991b) and the H3 chondrite Tieschitz (Nichols *et al.*, 1991) give very similar results, despite the fact that the meteorites come from at least two different chemical classes, and have petrographic grades ranging from 3.0 for Semarkona to 3.8 for Tieschitz.

Many of the chondrules give well-defined apparent ages. These ages span at least 10 Ma for each meteorite (for Chainpur, the span is 50 Ma). The highest initial $(^{129}I/^{127}I)_I$ ratios for both Chainpur and Semarkona are about 1.3×10^{-4}; the highest $(^{129}I/^{127}I)_I$ ratio observed in a Tieschitz chondrule is about 1.1×10^{-4} and corresponds to an apparent age about 4 Ma later. I-Xe ages are frequently given relative to the age of the L4 chondrite Bjurböle, which gives an apparent age reproducible to about 2 Ma or better, even for individual chondrules (Caffee *et al.*, 1982; Gilmour and Turner, 1994). The $(^{129}I/^{127}I)_I$ ratio in Bjurböle is roughly equal to the highest ratio seen in Tieschitz.

Of course, for comparison with the Al-Mg and Mn-Cr systems, we would like to compare the I-Xe ages to the ages of the oldest CAIs in Allende. Although I-Xe studies have been performed on many CAIs and chondrules from Allende (Nichols *et al.*, 1990; see also summary by Swindle *et al.*, 1988), there is a large variation in

apparent $(^{129}I/^{127}I)_I$ within most objects, with the apparent ages spanning much of the range of I-Xe ages of Semarkona, Chainpur and Tieschitz chondrules. Much of the iodine in Allende is contained within sodalite (Kirschbaum, 1988), one of the products of the secondary mineralization/alteration, so Swindle *et al.* (1988) suggested that the variable apparent ages reflect this secondary processing. Some other carbonaceous chondrites, however, appear to be less altered than Allende. In particular, a whole-rock sample of Vigarano, another CV3 whose CAIs exhibit substantially less secondary alteration than Allende CAIs, gave $(^{129}I/^{127}I)_I = 1.6 \times 10^{-4}$ (Crabb *et al.*, 1982), corresponding to the oldest well-defined apparent I-Xe age yet determined (Swindle and Podosek, 1988). It is more than 8 Ma older than Bjurböle, and 4 Ma older than the oldest chondrules from Type 3 ordinary chondrites. The implications of this age for the ages of CAIs will be discussed below.

Discussion

There are three possible interpretations for the large span in apparent ages of individual chondrules: isotopic inhomogeneity, chondrule formation over a 50 Ma interval, or postformation alteration.

The strongest argument that the apparent ages reflect chronology, and not isotopic inhomogeneity, is that in individual meteorites, $(^{129}I/^{127}I)_I$ is anticorrelated with initial (trapped) $(^{129}Xe/^{132}Xe)_I$ ratios (Fig. 7; Nichols *et al.*, 1991; Swindle *et al.*, 1991a). On this plot, which is analogous to a plot of age *vs.* initial isotopic composition for the Rb-Sr or Sm-Nd system, the data are consistent with closed-system evolution in a system with a chondritic bulk I/Xe ratio. On the other hand, it is difficult to imagine an inhomogeneity scenario in which materials in the nebula with more ^{129}I would have less ^{129}Xe.

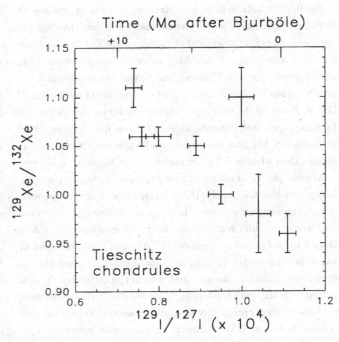

Fig. 7. $(^{129}I/^{127}I)_I$ *vs.* $(^{129}Xe/^{132}Xe)_I$ for Tieschitz chondrules (data from Nichols *et al.*, 1991). The data appear to fall along a single correlation line, suggesting closed-system chronological evolution in a single reservoir.

The possibility that the 50 Ma span in ages represents the duration of chondrule formation can not be ruled out on the basis of the I-Xe data alone, but we believe it is highly unlikely. The fact that the evolution is consistent with a chondritic, rather than a nebular, I/Xe ratio (*cf.,* evolution of the ^{53}Mn-^{53}Cr system) means that the material in these chondrules spent the time before their last resetting (as much as 50 Ma) in a system with chondritic composition rather than solar (*i.e.,* in a solid). If these ages represent chondrule formation, then chondrule formation would have to be a parent body, rather than nebular, process, but there is considerable evidence that such is not the case (see Taylor *et al.,* 1983). Hence, we believe these do not all represent formation ages.

If the apparent I-Xe ages are really ages, but they do not (or do not all) date chondrule formation, what could they be dating? The I-Xe system is still not well-enough understood to say for certain, but there are some hints in the data. For one thing, apparent ages in the chondrules correlate with bulk chemistry (*e.g.,* refractory compositions correspond to older ages) and with petrographic properties. Since bulk composition correlates with textural type, texture and ages might be expected to correlate. But apparent ages also correlate with more subtle petrographic features. For example, in Semarkona, chondrules with sulfides on the chondrule surface, and particularly those with sulfides penetrating between blades of radiating pyroxene chondrules, tend to have younger ages (*i.e.,* to be more easily reset). Since these sulfide textures are similar to those in which evidence of aqueous alteration have been found (Alexander *et al.,* 1989), Swindle *et al.* (1991b) suggested that the I-Xe ages in these chondrules reflect aqueous alteration. Also, since most iodine structural host phases are secondary minerals, the I-Xe system, in general, may reflect the chronology of secondary processes. .

But if some I-Xe ages reflect secondary processing, are any of them associated with primary chondrule formation? Despite the large number (45) of individual chondrules from Type 3 chondrites that have been analyzed, the oldest apparent ages are about 4 Ma after Vigarano for both Chainpur and Semarkona and about 8 Ma after Vigarano for the smaller sample (9 chondrules) from Tieschitz. Hence, if any of the ordinary chondrite chondrule ages represent formation ages, then some chondrule formation must have occurred at least a few Ma after the formation of Vigarano. Of course, it is also not clear what the I-Xe "formation" age of Vigarano represents.

Allende and Vigarano CAIs have the same Al-Mg systematics. Interpreted chronologically, this indicates that their CAI populations were contemporaneous. However, the I-Xe age of Vigarano does not necessarily represent the time of formation of its CAIs. Only a small fraction of Vigarano is CAIs, and since the siting of iodine in Vigarano (or virtually any other meteorite) is not known, there is no particular reason to think it would be preferentially sited in CAIs. On the other hand, if the CAIs are the first solids formed, the bulk I-Xe age presumably represents a younger limit to the time of formation of the CAIs. Furthermore, analyses of magnetite separates from the carbonaceous chondrites Orgueil (CI) and Murchison (CM) (Herzog *et al.,* 1973; Lewis and Anders, 1975) gave apparent ages that, while younger than Vigarano by about 2

Ma, are still older than the oldest type 3 ordinary chondrite chondrules by about 2 Ma. Except for Allende, where the I-Xe system is dominated by secondary alteration, there have been no I-Xe studies of individual chondrules from carbonaceous chondrites, so, unlike the case for Al-Mg, it is impossible to say how the ages of those chondrules compare with the whole rock ages of their hosts, or with the ages of type 3 ordinary chondrite chondrules. There is clearly a need for an I-Xe study of Vigarano chondrules.

SUMMARY AND CONCLUSIONS

Ages of significant numbers of individual chondrules and/or CAIs have been obtained using three different extinct radionuclide systems. Although the results differ in detail, all three systems seem to be indicating that chondrule formation occurred at least a few Ma after the formation of the earliest solids (CAIs).

In the Al-Mg system, this is based on the fact that the $(^{26}Al/^{27}Al)_I$ ratio of chondrules in carbonaceous chondrites or of other objects in ordinary chondrites are considerably lower than the ratios of the oldest (unaltered) CAIs in carbonaceous chondrites, implying age differences corresponding to several half-lives of ^{26}Al. In the Mn-Cr system, this is based on an apparent isochron for Chainpur chondrules that gives an age younger than that of an Allende CAI isochron. In the I-Xe system, the argument is based on the fact that the oldest apparent ages of chondrules from Type 3 ordinary chondrites are younger than ages from either whole rock or magnetite samples of some carbonaceous chondrites.

We have two caveats to the conclusion that chondrule formation occurred over an extended period of time. The first is that the three systems do not tell exactly the same story. The second is that because isotopic heterogeneity in the solar nebula must have existed, we can not rule it out as the cause for the variations in apparent ages in these systems.

There are several differences in the scenarios derived on the basis of different systems, but they may not be mutually contradictory. Although the numbers quoted for the apparent time scales are different, about 4 Ma for I- Xe, about 6 Ma for Mn-Cr, and at least 2–3 Ma for Al-Mg, these are not necessarily in disagreement, considering the uncertainty in the $(^{53}Mn/^{55}Mn)_I$ ratio for CAIs and the fact that the constraint from Al-Mg is only a lower limit. A more troublesome difference is the fact that the Al-Mg results suggest that chondrules (or at least plagioclase-rich chondrules) in carbonaceous chondrites are younger than CAIs, but the I-Xe argument is based on Chainpur chondrules being younger than the whole rock apparent age of the carbonaceous chondrite Vigarano, which might be dominated by chondrules (of course, CAIs in Vigarano might be older still). Finally, Nyquist *et al.* (1994a,b) have assumed that Chainpur chondrules all have the same $(^{53}Mn/^{55}Mn)_I$ ratio, while Swindle *et al.* (1991a) have shown that Chainpur chondrules have varying $(^{129}I/^{127}I)_I$ ratios. Of course, in many of these cases, the contradictory results for different systems might all be dating real events, since the three systems do not necessarily respond in the same way to later events. Also, part of the difficulty comes from the fact that no two comparisons are exactly the same—the Al-Mg results do not include the common ferromagnesian chondrules (because of their low Al/Mg ratios), the I-Xe results do not include any reliable formation ages for CAIs (because of the prevalent alteration in Allende, the most-studied carbonaceous chondrite), and the Mn-Cr results are for only one set of chondrules from one meteorite compared to data for one CAI to which no other chronometer has been applied. Furthermore, the Mn-Cr data are from an ongoing investigation.

Isotopic heterogeneity is always difficult to rule out. For both Al-Mg and I-Xe, for which extensive databases exist, there are reasons to think that many, if not all, of the apparent ages do truly reflect the passage of time (Swindle and Podosek, 1988; MacPherson *et al.*, 1995). It is more difficult to rule out heterogeneity in the case of the Mn-Cr system, because there are fewer data available. In addition, there is the hint in the Chainpur chondrule chromium data of a correlation between ε^{53}Cr and ε^{54}Cr. However, as pointed out in the discussion of the Mn-Cr results, that would imply that the difference in formation times is even longer.

Despite obvious difficulties with interpreting the results from the three isotopic systems in detail, the general conclusion can be made that chondrule formation was occurring several Ma after the formation of the first solids in the solar system.

We recognize that such a long timescale poses problems in some models, because, for example, it is far shorter than the typical orbital evolution time of small solids in the solar nebula (Cameron, 1995; Wood, this volume). We do not have a ready explanation for a way out of this dilemma. We merely point out that the isotopic evidence for an extended time scale seems to us to be getting stronger, and that it is possible to devise plausible scenarios in which such age differences might exist (Cameron, 1995).

REFERENCES

Alexander C. M. O'D., Barber D. J. and Hutchison R. (1989) The microstructure of Semarkona and Bishunpur. *Geochim. Cosmochim. Acta* **53**, 3045–3057.

Anders E. and Grevesse N. (1989) Abundances of the elements: Meteoritic and solar. *Geochim. Cosmochim. Acta* **53**, 197–214.

Birck J.-L. (1991) Isotopes in comets: Implications from Cr in carbonaceous chondrites. *Space Sci. Rev.* **56**, 141–146.

Birck J.-L. and Allègre C. J. (1985) Evidence for the presence of ^{53}Mn in the early solar system. *Geophys. Res. Lett.* **12**, 745–748.

Birck J.-L. and Allègre C. J. (1988) Manganese-chromium isotope systematics and the development of the early Solar System. *Nature* **331**, 579–584.

Birck J.-L., Rotaru M., Allègre C. J. (1990) ^{53}Mn in carbonaceous chondrites (abstract). *Meteoritics* **25**, 349–350.

Brazzle R. H., Kehm K., Hohenberg C. M., Göpel C., Swindle T. D., Davis A. M. and MacPherson G. (1995) I-Xe chronometry: crock or clock? A program to test and interpret the I-Xe system (abstract). *Lunar Planet. Sci.* **26**, 165–166.

Caffee M. W., Hohenberg C. M., Swindle T. D. and Hudson B. (1982) I-Xe ages of individual Bjurböle chondrules. *Proc. Lunar Planet. Sci. Conf. 13th*, *J. Geophys. Res.* **87**, A303–A317.

Caillet C., MacPherson G. J. and Zinner E. K. (1993) Petrologic and Al-Mg isotopic clues to the accretion of two refractory inclusions onto the Leoville parent body: One was hot, the other wasn't. *Geochim. Cosmochim. Acta* **57**, 4725–4743.

Cameron A. G. W. (1995) The first ten million years in the solar nebula. *Meteoritics* **30**, 133–161.

Clayton D. D. (1981) Some key issues in isotopic anomalies: Astrophysical history and aggregation. *Proc. Lunar Planet. Sci.* **12B**, 1781–1802.

Clayton R. N., Mayeda T. K., Goswami J. N. and Olsen E. J. (1991) Oxygen isotope studies of ordinary chondrites. *Geochim. Cosmochim. Acta* **55**, 2317–2337.

Crabb J., Lewis R. S. and Anders E. (1982) Extinct ^{129}I in C3 chondrites. *Geochim. Cosmochim. Acta* **46**, 2511–2526.

El Goresy A., Wadhwa M., Nagel H.-J., Zinner E. K., Janicke J. and Crozaz G. (1992) ^{53}Cr-^{53}Mn systematics of Mn-bearing sulfides in four enstatite chondrites (abstract). *Lunar Planet. Sci.* **23**, 331–332.

Gilmour J. D. and Turner G. (1994) Studies of extraterrestrial xenon samples using RELAX (abstract). *USGS Circular* **1107**, 111.

Gray C. M. and Compston W. (1974) Excess ^{26}Mg in the Allende meteorite. *Nature* **251**, 495–497.

Herzog G. F., Anders E., Alexander E. C. Jr., Davis P. K. and Lewis R. S. (1973) Iodine-129/xenon-129 age of magnetite from the Orgueil meteorite. *Science* **180**, 489–491.

Hinton R. W., and Bischoff A. (1984) Ion microprobe magnesium isotope analysis of plagioclase and hibonite from ordinary chondrites. *Nature* **308**, 169–172.

Hutcheon I. D. (1982) Ion probe magnesium isotopic measurements of Allende inclusions. *Amer. Chem. Soc. Symposium Ser.* **176**, 95–128.

Hutcheon I. D. and Olsen E. (1991) Cr isotopic composition of differentiated meteorites (abstract). *Lunar Planet. Sci.* **22**, 605–606.

Hutcheon I. D., Olsen E., Zipfel J. and Wasserburg G. J. (1992) Cr isotopes in differentiated meteorites: Evidence for ^{53}Mn (abstract). *Lunar Planet. Sci.* **23**, 565–566.

Hutchison R., Alexander C. M. O'D. and Barber D. J. (1987) The Semarkona meteorite: First recorded occurrence of smectite in an ordinary chondrite, and its implications. *Geochim. Cosmochim. Acta* **51**, 1875–1882.

Ireland T. R., Zinner E. K., Fahey A. J. and Esat T. M. (1992) Evidence for distillation in the formation of HAL and related hibonite inclusions. *Geochim. Cosmochim. Acta* **56**, 2503–2520.

Jeffery P. M., and Reynolds J. H. (1961) Origin of excess ^{129}Xe* in stone meteorites. *J. Geophys. Res.* **66**, 3582–3594.

Kirschbaum C. (1988) Carrier phases for iodine in the Allende meteorite and their associated ^{129}Xe/^{127}I ratios: A laser microprobe study. *Geochim. Cosmochim. Acta* **52**, 679–699.

Lee T. and Papanastassiou D. A. (1974) Mg isotopic anomalies in the Allende meteorite and correlation with O and Sr effects. *Geophys. Res. Lett.* **1**, 225–228.

Lee T. and Tera F. (1986) The meteoritic chromium isotopic composition and limits for radioactive ^{53}Mn in the early solar system. *Geochim. Cosmochim. Acta* **50**, 199–206.

Lee T., Papanastassiou D. A. and Wasserburg G. J. (1976) Demonstration of ^{26}Mg excess in Allende and evidence for ^{26}Al. *Geophys. Res. Lett.* **3**, 109–112.

Lewis R. S., and Anders E. (1975) Condensation time of the solar nebula from extinct ^{129}I in primitive meteorites. *Proc. Nat. Acad. Sci. USA* **72**, 268–273.

Loss R. D., Lugmair G. W., Davis A. M. and MacPherson G. J. (1994) Isotopically distinct reservoirs in the solar nebula: Isotope anomalies in Vigarano meteorite inclusions. *Astrophys. J.* **436**, L193–L196.

Lugmair G. W. and Galer S. J. G. (1992) Age and isotopic relationships among the angrites Lewis Cliff 86010 and Angra dos Reis. *Geochim. Cosmochim. Acta* **56**, 1673–1694.

Lugmair G. W., MacIsaac C. and Shukolyukov A. (1992) The ^{53}Mn-^{53}Cr isotopic system and early planetary evolution (abstract). *Lunar Planet. Sci.* **23**, 823–824.

Lugmair G. W., MacIsaac C. and Shukolyukov A. (1994a) Small time differences in differentiated meteorites recorded in the ^{53}Mn-^{53}Cr chronometer (abstract). *Lunar Planet. Sci.* **25**, 813–814.

Lugmair G. W., MacIsaac Ch. and Shukolyukov A. (1994b) Small time differences recorded in differentiated meteorites (abstract). *Meteoritics* **29**, 493–494.

MacPherson G. J., and Davis A. M. (1993) A petrologic and ion microprobe study of a Vigarano Type B refractory inclusion: Evolution by multiple stages of alteration and melting. *Geochim. Cosmochim. Acta* **57**, 231–243.

MacPherson G. J., Wark D. A. and Armstrong J. T. (1988), Primitive material surviving in chondrites: Refractory inclusions. In Meteorites and the Early Solar System (J. F. Kerridge and M. S. Matthews, eds.), 746–807. Univ. of Arizona Press, Tucson.

MacPherson G. J., Davis A. M. and Zinner E. K. (1995) The distribution of ^{26}Al in the early solar system—a reappraisal. *Meteoritics*, in press.

Nichols R. H. Jr., Hohenberg C. M., Olinger C. T. and Rubin A. E. (1990) Allende chondrules and rims: I-Xe systematics (abstract). *Lunar Planet. Sci.* **21**, 879–880.

Nichols R. H. Jr., Hagee B. E. and Hohenberg C. M. (1991) Tieschitz chondrules: I-Xe systematics. *Lunar Planet. Sci.* **22**, 975–976.

Nichols R. H. Jr., Hohenberg C. M., Kehm K., Kim Y. and Marti K. (1994) I-Xe studies of the Acapulco meteorite: Absolute I-Xe ages of individual phosphate grains and the Bjurböle standard. *Geochim. Cosmochim. Acta* **58**, 2553–2561.

Nyquist L. E., Lindstrom D., Wiesmann H., Martinez R., Bansal B., Mittlefehldt D., Shih C.-Y. and Wentworth S. (1994a) Mn-Cr isotopic systematics of individual Chainpur chondrules (abstract). *Meteoritics* **29**, 512.

Nyquist L. E., Lindstrom D., Wiesmann H., Bansal B., Shih C.-Y., Mittlefehldt D., Martinez R. and Wentworth S. (1994b) Mn-Cr isotopic systematics of Chainpur chondrules and bulk ordinary chondrites (abstract). In *Papers Presented to Chondrules and the Protoplanetary Disk*, Oct. 13–15, Albuquerque, New Mexico.

Nyquist L. E., Bansal B. M., Wiesmann H. and Shih C.-Y. (1994c) Neodymium, strontium, and chromium isotopic studies of the LEW86010 and Angra dos Reis meteorites and the chronology of the angrite parent body (abstract). *Meteoritics* **29**, 872–885.

Ott U., Podosek F. A., Brannon J. C., Bernatowicz T. J. and Neal C. J. (1994) Chromium isotopic anomalies in stepwise dissolution of Orgueil (abstract). *Lunar Planet. Sci.* **25**, 1033–1034.

Papanastassiou D. A. (1986) Chromium isotopic anomalies in the Allende meteorite. *Astrophys. J.* **308**, L27–L30.

Podosek F. A., and Cassen P. (1994) Theoretical, observational, and isotopic estimates of the lifetime of the solar nebula. *Meteoritics* **29**, 6–25.

Podosek F. A., Zinner E. K., MacPherson G. J., Lundberg L. L., Brannon J. C. and Fahey A. J. (1991) Correlated study of initial ^{87}Sr/^{86}Sr and Al/Mg isotopic systematics and petrologic properties in a suite of refractory inclusions from the Allende meteorite. *Geochim. Cosmochim. Acta* **55**, 1083–1110.

Ruzicka A. (1990) Deformation and thermal histories of chondrules in the Chainpur (LL3.4) chondrite. *Meteoritics* **25**, 101–113.

Rotaru M., Birck J.-L. and Allègre C. J. (1992) Clues to early Solar System history from chromium isotopes in carbonaceous chondrites. *Nature* **358**, 465–470.

Sheng Y. J., Hutcheon I. D. and Wasserburg G. J. (1991) Origin of plagioclase-olivine inclusions in carbonaceous chondrites. *Geochim. Cosmochim. Acta* **55**, 581–599.

Shih C.-Y., Nyquist L. E. and Wiesmann H. (1995) Iron isotopes in Chainpur chondrules (abstract). *Lunar Planet. Sci.* **26**, 1289–1290.

Swindle T. D. (1993) Extinct radionuclides and evolutionary time scales. In *Protostars and Planets III* (eds. E. H. Levy and J. I. Lunine), 867–881. Univ. of Arizona Press, Tucson.

Swindle T. D., and Podosek F. A. (1988) Iodine-xenon dating. In *Meteorites and the Early Solar System* (eds. J. F. Kerridge and M. S. Matthews), 1127–1146. Univ. of Arizona Press, Tucson.

Swindle T. D., Caffee M. W., Hohenberg C. M., and Lindstrom M. M. (1983) I-Xe studies of individual Allende chondrules. *Geochim. Cosmochim. Acta* **47**, 2157–2177.

Swindle T. D., Caffee M. W. and Hohenberg C. M. (1988) I-Xe studies of Allende inclusions: EGGs and the Pink Angel. *Geochim. Cosmochim. Acta* **52**, 2215–2227.

Swindle T. D., Caffee M. W., Hohenberg C. M., Lindstrom M. M. and Taylor G. J. (1991a) Iodine-xenon studies of petrographically and chemically characterized Chainpur chondrules. *Geochim. Cosmochim. Acta* **55**, 861–880.

Swindle T. D., Grossman J. N., Garrison D. H. and Olinger C. T. (1991b) Iodine-xenon, chemical and petrographic studies of Semarkona chondrules. *Geochim. Cosmochim. Acta* **55**, 3723–3734.

Taylor G. J., Scott E. R. D. and Keil K. (1983) Cosmic setting for chondrule formation. In *Chondrules and their Origins* (ed. E. A. King), pp. 262–278. Lunar and Planetary Institute, Houston.

Urey H. C. (1955) The cosmic abundances of potassium, uranium, and thorium and the heat balances of the Earth, the Moon and Mars. *Proc. Nat. Acad. Sci. USA* **41**, 127–144.

Walker F. W., Parrington J. R. and Feiner F. (1989) *Nuclides and Isotopes*. 14th Ed., General Electric Co.

Wasserburg G. J., Lee T. and Papanastassiou D. A. (1977) Correlated oxygen and magnesium isotopic anomalies in Allende inclusions: II. Magnesium. *Geophys. Res. Lett.* **4**, 299–302.

Williamson J. H. (1968) Least-squares fitting of a straight line. *Can. J. Phys.* **46**, 1845–1847.

Zinner E. K., and Göpel C. (1992) Evidence for ^{26}Al in feldspars from the H4 chondrite Ste. Marguerite (abstract). *Meteoritics* **27**, 311–312.

10: Formation of Chondrules and Chondrites in the Protoplanetary Nebula

EDWARD R.D. SCOTT, STANLEY G. LOVE* and ALEXANDER N. KROT

Hawai'i Institute of Geophysics and Planetology, School of Ocean and Earth Science and Technology, University of Hawai'i at Manoa, Honolulu, HI 96822, U.S.A. *Present address: California Institute of Technology, Division of Geological and Planetary Sciences, Mail Code 252–21, Pasadena, CA 91125, U.S.A.

ABSTRACT

Chemical, petrologic and isotopic data for chondrite groups and their components suggest that chondrule formation and accretion were complex and repetitive processes that may have operated simultaneously over a period of several Myr after Ca-, Al-rich inclusions (CAIs) formed. Chondrules formed in dusty regions from different reservoirs of fine-grained, matrix-like material, refractory CAI-like material and previous generations of chondrules. Evidence for recycling of chondrules is provided by relict grains, independent compound chondrules, and igneous rims. The thermal history of chondrules inferred from experiments, chondrule textures and mineral chemistry, is still rather uncertain. Realistic cooling paths which are consistent with plausible chondrule-forming mechanisms must be studied. The narrow range of estimated peak temperatures (1700–2100 K) of most chondrules may be due to dissociation of nebular hydrogen and the latent heat of melting. Two plausible chondrule-forming mechanisms, nebular lightning and gas drag heating, are discussed. Nebular lightning remains a possible chondrule-forming mechanism but faces serious challenges from the low ambient gas pressure and high electrical conductivity of the nebula.

Turbulent concentration of chondrules and other components between eddies (Cuzzi et al., this volume) appears consistent with several features of chondrites, e.g., matrix rims on chondrules and other chondritic components and characteristic chondrule size distributions in different groups. Systematic variations of chondrule sizes among CO3 and ordinary chondrites suggest that planetesimals accreted at the nebular midplane while turbulent concentration of chondrules continued elsewhere. Accretion of aggregations or planetesimals at the midplane may also have hampered fast migration of early-formed particles into the sun and allowed altered chondritic and igneous clasts to be accreted simultaneously with late-formed chondrules.

INTRODUCTION

We should not be surprised at the absence of a consensus regarding chondrule formation in view of the considerable disagreement about the nature of the solar nebula and the importance of various nebular processes. During the past 20 years, our views on this topic have changed drastically. Chondrites and their ingredients are no longer considered to be equilibrium condensates from a slowly cooled, quiescent nebula. Instead, most researchers believe they are products of a dynamic, energetic nebula in which solids were processed rapidly and repeatedly at temperatures high enough to cause melting and evaporation (Wood, 1988; Kerridge, 1993).

In this chapter we offer a broad overview of meteoritic and experimental constraints on the origins of chondrules and other components in chondrites and the accretion processes that transformed them into chondritic planetesimals. We discuss the timing of the formation of chondrules and other ingredients, and how formation and accretion of these objects might be related. We examine two mechanisms for chondrule formation, lightning and gas-drag heating resulting from nebular shock waves, that offer insights into the nature of chondrule formation and highlight critical data needed to elucidate chondrule formation.

CHONDRITE GROUPS AND THEIR COMPONENTS

There are currently 13 chondrite groups recognized; only 8 of these were known 20 years ago (Wasson, 1974). Each chondrite group is probably derived from a single parent body, but multiple bodies cannot be excluded. The main chondrite components and their approximate proportions in each group are given in Table 1.

Chondrules, which are the major component of most but not all groups, are µm-cm sized rounded or irregular objects, composed largely of ferromagnesian silicates (olivines and pyroxenes) with feldspathic interstitial material that crystallized from molten droplets during transient high-temperature events in the solar nebula (e.g., Wood, 1988; Wasson, 1993).

Ca-Al rich inclusions (CAIs) also formed during transient high-temperature events in the solar nebula but are enriched in the so called "refractory" minerals, melilite, spinel, fassaite, anorthite, perovskite (Davis and MacPherson, this volume). Fine-grained CAIs are largely nebular condensates: most have a highly distinctive rare-earth element (REE) abundance pattern (called group II) caused by condensation from a nebular reservoir that was depleted in the most refractory REE (e.g., Palme and Boynton, 1993). Most CAIs formed as a result of complex, multi-stage processes involving evaporation-condensation and melting-solidification; a few coarse-grained CAIs are simple evaporative refractory residues (Davis and MacPherson, this volume). Included with CAIs in Table 1 are amoeboid-olivine aggregates, which are fine-grained and contain refractory nodules but are closer in bulk composition to chondrules. There are also relatively rare objects with igneous textures that are compositionally intermediate between CAIs and ferromagnesian chondrules (Sheng *et al.*, 1991; Bischoff and Keil, 1984). Key isotopic, chemical and petrologic data for CAIs are reviewed by Clayton *et al.* (1988) and MacPherson *et al.* (1988).

Matrix material, which most workers consider to be the least

thermally processed component in chondrites, forms 20–200 µm thick rims on chondrules, CAIs and metal grains and also occurs in mm-sized lumps (Brearley, this volume). It is largely composed of FeO-rich, micron-sized silicates and contains fragments of chondrules and trace amounts of submicron, pre-solar dust (Huss and Lewis, 1995). Matrix material was probably heated but never molten.

The other major ingredient, metallic Fe, Ni, is commonly associated with troilite (FeS). These minerals are found inside, around and outside chondrules. Metallic nodules outside chondrules are typically smaller than silicate chondrules and may have been derived from them (Skinner and Leenhoutz, 1993).

RELATIONSHIPS AMONG CHONDRITE GROUPS

Twelve of the 13 chondrite groups constitute the carbonaceous (C), enstatite (E), and ordinary (O) classes of chondrites (Table 1). These terms now have only a minor genetic importance: CK carbonaceous chondrites, for example, are deficient in carbon. Chondrite groups are now defined largely in terms of refractory abundances and oxygen isotopic compositions (Clayton, 1993). Table 1 lists the mean abundances of the refractory elements, Ca, Al and Ti, relative to Mg after normalization to the CI chondrite ratios. C chondrites have refractory abundances of 1.0–1.4, whereas O and E chondrites have normalized refractory abundances of 0.8–1.0 (Larimer and Wasson, 1988). CI chondrites are used as a reference as they are closest in composition to that of the solar photosphere, presumably because they are largely composed of the most primitive chondritic component, matrix material. Ironically, CI chondrites are among the most heavily altered chondrites.

Each chondrite group (excluding CI) has a unique chondrule size distribution (Table 1; Grossman *et al.*, 1988a; Rubin, 1989). In CR chondrites, metal spherules and chondrules have size distributions consistent with aerodynamic sorting of initially unsorted objects (Skinner and Leenhouts, 1993). However, the sizes of metal particles may have been controlled also by the sizes of their parent chondrules. Sizes of chondrules and CAIs are not closely correlated. However, the two groups that have the largest and

Table 1. *Approximate abundances (vol%) of major components in chondrite groups and other key properties*

class	C (carbonaceous)							E (enstatite)		O(ordinary)			
	CK	CO	CV	CH#	CR	CM	CI	EH	EL	H	L	LL	R
CAI (+AOA)*	4	13	10	0.1	0.5	5	<0.0001	0.1–1?		0.1–1?			0.0
chondrules+	15	40	45	~70	50–60	20	0.001?	60–80		60–80			≥40
matrix material	75	30	40	5	30–50	70	99	<2–15?		10–15			35
Fe-Ni metal	<0.01	1–5	1–5	20	5–8	0.1	0.0	8	15	10	5	2	0.1
chondrule size (mm)^	0.7	0.15	1.0	<0.1	0.7	0.3	-	0.2	0.6	0.3	0.7	0.9	0.4
(refract. elem./Mg)CI	1.21	1.13	1.35	1.00	1.03	1.15	1.00	0.87	0.83	0.93	0.94	0.90	0.95

*CAI - Ca, Al-rich inclusions; AOA - amoeboid alivine aggregates; +including fragments; ^sizes from Grossman *et al.* (1988a,b), Rubin (1989), A. E. Rubin (private communication, 1994); other sources: see Scott and Newsom (1989); Kallemeyn *et al.* (1991, 1994); Huss and Lewis (1995); I. D. Hutcheon (private communication, 1994); Rubin and Kallemeyn (1994); Weisberg *et al.* (1991); Bischoff *et al.* (1993a,b); Wasson and Kallemeyn (1988, 1990). #Wasson and Kallemeyn (1990) suggest that CH chondrites were fractionated by asteroidal processes.

smallest mean chondrule sizes, CV (1.0 mm) and CH chondrites (<<0.1 mm), also have the largest and smallest CAIs: 1–10 mm and 0.01–0.1 mm, respectively (MacPherson *et al.*, 1988; Kimura *et al.*, 1993). We infer that all particles may have experienced aerodynamic size sorting after chondrule formation. Size sorting of chondrule precursors and CAIs before and during chondrule formation may also have occurred.

Isotopic and chemical data show that components in chondrite groups come from many different reservoirs. For example, chondrules in C, E and O chondrites have very different oxygen isotopic compositions (~2% overlap; Clayton, 1993). Since oxygen isotopic diffusion rates in silicate minerals are very slow, we can infer that the isotopic compositions of the chondrules have not changed since chondrule formation. Chondrules must therefore have been derived from at least three different nebular sources and not subsequently mixed.

Chemical differences among chondrules in carbonaceous chondrites suggest that each C chondrite group was probably derived from a separate nebular source (Grossman *et al.*, 1988a). However, chondrules in the three O chondrite groups, H, L and LL, may have been derived by size sorting from a single nebular reservoir (see Clayton, 1993). This sorting process may also have produced the relationship among H-L-LL groups between chondrule size and metal abundance (Haack and Scott, 1993). The EH and EL groups may be similarly derived from another nebular source. In E and O classes, the group with the highest metal abundance has the smallest chondrules.

Chondrules in C chondrites are commonly FeO-poor (Fe/(Fe+Mg) < 0.1) and metal-rich, whereas chondrules in O chondrites are mostly FeO-rich (Fe/(Fe+Mg) > 0.1), coarser grained, and metal-poor. These two chondrule varieties, which are also called types I and II, respectively, probably formed from rather different precursor materials and experienced different thermal histories (e.g., Jones, this volume). Although chondrites in a group typically have similar chemical compositions, mineralogy and chondrule sizes, there are notable exceptions, e.g., mean chondrule sizes in CO chondrites are correlated with the abundance of amoeboid-olivine aggregates (Rubin, 1989).

Many chondrites contain clasts of other chondrite types: e.g., dark inclusions of C-like material in CV (e.g., Kracher *et al.*, 1985) and CR chondrites (Endress *et al.*, 1994) and igneous clasts in H-L-LL3 chondrites (e.g., Hutchison, 1992; this volume). Some rare clasts, e.g., the heavily shocked LL clast in the H chondrite Dimmitt (Rubin *et al.*, 1983) may be projectile fragments that struck the asteroidal regolith at high speed after accretion ended. But others such as the dark chondritic clasts in CV3 chondrites (Fig. 1) that are more abundant (up to several vol. %), apparently not shocked and more closely related to the host chondrite, may have accreted at low speed together with the host chondrules.

To correctly infer nebular conditions from chondrites, we need to distinguish between nebular and asteroidal alteration effects which may be very complex (e.g., Krot *et al.*, 1995). All chondrites contain components that have been modified by heating or aqueous alteration in asteroids. Although there is now much agreement on the identification of secondary features, at least for CAIs and chondrules, it is unlikely that all such features have been recognized.

Fig. 1. A 9 cm wide slab of the Leoville CV chondrite showing a large dark clast that experienced aqueous alteration on a parent body prior to the acquisition of its dark rim (Kracher *et al.*, 1985). The similarity of the clast rim to the matrix rims around Leoville chondrules suggests that this clast was derived from a body that formed and was aqueously altered before the Leoville chondrules accreted. (From Kracher *et al.*, *J. Geophys. Res.* (1985) **90 Suppl.**, D123–D135.)

TIMESCALES FOR CAI AND CHONDRULE FORMATION

CAIs with canonical inferred initial $^{26}Al/^{27}Al$ ratios of 4.5×10^{-5} are found in CV and CM groups: they formed during a period of $<10^5$ yr (Hutcheon, 1982; Swindle *et al.*, this volume). Some CAIs were formed at this time but were partially remelted to form minerals with lower inferred initial $^{26}Al/^{27}Al$ ratios. Studies of these objects by Hutcheon (1982) and MacPherson and Davis (1993) suggest that CAIs had nebular lifetimes of 2–4 Myr. If ^{26}Al was uniformly distributed, some plagioclase-rich chondrules in O and C chondrites formed >2 Myr after CAIs formed (Sheng *et al*, 1991; Hutcheon *et al.*, 1994). (Wood (this volume) disputes these timescales.) The FUN and hibonite inclusions that formed without ^{26}Al but with many nuclear anomalies presumably formed in the solar nebula before ^{26}Al was present. From this chronology and the isotopic evidence that CAIs formed in the solar nebula, we infer that chondrules are unlikely to have formed during infall into the nebula as the timescale for gravitational collapse of the pre-solar molecular cloud to a protostar and disk is probably only 10^{5-6} yr.

The nebula lifetime of several Myr inferred from isotopic data is consistent with the duration of circumstellar disks of low-mass stars derived from astronomical observations (Podosek and Cassen, 1994). However, theoretical lifetimes for protosolar disks calculated on the basis of turbulent dissipation or drag on solids with Keplerian orbits moving through pressure-supported gas are at least 10 times shorter (Weidenschilling, 1988). The formidable complexity of the full disk evolution problem, including radial and convection-driven turbulence, ongoing grain and planetesimal growth, transport, and destruction, and other significant processes (e.g., Cuzzi *et al.*, 1993) suggests that meteoritic and astronomical evidence should prevail.

^{26}Al -based chronometry and textural evidence (CAIs inside chondrules, but not vice versa) (McSween, 1977; Misawa and Nakamura, this volume) suggest that CAIs formed before chondrules and that there was virtually no overlap in their formation periods. Although chondrules and CAIs both formed during brief high-temperature events and have some common features (e.g., presence of relict grains, evidence for recycling) we infer from the chronologic constraints that two different energy sources were responsible. Nevertheless, it seems unwise to investigate the origin of one chondritic component without regard for the nebular history of other components.

PETROLOGIC, CHEMICAL AND EXPERIMENTAL CONSTRAINTS ON CHONDRULE FORMATION

Chondrule precursor materials and chondrule recycling

Petrologic evidence indicates that chondrules formed by melting of coarse- and fine-grained precursors (Grossman *et al.*, 1988a; Wasson, 1993). The existence of fine-grained chondrule precursor material is indicated by the presence of irregular, fine-grained agglomerates that appear to represent lower degrees of melting than chondrules. In ordinary chondrites they are largely composed of FeO-rich olivine and contain abundant chondrule fragments (Weisberg and Prinz, this volume). In carbonaceous chondrites these agglomerates are typically FeO-poor (e.g., Nagahara and Kushiro, 1982). Direct evidence for formation of chondrules by melting of matrix material is rare. Krot and Rubin (this volume) found FeO-rich microchondrules in matrix rims of a few FeO-poor chondrules and inferred that these microchondrules had formed by melting of matrix-like material.

Coarse-grained chondrule precursors are observed in the form of relict grains in chondrules; these are grains that did not crystallize from host chondrule melts. Relict grains are largely derived from fragments of both FeO-poor and FeO-rich chondrules, indicating that chondrule formation was a multiple process and that chondrules of early generations were recycled (Jones, this volume). Additional evidence for chondrule recycling and for a dusty environment during chondrule formation is provided by igneous rims on chondrules. These rims formed by the accretion of fine-grained material and chondrule fragments to host chondrules that were subsequently partly or completely melted (Rubin and Krot, this volume). FeO concentrations in an igneous rim are similar to those of the enclosed chondrule (Rubin and Krot, this volume). This suggests that these two types of chondrules were made predominantly from separate types of material in separate environments. Relict grains indicate that a fraction were mixed and remelted.

CAI-like material was a common component in the chondrule precursor material, although direct evidence is rare. Rare relict CAIs are found inside chondrules; a few chondrules probably inherited type II REE patterns in this way (Misawa and Nakamura, this volume). In addition, bulk refractory abundances in chondrites, which are highly dependent on chondrule compositions, are correlated with the abundance of CAIs (Table 1).

Thermal history of chondrules

The thermal history of chondrules has been experimentally studied and reviewed in several papers (see Connolly and Hewins, this volume; Hewins and Connolly, this volume). Textures and mineral compositions of chondrules depend on the bulk composition, degree of melting, precursor grain size, and interactions with other particles as well as the thermal history. Early experiments to derive cooling rates from textures assumed linear cooling after some arbitrary period at a fixed temperature and did not include effects of precursor grain size and interactions with dust particles. These studies inferred cooling rates of 5–3000 K/hr during crystallization and peak temperatures of ~ 1700–2100 K. Later experiments that studied all of the variables suggest that we cannot confidently infer thermal histories for chondrules from textures and mineral chemistry. Moreover, if chondrules behaved as open systems during chondrule formation and lost significant amounts of Si, Fe, etc., any estimates of peak temperatures and cooling rates based on texture and mineral chemistry may be erroneous (Nagahara *et al.*, 1994). Realistic non-linear cooling paths which are consistent with plausible chondrule melting mechanisms must be studied.

The extent of loss of volatile elements like S and Na in chondrules is controlled by fewer variables and gives some measure of thermal history at peak temperatures. However, Na loss is also controlled by redox conditions and the extent of S loss or recondensation is controversial (e.g., Yu *et al.*, this volume; Grossman, this volume).

Mineral microstructures may also be used to constrain subsolidus cooling rates. Weinbruch and Müller (1994) inferred cooling rates for a few CV3 chondrules based solely on the microstructures of clinopyroxene and plagioclase. For the temperature range 1500–1650 K, they inferred rates of 1–10 K/hr and ~ 10,000 K/hr at ~1300 K. Further studies are needed to confirm this work.

FACTORS CONTROLLING CHONDRULE PEAK TEMPERATURES

Although some chondrule-forming events were mild enough to form agglomerates or intense enough to vaporize material, most chondrules appear to have been heated within a relatively narrow temperature range of 1750–2150 K without significant vaporization. Given the evidence for multiple chondrule-forming events in different chondrite groups, it is reasonable to expect a wider spread of heating intensities. In this section, we explore two mechanisms which might have restricted the range of peak chondrule temperatures: the chondrules' own heat of fusion, and dissociation of nebular hydrogen gas.

Both of these mechanisms are related to the degree of thermal contact between the chondrules and the surrounding gas, a parameter which is not well known. In a transparent environment there will be poor thermal contact between the nebular gas and solids. In an opaque environment, favored in several proposed chondrule formation mechanisms, gas and solids can share heat via conduction.

Heat of chondrules' fusion

The latent heat of chondrule melting (about 400 J/g) may have regulated chondrule peak temperatures. Assuming an initial nebular temperature of less than 700 K (e.g., Wasson, 1993), a peak chondrule temperature of 1800 K, and a heat capacity of ~0.8 J/g K (Wasson, 1974), energy inputs of 900 to 1300 J/g – a range of reasonable width, which widens further if the heat capacity between the solidus and liquidus is taken into account – produces the observed majority of incompletely melted chondrules.

The energy necessary to heat hydrogen gas from 700 to 1800 K is ~15,000 J/g, overwhelming compared to the 400 J/g wide "window" which allows an equal mass of chondrules to be partly melted. For chondrule melting to consume more energy than heating the gas, the solid:gas ratio must have been ~10^4 higher than the solar value of 0.01.

Dissociation of nebular hydrogen gas

If there was good thermal contact between the chondrules and the gas, dissociation of molecular hydrogen (H_2) into atomic hydrogen (H) could have limited chondrule peak temperatures. At low pressures, hydrogen dissociation is important at relatively modest temperatures: 1900–2400 K at 10^{-4} atm, 1700–2000 K at 10^{-6} atm, and 1500–1700 K at 10^{-8} atm (Liepmann and Roshko, 1957). The gas pressure during chondrule formation is not well known, and certainly depended on radial and vertical position in the solar nebula, but 10^{-6} atm is a reasonable estimate. Thus, the range of chondrule (and CAI) peak temperatures corresponds to the regime where H_2 dissociation may have been important. The dissociation energy of hydrogen is 214,000 J/g. It would have dominated the energetics of both phases unless the mass density ratio of solids to gas was 10^3 or higher. For a 300 K, 10^{-6} atm nebula, this limit corresponds to a chondrule number density of about 50 m^{-3}.

Summarizing, it appears that both the latent heat of fusion of chondrules and the dissociation of hydrogen may have restricted peak chondrule temperatures to a narrow range. At solid/gas mass density ratios above 10^3, or in heating events affecting primarily the solids while leaving the gas cool, the latent heat of fusion of chondrule material could have played a role in temperature regulation. At lower solid concentrations and/or in heating events controlled by non-radiative processes, dissociation of H_2 could have limited chondrule temperatures.

TWO MECHANISMS FOR CHONDRULE FORMATION

Having examined thermostats for chondrule heating mechanisms, we now briefly treat two such mechanisms themselves. Both broadly satisfy the constraints on chondrule heating discussed above and are currently the subjects of active study: electrical discharge ("lightning") heating (e.g., Horanyi and Robertson, this volume; Wasson 1993; Morfill *et al.*, 1993) and gas drag (e.g., Hood and Kring, this volume; Ruzmaikina and Ip, this volume). Both can create the multiple, high-temperature (> ~ 1900 K), short-duration (< ~ 10 s) thermal events indicated by the chondritic record.

Nebular lightning

Current theoretical and experimental work (e.g. Horanyi and Robertson, this volume) investigates whether chondrules could have been formed by nebular lightning bolts: giant (~0.01 AU long) columns of fully ionized gas with diameters of several thousand electron-molecule collision mean free paths. Gibbard and Levy (1994) highlight the difficulties of creating the large potential differences required for such bolts in the presence of the free charge abundance (mostly produced by the decay of natural radionuclides and by cosmic rays) thought to have existed in the solar nebula. Here, we address the intermediate issue of whether processes analogous to those operating in terrestrial thunderstorms could produce nebular lightning under ideal conditions. We try to predict the degree to which those discharges might resemble terrestrial lightning, and to gauge their efficacy at melting mm-sized stony chondrule precursors. A full discussion is presented elsewhere (Love *et al.*, 1995).

Lightning is sensitive to its environment. We treat nebular gas pressures of 10^{-7} to 10^{-3} bar, with 10^{-5} bar as a reference value (Cuzzi *et al.*, 1993). The ambient temperature of the nebula – of only minor importance – was probably less than 700 K (Wasson

1993); we choose 300 K. We treat dust:gas mass ratios of 0.01 (the "solar" value) to 10, with 1.0 as a reference value. We presume 10 m/s turbulent velocities, and two phases of 2 g/cm³ solid dust, present in equal masses: 1-mm chondrule precursor particles, and 1-μm pre-matrix grains.

This lightning model incorporates one of the many competing published mechanisms for accumulating opposite charges on particles of different sizes in terrestrial thunderstorms (Keith and Saunders, 1989), which has the advantage of having been studied well enough to be extended to the particle sizes and collision speeds believed to have existed in the nebula. Neglecting the effect of different materials (the Keith and Saunders model was developed for ice), we find a highly uncertain preferential transfer of ~ 1 electron per chondrule/matrix grain collision. Large-scale separation of the two particle types is assumed to occur via their different responses to turbulent motions in the gas: the 1-μm particles (with stopping distances of ~1 km under realistic conditions) are rapidly entrained in a "gust front" of turbulent moving gas, while the 1-mm chondrule precursors (with ~1000 km stopping distances) lag behind. Grain-grain collisions result in an accumulation of small grains with one charge polarity which are carried with the gust, leaving behind a volume filled with oppositely charged chondrule precursors. In a quiescent nebula, different gravitational settling speeds can also serve for large-scale charge separation, as in terrestrial thunderstorms.

As the charges in the gust front and in the swept volume behind it grow, so does the electrical field between them. When the field is strong enough that a free electron drifting with the field accumulates a significant fraction of a gas molecule's ionization energy between successive collisions, the resistance of the gas breaks down and a current flows. For a given gap between the charge centers, this threshold is called the breakdown voltage. It is customarily expressed in terms of the charge separation (d, in cm) and the gas pressure (p, in torr): V (volts) $\approx 30pd$ (Paschen's Law), an expression accurate within the uncertainties in the assumed nebular environment. Although Paschen's Law is well determined for many gases in clean, controlled laboratory experiments, it predicts breakdown voltages ~ 20 times larger than those observed in thunderstorms (Uman, 1987). This discrepancy is thought to arise because of field inhomogeneities in natural systems; it is the local maxima that trigger the discharge. Although Paschen's Law is probably no more accurate in the solar nebula, we retain it in order to obtain generous maximum values for the voltage – and hence the strength – of nebular lightning.

Using the geometry, assumptions, and reference values discussed above, it is possible to solve for the length and voltage of a discharge in the solar nebula: respectively, ~ 1500 m (comparable to the lengths of terrestrial lightning bolts) and 30 kV (much less than the ~ 100 MV typical for terrestrial lightning). The low voltage of nebular discharges is dictated primarily by the low pressure of the solar nebula.

We assume that the breakdown funnels charge from a region of the gust front with a radius comparable to the discharge gap. The energy thus released is ~ 40 J in the reference case mentioned

above. We assume a discharge channel radius of one molecular collision mean free path (as do Wasson and Rasmussen, 1994). It cannot be less than this distance, and, by analogy with terrestrial lightning, could be several thousand times greater. Again, we have chosen values providing an upper limit to the discharge strength. Under the nebular conditions discussed above, a 1-mm diameter particle with its ion collision cross-section increased by charging (e.g. Morfill *et al.*, 1993) in the channel intercepts ~ 0.02 J. This is insufficient for melting, which requires ~ 1 J if the particle is not allowed to reject any heat. Results for different conditions are illustrated in Fig. 2. At high pressures and low dust/gas ratios it is ideally possible to melt chondrules with lightning. Note that porous or fluffy particles with low mass:area ratios may melt more easily than the solids assumed here.

In spite of this possibly encouraging result, there are difficulties with lightning as a chondrule formation mechanism. First, if chondrule precursors were heated by ion collisions or radiation from lightning (or any other mechanism that delivers energy according to the chondrule's cross-sectional area), the energy they absorbed should have scaled as r^2, while the energy required to melt them was proportional to mass (scaling as r^3). Thus, there should be a size dependence of heating, with smaller bodies more intensely heated than larger ones. Such a dependence has not been found, but the issue should be reinvestigated. Micrometer-sized matrix grains subjected to an energy flux strong enough to melt mm-sized chondrule precursors should have experienced very intense heating, complete melting, and probably significant evaporation.

Second, we find under most likely conditions, a nebular discharge does not release enough energy to ionize the gas included in its channel. This is in strong contrast to terrestrial lightning, which

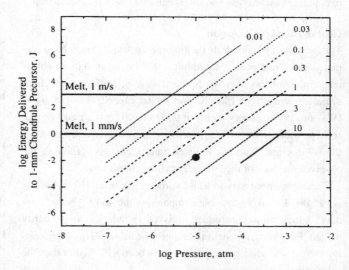

Fig. 2. Energy intercepted by a 1 mm diameter chondrule precursor in the discharge channel of an ideal nebular lightning bolt as a function of pressure (x axis) for various dust-gas mass density ratios (labeled curves), given the charge separation mechanism discussed in the text. The horizontal lines indicate limits on the drift speeds of chondrules if they experienced melting in a discharge lasting tens of seconds as predicted by Morfill *et al.* (1993). For the reference values of pressure and dust-gas ratio, indicated by the large spot, we infer that electrical discharge cannot form chondrules. (From Love *et al.*, *Icarus* (1995), **115**, 97–108.)

completely ionizes its path, and suggests that most nebular discharges were probably not analogous to terrestrial lightning.

Finally, additional difficulties arise if nebular discharge durations are 10–100 s, as predicted by Morfill *et al.* (1993). (This claim, however, remains in dispute: Wasson and Rasmussen, 1994, find millisecond durations.) If a discharge lasts many seconds, a particle in the discharge must not reject the energy incident upon it, or the losses (especially radiative) will hold its temperature below melting. Second, a particle must not move faster than ~ 1 mm/s, or it will escape from the ionized channel before melting. Random drift speeds in the nebula were ~ 10^3 times higher (e.g. Cuzzi *et al.*, 1993). Relaxing the lower-limit estimate of the bolt diameter worsens the problem. Only in the region indicated in the upper portion of Fig. 2 can a chondrule precursor moving at 1 m/s be melted.

In summary, we can construct a primitive model suggesting that electrical discharges could ideally have occurred in the solar nebula. Despite the model's uncertainty, it is possible to estimate an upper limit to the efficiency of nebular lightning at melting mm-sized stony objects. Even with the most advantageous assumptions, we find that these discharges were only marginally effective at melting chondrules. More realistic assumptions yield vastly poorer performance. Moreover, Gibbard and Levy (1994) suggest a concentration of free charges in the solar nebula high enough to "short-circuit" the charge separation required for nebular lightning, preventing it from occurring at all.

Nonetheless, we believe that nebular lightning is worthy of further investigation. A high concentration of fine-grained dust might facilitate charge separation (Wasson and Rasmussen, 1994). Study of the recently recognized phenomenon of energetic "upward lightning" in the low-pressure air above terrestrial thunderstorms (e.g. Vaughan and Vonnegut, 1989) may provide insight into the nature of electrical discharges in the solar nebula. Finally, laboratory investigations of charge separation mechanisms (analogous to those currently underway to investigate terrestrial thunderstorm lightning) could be fruitful in refining nebular lightning models, and would be doubly interesting in application to non-thunderstorm and planetary lightnings.

Gas drag heating

Another plausible method for chondrule formation is heating in shock waves or high-speed gas motions (e.g. Hood and Horanyi, 1993) of diverse nature: disk shocks triggered by outbursts from the young sun (Huss, 1994), packets of interstellar gas striking the solar nebula (e.g., Hood and Kring, this volume), a steady nebular accretion shock (e.g., Ruzmaikina and Ip, this volume), and stellar outflow jets (Liffman and Brown, this volume). The last two models may be challenged by the evidence for multiple episodes of heating and reprocessing, as discussed above.

Shocks can heat and alter chondrules in many different ways. A strong, hot shock can radiatively pre-heat particles as it approaches them. Particles overrun by the shock can be heated by radiation as before, and also by conduction with the shock-heated gas.

After the shock passes, an overtaken particle suddenly finds itself moving with significant speed relative to the gas. Here, μm-sized dust grains, with their short stopping distances, will match velocities with the moving gas and accumulate only a few km behind the shock front. The mm-size chondrule precursors, on the other hand, require thousands of km to come up to speed. (Note that this effect automatically sorts an initially heterogeneous assemblage of particles by size-density product. This sorting occurs on a distance scale of a few thousand km, much smaller than an AU, and may be blurred or erased by postshock turbulence. A chondrule precursor overtaken by a shock would immediately be pelted with tiny grains (themselves relatively undamaged by drag heating; see below) moving at the post-shock gas speed. This could impart significant collisional heating to the chondrule, and could explain the accretion of fine-grained chondrule rims. This model predicts that chondrule rims should have accreted smaller grains first, then increasingly larger grains whose larger stopping distances would have led to entrainment further behind the shock. The possibility of erosion rather than accretion (expected for faster shocks or weaker particles) complicates the picture somewhat.

Another post-shock effect is gas drag heating, upon which the following discussion will concentrate for several reasons. Unlike radiation and conduction, gas drag is relatively insensitive to the details of the environment, and is thus easier to model. Gas drag is also expected to dominate at low (< 10 km/s) speeds where radiation is less important. Furthermore, a brief treatment of gas drag illustrates predictions for the size dependence of chondrule heating and exemplifies a physical heating mechanism with a variable cooling rate. Gas drag heating theory has been successfully tested on micrometeorites. Note, however, that chondrules do not resemble mm-sized melted micrometeorites, which often show significant vesicularization and which show a much greater range of heating than do chondrules. It is not known whether these discrepancies can be explained by the differences in the composition and density structure between the solar nebula and the Earth's atmosphere, or by differences in the precursor particles.

During high-speed flight, a small particle's temperature is largely determined by the balance between the kinetic energy flux incident on the particle and thermal energy it radiates. For a spherical object compact enough that internal temperature gradients are unimportant on the time scale of the heating event, as is probably the case with drag heating of mm-sized stones, and neglecting thermal inertia and phase transitions, the energy balance equation is:

$$\varepsilon 4\pi r^2 \sigma T^4 = \pi r^2 \frac{1}{2}\rho_{gas}v^3 \qquad (1)$$

where ε is the object's emissivity, r its radius, T its temperature, and v its velocity; σ is the Stefan-Boltzmann constant, and ρ_{gas} is the density of the gas. If the particle is suddenly injected into a region of constant density gas (or overtaken by a shock) at an initial speed v_0 its velocity as a function of time (t) becomes:

$$v = \frac{1}{\dfrac{3\rho_{gas}}{4\rho r}t + \dfrac{1}{v_0}} \qquad (2)$$

where ρ is the particle's density, assuming molecular flow conditions and a drag coefficient of unity. We also assume that molecules striking the particle stop on impact, but do not stick. Combining Eqs. (1) and (2), neglecting background radiation and conduction heating, yields the temperature-time profile:

$$T = \left(\frac{\rho_{gas}}{8\varepsilon\rho} \left(\frac{3\rho_{gas}}{4\rho r} t + \frac{1}{v_0} \right)^{-3} \right)^{0.25} \qquad (3)$$

All else being equal, drag heating implies that larger, denser particles should suffer longer heating pulses – and thus more intense heating – than their smaller siblings. The opposite is predicted for heating by lightning, as pointed out above. We stress that evaluating the size dependence of chondrule heating may permit significant narrowing of the field of possible chondrule heating mechanisms.

Another feature of drag heating is the nonlinear nature of the temperature curve, which demonstrates that reported chondrule cooling rates can constrain the heating mechanism only if the temperatures at which those rates apply are also specified. Different physical heating and cooling mechanisms imply different temperature profiles. Current estimates of cooling rates, however, suffer from ambiguities in the effects of peak temperature, heat pulse duration, chondrule composition, and chondrule precursor grain size, and are thus probably too uncertain to rule out or verify any specific mechanism.

ACCRETION MECHANISMS AND RELATIONSHIP TO CHONDRULE FORMATION

To account for the extraordinary variations in the compositions of chondritic ingredients in the 13 groups (Table 1), chondrules and other ingredients must have accreted before they were mixed with material destined for other groups. Wood (1985) argues that to maintain the near-solar chondritic compositions, matrix, chondrules and CAIs must have accreted within a year of chondrule formation. We infer from Table 1 and other data, however, that the proportions of the various ingredients were only loosely constrained so that asteroids composed, for example, largely of CAIs, metal grains, or FeO-poor chondrules did not form. If any of these particles were ever preferentially aggregated in the nebula, the aggregations were not made into meteorites.

In the conventional model of accretion, particles gradually settled to the midplane of the solar nebula where they accreted into kilometer-sized planetesimals as a result of gravitational instabilities (Goldreich and Ward, 1973). Kerridge (1993) concludes that the meteorite record appears to favor this model because chondritic particles seem to have accreted without much regard for the characteristics of particle types. Metal particles, for example did not accrete into cm- or m-sized lumps as a result of magnetic forces. But Weidenschilling (1988) suggests that if such a dusty midplane had formed, differential shear at its edge caused by slower rotation of the gas would loft particles and prevent gravitational collapse. Cuzzi *et al.* (1993) further conclude that turbulent motions in the nebula would have enhanced random velocities of particles, pre-

venting the formation of a dusty midplane that was dense enough to collapse gravitationally into planetesimals.

Cuzzi *et al.* (this volume) propose instead that accretion began in a turbulent nebula when particles were concentrated by centrifugation between the smallest eddies. They find that mm-sized, compact particles would have been preferentially concentrated by this mechanism. According to this model, chondrule formation was an essential first step in converting porous aggregates into denser objects that were better able to decouple from the turbulent gas. Chondrules could have acquired their rims of matrix material during turbulent motion.

Several properties of chondrites appear to be much more consistent with primary accretion via turbulent concentration than via gravitational instabilities. The dependence of the thickness of matrix rims on chondrules (and CAIs) on chondrule size in CM2 chondrites (Metzler *et al.*, 1992) may have arisen during turbulent motions. Matrix grains that were not part of rims or compact lumps would have been largely excluded from chondrites, consistent with most studies (e.g., Metzler *et al.*, 1992; Scott *et al.*, 1984). The characteristic sizes of chondrules in the chondrite groups (Table 1) could have reflected variations in the gas pressure or turbulent speeds, for example.

Turbulent accretion may have allowed nebular solids at a given location to have accreted over an extended period to form closely related kinds of chondritic planetesimals. A necessary condition is that inter-eddy dust concentrations which were large and dense enough to partially decouple from the turbulence settled to the midplane and were incorporated into planetesimals while turbulent accretion of chondrules continued above. In addition the timescale for removal of the inter-eddy dust concentrations must have been shorter than the timescale for temporal changes in transport properties in the nebula to ensure that other chondrules having sizes outside the observed distributions were not remixed. Then, the parent bodies of the LL, L and H chondrites, for example, may have formed sequentially from a single reservoir of chondrules and metal grains in which turbulent velocities and the mean size of accreting chondrules gradually decreased. The correlation among CO chondrites between chondrule diameter and degree of metamorphism (Rubin, 1989) may have resulted from a decrease in turbulence speeds during the accretion of inter-eddy dust concentrations to a single body that was subsequently heated internally.

Finally, we discuss whether planetesimals accreted during the period of several Myr that appears to have elapsed between the times of CAI and chondrule formation. Hutchison (1992; this volume) argues that some planetesimals formed and were melted by ^{26}Al before chondrules formed. He cites rare inclusions in ordinary chondrites with fractionated REE patterns indicative of a planetary, igneous origin. The best candidate is an olivine-plagioclase clast in Semarkona with an initial inferred $^{26}Al/^{27}Al$ ratio of 8×10^{-6}. This clast formed 2 Myr after CAI formation when ^{26}Al would have been a viable heat source, assuming it was uniformly distributed (Hutcheon and Hutchison, 1989).

Many authors who have studied the abundant chondritic clasts in CV and CM chondrites infer that these clasts were derived from earlier generations of planetesimals, some of which suffered aqueous alteration (e.g., Endress *et al.*, 1994). We lack formation ages for these clasts, but the presence of a matrix-rimmed, chondritic fragment in Leoville, which was altered before incorporation (Fig. 1; Kracher *et al.*, 1985) suggests that fragments of early formed planetesimals were recycled back into the nebula to reaccrete with chondrules. To account for secondary hydrous minerals with a supposedly planetary origin inside fine-grained chondrule rims of nebular origin, Metzler *et al.* (1992) suggest that alteration may have occurred in small uncompacted precursor planetesimals that were dispersed prior to the formation of the CM body.

Skinner (1994) suggests that CAIs were preserved in the nebula for several Myr because they were temporarily stored in objects that were large enough to survive gas-drag. It may not have been necessary to form km-sized planetesimals, as Skinner inferred, as more numerous smaller aggregations may have caused mid-plane gas to rotate at Keplerian speeds (e.g., Cuzzi *et al.*, 1993).

Final comment

We cannot assume that nebular processes proceeded sequentially from CAI and chondrule formation to planetesimal accretion and must avoid prejudice against models invoking planetesimals to form chondrules or prevent chondrules and CAIs from spiraling into the Sun. We no longer believe in a slowly cooling, quiescent solar nebula and argue instead for a maelstrom where temperatures fluctuated through ~ 1000 K and solids experienced "multiple cycles of melting, evaporation, recondensation, crystallization, and aggregation" (Wood, 1988). Studies of the mineralogical, chemical and isotopic diversity of CAIs and chondrules have required us to revise our ideas about the thermal history of the nebula. The diversity of chondrites themselves may require comparable changes in our understanding of how chondritic ingredients accreted into planetesimals.

ACKNOWLEDGEMENTS

We thank many colleagues for generously discussing ideas and data, especially I. Hutcheon, A. Rubin, J. Cuzzi, W. Skinner, H. Haack, I. Sanders, A. Davis, G. MacPherson, J. Wasson, H. Nagahara, D. Kring, K. Liffman, M. Horanyi, P. Cassen. We also thank A. Kracher for kindly providing a photograph and A. Boss, J. Gooding, J. Wasson and A. Rubin for helpful reviews. This work was partially funded by NASA grant NAGW 3281 to K. Keil. This is Hawai'i Institute of Geophysics and Planetology publication no. 862 and School of Ocean and Earth Science and Technology no. 4017.

REFERENCES

Bischoff A. and Keil K. (1984) Al-rich objects in ordinary chondrites: Related origin of carbonaceous and ordinary chondrites and their constituents. *Geochim. Cosmochim. Acta* **48**, 693–709.

Bischoff A., Palme H., Ash R. D., Clayton R. N., Schultz L., Herpers U., Stöffler D., Grady M. M., Pillinger C. T., Spettel B., Weber H., Grund T., Endress M. and Weber D. (1993a) Paired Renazzo-type (CR) carbonaceous chondrites from the Sahara. *Geochim. Cosmochim. Acta* **57**, 1587–1603.

Bischoff A., Palme H., Schultz L., Weber D., Weber H. W. and Spettel B. (1993b) Acfer 182 and paired samples, an iron-rich carbonaceous chondrite: Similarities with ALH 85085 and relationship to CR chondrites. *Geochim. Cosmochim. Acta* **57**, 2631–2648.

Clayton R. N. (1993) Oxygen isotopes in meteorites. *Ann. Rev. Earth Planet. Sci.* **21**, 115–149.

Clayton R. N., Hinton R. W. and Davis A. M. (1988) Isotopic variations in the rock-forming elements in meteorites. *Phil. Trans. R. Soc. London* **A325**, 483–501.

Cuzzi J. N., Dobrolvoskis A. R. and Champney J. M. (1993) Particle-gas dynamics in the midplane of a protoplanetary nebula. *Icarus* **106**, 102–134.

Endress M., Keil K., Bischoff A., Spettel B., Clayton R.N. and Mayeda T. K. (1994) Origin of dark clasts in the Acfer 059/El Djouf 001 CR2 chondrite. *Meteoritics* **29**, 26–40.

Gibbard S. G. and Levy E. H. (1994) On the possibility of precipitation-induced vertical lightning in the protoplanetary nebula. In *Chondrules and the Protoplanetary Disk*. **LPI Contribution No. 844**, pp. 9–10. Lunar and Planetary Institute, Houston.

Goldreich P. and Ward W. R (1973) The formation of planetesimals. *Astrophys. J.* **183**, 1051–1061.

Grossman J. N., Rubin A. E., Nagahara H. and King E. A. (1988a) Properties of chondrules. In *Meteorites and the Early Solar System* (eds. J. F. Kerridge and M. S. Matthews), pp. 619–659. University of Arizona Press.

Grossman J. N., Rubin A. E. and MacPherson G. J. (1988b) ALH85085: A unique volatile-poor carbonaceous chondrite with possible implications for nebular fractionation processes. *Earth Planet. Sci. Lett.* **91**, 33–54.

Haack H. and Scott E. R. D. (1993) Nebula formation of the H, L, and LL parent bodies from a single batch of chondritic materials (abstract). *Meteoritics* **28**, 358–359.

Hood L. L. and Horanyi M. (1993). The nebular shock wave model for chondrule formation. *Icarus* **106**, 179–189.

Huss G. R. (1994) The early sun and the formation of chondrules. In *Chondrules and the Protoplanetary Disk*. **LPI Contribution No. 844**, pp. 14–15. Lunar and Planetary Institute, Houston.

Huss G. R. and Lewis R. S. (1995) Presolar diamond, SiC, and graphite in primitive chondrites: Abundances as a function of meteorite class and petrologic type. *Geochim. Cosmochim. Acta* **59**, 115–160.

Hutcheon I. D. (1982) Ion probe magnesium isotopic measurements of Allende inclusions. In *Nuclear and Chemical Dating Techniques: Interpreting the Environmental Record* (ed. L. A. Curie), pp. 95–128. Amer. Chem. Soc. Symp. Series 176.

Hutcheon I. D. and Hutchison R. (1989) Evidence from the Semarkona ordinary chondrite for [26]Al heating of small planets. *Nature* **337**, 238–241.

Hutcheon I. D., Huss G. R. and Wasserburg G. J. (1994) A search for [26]Al in chondrites: Chondrule formation time scale (abstract). *Lunar Planet. Sci.* **25**, 587–588.

Hutchison R. (1992) Earliest planetary melting – the view from meteorites. *Volcanol. Geotherm. Res.* **50**, 7–16.

Kallemeyn G. W., Rubin A. E. and Wasson J. T. (1991) The compositional classification of chondrites: V. The Karoonda (CK) group of carbonaceous chondrites. *Geochim. Cosmochim. Acta* **55**, 881–892.

Kallemeyn G. W., Rubin A. E. and Wasson J. T. (1994) The compositional classification of chondrites: VI. The CR carbonaceous chondrite group. *Geochim. Cosmochim. Acta* **58**, 2873–2888.

Keith W. D. and Saunders C. P. R. (1989) Charge transfer during multiple large ice crystal interactions with riming target. *J. Geophys. Res.* **94**, 13013–13106.

Kerridge J. F. (1993) What can meteorites tell us about nebular conditions and processes during planetesimal accretion? *Icarus* **106**, 135–150.

Kimura M., El Goresy A., Palme H. and Zinner E. (1993) Ca-, Al-rich inclusions in the unique chondrite ALH85085: Petrology, chemistry, and isotopic compositions. *Geochim. Cosmochim. Acta* **57**, 2329–2359.

Kracher A., Keil K., Kallemeyn G. W., Wasson J. T. and Clayton R. N.

(1985) The Leoville (CV3) accretionary breccia. *Proc. Lunar Planet. Sci. Conf.* **16**, D123–D135.

Krot A. N., Scott E. R. D. and Zolensky M. E. (1995) Mineralogical and chemical modification of components in CV3 chondrites: Nebular or asteroidal processing? *Meteoritics* **30**, 748–775 .

Larimer J. W. and Wasson J. T. (1988) Refractory lithophile elements. In *Meteorites and the Early Solar System* (eds. J. F. Kerridge and M. S. Matthews), pp. 416–435. University of Arizona Press.

Liepmann H. W. and Roshko A. (1957) *Elements of Gasdynamics*, John Wiley and Sons, 439pp.

Love S. G., Keil K., and Scott E. R. D. (1995) Electrical discharge heating of chondrules in the solar nebula. *Icarus* **115**, 97–108.

MacPherson G. J. and Davis A. M. (1993) A petrologic and ion microprobe study of a Vigarano Type B refractory inclusion: Evolution by multiple stages of alteration and melting. *Geochim. Cosmochim. Acta* **57**, 231–243.

MacPherson G. J., Wark D. A. and Armstrong J. T. (1988) Primitive material surviving in chondrites: Refractory inclusions. In *Meteorites and the Early Solar System* (eds. J. F. Kerridge and M. S. Matthews), pp. 746–807. University of Arizona Press.

McSween H. Y. (1977) Chemical and petrographic constraints on the origin of chondrules and inclusions in carbonaceous chondrites. *Geochim. Cosmochim. Acta* **41**, 1843–1860.

Metzler K., Bischoff A. and Stöffler D. (1992) Accretionary dust mantles in CM chondrites: Evidence for solar nebula processes. *Geochim. Cosmochim. Acta* **56**, 2873–2897.

Morfill G., Spruit H., and Levy E. H. (1993) Physical processes and conditions associated with the formation of protoplanetary disks. In *Protostars and Planets III* (eds. E. H. Levy and J. I. Lunine), pp. 939–978. University of Arizona Press.

Nagahara H. and Kushiro I. (1982) Petrology of chondrules, inclusions and isolated olivine grains in ALH-77307 (CO3) chondrite. *Proc. 7th Symp. Antarctic Meteorites*, 66–77

Nagahara H., Kushiro I., and Mysen B. O. (1994) Evaporation of olivine: Low pressure phase relations of the olivine system and its implication for the origin of chondritic components in the solar nebula. *Geochim. Cosmochim. Acta* **58**, 1951–1963.

Palme H. and Boynton W. V. (1993) Meteoritic constraints on conditions in the solar nebula. In *Protostars and Planets III* (eds. E. Levy and J. I. Lunine), pp. 979–1004. University of Arizona Press.

Podosek F. A. and Cassen P. (1994) Theoretical, observational, and isotopic estimates of the lifetime of the solar nebula. *Meteoritics* **29**, 6–25.

Rubin A. E. (1989) Size-frequency distributions of chondrules in CO3 chondrites. *Meteoritics* **24**, 179–189.

Rubin A. E. and Kallemeyn G. W. (1994) Pecora Escarpment 91002: A new Carlisle Lakes chondrite. *Meteoritics* **29**, 255–264.

Rubin A. E., Scott E. R. D., Taylor G. J., Keil K., Allen J. S. B., Mayeda T. K., Clayton R. N. and Bogard D. D. (1983) Nature of the H chondrite parent body regolith: Evidence from the Dimmitt breccia. *Proc. Lunar Sci. Conf. J. Geophys. Res.* **88**, A741–A754.

Scott E. R. D. and Newsom H. E. (1989) Planetary Compositions - Clues from Meteorites and Asteroids. *Zeitschr. Naturforsch.* **44a**, 924–934.

Scott E. R. D., Rubin A. E., Taylor G. J. and Keil K. (1984) Matrix material in type 3 chondrites - Occurrence, heterogeneity and relationship with chondrules. *Geochim. Cosmochim. Acta* **48**, 1741–1757.

Sheng Y. J., Hutcheon I. D. and Wasserburg G. J. (1991) Origin of plagioclase-olivine inclusions in carbonaceous chondrites. *Geochim. Cosmochim. Acta* **55**, 581–599.

Skinner W. R. (1994) Pre-Allende planetesimals with refractory compositions: The CAI connection (abstract). *Lunar Planet. Sci.* **25**, 1283–1284.

Skinner W. R. and Leenhoutz J. M. (1993) Size distributions and aerodynamic equivalence of metal chondrules and silicate chondrules in Acfer 059 (abstract). *Lunar Planet. Sci.* **24**, 1315–1316.

Uman M. A. (1987) *The Lightning Discharge*. Academic Press, Orlando. 377 pp.

Vaughan O. H. Jr. and Vonnegut B. (1989) Recent observations of lightning discharges from the top of a thundercloud into the clear air above. *J. Geophys. Res.* **94**, 13179–13182.

Wasson J. T. (1974) *Meteorites—Classification and Properties*, Springer, 316 pp.

Wasson J. T. (1993) Constraints on chondrule origins. *Meteoritics* **28**, 13–28.

Wasson J. T. and Kallemeyn G. W. (1988) Compositions of chondrites. *Phil. Trans. R. Soc. London* **A325**, 535–544.

Wasson J. T. and Kallemeyn G. W. (1990) Allan Hills 85085: a subchondritic meteorite of mixed nebular and regolithic heritage. *Earth Planet. Sci. Lett.* **101**, 148–161.

Wasson J. T. and Rasmussen K. L. (1994) The fine nebula dust component: A key to chondrule formation by lightning. In *Chondrules and the Protoplanetary Disk.* **LPI Contribution No. 844**, p. 43. Lunar and Planetary Institute, Houston.

Weidenschilling S. J. (1988) Formation processes and time scales for meteorite parent bodies. In *Meteorites and The Early Solar System* (eds. J. F. Kerridge and M. S. Matthews), pp. 348–375. University of Arizona Press.

Weinbruch S. and Müller W. F. (1994) Cooling rates of chondrules: A new approach (abstract). *Chondrules and the Protoplanetary Disk.* **LPI Contribution No. 844**, pp. 43–44. Lunar and Planetary Institute, Houston.

Weisberg M. K., Prinz M., Kojima H., Yanai K., Clayton R. N. and Mayeda T. K. (1991) The Carlisle Lakes-type chondrites: A new grouplet with high $\Delta^{17}O$ and evidence for nebular oxidation. *Geochim. Cosmochim. Acta* **55**, 2657–2669.

Wood J. A. (1985) Meteoritic constraints on processes in the solar nebula. In *Protostars and Planets II* (eds. D. C. Black and M. S. Matthews), pp. 687–702. University of Arizona Press.

Wood J. A. (1988) Chondritic meteorites and the solar nebula. *Ann. Rev. Earth Planet. Sci.* **16**, 53–72.

III. Chondrule Precursors and Multiple Melting

11: Origin of Refractory Precursor Components of Chondrules from Carbonaceous Chondrites

KEIJI MISAWA and NOBORU NAKAMURA

Department of Earth and Planetary Sciences, Faculty of Science, and Department of Nature of the Earth, Graduate School of Science and Technology, Kobe University, Nada, Kobe 657, Japan.

ABSTRACT

Trace element analyses of individual chondrules from CV and CO chondrites suggest that two major constituents of chondritic meteorites, chondrules and Ca, Al-rich inclusions (CAIs) are genetically related. One of the refractory precursor components of CV–CO chondrules is a high-temperature condensate from the nebular gas or a distillation residue of primitive dust clumps. This component is closely related to CAIs found in carbonaceous chondrites and may have been formed earlier than common ferromagnesian chondrules. Carbonaceous chondrite chondrules preserve refractory-element fractionations, suggesting that refractory precursors did not homogenize after their formation. In unequilibrated ordinary chondrite chondrules, on the basis of refractory lithophile element fractionation, no single variety of refractory precursor component can be invoked. In the region of the nebula where the refractory precursors of ordinary chondrite chondrules formed, the high-temperature process did not fractionate refractory elements or the refractory precursors were well mixed before chondrule formation.

INTRODUCTION

Chondrules and CAIs are important constituents of chondritic meteorites and may have preserved fractionations which occurred in the early solar system. Chemical, petrological, and isotope studies of chondrites reveal that CAIs were formed by condensation from the nebular gas or by high-temperature processing of primitive dusts in the nebula (e.g., MacPherson *et al.*, 1988), and that chondrules were formed by the melting of pre-existing solid precursor materials (e.g., Grossman *et al.*, 1988). Chondrules and CAIs are early solar system objects but they have been commonly considered separately and studied by separate research groups.

There exist Al-rich chondrules in carbonaceous chondrites whose bulk chemistry is Ca, Al-rich compared with typical magnesian silicate chondrules (McSween, 1977; Wark, 1987; Beckett and Grossman, 1988). Bischoff and Keil (1984) found Al-rich chondrules in unequilibrated ordinary chondrites and suggested that chondrules and CAIs in ordinary and carbonaceous chondrites formed by related processes in the solar nebula. Wark (1987) also suggested that Ca, Al-rich chondrules are chemically and mineralogically the most similar components shared by carbonaceous and ordinary chondrites. Kring and Holmèn (1988) described anor-

thite-rich chondrules that are compositionally intermediate between ferromagnesian chondrules and CAIs, and suggested that plagioclase-olivine inclusions identified by Sheng *et al.* (1991) are a subset of these anorthite-rich chondrules. Their bulk compositions plot in the same chemical fields as some Al-rich chondrules and barred olivine chondrules in unequilibrated ordinary chondrites (e.g., Bischoff and Keil, 1984; Weisberg, 1987).

One of the most important unresolved issues is the relationship between CAIs and ferromagnesian chondrules, which needs further clarification with respect to the time and place of their formation. Trace elements, especially rare-earth elements, may be used to constrain high-temperature processes as well as the redox conditions in the early solar system. Instrumental neutron activation analysis has been used for determination of abundances of rare-earth elements in individual chondrules from carbonaceous, unequilibrated ordinary and enstatite chondrites (Osborn *et al.*, 1974; Gooding *et al.*, 1980, 1983; Grossman and Wasson, 1983; Kurat *et al.*, 1984; Rubin and Wasson, 1987, 1988; Grossman *et al.*, 1985). Analyses of chondrule bulk compositions suggest that several chemically discrete precursor components were involved (Grossman and Wasson, 1983). Rare-earth abundances of individual chondrules were also obtained by mass spectrometric isotope dilution technique (Tanaka and Masuda, 1973; Tanaka *et al.*, 1975; Hamilton *et al.*, 1979). Improved analytical techniques were developed and precise trace element data on chondrules from carbonaceous chondrites as well as unequilibrated ordinary chondrites were obtained (Misawa and Nakamura, 1988a,b; Nakamura *et al.*, 1989; see review of Nakamura, 1993). In this paper we explore the relationship between CAIs and chondrules in CV and CO chondrites.

RARE-EARTH ELEMENT SIGNATURES OF CHONDRULES IN ORDINARY AND ENSTATITE CHONDRITES

Unequilibrated ordinary chondrites

The data obtained by mass spectrometric isotope dilution technique revealed that rare-earth patterns of chondrules in unequilibrated ordinary chondrites are generally unfractionated (Hamilton *et al.*, 1979; Nagamoto *et al.*, 1987; Nakamura *et al.*, 1990). There are large differences in trace element abundances between chondrules from carbonaceous chondrites and from unequilibrated ordinary chondrites. The former sometimes show fractionated rare-earth patterns but the latter show less fractionated and chondritic rare-earth patterns. This is consistent with the data obtained by instrumental neutron activation analysis (Gooding *et al.*, 1980, 1983; Grossman and Wasson; 1983, Kurat *et al.*, 1984). Chondrules from unequilibrated ordinary chondrites so far studied rarely exhibit highly fractionated rare-earth patterns (e.g., light/heavy rare-earth fractionation), so that no single variety of refractory precursor component can be invoked. Alexander (1994) measured trace elements in chondrule glass using an ion microprobe and suggested that there are no gas/solid or gas/liquid fractionation in precursors of unequilibrated ordinary chondrite chondrules. However, some chondrules possess minor but significant positive or negative anomalies of Ce, Eu, and/or Yb (Hamilton *et al.*, 1979; Nakamura *et al.*, 1990). Since no solid/liquid fractionation can produce Ce or Yb anomalies, the anomalies must be explained by gas/solid or gas/liquid fractionation processes of their precursor materials at high temperatures in the nebula.

Trace element data on Al-rich chondrules in unequilibrated ordinary chondrites are limited (Boynton *et al.*, 1983; Alexander, 1994). They show flat rare-earth patterns. Alexander (1994) suggested that these Al-rich chondrules could have formed almost entirely from an earlier generation of chondrule glass by splashing as earlier suggested by Bischoff *et al.* (1989).

Unequilibrated enstatite chondrites

Trace element data on individual chondrules from unequilibrated enstatite chondrules are limited. Most Qingzhen (EH3) chondrules show flat rare-earth patterns with negative Eu anomalies (Grossman *et al.*, 1985). The petrology and trace element chemistry of relict FeO-rich pyroxene grains in unequilibrated enstatite chondrite chondrules suggest that they are derived from an earlier generation of chondrules (Weisberg *et al.*, 1994). Bischoff *et al.* (1985) found Al-rich chondrules in several enstatite chondrites.

PETROGRAPHY AND CHEMISTRY OF CV AND CO CHONDRULES HAVING FRACTIONATED RARE-EARTH PATTERNS

In contrast to the unequilibrated ordinary chondrite chondrules with unfractionated rare-earth patterns at levels ~2× higher than CI abundances, the rare-earth elements in individual chondrules from CV-CO chondrites vary from 0.15 to 10 times CI-chondrite and are generally fractionated (Misawa and Nakamura, 1988a,b). Several chondrules from CM chondrites show fractionated rare-earth patterns distinctly different from those of CV and CO chondrules (Inoue and Nakamura, 1995). Because CM chondrules have been affected by aqueous alteration, we will not discuss CM chondrules in detail in this paper.

Most chondrules have rare-earth abundances that are uniform or vary smoothly with ionic radius. Specific anomalies of rare-earth elements are clearly resolved in mass spectrometric isotope dilution data. Chondrules from CV-CO chondrites sometimes show a light/heavy rare-earth fractionation along with positive or negative anomalies of Ce, Eu, and Yb (Fig. 1a). If elemental abundances are arranged in the order of elemental volatilities expected from a gas of solar composition, fractionation patterns of rare-earth elements in many chondrules appear to be smooth but not monotonic (Fig. 1b). These anomalies and fractionations appear to be due to nebular fractionation of refractory precursor materials and not due to fractionation during chondrule melting.

There are several chondrules whose bulk rare-earth abundances are highly fractionated. Allende 24 and Ornans 1 of Rubin and Wasson (1987, 1988), Allende 1, 2, 9, R-11 and Felix 8 of Misawa and Nakamura (1988a,b), Allende BG82CLII of Sheng *et al.* (1990), Kaba KI77 and KI88 of Liu *et al.* (1988), and Ningqiang N1 of Mayeda *et al.* (1988) belong to this group. The similarity of rare-earth fractionations in CAIs and in ferromagnesian chondrules (Fig. 2) suggests that refractory chondrule precursor materials and CAIs formed in related high-temperature processes in the nebula. Nine of the eleven are related to group-II CAIs (Boynton, 1975; Davis and Grossman, 1979). Two Allende barred olivine chondrules may be related to an ultra-refractory inclusion (Boynton *et al.*, 1980). These chondrules differ in mineral assemblages from typical Ca, Al-rich chondrules and CAIs found in carbonaceous chondrites (McSween, 1977). They mainly consist of olivine and pyroxene with glass, and sometimes contain anorthite and spinel but fassaite and melilite which are usually found in CAIs are absent.

Figure 3a shows a back-scattered electron image of Allende 2 barred olivine chondrule. The chondrule consists of olivine ($Fa_{0.5-4}$), Ca, Al-rich mesostasis, and Mg-spinel. The rare-earth pattern is heavy rare-earth enriched and shows large negative anomalies of Eu and Yb (Fig. 2a). This newly found rare-earth fractionation for Allende barred olivine chondrules may be explained as representing one kind of ultra-refractory component as inferred from group-II CAIs. Felix 8 consists of enstatite phenocrysts (Fs_1) and Al, Si-rich dendrites with an olivine (Fa_{4-9})-iron sulfides aggregate (Fig. 3b). The chondrule shows a highly frac-

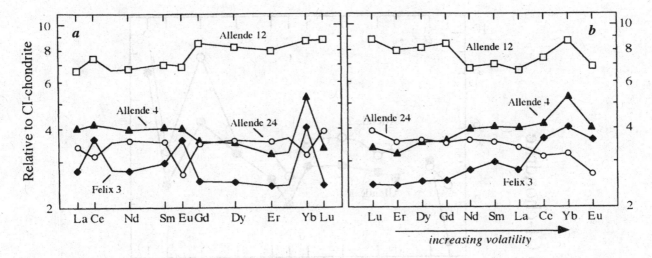

Fig. 1. CI-normalized (weight ratios) rare-earth abundance patterns of CV-CO chondrules. Orgueil data are from Anders and Grevesse (1989) and from Nakamura (1974). (a) Light/heavy rare-earth fractionation along with positive or negative anomalies of Ce, Eu, and Yb are recognized. (b) Rare-earth element abundances are arranged in the order of elemental volatilities expected from a gas of solar composition. Fractionation patterns appear to be smooth. Data are from Misawa and Nakamura (1988a,b).

tionated, light rare-earth-enriched, heavy rare-earth-depleted pattern with positive Ce and Yb anomalies, that is related to group-II rare-earth patterns (Fig. 2b). Allende 9 consists of low-Ca pyroxene ($Fs_{0.5-0.9}$), augite($Fs_{0.4-0.7}Wo_{30-43}$), olivine (Fa_{12-34}), and interstitial nepheline and sodalite along with abundant iron sulfides and minor metal (Fig. 3c). Although a light/heavy rare-earth fractionation is not observed, the general pattern and the positive anomalies of Ce and Yb are similar to those of Felix 8 (Fig. 2b).

The group-II rare-earth pattern found in the chondrules shown in Fig. 2b is different in detail from typical group-II patterns found in Allende CAIs. The pattern is characterized by a gradual increase of abundances from La to Sm with a positive anomaly of Ce, heavy rare-earth depletion with a positive anomaly of Yb (and Tm), but the absence of the negative Eu anomaly usually found in typical group-II CAIs. The pattern resembles the modified group-II pattern (Fig. 2c) described by Davis and Grossman (1979). The modified group-II rare-earth pattern is found in CAIs from Allende, Murchison, Leoville, Vigarano, Mokoia, Mighei, Acfer-El Djouf, Acfer 182, and ALH85085 (Conard, 1976; Davis and Grossman, 1979; Ireland *et al.*, 1988; Sylvester *et al.*, 1992; Mao *et al.*, 1990; Liu *et al.*, 1987; MacPherson and Davis, 1994; Weber *et al.*, 1995; Kimura *et al.*, 1993).

In the case of refractory chondrule precursors the removal temperature of the ultra-refractory component and the condensation temperature of the group-II component may be different from those of typical Allende group-II CAIs. Allende 9 and R-11 exhibit complementary Ce, Sm, and Eu abundances (Fig. 4). An ultra-refractory component may have been removed at relatively low temperature in the nebular gas and the refractory precursor material of Allende 9 isolated from the gas after the most volatile rare-earth Eu was nearly fully condensed. Alternatively, an ultra-refractory component may have been removed at high-temperature in the nebular gas and refractory precursor material of

Allende R-11 was isolated from the gas under relatively oxidizing condition before Eu was fully condensed. It is worth pointing out that there is little, if any, fractionation between Sr and Eu in chondrules possessing modified group-II patterns.

Fractionated rare-earth patterns similar to CAIs strongly suggest that the refractory precursor of those chondrules was CAI-like. However, there is little direct evidence that chondrule precursors and CAIs have formed in the same nebular region. Bischoff and Keil (1984) reported a relict hibonite-spinel inclusion (Sharps No. 9 of Bischoff and Keil, 1983) within a chondrule-like object from the Sharps (H3.4) chondrite. Allende chondrule R-11 (Fig. 5) is the first example of a CAI and a ferromagnesian chondrule mechanically joined together in carbonaceous chondrites (Misawa and Fujita, 1994). A 250 µm-sized Mg-spinel containing refractory silicate inclusions is associated with a circular void of ~5 µm and with abundant submicrometer-sized refractory platinum-group metal nuggets. The spinel core and silicate inclusions are iron-free. Silicate inclusions consist of Ti, Al-rich pyroxene (fassaite) and a Ca, Al-rich silicate phase. Refractory platinum-group metal abundances are comparable to those reported in Type A CAIs. The spinel appears to be a relict fragment of a high-temperature condensate from the nebular gas. The petrographic features of the spinel grain are similar to those of the FeO-poor spinel from the Murchison (CM) meteorite described by Kuehner and Grossman (1987).

Some refractory siderophile components may have been separated from the common siderophiles, Fe, Co, and Ni, during nebular processes and have been incorporated into chondrule precursors with refractory oxides and silicates. Because they were protected by surrounding oxides or silicates, they avoided low-temperature oxidization and sulfidization. Elemental fractionations among refractory siderophiles (Ir, Os) and common siderophiles (Fe, Co, Ni) observed in a metal-rich chondrule from Allende (Rubin and

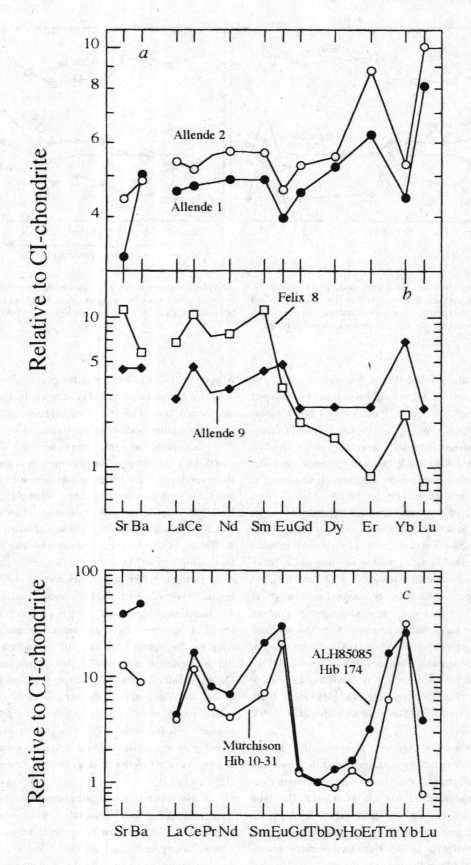

Fig. 2. Highly fractionated CI-normalized rare-earth abundance patterns are found in chondrules from carbonaceous chondrites. (a) Ultra-refractory rare-earth patterns found in Allende barred olivine chondrules (Misawa and Nakamura, 1988b). (b) Modified group-II rare-earth patterns in Felix (Misawa and Nakamura, 1988a) and Allende (Misawa and Nakamura, 1988b) chondrules. (c) For comparison, modified group-II rare-earth patterns of CAIs (hibonite grains) in Murchison (Ireland *et al.*, 1988) and ALH85085 (Kimura *et al.*, 1993) are shown.

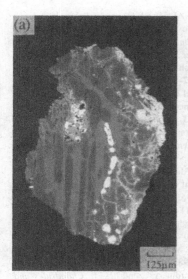
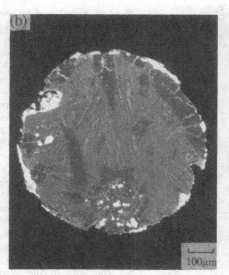
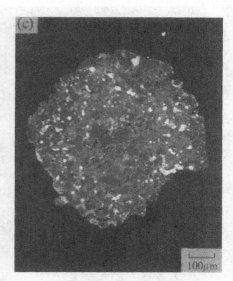

Fig. 3. Back-scattered electron images of Allende and Felix chondrules whose rare-earth patterns are highly fractionated. (a) Part of the Allende 2 chondrule consisting of olivine ($Fa_{0.5-4}$, dark gray) and spinel (FeO = 1.2wt%, Cr_2O_3 = 2.9wt%, not shown in this image) with Ca, Al-rich mesostasis (light gray). Bright blebs are iron-sulfides. Scale bar = 125 μm. (b) Felix 8 shows porphyritic pyroxene texture with enstatite phenocrysts (Fs_1) containing Al_2O_3 = ~3wt% and Si, Al-rich dendrites. The chondrule

contains an olivine (Fa_{4-9})-iron sulfide aggregate (lower center) which may represent one of the common ferromagnesian precursor components. A similar olivine aggregate is also found in Allende BG82CLII (Sheng *et al.*, 1991). Scale bar = 100 μm. (c) Allende 9 consists of low-Ca pyroxene ($Fs_{0.4-0.9}$, dark gray), augite ($Fs_{0.4-0.7}Wo_{30-43}$, gray), olivine (Fa_{12-34}, light gray), and interstitial nepheline and sodalite along with abundant iron sulfides (white). Scale bar = 100 μm.

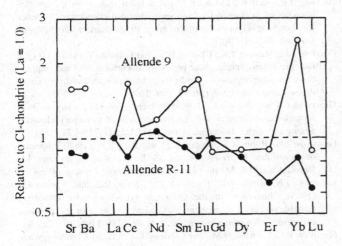

Fig. 4. CI-chondrite normalized (re-normalized to La ≡ 1.0) Sr, Ba, and rare-earth abundances of Allende 9 and R-11 are roughly complementary; CI-normalized La abundances are 2.87 and 7.35, respectively.

Oxygen isotopic compositions of individual chondrules in CV and CO chondrites plot below the terrestrial fractionation line and form a curved array roughly parallel to the Allende anhydrous mineral line. These features are interpreted to be the result of isotope exchange between an ${}^{16}O$-rich precursor solid and an ${}^{16}O$-poor ambient nebular gas during chondrule formational melting (Clayton *et al.*, 1983). The discovery of a relict CAI in a ferromagnesian chondrule also improves our understanding of the oxygen isotope variation in Allende chondrules. Since oxygen diffusion in spinel is very slow (Ryerson and McKeegan, 1994) and the grain size of spinel in Allende R-11 is large, its oxygen isotopic composition was probably preserved, even though the spinel was partly dissolved during chondrule formation. The refractory precursor components of Allende chondrules may be a ${}^{16}O$-rich end member ($\delta^{17}O$= –40‰ and $\delta^{18}O$= –40‰ relative to Standard Mean Ocean Water) as suggested by Rubin *et al.* (1990).

CONCLUSIONS

The refractory chondrule precursors of carbonaceous chondrite chondrules and CAIs are more closely related than previously thought. Formation of CAIs preceded that of common ferromagnesian chondrules. After CAIs formed in the nebula, most large CAIs were removed from the chondrule formation region. Flash-heating of nebular dust clumps produced chondrules. These chondrules preserved refractory-element fractionations even though diluted with common ferromagnesian precursors. In the region of the nebula where the refractory precursors of unequilibrated ordinary chondrite chondrules formed, high-temperature processes did not fractionate the refractory elements or the refractory precursor materials were well mixed before chondrule formation.

Wasson, 1987) can be explained by incorporation of this type of refractory precursors into the metal-rich common ferromagnesian precursors.

Thus, trace element signatures suggest that one of the refractory precursor components of carbonaceous chondrite chondrules is a high-temperature condensate from the nebular gas or a high-temperature distillation residue of dust clumps, and that the formation of CAIs preceded that of ferromagnesian chondrules. This is consistent with previous trace element data on carbonaceous chondrite chondrules along with ${}^{129}I$–${}^{129}Xe$, ${}^{53}Mn$–${}^{53}Cr$, and ${}^{26}Al$–${}^{26}Mg$-based chronometries of CAIs and chondrules (Hutcheon *et al.*, 1994; Swindle *et al.*, this volume).

Fig. 5. Back-scattered electron image of Allende R-11 chondrule which contains a relict CAI fragment. The 250 μm sized, rounded grain is $MgAl_2O_4$ spinel containing Ti, Al-pyroxene (fassaite), a Si, Al-rich inclusion, and abundant submicrometer-sized refractory platinum-group metal nuggets (see Figs. 1 and 2 of Misawa and Fujita, 1994). The spinel shows Fe-Mg zoning. Spinel core and silicate inclusions are almost iron free. The host chondrule consists of zoned olivine (Fa_{7-18}), low-Ca pyroxene (Fs_1), augite ($Fs_{0.9-1.4}Wo_{22-33}$), plagioclase (An_{78-83}), and glass along with iron sulfides. Scale bar = 200 μm.

Alternatively, chondrule formation was repeated enough times to mix the refractories. The most probable scenario at the location where CV and CO chondrules formed is that, before chondrule formation, several high-temperature fractionations occurred, and that the identity of different components formed by these nebular fractionations was not destroyed during the thermal processes that melted the chondrules. Recycling of CV and CO chondrules did not significantly alter the chemical memory of earlier generations.

ACKNOWLEDGEMENTS

We thank T. Fujita and S. Hayashi for their technical assistance. A. Bischoff, A.M. Davis, and J.T. Wasson provided thoughtful reviews that improved the manuscript. This work was supported in part by the cooperative program of the Institute of Cosmic Ray Research, University of Tokyo and by funds from the Inoue Foundation for Science.

REFERENCES

Alexander C.M.O'D. (1994) Trace element distributions within ordinary chondrite chondrules: Implications for chondrule formation condition and precursors. *Geochim. Cosmochim. Acta* **58**, 3451–3467.

Anders E. and Grevesse N. (1989) Abundances of the elements: Meteoritic and solar. *Geochim. Cosmochim. Acta* **53**, 197–214.

Beckett J.R. and Grossman L. (1988) The origin of type C inclusions from carbonaceous chondrites. *Earth Planet. Sci. Lett.* **89**, 1–14.

Bischoff A. and Keil K. (1983) Catalog of Al-rich chondrules, inclusions and fragments in ordinary chondrites. Spec. Publ. No. 22, University of New Mexico, Institute of Meteoritics, pp. 33.

Bischoff A. and Keil K. (1984) Al-rich objects in ordinary chondrites: Related origin of carbonaceous and ordinary chondrites and their constituents. *Geochim. Cosmochim. Acta* **48**, 693–709.

Bischoff A., Keil K. and Stöffler D. (1985) Perovskite-hibonite-spinel-bearing inclusions and Al-rich chondrules and fragments in enstatite chondrites. *Chem. Erde* **44**, 97–106.

Bischoff A., Palme H. and Spettel B. (1989) Al-rich chondrules from the Ybbsitz H4-chondrite: Evidence for formation by collision and splashing. *Earth Planet. Sci. Lett.* **93**, 170–180.

Boynton W.V. (1975) Fractionation in the solar nebula: Condensation of yttrium and the rare earth elements. *Geochim. Cosmochim. Acta* **39**, 569–584.

Boynton W.V., Frazier R.M. and MacDougall J.D. (1980) Identification of an ultra-refractory component in the Murchison meteorite (abstract). *Lunar Planet Sci.* **XI**, 103–105.

Boynton W.V., Hill D.H., Wark D.A. and Bischoff A. (1983) Trace elements on Ca, Al-rich chondrules in the Dhajala (H3) chondrite (abstract). *Meteoritics* **18**, 270–271.

Clayton R.N., Onuma N., Ikeda Y., Mayeda T.K., Hutcheon I.D., Olsen E.J. and Molini-Velsko C.A. (1983) Oxygen isotopic compositions of chondrules in Allende and ordinary chondrites. In *Chondrules and Their Origins* (ed. E.A. King), pp. 195–210. Lunar and Planetary Institute.

Conard R.L. (1976) A study of the chemical composition of Ca-Al-rich inclusions from the Allende meteorite. M.S. Thesis. Oregon State University.

Davis A.M. and Grossman L. (1979) Condensation and fractionation of rare earths in the solar nebula. *Geochim. Cosmochim. Acta* **43**, 1611–1632.

Gooding J.L., Keil K., Fukuoka T. and Schmitt R.A. (1980) Elemental abundances in chondrules from unequilibrated chondrites: Evidence for chondrule origin by melting of preexisting materials. *Earth Planet. Sci. Lett.* **50**, 171–180.

Gooding J.L., Mayeda T.K., Clayton R.N. and Fukuoka T. (1983) Oxygen isotopic heterogeneity, their petrological correlations, and implications for melt origins of chondrules in unequilibrated ordinary chondrites. *Earth Planet. Sci. Lett.* **63**, 209–224.

Grossman J.N. and Wasson J.T. (1983) Refractory precursor components of Semarkona chondrules and the fractionation of refractory elements among chondrites. *Geochim. Cosmochim. Acta* **47**, 759–771.

Grossman J.N., Rubin A.E., Nagahara H. and King E.A. (1988) Properties of chondrules. In *Meteorites and the Early Solar System* (eds. J.F. Kerridge and M.S. Matthews) pp. 619–659. Univ. of Arizona Press.

Grossman J.N., Rubin A.E., Rambaldi E.R., Rajan R.S. and Wasson J.T. (1985) Chondrules in the Qingzhen type-3 enstatite chondrite: Possible precursor components and comparison to ordinary chondrite chondrules. *Geochim. Cosmochim. Acta* **49**, 1781–1795.

Hamilton P.J., Evensen N.M. and O'Nions R.K. (1979) Chronology and chemistry of Parnallee (LL-3) chondrules (abstract). *Lunar Planet. Sci.* **X**, 494–495.

Hutcheon I.D., Huss G.R. and Wasserburg G.J. (1994) A search for ^{26}Al in chondrites: Chondrule formation time scales (abstract). *Lunar Planet. Sci.* **XXV**, 587–588.

Inoue M. and Nakamura N. (1995) REE abundances in chondrules from the Murchison and Yamato 793321 (CM) meteorites: Constraints on the formation of CM chondrules (abstract). *Lunar Planet. Sci.* **XXVI**, 653–654.

Ireland T.R., Fahey A.J. and Zinner E. (1988) Trace-element abundances in hibonites from the Murchison carbonaceous chondrite: Constraints on high-temperature processes in the solar nebula. *Geochim. Cosmochim. Acta* **52**, 2841–2854.

Kimura M., El Goresy A., Palme H. and Zinner E. (1993) Ca-,Al-rich inclusions in the unique chondrite ALH85085: Petrology, chemistry, and isotopic compositions. *Geochim. Cosmochim. Acta* **57**, 2329–2359.

Kring D.A. and Holmèn B.A. (1988) Petrology of anorthite-rich chondrules in CV3 and CO3 chondrites (abstract). *Meteoritics* **24**, 282–283.

Kurat G., Pernicka E. and Herrwerth I. (1984) Chondrules from Chainpur (LL-3): reduced parent rocks and vapor fractionation. *Earth Planet. Sci. Lett.* **68**, 43–56.

Kuehner S.M. and Grossman L. (1987) Petrography and mineral chemistry of spinel grains separated from the Murchison meteorite (abstract). *Lunar Planet. Sci.* **XVIII**, 519–520.

Liu Y.-G., Rajan R.S. and Schmitt R.A. (1987) Mokoia Ca-Al inclusions (CAIs) with negative and positive Ce anomalies – Interim report #2 (abstract). *Lunar Planet. Sci.* **XVIII**, 562–563.

Liu Y.-G., Schmitt R.A., Holmèn B.A., Wood J.A. and Kring D.A. (1988) A trace element/petrographic study of refractory inclusions in Kaba (CV3) (abstract). *Lunar Planet. Sci.* **XIX**, 686–678.

MacPherson G.J. and Davis A.M. (1994) Refractory inclusions in the prototypical CM chondrite, Mighei. *Geochim. Cosmochim. Acta* **58**, 5599–5625.

MacPherson G.J., Wark D.A. and Armstrong J.T. (1988) Primitive material surviving in chondrites: Refractory inclusions. In *Meteorites and the Early Solar System* (eds. J.F. Kerridge and M.S. Matthews) pp. 619–659. Univ. of Arizona Press.

Mao X.-Y., Ward B.J., Grossman L. and MacPherson G.J. (1990) Chemical composition of refractory inclusions from the Vigarano and Leoville carbonaceous chondrites. *Geochim. Cosmochim. Acta* **54**, 2121–2132.

Mayeda T.K., Clayton R.N., Kring D.A. and Davis A.M. (1988) Oxygen, silicon, and magnesium isotopes in Ningqiang chondrules (abstract). *Meteoritics* **23**, 288.

McSween H.Y.Jr. (1977) Chemical and petrographic constraints on the origin of chondrules and inclusions in carbonaceous chondrites. *Geochim. Cosmochim. Acta* **41**, 1843–1860.

Misawa K. and Fujita T. (1994) A relict refractory inclusion in a ferromagnesian chondrule from the Allende meteorite. *Nature* **368**, 723–726.

Misawa K. and Nakamura N. (1988a) Highly fractionated rare-earth elements in ferromagnesian chondrules from the Felix (CO3) meteorite. *Nature* **334**, 47–50.

Misawa K. and Nakamura N. (1988b) Demonstration of REE fractionation among individual chondrules from the Allende (CV3) chondrite. *Geochim. Cosmochim. Acta* **52**, 1699–1710.

Nagamoto H., Nishikawa Y., Misawa K. and Nakamura N. (1987) REE, Ba, Sr, Rb, and K characteristics of chondrules from the Tieschitz (H3) chondrite (abstract). *Lunar Planet. Sci.* **XVIII**, 696–697.

Nakamura N. (1974) Determination of REE, Ba, Fe, Mg, Na and K in carbonaceous and ordinary chondrites. *Geochim. Cosmochim. Acta* **38**, 757–775.

Nakamura N. (1993) Trace element fractionation during the formation of chondrules. In *Primitive Solar Nebula and Origin of Planets* (ed. H. Oya), pp. 409–425. Terra Sci. Publ.

Nakamura N., Misawa K., Kitamura M., Masuda A., Watanabe S. and K. Yamamoto (1990) Highly fractionated REE in the Hedjaz (L) chondrite: implications for nebular and planetary processes. *Earth Planet. Sci. Lett.* **99**, 290–302.

Nakamura N., Yamamoto K., Noda S., Nishikawa Y., Komi H., Nagamoto H., Nakayama T. and Misawa K. (1989) Determination of picogram quantities of rare-earth elements in meteoritic materials by direct-loading thermal ionization mass spectrometry. *Anal. Chem.* **61**, 755–762.

Osborn T.W., Warren R.G., Smith R.H., Wakita H., Zellmer D.L. and Schmitt R.A. (1974) Elemental composition of individual chondrules from carbonaceous chondrites, including Allende. *Geochim. Cosmochim. Acta* **38**, 1359–1378.

Rubin A.E. and Wasson J.T. (1987) Chondrules, matrix and coarse-grained chondrule rims in the Allende meteorite: Origin, interrelationship and possible precursor components. *Geochim. Cosmochim. Acta* **51**, 1923–1937.

Rubin A.E. and Wasson J.T. (1988) Chondrules and matrix in the Ornans CO3 meteorite: Possible precursor components. *Geochim. Cosmochim. Acta* **52**, 425–432.

Rubin A.E., Wasson J.T., Clayton R.N. and Mayeda T.K. (1990) Oxygen isotopes in chondrules and coarse-grained rims from the Allende meteorite. *Earth Planet. Sci. Lett.* **96**, 247–255.

Ryerson F.J. and McKeegan K.D. (1994) Determination of oxygen self-diffusion in åkermanite, anorthite, diopside, and spinel: Implications for oxygen isotopic anomalies and the thermal histories of Ca-Al-rich inclusions. *Geochim. Cosmochim. Acta* **58**, 3713–3734.

Sheng Y.J., Hutcheon I.D. and Wasserburg G.J. (1990) Magnesium isotope heterogeneity in plagioclase olivine inclusions (abstract). *Lunar Planet. Sci.* **XXI**, 1138–1139.

Sheng Y.J., Hutcheon I.D. and Wasserburg G.J. (1991) Origin of plagioclase-olivine inclusions in carbonaceous chondrites. *Geochim. Cosmochim. Acta* **55**, 581–599.

Sylvester P.J., Grossman L. and MacPherson G.J. (1992) Refractory inclusions with unusual chemical compositions from the Vigarano carbonaceous chondrite. *Geochim. Cosmochim. Acta* **56**, 1343–1363.

Tanaka T. and Masuda A. (1973) Rare earth elements in matrix, inclusions, and chondrules of the Allende meteorite. *Icarus* **19**, 523–530.

Tanaka T., Nakamura N., Masuda A. and Onuma N. (1975) Giant olivine chondrule as a possible later-stage product in the nebula. *Nature* **256**, 27–28.

Wark D.A. (1987) Plagioclase-rich inclusions in carbonaceous chondrite meteorites: Liquid condensates? *Geochim. Cosmochim. Acta* **51**, 221–242.

Weber D., Zinner E. and Bischoff A. (1995) Trace element abundances and magnesium, calcium, and titanium isotopic compositions of grossite-containing inclusions from the carbonaceous chondrite Acfer 182. *Geochim. Cosmochim. Acta* **59**, 803–824.

Weisberg M.K. (1987) Barred olivine chondrules in ordinary chondrites. *Proc. 17th Lunar Planet. Sci. Conf.* E663–E678.

Weisberg M.K., Prinz M. and Fogel R.A. (1994) The evolution of enstatite and chondrules in unequilibrated enstatite chondrites: Evidence from iron-rich pyroxene. *Meteoritics* **29**, 362–373.

12: Mass-Independent Isotopic Effects in Chondrites: The Role of Chemical Processes

MARK H. THIEMENS

Department of Chemistry, University of California San Diego, La Jolla, CA 92093–0356, U.S.A.

ABSTRACT

It has been experimentally demonstrated that there are at least three varieties of chemically produced mass independent isotopic fractionation processes. These include addition reactions, gas phase isotopic exchange and thermal decomposition, all of which give rise to $\delta^{17}O = \delta^{18}O$ isotopic compositions. In addition, it has recently been experimentally demonstrated that the simple isotopic exchange process between an oxygen atom and CO_2 gives rise to a $\delta^{17}O/\delta^{18}O = 1$ fractionation with magnitude of 40 ‰, strikingly similar in slope and magnitude to that observed in CAI's. Since many chondrules, such as those in ordinary chondrites, define a similar isotopic trend, the possibility exists that chemical processes gave rise to these components. This is particularly true now that mass independent compositions have been experimentally produced in solid silicon oxide condensates. If chondrule oxygen isotopic components arise from chemical processes, only one reservoir is required, with the isotopic compositions arising as a result of the chondrule formation process itself. The mechanism is simpler than requiring two interacting reservoirs, which additionally requires that they must isotopically align themselves to produce a $\delta^{17}O = \delta^{18}O$ composition. It is also possible that the oxygen isotopic composition of chondrule precursors or CAI arose from chemical processes, such as via molecular partitioning of oxygen isotopes in the nebula. In this case, subsequent formation of CAI or chondrules from this material would give rise to the observed isotopic compositions. This could also allow for subsequent nebular exchange between chondrules and the anomalous gas as a mechanism to produce the observed isotopic compositions.

The different types of chemical mass independent isotopic fractionation processes are discussed in the context of chondrule, CAI and nebular reservoir events. Relevant observations of the occurrence of these fractionation processes in the Earth's atmosphere as the chemical processes which produce the mass independent isotopic composition in at least 4 different molecules could be relevant to nebular chemistry.

INTRODUCTION

The observation of a $\delta^{17}O = \delta^{18}O$ isotopic composition in Allende calcium-aluminum rich inclusions (CAI) was suggested as resulting from a nuclear event (Clayton, *et al.*,1973), based upon the premise that chemical processes may not produce mass independent isotopic compositions. The basis of this assumption for the distinction of nuclear vs. chemical processes was first suggested by Hulston and Thode (1965) for the 4 stable isotopes of sulfur. With the subsequent demonstration by Thiemens and Heidenreich (1983) of a chemical reaction which produced isotopic components

with a $\delta^{17}O = \delta^{18}O$ fractionation, that assumption had to be abandoned. The question as to the source of the oxygen isotopic anomalies is of fundamental importance as nearly all of the bulk oxygen isotopic compositions of meteoritic material are anomalous with respect to the earth and moon, as seen in Fig. 1. Thus, the process responsible for the generation of these isotopic compositions represents a central event in solar system history.

An important component of the nucleosynthetic model is the requirement that the oxygen isotopic anomalies arise due to admixture of nucleosynthetic, ^{16}O rich components. As stated by Clayton *et al.*, (1973), "The effect is the result of nuclear rather than chemi-

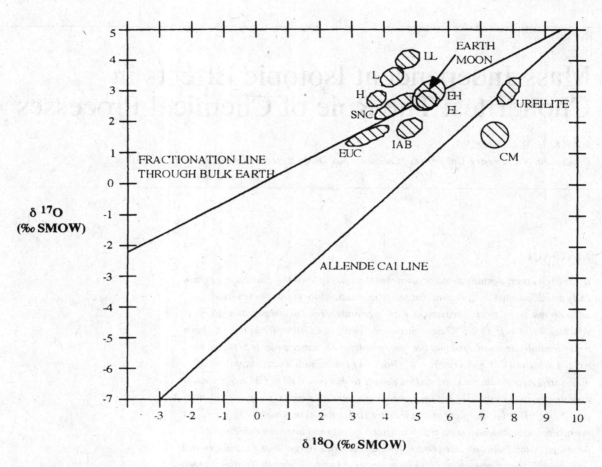

Fig. 1. A schematic diagram showing the oxygen isotopic composition of the meteorites. The slope one half line passes through the average mantle rock isotopic composition.

cal processes and probably results from the admixture of a component of almost pure ^{16}O." Recent isotopic investigations of the oxygen isotopic composition of individual pre-solar grains have not detected any evidence for ^{16}O carriers (Hutcheon *et al.*, 1994; Huss *et al.*, 1994). Corundum grains from Bishunpur, in fact, possess large ^{17}O enrichments (Huss *et al.*, 1994), as do corundum grains from Orgueil (Hutcheon *et al.*, 1994). Nittler *et al.*, (1994) have reported the oxygen isotopic composition of 21 interstellar oxide grains from the Tieschitz meteorite and concluded that they have not found any grains with a large ^{16}O enrichment which would be consistent with the notion of a ^{16}O carrier. Huss *et al.*, (1994) conclude "The presence of identifiable ^{16}O-rich carrier grains has not been established". Thus, while there is unmistakable evidence for interstellar grains carrying anomalous oxygen, their oxygen isotopic composition does not support the notion of admixture of pure ^{16}O components. At present, there is no evidence from the interstellar grain oxygen isotopic compositions for a pure ^{16}O carrier.

In the case of chondrules, mass independent isotopic compositions and variations are also observed. Chondrules from carbonaceous chondrites define a $\delta^{17}O = \delta^{18}O$ line in a three-isotope plot, offset, in the case of Allende porphyritic chondrules, from the $\delta^{17}O = \delta^{18}O$ line defined by the calcium aluminum rich inclusions. Barred chondrules in Allende deviate from this line and trend towards ordinary chondrite chondrules, as schematically shown in

Fig. 2. (Clayton *et al.*, 1983; Clayton and Mayeda, 1985; Clayton, 1993). The mean isotopic composition of chondrules from H, L, and LL chondrites define a line which fits the equation:

$$\delta^{17}O = 1.074\,\delta^{18}O - 1.53 \qquad (1)$$

(Clayton *et al*, 1991). This relation has been suggested as arising from isotopic exchange between solid and gas reservoirs (Clayton *et al.*, 1991) or solid reservoir exchange (Gooding *et al*, 1983).

The oxygen isotopic composition of size separated chondrules from ordinary chondrites also provide useful insight into chondrule formation processes (see e.g. Clayton, 1993). Size separated chondrules from Dhajala define a $\delta^{17}O = \delta^{18}O$ isotopic fractionation line (Clayton *et al.*, 1986). It was suggested that this trend results from differential isotopic exchange between a ^{16}O rich gas and ^{16}O poor solid reservoirs. The smallest chondrules, due to their greater surface to volume ratio, underwent the most extensive exchange with the gas reservoir. In Mezo-Madaras, however, the reverse trend is observed (Mayeda *et al.*, 1989). Analysis of six density fractions of the bulk meteorite also define a $\delta^{17}O = \delta^{18}O$ fractionation line, however, in this instance the lowest density fraction, consisting of mostly plagioclase and glass, is approximately 2‰ further up the fractionation line than any chondrule. This implies that the most easily exchanged mineral phases move up the slope one line, opposite to that suggested for Dhajala chondrules. There

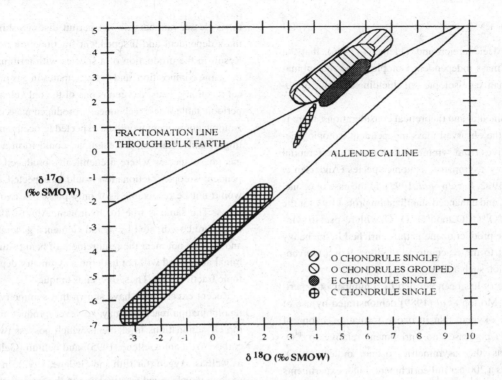

Fig. 2. The oxygen isotopic composition of different types of meteoritic chondrules.

is, at present, no consistent explanation for this paradox between the two meteorites.

This paper discusses the possible chemical mechanisms by which meteoritic oxygen isotopic compositions may be attained; in CAI/chondrule formation, nebular chemical reactions and during secondary exchange processes. There are recent observations regarding chemically produced mass independent isotopic fractionations which will be discussed within the context of meteoritic isotopic observations. This includes gas phase laboratory experiments, experiments demonstrating the production of a mass independent isotopic composition in solid SiO_x species and atmospheric observations. The state of the art with regard to understanding chemical, symmetry dependent isotope effects is sufficiently advanced that realistic consideration of their participation in early solar system processes may be evaluated.

OBSERVATIONS OF CHEMICALLY PRODUCED MASS-INDEPENDENT ISOTOPIC FRACTIONATIONS

It has been demonstrated that there are several processes which give rise to mass independent components from chemical reactions. Recent reviews are given by Thiemens (1992), Anderson *et al* (1992) and Larsen *et al* (1992). Several of these chemical mechanisms may occur, either individually, or simultaneously, during chondrule or CAI formation or nebular gas phase reactions. Along with the detection of mass independent isotopic fractionation processes in laboratory experiments, there are now atmospheric observations of isotopically anomalous components which are of relevance, and will be presented. At least four different molecules in the atmosphere possess anomalous oxygen isotopic compositions.

RECOMBINATION REACTIONS

The addition of two gas phase atomic or molecular species to form a new product is referred to as a recombination reaction. It was in the recombination reaction of molecular and atomic oxygen to form ozone that the first chemically produced mass independent isotopic fractionation process was observed (Thiemens and Heidenreich, 1983). Subsequently, there have been numerous experimental and theoretical studies directed towards understanding the mechanism by which this fractionation occurs (Anderson *et al*, 1992).

In a gas phase recombination reaction, such as ozone formation, the reaction consists of two steps, the initial attachment:

$$O + O_2 \longrightarrow O_3^* \tag{2}$$

where O_3^* represents a species possessing excess energy, generally vibrational, above the molecular bond strength. There must be a second reaction to remove this excess energy, otherwise the excited O_3^* species re-dissociates into atomic and molecular oxygen. The stabilization is normally accomplished via collisional quenching with a third body, M:

$$O_3^* + M \longrightarrow O_3 + M \tag{3}$$

with M carrying away the excess energy and permitting the excited species to stabilize. M is the predominant gas phase species. The quenching process is modestly efficient, most of the excited species break apart into fragments, by a factor of approximately a million to one. If there is collisional quenching within a nominal number of vibrations for the excited transition state, a stable product is formed. The overall reaction is then :

$$O + O_2 + M \longrightarrow O_3 + M \qquad (4)$$

It was suggested by Heidenreich and Thiemens (1986), that the source of the ozone mass independent isotopic is the recombination reaction rather than, e.g. isotopic self shielding or ozone dissociation.

Based upon experimental and theoretical considerations, there is now agreement that the observed mass independent isotopic fractionation process derives from symmetry factors, with preferential formation of the heavy, asymmetric isotopic species (Andersen *et al*, 1992, Thiemens, 1992, Larsen *et al*, 1992). In the case of ozone, this leads to an equal, and enhanced stabilization probability for the asymmetric species $^{16}O^{16}O^{17}O$ and $^{16}O^{16}O^{18}O$, with respect to symmetric $^{16}O^{16}O^{16}O$. The product ozone is thus enriched in the heavy isotopes with respect to the reactants, with slope 1 in a conventional three isotope plot, such as Fig. 1.

The role of symmetry has been verified by a number of experimental observations. Morton *et al* (1989) demonstrated by use of isotopically enriched oxygen, that in ozone formation, ozone of mass 54 ($^{18}O^{18}O^{18}O$) possesses no enrichment relative to $^{48}O_3$ ($^{16}O^{16}O^{16}O$), whereas the asymmetric ozone molecule $^{51}O_3$ ($^{16}O^{17}O^{18}O$) possesses a 200 per mil enrichment. Later experiments further confirmed and expanded upon these observations. It was demonstrated that symmetric ozone, $^{17}O^{17}O^{17}O$ and $^{18}O^{18}O^{18}O$ are depleted, whereas asymmetric ozone molecules , $^{16}O^{16}O^{17}O$, $^{16}O^{17}O^{17}O$, $^{16}O^{16}O^{18}O$, $^{16}O^{17}O^{18}O$, $^{16}O^{18}O^{18}O$, $^{17}O^{17}O^{18}O$ and $^{17}O^{18}O^{18}O$ are enriched with respect to their statistically expected abundances (Mauersberger *et al*, 1993). Tunable diode laser absorption studies by Anderson *et al* (1989) have directly demonstrated that the asymmetric ozone molecule possesses more than 80% of the isotopic enrichment. Thus, all experimental observations are consistent with the fractionation process deriving from symmetry factors during the formation process, as suggested by Heidenreich and Thiemens (1986).

There are several other experimental observations which confirm that symmetry dependent isotopic factors are responsible for the production of the $\delta^{17}O = \delta^{18}O$ fractionation. These observations are relevant with regard to their possible role in meteoritics, as they define the criteria for production of mass independent isotopic compositions. They are as follows.

1. **The effect has been observed in the production of molecules for which a symmetry dependent isotopic fractionation is expected.** This includes the reactions $O + O_2 \longrightarrow O_3$ (Thiemens and Jackson, 1987, 1988); $O + CO \longrightarrow CO_2$ (Bhattacharya and Thiemens 1989a,b) and $SF_5 + SF_5 \longrightarrow S_2F_{10}$ (Bains Sahota and Thiemens, 1987). Thus, the effect is general and not restricted to either ozone or oxygen. The observed isotopic fractionation is a feature of a gas phase reaction, occurring when terminal atoms of the product are the same species. There are numerous reactions which plausibly occur in the early solar system, and nature in general, which meet the criteria for production of a mass independent composition. Based upon oxygen's position on the periodic chart, It is one of few elements which a) has three or more stable iso-

topes (lesser numbers do not permit distinguishing between mass dependent and independent fractionation processes) b) Result in the production of a species with terminal atoms of the same composition, and c) Participates in gas phase chemical reactions. Thus, oxygen is one of the only elements of the periodic table where a chemically produced mass independent isotopic component would be expected to occur and to be distinguishable. Carbon, for example, could form a long chain gas phase species where a chemically produced mass independent isotopic fractionation would be expected. However, it would not be resolvable as it only possesses two stable isotopes. The same is true for hydrogen. Almost all other elements are coordinated by ligand elements, generally oxygen, and thus do not meet the requirement of being situated as terminal atoms and will not undergo a symmetry dependent isotopic fractionation. Thus, oxygen is unique.

Recent experiments have shown that symmetry dependent recombination reactions may, however, produce massive isotopic fractionations in elements which possess two isotopes, carbon (You and Gellene, 1995), and helium (Gellene, 1993) as well as oxygen (Griffith and Gellene, 1992). In the case of oxygen the observed fractionation in the formation of O_4^+ has been shown to be an equal ^{17}O, ^{18}O enrichment, with a magnitude of greater than 100,000 per mil (Griffith and Gellene, 1993). In the case of carbon, the ^{13}C enrichment as a result of a symmetry dependent isotopic fractionation exceeds 4000 per mil (You and Gellene, 1995). In both cases, the source of the anomalous fractionation is demonstrated to be molecular symmetry. Thus, symmetry dependent isotope effects may produce isotopic enrichments in atoms where there are only two stable isotopes, such as carbon and helium. It is plausible that in interstellar molecular clouds, where ion-molecule reaction readily occur, such isotopic fractionation processes could be important.

2. **There is no requirement for excitation of the reactants or photochemistry**. It has been experimentally demonstrated by electronic state-specific preparation of reactants, that the mass independent isotopic effect occurs in a strictly ground state reaction (Morton *et al*, 1990). This is an important observation, as solar activity or excitation of reactants is not required for a $\delta^{17}O = \delta^{18}O$ fractionation to occur. The fractionation is a consequence of gas phase reactions leading to a product.

ISOTOPIC EFFECTS FROM RECOMBINATION REACTIONS IN CHONDRULE FORMATION

Since chondrule formation events apparently result from short term heating and quenching events, it is highly unlikely that chondrules totally derive from gas-solid reaction. Rather, it is more likely that either some small fraction of the chemistry of the chondrule is processed through gas phase chemistry, or, alternatively, the precursor material to the chondrules was produced in a gas phase reaction. For example, in the case of both the $O + O_2$ and $O + CO$ reactions, the magnitude of the single stage fractionation factor is approximately 100 per mil in ^{17}O and ^{18}O. If the same is true in a reaction such as $O+SiO$, then only a small fraction of a chondrules oxygen isotopic composition would derive from such reactions in the enstatite and ordinary chondrite chondrules. Thus, the original isotopic composition would be somewhere near the terrestrial fractionation line, recombination reactions producing chondrules along a slope 1 line, with the magnitude proportional to the proportion of the chondrule which experiences gas phase chemistry. In the case of carbonaceous chondrite chondrules, it is more likely that they arose from material that was CAI-like in terms of the oxygen isotopes, with secondary exchange with nebular gas producing the observed spread in isotopic composition (Clayton, 1993). As will be discussed, the CAI isotopic composition may have developed from the process of isotopic exchange, which produces a steady state fractionation of 40 per mil in both ^{17}O and ^{18}O.

In nebular chemical processes, ground state gas phase reactions will partition oxygen isotopes into different molecules. This is the case, for example, in the Earth's the present atmosphere where CO_2, O_3, CO, and N_2O simultaneously acquire different mass independent oxygen isotopic compositions. In the nebula, it is possible that chemical interactions, such as formation/destruction and isotopic exchange, resulted in the production of distinct oxygen isotopic gas phase reservoirs (e.g. CO, CO_2, H_2O, SiO). Subsequent condensation and isotopic exchange processes reflect differing interactions with these molecules. In this case, the observed isotopic compositions would reflect nebular gas phase chemical reactions rather than the actual condensation/formation mechanisms associated with CAI or chondrule precursors. As regards recombination reactions, $O + CO$, $OH + CO$, $O + MO$ (where M is a metallic atom) are candidates for production of isotopically anomalous gas phase reservoirs. There are a number of possibly significant reactions which may have been important in the early solar system arising from recombination reaction effects. For example, the reaction $O + CO$ has been experimentally demonstrated to produce a ~100 per mil mass independent isotopic fractionation (Bhattacharya and Thiemens, 1989 a,b). This reaction then subsequently produces two reservoirs (CO and CO_2) of distinct oxygen isotopic composition in the same parcel of nebular gas. To account for the actual potential partitioning in the early solar system, as is the case for the contemporary atmosphere, a simultaneous kinetic solution of all chemical reactions and the relevant fractionation factors must be done to model the evolution of nebular materials.

While there has been considerable progress in determining the mechanism responsible for the generation of chemically produced mass independent isotopic components, one of the limits in modeling the early solar system, or the role in chondrule formation, is the level of understanding of the fundamental chondrule forming process itself. Specifically, the parameters which must be quantitatively known to construct a valid kinetic-isotopic numerical model include:

1. **The nature of the energy source**. Specifically, what process provided the energy for the chondrule producing event. This is particularly important, e.g., if the energy derives from solar

activity (either flares or a T Tauri phase) or nebular lightning, excited state chemistry and reactions must be included in kinetic-isotopic models. While the recombination reaction in ozone is strictly a ground state isotope effect, it is observed in atmospheric reactions that a $\delta^{17}O = \delta^{18}O$ fractionation may also arise from isotopic exchange involving electronically excited states, such as $O(^1D)$. The experimental and atmospheric observations for this type of isotopic fractionation process and the potential role in chondrule forming events will be discussed in a subsequent section.

If chondrule formation arises from a process such as frictional drag heating of infalling grains to the nebula, as suggested by Wood (1983, 1984) involvement by molecular excited states in the isotopic fractionation processes would be unlikely. In this instance, a $\delta^{17}O = \delta^{18}O$ fractionation may arise from recombination processes occurring during frictional heating with subsequent recondensation. In addition, there is a second type of chemically produced isotopic fractionation which is observed following a thermal dissociation which would be important in such heating events (Wen and Thiemens, 1990, 1991). As will be discussed, gas phase thermal dissociation produces a $\delta^{17}O = \delta^{18}O$ fractionation. This process is *not* the same fractionation mechanism observed in a recombination reaction. Thus, in a frictional heating event associated with chondrule formation, a second chemical mechanism exists which may produce the oxygen isotopic composition observed in many chondrules.

2. **The chemical composition of the reaction precursor species and the accompanying gas/dust number densities must be precisely known**. As discussed by Grossman *et al.* (1988), the chemical composition of the precursor material for chondrules is not known, at least to a precision necessary to construct a chemical, time evolutionary kinetic reaction model which incorporates isotopic fractionation processes such as utilized by Wen and Thiemens (1993). As discussed by Grossman *et al.* (1988), there is a wide range of possible precursor material, including infalling interstellar material, nebular condensates or chondrite matrix. The reaction mechanisms associated with evaporation and chondrule formation from any of these three sources would be quite different as the chemical speciation would vary significantly. The total and individual molecular number densities of the evaporated molecules, oxidation state, dust/gas ratio and temperature arising from each type of material could differ significantly.

POTENTIAL RECOMBINATION REACTIONS IN THE NEBULA

With respect to the production of chondrules which plot along a $\delta^{17}O = \delta^{18}O$ composition line, reactions such as:

$$O + SiO \longrightarrow OSiO \tag{5}$$

$$(\text{or } O + Si_2O_2 \longrightarrow Si_2O_3) \tag{6}$$

$$O + FeO \longrightarrow OFeO \tag{7}$$

fulfill the criteria for production of a $\delta^{17}O = \delta^{18}O$ fractionation and could occur during a condensation reaction associated with chondrule formation. Not all oxygen within a chondrule need be processed via reactions such as (5)–(7), only some proportion. The isotopic composition deriving from such reactions may not have been established during the chondrule formation event itself, but rather during the formation of the chondrule pre-cursor material. At present it is not possible to distinguish between the two possibilities. Measurements of the micro-oxygen isotopic distribution within different chondrules, particularly by ion-probe techniques, could be important in addressing the source of the oxygen isotopic composition of chondrules. For example, if there was no internal variation in the oxygen isotopic composition, this suggests that the relevant fractionation process effected the entire bulk chondrule. Alternatively, a gradient may suggest that an exchange process with an isotopically different gas-phase reservoir established the chondrules isotopic composition, especially if the exchange process between the chondrule and gas did not go to completion. Rubin *et al.* (1990) have observed that rims from coarse-grained chondrules in Allende are depleted in ^{16}O with respect to their enclosed chondrules, thus there is evidence for the importance of exchange processes, supporting the notion that measurements of the micro distribution of oxygen isotopes within chondrules will be of importance.

It has recently been shown that in the gas phase production of a SiOx species, which subsequently condenses to form a solid, a mass independent isotopic fractionation occurs (Thiemens *et al.*, 1994). As discussed by Thiemens *et al* (1994) this isotopic fractionation is associated with the formation of the product SiOx species, and not derived from secondary reactions, such as ozone formation. Though the reaction conditions are not nebular, the experiments do demonstrate that solid silicon oxides may be produced which possess mass independent isotopic compositions. Such processes may be associated with production of the oxygen isotopic components present in chondrules and CAI. It is also possible that gas phase reactions involving silicon oxide species initially provide the different gaseous nebular reservoirs associated with secondary chondrule events, such as isotopic exchange. This would be compatible with the suggestion for the production of the observed size dependent isotopic composition of chondrules (Clayton *et al*, 1983; Clayton, *et al*, 1986).

There are other recombination reactions which may have been of importance in the early solar system. For discussion, consider the hypothetical condensation of forsterite or corundum from a gas of solar composition:

$$2Mg_{(g)} + SiO_{(g)} + 3H_2O_{(g)} \longrightarrow Mg_2SiO_{4(s)} + 3H_{2(g)} \tag{8}$$

$$H_2O_{(g)} + 2AlO_{(g)} \longrightarrow Al_2O_{3(s)} + H_{2(g)} \tag{9}$$

The reaction products, forsterite and corundum, should be subject to symmetry dependent isotopic selective chemistry, resulting in an isotopic composition with $\delta^{17}O = \delta^{18}O$, and a fractionation line passing through the bulk oxygen isotopic composition. Since recombination reactions enrich the heavy isotopes in the products, the corundum and forsterite would be equally enriched in ^{17}O and ^{18}O. Reactions (8) and (9) however, are not stoichiometric reactions and do not occur in a concerted three-body reaction. At a more basic level, for example, in reaction (8) the fractionation may be established in the production of the precursor SiO_4 tetrahedron as it too would be subject to a symmetry selection fractionation. Such reactions fulfill the known requirements for production of a chemically produced mass independent isotopic composition. While forsterite and corundum were used for this particular example, most oxygen bearing minerals would have sufficed as it is the position of the oxygen atoms on the product mineral which gives rise to the isotopic fractionation trend. This is a reflection of the propensity of oxygen to serve as a ligand and be situated in a relevant symmetric site. If this process accounts for the observed chondrule isotopic composition, this removes the restrictions on the isotopic composition of the precursor reservoir for production of the chondrules or CAI. It would be the recombination reaction itself which produces the isotopic composition, not mixture of different reservoirs. This is consistent with the observation that the material which apparently gives rise to the chondrules is isotopically represented by the bulk isotopic composition of the meteorites which contain the chondrules. Formation of the chondrules out of this material would then give rise to a $\delta^{17}O = {}^{18}O$ composition line which approximately passes through the bulk ordinary chondrite isotopic composition, as observed (Clayton *et al*, 1991). With multiple chondrule forming events, chondrules both enriched and depleted in the heavy isotopes, with respect to the bulk initial composition could be produced. By material balance, as a chondrule is formed by a symmetry dependent isotopic fractionation, its isotopic composition would be equally enriched in ^{17}O and ^{18}O with respect to the precursor isotopic composition and the residual material depleted. This process is schematically shown in Fig. 3. The first chondrule population would be at point A in Fig. 3 and lie along a slope 1 fractionation line with respect to the original reservoir. If the source material for the chondrules was limited, the residual material would be depleted in heavy isotopes, also along a slope one fractionation line, and may be at point B, depending upon the magnitude of the depletion of the heavy isotopes. Generation of chondrules from a reservoir B, with a symmetry dependent isotopic fractionation would again produce chondrules enriched in the heavy isotopes with respect to the precursor material, and still be depleted with respect to the original isotopic reservoir. Thus, in this fashion, the observed range in fractionation along the $\delta^{17}O = \delta^{18}O$ fractionation line could be produced.

Size separated chondrules from the Dhajala meteorite define an isotopic composition line with a slope of approximately 1 (Clayton *et al.*, 1986). The isotopes of silicon from the same chondrules also exhibit a size dependent isotopic fractionation, however, the compositions are strictly mass fractionated. In terms of a chemical production mechanism for the chondrules, a mass independent fractionation is expected for oxygen but **not** for silicon, as it is coordinated by oxygen. To produce a mass independent isotopic composition by a chemical process, the silicon would have to be

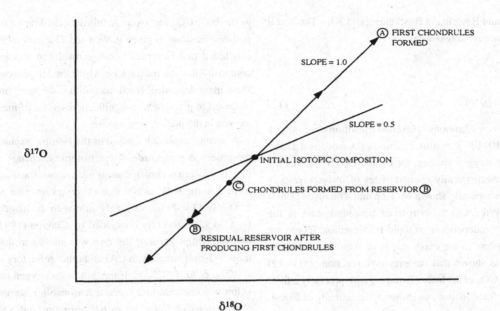

Fig. 3. The isotopic fractionation associated with chondrule formation with an initial bulk oxygen isotopic composition as labeled. The first chondrules produced by a symmetry dependent reaction will be at A. Depending upon extent of formation, the residual bulk composition will move along a slope 1 line towards B. If this material at B forms chondrules, depending upon extent of reaction, they might have an isotopic composition at C. The fractionation does not require all chondrule oxygen to react, only some proportion.

situated where oxygen resides. Thus, these observations are also consistent with a chemical process.

As discussed, the fractionation could have occurred at differing times in the evolution of meteoritic material, specifically during the formation of CAI's, chondrules or their progenitors. It is also possible that symmetry dependent isotopic fractionation processes during recombination reactions gave rise to reservoirs of isotopically distinct gaseous species, such as occurs in the earth's present atmosphere (Thiemens *et al*, 1991,1995). In this situation, the oxygen would be fractionated during gas phase nebular reactions, resulting in production of molecules possessing unique oxygen isotopic compositions, e.g. in CO, H_2O, CO_2 or SiO. Formation of chondrules or CAI from such reservoirs, or exchange between them, would also give rise to isotopically distinct meteoritic materials. In the next section, we will discuss how gas phase exchange processes simultaneously produce several distinct oxygen isotopic reservoirs in the terrestrial atmosphere.

ISOTOPE EXCHANGE PROCESSES AND OXYGEN ISOTOPIC FRACTIONATIONS

Many models which discuss the isotopic composition of chondrules and CAI, including chondrules from enstatite, ordinary and carbonaceous chondritic meteorites, invoke isotopic exchange or mixing between different reservoirs (Clayton *et al*, 1983; Clayton and Mayeda, 1985; Rubin *et al*, 1990); Clayton *et al*, 1991; McSween *et al*, 1985). These models do not address how these different ^{16}O rich and poor reservoirs, be they solid or gas, originated. Recently, experimental and atmospheric observations have shown that a second chemical process could have produced meteoritic

oxygen isotopic components or reservoirs, especially the $\delta^{17}O = \delta^{18}O$ compositions (Wen and Thiemens, 1993; Thiemens *et al*, 1991; Thiemens *et al*, 1995). This effect is not derived from a recombination reaction and is apparently a different type of symmetry dependent mass independent fractionation process. In addition, this process has now been verified as occurring in the atmospheres of Earth, and possibly Mars (Thiemens *et al*, 1991, 1995).

It was first observed by Thiemens *et al.*, (1991) that stratospheric CO_2 possesses a large mass independent isotopic composition. The isotopic enrichment of stratospheric carbon dioxide was observed to be as much as 12 per mil enriched in $\delta^{18}O$ and $\delta^{17}O$, with respect to tropospheric CO_2 (measured to be strictly mass fractionated). More recently, routine measurements from rocket borne whole air samples cryogenically collected at 6 altitudes in the stratosphere and mesosphere (to 60 km) also reveal a large, and variable mass independent isotopic composition in upper atmospheric carbon dioxide (unpublished data). The site for production of the anomalous component is the stratosphere, as the magnitude of the mass independent fractionation linearly correlates with, e.g. ^{14}CO, known to be of stratospheric origin (Thiemens *et al.*, 1995). The mechanism by which the mass independent isotopic fractionation occurs in stratospheric carbon dioxide may be of relevance in nebular chemistry.

It was experimentally demonstrated by Wen and Thiemens (1993), that the observed mass independent stratospheric isotopic composition derives from the isotopic exchange process:

$$O(^1D) + CO_2 \longrightarrow CO_2 + O(^3P) \tag{10}$$

This is a well known process, having first been studied by Katakis and Taube (1962) and Baulch and Breckenridge (1966). The actual exchange process occurs in two-steps:

$$O(^1D) + CO_2 \Leftrightarrow CO_3^* \tag{11}$$

and

$$CO_3^* \longrightarrow CO_2 + O(^3P) \tag{12}$$

The species CO_3^* is a vibrationally excited transition state, with a lifetime of some 10–100 vibrational periods (Demore and Dede, 1970). The atomic oxygen species, $O(^3P)$ and $O(^1D)$ refer to the ground state and electronically excited states of oxygen, respectively. It was experimentally shown by Wen and Thiemens (1993) that production of the CO_3^*, short-lived transition state is the source of the mass independent isotopic fractionation. There are two significant aspects to the experiments of Wen and Thiemens (1993). First, it was shown that the exchange reactions (11)–(12) enrich the product CO_2 in the heavy isotopes, with precisely a $\delta^{17}O = {}^{18}O$ (or slope 1) fractionation, as shown schematically in Fig. 4. Secondly, the exchange is independent of the oxygen atom isotopic composition, that is, a $\delta^{17}O = \delta^{18}O$ fractionation occurs **independent** of the oxygen atom isotopic composition, as schematically shown in Fig. 5. Oxygen atoms ranging in their initial isotopic composition by more than 70 per mil were utilized for the experiments, including oxygen atoms of both mass dependent and independent composition. When these oxygen atoms isotopically exchange with CO_2 via reactions (10)–(11), a steady state value is ultimately obtained, that is, further recycling and exchange does not alter the isotopic composition of the product oxygen, as shown in Fig. 5. It is observed that the final isotopic composition of the oxygen, following exchange, is for the oxygen atom to be equally depleted in ^{18}O and ^{17}O, or a slope one fractionation, with respect to the bulk CO_2 reservoir. A full detailed kinetic treatment of all possible reactions is given in Wen and Thiemens (1993). Thus, it is concluded that isotopic exchange results in a slope one isotopic composition, even for the case where the initial isotopic reservoirs show mass dependent fractionation. It is the same process which is observed to give rise to a significant reservoir of mass independent oxygen in the present atmosphere.

A second important feature of the isotopic exchange reaction is the observed **magnitude** of the isotopic exchange. It is observed that the oxygen is depleted along a slope one fractionation line with respect to the bulk carbon dioxide reservoir, with a magnitude of $\delta^{17}O = \delta^{18}O = 40‰$, strikingly similar to the magnitude observed for CAI's. As recently concluded by Clayton (1993) " The mean isotopic composition of the dust was similar to that now seen in most ^{16}O-rich minerals in CAI and other refractory minerals ($\delta^{18}O = -40‰$. $\delta^{17}O = -42‰$)". If the bulk solar oxygen isotopic composition is somewhat like earth, a reasonable assumption given the observations of lunar, terrestrial, enstatite and SNC meteorites, then **BOTH** the magnitude and the slope of the fractionation of the chemical processes is similar to that observed in CAI. Isotopic compositions greater than 40 per mil may be achieved by secondary, or continual material processing. For example, if there is secondary chemistry of an oxygen reservoir which has also moved in the ^{16}O rich direction along the slope 1 line, a ^{16}O enrichment beyond –40 per mil will be achieved when the 40 per mil fractionation factor is applied to that material.

In the case of chondrule or CAI formation, if, during the high temperature chondrule formation process, a gas phase isotopic exchange occurs, such as:

$$O^* + OSiO = O + OSiO^* \tag{13}$$

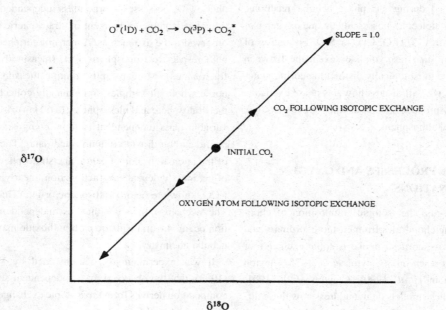

Fig. 4. The isotopic fractionation associated with isotopic exchange between atomic oxygen and carbon dioxide. The carbon dioxide becomes enriched in heavy isotopes along a slope 1 line while the atomic oxygen becomes depleted, also along a slope 1 line.

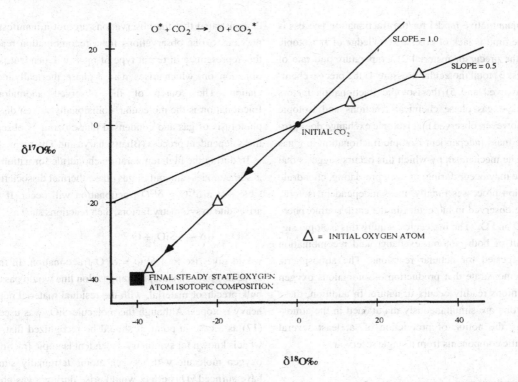

Fig. 5. The exchange between an oxygen atom of different initial isotopic composition with a large reservoir of carbon dioxide. The final isotopic composition of the exchanged oxygen atom is always depleted in ^{17}O and ^{18}O by 40 per mil at steady state, with respect to the carbon dioxide it has undergone exchange with.

$$O^* + SiO = O + SiO^* \quad (14)$$

or

$$O^* + CO = O + CO^* \quad (15)$$

a $\delta^{17}O = \delta^{18}O$ fractionation would be expected, as observed for the carbon dioxide experiments and stratospheric carbon dioxide. As discussed by Wen and Thiemens (1993), due to the structure of the CO_3^* transition state, the fractionation may be a manifestation of a symmetry dependent isotopic fractionation. In the case of SiO_2, there would be a short lived SiO_3^* transition state, thus, this mechanism could be a second process by which chondrules, their parental material (or CAI), or, the gaseous nebular reservoirs acquired their oxygen isotopic components. The final products in the nebular case, e.g. SiO, SiO_2 or CO would be the species responsible for the production of the initial isotopic reservoirs. Minerals formed subsequently by conventional geochemical reactions then inherit this composition. Exchange of condensed material with the gaseous reservoirs would also result in mass independent compositions being produced in meteoritic material. It is a requirement that the exchange which produces the isotopically anomalous reservoirs (molecular species) occur in the gas phase as other isotopic exchanges are mass dependent (Bains-Sahota and Thiemens, 1987). This accounts for the observation of strictly mass dependent isotopic fractionations in experimental studies of gas-solid, solid-solid, liquid-solid and liquid-liquid exchange processes. A mass independent isotopic fractionation is not expected in an evaporation/condensation processes as well.

For chondrules and CAI, the magnitude of the isotopic composition along a slope one line may, in part, be a measure of degree of isotopic exchange. As demonstrated by Wen and Thiemens (1993), when atomic oxygen and carbon dioxide exchange with each other, the product oxygen and carbon dioxide move along a slope one line (in opposite directions) until they achieve steady state (40 per mil fractionation). If steady state is not achieved, then the fractionation is less and varies somewhere between no fractionation and 40 per mil. Thus, the observed spread in the chondrule and CAI data may be a reflection of the degree of exchange. In an exchange process, for example with SiO, the product SiO (or relevant oxygen bearing species such as CO) would become enriched in the heavy isotopes and the atomic oxygen depleted. Subsequent reaction of the exchanged atom could then simultaneously form a chondrule depleted in the heavy isotopes with respect to the initial reservoir and the SiO, enriched in the heavy isotopes. Thus, heavy and light (with respect to bulk) chondrules could be formed simultaneously.

In the case of CAI, if there is exchange between an oxygen atom and, for example, nebular CO or CO_2 of isotopic composition essentially terrestrial, the exchanged atom would have an isotopic composition at –40 per mil for both $\delta^{17}O$ and $\delta^{18}O$. Subsequent reaction to form any mineral would then trap the isotopic composition of the exchanged atom in a stable phase. This would not require any fractionation during the formation of the first minerals as the precursor oxygen has already attained the anomalous isotopic composition by gas phase exchange.

At present, a quantitative model for the fractionation process is not realistic. The limit is lack of precise knowledge of 1) isotopic composition of the precursor material 2) temperature and rate of the overall process 3) total molecular pressure 4) the precise chemical reactions involved and 5) the isotopic fractionation factors associated with the gas phase chemical reactions and isotopic exchange. It is, however, observed that isotopic exchange does produce a slope one mass independent isotopic fractionation in a gas phase reaction. The mechanism by which this occurs suggests that such an exchange may occur during, or even pre-dating, chondrule and CAI formation processes Finally, mass independent isotopic compositions are observed in molecules in the earth's atmosphere in CO_2, CO, N_2O and O_3. The molecules acquire this isotopic signature as a result of both isotopic exchange and recombination reactions, as suggested for nebular reactions. The atmospheric measurements demonstrate that production of anomalous oxygen isotopic compositions readily occurs in nature. In addition, these molecular reservoirs are simultaneously maintained in the atmosphere, supporting the notion of production of, at least several observed meteoritic components from a single reservoir.

THERMAL DISSOCIATION ISOTOPIC EFFECTS IN CHONDRULE FORMATION

The process of chondrule formation undeniably involved extraordinarily high temperature events with relatively short cooling periods. Therefore, isotopic fractionations associated with high temperature reactions must also be considered, along with the recombination and isotopic exchange processes. Of particular importance may be gas phase, thermal molecular dissociation effects. It has been experimentally demonstrated in the case of thermal ozone decomposition, that an anomalous isotopic fractionation occurs (Wen and Thiemens, 1990, 1991). In the ozone thermolysis reaction, the product oxygen of the reaction:

$$2O_3 + \Delta \longrightarrow 3O_2 \tag{16}$$

is isotopically **enriched** in the heavy isotopes, with a fractionation of $\delta^{17}O = \delta^{18}O$, with respect to the initial ozone isotopic composition. The term Δ represents heat, in this case a temperature of less than 90 °C. The measured isotopic fractionation is the reverse in sign with respect to that expected for a conventional isotopic fractionation. The lighter isotopes are expected to preferentially dissociate due to their relatively weaker bond strength (Kaye, 1986). In addition, since bond strength differences arise from differing vibrational frequencies, the effect should be mass dependent. It has, for example, been shown that in the photo decomposition of ozone, the process is mass dependent, with a preferential dissociation of the light isotopes (Bhattacharya and Thiemens,1988). In the experiments reported by Wen and Thiemens (1990, 1991), it was demonstrated that the anomalous $\delta^{17}O = \delta^{18}O$ fractionation arises due to the gas phase thermal decomposition process. Secondary reactions, wall effects and excited state chemistry are ruled out, based upon the experimental observations. In addition, from kinetic considerations and numerical modeling of the experimental results, it was

demonstrated that the observations are not a manifestation of the previous ozone observations for a recombination reaction. Thus, this represents a different type of mass independent isotopic fractionation, one which arises in a gas phase, thermally induced dissociation. The source of the observed anomalous isotopic fractionation is the molecular, collisionally driven dissociation, as photolysis of gas and condensed phase ozone is observed to be a mass dependent process (Bhattacharya and Thiemens, 1988).

If, during the high temperature chondrule formation event, there is some modest amount of gas phase thermal dissociation, it is possible that a $\delta^{17}O = \delta^{18}O$ fractionation will occur. If that process arises due to symmetry factors, then reactions such as:

$$SiO_{4(g)} + \Delta \longrightarrow SiO_3 + O \tag{17}$$

would give rise to a $\delta^{17}O = \delta^{18}O$ fractionation. In this case, the slope one oxygen isotopic composition line would pass through the bulk precursor material, with the residual material depleted in the heavy isotopes. Although the molecule SiO_4 was used in reaction (17) as a case in point, it should be recognized that, based upon what is known for symmetry dependent isotopic fractionations, any oxygen molecule with oxygen atoms terminally situated would have sufficed. This effect would arise during a gas phase reaction and not from evaporation, a process found to be mass dependent for oxygen (Davis *et al.*, 1990) silicon (Molini-Velsko *et al.*, 1987) and magnesium (Uyeda *et al.*, 1991). Thus, based upon experimental observations, there is a third type of reaction which may give rise to a $\delta^{17}O = \delta^{18}O$ isotopic fractionation. This process is simple in the sense that it is purely gas phase and thermally driven. Such a process could plausibly occur during chondrule formation, a process requiring intense heat, which may have promoted some thermal decomposition. Unless 100% of the evaporated and thermally dissociated, material is re-condensed onto the chondrule, a slope one composition in the product chondrules would arise. This does not require that all oxygen in chondrules underwent dissociation via e.g. (17), which is highly unlikely. Recondensation, for example, of some proportion of the product SiO_3 onto the precursor silicate solid would make that material isotopically heavier with a $\delta^{17}O = \delta^{18}O$ fractionation line defined. Reaction of the precursor material with proportionately more of the O atom than SiO_3, would result in a chondrule depleted in the heavy isotopes with respect to bulk, initial. Thus, heavy and light chondrules would simultaneously be produced with magnitude and sign dependent upon relative rates of reaction and degree of dissociation.

It has recently been suggested that the mass independent isotopic composition observed in experimentally produced SiOx condensates may arise from some gas phase thermal decomposition of the precursor species (Thiemens *et al.*, 1994). This may indicate that thermal dissociative effects in SiO_x gas phase reactions are important. These reactions are complex, however, and the studies are in their infancy. Experiments are needed to further understand the potential role of thermal dissociation, especially for reactions with solid condensate products. The obvious difficulty in such experiments is providing the relevant experimental conditions. Such experiments require not only high temperatures, but also relatively low molecular number densities. If the gas phase number densities become too high, the reaction mechanism and order is altered and avalanche nucleation occurs. In order to kinetically resolve isotopic fractionation factors associated with the individual reactions, there must be suitable control of the reaction environment to permit controlled and characterizable chemical reactions.

SUMMARY AND CONCLUSIONS

There are at least three different types of mass independent isotopic fractionation processes that produce a slope one oxygen isotopic composition. These are: recombination reactions (observed in several different molecules and two different elements), isotopic exchange, and thermal dissociation. In addition, symmetry dependent isotopic fractionation processes are observed to occur in nature. Stratospheric ozone and carbon dioxide possess large mass independent isotopic fractionations. Atmospheric nitrous oxide (Cliff and Thiemens, 1994) and carbon monoxide (Hodder, Brenninkmeijer and Thiemens, 1994) also possess significant mass independent oxygen isotopic compositions. These observed fractionations arise due to different reactions, all presumably involving symmetry dependent gas phase reactions. Thus, the laboratory observations first observed by Thiemens and Heidenreich (1983) are not simply laboratory phenomena, they occur in nature and are observed in several of the most important molecules in the earth's atmosphere.

Based upon experimental observations , it has been suggested that there are three types of chemical isotopic fractionation processes which may have produced the oxygen isotopic compositions observed in some chondrules and CAI. If this is the case, only one reservoir is required to produce chondrule isotopic compositions. The observed isotopic components are then a result of the chemical reactions associated with chondrule and CAI formation processes. This obviates the need for interaction of multiple isotopic reservoirs, which have the requirement of possessing the precise isotopic compositions to produce a slope one composition, rather than essentially an infinite number of other possible isotopic compositions.

It is also possible that gas phase nebular reactions gave rise to isotopically distinct molecular reservoirs, as occurs in the Earth's atmosphere. Subsequent formation of CAI or chondrule parental material would then reflect the isotopic composition of this source oxygen. In this scenario, the observed meteoritic oxygen isotopic anomalies were produced in the nebula by chemical reactions and partitioning of oxygen isotopes among the molecular species, presumably CO, water, metallic oxides and perhaps CO_2. Subsequent condensation of CAI or chondrule material would produce their observed isotopic compositions. Secondary processes, such as isotopic exchange with such reservoirs could also account for the observed systematics of size separated chondrules.

The observed isotopic composition of meteoritic material thus may provide insight into the chemistry of the early solar system. As resolution of the parameters associated with chondrule and CAI formation advances, such as precise definition of the molecular

composition, temperature, molecular number density and dust/gas ratio, it will be possible to further develop the framework of this hypothesis. Further laboratory experiments, particularly studies of isotopic fractionations associated with gas phase high temperature reactions of geochemical material will be of fundamental importance, both for understanding chondrule or CAI formation processes, and nebular processes in general.

ACKNOWLEDGEMENTS

Jeff Johnston is thanked for his help in preparing illustrations. Andy Davis, Harmon Craig and Ed Scott are thanked for comments, which greatly helped in developing the paper. NASA (grants NAGW 3551 and NAGW 2251).

REFERENCES

Anderson S.M., Mauersberger K., Morton J. and Schueler B. (1992) Heavy ozone anomaly: evidence for a mysterious mechanism. In *Isotope Effects in Gas-Phase Chemistry* (ed. J.A. Kaye) pp. 155–166. American Chemical Society.

Bains-Sahota S. and Thiemens M.H. (1987) Mass independent isotopic fractionation in a microwave plasma. *J. Phys. Chem.* **91**, 4370–4374.

Bains-Sahota S. and Thiemens M.H. (1989) A mass-independent sulfur isotope effect in the non-thermal formation of S_2F_{10}. *J. Chem. Phys.* **90**, 6099–6109.

Baulch D.L. and Breckenridge W.H. (1966) Isotopic exchange of O(^{1}D) with carbon dioxide. *Trans. Faraday Soc.* **62**, 2768–2773.

Bhattacharya S.K. and Thiemens M.H. (1988) Isotopic fractionation in ozone decomposition. *Geophys. Res. Lett.* **15**, 9–12.

Bhattacharya S.K. and Thiemens M.H. (1989a) New evidence for symmetry dependent isotope effect: O + CO reaction. *Z. Naturforsch.* **44a**, 435–444.

Bhattacharya S.K. and Thiemens M.H.(1989b) Effect of isotopic exchange upon symmetry dependent fractionation in the O+CO —> CO_2 reaction. *Z. Naturforsch.* **44a**, 811–813.

Clayton R.N., Grossman L. and Mayeda T.K. (1973) A component of primitive nuclear composition in carbonaceous meteorites. *Science* **182**, 485–488.

Clayton R.N., Onuma N., Ikeda Y., Mayeda T.K., Hutcheon I.D., Olsen E.J. and Molini-Velsko C. (1983) Oxygen isotopic compositions of chondrules in Allende and ordinary chondrites. In *Chondrules and Their Origins* (ed. E.A. King), pp. 37–43. Houston: Lunar Planet. Inst.

Clayton R.N. and Mayeda T.K. (1985) Oxygen isotopes in chondrules from enstatite chondrites: possible identification of a major nebular reservoir. *Lunar Planet. Sci.* **XVI**, 142–3.

Clayton R.N., Mayeda T.K. Goswami J.N. and Olsen E.J. (1986) Oxygen and silicon isotopes in chondrules from ordinary chondrites. *Terra Cognita* **6**, 130.

Clayton R.N., Mayeda T.K., Goswami J., and Olsen E. J. (1986) Oxygen isotope studies of ordinary chondrites. *Geochim. Cosmochim. Acta* **55**, 2317–2137.

Clayton R.N. (1993) Oxygen isotopes in meteorites. *Ann. Rev. Earth Planet. Sci.* **21**, 114–49.

Cliff S.S. and Thiemens M.H. (1994) Isotopic analysis of both $\delta^{18}O$ and $\delta^{17}O$ in atmospheric nitrous oxide. New insights from an observed mass independent anomaly. *Eighth International Conference on Geochronology, Cosmochronology and Isotope Geology*, 63.

Davis A.M., Hashimoto A., Clayton R.N. and Mayeda T.K. (1990) Isotope fractionation during evaporation of Mg_2SiO_4. *Nature* **347**, 655–658.

DeMore W.B., Horibe Y. and Sowers T. (1970) Pressure dependence of carbon trioxide formation in the gas phase reaction O(^{1}D) with carbon dioxide. *J. Phys. Chem.* **74**, 2621–2625.

Gellene G.I. (1993) Symmetry-dependent isotopic fractionation in the formation of He_2^+. *J. Phys. Chem.* **97**, 34–39.

Gooding J.L., Mayeda T.K., Clayton R.N. and Fukuoka T (1983) Oxygen isotopic heterogeneities, their petrological correlations, and implications for melt origins of chondrules in unequilibrated ordinary chondrites. *Earth Planet. Sci. Lett.* **65**, 209–24.

Griffith K.S. and Gellene, G.I. (1992) Symmetry restrictions in diatom/diatom reactions. II. Nonmass-dependent isotope effects in the formation of O_4^+. *J. Chem. Phys.* **96**, 4403–11.

Grossman J.N., Rubin A.E., Nagahara H. and King E.A. (1988) Properties of Chondrules. In *Meteorites and the Early Solar System*. (ed. J.F. Kerridge and M.S. Matthews), pp. 619–659. University of Arizona Press.

Heidenreich III J.E. and Thiemens M.H. (1986) A non-mass dependent isotope effect in the production of ozone from molecular oxygen: the role of molecular symmetry in isotope chemistry. *J. Chem. Phys.* **84**, 2129–36.

Hodder P.S., Brenninkmeijer C.A.M. and Thiemens M.H. (1994) Mass independent fractionation in tropospheric carbon monoxide. *Eighth International Conference On Geochronology, Cosmochronology and Isotope Geology*, p.138

Hulston J.R. and Thode H.G. (1965) Variations in the S^{33}, S^{34} and S^{36} contents of meteorites and their relation to chemical and nuclear effects. *J. Geophys. Res.* **70**, 3475–84.

Huss G.R., Fahey A.J., Gallino R and Wasserburg G.J. (1994) Oxygen isotopes in circumstellar Al_2O_3 grains from meteorites and stellar nucleosynthesis. *Astrophys. J.* **430**, L81–L84.

Hutcheon I.D. Huss G.R. Fahey A.J. and Wasserburg G.J. (1994) Extreme ^{26}Mg and ^{17}O enrichments in an Orgueil corundum: identification of a presolar oxide grain. *Astrophys. J.* **425**, L97–L100.

Katakis D. and Taube H. (1962) Some photochemical reactions of O_3 in the gas phase. *J. Chem. Phys.* **36**, 416–422.

Kaye J.A. (1987) Mechanisms and observations for isotope fractionation of molecular species in planetary atmospheres. *Rev. Geophys.* **25**, 1609–58.

Larsen N.W., Pedersen T. and Sehested J. (1992) Isotopic study of the mechanism of ozone formation. In *Isotope Effects In Gas-Phase Chemistry* (J.A. Kaye, Ed.), American Chemical Society.

Mauersberger K., Morton J., Schueler B.,Stehr J. and Anderson S.M. (1993) Multi-isotope study of ozone: implications for the heavy ozone anomaly. *Geophys. Res. Lett.* **20**, 1031–34.

Mayeda T.K., Clayton R.N. and Sodonis A. (1989) Internal oxygen isotope variations in two unequilibrated chondrites. *Meteoritics* **24**, 301.

Morton J., Schueler B., and Mauersberger K (1989) Oxygen fractionation of ozone isotopes $^{48}O_3$ through $^{54}O_3$, *Chem. Phys. Lett.* **154**, 143–45.

Morton J., Barnes B., Schueler B. and Mauersberger K. (1990) Laboratory studies of heavy ozone. *J. Geophys. Res.* **95**, 901–8.

Molini-Velsko C., Mayeda T.K. and Clayton R.N. (1987) Silicon isotope systematics during distillation. *Lunar and Planetary Science* **XVIII**, 657–8.

Nittler L.R., O'D Alexander C.M. Gao X., Walker R.M. and Zinner E.K. (1994) Interstellar oxide grains from the Tieschitz ordinary chondrite. *Nature* **370**, 443–6.

Rubin A.E., Wasson J.T., Clayton R.N., and Mayeda T.K. (1990) Oxygen isotopes in chondrules and coarse-grained chondrule rims from the Allende meteorite. *Earth Planet. Sci. Lett.* **96**, 247–55.

Thiemens M.H. and Heidenreich III J.E. (1983) The mass-independent fractionation of oxygen: a novel isotope effect and it's possible cosmochemical implications. *Science* **219**, 1073–75.

Thiemens M.H. and Jackson T. (1987) Production of isotopically heavy ozone by ultra-violet light photolysis of ozone. *Geophys. Res. Lett.* **14**, 624–7.

Thiemens M.H. and Jackson T.(1988) New experimental evidence for the mechanism for production of isotopically heavy O_3 *Geophys. Res. Lett.* **18**, 639–672.

Thiemens M.H., Jackson T., Mauersberger K., Schueler B., and Morton J. (1991) Oxygen isotope fractionation in stratospheric CO_2. *Geophys. Res. Lett.* **18**, 669–672.

Thiemens M.H. (1992) Mass-independent isotopic fractionations and their applications. In *Isotope Effects in Gas-phase Chemistry* (ed. J.A. Kaye), pp. 138–154. American Chemical Society.

Thiemens M.H., Nelson R., Dong Q.W. and Nuth III J.A. (1994) First observation of a mass independent isotopic fractionation in a condensation reaction (1994). *Lunar. Planet. Sci.* **XXV**, 1389–90.

Thiemens M.H., Jackson T.L. and Brenninkmeijer C.A.M. (1995) Observation of a mass independent oxygen isotopic composition in stratospheric CO_2, the link to ozone chemistry, and the possible occurrence in the Martian atmosphere. *Geophys. Res. Lett.* **22**, 255–7.

Uyeda C., Tsuchiyama A. and Okano T. (1991) Magnesium isotopic fractionation of silicates produced in condensation experiments. *Earth Planet. Sci. Lett.* **107**, 138–47.

Wen J. and Thiemens M.H. (1990) An apparent new isotope effect in a molecular decomposition and its implications for nature. *Chem. Phys. Lett.* **172**, 416–20.

Wen J. and Thiemens M.H. (1991) Experimental and theoretical study of kinetic isotopic effect on ozone decomposition. *J. Geophys. Res.* **96**, 10911–22.

Wen J. and Thiemens M.H.(1993) Multi-isotope study of the $O(^1D)$ + CO_2 exchange and stratospheric consequences. *J. Geophys. Res.* **98**, 12,801–12,808.

Wood J.A. (1983) Formation of chondrules and CAI's from interstellar grains accreting to the solar nebula. *Mem. Natl.Inst. Polar. Res. Special Issue* **30**, 84–92.

Wood J.A. (1984). On the formation of meteoritic chondrules by aerodynamic drag heating in the solar nebula. *Earth Planet. Sci. Lett.* **70**, 11–26.

You R.K. and G.I. Gellene (1995). Symmetry induced kinetic isotope effects in the formation of $(CO_2)^{2+}$. *J. Chem. Phys.* **102**, 3227–37.

13: Agglomeratic Chondrules, Chondrule Precursors, and Incomplete Melting

MICHAEL K. WEISBERG and *MARTIN PRINZ*

Department of Earth and Planetary Sciences, American Museum of Natural History, New York, NY 10024, U.S.A.

ABSTRACT

Chondrules form as a result of heating processes that range from sintering, to varying degrees of melting, of agglomerations of precursor solids. The abundances of aggregational objects, and incompletely melted chondrules in chondrites may be much greater than previously recognized. Agglomeratic olivine chondrules are dominated by fine-grained olivine, but also contain pyroxene crystals, chondrule fragments and refractory inclusions. These chondrules were heated to lesser degrees than most other chondrules, and provide information about the materials present in the nebula prior to and during chondrule formation. They have refractory element (Si/Al, Mg/Al) ratios similar to the common porphyritic chondrules and are probably porphyritic chondrule precursor materials that were heated to temperatures (~1200°C) high enough to result in sintering and only minor degrees of melting. Their relatively narrow compositional range suggests that most chondrules formed from a fairly homogeneous mix of materials that were greatly dominated by olivine, but many were later modified by open system behavior (evaporation of volatile elements and reduction of Fe) during chondrule formation. Based on the characteristics of agglomeratic chondrules, the grain sizes of the precursor materials may have been in the 2–5 μm range or smaller, but larger olivine and/or pyroxene grains (up to 400 μm) were also present. Microchondrules and refractory inclusions were recycled into the chondrule precursor mix as minor components. Layered chondrules show textural, mineralogical, compositional and oxygen isotopic differences between core and rim layers that suggest a spatial or temporal progressive accretionary growth process of nebular materials. Some porphyritic chondrules may be incompletely melted aggregates of crystals and dust, rather than crystals grown from a complete melt.

INTRODUCTION

There is considerable evidence that chondrules formed by the melting and incomplete melting of solid precursor materials (Gooding *et al.*, 1980; Grossman and Wasson, 1982; Nagahara, 1983a) and, by default, the early solar nebula is the preferred location for their formation (Taylor *et al.*, 1983). The degree of melting that any single chondrule has undergone is, however, not always clear. Some, such as the cryptocrystalline, radial pyroxene and barred olivine chondrules, appear to have formed through the crystallization of completely molten droplets, as suggested by their textures (Nagahara, 1983b) and by experimental simulations (Radomsky, 1988). Others, such as agglomeratic olivine chondrules, are consid-

ered to be aggregates of material that experienced very low degrees of melting (Weisberg and Prinz, 1994). In the case of the more common "porphyritic" chondrules, some have argued that the majority (both FeO-poor and FeO-rich) formed through complete melting and that the olivine and pyroxene crystals in them crystallized *in situ* from the melt (Jones, 1990; Jones and Scott, 1989). However, identification of relict crystals in chondrules that did not crystallize *in situ* from the chondrule melt (e.g., Rambaldi, 1981; Nagahara, 1981; Steele, 1986a,b; Jones, this volume), as well as less easily identified relict crystals, have led to the conclusion that most chondrules form through incomplete melting (Nagahara, 1983a).

The matrix in type 3 unequilibrated chondrites is an aggregate of fine-grained minerals that is compositionally and texturally distinct from chondrules, and may include the products of condensation and solid-solid or solid-gas reactions in the nebula, pre-solar grains, and finely ground fragments of chondrules and inclusions (Huss *et al.*, 1981; Scott *et al.*, 1984; 1988; Nagahara, 1984; Brearley, 1989; Alexander *et al.*, 1989; Brearley *et al.* 1989). Most mineral grains in the fine-grained portions of dark inclusions are similar to those in the matrix, suggesting that dark inclusions are consolidated lithic clasts of matrix material and, thus, share a similar origin. Fluffy type A refractory inclusions consist of aggregates of tiny nodules, rich in the mineral melilite, most of which do not appear to have experienced melting during their formational histories (MacPherson and Grossman, 1984). Amoeboid olivine aggregates, as their name implies, are irregular-shaped olivine-rich aggregates that contain feldspathoids, as well as nodules consisting of the refractory element-rich minerals spinel, perovskite and/or melilite. Spinel-pyroxene-rich aggregates are unmelted aggregations of nodules consisting of spinel rimmed by pyroxene minerals. Agglomeratic olivine chondrules in ordinary chondrites, the main focus of this paper, are olivine-rich chondrules which are mechanical mixtures of mineral grains and fragmental components that did not experience high degrees of melting. All of the aggregational objects discussed here contain primitive mineral components that appear to have originated in the nebula and may therefore be the best source of information on the materials available in the nebula during chondrule formation.

We suggest that agglomeratic olivine chondrules furnish the most compelling, direct evidence for the formation of chondrules from the agglomeration and incomplete melting of solids. Agglomeratic chondrules have been defined by Van Schmus (1969) as "a type of chondrule that has internal textures that suggest that they are mechanical mixtures of individual small crystal fragments of silicates, oxides, sulfides, and metal rather than a crystalline assemblage that is the result of crystallization from a melt." They appear to be chondrules that were heated to temperatures too low (<1200°C) to initiate much melting of the olivine and pyroxene, and as a result they are bound together by a minor feldspathic mesostasis and do not have the igneous textures commonly associated with chondrules (Weisberg and Prinz, 1994). They therefore provide insights into the solids that were available in the nebula during chondrule formation. Since all models for chondrule formation must consider the nature of the chondrule precursors, it is important to understand the physical, mineralogical and chemical characteristics of the materials that were melted or incompletely melted to form chondrules.

Here, we discuss the characteristics of agglomeratic chondrules and discuss their implications for understanding chondrule precursors and chondrule evolution. Additionally, we review some of the other aggregational components in chondrites in order to emphasize their ubiquity in many chondrites. Finally, we evaluate the evidence for chondrule formation by complete vs. incomplete melting and try to assess which of these processes was dominant.

AGGREGATIONAL COMPONENTS IN CHONDRITES

Chondrites are breccias of several components, many of which are clearly aggregates of unmelted, pre-existing crystals and fragments. These components include matrix, dark inclusions (with matrix-like characteristics), some refractory inclusions (including amoeboid olivine aggregates, fluffy type A inclusions and spinel-pyroxene aggregates), and agglomeratic chondrules. Some of the chondrules that appear to have igneous textures, such as the porphyritic chondrules, may also be aggregates of olivine and/or pyroxene crystals that are cemented together by an amorphous to microcrystalline feldspathic mesostasis or "glue".

The abundance of aggregational components in carbonaceous chondrites is greater than in ordinary or enstatite chondrites, as compared to completely or incompletely melted objects. Many of the chondrules in the ordinary chondrites appear to have crystallized from molten or incompletely molten droplets, with spherical to near-spherical shapes, glass-rich mesostases, and textures that appear to be igneous (Fig. 1a).

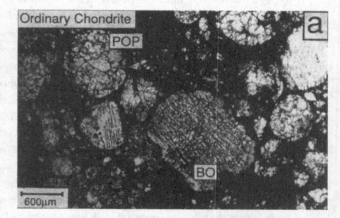

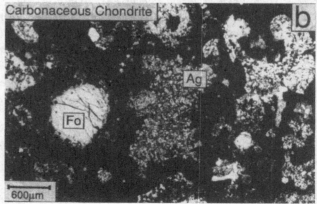

Fig. 1. Photomicrographs of: (a) The Krymka LL3.0 chondrite showing a barred olivine chondrule (BO) and porphyritic olivine-pyroxene chondrule (POP). Most chondrules are spherical and contain a feldspathic glassy mesostasis, suggesting they crystallized from melts; however, the texture of the POP chondrule is ambiguous in that some or all of the crystals may be from a precursor aggregate that did not crystallize in situ from this melt. (b) The Ningqiang CK3-anomalous chondrite showing that many of its chondrules are irregular-shaped olivine-rich aggregates (Ag) in contrast with the ordinary chondrite (a). The aggregate (Ag) consists of angular olivine fragments and olivine clusters and interstitial melt is essentially absent. Another chondrule consists of a single near-spherical forsterite crystal (Fo) with an accretionary dusty rim.

However, in carbonaceous chondrites many of the chondrules appear to be aggregational objects, with highly irregular shapes, non-igneous textures, and consist mainly of mineral fragments (Fig. 1b). The Ningqiang meteorite is an example of a carbonaceous chondrite that contains a relatively large number (>13vol.%, more than half of its chondrules and inclusions) of unmelted aggregational objects formed by the cementation of mineral grains in the nebula (Rubin *et al.*, 1988; Fig. 1b). Some Ningqiang chondrules consist of single near-spherical crystals of olivine surrounded by an accretionary rim (Fig. 1b). Such chondrules contain no evidence of an igneous origin and are aggregates of olivine and dust. Cohen *et al.* (1983) have also shown that CV carbonaceous chondrites (e.g., Mokoia) have a high abundance of aggregational chondrules which consist of aggregates of olivine cemented together, as well as chondrules formed by the recrystallization of fine-grained dust aggregates due to sintering, all in the nebula. They determined that aggregational chondrules make up ~20 vol.% of Mokoia, nearly half of its chondrules and inclusions.

There are a number of other types of chondrules of aggregational origin. Fig. 2a shows a chondrule from Semarkona that is an aggregate of Fe-rich crystals incorporated within a chondrule melt, which survived melting, and on which newly formed FeO-poor olivine crystallized. The Fe-rich relict crystals are easily identifiable because they contain numerous blebs of Fe-metal which formed as a result of solid-state reduction of Fe, in response to incorporation into a more reducing environment. The chondrule shown in Fig. 2b is a fine-grained olivine-rich aggregate that contains two larger olivine crystals within its accretionary rim. The larger crystals are more MgO-rich than the finer olivine grains in the core, and also have higher Al_2O_3 contents (up to 0.3 wt. %). The grain size and compositional differences between the core and rim olivine grains indicate that they did not crystallize from the same melt. High Al in the rim olivine is inconsistent with the distribution coefficients for Al between olivine and melt, and suggests that this olivine did not crystallize from a melt. It has been shown that several percent of the olivine in most chondrites has relatively high Al contents and probably did not form through crystallization from a melt, and instead may have condensed from a nebular vapor (Steele, 1986a, b), although the possibility of Al-bearing olivine forming from a melt has also been discussed (Kring, 1988; Alexander *et al.*, 1994). In conclusion, a large number of the components in both ordinary and carbonaceous chondrites, including many chondrules, appear to be aggregates of materials that were never completely molten.

AGGLOMERATIC OLIVINE CHONDRULES

Agglomeratic chondrules in ordinary chondrites were first described by Van Schmus (1969), and similar objects have been referred to as "dark zoned" chondrules (Dodd, 1971), and "fine-grained lumps" (Rubin, 1984). Weisberg and Prinz (1994) used the term "agglomeratic olivine" chondrules because this expresses their aggregational nature, as well as their high modal abundance of olivine. The following is based on obervations of 27 agglomeratic olivine chondrules in four unequilibrated ordinary chondrites (Krymka, Semarkona, Bishunpur, Chainpur). They are fine-grained (>50 vol.% of grains are <5 μm), olivine-rich (>50 vol.% of grains are olivine) objects that are similar in size to coexisting chondrules in the same chondrite and, on average, make up 2 vol.% of unequilibrated ordinary chondrites. Their shapes range from irregular to round and they have a dark to opaque appearance in thin section, with a few larger transparent olivine crystals in many cases (Fig. 3a, b). Olivine is, by far, the dominant phase (Fig. 3c, d), ranging from 57 to 94 vol.%. Pyroxene and a feldspathic-glassy mesostasis are much less abundant, generally making up less than 20 vol.% each. Other phases are present in minor to trace amounts (<5 vol.%) and include chromite, metal, and sulfide. In addition to mineral fragments, some agglomeratic chondrules contain broken chondrules, microchondrules, and refractory inclusions (Fig. 3c, e).

The fine-grained olivine ranges from subhedral to anhedral in shape. Many are broken crystal fragments, and a few occur as blades, similar to some of the olivine in chondrule rims and in the matrix. Some agglomeratic chondrules have porous areas consisting of submicrometer-sized (<1 μm) olivine which, in some cases, occurs interstitially to the fine-grained (2–5 μm) olivine (Fig. 3c,

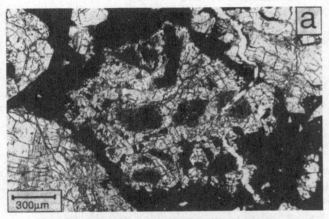

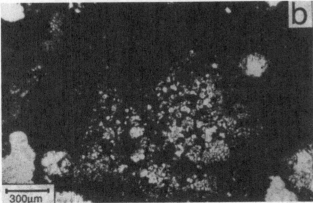

Fig. 2. Photomicrographs of: (a) An aggregational chondrule from the Semarkona LL3.0 chondrite consisting of relict crystals with darkened cores, due to numerous tiny blebs of Fe-metal formed as a result of solid-state reduction reactions. (b) An aggregational chondrule from the ALH84028 CV3 chondrite consisting of olivine crystals, lithic fragments and two large olivine grains in its outer rim. The large olivines have relatively high Al_2O_3 (0.3 wt.%), whereas the smaller olivine grains have low Al_2O_3 (<0.03%) more typical of olivine.

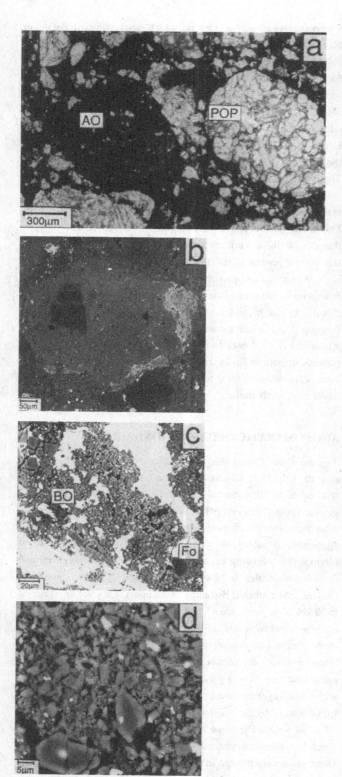

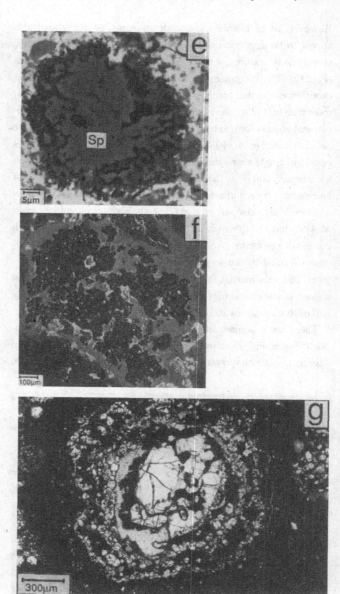

(dark gray) olivine fragment sharply rimmed by FeO-rich olivine compositionally similar to the fine-grained olivine. (c) BSE image of an agglomeratic olivine chondrule from Bishunpur that consists of abundant fine-grained olivine (light gray), FeS (white), a barred olivine microchondrule (BO), and an MgO-rich olivine (Fo) with a sharply bound (light gray) FeO-rich olivine rim compositionally similar to the surrounding fine-grained olivine. (d) Enlargement of (c) showing three types of olivine: zoned MgO-rich olivine 5 μm in size, finer grained olivine (2–3 μm), and interstitial submicrometer-sized olivine (<1 μm). (e) BSE image of a refractory element-rich nodule from an agglomeratic olivine chondrule in the Chainpur L3.4 chondrite that consists of a spinel-rich (Sp) core surrounded by a Ca pyroxene-rich rim. The inclusion is texturally similar to the spinel-pyroxene aggregates that occur in carbonaceous chondrites (MacPherson and Grossman, 1988). (f) BSE image of a pyroxene-rich inclusion in Krymka, with small inclusions of olivine in the pyroxene. (g) Multi-layered chondrule from the Al Rais CR chondrite which consists of six units: a core (1) made up of a large single crystal of olivine, surrounded by layer (2) of FeNi-metal (black), a finer-grained olivine-rich layer (3), another FeNi-metal-rich layer (4), an olivine-rich layer (5) even finer-grained than layer 3, and an accretionary dust layer (6). Some layered chondrules in CR chondrites have an outer layer of carbonate (Weisberg *et al.*, 1993).

Fig. 3. Photomicrographs of: (a) Agglomeratic olivine (AO) chondrule from the Bishunpur L3.2 chondrite shown in comparison to a porphyritic (POP) chondrule. Agglomeratic olivine chondrules in ordinary chondrites are very fine-grained and are dark to near-opaque in thin section. (b) Backscattered electron (BSE) image of an agglomeratic olivine chondrule from Bishunpur showing that it consists mainly of fine-grained (<5 μm) olivine, some coarser olivine (10–20 μm), and a large (100 μm) MgO-rich

d). Many agglomeratic olivine chondrules also contain large (up to 400 μm) transparent olivine crystals as well as large low-Ca pyroxene (Fig. 3b, c).

In Krymka, one of the least equilibrated ordinary chondrites, the dominant fine-grained olivine is Fe-rich (Fa_{32-53}), similar in composition to olivine in chondrule rims (Rubin, 1984). This is in contrast to the fine-grained olivine (blades and anhedral crystals) in Krymka matrix which is generally more Fe-rich (>Fa_{62}), as shown in Fig. 4. In some agglomeratic olivine chondrules, the olivine near the chondrule periphery is finer grained and more Fe-rich than the fine olivine in the chondrule interior (Table 1). The larger olivines (>5 μm) are much more magnesian (Fa_{2-20}) than the fine olivine, but in some cases they are rimmed (up to 5 μm) by overgrowths of olivine similar in composition to the surrounding fine-grained olivine (Fig. 3b, c). The cores of some of the larger olivine grains (Table 1) have compositions similar to relict forsterite grains in FeO-rich chondrules described by Jones (this volume). Low-Ca pyroxene is generally MgO-rich (<Fs_{10}) and is not in chemical equilibrium with the surrounding fine-grained olivine. Chromites occur in clusters of euhedral crystals and are essentially $FeCr_2O_4$. The mesostasis is generally similar in all agglomeratic olivine chondrules from all ordinary chondrites and has an average (in wt.%) of 63% SiO_2, 17% Al_2O_3, 4% CaO, 9% Na_2O, 0.1% K_2O, with up to 5% FeO and 3% MgO. In some cases the mesostasis is heterogeneous and has a higher K_2O content (1.5%) near the chondrule periphery.

Some of the other components that occur within agglomeratic chondrules include barred olivine microchondrules (Fig. 3c). One agglomeratic chondrule has spinel-rich nodules in its rim (Fig. 3e), and has textural similarities to spinel-rich inclusions in carbona-

ceous chondrites (MacPherson et al., 1983). The nodules consist of spinel-rich cores surrounded by a Ti-, Al-rich Ca-pyroxene. Additionally, some agglomeratic chondrules contain irregular-shaped pyroxene-rich fragments which consist of pyroxene with small rounded inclusions of olivine. Similar pyroxene-rich inclusions occur as smaller (<10 μm) inclusions in the chondrite matrix and as large (~1mm) inclusions in ordinary chondrites (Fig. 3f). The pyroxene in pyroxene-rich inclusions is Mg-rich enstatite (Fs_2) and the rounded olivine within it is generally more FeO-rich (Fa_6) and is not in chemical equilibrium with the pyroxene host.

Bulk compositions of agglomeratic olivine chondrules generally have compositions similar to other chondrules, but have higher Fe and S contents (Table 2; Rubin, 1984). Their refractory lithophile ratios (Si/Al, Mg/Al) are similar to those of the porphyritic chondrules that dominate ordinary chondrite chondrule populations suggesting that agglomeratic chondrules formed from the same mix of precursors that formed porphyritic chondrules (Fig. 5).

Table 1. *Electron microprobe analyses of olivine (wt.%) in representative agglomeratic olivine chondrules in ordinary chondrites*

	Krymka (48-1)			Bishunpur (418-11)		Chainpur (40-3)	
	Large	Fine[1]	Fine[2]	Large	Fine	Large	Fine
SiO_2	42.4	37.6	35.4	42.0	37.2	40.6	35.9
Al_2O_3	<0.03	<0.03	<0.03	0.3	<0.03	<0.03	<0.03
Cr_2O_3	0.06	0.1	0.2	0.2	0.3	0.04	0.06
FeO	0.7	27.2	37.5	0.7	26.8	6.1	31.5
MnO	<0.03	0.4	0.3	<0.03	0.4	0.1	0.4
MgO	56.1	35.1	26.9	55.4	34.3	53.0	31.9
CaO	0.8	0.3	0.3	0.4	0.3	0.2	0.2
Total	100.1	100.7	100.6	99.0	99.3	100.0	100.0
Fa mol.%	0.7	30.2	43.9	0.7	30.5	6.1	35.6

1-Chondrule interior. 2-Chondrule periphery.

Table 2. *Electron microprobe analyses of bulk compositions (wt.%) of representative agglomeratic olivine chondrules in ordinary chondrites*

	Krymka 48–1	Krymka 48–18	Bishunpur 418–11	Chainpur 40–12
SiO_2	38.0	35.9	38.7	38.7
TiO_2	0.1	0.1	0.1	0.1
Al_2O_3	3.1	1.7	2.0	0.7
Cr_2O_3	0.4	0.5	0.5	0.5
FeO	28.1	33.9	20.1	30.5
MnO	0.3	0.2	0.3	0.4
MgO	25.9	25.3	35.8	26.0
CaO	1.9	0.6	1.3	2.3
Na_2O	1.4	1.1	0.8	0.3
K_2O	0.1	0.1	0.1	0.1
P_2O_5	0.7	0.6	0.3	0.4

Recalculated to 100% metal- and sulfide-free.

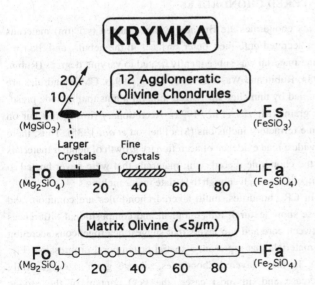

Fig. 4. Endmember compositions of pyroxenes and olivines (solid solution series) from 12 agglomeratic olivine chondrules in the Krymka LL3.0 chondrite, shown in comparison to the matrix olivines of the same meteorite. The fine-grained olivine that dominates these chondrules is FeO-rich (Fa_{32-53}), but larger olivines and pyroxenes are more MgO-rich (Fa_{80-99}; Fs_{90-99}). Olivine in the Krymka matrix has a wide range of compositions, mostly more FeO-rich (> Fa_{62}) than that in the agglomeratic olivine chondrules.

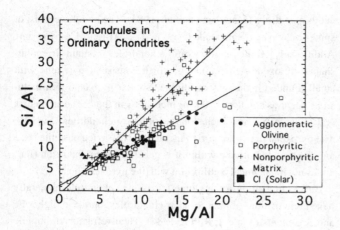

Fig. 5. Bulk compositional, refractory lithophile element ratios (determined by broad beam electron microprobe analysis) of chondrules from many ordinary chondrites shown on a Si/Al vs. Mg/Al diagram. The agglomeratic olivine chondrules (solid circles) have solar compositions for these ratios and overlap with the porphyritic chondrules (open squares) suggesting that both formed from a similar mix of precursor components. The nonporphyritic (radial pyroxene and cryptocrystalline) chondrules (crosses) generally have higher Si/Al ratios suggesting a somewhat different mix of precursor components with more pyroxene or silica. Also shown on this diagram are compositions of matrices (open triangles) from many ordinary chondrites which do not have solar ratios and are not likely to be the precursor to chondrules; however, matrix may be one component of a complex mix of precursors.

However, the nonporphyritic (cryptocrystalline and radial pyroxene) chondrules differ from the porphyritic chondrules in their bulk chemical compositions in having higher Si/Al ratios and they cannot be derived by closed system melting of the mix of precursors that formed the agglomeratic olivine and porphyritic chondrules.

Formation of agglomeratic olivine chondrules can be summarized in a three stage model. STAGE 1: Formation of free-floating or loose agglomerates of fine olivine crystals in the nebula. STAGE 2: Agglomeration of the fine olivines with other nebular components that include larger olivine and pyroxene mineral fragments, microchondrules, pyroxene-rich inclusions, and refractory inclusions. STAGE 3: Sintering of these agglomerations by a transient heat source resulted in small degrees of melting and solid state recrystallization. More intense heating and incomplete melting of these agglomerations to subliquidus temperatures may have resulted in the formation of porphyritic chondrules and complete melting to liquidus temperatures may have resulted in the barred olivine chondrules. Most of the FeS within the chondrules at their time of formation may have been expelled and redeposited as FeS-rich rims (e.g., Zanda *et al.*, in press).

Since agglomeratic olivine chondrules appear to have been heated to a lesser degree than the other chondrules they offer a unique opportunity to have available the solids that were present in the nebula during chondrule formation. Although mild heating and subsolidus reactions may have altered the compositions of agglomeratic olivine chondrules somewhat, their mineralogical and compositional characteristics may be representative of the precursor mix of materials that were melted to form chondrules. It is clear

that fine-grained olivine was the major precursor component. The presence of microchondrules and refractory inclusions, in minor abundances, in agglomeratic olivine chondrules suggests recycling of materials from earlier generations of chondrule formation; however, this was not common. The large olivine crystals commonly present in agglomeratic olivine chondrules may also be recycled material from an earlier generation of broken chondrules or, in some cases, may represent nebular condensation products that never resided in chondrules.

There is considerable controversy as to whether the variations in chondrule compositions are mainly controlled by the compositions of the precursors or by open system behavior during chondrule formation. It is perplexing that AO chondrules have a relatively narrow range of compositions that cannot easily account for the compositions of all chondrules. For example, why are there no agglomeratic chondrules compositionally equivalent to the FeO-poor chondrules? Either we have not yet sampled these agglomerates, none survived chondrule formation, or most chondrules formed from a relatively homogeneous mix of precursors which were later modified by open system behavior (evaporation of volatile elements and reduction of Fe) during chondrule formation. Nonporphyritic chondrules (e.g. radial pyroxene, cryptocrystalline) require a somewhat different set of precursors that were enriched in a pyroxene and/or silica component relative to the agglomeratic olivine chondrules or were also modified by open system behavior. They could have been derived from AO chondrule-like materials by later incorporation of free SiO_2 into the precursor mix, or were formed by later liquid/gas exchange with a Si-rich nebular gas.

LAYERED CHONDRULES

Many chondrules are rimmed by fine-grained (<5 μm) materials that accreted onto their outer surfaces in the nebula, and the rimming material was subsequently heated to varying degrees (Rubin, 1984; Rubin and Wasson, 1987; Kring, 1991). CM chondrules are rimmed by unmelted dust-sized mineral grains that include presolar grains (Metzler *et al.*, 1992). Accretionary rims also occur on some refractory inclusions (MacPherson *et al.*, 1985). Thus, there is widespread evidence of accretionary growth of nebular materials in the chondrule record, with materials that were never heated to temperatures high enough to initiate major melting.

In CR chondrites, multi-layered chondrules are common, and these show textural, mineralogical and compositional differences between core and rim layers that suggest heterogeneous accretion of materials that represent a spatial or temporal progression (Fig. 3g). In these layered chondrules, silicate grain sizes generally decrease and, in most cases, the FeO content of the olivine increases from the core to rim layers. Some cores of layered chondrules are essentially single crystals of olivine (Fig. 3g). In most layered chondrules the composition of the metal in the rim layers is lower in Ni and Co than is the core metal, while maintaining approximately solar Ni:Co ratios (Fig. 6). Oxygen isotopic compositions of rim layers are always ^{16}O-poor relative to their cores

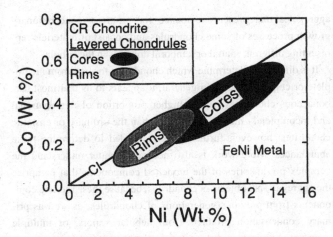

Fig. 6. Compositions of FeNi metal in the cores and rims of layered chondrules in the CR chondrites. This shows that metal in these primitive chondrules have near-solar Co-Ni ratios, with cores having higher Ni and Co than that in their associated rims. (Data from Weisberg *et al.*, 1993.)

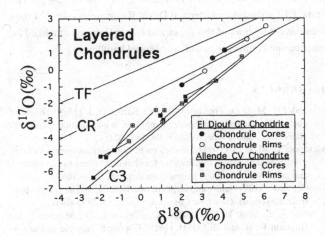

Fig. 7. Oxygen 3-isotope diagram showing compositions of the cores and associated rims of layered chondrules in the El Djouf 001 CR and Allende CV chondrites. Chondrule rims are always ^{16}O-poor than their cores. (Data from Rubin *et al.*, 1990; Weisberg *et al.*, 1992; Weisberg *et al.*, 1995.)

(Weisberg *et al.*, 1992; Fig. 7). A similar trend was observed in CV3 chondrules, in which chondrule rims are always ^{16}O-poor relative to their cores (Rubin *et al.*, 1990; Fig. 7). Thus, the layered chondrules are aggregates that were incompletely melted and preserve evidence of a progressive accretionary growth process of nebular materials that were heated to varying degrees during chondrule formation.

COMPLETE VS INCOMPLETE MELTING DURING CHONDRULE FORMATION

From the study of FeO-poor and FeO-rich porphyritic chondrules in ordinary chondrites, Jones and Scott (1989) and Jones (1990) suggested that most chondrules (~85 vol.%) form through near-complete melting of mineral dust and that the crystals within these

chondrules are phenocrysts that crystallized *in situ* from their chondrule melts. Relict crystals that survived melting, if present, are submicroscopic in size. By contrast, the recognition of relict crystals (e.g., Rambaldi, 1981; Nagahara, 1981; Steele, 1986a, b; Jones, this volume), with compositional and textural characteristics that preclude *in situ* crystallization in chondrules, led workers to conclude that most chondrules were incompletely molten and that a high proportion of olivine grains in chondrules are precursors that did not crystallize from the chondrule melt (e.g., Nagahara, 1983a).

The evidence presented for the formation of most chondrules through complete melting include: (1) Relict crystals appear to be absent in most chondrules. (2) The textures of most chondrules appear to be igneous in that they contain olivine and pyroxene crystals in an amorphous to microcrystalline mesostasis. (3) Olivine is enclosed in pyroxene, as in many igneous rocks. (4) The olivine enclosed within pyroxene is compositionally similar to the olivine in contact with the mesostasis, suggesting that both share a similar origin, presumably crystallization from the melt. (5) Minor element (Al and Ti) abundances in olivine are consistent with crystal/liquid partition coefficients. (6) Chemical zoning in olivine is similar to the zoning patterns expected of olivine crystallization from a melt.

The arguments for the complete or near-complete melting of most chondrules could be interpreted differently, and are not conclusive evidence for their formation in this manner. In addition, the arguments may apply mainly to selected examples. Counter-arguments for the formation of most chondrules by incomplete melting include: (1) The presence of easily identifiable relict crystals in some chondrules may be indicative that many crystals in chondrules are unmelted relict crystals that are not so easily recognizable. For example, many relict crystals are recognized because they differ compositionally or texturally from other crystals in the same chondrule. FeO-rich olivine in a chondrule consisting predominantly of MgO-rich olivine, or vice versa is easily recognized, but MgO-rich olivine relicts in a chondrule dominated by MgO-rich olivine may be easily overlooked. Careful minor element analyses of numerous crystals in a wider range of less obvious igneous chondrules may help identify relicts (e.g., Steele, 1986a, b). (2) The textures of many chondrules often assumed to be igneous may be aggregational on more careful examination. The common "porphyritic" textures of many chondrules have a wide range of grains sizes and shapes that are more indicative of incompletely melted aggregates of crystals and dust, rather than crystals grown from complete melts. (3) Nagahara (1983a) showed that olivine within pyroxene is not always compositionally similar to olivine in contact with the melt and these olivines may have a different origin. In the agglomeratic chondrules the olivine enclosed in pyroxene inclusions is not in equilibrium with other olivine in the same chondrule. (4) A poikilitic texture (olivine enclosed within pyroxene) is not conclusive evidence of igneous growth. Irregular-shaped pyroxene aggregates with olivine enclosed within the pyroxene occur in the host chondrite matrix, in dark inclusions, and in agglomeratic chondrules. This suggests that these materials were available as precursors in the nebula, and they were not necessarily from pre-existing chondrules. (5) Some olivine in chondrules has relatively high concentrations of Al, especially in carbonaceous chondrites, and these are clearly inconsistent with the distribution of Al between olivine and liquid that would be expected from olivine that crystallized in a melt. Steele (1986) showed that several percent of all olivine present in carbonaceous and ordinary chondrites have minor and trace element compositions that suggest growth of the olivine from a nebular vapor. Additionally, the presence of low levels of Al_2O_3 in olivine considered consistent with igneous partitioning does not preclude the possibility that it is relict. (6) Some chemical zoning of FeO-rich rims of olivine onto FeO-poor olivine, in some chondrules, is the result of solid-gas exchange reaction as well as condensation (Peck and Wood, 1987; Hua *et al.*, 1988; Weinbruch *et al.*, 1990). This process can occur both within, and on the outer margins of these chondrules. Zoning that appears to be the result of igneous processes, within a chondrule, does not necessarily imply complete melting of the entire chondrule.

On balance, it appears to us that porphyritic chondrules formed from both complete and incomplete melting processes. We suggest that chondrules that formed from incomplete melts of agglomerations, heated to below liquidus (<1900°C) and slightly above solidus (1200°C) temperatures, may be more common than presently recognized. Chondrules represent a continuum that ranges from the agglomeratic chondrules which formed from sintered aggregates, to porphyritic chondrules which formed from incompletely melted to completely melted aggregates, to the nonporphyritic chondrules formed mainly by complete melting of aggregates that differed from most other chondrules in being enriched in pyroxene and/or silica components.

CONCLUSIONS

Chondrules form as a result of varying degrees of melting of mixtures of pre-existing mineral crystals, fragments and dust. Agglomeratic olivine chondrules clearly exerienced very low degrees of melting and make up ~2 vol.% of the ordinary chondrites. In carbonaceous chondrites the abundance of aggregational objects is much higher. It is likely that most chondrules formed from incomplete melts that were heated to below liquidus (<1900°C) and slightly above solidus (1200°C) temperatures.

The relatively narrow range of compositions of agglomeratic olivine chondrules suggests that they, as well as most chondrules, formed from a fairly homogeneous mix of precursor materials that were greatly dominated by olivine. Open system behavior may therefore account for differences in the Fe and volatile contents between the FeO-rich and FeO-poor chondrules. Grain sizes of the precursor materials that are represented in agglomeratic chondrules are in the 2–5 μm range or smaller; larger olivine and/or pyroxene grains (up to 400 μm) are also present. The microchondrules and refractory inclusions found in some agglomeratic chondrules, suggest that these components were recycled into the chondrule precursor mix as minor components. Layered chondrules are

aggregates that preserve evidence for progressive accretionary growth processes of some chondrules and consist of materials representing different spatial or temporal domains in the nebula.

It is difficult to determine which chondrules formed from incomplete or complete melts. In general, it appears to us that most carbonaceous chondrites have a higher proportion of aggregational and incompletely melted chondrules than the ordinary or enstatite chondrites; however, further work is needed to determine these abundances. More work is also needed to better understand the complex pre-histories of the unmelted components that comprise the aggregational objects in chondrites, because these include components from previous generations of chondrules, as well as primary condensation products from nebular vapors, or multiple stages of vaporization and recondensation.

ACKNOWLEDGEMENTS

This work was supported by NASA grant NAGW 34–90 (M. Prinz, P.I.). Conel Alexander and David Kring are thanked for their constructive reviews of this paper and Roger Hewins is thanked for contributing comments as well as editorial handling.

REFERENCES

Alexander C. M. O'D., Hutchison R., and Barber D. J. (1989) Origin of chondrule rims and interchondrule matrices in unequilibrated ordinary chondrites. *Earth Planet Sci. Lett.* **95**, 187–207.

Alexander C. M. O'D. (1994) Trace element distributions within ordinary chondrite chondrules: Implications for chondrule formation conditions and precursors. *Geochim. Cosmochim. Acta* **58**, 3451–3467.

Brearley A. J. (1989) Nature and origin of matrix in the unique type 3 chondrite, Kakangari. *Geochim. Cosmochim. Acta* **53**, 2395–2411.

Brearley A. J., Scott E. R. D., Keil K., Clayton R. N., Mayeda T. K., Boynton W. B. and Hill D. H. (1989) Chemical, isotopic and mineralogical evidence for the origin of matrix in ordinary chondrites. *Geochim. Cosmochim. Acta* **53**, 2081–2094.

Cohen R. E., Kornacki A. S. and Wood J. A. (1983) Mineralogy and petrology of chondrules and inclusions in the Mokoia CV3 chondrite. *Geochim. Cosmochim. Acta* **47**, 1739–1757.

Dodd R. T. and Van Schmus W. R. (1971) Dark-zoned chondrules. *Chem. Erde* **30**, 49–69.

Gooding J. L., Keil K., Fukuoka T. and Schmitt R. A. (1980) Elemental abundances in chondrules from unequilibrated chondrites: Evidence for chondrule origin by melting of pre-existing materials. *Earth Planet. Sci. Lett.* **50**, 171–180.

Grossman J. N. and Wasson J. T. (1982) Evidence for primitive nebular components in chondrules from the Chainpur chondrite. *Geochim. Cosmochim. Acta* **46**, 1081–1099.

Hua X., Adam J., Palme H. and El Goresy A. (1988) Fayalite-rich rims, veins, and halos around and in forsterite olivines in CAIs and chondrules in carbonaceous chondrites: types, compositional profiles and constraints for their formation. *Geochim. Cosmochim. Acta* **52**, 1389–1408.

Huss G. R., Keil K. and Taylor G. J. (1981) The matrices of unequilibrated ordinary chondrites: implications for the origin and history of chondrites. *Geochim. Cosmochim. Acta* **45**, 33–51.

Jones R. H. and Scott E. R. D. (1989) Petrology and thermal history of type IA chondrules in the Semarkona (LL3.0) chondrite. *Proc. Lunar Planet. Sci. Conf. 19th*, 523–536.

Jones R. H. (1990) Petrology and mineralogy of Type II, FeO-rich chondrules in Semarkona (LL3.0): Origin by closed-system fractional crystallization with evidence for supercooling. *Geochim. Cosmochim. Acta* **54**, 1785–1802.

Kring D. A. (1988) *The petrology of meteoritic chondrules: Evidence for fluctuating conditions in the solar nebula.* Ph.D. Thesis, 346pp. Harvard University, Cambridge.

Kring D. A. (1991) High temperature rims around chondrules in primitive chondrites: Evidence for fluctuating conditions in the solar nebula. *Earth Planet. Sci. Lett.* **105**, 65–80.

MacPherson G. J., Bar-Matthews M., Tanaka T., Olsen E. and Grossman L. (1983) Refractory inclusions in the Murchison meteorite. *Geochim. Cosmochim. Acta* **47**, 823–839.

MacPherson G. J. and Grossman L. (1984) "Fluffy" Type A Ca-Al-rich inclusions in the Allende meteorite. *Geochim. Cosmochim. Acta* **48**, 29–46.

MacPherson G. J., Hashimoto A. and Grossman L. (1985) Accretionary rims on inclusions in the Allende meteorite. *Geochim. Cosmochim. Acta* **49**, 2267–2279.

Metzler K., Bischoff A. and Stöffler D. (1992) Accretionary dust mantles in CM chondrites: Evidence for solar nebula processes. *Geochim. Cosmochim. Acta* **31**, 661–664.

Nagahara H. (1981) Evidence for secondary origin of chondrules. *Nature* **292**, 135–136.

Nagahara H. (1983a) Chondrules formed through incomplete melting of the pre-existing mineral clusters and the origin of chondrules. In *Chondrules and their Origins.* (ed. E. A. King), pp. 211–222. Lunar and Planetary Institute, Houston.

Nagahara H. (1983b) Texture of chondrules. *Mem. Natl. Inst. Polar Res.* Special Issue **30** 61–83.

Nagahara H. (1984) Matrices of type 3 ordinary chondrites-primitive nebular records. *Geochim. Cosmochim. Acta* **48**, 2581–2595.

Peck J. A. and Wood J. A. (1987) The origin of ferrous zoning in Allende chondrule olivines. *Geochim. Cosmochim. Acta* **51**, 1503–1510.

Radomsky P. M. (1988) *Dynamic crystallization experiments on magnesian olivine-rich and pyroxene-olivine chondrule compositions.* M. S. thesis, Rutgers University.

Rambaldi E. R. (1981) Relict grains in chondrules. *Nature* **293**, 558–561.

Rubin A. E. (1984) Coarse-grained chondrule rims in type 3 chondrites. *Geochim. Cosmochim. Acta* **48**, 1779–1789.

Rubin A. E. and Wasson J. T. (1987) Chondrules, matrix and coarse-grained rims in the Allende meteorite: Origin, interelationships and possible precursor components. *Geochim. Cosmochim. Acta* **51**, 1923–1937.

Rubin A. E., Daode W., Kallemeyn G. W. and Wasson J. T. (1988) The Ningqiang meteorite: Classification and petrology of an anomalous CV chondrite. *Meteoritics* **23**, 12–23.

Rubin A. E., Wasson J. T., Clayton R. N. and Mayeda T. K. (1990) Oxygen isotopes in chondrules and coarse-grained chondrule rims from the Allende meteorite. *Earth Planet. Sci. Lett.* **96**, 247–255.

Scott E. R. D., Rubin A. E., Taylor G. J. and Keil K. (1984) Matrix material in type 3 chondrites-Occurrence, heterogeneity and relationship with chondrules. *Geochim. Cosmochim. Acta* **48**, 1741–1757.

Scott E. R. D., Barber D. J., Alexander C. M. O'D., Hutchison R. and Peck J. (1988) Primitive material surviving in chondrites: Matrix. In *Meteorites and the Early Solar System* (eds. J. F. Kerridge and M. S. Matthews), University of Arizona Press, Tucson, 718–745.

Steele I. M. (1986a) Compositions and textures of relic forsterite in carbonaceous and ordinary unequilibrated chondrites. *Geochim. Cosmochim. Acta* **50**, 1379–1395.

Steele I. M. (1986b) Cathodoluminescence and minor elements in forsterites from extraterrestial samples. *Amer. Mineral.* **71**, 966–970.

Taylor G. J., Scott E. R. D. and Keil K. (1983) Cosmic setting for chondrule formation. In *Chondrules and Their Origin.* (ed. E. A. King), pp. 262–278. Lunar and Planetary Institute, Houston.

Van Schmus W. R. (1969) The mineralogy and petrology of chondritic meteorites. *Earth-Science Rev.* **5**, 145–184.

Weinbruch S., Palme H., Müller W. F. and El Goresy A. (1990) FeO-rich rims and veins in Allende forsterite: Evidence for high temperature condensation at oxidizing conditions. *Meteoritics* **25**, 115–125.

Weisberg M. K. and Prinz M. (1994) Agglomeratic olivine (AO) chondrules in ordinary chondrites (abstract). *Lunar Planet. Sci.* **XXV**, 1481–1482.

Weisberg M. K., Prinz M., Clayton R. N. and Mayeda T. K. (1992) Formation of layered chondrules in CR2 chondrites: A petrologic and oxygen isotope study (abstract). *Meteoritics* **27**, 306.

Weisberg M. K., Prinz M., Clayton R. N. and Mayeda T. K. (1993) The CR (Renazzo-type) carbonaceous chondrite group and its implications. *Geochim. Cosmochim. Acta* **57**, 1567–1586.

Weisberg M. K., Prinz M., Clayton R. N., Mayeda T. K., Grady M. M. and Pillinger C. T. (1995) The CR chondrite clan. *Proc. Symposium on Antarctic Meteorites*, **8**, 11–32.

Zanda B., Yang Y., Bourot-Denise M., Hewins R. H. and Connolly H. C. (1996) Sulfur behavior in chondrule formation and metamorphism. *Geochim. Cosmochim. Acta*, submitted.

14: Constraints on Chondrule Precursors from Experimental Data

HAROLD C. CONNOLLY JR and ROGER H. HEWINS

Department of Geological Sciences, Rutgers University, Piscataway, NJ 08855, U.S.A.

ABSTRACT

Few experiments have attempted to determine the nature of chondrule precursors. What data are available suggest that microporphyritic chondrule textures form by the incomplete melting of fine-grained starting materials (<63 μm). Relict grains in chondrules suggest that at least some precursor grains were large. However, chondrules are rarely composed mainly of large relict grains suggesting that chondrules with relict grains formed from precursors with a range of grain sizes. Incorporation of a solid reducing agent in starting material reproduces dusty relict olivine grains and Fe metal with SiO_2 inclusions that are found in many reduced chondrules. Such chondrules probably incorporated C-rich material in their precursors. With the exception of microporphyritic chondrule textures, other chondrule textures can be produced from precursor grains of various sizes depending on the intensity of the melting event. Therefore, the precursors of chondrules that lack relict grains are difficult to determine with certainty. Experimental petrology studies of chondrule precursors are clearly in an infancy stage and many more experiments are needed.

INTRODUCTION

The experimental investigation of chondrule origins has concentrated on determining how chondrules were melted and what type of cooling they experienced, thus establishing some insight into potential processes occurring within the chondrule-forming region(s) of the nebula. To learn what constraints these experiments have placed on chondrule production we refer the reader to Hewins and Connolly (this volume). However, chondrules not only provide insight into processes that occurred within the solar nebula but their very existence provides clues as to what types of materials were present within the solar nebula.

Experiments have been performed under the assumption, often not formally stated, that chondrules were formed from crystalline precursors. As long as crystalline materials were present, nucleation and crystal growth could have occurred in chondrule melts. Experiments have not been designed that specifically address the original source of chondrule precursors (i.e. condensates or recycled chondrule materials). Therefore, in the context of this paper, the word "precursors" will denote any type of phase present that could have been used to form chondrules regardless of its original source.

The focus of this paper is to present a brief summary of experiments that were designed to study some aspect of chondrule precursors and the constraints that can be derived from them. Furthermore, this paper will also present the reader with knowledge of what we cannot say about chondrule precursors based on experiments.

Petrologic observations

The inference that chondrules were formed from crystalline minerals is based on the existence of relict grains (Nagahara, 1981; Rambaldi, 1981; Kracher *et al.*, 1983). A relict grain is a mineral grain within a chondrule that was a part of a chondrule's precursor assemblage that was not fully melted during the chondrule-forming process. Relict grains vary in grain size and mineralogy(Grossman *et al.*, 1988; Jones, this volume). The most common types of relict grains are olivine and pyroxene, but phases such as spinel and fassaite which are characteristic of CAI, are occasionally observed (Misawa and Nakamura, this volume). The origins of relict grains remain a debated issue (Grossman *et al.*, 1988; Steele, 1986; 1989; Alexander, this volume; Jones, this volume) but the presence of these grains provides the experimentalist with a basis for studying the consequences of physical and mineralogical variations in chondrule precursors.

Dusty olivine grains, one specific type of relict, are not just simply unmelted precursor grains. They have tiny metal inclusions, indicative of reduction of the host grain either before or during the chondrule-forming process. Another major issue in the formation of chondrules is whether chondrule redox was controlled by nebular gases or by the precursor minerals. Dusty relict grains and chondrules may have recorded fluctuations in the redox state of nebular gases within the chondrule-forming region(s) (Nagahara, 1981) or the presence of a reducing agent such as carbon within the precursors (Rambaldi, 1981). Therefore, understanding the formation of dusty olivine grains has implications for understanding nebular processes in general.

The bulk composition of chondrules provides another major clue to the possible mineralogy of chondrule precursors. It has been suggested that chondrule minerals are the products of *in situ* crystallization from a melt that corresponds to the bulk composition of the chondrules (Jones and Scott, 1988; Jones, 1990; Jones, 1994). Therefore, studying what possible combinations of minerals can produce observed chondrule bulk compositions may provide clues to chondrule precursors.

From the above observations, experimentalists are provided with a range of possible chondrule-precursor issues to test experimentally: (1) consequences of variations in grain size and in mineralogy (silicates and non-silicates) of chondrule precursors and (2) ways to reproduce relict grains. Below we review the experiments that emphasized some aspects of these issues and the limited constraints these experiments have placed on our knowledge of chondrule precursors.

EXPERIMENTS

The effects of the grain size of starting compositions on chondrule textures

The experiments of Radomsky and Hewins (1990) were the first to show how differences in the grain size of a starting composition affect the production of chondrule textures. They placed millimeter-sized olivine grains into their starting pellets of a FeO, SiO_2-poor (Type I) composition (liquidus temperature of 1570°C), that was ground to a fine, unsorted powder. These pellets were melted isothermally for 30 minutes at a superliquidus temperature of 1601°C and then cooled at various cooling rates. In all these runs the millimeter sized olivine grains survived as relict grains with overgrowths of melt-derived olivine. These charges had igneous textures (Fig. 1) but in the equivalent experiments that did not have large olivines added to the pellets, melting was always complete and resulted in glassy spheres. These experiments clearly showed,

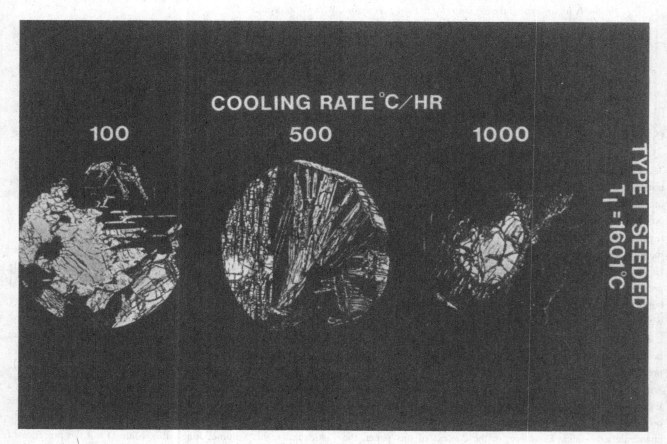

Fig. 1. A series of experiments where 1.5 mm olivine (Fo_{89}) grains were added to the pellets before melting at superliquidus temperatures. Pellets without these olivine seeds that experienced identical thermal histories produced glass, but the seeded charges are crystalline, indicating that the grain size of the starting material affects texture production. The added olivine grains survived as relict grains in all the experiments (white arrows). This figure is from Radomsky and Hewins (1990).

for the first time, that the grain size of chondrule precursors could have a profound effect on the dynamic melting process and hence on the production of chondrule textures.

The results of Radomsky and Hewins (1990) demonstrated the need for a systematic study of how the grain size of starting material affects the final texture of melt spheres. In just such a study, Connolly *et al.* (1991) and Connolly (unpublished data, 1995) used a very FeO-rich composition (liquidus temperature of 1211°C) made from ground minerals that was separated into four different initial size fractions (20–44 μm; 45–62 μm; 63–124 μm; 125–249 μm). They subjected pellets of each grain size to identical thermal conditions similar to a type of flash melting (a rapid, nonisothermal, superliquidus melting). The pellets were placed into a furnace that was at a specific, superliquidus temperature and then immediately cooled linearly at 2800°C/hr to the liquidus temperature of 1211°C. From 1211°C to 1000°C charges were cooled at 500°C/hr. At 1000°C charges were removed (quenched) from the furnace.

A highlight of these experiments is that microporphyritic olivine texture (Fig. 2) similar to that of MgO-rich (Type IA) chondrules could *only* be reproduced using a starting material that had a grain size less than 63 μm and a melting event that produced a high degree of incomplete melting (i.e. a melting event that was not intense). Furthermore, their results also showed that porphyritic olivine (PO) textures typically found in FeO-rich (Type II) chondrules and barred olivine (BO) textures could be produced from any grain size starting materials providing they experienced the proper melting conditions. However, in none of their experiments did they observe any relict grains. Previous experiments that used isothermal, long duration melting, produced microporphyritic olivine textures when materials of an uncontrolled or random grain size were melted at temperatures far below the liquidus (Lofgren and Russell, 1988; Radomsky and Hewins, 1990). Therefore, with the exception of the Radomsky and Hewins (1990) experiments, it was unknown if the grain size had any effect on producing the texture.

In another series of experiments that investigated the effect of the grain size of starting material on texture production, Connolly *et al.* (1993a,b; 1994) used FeO-rich (liquidus temperature of 1556°C) and MgO-rich (liquidus temperature of 1692°C) chondrule compositions made from ground minerals with the same grain-size ranges as above. Pellets were placed into a furnace that was preheated to a specific, superliquidus temperature and cooled, non-linearly to 1000°C where the charges were quenched. The results of these experiments were similar to the results of Connolly *et al.* (1991) with the exception that many visible relict olivine grains were produced (Fig. 3). Relict olivine grains are identical to the starting olivines and as many as 60% of the phenocrysts within a charge contained visible relict olivine grains.

The role of non-silicates in the formation of synthetic chondrules

In most experiments it is assumed that the presence of relict grains indicates that chondrule precursors were composed of anhydrous silicates. However, phases such as troilite and serpentine are found within chondrules (Grossman *et al.*, 1988; Metzler *et al.*, 1994)

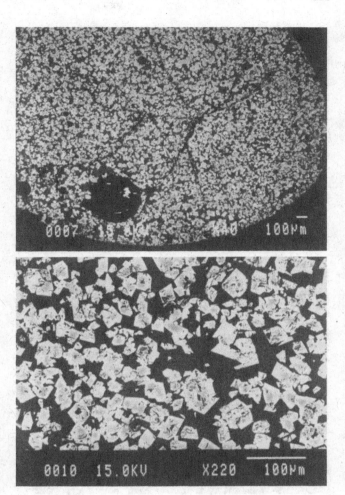

Fig. 2. Backscatter electron images of an experimental charge with a microporphyritic olivine (MPO) texture from an experiment of Connolly *et al.* (1991) who used a very FeO-rich starting composition. This texture is similar to that of Type IA (MgO-rich) chondrules. The lower image is a magnified (220×) area of the upper image that shows the small size of the phenocrysts. This type of MPO texture can only be produced when the starting material is very fine grained (< 63 μm).

although their exact origins may be debated. A few experiments have been performed that attempt to understand the effect of phyllosilicates or non-silicates within chondrule precursors on chondrule production.

Maharaj and Hewins (1994) have studied the potential role of serpentine as a chondrule precursor. They added 50 wt% serpentine to an FeO–SiO$_2$-rich (Type IIAB) composition as used by Radomsky and Hewins (1990). Pellets were placed in a furnace heated to 1500°C and immediately cooled using a linear cooling rate of 500°C/hr. They showed that upon heating, the serpentine dehydrated producing voids as large as 3 mm with 88% void spaces in the synthetic chondrule. Because of the absence of such voids in natural chondrules they suggested that chondrule precursors were anhydrous phases. However, more research is needed to determine if smaller quantities of added serpentine and different conditions would produce the same results.

In another series of experiments, Yu *et al.* (this volume) added

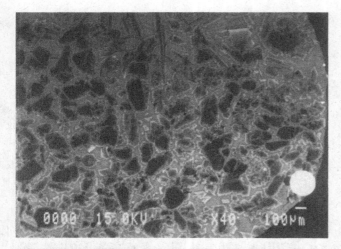

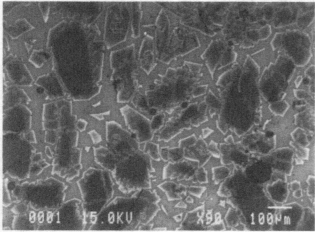

Fig. 3. The upper image is a backscatter electron image of an experimental charge from Connolly *et al.* (1993a) showing a porphyritic olivine (PO) texture similar to that of Type IIA (FeO-rich) chondrules. This charge was made from the flash melting of a Type II chondrule composition with a grain size of 125–250 μm. The lower image is a backscatter electron image of a magnified (45×) area of the upper figure showing the large size of the phenocrysts. The dark cores within the phenocrysts are relict olivine grains.

and then immediately cooled using a nonlinear cooling rate (Connolly *et al.*, 1994a). The most striking result of their experiments was the production of relict olivines with numerous Fe-rich metal inclusions, or dusty olivines (Fig. 4). The use of graphite and diamond was as a proxy for organic carbon which was probably more abundant in the nebula. However, graphite does occur in chondrites and can be found in large metal grains. It has been suggested (Mostefaoui and Perron, 1994) that graphite in metal apparently survived reduction reactions, by separating from silicate early in chondrule melting. This suggests that graphite was present in chondrule precursors.

In addition to the production of dusty olivines, the original starting olivine ($Fo_{90.5}$) was reduced by the addition of carbon to produce phenocrysts with forsterite contents that were almost Fo_{100}. The reduction also produced metallic Fe and FeNi grains during melting and through part of the crystallization process. The metal grains range in size from micrometer to several hundreds of micrometers in diameter, and are visible throughout the interior and on the exterior of charges. The large metal grains are Fe-rich and contain 1–2 wt% Ni (derived from the olivine in the starting

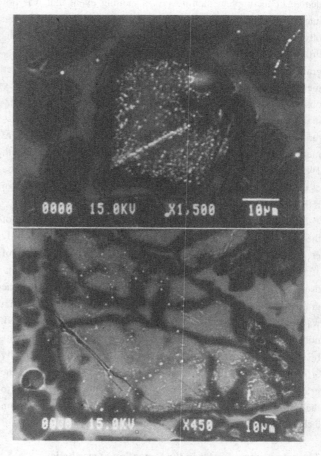

Fig. 4. Backscatter electron images of synthetic dusty olivine grains produced when graphite was added to an analog Type IA chondrule composition. The upper image shows a phenocryst with a relict dusty olivine core that is overgrown with more reduced olivine. The lower image shows a large relict dusty olivine with more reduced olivine that has formed along cracks in the relict dusty olivine. Note that the scale is different for each image. Image from Connolly *et al.* (1995).

troilite (5 wt%) to 60 mg pellets of both the FeO, SiO_2-poor (Type I) and the FeO, SiO_2-rich (Type IIAB) starting compositions of Radomsky and Hewins (1990). Charges were subjected to a variety of melting, cooling, and fO_2 conditions. Their results showed that troilite added to the starting material often survived the imposed thermal/fO_2 conditions. The occurrence of troilite in natural chondrules from Semarkona and Renazzo with textures similar to those in experimental chondrules motivated Yu *et al.* to suggest that troilite was present as a precursor in these types of chondrules.

Another potential nonsilicate chondrule precursor is carbon (in many forms). In a series of experiments designed to understand the potential role of nonsilicate precursor material, Connolly *et al.* (1994) added from 1 to 10 wt.% carbon in the form of graphite and diamond to a MgO-rich chondrule analog starting composition (liquidus temperature of 1692°C) that was made into two different grain size ranges of 20–44 μm and 125–249 μm. Pellets of the carbon-rich composition were placed into a furnace heated to 1750°C

material) and approximately 0.2 wt% Co with significant amounts of dissolved Cr and Si. These large metal grains also contained numerous inclusions of silica and fayalitic glass. The smaller grains are taenite with 25–30 wt% Ni and less than 1 wt% Co. These small grains had neither dissolved Cr nor Si and did not contain any detectable inclusions.

Dusty olivines, large metal grains that contain inclusions of silica and fayalitic glass, and reduced olivines within a single experimental chondrule are all features found in many reduced, natural chondrules. Because of the great similarity between the synthetic, reduced chondrules and natural, reduced chondrules, Connolly *et al.* (1994) suggested that the latter contained a carbon phase within the precursor material that acted as a reducing agent during formation.

DISCUSSION

It is clear from a review of the experimental data that few experiments have been designed to determine the nature of chondrule precursors. Before we discuss the implications of the experimental data it is necessary to emphasize the complex nature of chondrule formation and our ignorance of the circumstances of chondrule melting. Unfortunately, chondrules could have experienced more than one type of melting (Hewins and Connolly, this volume; Wood, this volume). Traditionally, chondrule textures have been thought to be the product of incomplete melting. However, Connolly and Hewins (1995) have shown that some chondrules could have been totally melted and that their textures could be the product of collisions between the melt droplet and dust grains. If this is true, then relict grains would be injected grains (i.e. projectiles) that caused the production of textures and could have no relationship to the immediate precursors of the liquid droplet. However, the research of Connolly *et al.* (1991; 1993a,b) has shown that relict grains and most porphyritic chondrule textures are easily produced by incomplete melting. Therefore, for the purpose of our discussion we will *assume* that *most* porphyritic textures are the product of incomplete melting from a flash-melting event. As we cannot exactly constrain the nature or intensity of the melting event, our interpretations of chondrule precursors based on experiments to date could be revised with future research.

Texture production and precursors

It is possible to produce classic porphyritic textures similar to those of FeO-rich (Type II) chondrules from starting material of any grain size provided that the precursors are heated sufficiently. Coarse-grained starting material (> 63 μm) easily produced porphyritic and barred textures when melted. However, microporphyritic textures were never produced from starting material with a grain size greater than 63 μm for any of the initial melting conditions. Textures similar to those characteristic of MgO-rich (type IA) chondrules can be produced only by using a fine-grained starting material (< 63 μm) with only slight initial melting. The main difference between MPO and PO textures is the number and size of phenocrysts. MPO texture requires that a larger number of relict

silicate grains (or nuclei) remain in the melt after melting and therefore the grain size of the individual precursor particles must be small (cf. Lofgren, this volume). We suggest that the average grain size of MPO chondrule precursors must have been finer than 63 μm and the intensity of the melting event they experienced produced only slight melting of the precursors. Finer-grained chondrules such as dark-zoned chondrules (Hewins and Connolly, this volume) must have had an average precursor grain size less than that of typical Type IA chondrules suggesting that very fine grained (a few micrometers) precursor material existed at some stage of the chondrule-forming process.

Unlike MPO textures, textures characteristic of FeO-rich chondrules (i.e. porphyritic, barred) are easily produced from any grain size starting material if melting is significant. Therefore, unless relict grains exist within porphyritic or barred textures little can be said as to the physical nature of their precursors.

Relict grains

Relict grains are easily produced by the incomplete melting of crystalline materials. For any relict to have survived the melting process it is obvious that the parent grain must have initially been at least as large as it is now, and melting was limited in intensity. Therefore, chondrules with relict grains had at least some crystalline precursor material that was at least as large as the relict grains. Because experiments produce abundant relict grains by incomplete melting and natural chondrules generally contain only a few relict grains at most, we suggest that the average grain size of the precursors was not the size of the commonly reported relict grains. It is possible that chondrule precursors could have resembled the aggregational chondrules of Weisberg and Prinz (this volume), which contain mainly tiny silicate grains with a few larger ones. We cannot, however, say that all of the precursors were crystalline or even that all were silicates. Furthermore, only those coarse-grained (i.e., porphyritic) chondrules that have relict grains permit any constraints to be placed on the physical properties of chondrule precursors.

The production of dusty relict grains, large Fe-rich, metal grains with SiO_2 inclusions, and reduced olivines in synthetic chondrules, by the addition of carbon phases to a starting composition strongly suggests that carbon phases were a part of reduced chondrule precursors. Carbon in several forms is well known to have been present within the early solar nebula, and therefore it is plausible for carbon to have been incorporated into chondrule precursors, even as a type of glue that bonded the precursors together (Wood, 1985; Connolly *et al.*, 1994). Furthermore, the redox conditions of chondrules may therefore have been dependent on their precursors and not solely on the composition and redox properties of nebular gas. It is important to point out that the experiments of Connolly *et al.* (1994) are the first to study the role of carbon and silicate melting with chondrule compositions. Recent preliminary experiments using isolated olivines (Danielson and Jones, 1995; Libourel and Chaussidon, 1995) show that it may also be possible to produce isolated dusty olivine grains in reducing gases without the aid of carbon. As these experiments are preliminary, it is too early to use them to place substantial constraints on chondrule precursors. However, they do reinforce the possibility that nebular gases played an important role in the formation of chondrule precursors. What the experiments do not demonstrate is the importance of nebular gases in the production of chondrules. These experiments suggest that dusty olivines were produced by reaction with nebular gases prior to the formation of chondrules. Therefore, dusty olivines were formed either as (1) isolated grains that reacted with nebular gases to cause reduction prior to melting within a chondrule precursor clump or (2) during the chondrule melting process by reduction from a carbon phase. Because experiments that use only reducing gases have not reproduced dusty olivine and metal grains with SiO_2 inclusions that were produced by Connolly *et al.* (1994) using carbon in the starting material, we suggest that carbon seems to have been an important component of reduced chondrule precursors.

OTHER QUESTIONS ON CHONDRULE PRECURSORS

Many more questions exist concerning the nature of chondrule precursors. It is possible that the observed population of chondrules was formed from recycled chondrule materials (Alexander, this volume). Future research must study the issue of chondrule precursors being chondrule material recycled either by the remelting of whole chondrules or by the remixing of chondrule debris to make new chondrule-precursor clumps. In addition, most experiments have only concentrated on the role of silicates in chondrule formation. From our review, it is clear that nonsilicates such as carbon and troilite could have been present as chondrule precursors. Therefore, further research needs to be conducted on the potential role of nonsilicates in the production of chondrules.

An additional concern with the issue of chondrule precursors is that Wood (this volume) has pointed out that the assumption that chondrules formed from dustballs is unproved. In the current models of our nebula the formation of dustballs may take thousands of years to form and chondrules could have been formed by coalescing melted dust grains (Skinner, 1991). As experimentalists, we have assumed that chondrules were formed from compact dustballs composed of various minerals and Wood's suggestion needs attention. It is clear that our understanding of chondrule precursors is in its infancy.

CONCLUSIONS

The constraints that experiments have placed on chondrules are as follows:

1. Fine-grained, porphyritic textures typical of MgO-rich chondrules imply fine-grained (< 63 μm) minerals as precursors and ultra-fine-grained chondrules suggest the existence of a micrometer-sized dust precursor.

2. If a porphyritic chondrule texture contains relict grains, then the precursors for that chondrule had *at least some* crystalline olivine and/or pyroxene precursors that were at least as large (up to a few hundred micrometers) as the relict grain(s). Such

chondrules could, however, have been generated from fine material with only a few large grains. It is difficult to state any major constraints on the nature of FeO-rich chondrule precursors based on the variety of melting and cooling conditions that they could have experienced.

3. It is likely that chondrule precursors were anhydrous and that FeS was generally present.

4. Reduced chondrules with dusty relict grains probably had carbon phase(s) as part of their precursors which may have been in part or totally responsible for the reduction. Therefore, chondrules may actually provide few clues to the composition and redox properties of nebular gases. However, the production of dusty olivines from reactions with nebular gases before assemblage of chondrule precursor clumps is another possibility that cannot be ignored.

5. Experiments have not studied the origins of chondrule precursors and the reader is reminded that we do not know if the chondrules we observe were made from recycled chondrule material, nebular condensates, or some combination of the two.

6. Experimentalists can now produce a wide variety of chondrule textures from many different melting and cooling histories. Exact conditions that formed chondrules cannot be unequivocally defined. Therefore, it is difficult to place exact constraints on the nature of chondrule precursors until we learn more about chondrule melting and cooling.

ACKNOWLEDGEMENTS

We thank J. S. Delaney and G. E. Lofgren for numerous discussions and assistance. We also thank the Johnson Space Center Experimental Petrology Laboratories where many of these experiments were performed. Constructive reviews were provided by D. Kring and M. Weisberg and editorial help by A. E. Rubin. This study was supported by NASA PMG and the NASA Graduate Student Researcher Program NGT-50836.

REFERENCES

Connolly H. C. Jr. and Hewins R. H. (1995) Chondrules as products of dust collisions with totally molten droplets within a dust-rich nebular environment: An experimental investigation. *Geochim. Cosmochim. Acta* **59**, 3231–3246.

Connolly H. C. Jr., Jones B. D., and Hewins R. H. (1991) The effect of precursor grain size on chondrule textures (abstr). *Meteoritics* **26**, 329.

Connolly H.C.Jr., Hewins R.H. and Lofgren G. E. (1993a) Flash melting of chondrule precursors in excess of 1600C: series I: Type II (B1) chondrule composition experiments (abstr). *Lunar Planet Sci. Conf.* **XXIV**, 329–330.

Connolly H. C. Jr., Hewins R. H. and Lofgren G. E. (1993b) Possible clues to the physical nature of chondrule precursors: An experimental study using flash melting conditions (abstr). *Lunar Planet. Sci. Conf.* **XXIV**, 329–330.

Connolly H. C. Jr., Hewins R. H., Ash R. D., Zanda B., Lofgren G. E. and Bourot-Denise M. (1994) Carbon and the formation of reduced chondrules. *Nature* **371**, 136–139.

Danielson L. R. and Jones R. H. (1995) Experimental reduction of olivine: constraints on formation of dusty relict olivine in chondrules (abstr). *Lunar and Planet. Sci. Conf.* **XXIV**, 309–310.

Grossman J. N., Rubin A. E., Nagahara H., and King E. A. (1988) Properties of chondrules. In *Meteorites and the Early Solar System.* (eds. J. F.Kerridge and M. S. Mathews), pp 619– 659. University of Arizona Press.

Jones R.H. (1990) Petrology and mineralogy of Type II, FeO-rich chondrules in Semarkona (LL3.0): Origin by closed-system fractional crystallization with evidence for supercooling. *Geochim. Cosmochim. Acta* **54**, 1785–1802.

Jones R.H. and Scott E.R.D. (1988) Petrology and thermal history of chondrules in the Semarkona (LL3.0) chondrite. *Proc. Lunar Planet Sci. Conf. 19th*, 523–536.

Jones R. H. (1994) Petrology of FeO-poor, porphyritic pyroxene chondrules in the Semarkona chondrite. *Geochim. Cosmochim. Acta* **54**, 5325–5340.

Kracher A., Scott E. R. D. and Keil K. (1983) Relict and other anomalous grains in chondrules: implications for chondrule formation. Proc. Lunar Planet. Sci. Conf. 14th (Part 2). *J. Geophys. Res. Suppl.* **89**, B559–B566.

Libourel G. and Chaussidon M. (1995) Experimental constraints on chondrule reduction (abst). *Meteoritics* **30**, 536–537.

Lofgren G. E. and Russell W. J. (1986) Dynamic crystallization of chondrule melts of porphyritic and radial pyroxene composition. *Geochim. Cosmochim. Acta* **50**, 1715–1726.

Maharaj S. V. and Hewins R. H. (1994) Clues to chondrule precursors: An investigation of vesicle formation in experimental chondrules. *Geochim. Cosmochim. Acta* **58**, 1335–1342.

Metzler K., Bischoff, A. and Stoffler D. (1992) Accretionary dust mantles in CM chondrites: Evidence for solar nebula processes. *Geochim. Cosmochim. Acta* **56**, 2873–2897.

Mostefaoui S. and Perron C. (1994) Redox processes in chondrules recorded in metal (abstr). *Meteoritics* **29**, 506.

Nagahara H. (1981) Evidence for secondary origin of chondrules. *Nature* **292**, 135–136.

Radomsky P. M. and Hewins R. H. (1990) Formation conditions of pyroxene-olivine and magnesian olivine chondrules. *Geochim. Cosmochim. Acta* **54**, 3475–3490.

Rambaldi E. R. (1981) Relict grains in chondrules. *Nature* **293**, 558–561.

Skinner W. R. (1991) Origin of chondrules by droplet coalescence: An alternative to the dust- ball hypothesis (abst). *Lunar Planet. Sci.* **XXII**, 1269–1270.

Steele I. M. (1986) Compositions and textures of relic forsterite in carbonaceous and unequilibrated ordinary chondrites. *Geochim. Cosmochim. Acta* **50**, 1379–1395.

Steele I. M. (1989) Compositions of isolated forsterites in Ornans (C3O). *Geochim. Cosmochim. Acta* **53**, 2069–2079.

Wood J. A. (1985) Meteoritic constraints on processes in the solar nebula: An overview. In *Protostars and Planets II* (eds. D. C Black . and M. S Matthews .) pp. 687–701, University of Arizona Press, Tucson.

15: Nature of Matrix in Unequilibrated Chondrites and its Possible Relationship to Chondrules

ADRIAN J. BREARLEY

Institute of Meteoritics, Department of Earth and Planetary Sciences, University of New Mexico, Albuquerque, NM 87131, U.S.A.

ABSTRACT

The petrology, mineralogy and composition of fine-grained matrix materials in unequilibrated chondritic meteorites (ordinary, CV, CO and the unique chondrite, Kakangari) are reviewed. Matrices consist of a complex mixture of material from different sources, including presolar grains, chondrule fragments and condensate material, which may have been formed by a variety of different mechanisms. The relative proportions of these components is currently poorly understood. In comparison with chondrules, matrices are FeO-rich and have refractory element ratios which are fractionated from chondrules. The volatile lithophiles, Na and K, and moderately volatile siderophile and chalcophile elements are also typically enriched in matrices. A direct relationship between chondrules and matrix (either as the precursor or a product) based on their respective refractory element ratios is difficult unless a further stage of fractionation is invoked, prior to or after chondrule formation. Although fine-grained matrix may be a suitable site for the recondensation of volatiles lost from chondrules during chondrule formation, the current data indicate that the volatile content of matrix cannot be simply related to recondensation of volatiles alone.

INTRODUCTION

One of the major questions facing chondrule researchers has been, and continues to be, the nature of the precursor materials from which chondrules themselves form. Many, if not most, of the theories for the origin of chondrules involve the accretion of so called "dustballs", aggregates of fine-grained dust which were melted by episodic, excursive high temperature events within the solar nebula. Although recycling of chondrules as a result of collisional fragmentation, etc., is indicated by the presence of relict grains in chondrules, it is reasonable to assume that a major component of the chondrule precursors was probably fine-grained. This argument would be consistent with the observations that dust in the interstellar medium is extremely fine-grained and it is this type of material which constitutes the raw material from which the solar nebula formed.

A possible candidate for a fine-grained precursor for chondrules is represented by the matrix of chondritic meteorites. Although the abundance of this fine-grained component varies between chondrite groups it is present in all except the enstatite chondrites. The intimate association of matrix and chondrules, including the frequent presence of rims of fine-grained matrix on chondrules, has led to the view that fine-grained dust (as represented by matrix) was present within the solar nebula during chondrule formation and hence could have been involved in chondrule-forming events.

This paper reviews our current knowledge of the properties of matrix (mineralogy, chemistry and isotopic composition) and examines the possible relationships of chondrules to fine-grained matrix. For the purposes of this paper, the discussion will be limited to the most primitive chondrites in chondrite groups which have escaped extensive aqueous alteration and metamorphism and contain nominally anhydrous matrices, i.e., the unequilibrated ordinary chondrites (UOCs), the CO and CV chondrites and the unusual carbonaceous chondrite, Kakangari (for a broader discussion of matrix see Scott *et al*, 1988). However, it should be noted that some members of these chondrite groups have experienced some aqueous alteration and it is extremely important to understand the effects of these processes on the mineralogy and composition of matrix in order to fully examine the possible relationships between chondrules and matrix.

GENERAL CHARACTERISTICS OF MATRIX

Scott *et al.* (1988) defined matrix in primitive, unequilibrated chondrites as "the fine-grained, predominantly silicate material, interstitial to macroscopic, whole or fragmented, entities such as chondrules, inclusions and large isolated mineral (i.e., silicate, metal, sulfide and oxide) grains". This is the definition that is adopted in this paper. This is an important distinction from the older definition in which matrix was regarded as everything external to optically-definable chondrules.

There have been considerable advances in the compositional and mineralogical characterization of matrices in the last 15 years, largely as a result of the use of electron microprobe analysis and transmission electron microscopy (TEM) to study their fine-grained mineralogy. In the last 5 years, the application of microbeam techniques, such as the ion microprobe and synchrotron X-ray fluorescence (SXRF) microprobe has begun to provide much needed trace element data for matrices. These studies show that matrix materials are diverse in character and vary from one chondrite group to another (Table 1). It is apparent that matrix is an extremely complex assemblage of unequilibrated material consisting of silicates (dominantly FeO-rich olivine), oxides, sulfides, sulfates, carbonates, Fe,Ni metal and carbonaceous materials. Matrix in chondritic meteorites is also the host of interstellar grains, such as diamond, silicon carbide and graphite which have now been isolated in almost all the chondrite groups (e.g., Anders and Zinner, 1993). Many of the components of matrix appear to have experienced complex and diverse formational and thermal histories.

This diversity in the properties of matrix from one chondrite group to another has led to a plethora of models for the origin of matrix and no consensus currently exists for the origin of these materials, even within one chondrite group. Thus the situation for matrix differs considerably from that of chondrules, where most workers would regard the mechanism for chondrule formation (whatever that may be) in the different chondrite types as being the same.

ORDINARY CHONDRITES

Mineralogy and petrology

Two distinct occurrences of matrix have been widely recognized in the ordinary chondrites (e.g., Ashworth, 1977; Wlotzka, 1981; Nagahara, 1984; Scott *et al.*, 1984; Alexander *et al.*, 1989a; Matsunami *et al.*, 1990a), one being a porous or clastic matrix which occurs interstitially to chondrules (interchondrule matrix) (Fig. 1a) and the second, finer-grained material which rims chondrules and other macroscopic components (Fig. 1b). The mineralogy and petrology of rims in particular has been documented extensively (e.g., Allen *et al.*, 1980; King and King, 1981; Wilkening *et al.*, 1981; Grossman and Wasson, 1987; Matsunami, 1984; Alexander *et al.*, 1989a).

The mineralogy of matrix and rims is complex (Table 1). Electron microprobe studies have identified (in decreasing order of abundance), olivine (Fo$_{99}$ to Fo$_9$), low-Ca pyroxenes, augite, albite,

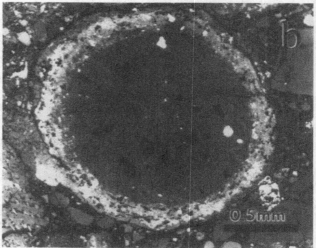

Fig. 1. Electron micrographs showing the typical and most common occurrences of fine-grained matrix materials in unequilibrated ordinary chondrites. a) Region of interchondrule matrix in Bishunpur (LL3.1). The area is dominated by fine-grained material, but larger, angular, clastic fragments of olivine and pyroxene which are probably derived from chondrules are also present. b) Well-developed, fine-grained rim on a porphyritic olivine chondrule in Chainpur (LL3.4). The rim is very fine-grained and is continuous around the periphery of the chondrule.

Fe, Ni metal, troilite, magnetite, spinel, chromite and calcite (Nagahara, 1984). Matrix in the ordinary chondrites is also the major carrier of carbonaceous material (Fredriksson *et al.*, 1969; Kurat, 1970). which consists of both poorly graphitized carbon (Christophe Michel-Lévy and Lautie, 1981) and presolar organic material, which is highly enriched in deuterium (Yang and Epstein, 1983; Alexander *et al.*, 1990).

TEM studies show that in the matrix of one of the least metamorphosed UOCs (Bishunpur, LL3.1) the very fine-grained component (< 1 μm) consists of an amorphous material rich in normative albite, which acts as a groundmass to clastic olivines and pyroxenes (Ashworth, 1977; Alexander *et al.*, 1989a,b). In other UOCs, this very fine-grained component is dominated by a densely-packed groundmass of FeO-rich olivine with a grain size < 0.1 μm

(Alexander *et al.*, 1989a; Brearley *et al.*, 1989). In Semarkona and Bishunpur, evidence for varying degrees of aqueous alteration in the matrix have been reported (Hutchison *et al.*, 1987).

Major and minor element composition of matrix

Data from several broad beam electron microprobe studies of matrix and rims (e.g., Ikeda *et al.*, 1981; Huss *et al.*, 1981; Wlotzka, 1983; Matsunami, 1984; Scott *et al.*, 1984; Brearley *et al.*, 1989; Alexander *et al.*, 1989a; Matsunami *et al*, 1990a,b) show that matrix is compositionally heterogeneous within an individual chondrite and from one chondrite to another. Matrix and rims are fractionated relative to the bulk chondrite with significant enrichments in Al, Na and K (Fig. 2). In comparison with chondrules, matrix is also very enriched in Fe (Fig. 3) based on INAA and electron microprobe data for Semarkona and Chainpur (Grossman and Wasson, 1983; Jones, 1990, 1994; Jones and Scott, 1989). The compositional fields of matrix and rims are quite distinct, although the overall trend of the matrix compositions appears to extend towards that of chondrules. It is apparent, however, that the compositional spread of chondrules in terms of their Si/Mg ratio is significantly larger than that for matrix.

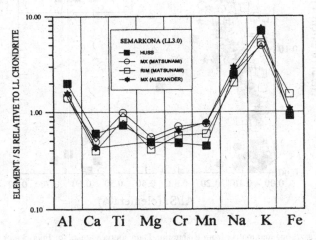

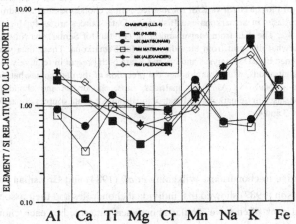

Fig. 2. Abundance diagrams for fine-grained matrices and rims from the unequilibrated ordinary chondrites, Semarkona and Chainpur normalized to the average bulk chondrite composition. The data are from a variety of sources (Huss *et al.*, 1981; Matsunami,1984; Alexander *et al.*, 1989).

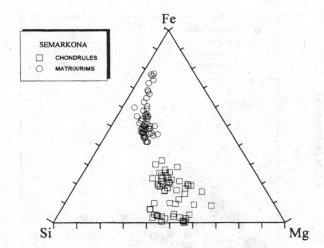

Fig. 3. Ternary Si-Fe-Mg (element wt%) diagram showing the compositional fields of chondrules and matrix in the unequilibrated ordinary chondrite, Semarkona (LL3.0). Matrix data are individual electron microprobe analyses (Brearley, unpubl. data) and the chondrule data are from Grossman and Wasson, 1983, Jones and Scott (1989) and Jones, (1990, 1994).

Refractory elements

It is well-established that chondrules have Ca/Al ratios close to CI, but this ratio in matrix and rims in ordinary chondrites is highly variable. For example, matrix in Semarkona has a relatively well-defined correlation between Ca and Al (Fig. 4a), but at a much lower Ca/Al ratio than CI and Bishunpur matrix Ca/Al ratios show considerable scatter. Data for other ordinary chondrites (e.g. Ikeda *et al.*, 1981) show similar features for their Ca/Al ratios as Semarkona and Bishunpur. Bulk chondrules and mesostasis in several different types of chondrules have correlated Ti/Al, with the CI ratio (Fig. 4b) but no such correlation exists for either matrix or rims and most compositions lie below the CI ratio.

Alkalis

The alkali elements, Na and K, are both moderately volatile elements and are enriched in rims and matrix in comparison with chondrules and bulk chondrites, although their behavior is variable (Fig. 4c). For example, Bishunpur matrix and rims have Na/Si vs K/Si ratios which define a very strong positive correlation, but Semarkona matrix and rims show considerably more scatter. Na can also exhibit significant correlations with Al in chondrules and matrix (Fig. 4d). This relationship is especially strong in Bishunpur matrix, but is less well-defined in Semarkona. Some chondrules also have this ratio but none lie above the CI ratio line (Grossman *et al.*, 1988; Alexander, 1994) and mesostasis compositions show variable behavior with both positive and inverse correlations between Na and Al.

Trace element composition of matrix

Trace element data for matrix and rim materials in the ordinary chondrites are scarce, a result of the difficulty of separating small, uncontaminated, samples of these materials for analysis. The majority of the data have been obtained by INAA techniques (e.g.

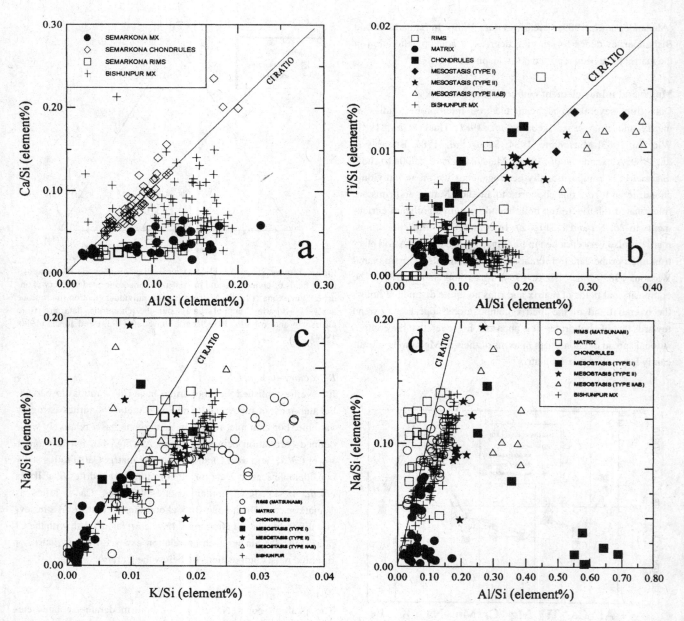

Fig. 4. (a) Ca/Si vs Al/Si plot for matrix, fine-grained rims and chondrules from Semarkona and matrix from Bishunpur. The matrix data from Bishunpur also include data for rims, but no differentiation is made between them on this plot. The data for Bishunpur and Semarkona matrix are individual electron microprobe analyses (Brearley, unpubl. data) whereas the Semarkona rim data are typically the average of several analyses in individual occurrences of matrix from Matsunami et al. (1990b). The chondrule data are from the same sources as Fig. 3. The CI abundance ratio is plotted for reference. (b) Ti/Si vs Al/Si for matrix, fine-grained rims, chondrules and chondrule mesostasis in type I, II and IIB chondrules from Semarkona and matrix from Bishunpur. Data sources are as for Figs. 3 and 4a. (c) Plot of Na/Si vs K/Si for matrices, rims, chondrules and chondrule mesostases in Semarkona and Bishunpur. Data sources are as for Figs. 3 and 4a.. The data from Matsunami et al. (1990b) for Semarkona rims are somewhat different from the other data for Semarkona. These data were obtained by EDS analyses and Na and K are both present in relatively low concentrations, which may reduce the precision of the data somewhat. (d) Plot of Na/Si vs Al/Si for matrices, rims, chondrules and chondrule mesostases in Semarkona and matrix in Bishunpur. Data sources are as for Figs. 3 and 4a.

Rambaldi et al., 1981; Wilkening et al. 1984; Grossman, 1985; Grossman and Wasson, 1987; Nagamoto et al, 1987; Brearley et al., 1989), but recently ion microprobe analyses of ordinary chondrite matrix materials have also become available (Alexander, 1991, 1995).

The INAA data are complex, but basically show that siderophile and volatile lithophile elements are enriched in rims and matrix, relative to chondrules. Wilkening et al. (1984) and Grossman and Wasson (1987) showed that individual chondrules and their associated rims often have similar abundance patterns, but each chondrule/rim pair appears to have a unique abundance pattern. The refractory element abundances are relatively flat and unfractionated, but siderophile and chalcophile element abundances are higher in rims. Many of these elements are volatile to moderately volatile (e.g. Au, As, Ga, Se, Zn) indicating that volatility may have played a role in the rim enrichments.

INAA data have been reported for matrix separates from Semarkona (Grossman, 1985) and for a large (800 μm) dark, fine-grained matrix lump from ALH A77299 (H3.7) (Brearley *et al.*, 1989). Grossman (1985) found that, in general, volatile elements (K, Se, Zn and Br) and siderophile elements were all enriched relative to the chondrules. In comparison, the matrix clast in ALH A77299 has an unfractionated pattern with refractory and volatile lithophiles and siderophile elements all having abundances within 0.3 × CI. Alexander (1991, 1995) has reported ion microprobe data for a variety of trace elements in rims and matrix in several UOCs, including the REE and refractory lithophiles such as Zr. The rims and matrix have essentially flat REE abundance patterns with a slight positive Eu anomaly and are only slightly depleted relative to CI (Fig. 5), whereas chondrule glasses are highly enriched in REE (10–70 × CI).

Oxygen isotopic composition

Clayton *et al.* (1991) have reviewed all the available oxygen isotopic data for ordinary chondrites (including matrix), so the subject will not be dealt with in detail here. The oxygen isotopic composition of matrix in ordinary chondrites is poorly known and not well understood, again because of the significant difficulties of separating samples for isotopic analysis.

CO CARBONACEOUS CHONDRITES

Mineralogy and petrology

Matrix in the CO3 chondrites constitutes between 30–50 volume percent of the meteorite (McSween, 1977a, Scott and Jones, 1990). Matrix occurs as a groundmass to chondrules, isolated mineral grains and CAIs and is frequently present as well-defined, fine-grained rims (Fig. 6a) surrounding all the macroscopic components

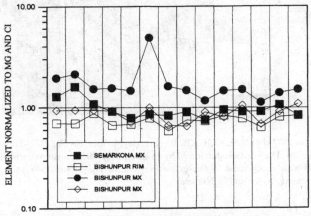

Fig. 5. Abundance diagrams showing typical rare earth element patterns for matrix from Semarkona (LL3.0) and matrix and rims from Bishunpur (LL3.1) ordinary chondrite, normalized to Mg and CI. REE in rims and matrix are typically relatively unfractionated relative to CI and have abundances which are very close to CI. Eu may show a slight positive anomaly. Data from Alexander (1955).

of the meteorite (Metzler *et al.*, 1988; Brearley, 1993). The matrix is extremely fine-grained, with a very low abundance of clastic mineral fragments above 1–2 μm in size.

The least equilibrated CO chondrite, ALH A77307 (3.0) has a matrix which consists of an unequilibrated assemblage of amorphous silicate material, olivine, low-Ca pyroxene, Fe,Ni metal, oxides, sulfides and sulfates (Table 1; Fig. 6b) (Brearley, 1993). In comparison, the matrices of type 3.1 to 3.7 CO chondrites are dominated by fine-grained (< 0.5 μm) FeO-rich olivine (Christophe Michel-Lévy, 1969; Keller and Buseck, 1990a; Brearley, 1994), which surrounds more angular and coarser-grained crystals of olivine.

Major and minor element composition of matrix

McSween (1977a), McSween and Richardson (1977), Ikeda *et al.* (1981), Scott and Jones (1990), Brearley (1993, 1994) and Zolensky *et al.* (1993) have all reported data on the average bulk compositions of matrices in CO3 chondrites. In comparison with bulk chondrule data for Ornans (Rubin and Wasson, 1988), matrices in the CO chondrites are FeO-rich (Fig. 7). However, the full range of chondrule compositions may not be represented in the Rubin and Wasson (1988) data, because more FeO-rich chondrule compositions, such as would be expected for type II, FeO-rich porphyritic olivine compositions are not apparent. Like the ordinary chondrites the matrices of the CO chondrites show no correlations between Ca and Al. This contrasts sharply with data for Ornans chondrules, which show very strongly correlated Ca and Al, with the CI ratio (Fig. 8).

Trace element data

Rubin and Wasson (1988) analyzed a sample of Ornans matrix by INAA techniques and Brearley *et al.* (1993, 1994, 1995) have determined the concentrations of a suite of moderately volatile elements (Cu, Zn, Ga, Ge, Se) in chondrule rims from several CO chondrites, using SXRF microprobe. The data obtained for CO chondrites using bulk and microbeam techniques show considerable discrepancies (Fig.9). The Ornans bulk matrix analysis has an abundance pattern which is essentially unfractionated relative to the bulk chondrite. In comparison, the SXRF data show that moderately volatile elements are significantly enriched in fine-grained rims in the four CO chondrites studied (ALH A77307, Kainsaz, Ornans and Warrenton).

CV CARBONACEOUS CHONDRITES

Mineralogy and petrology

There have been numerous studies of the matrices of the CV chondrites (e.g., Green *et al.*, 1971; Housley and Cirlin, 1983; Peck, 1984; Kornacki and Wood, 1984; Toriumi, 1989; Nakamura *et al.*, 1992; Keller and Buseck, 1990b; Keller *et al.*, 1994). The fine-grained material in the CV3 chondrites constitutes ~ 30–50 vol% (McSween, 1979) and occurs as both chondrule rims (often layered) and as interchondrule matrix. Typical CV3 matrix is somewhat coarser-grained than in the other chondrite groups, with an

Table 1. *Characteristics of matrices in unequilibrated chondrites (H, L, LL, CV, CO and Kakangari)*

Chondrite	% matrix	References	Mineralogy	References
CO3	30–50	1,2,3	amorphous silicate material, olivine (Fo_{0-70}) low-Ca pyroxene, augite, magnetite, Fe,Ni metal, pyrrhotite, pentlandite, Cr-spinel, hercynite, awaruite, albite, (anhydrite, serpentine, Fe^{3+} oxides).	3,4
CV3	35–50	2,5	olivine (Fa_{10}-Fa_{88}), high-Ca pyroxene (Fs_{10-50}, Wo_{45-50}), nepheline, sodalite, andradite, awaruite, pentlandite, troilite, magnetite, phyllosilicates, Ca phosphates,	6,7,8,9,
H, L, LL	5–15	10	olivine (Fa_{0-91}), low-Ca pyroxene (F_{1-20}), augite, amorphous feldspathic material, troilite, Fe,Ni metal, magnetite, chromite maghemite, smectite, calcite, pyrrhotite	11–15
Kakangari	30–50	16,17	low-Ca pyroxene (Fs_{2-5}), olivine (Fo_{0-3}), albite, anorthite, troilite, Fe,Ni metal.	18

Phases which are probably produced by secondary, aqueous alteration are indicated in italics.
References. 1) McSween, 1979; 2) Scott and Jones, 1990; 3) Brearley, 1993; 4) Keller and Buseck, 1990a; 5) McSween, 1977; 6) Scott *et al.*, 1988; 7) Toriumi, 1989; 8): Tomeoka and Buseck, 1990; 9) Keller *et al.*, 1994; 10: Huss *et al.*, 1981; 11) Ashworth, 1977; 12) Nagahara, 1984; 13) Alexander *et al.*, 1989; 14) Brearley *et al.*, 1989; 15) Matsunami *et al.*, 1990a; 16) Mason and Wiik, 1966; 17) Prinz *et al.*, 1989; 18) Brearley, 1989.

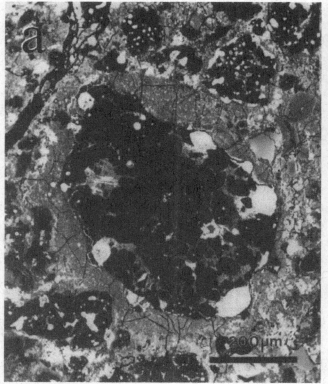
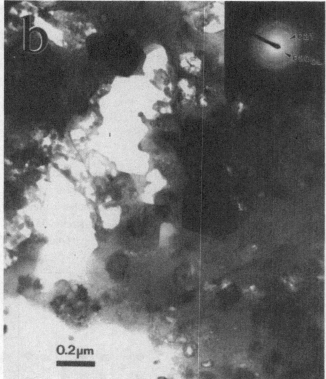

Fig. 6. (a) Backscattered electron (BSE) image of a well-defined chondrule rim surrounding a porphyritic MgO-rich olivine chondrule in the CO3 chondrite, ALH A77307 (Brearley, 1993. *GCA* **57**, 1521–1150, Fig. 1b). (b) Transmission electron micrograph of a typical region of fine-grained matrix in the CO3 chondrite, ALH A77307 (Brearley, 1993. *GCA* **57**, 1521–1150, Fig. 6b). The matrix is dominated by an amorphous silicate material, which acts as a groundmass to fine-grained crystalline phases such as olivine, pyroxene, troilite and Fe,Ni metal.

average grain size of ~5 μm (Scott *et al.*, 1988; Toriumi, 1989), but is also dominated by FeO-rich olivine (Peck, 1984) with a grain size which varies between 0.01 μm and 5 μm (Fig. 10). Two distinct types of olivine occur in CV chondrites matrices: elongate, euhedral grains and subhedral, rounded grains which are often finer

grained (Toriumi, 1989). A wide variety of other silicates, oxides and sulfides have also been identified in CV matrices (Table 1) and evidence for low temperature alteration is present in several CV chondrites including Mokoia, Kaba, Bali, Vigarano (Tomeoka and Buseck, 1990; Keller *et al.*, 1994).

Major and minor element matrix compositions.

Broad beam electron microprobe analyses of CV3 chondrite matrices have been reported by McSween and Richardson (1977), Scott *et al.* (1984, 1988) and Zolensky *et al.* (1993). These data are shown in Fig. 11 in comparison with data for chondrules in Allende determined by Rubin and Wasson (1987). The average matrix compositions of CV chondrites are very similar. In terms of the relationships between chondrules and matrix, Allende obviously differs from CO3 chondrites such as Ornans. Chondrules in Allende extend from very magnesian to iron-rich compositions which overlap the range for the matrix data. The situation in the CV3 chondrites is closer to that found in the ordinary chondrites, where matrix compositions extend towards those of chondrules, at least in terms of their major element (Si-Fe-Mg) composition.

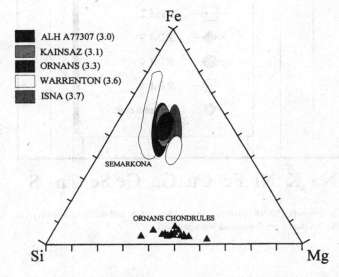

Fig. 7. Ternary Si-Fe-Mg (element wt%) plot showing the compositional fields of matrix in four CO chondrites and Semarkona, (Brearley, 1993, 1994; Brearley, unpubl. data) compared with data for Ornans chondrules from Rubin and Wasson (1988).

Chondrules in Allende (Rubin and Wasson, 1987) appear to have close to the CI Ca/Al ratio (Fig. 12), although Palme (1992, 1994) has reported additional data which suggest that on average chondrules have Ca/Al ratios somewhat lower than CI. Matrix in the different CV chondrites exhibits a range of Ca/Al ratios with no clear relationship between these two elements. There is also considerable variation in the published bulk Ca/Al ratios for Allende matrix. For example, individual broad beam electron microprobe analyses from this study span a range of Ca/Al ratios from below the CI ratio, to above it and Palme (1994) reported INAA data with matrix Ca/Al typically higher than CI.

Trace element data

Rubin and Wasson (1987) used INAA techniques to analyze two samples of matrix from Allende, in addition to several chondrules and coarse-grained rims. These data show that matrix and coarse-grained rims are not very fractionated relative to the bulk chondrite, although the refractory element abundances in matrix are somewhat lower than the bulk. The situation for refractory element abundances in chondrules appears to be rather different and Rubin and Wasson (1987) reported several chondrules which have highly variable refractory element abundances, often significantly fractionated from the bulk chondrite. For the Allende data it is also very apparent that the matrix does not appear to be enriched in volatiles relative to the bulk chondrite. In this respect Allende is very different from the situation in the UOC and CO chondrites.

Oxygen isotopic data

Allende has been analyzed extensively to determine the oxygen isotopic composition of its chondrules, refractory inclusions, coarse-grained rims and matrix (Clayton *et al.*, 1977, 1983; Rubin *et al.*, 1990). Chondrules in Allende span a broad range of oxygen isotopic compositions which lie parallel to the line defined by anhydrous minerals in refractory inclusions from the same meteorite. Matrix, coarse-grained rims and dark inclusions have compositions which lie on this line, but extend to much more ^{16}O poor

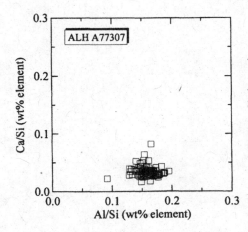

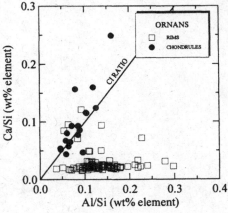

Fig. 8. Compositional variations of Ca and Al in rims from the CO3 chondrites, ALH A77307 (3.0) and Ornans (3.4). Data points represent individual 10 μm beam analyses. Data for Ornans chondrules from Rubin and Wasson (1988) are also shown.

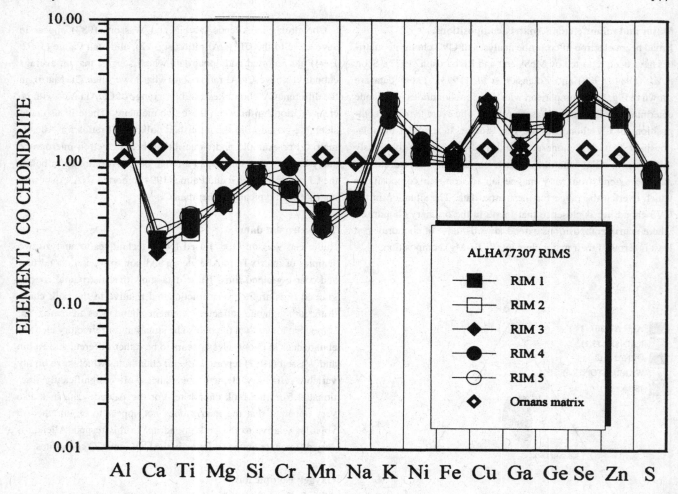

Fig. 9. Abundance patterns for major, minor and trace elements for several chondrule rims in ALHA 77307 and for one matrix separate from Ornans. The ALH A77307 data were obtained by synchrotron X-ray fluorescence microprobe (SXRF) (Brearley *et al.*, 1993, 1995) and the Ornans analysis is INAA data from Rubin and Wasson (1988).

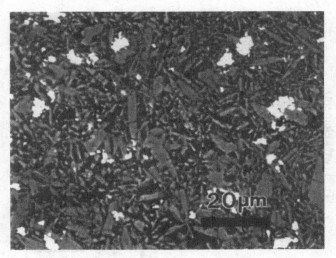

Fig. 10. Backscattered electron image of Allende matrix showing the presence of abundant, platy and elongate olivines with a grain size of < 10 μm. The porosity of the matrix appears to be quite high (~5 vol%).

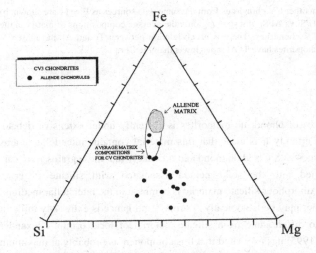

Fig. 11. Ternary Si-Fe-Mg (element wt%) plot showing the average compositions of matrix in several CV chondrites. The compositional field of matrix in Allende is shown based on 10 μm beam electron microprobe analyses (Brearley, unpubl. data). Also shown is the field defined by the average compositions of matrix in several other CV chondrites from McSween and Richardson (1977), Scott et al. (1984, 1988) and Zolensky et al. (1993). The Allende chondrule compositions are INAA analyses taken from Rubin and Wasson (1987).

compositions than porphyritic chondrules. The marked differences between matrix, chondrules and their enclosing coarse-grained rims indicate that they could not have been derived from one another by any simple, direct mechanism.

KAKANGARI (UNGROUPED CARBONACEOUS CHONDRITE)

Kakangari is an ungrouped, unequilibrated carbonaceous chondrite (e.g., Davis et al., 1977), rich in fine-grained matrix material (30 vol%; Prinz et al., 1989) that has several features which are of significance in addressing the relationship between chondrules and matrix.

Petrology and mineralogy

Kakangari matrix has a diverse primary mineralogy dominated by enstatite and olivine, with minor albite, anorthite, Cr-spinel, troilite and Fe,Ni metal (Brearley, 1989). The matrix consists of an extremely fine-grained (< 1 μm) non-clastic component dominated by MgO-rich olivine and pyroxene and a rarer, coarser-grained, clastic component consisting of angular mineral fragments. The non-clastic component of the matrix is characterized by the presence of irregular, distinct clusters or aggregates of crystals, 2–8 μm in diameter, distinguished by irregular boundaries and/or distinct mineralogies from adjacent mineralogical aggregates.

Major and minor element composition

McSween and Richardson (1977) showed that Kakangari has a matrix which is Fe-poor in comparison with matrices in the carbonaceous and unequilibrated ordinary chondrites. Kakangari matrix is also enriched in Na and K relative to CV3 chondrites, a reflection of the presence of albitic feldspar in the matrix, but the other moderately volatile elements, Cr and Mn, as well as the refractory elements, are very similar in abundance to CV3 matrix. Ca and Al are somewhat fractionated from the CI ratio with a Ca/Al ratio = 0.9 (cf., 1.069 for CI). No data are available for the trace element abundances in Kakangari matrix.

Oxygen isotopic composition

Prinz et al. (1989) measured several separated chondrules, one matrix sample and two whole rock samples from Kakangari. Chondrules have ^{16}O-poor compositions, which overlap the oxygen isotopic composition of chondrules in enstatite chondrites. In contrast, matrix lies at the opposite end of a line extrapolated from the chondrules through the whole rock data at more ^{16}O-rich compositions. This relationship indicates that simple fragmentation of chondrules could not have resulted in the formation of matrix.

DISCUSSION

Any discussion of the relationship between chondrules and matrix is closely linked to the origin of matrix itself. Several models for the origin of matrix materials have been proposed and the current situation is extremely complex. It is my view, however, that a general consensus about many aspects of matrix is emerging, although there is still disagreement on many of the details. For example, most workers probably now accept that matrix consists of a variety of different components which formed in different locations and at different times (e.g. interstellar grains, chondrule fragments and condensate material). The main arguments now essentially concern the relative importance of each of these components in matrix materials. A detailed discussion of the formation mechanisms of matrix is beyond the scope of this paper. However, for the purposes of this discussion three possible endmembers can be defined: 1) matrix was a precursor to chondrules, 2) matrix was not a precursor to chondrules and 3) matrix was not a precursor, but actually formed as a result of fragmentation of chondrules. In the context of model 3, if chondrules are viewed as the products of recycling of

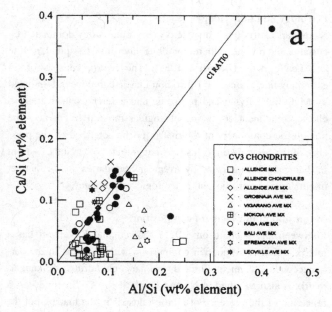

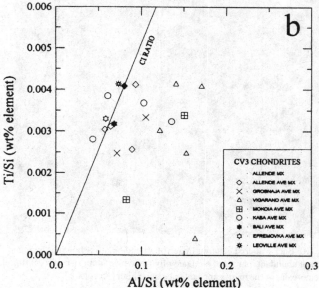

Fig. 12. Refractory element relationships in matrix and chondrules in the CV carbonaceous chondrites. a) Ca/Si vs Al/Si plot showing that chondrules lie close to the CI ratio, but average matrix compositions lie both above and below the CI line. Individual 10 μm beam analyses of matrix from Allende (Brearley, unpubl. data) and average compositions of matrix in other CV chondrites from the same data sources as Fig. 11 are shown. b) Ti/Si vs Al/Si plot showing only the average compositions of matrix in the CV chondrites. There is no correlation between Ti and Al and most CV chondrites have Ti/Al ratios lower than the CI ratio.

chondrules as suggested by Alexander (1994), then matrix could still be considered as a chondrule precursor (i.e., model 1). An additional way that matrix may be related to chondrules, exclusive of the above three models, is that it represents a sink for volatiles which were lost during chondrule formation. As discussed elsewhere in this volume, this is a question which is currently undergoing extensive debate, but clearly data from matrix materials can play an important role in addressing this question.

Having presented mineralogical and compositional data for matrix materials in a variety of chondrite groups above, I will now consider the constraints that these data place on the different models for the relationship between chondrules and matrix.

Mineralogical and petrological constraints

In the UOCs, relatively coarse-grained fragmental olivines and pyroxenes (clastic grains) are quite common within the interchondrule matrix (Alexander *et al.*, 1989a; Brearley *et al.*, 1989). These clastic grains often have compositions which overlap the field of compositions of chondrule olivines and pyroxenes, arguing for a link between these two components. In addition, the power law size distribution found for these fragments in the UOCs, reported by Ashworth (1977) and Alexander *et al.* (1989a), is consistent with an origin by fragmentation of chondrules. Jones (1992) has presented arguments that most isolated grains in the matrices of CO3 chondrites are probably debris from chondrules which were disaggregated by collisions within a nebular environment. Thus there is probably a general consensus that this type of fragmental grain represents chondrule fragments.

In comparison, the origin of the finest grain fraction of the matri-

ces of chondritic meteorites is currently under extensive debate. Certainly it is clear that this material contains interstellar grains such as SiC and diamond and it is likely that these grains were carried into the solar nebula associated with silicate material. Amorphous silicate material is common in the interstellar medium, but may be isotopically "normal" and hence is extremely difficult to differentiate from silicate material of local origin. Alexander (1995) has estimated that this component is probably at maximum < 10 vol% of the finest fraction of matrix material.

Mineralogically, the major difference between chondrules and matrix is the ubiquitous presence of FeO-rich olivine in chondrite matrices, which is absent from chondrules. The presence of this phase precludes direct derivation of matrix from chondrules by a simple, single step process. Several authors have presented the view that this component of matrix represents nebular condensates of one form or another (e.g., Nagahara, 1984; Matsunami, 1984; Brearley *et al.*, 1989; Brearley, 1993; Kornacki and Wood, 1984). In comparison, Alexander *et al.*, (1989a) and Alexander (1995) have argued that in the UOCs, the fine-grained fraction of matrix and rims largely represents fragments of mesostasis released from breakup of chondrules. In this model, the amorphous silicate material in the matrix of the UOC Bishunpur represents fragmented glassy mesostasis from chondrules and the FeO-rich olivine is formed by solid state oxidation and reaction between metal and this silica-rich chondrule glass, probably within a parent body environment. Brearley *et al.* (1989) and Brearley (1993) have suggested, alternatively, that such amorphous material, which is also present in abundance in the CO3 chondrite, ALH A77307 represents the products of disequilibrium condensation processes, which are widely known to produce amorphous, non stoichiometric products, rather than crystalline phases (e.g., Stephens and Kothari, 1979). Such material could represent interstellar material or could have formed within several nebular environments including the high temperature events which formed chondrules themselves. During such energetic events melting of chondrules obviously occurred, but higher energy events could certainly have caused the evaporation and rapid recondensation of dust, far from equilibrium.

In the CV and CO chondrites and Kakangari, there is little mineralogical evidence suggesting that a significant chondrule component is present in the fine-grained matrix. Many phases which are not present in chondrules occur in matrix and for the cases where both phases are present in chondrules and matrix, their compositions, microstructures and mode of occurrence are quite different from those found in chondrules. For example, low-Ca pyroxene in the matrices of the CO chondrites typically contains a high abundance of orthopyroxene, whereas chondrule low-Ca pyroxene is dominated by twinned clinopyroxene (Brearley, 1993). The highly unequilibrated nature of the matrix mineralogical assemblages in the least equilibrated CO chondrites and the diversity of the mineralogical aggregates found in this group of meteorites and in Kakangari also argue against any major contribution from chondrules.

In general, unlike the UOCs, the evidence to support a significant chondrule component within matrix in CV and CO chondrites, as well as Kakangari, appears weak and at present most workers regard the chondrule component to be low (< 10 vol%). This component is probably restricted to the clastic component of matrix (i.e., angular grains, which appear to be of fragmental origin), which have compositions consistent with derivation from chondrules. At present, formation of the bulk of the material in these meteorites within a nebular or possibly presolar environment, involving evaporation and condensation processes appears to be favored.

Compositional constraints

Matrix in all the chondrite groups is fractionated relative to both the bulk chondrite and to CI. It can certainly not be considered as representing samples of unfractionated nebular or presolar dust with a CI composition. The fractionations observed in the different chondrite groups are variable and indicate that the processes which affected matrix materials varied to some degree from one chondrite group to another. The chemical fractionations experienced by matrix materials cannot be simply related to equilibrium condensation, where the chemical signatures would be largely controlled by the relative volatilities of the elements. Other processes, such as aqueous alteration and mild metamorphism may have also influenced the final chemical composition of matrices, especially so in view of their very fine-grained characteristics.

Refractory elements

Many of the arguments for the relationship between chondrules and matrix have centered around the behavior of the refractory elements, especially Ca and Al. It has been argued that if a significant

component of matrix was a precursor to chondrules (or the reverse) then the Ca/Al ratios of matrix and chondrules should be the same. From the data presented above, it is apparent that for the UOCs, the CO3 and CV3 chondrites there are significant discrepancies between chondrule Ca/Al ratios (which are typically close to CI values) and matrix. In most meteorites, except the CV chondrites, Al is enriched significantly in matrix relative to Ca and CI abundances. These observations would appear to argue against any relationship between chondrules and matrix. However, secondary processes may have altered the Ca/Al ratio of matrix, without changing the ratio observed in chondrules. In addition, although data from one chondrite group may indicate that one type of process is important, comparison with data from other chondrite groups can often provide other important constraints.

For the case of the UOCs, the behavior of the refractory elements, Ca and Al, is highly variable from one chondrite to another (Fig. 4a). The data for Semarkona suggest that a relationship between chondrules and matrix is extremely unlikely, because of the differences in the Ca/Al ratios of chondrule and matrix. However, it has been shown by Hutchison *et al.* (1987) that Semarkona matrix has suffered from significant aqueous alteration, which has resulted in the remobilization of Ca to form calcite. Hence, Ca could be fractionated from Al and this obscures the original relationship between these two elements.

Although Ca may be fractionated from Al by aqueous alteration, the same process is not likely to affect Ti, which is widely recognized as being a geochemically immobile element (along with Al). The Ti/Al ratios (Fig. 4b) of matrix and rims in Semarkona are quite variable. In matrix, Ti appears to be uncorrelated with Al, and perhaps weakly correlated in rims, whereas chondrules have strongly correlated Ti and Al with close to the CI value. Ti and Al are both incompatible in chondrules and are enriched in the mesostases which have a Ti/Al ratio close to CI. Therefore, it is difficult to explain the fact that the Ti/Al ratios of chondrules (and chondrule mesostasis) and matrix are quite distinct if the fine-grained fraction of matrix is dominated by fragments of chondrule glass, as suggested by Alexander *et al.* (1989a).

An additional mechanism for fractionating Ca from Al can occur as a result of mild metamorphism. It has been recognized that P is lost rapidly from Fe,Ni metal in slightly metamorphosed chondrites and reacts with Ca to produce phosphates such as apatite or merrillite (Hutchison and Bevan, 1983). Such a reaction would also fractionate REE from the matrix as well. This hypothesis remains to be tested rigorously but it is currently a viable mechanism for fractionating Ca from Al. However, the problem with the discrepant Ti/Al ratio of chondrule mesostasis and matrix still exists for this model.

The fractionation of Ca from Al is also strongly apparent in the CO3 chondrites (Fig. 10). None of these meteorites have suffered extensive aqueous alteration, so it is not possible to invoke this mechanism for fractionating these two elements. In addition, the metal content of CO3 chondrites is also low (< 5 vol%, McSween, 1977a), so that the loss of P from metal to form Ca phosphates is implausible based on mass balance calculations. This suggests that the fractionation of Ca from Al may be a primary, rather than secondary feature, in the CO3 chondrites. If this is the case then it is highly improbable that matrix could have been a significant precursor of chondrules (or vice versa). Palme (1994) has also noted that the discrepant Ca/Al ratios of matrix and chondrules in Allende argue against chondrules being related to matrix as either a precursor or product phase.

Although the fractionation of Ca from Al and the enrichment of Al in matrix may be primary in some chondrite groups the mechanism for producing this fractionation is no less problematical than by a parent body process. It could be achieved if the dominant carriers of Al in matrix are phases such as corundum or spinel. These minerals would certainly have survived processing within a nebular environment if they were ever present within fine-grained nebular dust. However, neither of these phases has been found in any abundance in the fine-grained component of matrices, so such an explanation seems to be implausible. An additional possibility is that a component rich in albite or a feldspathoid such as nepheline might be incorporated into matrix materials, which would also increase the Na content of matrix, without appealing to loss of volatiles from chondrules. However, data for rims and matrix for the ordinary chondrites (Alexander, 1995) suggest that albite is not likely to be an important component in matrix. On the other hand, in Kakangari, albite is definitely present as inclusions within low-Ca pyroxene, although Al is actually depleted relative to Ca in the matrix of this meteorite.

Clearly, there are significant problems in establishing a direct relationship between chondrules and matrix in chondritic meteorites, based on refractory element abundances. Although it is possible to fractionate Ca from Al within a parent body environment, most reactions are selective and should not affect all refractory element ratios. In particular the excess Al in matrices is problematic for both nebular models and models which invoke formation of matrices from chondrules.

Volatile and siderophile enrichments in matrix

The origin of the siderophile and volatile element depletions observed in chondrules is currently an area of extensive discussion. Some workers (e.g., Grossman *et al.*, 1988; Rubin and Wasson, 1988) have argued that siderophile and volatile element fractionation occurred before chondrule formation, whereas other workers believe that this process occurred during (e.g., Alexander, 1994) or indeed any time before, during or after chondrule formation (e.g., Newsom, 1994). These questions are addressed in detail elsewhere in this volume. However, because matrix and rims may have been a sink for volatile and siderophile elements if they were lost from chondrules, it is important to examine the evidence from matrix.

Loss of siderophile elements could have occurred by physical separation of metal beads from molten chondrules (Grossman and Wasson, 1985) or by evaporation of volatile elements at high temperatures. Recondensation of these elements onto fine-grained dust (as represented by matrix) would be expected in dusty regions of the nebula. Superficially, at least, it is possible that the enrichment of Fe observed in matrix could result from metal loss from chon-

drules. This would, however, require extensive oxidation and reaction between metal and silicate phases to produce the FeO-rich olivines which dominate the mineralogy of matrix as proposed by Alexander *et al.*, (1989a). Fe could also have been lost as a result of volatilization from metal and silicates and recondensed onto matrix materials. For the case of the ordinary chondrites it is difficult to assess whether volatiles were lost from chondrules into matrix, because of a) the limited amount of data for minor and trace volatile elements and b) the complexity of the fractionation patterns obtained by INAA techniques. For the ordinary chondrites, the best data are for the alkalis, Na and K (e.g. Fig. 4c). In chondrules, rims and matrix, Na and K are strongly correlated, but chondrules are depleted and matrix and rims enriched in these elements. Chondrules appear to have close to the CI Na/K ratio, but Bishunpur matrix and rims, for example, have a lower Na/K ratio, the reverse to what would be expected based on volatility alone. Na is more volatile than K and would be expected to be enriched in matrix. Alexander (1995) has suggested that Na could have diffused back into chondrules if fine-grained rim material accreted onto chondrules while they were still hot. This process could fractionate Na from K, but it is not clear whether the excellent correlation between the two elements would be preserved.

If volatility played a major role in introducing volatile elements from chondrules into matrix, an important question which must be addressed is why the behavior of these elements varies from one chondrite group to another. In the matrices of ordinary and CO chondrites, as well as Kakangari, K is always enriched relative to Na, whereas the reverse is true for the CV chondrites. In addition the abundances of these elements in matrix in the different chondrite groups varies widely. CI normalized abundances for Na and K are enriched in the ordinary chondrites (2 and $6 \times$ CI respectively); in the CO chondrites and Kakangari, these values are 0.5–0.8 and 0.5–1.1 $\times$ CI respectively and for the CV chondrites Na = 0.2–0.8 $\times$ CI and K = 0.1–0.2 $\times$ CI. The process of volatile loss and back diffusion of Na into chondrules would be expected to be similar in all chondrites but this is obviously not the case. Assuming that chondrules did lose volatiles, MgO-rich chondrules with high liquidus temperatures should have lost the most mass. However, CO chondrites which have a very high abundance of MgO-rich chondrules, have matrices and rims which are not especially enriched in Na or K relative to either CI or bulk chondrite abundances.

Based on rather limited data it appears that the behavior of other moderately volatile elements is also complex (e.g., Brearley *et al.*, 1993, 1995) and is not simply related to volatility (e.g., Fig. 9). For example, Cu in CO chondrite rims is highly enriched, whereas Zn, which is significantly more volatile, is depleted. This type of behavior is the reverse of what would be anticipated for volatile enrichments produced as a result of recondensation of volatiles lost during chondrule formation alone. It is also evident that the behavior of moderately volatile lithophile elements is somewhat different from the moderately volatile siderophile and chalcophile elements. Cu (chalcophile) with a condensation temperature of 1037 K is enriched by a factor of two compared with K, which condenses at 1000 K (Wasson, 1985).

In conclusion, the current evidence linking volatile enrichments in matrix with depletions in chondrules is equivocal and further work is obviously required to clarify this situation. The lack of data on a significant suite of volatile elements for matrix and rims is the crucial factor in limiting constraints on such models.

Oxygen isotopic constraints

The current data for the ordinary chondrites do not provide any additional constraints on the relationship between chondrules and matrix. Chondrules could have formed from matrix or vice versa, based on these data alone. In contrast, the differences in isotopic composition of matrix and chondrules in Allende and Kakangari argue that matrix did not form from fragmentation of chondrules, at least not by any simple mechanism. If matrix was formed by fragmentation of chondrules, a change in oxygen isotopic composition is required in which the fine-grained component exchanged oxygen with a reservoir with a distinct oxygen isotopic composition, without affecting the chondrules. Such a process appears to be unlikely, although not completely implausible in a nebular environment.

CONCLUSIONS

Although many aspects of the relationships between chondrules and matrix are still unclear, there is a growing consensus about the types of materials that are present in matrix materials. Certainly some of these materials, such as interstellar grains and carbonaceous materials, which are concentrated in matrix, were not derived from chondrules. However, clastic mineral fragments certainly represent chondrule debris produced by collisions within the solar nebula. Arguments can be presented which support the idea that the finest grain fraction of ordinary chondrite matrices represent chondrule mesostasis glass released during chondrule fragmentation whereas for the CV and CO chondrites and Kakangari such an origin seems unlikely. In the latter case, an origin for matrix materials by nebular processes, involving evaporation and condensation, appears most likely. Neither of these models can entirely explain the mineralogical and compositional characteristics of matrix and several problems for both types of model remain. However, for some chondrites, the evidence from refractory elements argues that matrix could not have been a major precursor for most chondrules and probably evolved separately from chondrules. Finally, the question of whether volatiles lost from chondrules recondensed onto matrix cannot be resolved unequivocally at this point. Further detailed studies to examine the distribution of moderately volatile elements between chondrules and matrix materials are required to clarify this relationship.

ACKNOWLEDGEMENTS

I am grateful to Ed Scott, Rhian Jones, Conel Alexander, Frans Rietmeijer and Ian Mackinnon for many stimulating discussions about fine-grained materials over recent years and to Mike Zolensky and Hiroko Nagahara for their formal reviews. Funding was provided by the Institute of Meteoritics, University of New Mexico and NASA grant, NAGW 3347 to J.J. Papike (P.I.). Electron microprobe analyses and transmission electron microscopy were carried out in the Electron Microbeam Facility, Department of Earth and Planetary Sciences and Institute of Meteoritics, University of New Mexico.

REFERENCES

Alexander C.M.O'D. (1991) The origin of matrix and rims in Bishunpur (L/LL3.1): an ion microprobe trace element study. *Meteoritics* **26**, 312–313.

Alexander C.M.O'D. (1994) Trace element distributions within ordinary chondrite chondrules: Implications for chondrule formation conditions and precursors. *Geochim. Cosmochim. Acta* **58**, 3451–3467.

Alexander C.M.O'D. (1995) Trace element contents of chondrule rims and interchondrule matrix in ordinary chondrites. *Geochim. Cosmochim. Acta* (submitted).

Alexander C.M.O'D, Hutchison R. and Barber D.J. (1989a) Origin of chondrule rims and interchondrule matrices in unequilibrated ordinary chondrites. *Earth Planet. Sci. Lett.* **95**, 187–207.

Alexander C.M.O'D, Barber D.J. and Hutchison R. (1989b) The microstructure of Semarkona and Bishunpur. *Geochim. Cosmochim. Acta* **53**, 3045–3057.

Alexander C.M.O'D., Arden J.W., Ash R.D. and Pillinger C.T. (1990) Presolar components in the ordinary chondrites. *Earth Planet. Sci. Lett.* **99**, 220–229.

Allen J. S., Nozette S. and Wilkening L. L. (1980) A study of chondrule rims and chondrule irradiation records in unequilibrated ordinary chondrites. *Geochim. Cosmochim. Acta* **44**, 1161–1175.

Anders E. and Zinner E. (1993) Interstellar grains in primitive meteorites: diamond, silicon carbide and graphite. *Meteoritics* **28**, 490–514.

Ashworth J. R. (1977) Matrix textures in unequilibrated ordinary chondrites. *Earth Planet. Sci. Lett.* **35**, 25–34.

Brearley A.J. (1989) Nature and origin of matrix in the unique chondrite, Kakangari. *Geochim. Cosmochim. Acta* **53**, 2395–2411.

Brearley A.J. (1993) Matrix and fine-grained rims in the unequilibrated CO3 chondrite, ALH 77307: Origins and evidence for diverse, primitive nebular dust components. *Geochim. Cosmochim. Acta* **57**, 1521–1150.

Brearley A.J. (1994) Metamorphic effects in the matrices of CO3 chondrites: compositional and mineralogical variations. *LPS* **XXV**, 165–166.

Brearley A.J., Bajt S. and Sutton S.R. (1993) SXRF determination of trace elements in the chondrule rims in the unequilibrated CO3 chondrite, ALH A77307. *LPS* **XXIV** 187–188.

Brearley A.J., Bajt S. and Sutton S.R. (1994) Metamorphism in the CO3 chondrites: trace element behavior in matrices and rims. *LPS* **XXV** 167–168.

Brearley A.J., Bajt S. and Sutton S.R. (1995) Distribution of moderately volatile trace elements in chondrule rims in the unequilibrated CO3 chondrite, ALH A77307. *Geochim. Cosmochim. Acta* **59**, 4307–4316.

Brearley A.J., Scott E.R.D., Keil K., Clayton R.N., Mayeda T.K., Boynton W.V. and Hill D.H. (1989) Chemical, isotopic and mineralogical evidence for the origin of matrix in ordinary chondrites. *Geochim. Cosmochim. Acta* **53**, 2081–2093.

Christophe Michel-Lévy, M. (1969) Etude minéralogique de la chondrite CIII de Lancé. In *Meteorite Research* (ed. P. M. Millman) pp. 492–499. D. Reidel, Dordrecht.

Christophe Michel-Lévy M. and Lautie A. (1981) Microanalysis by Raman spectroscopy of carbon in the Tieschitz chondrite. *Nature* **292**, 321–322.

Clayton R. N. and Mayeda T. K. (1977) Anomalous anomalies in carbonaceous chondrites. *Lunar Sci.* **VIII**, 193–195 (abstract).

Clayton R. N., Onuma N., Ikeda Y., Mayeda T. K., Hutcheon I., Olsen E. J. and Molini-Velsko C. (1983) Oxygen isotopic compositions of chondrules in Allende and ordinary chondrites. In *Chondrules and Their Origins* (ed. E. A. King) pp. 37-43. Lunar and Planetary Inst., Houston.

Clayton R.N., Mayeda T.K., Goswami J.N. and Olsen E.J. (1991) Oxygen isotopic studies of ordinary chondrites. *Geochim. Cosmochim. Acta* **55**, 2317–2337.

Davis A.M., Grossman L. and Ganapathy R. (1977) Yes, Kakangari is a unique chondrite. *Nature* **265**, 230–232.

Fredriksson K., Jarosewich E. and Nelen J. (1969) The Sharps chondrite-new evidence for the origin of chondrules and chondrites. In *Meteorite Research* (ed. P.M. Millman), pp. 155–165. Springer-Verlag.

Green H. W., Radcliffe S. V. and Heuer A. H. (1971) Allende meteorite: A high voltage electron petrographic study. *Science* **172**, 936–939.

Grossman J. N. (1985) Chemical evolution of the matrix of Semarkona. *Lunar Planet. Sci.* **XVI**, 302–303 (abstract).

Grossman J. N. and Wasson J. T. (1983) Refractory precursor components of Semarkona chondrules and fractionation of refractory elements among chondrites. *Geochim. Cosmochim. Acta* **47**, 759–771.

Grossman J. N. and Wasson J. T. (1985) The origin and history of the metal and sulfide components of chondrules. *Geochim. Cosmochim. Acta* **49**, 925–939.

Grossman J.N. and Wasson J.T. (1987) Compositional evidence regarding the origin of rims on Semarkona chondrules. *Geochim. Cosmochim. Acta* **51**, 3003–3011.

Grossman J. N. Clayton R. N. and Mayeda T. K. (1987) Oxygen isotopes in the matrix of the Semarkona (LL3.0) chondrite. *Meteoritics* **22**, 395–396.

Grossman J.N. Rubin A.E., Nagahara H. and King E.A. (1988) Properties of chondrules. In *"Meteorites and the Early Solar System"*, (eds., J.F. Kerridge and M.S. Matthews), University of Arizona Press, pp. 619–659.

Housley R.M. and Cirlin E.H. (1983) On the alteration of Allende chondrules and the formation of matrix. In *Chondrules and Their Origins* (ed. E.A. King), pp. 145–161. Lunar Planet. Inst., Houston.

Huss G. R., Keil K. and Taylor G. J. (1981) The matrices of unequilibrated ordinary chondrites: Implications for the origin and history of chondrites. *Geochim. Cosmochim. Acta* **45**, 33–51.

Hutchison R. and Bevan A.W.R. (1983) Conditions and time of chondrule accretion. In *Chondrules and Their Origins* (ed. E.A. King), pp. 162–179. Lunar Planet. Inst., Houston.

Hutchison R. Alexander C. M. O. and Barber D. J. (1987) The Semarkona meteorite: First recorded occurrence of smectite in an ordinary chondrite, and its implications. *Geochim. Cosmochim. Acta* **51**, 1875–1882.

Ikeda Y., Kimura M., Mori H. and Takeda H. (1981) Chemical compositions of matrices of unequilibrated ordinary chondrites. In *Proc. 6th Symp. Antarctic Meteorites, Mem. Natl. Inst. Polar Res.*, Special Issue **20**, 124–144.

Jones R.H. and Scott E.R.D. (1989) Petrology and thermal history of type IA chondrules in the Semarkona (LL3.0) chondrite. *Proc. 19th Lunar Planet. Sci. Conf.* 523–536.

Jones R.H. (1990) Petrology and mineralogy of type II, FeO-rich, chondrules in Semarkona (LL3.0): Origin by closed-system fractional crystallization, with evidence for supercooling. *Geochim. Cosmochim. Acta* **54**, 1785–1802.

Jones R.H. (1992) On the relationship between isolated and chondrule olivine grains in the carbonaceous chondrite ALHA 77307. *Geochim. Cosmochim. Acta* **56**, 467–482.

Jones R.H. (1994) Petrology of FeO-poor, porphyritic pyroxene chondrules in the Semarkona chondrite. *Geochim. Cosmochim. Acta* **58**, 5325–5340.

Keller L.P. and Buseck P.R. (1990a) Matrix mineralogy of the Lancé CO3 carbonaceous chondrite: A transmission electron microscope study. *Geochim. Cosmochim. Acta* **54**, 1155–1163.

Keller L.P. and Buseck P.R. (1990b) Aqueous alteration in the Kaba CV3 carbonaceous chondrite. *Geochim. Cosmochim. Acta* **54**, 2113–2110.

Keller L.P., Thomas K.L., Clayton R.N., Mayeda T.K., DeHart, J.M. and McKay D.S. (1994) Aqueous alteration of the Bali CV3 chondrite: Evidence from mineralogy, mineral chemistry, and oxygen isotopic compositions. *Geochim. Cosmochim. Acta* **58**, 5589–5598.

King T. V. V. and King E. A. (1981) Accretionary dark rims in unequilibrated ordinary chondrites. *Icarus* **48**, 460–472.

Kornacki A.S. and Wood J.A. (1984) The mineral chemistry and origin of inclusion matrix and meteorite matrix in the Allende CV3 chondrite. *Geochim. Cosmochim. Acta* **48**, 1663–1676.

Kurat G. (1970) Zur genese des köhligen materials im meteoriten von Tieschitz. *Earth Planet. Sci. Lett.* **7**, 317–324.

Mason B. and Wiik H.B. (1966) The composition of the Bath, Frankfort, Kakangari, Rose City and Tadjera meteorites. *Am. Mus. Novitates* **2272**, 23 pp.

Matsunami S. (1984) The chemical compositions and textures of matrices and chondrule rims of eight unequilibrated ordinary chondrites. In *Proc. 9th Symp. Antarctic Meteorites. Natl. Inst. Polar Res.* **35**, 126–148.

Matsunami S., Nishimura H. and Takeshi H. (1990a) The chemical compositions and textures of matrices and chondrule rims of unequilibrated ordinary chondrites-II. Their constituents and implications for the formation of matrix olivine. In *Proc. NIPR Symp. Antarct. Meteorites* no. **3**, 126–148.

Matsunami S., Nishimura H. and Takeshi H. (1990b) Compositional heterogeneity of fine-grained rims in the Semarkona (LL3) chondrite. In *Proc. NIPR Symp. Antarct. Meteorites* no. **3**, 181–193.

McSween H. Y., Jr. (1977a) Carbonaceous chondrites of the Ornans type: a metamorphic sequence. *Geochim. Cosmochim. Acta* **44**, 477–491.

McSween H. Y., Jr. (1977b) Petrographic variations among carbonaceous chondrites of the Vigarano group. *Geochim. Cosmochim. Acta* **47**, 1777–1790.

McSween, H. Y., Jr. (1979). Are carbonaceous chondrites primitive or processed? A review. *Rev. Geophys. Space Phys.* **17**, 1059–1078.

McSween H. Y., Jr. and Richardson S. M. (1977) The composition of carbonaceous chondrite matrix. *Geochim. Cosmochim. Acta* **41**, 1145–1161.

Metzler K., Bischoff A. and Stöffler D. (1988) Characteristics of accretionary dark rims in carbonaceous chondrites. *Lunar Planet. Sci.* **XIX**, 772–773.

Nagahara H. (1984) Matrices of type 3 ordinary chondrites—Primitive nebular records. *Geochim. Cosmochim. Acta* **48**, 2581–2595.

Nagamoto H., Nakamura N., Nishikawa Y., Misawa K. and Noda S. (1987) Distribution of trace elements in the rim-core of Tieschitz (H3) chondrules and matrix. *Abstracts 12th Symp. Antartic. Meteorites*, 84–86.

Nakamura T., Tomoeka K. and Takeda H. (1992) Shock effects of the Leoville CV carbonaceous chondrite: a transmission electron microscope study. *Earth Planet. Sci. Lett.* **114**, 159–170.

Newsom H.E. (1994) Siderophile elements and metal-silicate fractionation in the solar nebula. In *Chondrules and the Protoplanetary Disk*. LPI Contribution No. 844, Lunar and Planetary Institute, Houston. 50 pp. (abstract).

Palme H. (1992) Formation of Allende chondrules and matrix. *NIPR 17th Symp. Antarctic. Meteor.* 193–195.

Palme H. (1994) Formation of chondrules and CAIs by nebular processes. In *Chondrules and the Protoplanetary Disk*, LPI Contribution No. 844, Lunar and Planetary Institute, Houston. 50 pp. (abstract).

Peck J. A. (1984) Origin of the variation in properties of CV3 meteorite matrix and matrix clasts. Lunar *Planet. Sci.* **XVI** 635–636 (abstract).

Prinz M., Weisberg M.K., Nehru C.E., MacPherson G.J., Clayton R.N. and Mayeda T.K. (1989) Petrologic and stable isotope study of the Kakangari, (K-group) chondrite: Chondrules, matrix and CAI's (abstr.) *Lunar Planet Sci.* **XX**, 870–871.

Rambaldi E. R., Fredriksson B. I. and Fredriksson K. (1981) Primitive ultrafine matrix in ordinary chondrites. *Earth Planet. Sci. Lett.* **56**, 107–126.

Rubin A. E. and Wasson J. T. (1987) Chondrules, matrix and coarse-grained chondrule rims in the Allende meteorite. *Geochim. Cosmochim . Acta* **51**, 1923–1937.

Rubin A. E. and Wasson J. T. (1988) Chondrules in the Ornans CO3 meteorite and the timing of chondrule formation relative to nebular fractionation events. *Geochim. Cosmochim. Acta*, **52**, 425–432.

Rubin A. E., Wasson J. T. Clayton R. N. and Mayeda T. K. (1990) Oxygen isotopes in chondrules and coarse-grained chondrule rims from Allende. *Earth Planet. Sci. Lett.* **96**, 247–255.

Scott E.R.D. and Jones R.H. (1990) Disentangling nebular and asteroidal features of CO3 carbonaceous chondrites. *Geochim. Cosmochim. Acta* **54**, 2485–2502.

Scott E. R. D., Rubin A. E., Taylor G. J. and Keil K. (1984) Matrix material in type 3 chondrites-Occurrence, heterogeneity and relationship with chondrules. *Geochim. Cosmochim. Acta* **48**, 1741–1757.

Scott E.R.D., Barber D.J., Alexander C.M., Hutchison R. and Peck J.A. (1988) Primitive material surviving in chondrites: matrix. In *"Meteorites and the Early Solar System"*, (eds., J.F. Kerridge and M.S. Matthews), University of Arizona Press, pp. 718–745.

Stephens J.R. and Kothari B.K. (1978) Laboratory analogues to cosmic dust. *Moon and Planets* **19**, 139–152.

Tomeoka K. and Buseck P.R. (1990) Phyllosilicates in the Mokoia CV carbonaceous chondrite: Evidence for aqueous alteration in an oxidizing condition. *Geochim. Cosmochim. Acta* **54**, 1787–1796.

Toriumi M. (1989) Grain size distribution of the matrix in the Allende chondrite. *Earth Planet. Sci. Lett.* **92**, 265–273.

Wasson J.T. (1985) *Meteorites.* (New York, W.H. Freeman and Company).

Wilkening L. L., Boynton W. V. and Hill D. H. 1984. Trace elements in rims and interiors of Chainpur chondrules. *Geochim. Cosmochim. Acta* **48**, 1071–1080.

Wlotzka F. 1983. Composition of chondrules, fragments and matrix in the unequilibrated ordinary chondrites Tieschitz and Sharps. In *Chondrules and Their Origins.* (ed. E. A. King), pp. 296–318. Lunar and Planetary Inst., Houston.

Yang J. and Epstein S. (1983) Interstellar organic matter in meteorites. *Geochim. Cosmochim. Acta* **47**, 2199–2216.

Zolensky M.E., Barrett R. and Browning L. (1993) Mineralogy and composition of matrix and chondrule rims in carbonaceous chondrites. *Geochim. Cosmochim. Acta* **57**, 3123–3148.

16: Constraints on Chondrite Agglomeration from Fine-Grained Chondrule Rims

KNUT METZLER [1] and *ADOLF BISCHOFF* [2]

[1]*Institut für Mineralogie, Museum für Naturkunde der Humboldt-Universität, Invalidenstr. 43, 10115 Berlin, Germany.* [2]*Institut für Planetologie, Universität Münster, Wilhelm-Klemm-Str. 10, 48149 Münster, Germany.*

ABSTRACT

Fine-grained rims (dust mantles) around chondrules, CAIs, etc. occur in most types of unequilibrated chondrites and share many common characteristics. We have summarized literature data on fine-grained rims from various chondrite classes and focus in detail on rims from CM chondrites. These are characterized as follows: 1) they occur around all types of coarse chondritic components, i.e. chondrules, CAIs, etc.; 2) they exhibit sharp contacts with the central objects; 3) their overall shapes are smooth and subrounded, despite angular and irregularly shaped central objects; 4) they show sedimentary textures, consisting of mineral grains < 1 to ~50 micrometer in size, embedded in a submicron-sized groundmass; 5) they frequently show multi-layering consisting of two or more concentric dust layers of different composition; 6) their bulk chemical compositions vary considerably in a given meteorite; 7) their chemical composition is not correlated to that of the central object; 8) they lack solar wind implanted noble gases; 9) they contain presolar SiC grains; 10) their thickness is positively correlated to the size of the central object. We suggest that fine-grained rims formed in the solar nebula by accretion of dust onto the surfaces of chondrules and other coarse components during their passage through dust-rich regions of the solar nebula.

The CM chondrite Y-791198 and many lithic clasts from brecciated CM chondrites consist almost entirely of dust-mantled chondrules, CAIs, etc. This rock type seems to represent a mechanically unaltered "cosmic sediment" which has preserved its accretionary texture over the last 4.5 Ga and may represent the first rock generation of the CM parent body.

Since the fragile dust mantle structures were not destroyed during parent body formation, we conclude that accretion took place in a regime of low relative velocities (< 5 ms⁻¹). Noble gas measurements and nuclear track studies indicate that the CM parent body formed in a nebula region that was shielded from solar and galactic radiation.

INTRODUCTION

Fine-grained rims around chondrules, Ca, Al-rich inclusions, and other coarse components occur in most types of unequilibrated chondrites, most prominently in carbonaceous chondrites of the CM group. Kurat (1970) observed that all chondrules, chondrule and mineral fragments in the H3 chondrite Tieschitz are covered with thin layers of "carbonaceous" material. He concluded that these layers condensed onto the surfaces of these objects in impact generated gas and dust clouds, whereas Allen *et al.* (1979, 1980) suggested that rims could have formed in the solar nebula by accretion of dust. Since then, there is a growing acceptance that fine-grained rims may represent the result of chondrule/dust interactions in low temperature regions of the solar nebula (e.g. MacPherson and Grossman, 1981; King and King, 1981; Bunch and Chang, 1984; Scott *et al.*, 1984; MacPherson *et al.*, 1985; Metzler and Bischoff, 1987; Kring, 1988; Brearley and Geiger,1991; Tomeoka *et al.*, 1991; Nakamura *et al.*, 1991; Metzler *et al.*, 1992). There is a large amount of detailed work in the literature concerning the texture, mineralogy and chemistry of fine-grained rims in chondrites. In Table 1 we have summarized most of these publications for several chondrite classes. An overview about the mineralogical compositions of rims from carbonaceous chondrites is given in Zolensky *et al.* (1993; their Table 2).

Several other rim types have been observed in chondritic meteorites. These include coarse-grained rims (e.g. Rubin, 1984; Rubin and Wasson, 1987), rims with igneous textures (e.g. Kring, 1991) and layers around CAIs (e.g. Wark and Lovering, 1977). These rim types were formed when rim material was molten or processed in hot nebular regimes (Rubin and Krot, this volume). Rims of this type should not be mistaken for the fine-grained rims that were probably formed in low temperature environments.

In the following we will focus in detail on the occurrence, mineralogy and chemistry of fine-grained rims in CM chondrites, followed by a brief summary of descriptions of this rim type in various other chondrite classes. Finally, it will be discussed what can be learned from fine-grained rims and their host rock about solar nebula conditions and chondrule agglomeration processes that led to the formation of primitive chondritic planetesimals.

FINE-GRAINED RIMS IN CM CHONDRITES

Several investigators have recognized that fine-grained dust-like materials surround various components (chondrules, fragments,

etc.) in CM chondrites. These structures have been described as "narrow bands of fine-grained, inclusion-free matrix" (Fuchs *et al.*, 1973), "accretionary dark rims" (King and King, 1981), "black rinds" (Kinnunen and Saikkonen, 1983), "dust balls" (Bunch and Chang, 1984) and "accretionary dust mantles" (Metzler *et al.*, 1992). In Fig. 1 a low-FeO chondrule fragment from the Antarctic CM chondrite Y791198 with its dust mantle is shown as an example.

Concentration of components with fine-grained rims in "primary rock"

Although chondrules, CAIs, etc. with fine-grained rims (dust mantles) can be found in most CM chondrites, there are significant differences in the abundance and textural settings of these objects. The CM chondrite Y-791198, that seems to have escaped impact brecciation on its parent body, represents a densely packed agglomerate of those components, without significant amounts of fine-grained interchondrule matrix. This texture (see Fig. 6) is interpreted as accretional and this meteorite was defined as "primary accretionary rock" or "primary rock" for short (Metzler *et al.*, 1992; see discussion). All other CM chondrites are brecciated rocks containing up to cm-sized clasts of primary rock embedded in a fine-grained clastic matrix. Both these lithic clasts and the fine-grained clastic matrix probably originated from coherent primary rocks by impact comminution on the parent body. An example of a primary rock from the CM chondrite Y74662 is shown in Fig. 2.

Table 1. *Mineralogy and thickness of fine-grained rims from unequilibrated chondrites of various classes*

	Minerals in fine-grained rims	Thickness	References
Ordinary chondrites			
H3	ol,px,pl,m,asm,tr	<60 μm	1,2,3,7,9,11,17,18,19,20,23,26,31,54
L3	ol,px,pl,m,asm,tr	<70 μm	1,2,3,9,11,23,31
LL3	ol,px,pl,m,asm,tr	<60 μm	1,2,3,6,7,10,21,23,31
Carbonacceous chondrites			
CM2	serp,toch,mlp,ol,px,cb,tr,pt	<400 μm	4,5,16,19,24,25,27,28,29,30,32,37,38,39,40,41,42, 43,44,46,47,48,50,51,52,53,55
CO3	ol,px,m,mag,pt,pyrr,anhy,mlp	<150 μm	8,12,37,38
CV3	ol,px,and,ne,tr,m,sap	<300 μm	13,14,15,16,19,25,27,29,30,33,34
CR2	serp,smc,ol,px,cb,sfi	<100 μm	14,35,36
ungrouped	?	<100 μm	22,30,45,49

and: andradite; anhy: anhydrite; asm: amophous Si-rich material; cb: carbonates; sfi: sulfides; m: metal; mag: magnetite; mlp: mixed layer phyllosilicates; ne: nepheline; ol: olivine; pt: pentlandite; pl: plagioclase; px: pyroxene; pyrr: pyrrhotite; sap: saponite; serp: serpentines; smc: smectites; toch: tochilinite; tr: troilite.

1) Allen et al., 1980; 2) Nagahara, 1984; 3) Matsunami; 1984 4) Metzler et al., 1992; 5) Kinnunen and Saikkonen, 1983; 6) Grossman and Wasson, 1987; 7) Alexander et al., 1989; 8) Brearley, 1993; 9) Kitamura and Watanabe, 1985; 10) Ashworth, 1977; 11) Ikeda et al., 1981; 12) Kurat, 1975; 13) MacPherson et al., 1985; 14) Prinz et al., 1985; 15) MacPherson and Grossman 1981; 16) Bunch and Chang, 1984; 17) Kurat, 1970; 18) Christophe Michel-Levy, 1976; 19) King and King, 1981; 20) Wlotzka, 1983; 21) Wilkening et al., 1984; 22) Kimura and Ikeda; 1992; 23) Scott et al., 1984; 24) Fuchs et al., 1973, 25) Bunch and Chang, 1980; 26) Bunch et al., 1991; 27) Zolensky et al., 1993; 28) Nakamura et al., 1991; 29) Zolensky et al., 1988; 30) Zolensky et al., 1989; 31) Allen et al., 1979; 32) Brearley and Geiger, 1991; 33) Fruland et al., 1978; 34) Bischoff, 1989; 35) Bischoff et al., 1993; 36) Weisberg et al., 1993; 37) Metzler, 1990; 38) Metzler et al., 1988; 39) Metzler and Bischoff, 1989a; 40) Metzler and Bischoff, 1989b; 41) Metzler et al., 1991; 42) Brearley and Geiger, 1991; 43) Metzler and Bischoff, 1987; 44) Zolensky et al.,(1990); 45) Bischoff and Metzler, 1990; 46) Metzler and Bischoff, 1990; 47) Sears et al., 1993; 48) Tomeoka et al., 1991; 49) Bischoff and Metzler, 1991; 50) Kring, 1988; 51) Kring, 1991; 52) Nagao, 1991; 53) Metzler and Bischoff, 1991; 54) Hutchison and Bevan (1983); 55) Brearley and Geiger (1993)

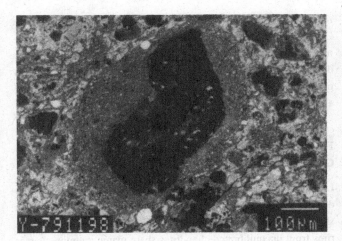

Fig. 1. Low-FeO chondrule fragment with fine-grained rim in the CM chondrite Y-791198 showing a sharp chondrule/rim boundary. The outer shape of the aggregate is smooth and subrounded, despite the angular and irregularly shaped core (back scattered electron image).

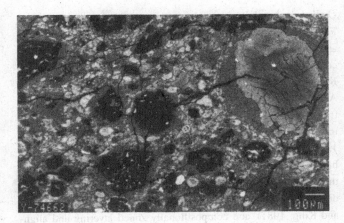

Fig. 2. Texture of a clast of primary rock in the CM chondrite Y-74662. Almost all chondrules and chondrule fragments are mantled by dust layers. Low-FeO chondrules appear dark; the zoned, light-grey object on the upper right is a high-FeO olivine fragment with dust mantle (back scattered electron image).

The occurrence of components with fine-grained rims is restricted to the clasts of primary rock. Nevertheless, in the case of Murchison, Murray, and Banten, dust-mantled chondrules, CAIs, etc. can be found within the clastic matrix. For an explanation of this texture see Metzler *et al.* (1992).

Rim occurrences, textures and compositions

Fine-grained rims in CM chondrites occur around all types of coarse-grained chondritic components, i.e. chondrules, chondrule fragments, CAIs, mineral fragments, and PCP-rich objects (for the latter see e.g. Fuchs *et al.*, 1973; Tomeoka and Buseck, 1985; Metzler *et al.*, 1992). The contacts between dust mantles and mantled constituents are always very sharp, whereas the transitions between rims and the surrounding material tend to be less well defined (Figs. 1,2,6). The total thickness of fine-grained rims varies between a few and ~400 microns and their outer shape is always very smooth and rounded, in contrast to the angular and irregular

shape of the mantled cores (Figs. 1, 2, 6). They are characterized by sedimentary fabrics; igneous textures were never observed. The groundmass with grain sizes in the submicron range consists mainly of hydrous silicates like serpentines (e.g. antigorite, cronstedtite) and the S- and OH-bearing phase tochilinite (e.g. Zolensky *et al.*, 1990; Brearley and Geiger, 1991), and contains a great number of inclusions (olivine, pyroxene, sulfides, metal, magnetite, etc.) with sizes of up to ~50 microns. The mineral assemblages found in the rims underscore the highly unequilibrated nature of this material. It seems to represent a mechanical mixture of components with totally different origins and geneses (Metzler *et al.*, 1992). The rims frequently consist of two or more concentric dust layers of different composition (e.g. Bunch and Chang, 1984; Kring, 1988), where the outer layers tend to be more magnesian than the inner ones. The mean chemical composition of fine-grained rims is remarkably similar to the mean bulk chemistry of CM chondrites, but shows some characteristic chemical deviations. For example, Ni is enriched in the rims and Ca is decoupled from Al, with low Ca/Al ratios compared to mean CM composition (Metzler *et al.*, 1992). Although bulk chemical compositions of individual dust mantles in a given meteorite vary significantly, there is no mineralogical and major element correlation between dust mantles and their central objects (Metzler *et al.*, 1992). It remains unclear whether the rim-forming dust in CM chondrites was partly or totally altered to hydrous minerals prior to its accretion onto chondrules. This topic is beyond the scope of the present paper, and the reader is referred to Metzler *et al.* (1992) and Kerridge *et al.* (1994).

Noble gas signatures

Nagao (1991) measured the noble gas content of a dust mantle in Murchison, using a mass spectrometer with a laser extraction system. He found that the Ne isotopic composition in this dust mantle represents a mixture of planetary and spallogenic Ne, without any indication for solar wind implanted noble gases. This observation is of special interest, since Murchison is a *solar-gas-rich* CM chondrite. It implies that the solar gases must reside in other lithologies than in dust mantles. This conclusion is supported by measurements of the bulk isotopic compositions of Ne for the CM chondrites Y791198 and Y74662 (Nagao, 1989; Pedroni, 1991; personal communications). Although both meteorites consist essentially of primary rock composed of about 50 vol% fine-grained rims (Metzler *et al.*, 1992), they do not contain solar noble gases. It follows that fine-grained rims in CM chondrites in general are free of solar noble gases. In analogy to other chondritic regolithic breccias, the fine-grained clastic matrix (rock debris) is the most favorable host lithology for solar gases, as also indicated by nuclear track data (Metzler, 1993).

Occurrence of presolar grains

Presolar SiC has been found in CM chondrites (see Alexander, 1993 and references therein) and these grains are located outside chondrules in the fine-grained matrix portions of these meteorites. Using an automated SEM, especially designed for detection of SiC grains in meteoritic thin sections, R.M. Walker from Washington University, St. Louis (pers. com., 1994) found that many of the SiC grains in Cold Bokkeveld are located in the fine-grained chondrule rims. The presolar origin of these grains was proven by ion microprobe measurements of their isotopic composition.

Rim thickness/core size correlation

The thickness of dust mantles and the diameter of the mantled cores have been measured in SEM images of thin sections. The data for mantled components in Kivesvaara, Y-791198, Y-74662, and Murray are shown in Fig. 3. A positive correlation can be observed between the thickness of dust mantles and the diameters of the corresponding cores (Metzler and Bischoff, 1989a; Metzler *et al.*, 1992). This means that small chondrules are surrounded by thin dust layers, large chondrules by thick layers. Although effects of sectioning have to be considered in the interpretation of these data, a distinct correlation is obvious. The slopes of the calculated linear regression lines vary between 0.17 and 0.21. The measurements reveal that the average thickness of the dust mantles reaches roughly 19 % of the core diameters. Recently, Sears *et al.* (1993)

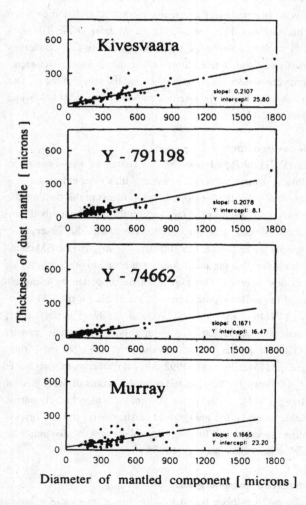

Fig. 3. Correlation between the apparent thickness of dust mantles and the apparent diameter of the mantled cores (e.g. chondrules, CAI's, etc.) as measured in thin sections of the CM chondrites Kivesvaara, Y-791198, Y-74662, and Murray.

reported two chondrule groups each with distinctive rims in the CM chondrite Murchison. Using cathodoluminescence (CL) techniques they claimed that low-FeO chondrules in Murchison have developed consistently thicker rims than high-FeO chondrules. In our view, this observation remains to be confirmed using scanning electron microscopy.

FINE-GRAINED RIMS FROM OTHER CHONDRITE GROUPS

Literature studies reveal that dark fine-grained rims have been described from many other chondrite classes. In the following a short and incomplete summary is given, to show that fine-grained rims from unequilibrated chondrites share many common characteristics and may have formed by a similar process. Most authors come to the conclusion that these rims are products of dust accretion onto larger objects in a nebular environment. For further information and references see Table 1.

Ordinary chondrites

Fine-grained rims in ordinary chondrites show a dark to opaque appearance in transmitted light and occur around all types of coarse components like chondrules, chondrule fragments, etc. (e.g. Kurat, 1970; Christophe Michel-Levy, 1976; Ashworth, 1977; Allen *et al.*, 1980; King and King, 1981; Wlotzka, 1983). Rims from H3 and L3 chondrites are enriched in moderately volatile trace elements relative to matrix and chondrules (King and King, 1981). Chemically, fine-grained rims are variable in composition, but like CM chondrites, there is no chemical correlation between the dark rims and the central objects (Wlotzka, 1983). Rim/chondrules and rim/matrix boundaries are mostly well defined (King and King, 1981), and compositionally zoned layering and alignment of mineral grains parallel to the surface of chondrules is observed (Nagahara, 1984). TEM studies on fine-grained rims revealed that they always contain some clastic grains, usually cemented by a non-clastic groundmass (Alexander *et al.*, 1989).

Ordinary chondrites are frequently internally disrupted and mechanically altered (Scott, 1984; Scott *et al.*, 1985; Romstedt and Metzler, 1994). Only those samples that escaped these brecciation processes (e.g. Tieschitz) show fine-grained rims around all coarse components (Romstedt and Metzler, 1994).

CO chondrites

Fine-grained rims in Lancé were first studied by Kurat (1975), who found that this dark material is chemically similar to CI material, despite depletions in Al, Ca, and Na. Metzler *et al.* (1988) investigated two fine-grained chondrule rims from the least equilibrated CO chondrite, ALHA 77307, using electron microprobe and find considerable differences in grain size and composition. On the other hand, Brearley (1993) states that rims in this meteorite are remarkably homogeneous. TEM investigations revealed that these rims consist of a highly unequilibrated assemblage of Si- and Fe-rich amorphous materials, olivine, pyroxene, Fe,Ni metal, magnetite, pentlandite, pyrrhotite, anhydrite, and mixed layer phyllosilicate phases (Brearley, 1993). In contrast to CM chondrites, there is clear evidence for an isolated fine-grained matrix component between dust-mantled objects in CO chondrites.

CV chondrites

In CV3 chondrites many chondrules, Ca,Al-rich inclusions and dark inclusions are surrounded by distinct rims of fine-grained materials (King and King, 1981; MacPherson *et al.*, 1985; Bischoff, 1989). Rims in Allende have a zonal sequence consisting of up to four distinct layers that can be distinguished by their different mineralogical and chemical compositions (MacPherson *et al.*, 1985). These authors found that rims consist of olivine, various pyroxenes, andradite, nepheline, and iron sulfides. Besides fine-grained rims, CV chondrites contain a considerable fraction of fine-grained interchondrule matrix. In a given CV chondrite, rim fabrics always resemble the fabric of the interchondrule matrix, indicating a close relationship between both components. For example, fine-grained rims in Allende mainly show a porous texture of euhedral silicate grains, similar to Allende matrix, whereas the rim material and matrix of Efremovka consist of densely aggregated and irregularly shaped silicate materials (Metzler *et al.*, 1988).

CR chondrites

The occurrence of fine-grained chondrule rims is also known from CR chondrites (Prinz *et al.*, 1985; Weisberg *et al.*, 1993; Bischoff *et al.*, 1993, Zolensky *et al.*, 1993). The latter authors found that most chondrules in the Acfer 059/El Djouf 001 CR2 chondrite are surrounded by several mineralogically distinct layers and that the outermost rim consists of fine-grained material similar in appearence to the "accretionary dust mantles" found in CM chondrites.

DISCUSSION AND CONCLUSIONS

Major parts of the following discussion are based on observations on fine-grained rims from CM, CO, and CV carbonaceous chondrites, representing only ~4 % of all meteorite falls (Sears and Dodd, 1988). Nevertheless, their potential parent bodies (C-type asteroids) contribute more than 30 % to the present asteroid population (Gaffey *et al.*, 1989). Hence, conclusions on solar nebula processes drawn from textures and compositions of these meteorites may hold for a significant volume of the solar nebula, represented within the asteroid belt. Furthermore, rims from mechanically unaltered ordinary chondrites of type 3 share many similarities with fine-grained rims from carbonaceous chondrites, indicating similar rim-forming mechanisms for many chondrite classes.

Common characteristics of rims from various chondrite classes

Fine-grained rims from different chondrite groups generally share the following characteristics:

1) sedimentary textures; 2) multi-layering (especially in CM, CV, and CR chondrites; 3) rim material and interchondrule matrix are mineralogically and chemically similar; and 4) rims are finer grained and less porous than matrix.

The origin of matrix (including rim material) is still the subject of considerable controversy. The entire spectrum of models exists from matrix as a nebular product to derivation entirely from chondrules (see Scott *et al.*, 1988; Brearley, this volume), indicating that there are several distinct types of fine-grained material. Probably, the fine-grained material that forms rims and matrix is "..best viewed as a complex mixture of interstellar material, nebular condensates, and fragments that may have been derived from chondrules" (Brearley, 1994). The question of chondrule-matrix and chondrule-rim relationships is addressed in detail by Brearley (this volume).

Rim formation by chondrule/dust interaction in the solar nebula

Many authors have suggested that rim formation occurred by chondrule/dust interaction in free space. Kurat (1970, 1975) concluded that this took place in gas and dust clouds, produced by large-scale impact events. Others conclude that fine-grained rims may represent the result of chondrule/dust interactions in low temperature regions of the solar nebula (e.g. Ashworth, 1977; Allen *et al.*, 1979, 1980; MacPherson and Grossman, 1981; King and King, 1981; Nagahara, 1984; Bunch and Chang, 1984; Scott *et al.*, 1984; MacPherson *et al.*, 1985; Prinz *et al.*, 1986; Grossman and Wasson, 1987; Metzler and Bischoff, 1987; Kring, 1988, 1991; Brearley and Geiger,1991; Tomeoka *et al.*, 1991; Nakamura *et al.*, 1991; Metzler *et al.*, 1988, 1992; Brearley, 1993).

Model for rim formation in CM chondrites

A detailed model for the formation of fine-grained rims in CM chondrites was proposed by Metzler (1990) and Metzler *et al.* (1992). This model (Fig. 4) starts with preexisting chondrules, CAI`s and other objects that developed dust mantles in the solar nebula by sticking of dust onto their surface. Concentric dust layers may have formed by encountering various dust reservoirs of different chemical and mineralogical composition (Kring, 1988), a conclusion similar to that of MacPherson *et al.* (1985) for the CV chondrite Allende. Subsequently, the agglomeration of these mantled components with low relative velocities led to the formation of CM planetesimals, consisting of "primary accretionary rock".

Sticking mechanisms.

Blum and Münch (1993) concluded from experiments and calculations that collisions between single submicon-sized dust grains and larger aggregates generally result in coagulation by van der Waals bonding, provided the relative velocity is ≤ 1 ms^{-1}. A summary of other potential sticking mechanisms was given by Weidenschilling (1988). This includes electrostatic attraction, ferromagnetism, chemical reactions and sticky coatings of dust grains. From observations on fine-grained rims it is not possible to deduce which of these processes dominated rim formation and which can be ruled out.

Time scales of rim formation

The time scale of dust mantle formation for CM chondrites was first calculated by Kring (1988). Depending on the collision veloc-

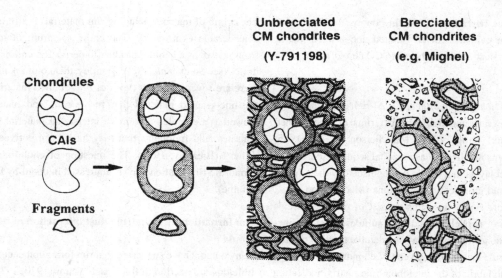

SOLAR NEBULA   **PARENT BODY**

Fig. 4. Model for the formation of fine-grained rims and primary rock in CM chondrites (Metzler *et al.*, 1992). Starting with already formed chondrules, CAI's, and fragments, rims formed by sticking of dust onto the surfaces of these components in the solar nebula. Subsequently, the agglomeration of these mantled components led to the formation of CM planetesimals, consisting of primary rock with accretionary texture. In a last phase, this rock was brecciated by high velocity impacts on the planetesimal surface, producing clasts of primary rock embedded in a fine-grained clastic matrix (rock debris).

ity and the dust/gas ratio, he concluded that rims were formed within minutes to some tens of years. Metzler *et al.* (1991) and Morfill (pers. com., 1993) found that the rim thickness/core size correlation is best compatible with a scenario where chondrules spend a certain "residence time" in a turbulent protoplanetary disk with cosmic dust/gas ratio, developing their dust mantles within several thousand years. An increase in the dust/gas ratio certainly would have reduced these formation times. Both calculations revealed that fine-grained rims can be formed during plausible periods of time in a nebular environment.

Rim formation; Alternative scenarios
Formation in a planetary regolith
The possibility of rim formation in an active regolith on the CM parent body was often considered in discussions among meteoriticists. Although a self-consistent model for rim-forming regolith processes has not been developed yet, it is assumed by some authors that rims possibly formed by the "objects sloshing about in the mud" of the "apparently very wet regolith" (from Sears *et al.*, 1993). In this case one should expect spiral-like features in the rims due to wrapping up of fine-grained material around the chondrules and other components, but those features have never been observed. Furthermore, if the wetness of a regolith layer is considered as an important parameter for rim formation in CM chondrites, it cannot explain the striking textural similarities between rims from CM chondrites and those from the anhydrous CV and CO chondrites. If rims formed in a regolith, lithic clasts, consisting of several chondrules, CAIs, etc., should be mantled by dust, as well. A list of additional arguments against a regolith origin of rims is given by Metzler *et al.* (1992).

Formation from host chondrules by aqueous alteration
Some authors propose that fine-grained rims are products of *in situ* aqueous alteration of their host chondrules (e.g. Richardson, 1981; Sears *et al.*, 1991, 1993). Although there is no doubt that most rims were *altered* by aqueous alteration on the parent body (see below), we strictly rule out the possibility that rims *formed* by this process and "that they reflect the outline of the chondrules prior to attack" (Sears *et al.*, 1991). The following arguments point against this possibility. 1) There is no chemical relationship between fine-grained rims and their host chondrules (Metzler *et al.*, 1988). 2) There is no evidence in the rim textures for pseudomorphic replacement of chondrule minerals. 3) CAIs with perfectly preserved high temperature rims (Wark and Lovering, 1977) are mantled by dust layers. The Wark-Lovering rims unquestionably represent the former outline of the CAIs and their preservation clearly indicates the addition of a dust layer rather than formation of fine-grained rims by in-situ aqueous reactions at the expense of the enclosed CAIs. 4) The striking correlation between core size and rim thickness cannot be explained by an in-situ aqueous alteration model; even the contrary correlation should be expected. 5) Chondrules that were unquestionably altered and pseudomorphed *in-situ*, preserve their sharp core-rim boundary (see below). 6) Fine-grained rims in Cold Bokkeveld contain presolar SiC grains that would have been destroyed by melting. 7) Fine-grained rims with striking similarities are present in most other (anhydrous) chondrite classes.

Formation by quenching of liquids
Hutchison and Bevan (1993) suggested that the chondrule rim material from Tieschitz was hot (probably >900°C) and liquid

when added to chondrules and clasts. This view, where dark chondrule rims were formed by quenching of liquids, is in strong contrast to conclusions drawn by Allen *et al.* (1980) from observations on rims from the same meteorite.

Agglomeration of dust-mantled components during planetesimal formation: Primary rock in CM chondrites

Formation of primary rock

The concentration of dust-mantled components in the CM chondrite Y-791198 (see above) led to the idea that this meteorite may be a cosmic sediment formed by agglomeration of dust-mantled components during accretion (Metzler and Bischoff, 1989a; Metzler *et al.*, 1992). In our view it represents a mechanically unaltered sample of the first rock generation of the CM parent body. The occurrence of lithic clasts with similar textures in many brecciated CM chondrites shows that this primary rock type was widely distributed on the CM parent body.

Alteration by compaction, brecciation and aqueous alteration on the parent body

It has to be taken into account that most samples of primary rock were altered by parent body processes. First, in some meteorites, clasts of primary rock occasionally show preferred orientations of elongated dust-mantled particles (e.g. Cold Bokkeveld), indicative of compaction by overlying material. Second, most CM chondrites were mechanically altered after their formation by brecciation processes. Third, many samples of primary rock show evidence of intensive aqueous alteration on the parent body. The fact that many CM chondrites show evidence for in-situ aqueous alteration has long been accepted (see Kerridge and Bunch, 1979; Zolensky and McSween, 1988, and references therein). Many of these authors described chondrules that were pseudomorphed by calcite and phyllosilicate, but references to the bulk texture of the host meteorites are scarce. Recently, it was described for the first time that clasts of primary rock can be found, that are extensively altered *in-situ*, but without affecting their overall texture (Metzler, 1995). In Fig. 5 a

Fig. 5. Clast of primary rock that was altered *in-situ* by aqueous fluids in the CM parent body. Almost all primary silicates are pseudomorphed by hydrous minerals like serpentines (back scattered electron image). Despite this extensive parent body processing, the overall primary texture is still preserved.

clast of primary rock from Nogoya is shown, where almost all primary silicates are replaced by hydrous phases. Nevertheless, the "accretionary" texture of this rock is still preserved; even the typical rim thickness/core size correlation (see above) is still visible.

In general, it becomes clear that careful investigations of bulk meteorite textures can help to discern between primary features and effects that were introduced into the primary material by compaction, brecciation and *in-situ* aqueous alteration.

Constraints on chondrule agglomeration and conditions in the solar nebula

Based on the model of rim formation in the solar nebula, the following constraints on chondrule agglomeration and solar nebula conditions can be obtained.

Formation of fine-grained rims was a cosmically significant event

Fine-grained rims from various chondrite groups show striking textural similarities. On the other hand, compositional differences indicate that the rim-forming dust in various chondrite classes may have had many different origins. Obviously, rim-forming processes were active whenever a dust component came in contact with chondrules and other coarse objects in the nebula. During the sticking process the finest grain size fraction tends to form layers around the coarse components. In the case of CM chondrites almost all fine-grained material became attached to chondrule surfaces and was accreted to the parent body in the form of dust mantles. In the case of CO and CV chondrites considerable amounts of dust accreted independently of chondrules and fills the interstices between dust-mantled chondrules.

Existence of several dust reservoirs of different composition

Concentrically zoned rims with compositionally different layers may indicate the existence of various distinct dust reservoirs that were subsequently encountered by chondrules on their pathway from their source to the accreting parent body (MacPherson *et al.*, 1985; Kring, 1988; Metzler, 1990).

Rim formation enhanced the sticking efficiency

The formation of "fuzzy" balls may have had the effect of promoting the accretion process by increasing the sticking probability between colliding chondrules and dust grains (MacPherson and Grossman, 1981).

Low velocity agglomeration of chondrules during accretion

This constraint simply comes from the fact that fine-grained rims in chondrites are preserved. Relative velocities between mantled objects during accretion must have been below ~5 ms^{-1} (Hartmann, 1978), otherwise the rims would have been shattered by mutual collisions.

CM planetesimals accreted under shielding conditions

Dust mantles and the primary rock do not contain solar gases (Metzler and Bischoff, 1989b) or preirradiated grains (Metzler, 1993). This can be explained by a solar nebula that was opaque to

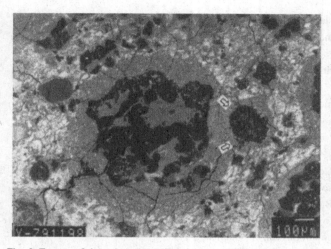

Fig. 6. Texture of the unbrecciated CM chondrite Y-791198 that consists entirely of primary rock. In the center a contact boundary between two dust-mantled chondrules is visible (arrows; back scattered electron image). It is not clear from the texture whether these chondrules encountered each other in space and stuck together or whether they came into contact only during their incorporation into the parent body.

solar emissions at the time of CM planetesimal accretion due to high concentrations of dust.

It remains unclear whether dust-mantled objects accreted to cm- or even m-sized objects in space or remained isolated objects until they reached the surface of the growing planetesimal. The question is whether the dust mantled chondrules shown in Fig. 6 encountered each other in space and stuck together, or whether they developed their common boundary only during incorporation into the planetesimal surface? Fine-grained rims (dust mantles) seem to have formed chronologically between chondrule-producing transient heating events and the accretion of chondritic parent bodies. For this reason the investigation of these structures may help to answer the question of how a dusty solar nebula was transformed into a planetary system.

ACKNOWLEDGEMENTS

We thank J. Wasson, A. Brearley, A. Krot for their helpful and constructive reviews.

Figs. 3 and 4 are reprinted from *Geochimica et Cosmochimica Acta* **56**, Metzler K., Bischoff A., and Stöffler D., pp.2873–2897, copyright by Pergamon Press, 1992.

REFERENCES

Alexander C.M.O., Hutchison R., and Barber D.J. (1989) Origin of chondrule rims and interchondrule matrices in unequilibrated ordinary chondrites. *Earth Planet. Sci. Lett.* **95**, 187–207.

Alexander C.M.O. (1993) Presolar SiC in chondrites: How variable and how many sources? *Geochim. Cosmochim. Acta* **57**, 2869–2888.

Allen J.S., Nozette S., and Wilkening L.L. (1979) Chondrule rims: Composition and texture (abstract). *Lunar Planet. Sci.* **X**, 27–29.

Allen J.S., Nozette S., and Wilkening L.L. (1980) A study of chondrule rims and chondrule irradiation records in unequilibrated ordinary chondrites. *Geochim. Cosmochim. Acta* **44**, 1161–1175.

Ashworth J.R. (1977) Matrix textures in unequilibrated ordinary chondrites. *Earth Planet. Sci. Lett.* **35**, 25–34.

Bischoff A. (1989) Mineralogische und chemische Untersuchungen an chondritischen Meteoriten: Folgerungen für die Entstehung fester Materie im Solarnebel und die Entwicklung der Meteoritenmutterkörper. *Habilitationsschrift*, Westf. Wilhelms-Universität, Münster, pp. 264.

Bischoff A. and Metzler K. (1990) Petrography and chemistry of the three carbonaceous chondrites Y-86720, Y-82162 and B-7904. *Proc. NIPR Symp. Antarc. Meteorites* **3**, Natl. Inst. Polar Res., Tokyo, 185–187.

Bischoff A. and Metzler K. (1991) Mineralogy and petrography of the anomalous carbonaceous chondrites Y-86720, Y-82162 and B-7904. *Proc. NIPR Symp. Antarc. Meteorites.* **4**, Natl. Inst. Polar Res., Tokyo, 226–246.

Bischoff A., Palme H., Ash R.D., Clayton R.N., Schultz L., Herpers U., Stöffler D., Grady M.M., Pillinger C.T., Spettel B., Weber H., Grund T., Endreß M., and Weber D. (1993) Paired Renazzo-type (CR) carbonaceous chondrites from the Sahara. *Geochim. Cosmochim. Acta* **57**, 1587–1603.

Blum J. and Münch M. (1993) Experimental investigations on aggregate-aggregate collisions in the early solar nebula. *Icarus* **106**, 151–167.

Brearley A.J. (1993) Matrix and fine-grained rims in the unequilibrated CO3 chondrite, ALHA 77307: Origins and evidence for diverse, primitive nebular dust components. *Geochim. Cosmochim. Acta* **57**, 1521–1550.

Brearley A.J. and Geiger T. (1991) Mineralogical and chemical studies bearing on the origin of accretionary rims in the Murchison CM2 carbonaceous chondrite (abstract). *Meteoritics* **26**, 323.

Brearley A.J. and Geiger T. (1993) Fine-grained chondrule rims in the Murchison CM2 chondrite: compositional and mineralogical systematics. *Meteoritics* **28**, 328–329.

Bunch T.E. and Chang S. (1980) Carbonaceous chondrites – II. Carbonaceous chondrite phyllosilicates and light element geochemistry as indicators of parent body processes and surface conditions. *Geochim. Cosmochim. Acta* **44**, 1543–1577.

Bunch T.E. and Chang S. (1984) CAI rims and CM2 dust balls: products of gas-grain interactions, mass transport, grain aggregation and accretion in the nebula? (abstract). *Lunar Planet. Sci.* **XV**, 100–101.

Bunch T.E., Schultz P., Cassen P., Brownlee D., Podolak M., Lissauer J., Reynolds R. and Chang S. (1991) Are some chondrule rims formed by impact processes? Observations and experiments. *Icarus* **91**, 76–92.

Christophe Michel-Lévy M. (1976) La matrice noire et blanche de la chondrite de Tieschitz (H3). *Earth Planet. Sci. Lett.* **30**, 143–150.

Fruland R.M., King E.A., and McKay D.S. (1978) Allende dark inclusions. *Proc. Lunar Planet. Sci. Conf.* **9th**, 1305–1329.

Fuchs L.H., Olsen E., and Jensen K.J. (1973) Mineralogy, mineral chemistry and composition of the Murchison (C2) meteorite. *Smithson. Contrib. Earth Sci.* **10**, 1–39.

Gaffey M.J., Bell J.F., and Cruikshank D.P. (1989) Reflectance spectroscopy and asteroid surface mineralogy. In *Asteroids II* (eds. R.P. Binzel, T. Gehrels, and M.S. Matthews) pp. 98–127.

Grossman J.N. and Wasson J.T. (1987) Compositional evidence regarding the origins of rims on Semarkona chondrules. *Geochim. Cosmochim. Acta* **51**, 3003–3011.

Hartmann W.K. (1978) Planet formation: Mechanisms of early growth. *Icarus* **33**, 50–61.

Hutchison R. and Bevan A.W.R. (1983) Conditions and time of chondrule accretion. In *Chondrules and their origins* (ed. E.A. King) pp.162–179.

Ikeda Y., Kimura M., Mori H. and Takeda H. (1981) Chemical compositions of matrices of unequilibrated ordinary chondrites. *Proc. 6th Symp. Antarct. Meteorites*, Natl. Inst. Polar Res., Tokyo, 124–144.

Kerridge J.F. and Bunch T.E. (1979) Aqueous activity on asteroids: Evidence from carbonaceous meteorites. In Asteroids (ed. T. Gehrels), pp. 745–764, University of Arizona Press, Tucson.

Kerridge J.F., McSween H.Y., and Bunch, T.E. (1994) Petrologic evolution of CM chondrites: The difficulty of discriminating between nebular and parent body effects (abstract) *Meteoritics* **29**, 481.

Kimura M and Ikeda Y. (1992) Mineralogy and petrology of an unusual Belgica-7904 carbonaceous chondrite: Genetic relationships among components. *Proc. NIPR Symp. Antarct. Meteorites* **5**, 74–119.

King T.V.V. and King E.A. (1981) Accretionary dark rims in unequilibrated ordinary chondrites. *Icarus* **48**, 460–472.

Kinnunen K.A. and Saikkonen R. (1983) Kivesvaara C2 chondrite: silicate petrography and chemical composition. *Bull. Geol. Soc. Finland* **55**, 35–49.

Kitamura M. and Watanabe S. (1985) Adhesive growth and abrasion of chondrules during the accretion process. *Proc. 10th. Symp. Antarct. Meteorites,* Natl. Inst. Polar Res., Tokyo, 222–234.

Kring D.A. (1988) The petrology of meteoritic chondrules: Evidence for fluctuating conditions in the solar nebula. *Ph.D. Thesis, Harvard University*, Cambridge, 346 pp.

Kring D.A. (1991) High temperature rims around chondrules in primitive chondrites: evidence for fluctuating conditions in the solar nebula. *Earth Planet. Sci. Lett.* **105**, 65–80.

Kurat G. (1970) Zur Genese des kohligen Materials im Meteoriten von Tieschitz. *Earth Planet. Sci. Lett.* **7**, 317–324.

Kurat G. (1975) Der kohlige Chondrit Lancé: Eine petrologische Analyse der komplexen Genese eines Chondriten. *Tschermaks Min. Petr. Mitt.* **22**, 38–78.

MacPherson G.J. and Grossman L. (1981) Clastic rims on inclusions: Clues to the accretion of the Allende parent body (abstract). *Lunar Planet. Sci.* **XII**, 646–647.

MacPherson G.J., Hashimoto, A., and Grossman L. (1985) Accretionary rims on inclusions in the Allende meteorite. *Geochim. Cosmochim. Acta* **49**, 2267–2279.

Matsunami S. (1984) The chemical compositions and textures of matrices and chondrule rims of eight unequilibrated ordinary chondrites: a preliminary report. *Proc. Symp. 9th Antarc. Meteorites,* Natl. Inst. Polar Res., Tokyo, 126–148.

Metzler K. (1990) Petrographische und mikrochemische Untersuchungen zur Akkretions- und Entwicklungsgeschichte chondritischer Mutterkörper am Beispiel der CM-Chondrite. *Ph.D. dissertation,* Univ. of Münster, 183 pp.

Metzler K. (1993) In situ investigation of preirradiated olivines in CM chondrites (abstract) *Meteoritics,* **28**, 398–399.

Metzler K. (1995) Aqueous alteration of primary rock on the CM parent body (abstract) *Lunar Planet. Sci.* **XXVI**, 961–962.

Metzler K. and Bischoff A. (1987) Accretionary dark rims in CM chondrites (abstract). *Meteoritics* **22**, 458–459.

Metzler K. and Bischoff A. (1989a) Accretionary dust mantles in CM chondrites as indicators for processes prior to parent body formation (abstract). *Lunar Planet. Sci.* **XX**, 689–690.

Metzler K. and Bischoff A. (1989b) Formation of accretionary dust mantles in the solar nebula as confirmed by noble gas data of CM chondrites (abstract). *Meteoritics* **24**, 303–304.

Metzler K. and Bischoff A. (1990) Petrography and chemistry of accretionary dust mantles in the CM-chondrites Y-791198, Y-793321, Y-74662, and ALHA 83100 – indications for nebula processes (abstract) *Symp. Antarc. Meteor. 15th,* 198–200.

Metzler K. and Bischoff A. (1991) Evidence for aqueous alteration prior to parent body formation; petrographic observations in CM chondrites (abstract). *Lunar Planet. Sci.* **XXII**, 893–894.

Metzler K., Bischoff A., and Morfill G. E. (1991) Accretionary dust mantles in CM chondrites: Chemical variations and calculated time scales of formation (abstract). *Meteoritics* **26**, 372.

Metzler K., Bischoff A., and Stöffler D. (1988) Characteristics of accretionary dark rims in carbonaceous chondrites (abstract). *Lunar Planet. Sci.* **XIX**, 772–773.

Metzler K, Bischoff A. and Stöffler D. (1992) Accretionary dust mantles in CM chondrites: Evidence for solar nebula processes. *Geochim. Cosmochim. Acta* **56**, 2873–2897.

Nagao K. (1991) Studies on the primordial noble gases and the carrier phases in carbonaceous chondrites using laser microprobe. *Report of the 1989–1990 Monbusyo Scientific Research program.*

Nagahara H. (1984) Matrices in type 3 ordinary chondrites – primitive nebular records. *Geochim. Cosmochim. Acta* **48**, 2581–2595.

Nakamura T., Tomeoka K., and Takeda H. (1991) Mineralogy of chondrule rims in the Yamato-791198 CM chondrite: comparison to Murchison (abstract). *Symp. Antarc. Meteor.* **16th**, 40–41.

Prinz M., Weisberg M.K., Nehru C.E., and Delaney J.S. (1985) Layered

chondrules: Evidence for multistage histories during chondrule formation. *Meteoritics* **20**, 732–733.

Prinz M., Weisberg M.K., Nehru C.E., and Delaney J.S. (1986) Layered chondrules in carbonaceous chondrites (abstract). *Meteoritics* **21**, 485–486.

Richardson S.M. (1981) Alteration of mesostasis in chondrules and aggregates from three C2 carbonaceous chondrites. *Earth Planet. Sci. Lett.* **52**, 67–75.

Romstedt J. and Metzler K. (1994) Brecciation and preirradiation of unequilibrated H chondrites (abstract) *Meteoritics* **29**, 523–524.

Rubin A.E. (1984) Coarse-grained chondrule rims in type 3 chondrites. *Geochim. Cosmochim. Acta* **48**, 1779–1789.

Rubin A.E. and Wasson J.T. (1987) Chondrules, matrix and coarse-grained chondrule rims in the Allende meteorite: Origin, interrelationships and possible precursor components. *Geochim. Cosmochim. Acta* **51**, 1923–1937.

Scott E.R.D. (1984) Classification, metamorphism and brecciation of type 3 chondrites from Antarctica. *Smithsonian Contrib. Earth Sci.* **26**, 73–94.

Scott E.R.D., Rubin A.E., Taylor G.J. and Keil K. (1984) Matrix material in type 3 chondrites-occurrence, heterogeneity and relationship with chondrules. *Geochim. Cosmochim. Acta* **48**, 1741–1757.

Scott E.R.D., Lusby D., and Keil K. (1985) Ubiquitous brecciation after metamorphism in equilibrated ordinary chondrites. Proc. Lunar Planet. Sci. Conf. 16th, *Journal Geophys. Res.* **90**, D137–D148.

Sears D.W.G. and Dodd R.T. (1988) Overview and classification of meteorites. In *Meteorites and the early Solar System* (eds. J.F. Kerridge and M.S. Matthews) pp. 3–31, University of Arizona Press, Tucson.

Sears D.W.G., Batchelor J.D., Lu J., and Keck B.D. (1991) Metamorphism of CO and CO-like chondrites and comparisons with type 3 ordinary chondrites. *Proc. NIPR Symp. Antarct. Meteorites* **4**, 319–343.

Sears D.W.G., Benoit P.H., and Jie L. (1993) Two chondrule groups each with distinctive rims in Murchison recognized by cathodoluminescence. *Meteoritics* **28**, 669–675.

Tomeoka K. and Busek P.R. (1985) Indicators of aqueous alteration in CM chondrites; Microtextures of a layered mineral containing Fe, S, O, and Ni. *Geochim. Cosmochim. Acta* **49**, 2149–2163.

Tomeoka K., Hatakeyama K., Nakamura T., and Takeda H. (1991) Evidence for pre-accretionary aqueous alteration in the Y-793321 CM carbonaceous chondrite (abstract) *Symp. Antarc. Meteor.* **16th**, 37–39.

Wark D.A. and Lovering J.F. (1977) Marker events in the early evolution of the solar system: evidence from rims on calcium-aluminum-rich inclusions in carbonaceous chondrites. *Proc.Lunar Sci. Conf. 8th.*, 95–112.

Weidenschilling S.J. (1988) Formation processes and time scales for meteorite parent bodies. In *Meteorites and the early Solar System* (eds. J.F. Kerridge and M.S. Matthews) pp. 348–371, University of Arizona Press, Tucson.

Weisberg M.K., Prinz M., Clayton R.T., and Mayeda T.K. (1993) The CR (Renazzo-type) carbonaceous chondrite group and its implications. *Geochim. Cosmochim. Acta* **57**, 1567–1586.

Wilkening L.L., Boynton W.V. and Hill D.H. (1984) Trace elements in rims and interiors of Chainpur chondrules. *Geochim. Cosmochim. Acta* **48**, 1071–1080.

Wlotzka F. (1983) Composition of chondrules, fragments, and matrix in the unequilibrated ordinary chondrites Tieschitz and Sharps. In *Chondrules and their Origins* (ed. E.A. King), 296–318. Lunar and Planetary Institute, Houston.

Zolensky M. and McSween H.Y (1988) Aqueous alteration. In *Meteorites and the early Solar System* (eds. J.F. Kerridge and M.S. Matthews) pp. 114–143, University of Arizona Press, Tucson.

Zolensky M.E., Barrett R.A. and Gooding J.L. (1989) Matrix and rim compositions compared for 13 carbonaceous chondrite meteorites and clasts. *Lunar Planet Sci.* **XX**, 1249–1250.

Zolensky M.E., Barrett R.A., Klöck W. and Gooding J.L. (1990) The mineralogy of matrix and chondrule rims in CM chondrites. *Lunar Planet. Sci.* **XXI**, 1383–1384.

Zolensky M., Barrett, R, and Browning L. (1993) Mineralogy and composition of matrix and chondrule rims in carbonaceous chondrites. *Geochim Cosmochim Acta* **57**, 3123–3148.

17: Relict Grains in Chondrules: Evidence for Chondrule Recycling

RHIAN H. JONES

Institute of Meteoritics, Department of Earth and Planetary Sciences, University of New Mexico, Albuquerque, NM 87131, U.S.A.

ABSTRACT

Relicts are grains which did not crystallize in situ in the host chondrule. They represent coarse-grained precursor material that did not melt during chondrule formation, and provide a tangible record of chondrule precursor grains. The compositions of most relicts are very similar to comparable grains in chondrules, providing evidence that they were derived from previous generations of chondrules. Relict forsterite and enstatite grains in FeO-rich chondrules are likely to be derived from FeO-poor chondrules, and "dusty" olivine grains in FeO-poor chondrules may be derived from FeO-rich chondrules. This evidence for recycling of chondrule material places important constraints on chondrule formation models. It appears that at least 15% of chondrules contain material that experienced at least two chondrule-forming events, with disruptive collisions probably occurring between the events. Chondrules with different oxidation states (FeO-poor and FeO-rich) were intimately mixed in chondrule-forming regions. This indicates either, i) that both kinds of chondrules formed in the same region, and that the oxidation state of each chondrule is determined by the intrinsic oxygen fugacity of the precursor assemblage, or ii) that regions of different oxidation state existed, but were close enough together for mixing to occur. Relict grains are also present in a limited number of CAI, indicating that recycling occurred during CAI as well as chondrule formation.

INTRODUCTION

Relict grains are defined as grains that did not crystallize *in situ* from the host chondrule melt as the chondrule cooled. They are therefore chondrule precursor grains, and constitute the only tangible remnants of the precursor assemblage. The presence of relict grains in chondrules has been recognized for some time, but there are conflicting opinions concerning which grains are true relicts, and the origins of grains that are considered to be relicts. Certain types of relicts appear to be derived from previous generations of chondrules, and hence provide evidence for recycling of material in the chondrule-forming region.

Relicts are usually identified by differences in size, texture and/or composition relative to normal grains in the host chondrule. Normal grains are those which crystallize from the chondrule melt during cooling. A common feature of relicts is the presence of an overgrowth with the composition of normal grains which grew dur-

ing the cooling episode. Relict grains usually show a somewhat rounded morphology indicating that they underwent partial resorption into the melt during chondrule formation. The dissolution rate of silicate minerals in chondrule melts is discussed by Greenwood and Hess (this volume).

Relict grains in chondrules may form by similar processes to coarse-grained rims and compound chondrules (Rubin and Krot, this volume). Two mechanisms are possible: 1) The solid relict grain or primary chondrule collides with a molten or partially molten secondary chondrule and becomes incorporated into the secondary chondrule (Lux *et al.*, 1981; Kracher *et al.*, 1984). 2) The solid relict grain or primary chondrule becomes surrounded by a layer of fine-grained dust. A subsequent heating event partially or completely melts the fine-grained dust, preserving a relict grain or primary chondrule (Wasson, 1993; Wasson *et al.*, 1995). For individual relict grains, it is difficult to distinguish between these two possibilities. In both models, relict grains may originate as individual grains that occur as primitive solar nebular material (nebular condensates or interstellar grains), grains produced from disruptive collisions between solid chondrules, or grains that are the remnants of primary chondrules that were almost completely resorbed into the secondary chondrule melt. One way to distinguish sources of relict grains may be by determination of oxygen isotopic compositions of relict and chondrule grains (e.g. Weinbruch *et al.*, 1993). Studies of oxygen isotopic compositions of relicts in chondrules have not been made to date but might provide an interesting insight into the nature of relicts. The discussion below is based entirely on textural and compositional comparisons between relict and chondrule grains.

A variety of chondrule grains has been described as relicts. These include "dusty", metal-rich, olivine (Rambaldi, 1981; Nagahara, 1981; Rambaldi and Wasson, 1982), and forsterite grains occurring in FeO-rich chondrules (Jones, 1990, 1992). Both of these types appear to be derived from previous generations of chondrules and are discussed in more detail in this chapter. In addition, an example of relict enstatite in FeO-rich chondrules is described. There is some evidence that relict grains of hibonite occur in some CAI (MacPherson *et al.*, 1984; Hinton *et al.*, 1988; Steele, 1995). Two other types of grains that have been described as relicts previously are olivine grains enclosed in low-Ca pyroxene in a poikilitic texture (Nagahara, 1983), and forsterite that exhibits blue cathodoluminescence (CL) under an electron beam (Steele, 1986). Arguments questioning whether these latter types of grains are truly relicts are summarized briefly below.

Experimental studies have shown that in order to produce a porphyritic texture in a chondrule, heterogeneous nucleation sites must be preserved in the chondrule during the heating event (e.g. Tsuchiyama and Nagahara, 1981; Lofgren and Russell, 1986; Hewins, 1988). Such nucleation sites are likely to be submicroscopic and are not necessarily equivalent to observable relict grains. Clearly, where relict grains are observed as the cores of grains, the relicts acted as heterogeneous nucleation sites for the normal overgrowth rim. However, this is not to say that every grain in a porphyritic chondrule has an identifiable relict core. In general, the variation in composition from grain to grain within individual chondrules is significantly less than the variation between chondrules (e.g. Jones, 1992). This indicates that for most porphyritic chondrules, cores of individual grains crystallized *in situ* during cooling, and heterogeneous nucleation sites were submicroscopic.

In order to examine the relationships between relict grains and normal grains in chondrules, it is necessary to study chondrites that have suffered minimal effects of thermal metamorphism. The effect of metamorphism is to equilibrate the compositions of all phases in a chondrite. Some elements are mobilized at lower temperature than others, and redistributed throughout existing or newly nucleated phases. This complicates the interpretation of comparisons between compositional data. The discussion in this chapter is restricted entirely to the least equilibrated chondrites (petrologic subtypes 3.0 and 3.1). In such chondrites, two types of chondrules are present, FeO-poor and FeO-rich. These types of chondrules are alternatively described as types I and II, or groups

A and B, respectively, but for clarity these terms will not be used in this paper. The terms FeO-poor and FeO-rich refer to the bulk compositions of the chondrules, as well as the compositions of their silicate minerals. FeO-poor and FeO-rich chondrules contain olivine and pyroxene with FeO/(FeO+MgO) ratios < 10 mole% and > 10 mole% respectively. In addition, FeO-poor chondrules are more reduced than FeO-rich chondrules, and commonly contain Fe metal.

RELICTS DERIVED FROM CHONDRULES

Forsterite relicts in FeO-rich chondrules

Forsterite relicts occur as the cores of FeO-rich olivine grains in FeO-rich chondrules. They are easily distinguished in back-scattered electron images because they are very dark in comparison with the lighter gray, more FeO-rich overgrowths (Fig. 1). Overgrowth olivines are usually zoned in FeO and minor elements (CaO, MnO, Cr_2O_3), similar to normal grains in the host chondrule. The outlines of the forsteritic cores are commonly rounded, indicating that they were partially resorbed into the chondrule melt during the chondrule heating event. Resorption occurs because forsterite is not in equilibrium with the melt composition (e.g. Greenwood and Hess, this volume). Some forsterite relicts with very low FeO contents (< 1 wt% FeO) show blue cathodoluminescence under an electron beam (Steele, 1986). Forsterite relicts have been described in FeO-rich chondrules in ordinary chondrites (Jones, 1990) and CO3 carbonaceous chondrites (Steele, 1989; Jones, 1992, 1993), and are also present in CM chondrites (Metzler *et al.*, 1992), although occurrences in the latter have not been described in detail.

Compositions of forsterite relicts are very similar to compositions of normal olivine grains in FeO-poor chondrules. For those

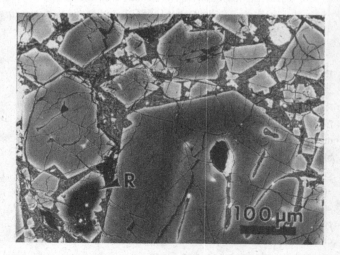

Fig. 1. Back-scattered electron image of a portion of an FeO-rich, porphyritic olivine chondrule in the Semarkona ordinary chondrite. Olivine grains are FeO-rich and zoned, with increasing FeO (lighter gray) towards the edges. One olivine grain contains a relict forsterite core (R) which appears very dark in the image, because it is MgO-rich. The relict is rounded, showing evidence for partial resorption, and overgrown with FeO-rich olivine.

grains that have been studied, FeO contents of relicts vary in the range 0.5-6 wt%, and minor element contents vary in the ranges: CaO 0.0–0.7, MnO 0.0–0.16, Cr_2O_3 0.1–0.4, Al_2O_3 0.0–0.2, TiO_2 0.0–0.1 wt% (Table 1; Jones, 1990, 1992). Olivine compositions in FeO-poor chondrules in the ordinary chondrite, Semarkona, and the CO3 chondrite, ALH A77307 lie within similar ranges and show similar interelement variations (Fig. 2; Jones and Scott,

1989; Jones, 1992). These very clear relationships argue strongly that the relicts are derived from FeO-poor chondrules.

FeO contents of two of the relicts are significantly higher than the chondrule data (Fig. 2a). This is probably attributable to partial equilibration of the relict with the overgrown olivine during chondrule formation. Diffusion rates of Fe in olivine are higher than diffusion rates of most minor elements (e.g. Jurewicz and Watson,

Table 1. *Electron microprobe analyses of relict forsterite grains in FeO-rich chondrules in unequilibrated chondrites*

Chondrule	SiO_2	TiO_2	Al_2O_3	Cr_2O_3	FeO	MnO	MgO	CaO	Total	Fa
Sem 41	41.7	0.09	0.15	0.09	0.71	0.04	56.2	0.72	99.7	0.7
Sem 51B	41.7	nd	nd	0.15	3.6	0.08	54.4	0.04	99.9	3.6
Sem C71	41.8	0.05	0.11	0.13	0.43	nd	56.2	0.52	99.2	0.4
A307 1	42.3	0.06	0.19	0.38	1.3	0.08	55.0	0.31	99.6	1.3
A307 5	42.1	0.05	0.10	0.18	2.2	0.06	54.7	0.46	99.9	2.2
A307 5	41.6	nd	0.06	0.36	5.3	0.16	52.6	0.21	100.3	5.3
A307 18	41.1	0.04	0.10	0.28	5.1	0.08	53.6	0.51	100.8	5.1
A307 21	42.1	nd	0.09	0.28	1.0	nd	56.3	0.29	100.1	1.0

Sem = Semarkona (LL3.0): Jones (1990).
A307 = ALHA77307 (CO3.0): Jones (1992) and unpublished data.
nd = not detected.

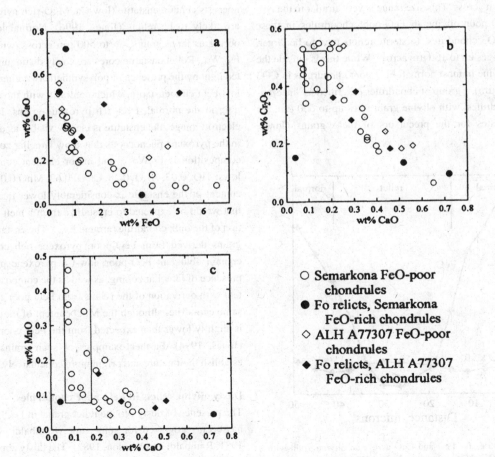

○ Semarkona FeO-poor chondrules
● Fo relicts, Semarkona FeO-rich chondrules
◇ ALH A77307 FeO-poor chondrules
◆ Fo relicts, ALH A77307 FeO-rich chondrules

Fig. 2. Comparison of the compositions of forsterite relicts in FeO-rich chondrules with mean olivine compositions in FeO-poor chondrules in the unequilibrated chondrites, Semarkona (LL3.0) and ALH A77307 (CO3.0). Chondrule compositions are from Jones and Scott (1989) and Jones (1992); relict compositions are from Jones (1990, 1992, and unpublished data). Boxes show the compositional variation of minor elements in FeO-rich chondrules (see Fig. 6), for comparison of different chondrule types.

1988) and FeO is likely to equilibrate quickly at the high temperatures reached during chondrule formation. FeO contents of olivine in FeO-rich chondrules are typically 15–20 wt% for ordinary chondrites (Jones, 1990) and 25–30 wt% in CO3 chondrites (Jones, 1992). The two relicts in ALH A77307 with high FeO and approximately 0.5 wt% CaO probably had FeO contents < 1 wt% originally, but partially equilibrated with the FeO-rich overgrowth at high temperatures. The effect of diffusion between forsterite relicts and FeO-rich overgrowths is illustrated in Fig. 3, which shows zoning profiles for FeO and CaO across a grain with a relict core. The central relict forsterite has an FeO content as low as 1 wt%. Normal olivine in this chondrule is zoned from FeO 18–28 wt%. In the intermediate region between the relict core and the overgrowth, the profile is not a sharp step, but is sloped, indicating that some diffusion has occurred between the two regions, blurring the original interface. The CaO content of the first olivine overgrowth (0.30 wt%) is lower than that in the relict core (0.43 wt%), resulting in a more complex profile for CaO than for FeO. However, diffusion does not appear to have modified the CaO content of the core of the relict significantly, and CaO content can thus be used to estimate the original FeO content of the relict on plots such as Fig. 2a.

The sizes of forsterite relicts are also consistent with derivation from FeO-poor chondrules. Forsterite relicts in ordinary chondrites are small, < 40 μm across. This size range is very similar to the typical size of FeO-poor olivine in FeO-poor chondrules in these chondrites. In CO3 chondrites, forsterite relicts tend to be larger, and have size ranges up to 100 μm across. While this is outside the size range of olivine in most normal, FeO-poor chondrules in CO3 chondrites (< 50 μm), a group of chondrules described as "macroporphyritic" chondrules, with olivine grain sizes up to 600 μm, are potential candidates for the precursors of these grains (Jones,

1992). Compositions of forsterite in macroporphyritic chondrules are very similar to those in normal, FeO-poor chondrules in CO3 chondrites.

Abundances of forsterite relicts in FeO-rich chondrules are different in ordinary and CO3 chondrites. In the ordinary chondrite Semarkona, out of eleven FeO-rich chondrules that were studied in detail, three contained forsterite relicts in the sections examined (Jones, 1990). Since FeO-rich, porphyritic olivine chondrules represent about 20% of the total chondrule population in ordinary chondrites (Jones, 1990), forsterite relicts are present in at least 5% of all chondrules. This is a minimum estimate, because the small size of the relicts means that they may be present in more chondrules, but not apparent in the particular cross-section of the chondrule observed in thin section. In the CO3 chondrite ALH A77307, five out of six FeO-rich chondrules contain forsterite relicts (Jones, 1992). FeO-rich chondrules are much less abundant in CO chondrites than ordinary chondrites, representing about 5% of all chondrules, so forsterite relicts are present in about 4% of all chondrules in these chondrites.

Enstatite relicts in FeO-rich chondrules

In one of ten FeO-rich, porphyritic pyroxene chondrules in the Semarkona LL3.0 ordinary chondrite, several of the pyroxene phenocrysts contain enstatite (low-Ca, MgO-rich pyroxene) cores that are likely to be relicts (Jones, 1996). Normal low-Ca pyroxene occurs as large grains, up to 500 μm across, with a composition $Fs_{16}Wo_4$. Relict enstatite cores are easily distinguished in transmitted light by the presence of polysynthetic twinning characteristic of low-FeO clinoenstatite, which contrasts with the absence of twinning in the normal, FeO-rich pyroxene grains. In back-scattered electron image, the enstatite is clearly visible as dark core regions in the pyroxene phenocrysts that show lamellar zoning (Fig. 4). Its composition is $Fs_7Wo_{0.3}$, and minor element contents are as follows: TiO_2 0.03, Al_2O_3 0.28, Cr_2O_3 0.62, MnO 0.06 wt%. The FeO content of the enstatite is considerably lower than a composition that would be expected to crystallize from a melt with a composition of the bulk chondrule (around Fs_{16}). The enstatite may be relict grains derived from FeO-poor, pyroxene-rich chondrules which contain abundant FeO-poor, low-Ca pyroxene phenocrysts. The presence of lamellar zoning, as well as the composition, are consistent with derivation of the relicts from FeO-poor chondrules in the same chondrite, although the MnO content of the enstatite regions is slightly lower than expected from the FeO-poor chondrule suite (Jones, 1994). Further examples of such grains would help to establish their nature and origin more conclusively.

Dusty olivine relicts in FeO-poor chondrules

The presence of dusty olivine relict grains in FeO-poor chondrules has been recognized for some time (Rambaldi, 1981; Nagahara, 1981; Rambaldi and Wasson, 1982). The dusty appearance of these grains in transmitted light is caused by the presence of a host of small, micron-sized grains of essentially Ni-free Fe metal (Fig. 5). This texture is interpreted to be the result of solid-state reduction of a more FeO-rich olivine. Most dusty grains are clearly relicts,

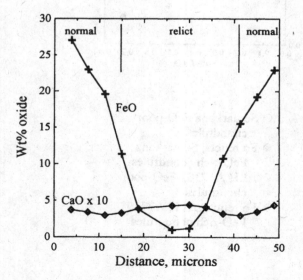

Fig. 3. Zoning profiles for FeO and CaO across an olivine grain with a relict core in an FeO-rich chondrule, in ALH A77307 (CO3.0). For FeO, the boundary between the relict core and the normal overgrowth is not a sharp step, showing that some diffusion has occurred between the two during chondrule formation. This chondrule is illustrated in Jones (1992), Fig. 2a.

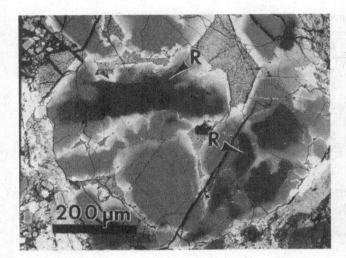

Fig. 4. Back-scattered electron image of a portion of an FeO-rich, por-phyritic pyroxene chondrule in the Semarkona ordinary chondrite. Pyroxene grains are FeO-rich and zoned, with increasing FeO (lighter gray) towards the edges and with overgrowth rims of Ca-rich pyroxene (augite). Several relict enstatite grains (R) are present in the cores of Fe-rich pyrox-ene: they appear dark gray in the image, because they are MgO-rich, and show lamellar zoning. The relicts have rounded outlines, indicating that they underwent partial resorption during chondrule formation.

although the source of the FeO-rich precursors, prior to reduction, has not been established previously. Reduction could have taken place either before or during chondrule formation. In some chon-drules, most olivine grains are dusty, indicating that reduction prob-ably took place after chondrule formation (Kracher *et al.*, 1984).

Dusty olivine grains are typically larger than the normal grains in their host chondrules: grains up to 1 mm across are observed in ordinary chondrites. They are commonly overgrown by a clear rim of olivine which has compositional and zoning characteristics sim-ilar to normal olivine in the host chondrule, and which is inter-preted as an overgrowth that grew during chondrule formation. The dusty region is commonly rounded and shows evidence for partial resorption into the chondrule melt, but the morphology of the relict can also be quite euhedral in some cases.

Compositions of dusty olivines may be used to interpret the ori-gin of their precursors. FeO contents of the relicts are, in most cases, similar to or slightly higher than the FeO contents of normal olivines in the host chondrules (Table 2; Rambaldi, 1981; Rambaldi and Wasson, 1982; Nagahara, 1981; Kracher *et al.*, 1984; Jones 1994). Abundances of metal vary from grain to grain, and lie in the range 2–8 vol% metal (Jones and Danielson, 1995). Calculated compositions of the olivine grains before reduction lie in the range Fa_{5-28}, with the most typical values Fa_{10-20}, consistent with deriva-tion of the precursors from FeO-rich chondrules in ordinary chon-drites. FeO-rich chondrules in carbonaceous chondrites are more FeO-rich than those in ordinary chondrites, around Fa_{30}, so these chondrules are potential candidates for dusty olivine precursors with high FeO contents. Some relict, FeO-rich grains that are only partially reduced have been observed in chondrules in ordinary chondrites (Rambaldi and Wasson, 1982; Jones and Danielson, 1995). The clear core of one such grain in a Semarkona chondrule

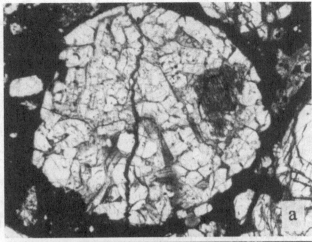

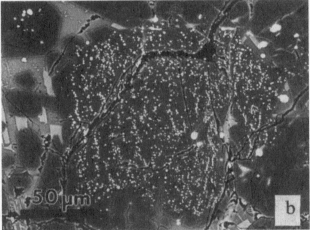

Fig. 5. a) Transmitted light photomicrograph of a chondrule containing a dusty relict olivine grain in the Semarkona chondrite. Chondrule is about 750 µm across. b) Back-scattered electron image of the same dusty olivine grain, showing that the dusty appearance is due to the presence of many micron-sized blebs of Fe metal. The metal abundance is estimated to be approximately 7 vol% in this grain.

has a composition Fa_{16} (Table 2). The FeO content, as well as minor element contents, of this grain are indistinguishable from the com-position of olivine in FeO-rich chondrules in the same chondrite (Fig. 6). Such grains provide direct evidence that compositions of dusty relicts are consistent with a derivation from chondrules.

Differences between minor element contents of dusty olivines and normal olivines in the host chondrules have been used to argue that the relict grains cannot be derived from chondrules, and may be nebular condensates or interstellar grains (e.g. Rambaldi, 1981; Grossman *et al.*, 1988). However, the minor element contents of dusty olivines (Table 2) actually compare very closely with minor element contents of FeO-rich olivines in FeO-rich chondrules from unequilibrated ordinary and CO3 chondrites, as illustrated in Fig. 6. This relationship argues strongly that the precursors of dusty olivines could indeed have a chondrule source. Reduction reactions would not be expected to change the minor element contents of olivines significantly, because CaO, MnO and Cr_2O_3 are not as eas-ily reduced as FeO.

Table 2. *Electron microprobe analyses of dusty olivine grains in FeO-poor chondrules in unequilibrated chondrites*

Chondrule	SiO$_2$	Cr$_2$O$_3$	FeO	MnO	MgO	CaO	Total	Fa	Fa(host)
Sem C70	40.7	0.48	4.0	0.38	52.0	0.12	97.7	4.1	2.4
Sem C90	41.0	0.49	6.1	0.32	51.9	0.11	99.9	6.1	2.9
Sem C111*	39.5	0.45	15.4	0.34	43.8	0.08	99.6	16.5	10.1
Murch A	43.5	0.42	0.81	0.30	55.4	0.14	100.6	0.8	
Murch D	43.1	0.44	1.1	0.18	56.2	0.16	101.2	1.1	1.0
Krymka B^R	41.8	0.15	3.4	0.25	55.6	0.10	101.3	3.3	1.0
Bish C^R	40.9	0.19	1.5	0.27	55.0	0.20	98.1	1.5	0.8
Bish B^{RW}	43.1	0.42	6.4	0.30	53.5	0.10	103.9	6.3	4.9
Bish C^{RW}	41.4	0.46	5.3	0.10	51.9	0.24	99.5	5.4	1.0

Sem = Semarkona (LL3.0): Jones (1994, 1996).
Murch = Murchison (CM): Jones and Danielson (1995).
Krymka (LL3.1), Bish = Bishunpur (LL3.1): R = Rambaldi (1981), RW = Rambaldi and Wasson (1982).
* = clear core of a partially reduced grain.

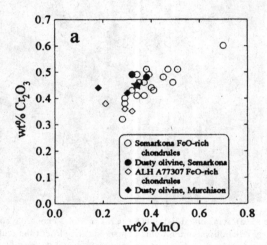

Fig. 6. Comparison of the minor element contents of dusty olivines in FeO-poor chondrules with mean olivine compositions in FeO-rich chondrules in the unequilibrated chondrites, Semarkona (LL3.0) and ALH A77307 (CO3.0). The star symbol represents the clear core of a partially reduced olivine grain in Semarkona (Jones, 1996).

Dusty olivines are common in ordinary chondrites, occurring in about 10% of all chondrules (Nagahara, 1983; Grossman *et al.*, 1988). They have also been observed in CM and CV chondrites (Kracher *et al.*, 1984), although they are considerably less common in carbonaceous chondrites than in ordinary chondrites. This observation is consistent with the fact that FeO-rich chondrules are also less abundant in carbonaceous than in ordinary chondrites, so that if dusty relicts are indeed derived from chondrules, there is a more limited source in carbonaceous relative to ordinary chondrite reservoirs. In enstatite chondrites, dusty textures are observed in pyroxene as well as olivine (Lusby *et al.*, 1987), but the abundance of dusty relict grains has not been estimated.

RELICTS DERIVED FROM CAI

In addition to the evidence for chondrule recycling, rare evidence for recycling of CAI in the solar nebula is provided by the presence of relict grains of hibonite in a limited number of refractory inclusions. Inclusion GR-1 in Murchison has two generations of hibonite that are texturally distinct, and show significant differences in trace element compositions (MacPherson *et al.*, 1984; Hinton *et al.*, 1988). Steele (1995) described an unusual CAI in Allende, in which a large hibonite/perovskite cluster in the core of the inclusion has exsolution textures indicating equilibration prior to formation of the finer-grained hibonite/spinel mantle. Observation of these hibonite relicts suggests that some recycling of CAI material occurred, but that the occurrence is much less common than that in chondrules.

One example of a CAI relict in a ferromagnesian chondrule has been reported (Misawa and Fujita, 1994). A relict grain of spinel, containing inclusions of fassaite, a Ca,Al-rich phase, and refractory, platinum-group metal grains is present in a barred olivine chondrule in the Allende chondrite. This is the only example reported to date that provides direct evidence of the presence of solid CAI material in the chondrule forming region.

DEBATABLE RELICTS

In previous discussions of relict grains, two additional types of olivine have been suggested to be relicts: olivine in poikilitic chondrules, and forsterite showing blue cathodoluminescence (CL). However, the evidence that these types of grains are truly relicts is equivocal. The following brief discussion summarizes arguments against the suggestion that they are relicts. It is important to establish the nature of these grains, so that estimates of the frequency of chondrule recycling events can be made. Poikilitic chondrules constitute about 30% of the total chondrule populations in ordinary and carbonaceous chondrites. If the olivine in these chondrules was relict in nature, and if it was derived from a previous generation of chondrules, this would increase significantly the number of collisional events inferred for the chondrule-forming region.

Olivine in poikilitic chondrules

A poikilitic texture is one in which relatively large crystals of one mineral enclose numerous smaller crystals of another mineral which are randomly oriented and generally uniformly distributed. This texture indicates that the larger grains grew after the smaller ones. A common texture in FeO-poor chondrules is one in which small grains of olivine are completely enclosed in large grains of low-Ca pyroxene. Nagahara (1983) suggested that the olivine grains observed in this textural association are relicts that did not melt in the chondrule-forming event. The following points can be made to argue against this suggestion. i) Poikilitic textures are reproduced experimentally in samples in which no olivine was present in the starting material (Lofgren and Russell, 1986), so it is not necessary to have relict olivine crystals to produce such a texture (Nagahara, 1983). ii) There is little compositional variation between poikilitic olivine grains in individual chondrules, and this would seem unlikely if the olivines in the poikilitic texture included random aggregates of relicts. iii) In unequilibrated chondrites, the composition of olivine poikilitically enclosed in low-Ca pyroxene is very similar to compositions of olivine phenocrysts in the same chondrule (Jones, 1994). Nagahara (1983) showed that compositions of these two types of olivines are different (poikilitically enclosed olivine has higher FeO and lower CaO than phenocrysts), but the chondrite these data were obtained from, ALH A77278, is partially equilibrated (petrologic subtype 3.7). For FeO-poor chondrules, the effect of metamorphism is to increase the FeO content of olivine, and to reduce the CaO content (e.g. McCoy *et al.*, 1991). It is likely that the poikilitic olivines were more altered than the phenocrysts in the chondrules described by Nagahara (1983), because such grains tend to be smaller than the phenocrysts, and are usually closer to the edge of the chondrule. The pyroxenes in which they are embedded contain numerous cracks which facilitate Fe-Mg and Ca exchange (e.g. Jones and Rubie, 1991). Hence, it is important to assess the compositional evidence from unequilibrated chondrites, and this evidence does not indicate that the poikilitic olivines are relicts.

Kurat (1988) interpreted this poikilitic texture as the result of reaction of silica-rich vapor with olivine aggregates to produce pyroxene. In this sense, olivine would be a relict phase. However, the rounded shape of chondrules and presence of glassy mesostasis imply that the chondrule was molten, and since the texture has been shown to occur by crystallization from a liquid, it seems reasonable to assume a liquid origin. Also, the minor element contents of olivine and pyroxene are consistent with crystallization from a melt (Jones, 1994). Low-Ca pyroxene typically contains about 0.4 wt% Al_2O_3 whereas olivine contains < 0.03 wt%. If the pyroxene formed by vapor-solid reaction, the Al_2O_3 in the pyroxene would have to be derived from the vapor as it would not be available from the olivine. It is not clear that a silica-rich vapor would have the appropriate composition.

Forsterite exhibiting blue cathodoluminescence (CL)

Forsterite with low FeO content ($Fa_{<1.5}$) exhibits blue CL under an electron beam (Steele, 1986). Chondritic olivine of this composition tends to have relatively high contents of refractory minor elements, including Ca, Ti, Al and V, and Steele (1986) has argued that such high refractory element contents are inconsistent with equilibrium crystallization from a melt of chondrule composition. As a result, cores of olivine grains in low-FeO chondrules that exhibit blue CL have been interpreted as relicts that originally crystallized by condensation from a vapor. However, olivine grains showing blue CL are produced in the laboratory during cooling of chondrule analog melts with low FeO contents (Lofgren and DeHart, 1992), so the exhibition of blue CL is not in itself an indication of a condensate origin. It is also important to recognize that crystallization in chondrule melts occurs under strongly disequilibrium conditions, because cooling rates are too high to allow equilibrium to be maintained (e.g. Lofgren, this volume), and the partitioning behavior of minor elements in forsteritic olivine is not well understood under disequilibrium conditions. In addition, the evidence that these "blue" forsterites are relict grains when they occur in FeO-poor chondrules is ambiguous. FeO and minor element contents vary continuously from the "blue" cores of grains towards their more FeO-rich rims (e.g. Jones and Scott, 1989; Jones, 1992), and there is generally little evidence for a step in the zoning pattern which would indicate a more FeO-rich overgrowth on a relict core. Such a step would be difficult to observe if the relicts and overgrowths are similar in composition, and if diffusion has occurred across the relict / overgrowth interface. While it is not possible to rule out the presence of relict cores of FeO-poor olivine grains in low-FeO, porphyritic olivine chondrules, this suggestion has not been demonstrated conclusively, and forsterites with blue CL may not represent ubiquitous relicts in this association.

DISCUSSION

The textural and compositional data presented above argue strongly that relict forsterite and enstatite in FeO-rich chondrules, and dusty relict olivine in FeO-poor chondrules, are likely to be derived from previous generations of chondrules. Since an obvious source (i.e. chondrules) is present in chondrites, and since relicts show compositional variations consistent with derivation from this source, it seems unnecessary to invoke an alternative origin. A derivation from chondrules can place important constraints on chondrule formation processes. It provides estimates of the frequency of repeated episodes of chondrule formation on the same material, as well as offering insights into the proximity of material of different compositions and oxidation states in the chondrule-forming region. If individual relict grains are derived from collisions between chondrules, the frequency of chondrule collisions can also be estimated.

Chondrule recycling

In type 3 ordinary chondrites, approximately 10% of all chondrules contain dusty relict grains, and approximately 5% of all chondrules contain relict forsterite. Thus, it is possible to state that at least 15% of chondrules contain material that experienced at least two episodes of chondrule formation. This estimate is likely to be a minimum value, because it does not include recycled material that may have been completely resorbed into the host chondrule, or relicts with compositions very similar to those of the host chondrule grains, which may be overlooked. In addition to relicts, enveloping (and possibly adhering) independent compound chondrules and coarse-grained rims are thought to provide evidence for recycling (Wasson *et al.*, 1995; Rubin and Krot, this volume). In ordinary chondrites, about 1% of chondrules are independent compound chondrules, and 10% have coarse-grained rims (Rubin and Krot, this volume). In CV chondrites, 50% of chondrules have coarse-grained rims (Rubin, 1984), indicating a higher incidence of multiple heating events for CV chondrites.

Wasson (1993) calculated the mean free path length, and time between collisions, for chondrules in the chondrule-forming region of the nebula. His equation for mean free path contained an error: the correct version is

$$\text{mean free path} = (\sqrt{2}.\pi.d^2.n)^{-1},$$

where d is the diameter of the chondrule and n is the space density of chondrules in the nebula.

The following parameters were chosen by Wasson (1993) to maximize the frequency of chondrule collisions. An Earth's mass of material is assumed to be present in the asteroid belt (2.1 to 3.5 AU), uniformly distributed in a relatively thin dusty midplane 200 km thick. For a mean chondrule radius of 0.1 mm, and assuming that all the solid matter is present as chondrules, the space density of chondrules is then 4.0×10^{-3} cm^{-3}. Using a mean interparticle velocity of 2 ms^{-1}, the correct mean free path length is 1.4 km, and the time between chondrule collisions (= mean free path length / velocity) is 0.2 hours. For a chondrule radius more appropriate for ordinary chondrites, 0.3 mm, the space density of chondrules becomes 1.5×10^{-4} cm^{-3}, and the time between collisions is 0.6 hours.

Individual relict grains derived from chondrules may provide evidence that disruptive collisions occurred frequently in the chondrule-forming region. For relative velocities of 2 ms^{-1}, the kinetic energy of chondrules of radius 0.1–0.3 mm lies in the range 0.3–7 erg. This energy is probably sufficient to disrupt solid chondrules and release free relict grains into the nebular environment (e.g.

Fujiwara *et al.*, 1977; Fujiwara, 1980). Thus, it is likely that disruptive encounters between chondrules were common, and derivation of relict grains from chondrules is not unreasonable. Ruzicka (1990) showed that dusty olivine grains contain higher dislocation densities than normal olivine in the same chondrules. This may be interpreted as evidence for impact deformation of the dusty grains, although the high dislocation density may also be the result of the reduction reaction itself which strains the olivine crystal lattice.

Lower relative velocities than 2 ms^{-1} would probably be required to incorporate a solid relict grain into a molten or partially molten chondrule, because molten chondrules are likely to be disrupted by splashing in an energetic collision. Wasson (1993) argued that the time between collisions is longer than the time that a chondrule is likely to be at least partially molten (e.g. Lofgren, this volume), hence that a solid / melt collision origin for relict grains and compound chondrules is unlikely. However, the corrected values for collision times do not allow this mechanism to be discounted easily.

The collision rates calculated above suggest that few chondrules are likely to survive intact for extended periods of time in the midplane of the nebula. At a rate of one collision per hour, each chondrule would encounter about 9000 collisions per year. In order to estimate a possible lower limit on collision rates, I carried out the same calculation, but assuming that the total material in the asteroid region is 0.5 of an Earth mass, the midplane is 400 km thick, and only half the material is present as chondrules. For a chondrule radius of 0.3 mm, the time between collisions is about 19 hours and 9 hours, for relative velocities of 1 and 2 ms^{-1} respectively. Such collision rates must still be regarded as causing a problem for extended survival times of chondrules in the midplane.

The evidence from relict grains suggests a limited number of recycling events. If there were multiple episodes, there would be an increasing likelihood of observing multiple generations of relicts. Although such objects may be hard to identify, to date none of the relict grains observed in chondrules have been interpreted as having undergone more than one recycling episode. One of the dusty grains observed in Murchison (grain A, Table 2; Kracher *et al.*, 1984; Jones and Danielson, 1995) is an isolated olivine grain. This grain could have been reduced as a relict in a chondrule, followed by collisional disruption of the chondrule, providing evidence for multiple disruptive collisions acting on a single grain. Similar evidence for multiple collisions is observed in CO3 chondrites, in which isolated olivine and pyroxene grains are common. Several isolated olivine grains show complex zoning, and are likely to be derived from FeO-rich olivine with relict forsterite cores in FeO-rich chondrules (Jones, 1992, 1993). However, the fragmentation that produced isolated olivines is not necessarily contemporaneous with chondrule formation, and does not provide direct evidence of multiple chondrule recycling events.

Of the relict grains described in this chapter, it is curious that no pyroxene relicts have been observed in olivine-rich chondrules, whereas dusty olivine relicts are commonly present in pyroxene-rich chondrules. This observation cannot be attributed easily to sampling bias, because pyroxene-rich chondrules are common constituents of most chondrite types (60% of all chondrules in ordinary chondrites: Gooding and Keil, 1980). One possible explanation is that pyroxene-rich chondrules were formed at a later stage in solar nebula evolution than olivine-rich chondrules, when solids were more silica-rich. Another possibility is that relict pyroxene grains are resorbed more rapidly into a chondrule melt than olivine. However, at present it is difficult to evaluate the likely reason for this observation.

Mixing of material of different oxidation states.

The examples of relict grains described above, relict forsterite and enstatite in FeO-rich chondrules, and dusty olivine in FeO-poor chondrules, represent major differences in redox conditions between relicts and host chondrules. This evidence shows that there was intimate mixing between different chondrule types. Two scenarios accounting for this are possible: i) the oxygen fugacity of each chondrule is defined by the intrinsic oxygen fugacity of the precursor dustball, and chondrules with a range of oxygen fugacities may form in one region, or ii) chondrules with different intrinsic oxygen fugacities formed in different regions, but close enough in time and space that there was mixing between these regions. Experimental evidence suggests that during the brief heating events of chondrule formation, a chondrule melt does not have sufficient time to equilibrate with the ambient atmosphere (Tsuchiyama and Miyamoto, 1984). Chondrule precursor assemblages may play a more important role in defining the intrinsic oxygen fugacity of each individual chondrule (Connolly *et al.*, 1994). These arguments suggest that the former scenario is more likely to be realistic, and that chondrules with different intrinsic oxygen fugacities formed in close proximity in the chondrule-forming region.

ACKNOWLEDGEMENTS

I would like to thank T.J. McCoy, H. Nagahara E.R.D. Scott and A.E. Rubin for reviews and discussions that have improved this manuscript. This work was supported by NASA grant NAG-3347 (J.J. Papike, P.I.) and the Institute of Meteoritics at the University of New Mexico. Electron microprobe analyses were performed at the Electron Microbeam Analysis Facility, Institute of Meteoritics and Department of Earth and Planetary Sciences, University of New Mexico.

REFERENCES

Connolly H.C.Jr., Hewins R.H., Ash R.D., Zanda B., Lofgren G.E. and Bourot-Denise M. (1994) Carbon and the formation of chondrules. *Nature* **371**, 136–139.

Fujiwara A. (1980) On the mechanism of catastrophic destruction of minor planets by high-velocity impact. *Icarus* **41**, 356–364.

Fujiwara A., Kamimoto G. and Tsukamoto A. (1977) Destruction of basaltic bodies by high-velocity impact. *Icarus* **31**, 277–288.

Gooding J.L. and Keil K. (1980) Relative abundances of chondrule primary textural types in ordinary chondrites and their bearing on conditions of chondrule formation. *Meteoritics* **16**, 17–43.

Grossman J.N., Rubin A.E., Nagahara H. and King E. (1988) Properties of chondrules. In *Meteorites and the Early Solar System* (eds. J.F. Kerridge and M.S. Matthews), 619–659. University of Arizona Press.

Hewins R.H. (1988) Experimental studies of chondrules. In *Meteorites and the Early Solar System*, (eds. J.F. Kerridge and M.S. Matthews) pp. 660–679. University of Arizona Press.

Hinton R.W., Davis A.M., Scatena-Wachel D.E., Grossman L. and Draus R.J. (1988) A chemical and isotopic study of hibonite-rich refractory inclusions in primitive meteorites. *Geochim. Cosmochim. Acta* **52**, 2573–2598.

Jones R.H. (1990) Petrology and mineralogy of type II, FeO-rich chondrules in Semarkona (LL3.0): Origin by closed-system fractional crystallization, with evidence for supercooling. *Geochim. Cosmochim. Acta* **54**, 1785–1802.

Jones R.H. (1992) On the relationship between isolated and chondrule olivine grains in the carbonaceous chondrite ALHA77307. *Geochim. Cosmochim. Acta* **56**, 467–482.

Jones R.H. (1993) Effect of metamorphism on isolated olivine grains in CO3 chondrites. *Geochim. Cosmochim. Acta* **57**, 2853–2867.

Jones R.H. (1994) Petrology of FeO-poor, porphyritic pyroxene chondrules in the Semarkona chondrite. *Geochim. Cosmochim. Acta* **58**, 5325–5340.

Jones R.H. (1996) FeO-rich, porphyritic pyroxene chondrules in unequilibrated chondrites. Submitted to *Geochim. Cosmochim. Acta*.

Jones R.H. and Danielson L.R. (1995) A chondrule origin for dusty, relict olivine grains. *Lunar Planet. Sci. XXVI*, 699–700. (Abstract).

Jones R.H. and Rubie D.C. (1991) Thermal histories of CO3 chondrites: Application of olivine diffusion modelling to parent body metamorphism. *Earth Planet. Sci. Lett.* **106**, 73–86.

Jones R.H. and Scott E.R.D. (1989) Petrology and thermal history of Type IA chondrules in the Semarkona (LL3.0) chondrite. *Proc. 19th Lunar Planet. Sci. Conf.*, 523–536. Lunar and Planetary Institute, Houston.

Jurewicz A.J.G. and Watson E.B. (1988) Cations in olivine, Part 2: Diffusion in olivine xenocrysts, with applications to petrology and mineral physics. *Contrib. Mineral. Petrol.* **99**, 186–201.

Kracher A., Scott E.R.D. and Keil K. (1984) Relict and other anomalous grains in chondrules: Implications for chondrule formation. *J. Geophys. Res.* **89**, Supp., B559–B566.

Kurat G. (1988) Primitive meteorites: An attempt towards unification. *Phil. Trans. R. Soc. Lond.* A **325**, 459–482.

Lofgren G.E. and DeHart J.M. (1992) Dynamic crystallization studies of enstatite chondrite chondrules: Cathodoluminescence properties of enstatite. *Lunar Planet. Sci. XXIII*, 799–800. (Abstract).

Lofgren G.E. and Russell W.J. (1986) Dynamic crystallization of chondrule melts of porphyritic and radial pyroxene composition. *Geochim. Cosmochim. Acta* **50**, 1715–1726.

Lusby D., Scott E.R.D. and Keil K. (1987) Ubiquitous high-FeO silicates in enstatite chondrites. *J. Geophys. Res.* **92**, B4, E679–E695.

Lux G., Keil K. and Taylor G.J. (1981) Chondrules in H3 chondrites: textures, compositions and origins. *Geochim. Cosmochim. Acta* **45**, 675–685.

MacPherson G.J., Grossman L., Hashimoto A., Bar-Matthews M. and Tanaka T. (1984) Petrographic studies of refractory inclusions from the Murchison meteorite. *J. Geophys. Res.* **89**, Supp., C299–C312.

McCoy T.J., Scott E.R.D., Jones R.H., Keil K. and Taylor G.J. (1991) Composition of chondrule silicates in LL3–5 chondrites and implications for their nebular history and parent body metamorphism. *Geochim. Cosmochim. Acta* **55**, 601–619.

Metzler K. Bischoff A. and Stöffler D. (1992) Accretionary dust mantles in CM chondrites: Evidence for solar nebula processes. *Geochim. Cosmochim. Acta* **56**, 2873–2897.

Misawa K. and Fujita T. (1994) A relict refractory inclusion in a ferromagnesian chondrule from the Allende meteorite. *Nature* **368**, 723–726.

Nagahara H. (1981) Evidence for secondary origin of chondrules. *Nature* **292**, 135–136.

Nagahara H. (1983) Chondrules formed through incomplete melting of the pre-existing mineral clusters and the origin of chondrules. In *Chondrules and Their Origins* (ed. E.A. King), pp. 211–222. Lunar and Planetary Institute, Houston.

Rambaldi E.R. (1981) Relict grains in chondrules. *Nature* **293**, 558–561.

Rambaldi E.R. and Wasson J.T. (1982) Fine, nickel-poor Fe-Ni grains in the olivine of unequilibrated ordinary chondrites. *Geochim. Cosmochim. Acta* **46**, 929–939.

Rubin A.E. (1984) Coarse-grained chondrule rims in type 3 chondrites. *Geochim. Cosmochim. Acta* **48**, 1779–1789.

Ruzicka A. (1990) Deformation and thermal histories of chondrules in the Chainpur (LL3.4) chondrite. *Meteoritics* **25**, 101–113.

Steele I.M. (1986) Compositions and textures of relic forsterite in carbonaceous and unequilibrated ordinary chondrites. *Geochim. Cosmochim. Acta* **50**, 1379–1395.

Steele I.M. (1989) Compositions of isolated forsterites in Ornans (C3O). *Geochim. Cosmochim. Acta* **53**, 2069–2079.

Steele I.M. (1995) Mineralogy of a refractory inclusion in the Allende (C3V) meteorite. *Meteoritics* **30**, 9–14.

Tsuchiyama A. and Miyamoto M. (1984) Metal grains in chondrules: an experimental study. *Lunar Planet. Sci. XV*, 868–869. (Abstract).

Tsuchiyama A. and Nagahara H. (1981) Effects of precooling thermal history and cooling rate on the textures of chondrules: A preliminary report. *Mem. Nat. Inst. Polar Res. Spec. Issue No. 20*, 175–192. NIPR, Tokyo.

Wasson J.T. (1993) Constraints on chondrule origins. *Meteoritics* **28**, 14–28.

Wasson J.T., Krot A.N., Lee M.-S. and Rubin A.E. (1995) Compound chondrules. *Geochim. Cosmochim. Acta*, in press.

Weinbruch S., Zinner E.K., El Goresy A., Steele I.M. and Palme H. (1993) Oxygen isotopic composition of individual olivine grains from the Allende meteorite. *Geochim. Cosmochim. Acta* **57**, 2649–2661.

18: Multiple Heating of Chondrules

ALAN E. RUBIN[1] and ALEXANDER N. KROT[2]

[1] Institute of Geophysics and Planetary Physics, University of California, Los Angeles, CA 90095–1567, U.S.A. [2] Hawai'i Institute of Geophysics and Planetology, School of Ocean and Earth Science and Technology, University of Hawai'i at Manoa, Honolulu, HI 96822, U.S.A.

ABSTRACT

Constraints can be placed on the nature of the chondrule-formation mechanism by the occurrence of coarse-grained chondrule rims, independent enveloping compound chondrules and relict grains inside chondrules. All three types of objects are interrelated and formed in the nebula by partly melting dust clumps that had entrained previously formed chondrules or chondrule fragments. If the primary object entrained in a large dust clump was an intact chondrule, large-scale melting of the surrounding dust (but not the primary object) resulted in the formation of an enveloping compound chondrule. If the primary object was a phenocryst derived from a disrupted chondrule, then the resultant object was a chondrule with a relict grain. If the dust clump was small and incompletely melted, then a coarse-grained rim formed around the primary object whether it was a chondrule, chondrule fragment or isolated mineral grain. It is probable that the heating mechanism that formed all of these objects is the same one responsible for chondrule formation. The large percentage of chondrules that contain relict grains or coarse-grained rims or are part of compound chondrule pairs (> 25% in type-3 ordinary chondrites; > 50% in CV3 chondrites) indicates that chondrules were commonly reheated to temperatures above the solidus by the chondrule-formation mechanism. Differences among the fractions of reheated chondrules in different chondrite groups may reflect differences in the intensity of the heating mechanism in different regions of the solar nebula.

INTRODUCTION

The number of times that individual chondrules have been heated is an important constraint on chondrule-formation mechanisms. Some proposed heating mechanisms (e.g., supernova shock waves; aerodynamic drag heating; Wark, 1979; Wood, 1984) are unlikely to heat chondrules more than once without additional *ad hoc* modifications. Other proposed mechanisms (e.g., nebular lightning; reconnecting magnetic field lines; Morfill *et al.*, 1993; Sonett, 1979) can accommodate multiple heating of chondrules even though they may suffer from unrelated drawbacks (e.g., Levy and Araki, 1989; Love *et al.*, 1994). Although rocks on asteroidal surfaces can be heated multiple times by impacts (e.g., Mittlefehldt *et al.*, 1992), the arguments against chondrule formation on parent bodies are compelling (Taylor *et al.*, 1983); therefore, a nebular setting for chondrule formation is assumed throughout this paper.

The evidence for multiple heating of chondrules is largely petrographic, deriving from three interrelated types of objects: (1) chondrules with coarse-grained rims, (2) independent enveloping compound chondrules, and (3) relict grains inside chondrules. The characteristics of each of these objects are described below. All are best studied in those meteorites least affected by thermal metamorphism and aqueous alteration – the type-3 ordinary and carbonaceous chondrites. Although EH3 chondrites also contain these objects (e.g., Rubin, 1984; Rubin and Grossman, 1987; Rambaldi *et al.*, 1983), members of this group are typically more metamorphosed than many type-3 ordinary and carbonaceous chondrites (and much less well studied) and will not be considered further here.

CHONDRULE RIMS

There are two major types of rims which surround chondrules: *fine-grained rims*, which contain little (< 10%) or no melted material, and *coarse-grained rims* (also known as *igneous rims*), which have been substantially (~20 to ≥80%) melted.

Fine-grained rims in ordinary chondrites (OC) are typically 10–30-μm thick (Scott *et al.*, 1988). They are nearly identical in texture, mineralogy and composition to interchondrule silicate matrix material (mainly matrix clumps), but there is some evidence that the rims are finer grained and somewhat less porous (Ashworth, 1977; Allen *et al.*, 1980, Matsunami, 1984; Alexander *et al.*, 1989). Fine-grained chondrule rims and matrix material both consist mainly of highly disequilibrated, irregularly shaped micrometer-to-submicrometer-size grains of olivine (with a variable but generally high Fa content), low-Ca pyroxene, Ca pyroxene, amorphous feldspathic material and variable amounts of troilite and metallic Fe-Ni (Christophe Michel-Lévy, 1976; Ashworth, 1977; Allen *et al.*, 1980; Huss *et al.*, 1981; Ikeda *et al.*, 1981; King and King, 1981; Matsunami, 1984; Scott *et al.*, 1984; Nagahara, 1984; Alexander *et al.*, 1989; Brearley *et al.*, 1989; Brearley, this volume). Microchondrules and grain fragments of magnesian olivine and low-Ca pyroxene (probably derived mainly

from disrupted chondrules) also occur (Nagahara, 1984; Alexander *et al.*, 1989; Brearley *et al.*, 1989; Krot and Wasson, 1995). Because of their fine grain size these rims appear opaque in thin sections of standard, 30-μm thickness.

Coarse-grained chondrule rims (Fig. 1) are considerably coarser grained than fine-grained rims: e.g., 5–40 μm vs. ~0.1 μm in type-3 OC, and ~10 μm vs. ~2 μm in CV3 chondrites (Rubin, 1984; Ashworth, 1977; Wark, 1979; Housley and Cirlin, 1983; Krot and Wasson, 1995). Such rims surround ~10% of the chondrules in type-3 OC (Fig. 1a,b) and ~50% of the chondrules in CV3 chondrites (Rubin, 1984); they also surround chondrule fragments (e.g., Fig. 1c) and rare isolated mineral grains. Rim thicknesses average ~160 ± 100 μm in OC (Krot and Wasson, 1995) and ~400 μm in CV3 chondrites (Rubin, 1984). Coarse-grained rims in OC (Fig. 1a,b) consist primarily of olivine, low-Ca pyroxene, sodic plagioclase or feldspathic glass, troilite and metallic Fe-Ni. CV3 coarse-grained rims (Fig. 1c,d) are similar in mineralogy to CV matrix material: both sets of materials consist mainly of olivine, Ca-pyroxene, calcic plagioclase, pentlandite, pyrrhotite and metallic

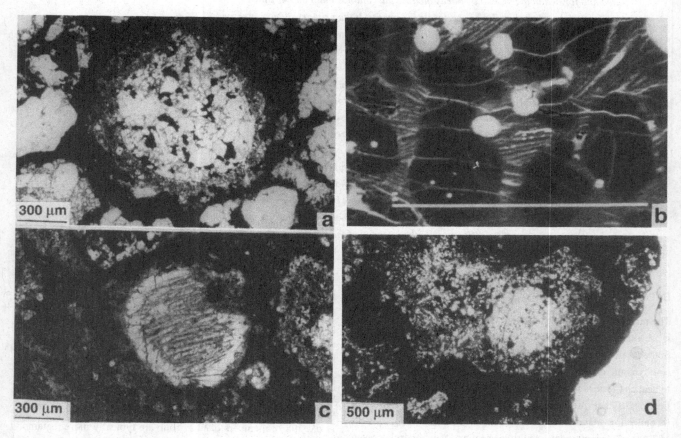

Fig. 1. Coarse-grained chondrule rims in ordinary and CV3 chondrites. (a) Symmetric coarse-grained rim around a low-FeO porphyritic olivine-pyroxene chondrule containing large patches of troilite (black) in the Semarkona LL3.0 chondrite. The coarse grain size of the rim renders it translucent in transmitted light. (b) Scanning electron microscope image of a portion of a coarse-grained rim with an obvious igneous texture around a barred olivine chondrule in the LEW86158 L3.2 chondrite. Dark crystals are subhedral olivine grains (mean size, ~30 μm). These are surrounded by partly recrystallized feldspathic mesostasis and numerous droplets (white) of metallic Fe-Ni and troilite. Scale bar is 100 μm in length. (c) Thin coarse-grained rim surrounding a broken barred olivine chondrule in the Allende CV3 carbonaceous chondrite. The rim lies outside of the olivine shell that encloses the bars. Rim formation must have occurred after the chondrule crystallized and fragmented; this indicates that the rim-formation event followed the chondrule-formation event. (d) Large, asymmetric coarse-grained rim enclosing a low-FeO porphyritic olivine chondrule in Allende. Black areas within the rim are composed of iron-rich sulfide (pyrrhotite and pentlandite); the sulfide forms a discontinuous ring partially surrounding the chondrule. a, c and d in transmitted light; b in back-scattered electrons.

Fe-Ni (Rubin, 1984); in oxidized CV3 chondrites, many coarse-grained rims and matrix regions also contain nepheline and sodalite. In many cases, discontinuous rings of sulfide occur between the chondrule and the coarse-grained rim, and between the rim and the matrix.

Rubin and Wasson (1987, 1988) analyzed coarse-grained rims in CV3 Allende by instrumental neutron activation analysis (INAA) and found that, in general, the rims are compositionally unrelated to the chondrules they enclose. As a group the rims are compositionally more homogeneous than chondrules, suggesting that they formed from finer-grained, well-mixed precursors. The INAA data on Allende matrix indicate that it is similar to some coarse-grained rims (Fig. 2); this supports the notion that these rims were derived in large part from matrix-like material. Oxygen isotope analysis of Allende chondrules and surrounding coarse-grained chondrule rims indicates that all the objects fall on a slope-1 mixing line and that the rims have higher $\delta^{18}O$ and $\delta^{17}O$ than their enclosed chondrules (Rubin et al., 1990). The rims which contain the heaviest oxygen (i.e., those with the highest $\delta^{18}O$ and $\delta^{17}O$ values) are most similar in O-isotopic composition to matrix material (Rubin et al., 1990; Clayton et al., 1983).

Krot and Wasson (1995) studied coarse-grained chondrule rims in type-3 OC by scanning electron microscopy and found that most rims around low-FeO chondrules (e.g., Fig. 1b) show evidence of extensive to complete melting (i.e., $\geq 80\%$ melting): (a) chondrule surfaces adjacent to the rims show resorption, (b) olivine and pyroxene grains in the rims are euhedral to subhedral, consistent with crystallization from a melt, (c) feldspathic mesostasis (i.e., solidified residual liquid) occurs between the mafic silicate grains in the rims, (d) metallic Fe-Ni and troilite occur in spheroidal blebs and discontinuous rings, and (e) relict (i.e., unmelted) grains are absent. The evidence for melting is so obvious that Krot and

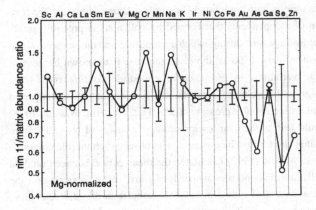

Fig. 2. Bulk composition of Allende coarse-grained rim 11 (open circles) normalized to mean matrix and to Mg (data from Rubin and Wasson, 1987, 1988). Elements are arranged from left to right in order of decreasing volatility except for the rare-earth elements which are listed in order of increasing atomic number. The rim is very similar in composition to the mean matrix, deviating less than 15% for most elements. This compositional similarity supports the inference that coarse-grained rims formed in part from matrix-like material. Error bars show the range in composition of matrix samples.

Wasson (1995) recommended renaming coarse-grained rims "igneous rims." In general, the coarse-grained rims surrounding high-FeO chondrules showed lower degrees of melting (but still generally > 50%): many of these rims contain chondrule (or microchondrule) fragments and unmelted magnesian olivine and pyroxene grains (in some cases derived from the host chondrule) (Krot and Wasson, 1995).

Kring (1991) described Fe,Ca-rich rims consisting mainly of Ca pyroxene, FeO-rich olivine, metallic Fe-Ni and sulfide surrounding a few percent of the chondrules in CO3 Kainsaz. Although the relationship between the Fe,Ca-rich rims and coarse-grained rims in OC and CV3 chondrites is unclear, some of the Fe,Ca-rich rims have unequivocal igneous textures. This indicates that their primary chondrules also experienced multiple heating events.

The resorption of mafic silicate grains at the outer surfaces of low-FeO OC chondrules with coarse-grained rims shows that these chondrules preserve evidence of the reheating event that formed their rims; it also shows that the coarse-grained rims might have been derived in part from their enclosed chondrules. In contrast, the much thicker coarse-grained rims around CV chondrules could not have been derived mainly from the outer portions of their enclosed chondrules. The similarities in CV chondrites between coarse-grained chondrule rims and matrix material in mineralogy, bulk chemistry and O-isotopic composition indicate that these rims consist largely of matrix-like material. New observations (A. N. Krot, unpublished data) indicate that CV3 coarse-grained chondrule rims contain low-FeO mafic silicate phenocrysts admixed with unmelted matrix-like material. Because many coarse-grained rims around low-FeO OC and CV chondrules are themselves surrounded by fine-grained rims (Rubin, 1984; Krot and Wasson, 1995), for which a nebular origin has previously been inferred (e.g., Rubin and Wasson, 1987), it is clear that melting of OC and CV coarse-grained rims occurred in the solar nebula. In addition, parent-body heating processes (e.g., impact; thermal metamorphism, igneous activity) are unlikely to have formed coarse-grained rims without significantly affecting the enclosed chondrules. The most likely nebular mechanism capable of melting clumps of matrix-like material to form coarse-grained rims is the same mechanism responsible for chondrule formation itself.

Connolly and Hewins (1991) produced synthetic coarse-grained rims with igneous textures by gluing fayalitic dust to cold chondrules and heating the assemblages in a furnace to ≥ 1000 °C. In one experiment metallic Fe-Ni and sulfide migrated from the chondrule and filled pore spaces in the rim. The moderate resemblance in texture of these experimental charges to OC chondrules with coarse-grained rims supports the idea that the latter objects formed by multiple episodes of heating.

Krot and Wasson (1995) found that in both low-FeO and high-FeO OC chondrules, the olivine Fa and low-Ca pyroxene Fs contents of many coarse-grained rims and their enclosed chondrules are similar (Fig. 3a). They suggested that the mean composition of nebular solids became more oxidized with time (e.g., as more metallic Fe reacted with H_2O in the gas to produce FeO as temperatures decreased); hence, the similarity in oxidation state between chondrules and their

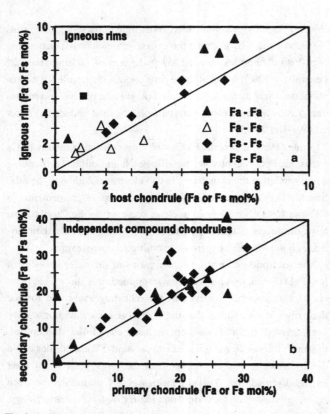

Fig. 3. (a) Diagram of the fayalite (Fa) content of olivine and ferrosilite (Fs) content of low-Ca pyroxene in igneous rims (i.e., coarse-grained rims) in ordinary chondrites versus Fa and Fs in the enclosed host low-FeO chondrule (data from Krot and Wasson, 1995). Line shows a one-to-one correspondence. Rims are generally more ferroan than chondrules, but many rims are close in Fa or Fs content to that of their enclosed chondrules. This correspondence suggests that rims formed not too long after their host chondrules, but in any case, before there was a substantial drift in the mean FeO content of nebular solids. (b) Diagram of the fayalite (Fa) content of olivine and ferrosilite (Fs) content of low-Ca pyroxene in independent compound chondrule secondaries in ordinary chondrites versus Fa and Fs in the corresponding primaries (data from Wasson *et al.*, 1995). The line shows a one-to-one correspondence. Approximately 40% of the pairs differ in Fa or Fs by ≥ 4 mol%, indicating that these objects formed from materials with significantly different oxidation states.

coarse-grained rims suggests that the time interval between these two heating events was short compared to the time required for a significant drift in the mean FeO content of nebular solids.

We conclude that chondrules with coarse-grained rims were heated to temperatures above the solidus more than once. During some reheating events, the primary chondrule itself may have been melted, obliterating petrographic evidence of reheating. Thus, in each chondrite group, the modal abundance of chondrules with coarse-grained rims is only a lower limit on the number of chondrules that experienced multiple heating events.

COMPOUND CHONDRULES

In the most comprehensive study of compound chondrules to date, Wasson *et al.* (1995) divided these conjoined objects into two major groups: *independent compound chondrules* (which differ

significantly in texture and/or mineral composition (e.g., by 4 mol% Fa or Fs; Fig. 3b) and most likely formed during separate heating events) and *sibling compound chondrules* (whose constituent objects have similar textures and mineral compositions and probably formed from the same precursor dust clump during a single heating event). Wasson *et al.* (1995) estimated that 1.4% of all OC chondrules are sibling compound chondrules and that 1.0% are independent compound chondrules. These authors distinguished primary chondrules (which formed first) from secondary chondrules (which formed later). They recognized three basic structural types of compound chondrules (which can occur either as siblings or independents): (1) *enveloping chondrules* (Fig. 4a,b), wherein a secondary chondrule encloses the primary (in many cases igneous mafic silicate overgrowths occur on the primary; these overgrowths are essentially identical in composition to phenocrysts in the secondary), (2) *adhering chondrules* (Fig. 4c), wherein a small secondary chondrule forms a quasi-hemispherical bump on the surface of a larger, primary chondrule, and (3) *consorting chondrules* (Fig. 4d), a special case of adhering chondrules, wherein the conjoined chondrules are of similar size.

Because sibling compound chondrules are assumed to have formed in a single flash-heating event, they will not be considered further here. However, independent compound chondrules could in principle record evidence of multiple flash-heating events, and they are the focus of this section.

There are two major models for the formation of independent compound chondrules:

Random collisions

Lux *et al.* (1981) and Gooding and Keil (1981) proposed that compound chondrules formed by random collisions during the period in which at least one of the chondrules was molten or partly molten. It is clear that siblings formed in this manner, and the experimental production of siblings (Connolly *et al.*, 1994a) creates objects with compound chondrule textures. However, the fraction of independents that formed by collisions is probably small. The average time during which a chondrule was molten was probably brief (say, 1–60 s); the exact duration is unknown because of large variations in estimates of chondrule cooling rates (e.g., Lofgren, this volume). Nevertheless, a much longer molten period at plausible nebular pressures (i.e., $pH_2 \leq 10^{-5}$ atm; Wasson, 1978) would lead to the evaporation of large amounts of Na and S from molten chondrules, inconsistent with observed chondrule compositions. Wasson (1993) calculated the mean period between random collisions of chondrules at speeds of 1 ms⁻¹; after correcting for the mistaken use of radius instead of diameter in this determination, Wasson *et al.* (1995) recalculated that the shortest mean time between collisions is ~1500 s. This greatly exceeds reasonable estimates for the upper limit of the period during which chondrules were molten or plastic (i.e., no more than a few tens of seconds; Wasson, 1993). Significantly higher speeds would result in chondrule fragmentation. In order to achieve a likelihood of forming compound chondrules via random collisions, the mean density of matter in the nebula would have to be implausibly high (i.e., much

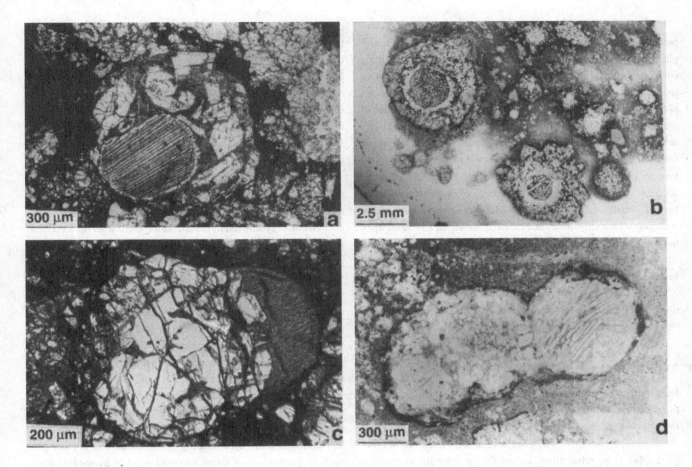

Fig. 4. Compound chondrules in ordinary and CV3 chondrites. (a) Independent enveloping compound chondrule in the Inman L/LL3.4 chondrite composed of a fragmented barred olivine primary and a large porphyritic olivine secondary. (b) Two neighboring independent enveloping compound chondrules with coarse-grained rims in Allende; both chondrules consist of a barred olivine primary and a porphyritic olivine-pyroxene secondary and are surrounded by symmetric coarse-grained rims. The great thickness of the olivine rim encircling the barred olivine primary chondrule in the compound object at lower right is due to igneous overgrowths in the enveloping secondary; the rim served as a nucleation site for olivine crystallizing from the melt of the secondary. The coarse-grained rim around this object is relatively narrow. In contrast the coarse-grained rim around the chondrule at upper left exceeds 1 mm in thickness in some areas; it is clearly separated from the secondary by a thin ring of troilite (black). (c) Adhering independent compound chondrule in the LEW86144 L3.2 chondrite consisting of a broken porphyritic olivine-pyroxene primary (center) with a radial pyroxene secondary (right) forming a quasi-hemispherical bump at the primary surface. (d) Three conjoined chondrules surrounded by a fine-grained rim in Allende. The large chondrule at right is barred olivine; the other two chondrules (which have a boundary that is not easily distinguished) are porphyritic olivine. Although the two porphyritic olivine chondrules may be siblings, the barred olivine chondrule probably formed in a separate event, making this consorting trio independent. a - d in transmitted light.

higher than that required to form the planets in minimum-mass solar-nebula models; Wasson, 1993; Wasson *et al.*, 1995).

Melting of dust around a primary chondrule

Wasson (1993) suggested that independent enveloping compound chondrules form a continuum with coarse-grained chondrule rims; independent petrographic studies by both of the present authors turned up many intermediate cases. It seems likely that independent enveloping compound chondrules and coarse-grained chondrule rims both formed by flash melting a porous dust clump containing a primary chondrule (which was not itself melted). Enveloping compound chondrules formed from large dust clumps that underwent large-scale melting; coarse-grained chondrule rims formed from small dust clumps that may have been less extensively melted. Flash-heating mechanisms such as lightning

(Rasmussen and Wasson, 1982) or magnetic-field reconnection (Levy and Araki, 1989) mainly heat grain surfaces by the impact of charged particles; they are more likely to melt fine-grained, porous materials such as clumps of matrix-like dust than solid chondrules (Wasson *et al.*, 1995).

Below we assume that most independent enveloping compound chondrules formed by the melting and collapse of a fine-grained dust clump in which a previously formed chondrule was entrained. These compound chondrules are the ones most analogous to coarse-grained chondrule rims. Random positioning of the primary chondrule in the dust clump could account for both central and off-center positioning of the primary chondrule in different enveloping compound chondrules (e.g., Fig. 1d of Rubin, 1984; Fig. 1 of Keil *et al.*, 1978). Analogously, there are symmetric (Figs. 1a, 4b) and asymmetric (Fig. 1d) coarse-grained rims.

It is less clear how independent adhering and consorting compound chondrules formed. The evidence that they formed by multiple heating events is less strong than for independent enveloping compound chondrules. There are several possibilities for forming independent adhering compound chondrules including: (1) complete melting of a small, dust-rich clump adhering to the surface of the primary chondrule, (2) melting of a portion (or portions) of a dust clump enclosing the primary chondrule, and (3) low-speed collision of a molten droplet formed by melting a dust clump in the vicinity of the primary chondrule. This last possibility probably accounts for few independent adhering compound chondrules because small molten droplets are likely to solidify long before colliding with a chondrule, as discussed above. In any case, upon melting, surface tension molded the adhesions into quasi-hemispherical bumps protruding from the surface of the primary. Independent consorting compound chondrules probably formed in a manner analogous to independent adhering compound chondrules except that, in the case of consorting chondrules, the dust-rich clumps were large relative to the primary chondrules.

RELICT GRAINS

Relict grains are solid precursors inside chondrules that survived the chondrule melting event. They are reviewed elsewhere (Jones, this volume). Such grains are abundant, occurring in $\geq 15\%$ of all chondrules in type-3 OC. There are several different kinds of relict grains: (a) Some relict grains in low-FeO porphyritic chondrules contain numerous small blebs of low-Ni metallic Fe formed by solid-state reduction of FeO during the chondrule melting event (e.g., Nagahara, 1981; Rambaldi, 1981); such grains appear dusty in transmitted light (Fig. 5). The reducing agent may have been C, which was itself oxidized and lost as CO during melting (Rambaldi and Wasson, 1982; Connolly *et al.*, 1994b). (b) Magnesian relict

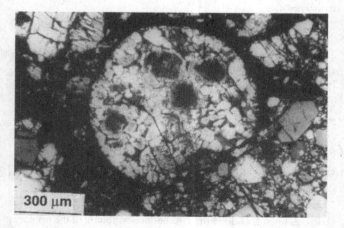

Fig. 5. Low-FeO porphyritic olivine chondrule in the Chainpur L/LL3.4 chondrite containing at least seven relict "dusty" olivine grains with numerous small inclusions of low-Ni metallic Fe formed by reduction of FeO in silicate during the chondrule melting event. Also present are several zoned, inclusion-free relict olivine grains with overgrowths produced in the chondrule melt. Transmitted light with crossed polarizers.

grains also occur; these grains are not dusty but can be identified in high-FeO chondrules, wherein the phenocrysts (grains which crystallized from the melt) have appreciably higher FeO contents (e.g., Fig. 4b of Jones, 1990). These relict grains typically have ferroan overgrowths similar in composition to the phenocrysts. (c) Other relict grains are recognized because they are very large relative to the majority of grains in an individual chondrule (Kracher *et al.*, 1984); in some cases, these grains constitute ≥ 90 vol.% of the chondrule (e.g., Fig. 5a of Rubin, 1989).

Both the high-FeO relict grains inside low-FeO chondrules and the low-FeO relict grains inside high-FeO chondrules are easily recognized because they are far out of equilibrium with the phenocrysts that crystallized in the chondrule melt. It seems likely that many, perhaps most, relict grains go undetected because their FeO contents are similar to those of the surrounding phenocrysts. We speculate that a hypothetical comprehensive survey of relict grains would show that their FeO contents are, on the whole, similar to those of their host chondrules; the situation could resemble that of independent enveloping compound chondrules wherein the FeO contents of a majority of olivine and low-Ca pyroxene grains in the primary and secondary objects are fairly similar (Fig. 6 of Wasson *et al.*, 1995).

Relict grains were probably derived from previous generations of chondrules as isolated phenocrysts liberated from collisionally disrupted chondrules. The liberated phenocrysts became entrained in clumps of dust; upon heating, the dust melted but the phenocrysts (or portions of them) survived as relict grains. It is possible that some relict grains are remnants of intact chondrules or chondrule fragments that were largely (but incompletely) melted during reheating. In either case, it is clear that relict grains constitute *prima facie* evidence for chondrule reheating.

CONSTRAINTS ON CHONDRULE ORIGIN

There is a close relationship between relict grains, independent enveloping compound chondrules and coarse-grained rims. If an intact chondrule gets entrained in a large dust clump and the clump undergoes large-scale melting during a reheating event that does not also melt the primary chondrule, the resultant object is an independent enveloping compound chondrule. If the primary object is not an entire chondrule but rather a phenocryst liberated from a previously disrupted chondrule, then the resultant object is a chondrule with a relict grain. In those cases wherein the dust clump is small and experiences ~20 to ~80% melting, a coarse-grained rim forms irrespective of the nature of the primary object (be it a chondrule, chondrule fragment or isolated mineral grain).

It seems very likely that the same heating mechanism is responsible for forming all of these interrelated objects; it is probably the same mechanism that formed chondrules in the first place. This indicates that the chondrule-formation mechanism is repeatable (**constraint 1**).

The large percentage of chondrules that contain relict grains or coarse-grained rims or belong to independent compound chondrules (> 25% in OC; > 50% in CV3 chondrites) indicates that the chondrule formation mechanism commonly reheated nebular solids. Because different chondrite groups may have formed at significantly different heliocentric distances (e.g., Wasson, 1985, 1988; Rubin and Wasson, 1995), the differences among the fractions of reheated chondrules in different groups may indicate that the intensity of the heating mechanism varied with nebular region (**constraint 2**). This is consistent with observed differences in the percentage of (completely melted) droplet chondrules in different chondrite groups (Rubin and Wasson, 1995).

The low degree of volatile (Na, S) loss in chondrules during melting indicates that chondrules were molten for short periods of time (e.g., Grossman, 1988; Wasson, 1993; Yu *et al.*, this volume), perhaps only seconds to tens of seconds. Although not a constraint developed in the present study, this observation strongly indicates that the chondrule-formation mechanism involved flash-heating (**constraint 3**).

These three constraints on chondrule formation can help identify the specific mechanism involved. Plausible, recurrent, flash-heating mechanisms that could be expected to vary in intensity in different nebular regions include nebular lightning (Rasmussen and Wasson, 1982; Gibbard and Levy, 1994; Horanyi *et al.*, 1995; Horanyi and Robertson, this volume; Wasson and Rasmussen, 1994), solar flares (Levy and Araki, 1989) and shock waves within the nebular midplane (Hood and Kring, this volume).

ACKNOWLEDGEMENTS

We thank J.T. Wasson for valuable discussions and comments on the manuscript. Reviews by K. Metzler and H. Connolly and suggestions by R.H. Jones are greatly appreciated. This work was supported in part by NASA grants NAGW-3535 (J.T. Wasson) and NAGW-3281 (K. Keil). This is Hawai'i Institute of Geophysics and Planetology Publication number 863 and School of Ocean and Earth Science and Technology, University of Hawai'i, Publication number 4018.

REFERENCES

Alexander C. M. O., Hutchison R. and Barber D. J. (1989) Origin of chondrule rims and interchondrule matrices in unequilibrated ordinary chondrites. *Earth Planet. Sci. Lett.* **95**, 187–207.

Allen J. S., Nozette S. and Wilkening L. L. (1980) A study of chondrule rims and chondrule irradiation records in unequilibrated ordinary chondrites. *Geochim. Cosmochim. Acta* **44**, 1161–1175.

Ashworth J. R. (1977) Matrix textures in unequilibrated ordinary chondrites. *Earth Planet. Sci. Lett.* **35**, 25–34.

Brearley A. J., Scott E. R. D., Keil K., Clayton R. N., Mayeda T. K., Boynton W. V. and Hill D. H. (1989) Chemical, isotopic, and mineralogical evidence for the origin of matrix in ordinary chondrites. *Geochim. Cosmochim. Acta* **53**, 2081–2093.

Christophe Michel-Lévy M. (1976) La matrice noire et blanche de la chondrite de Tieschitz (H3). *Earth Planet. Sci. Lett.* **30**, 143–150.

Clayton R. N., Onuma N., Ikeda Y., Mayeda T., Hutcheon I. D., Olsen E. J. and Molini-Velsko C. (1983) Oxygen isotopic compositions of chondrules in Allende and ordinary chondrites. In *Chondrules and their Origins* (ed. E. A. King), pp. 37–43. Lunar Planet. Inst.

Connolly H. C. and Hewins R. H. (1991) The experimental production of chondrule rims: Constraints on chondrule rim origins (abstract). *Lunar Planet. Sci.* **22**, 233–234.

Connolly H. C., Hewins R. H., Atre N. and Lofgren G. E. (1994a) Compound chondrules: An experimental investigation (abstract). *Meteoritics* **29**, 458.

Connolly H. C., Hewins R. H., Ash R. D., Zanda B., Lofgren G. E. and Bourot-Denise M. (1994b) Carbon and the formation of reduced chondrules. *Nature* **371**, 136–139.

Gibbard S. G. and Levy E. H. (1994) On the possibility of precipitation-induced vertical lightning in the protoplanetary nebula (abstract). *Papers Presented to Chondrules and the Protoplanetary Disk* LPI Contribution No. 844, 9–10.

Gooding J. L. and Keil K. (1981) Relative abundances of chondrule primary textural types in ordinary chondrites and their bearing on conditions of chondrule formation. *Meteoritics* **16**, 17–43.

Grossman J. N. (1988) Formation of chondrules. In *Meteorites and the Early Solar System* (eds. J. F. Kerridge and M. S. Matthews), pp. 680–696. Univ. Arizona Press.

Horanyi M., Morfill G., Goertz C. K. and Levy E. H. (1995) Chondrule formation in lightning discharges. *Icarus* **114**, 174–185.

Housley R. M. and Cirlin E. H. (1983) On the alteration of Allende chondrules and the formation of matrix. In *Chondrules and Their Origins* (ed. E. A. King), pp. 145–161. Lunar and Planetary Institute.

Huss G. R., Keil K. and Taylor G. J. (1981) The matrices of unequilibrated ordinary chondrites: Implications for the origin and history of chondrites. *Geochim. Cosmochim. Acta* **45**, 33–51.

Ikeda Y., Kimura M., Mori H. and Takeda H. (1981) Chemical composition of matrices of unequilibrated ordinary chondrites. *Mem. Natl. Inst. Polar Res. Special Issue* **20**, 124–144.

Jones R. H. (1990) Petrology and mineralogy of type II, FeO-rich chondrules in Semarkona (LL3.0): Origin by closed-system fractional crystallization, with evidence for supercooling. *Geochim. Cosmochim. Acta* **54**, 1785–1802.

Keil K., Lux G., Brookins D. G., King E. A. and King T. V. V. (1978) The Inman, McPherson County, Kansas meteorite. *Meteoritics* **13**, 11–22.

King T. V. V. and King E. A. (1981) Accretionary dark rims in unequilibrated chondrites. *Icarus* **48**, 460–472.

Kracher A., Scott E. R. D. and Keil K. (1984) Relict and other anomalous grains in chondrules: Implications for chondrule formation. *Proc. Lunar Planet. Sci. Conf.* **14**, B559–B566.

Kring D. A. (1991) High temperature rims around chondrules in primitive chondrites: Evidence for fluctuating conditions in the solar nebula. *Earth Planet. Sci. Lett.* **105**, 65–80.

Krot A. N. and Wasson J. T. (1995) Igneous rims on low-FeO and high-FeO chondrules in ordinary chondrites. *Geochim. Cosmochim. Acta* **59**, 4951–4966.

Levy E. H. and Araki S. (1989) Magnetic reconnection flares in the protoplanetary nebula and the possible origin of meteoritic chondrules. *Icarus* **81**, 74–91.

Love S. G., Keil K. and Scott E. R. D. (1994) Formation of chondrules by electrical discharge heating (abstract). *Papers Presented to Chondrules and the Protoplanetary Disk* LPI Contribution No. 844, 21–22.

Lux G., Keil K. and Taylor G. J. (1981) Chondrules in H3 chondrites: textures, compositions and origins. *Geochim. Cosmochim. Acta* **45**, 675–685.

Matsunami S. (1984) The chemical compositions and textures of matrices and chondrule rims of eight unequilibrated ordinary chondrites: A preliminary report. *Mem. Natl. Inst. Polar Res. Special Issue* **35**, 126–148.

Mittlefehldt D. W., Rubin A. E. and Davis A. M. (1992) Mesosiderite clasts with the most extreme positive Eu anomalies among solar system rocks. *Science* **257**, 1096–1099.

Morfill G., Spruit H. and Levy E. H. (1993) Physical processes and conditions associated with the formation of protoplanetary disks. In *Protostars and Planets III* (eds. E. H. Levy and J. I. Lunine), pp. 939–978. Univ. Arizona Press.

Nagahara H. (1981) Evidence for secondary origin of chondrules. *Nature* **292**, 135–136.

Nagahara H. (1984) Matrices of type 3 ordinary chondrites – primitive nebular records. *Geochim. Cosmochim. Acta* **48**, 2581–2595.

Rambaldi E. R. (1981) Relict grains in chondrules. *Nature* **293**, 558–561.

Rambaldi E. R. and Wasson J. T. (1982) Fine, nickel-poor Fe-Ni grains in the olivine of unequilibrated ordinary chondrites. *Geochim. Cosmochim. Acta* **46**, 929–939.

Rambaldi E. R., Rajan R. S., Wang D. and Housley R. M. (1983) Evidence for relict grains in chondrules of Qingzhen, an EH3 type enstatite chondrite. *Earth Planet. Sci. Lett.* **66**, 11–24.

Rasmussen K. L. and Wasson J. T. (1982) A new lightning model for chondrule formation (abstract). In *Papers Presented to the Conference on Chondrules and Their Origins*, p. 53. Lunar and Planetary Institute.

Rubin A. E. (1984) Coarse-grained chondrule rims in type 3 chondrites. *Geochim. Cosmochim. Acta* **48**, 1779–1789.

Rubin A. E. (1989) Size-frequency distributions of chondrules in CO3 chondrites. *Meteoritics* **24**, 179–189.

Rubin A. E. and Grossman J. N. (1987) Size frequency distributions of EH3 chondrules. *Meteoritics* **22**, 237–251.

Rubin A. E. and Wasson J. T. (1987) Chondrules, matrix and coarse-grained chondrule rims in the Allende meteorite: Origin, interrelationships and possible precursor components. *Geochim. Cosmochim. Acta* **51**, 1923–1937.

Rubin A. E. and Wasson J. T. (1988) Errata. *Geochim. Cosmochim. Acta* **52**, 2549.

Rubin A. E. and Wasson J. T. (1995) Variations of chondrite properties with heliocentric distance (abstract). *Meteoritics* **30**, 569.

Rubin A. E., Wasson J. T., Clayton R. N. and Mayeda T. K. (1990) Oxygen isotopes in chondrules and coarse-grained chondrule rims from the Allende meteorite. *Earth Planet. Sci. Lett.* **96**, 247–255.

Scott E. R. D., Rubin A. E., Taylor G. J. and Keil K. (1984) Matrix material in type 3 chondrites - Occurrence, heterogeneity and relationship with chondrules. *Geochim. Cosmochim. Acta* **48**, 1741–1757.

Scott E. R. D., Barber D. J., Alexander C. M. O., Hutchison R. and Peck J. A. (1988) Primitive material surviving in chondrites; Matrix. In *Meteorites and the Early Solar System* (eds. J. F. Kerridge and M. S. Matthews), pp. 718–745. University of Arizona Press.

Sonett C. P. (1979) On the origin of chondrules. *Geophys. Res. Lett.* **6**, 677–680.

Taylor G. J., Scott E. R. D. and Keil K. (1983) Cosmic setting for chondrule formation. In *Chondrules and their origins* (ed. E. A. King), pp. 262–268. Lunar and Planetary Institute, Houston.

Wark D. A. (1979) Birth of the presolar nebula: The sequence of condensation revealed in the Allende meteorite. *Astrophys. Space Sci.* **65**, 275–295.

Wasson J. T. (1978) Maximum temperatures during the formation of the solar nebula. In *Protostars and Planets* (ed. T. Gehrels), pp. 488–501, University of Arizona Press.

Wasson J. T. (1985) *Meteorites: Their Record of Early Solar System History*, Freeman, 267p.

Wasson J. T. (1988) The building stones of the planets. In *Mercury* (ed. F. Vilas, C. R. Chapman and M. S. Matthews), pp. 622–650, University of Arizona Press.

Wasson J. T. (1993) Constraints on chondrule origins. *Meteoritics* **28**, 13–28.

Wasson J. T. and Rasmussen K. L. (1994) The fine nebula dust component: A key to chondrule formation by lightning (abstract). *Papers Presented to Chondrules and the Protoplanetary Disk* LPI Contribution No. 844, 43.

Wasson J. T., Krot A. N., Lee M. S. and Rubin A. E. (1995) Compound chondrules. *Geochim. Cosmochim. Acta* **59**, 1847–1869.

Wood J. A. (1984) On the formation of meteoritic chondrules by aerodynamic drag heating in the solar nebula. *Earth Planet. Sci. Lett.* **70**, 11–26.

19: Microchondrule-Bearing Chondrule Rims: Constraints on Chondrule Formation

ALEXANDER N. KROT[1] and ALAN E. RUBIN[2]

[1]*Hawai'i Institute of Geophysics and Planetology, School of Ocean and Earth Science and Technology, University of Hawai'i at Manoa, Honolulu, HI 96822, U.S.A.* [2]*Institute of Geophysics and Planetary Physics, University of California, Los Angeles, CA 90095–1567, U.S.A.*

ABSTRACT

Microchondrules, arbitrarily defined as chondrules less than 40 μm in apparent diameter, have been found in fine-grained rims around five normal-size low-FeO chondrules in LL3.1 Bishunpur, L3.4 EET90161, L3.4 EET90261, LL3.4 Piancaldoli and LL3.0 Semarkona. Two kinds of microchondrules are observed in these rims: abundant low-FeO microchondrules consisting of low-Ca pyroxene and relatively rare high-FeO microchondrules consisting of fayalitic olivine. Both types of microchondrules are embedded in fine-grained FeO-rich matrix-like material; they typically are accompanied by irregularly shaped pyroxene fragments having compositions similar to the low-FeO microchondrules. The pyroxene-rich surfaces of the host chondrules are very irregular and show evidence of remelting; these surfaces appear to have been the main source of the low-FeO pyroxene fragments and microchondrules. We suggest that the microchondrules within each fine-grained chondrule rim formed during a single flash-heating event which melted pyroxene-rich chondrule exteriors and small amounts of FeO-rich matrix material.

INTRODUCTION

Chondrules in different chondrite groups have average apparent diameters ranging from ~200–1000 μm (Grossman *et al.*, 1988a). In addition to these "normal-size" chondrules there are macrochondrules ranging up to 5 cm in diameter (Prinz *et al.*, 1988; Weisberg *et al.*, 1988a) and microchondrules (arbitrarily defined as chondrules < 40 μm in diameter) ranging down to submicrometer sizes (Rubin *et al.*, 1982). One group of carbonaceous chondrites (sometimes called CH), typified by ALH85085, contain microchondrules as the predominant chondrule size (e.g., Grossman *et al.*, 1988b; Scott, 1988; Weisberg *et al.*, 1988b; Bischoff *et al.*, 1993).

Microchondrules in ordinary chondrites (OC) have been described in two principal settings: (1) as rare individuals in the fine-grained matrix of many type-3 chondrites (Rubin *et al.*, 1982; Nagahara, 1984), and (2) as the predominant chondrule size in a few unequilibrated clasts in the Rio Negro L regolith breccia (Fodor *et al.*, 1977; Rubin *et al.*, 1982), L3.7 Mezö-Madaras (Christophe Michel-Lévy, 1988) and LL3.1 Krymka (Rubin, 1989). In most cases the microchondrules are embedded in fine-grained FeO-rich matrix-like material and have similar textures and compositions within individual clasts. Here we report new occurrences of microchondrules in type-3 ordinary chondrites: five normal-size low-FeO chondrules are surrounded by fine-grained matrix-like rims containing numerous microchondrules. This discovery sheds light on the origin of microchondrules and on the general question of chondrule formation.

RESULTS

Microchondrules in fine-grained rims were found around three normal-size chondrules in L3.4 EET90161, L3.4 EET90261 and LL3.1 Bishunpur. The microchondrule-bearing object in LL3.4 Piancaldoli that Rubin *et al.* (1982) called a clast was also found to be a fine-grained chondrule rim – one adjacent to a pyroxene grain at the edge of the thin section (Fig. 1a). In addition, D.A. Kring (pers. commun., 1995) found abundant microchondrules within the fine-grained rim surrounding a low-FeO barred olivine chondrule in LL3.0 Semarkona.

The 850 × 1500 μm porphyritic olivine (PO) chondrule in

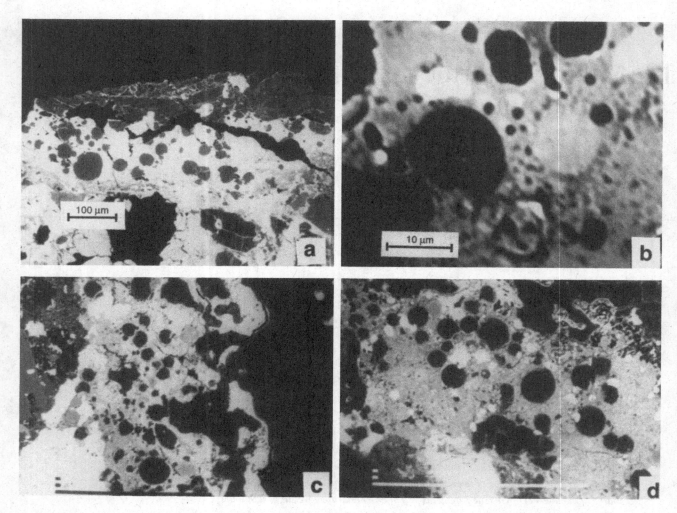

Fig. 1. (a) Fine-grained matrix-like material (white) containing numerous low-FeO, low-Ca pyroxene microchondrules (dark gray) adjacent to a pyroxene grain in Piancaldoli. The pyroxene grain is probably part of a porphyritic chondrule that was lost during thin section preparation; the matrix-like material probably represents a rim that surrounded the entire chondrule. (b) Along with the abundant FeO-poor pyroxene microchondrules (dark gray) in the Piancaldoli rim, there is a single FeO-rich olivine microchondrule (light gray). (c) A fine-grained matrix-like rim (light gray) surrounding a porphyritic olivine chondrule (right) in EET90161. The chondrule surface is very irregular and embayed, suggestive of remelting. Two kinds of microchondrules occur in the rim: abundant low-FeO pyroxene (dark gray) and relatively rare FeO-rich olivine (light gray); also present are irregularly shaped pyroxene fragments (dark gray). (d) A microchondrule-bearing rim (light gray) around a compound porphyritic olivine chondrule (dark gray) in Bishunpur. The surface of the host chondrule (top right) is very irregular and protrudes into the rim. Two kinds of microchondrules occur: abundant low-FeO pyroxene (dark gray) and relatively rare FeO-rich olivine (light gray). Also present are irregularly shaped pyroxene fragments. The long scale bars in (c) and (d) are 100 μm. Backscattered electron images.

EET90161 consists of euhedral and subhedral grains of forsterite and numerous metal-rich opaque nodules embedded in Ca-rich feldspathic glass. The chondrule surface is irregular, but not angular or fragmented; it shows rounded embayments suggestive of remelting and flowing. The chondrule is surrounded by a 50–70-μm-thick fine-grained FeO-rich matrix-like rim containing abundant (n > 100) microchondrules as well as irregularly shaped pyroxene (Fs_2Wo_8) fragments. The microchondrules, which are homogeneously distributed in the rim, occur in two varieties: abundant low-FeO cryptocrystalline microchondrules consisting of low-Ca pyroxene ($Fs_{1-22}Wo_{1.4\pm0.9}$) and rare high-FeO microchondrules consisting of fayalitic olivine ($Fa_{58\pm10}$). The Fs values in the low-Ca pyroxene in the low-FeO microchondrules overlap with those of the pyroxene in the outermost part of the host chondrule;

there is also marginal overlap in the Wo values, but, in general, the microchondrules are significantly less calcic.

The 500-μm-diameter porphyritic olivine-pyroxene (POP) chondrule in EET90261 has an irregular shape; it consists of large laths of forsterite, calcic plagioclase, sub-calcic augite, diopside and Ca-rich feldspathic mesostasis. The chondrule is surrounded by a 50–70-μm-thick fine-grained FeO-rich matrix-like rim containing ~30–50 microchondrules, < 20 μm in apparent diameter, and irregularly shaped pyroxene fragments. The microchondrules are homogeneously distributed in the rim; they consist of cryptocrystalline low-Ca pyroxene of fairly uniform composition ($Fs_{3.6\pm1.4}Wo_{1.7\pm0.9}$).

The 1250×1850 μm chondrule in LL3.1 Bishunpur is a compound object consisting of a PO primary and two PPO (porphyritic pyroxene-olivine) adhering secondaries (Fig. 3d of Wasson *et al.*, 1995). The primary consists of forsterite and anorthitic glass; it also contains several relict grains of forsterite with abundant inclusions of tiny "dusty" metallic Fe-Ni grains. Low-Ca pyroxene (Fs_2Wo_1) occurs only in the outermost part of the primary chondrule. The adhering secondaries consist of major forsterite and accessory low-Ca pyroxene, high-Ca pyroxene and moderately sodic feldspathic mesostasis. The surfaces of the host chondrule and the adhering secondaries are fragmented and irregular; they commonly form irregularly shaped peninsulas (many of which have rounded edges) protruding into the fine-grained rim. We interpret these textures as evidence of remelting. The rim around the entire compound object is 100–150 µm thick and contains a high abundance (n > 100) of homogeneously distributed, irregularly shaped pyroxene fragments and microchondrules. Two types of microchondrules are present: abundant low-FeO cryptocrystalline low-Ca pyroxene microchondrules and relatively rare high-FeO olivine ($Fa_{52 \pm 9}$) microchondrules. The pyroxene in the low-FeO microchondrules overlaps in composition with the pyroxene peninsulas protruding from the host chondrule surface, but, on average, the microchondrule pyroxene has more FeO and less CaO.

Reexamination of Piancaldoli revealed that the microchondrule-bearing object described by Rubin *et al.* (1982) is actually a fine-grained FeO-rich matrix-like rim adjacent to a pyroxene grain. The grain itself is probably a phenocryst of a porphyritic chondrule that was lost during thin section preparation; the matrix-like rim probably surrounded the entire chondrule. The surface of the Piancaldoli pyroxene grain is irregular and embayed, suggestive of remelting. Many of the microchondrules in the rim are compound. The microchondrules consist of cryptocrystalline or radial low-Ca pyroxene ($Fs_{3.6 \pm 2.4}Wo_{1.0 \pm 0.5}$); one high-FeO olivine microchondrule (Fa_{77}) was also found in the rim.

Another microchondrule-bearing rim occurs around a low-FeO barred olivine (BO) chondrule in LL3.0 Semarkona (D.A. Kring, pers. commun., 1995). It is essentially identical to those described above.

DISCUSSION

The occurrence in each of these rims of numerous microchondrules (including many compound objects) with very similar compositions suggests that they are analogous to "sibling" compound chondrules (Wasson *et al.*, 1995), which are believed to have formed simultaneously from the same precursor dustball. We suggest that the microchondrules in each rim formed from a "cloud" of droplets around the host chondrule during a single heating event. Because such a microchondrule cloud would dissipate quickly due to random motions, a plausible model seems to require that the fine-grained material which now composes the rims was in the immediate vicinity of the microchondrule droplets and entrained them shortly after melting. The microchondrules may have existed for 1 s as independent free-floating droplets within voids in the matrix-like rims before colliding to form compound objects or getting trapped in the rim.

The compositional similarity between the large pyroxene phe-

nocryst and the adjacent pyroxene microchondrules in Piancaldoli suggests that the microchondrules may have been produced by remelting of the host chondrule surface. Although the other two chondrules with microchondrule-bearing rims are PO, it is plausible that their remelted surfaces were pyroxene-rich as well: Jones and Scott (1989) showed that some low-FeO PO chondrules consist essentially of monomineralic pyroxene shells surrounding an olivine-rich core. Alternatively, the pyroxene could have been derived from a coarse-grained (or igneous) rim; Krot and Wasson (1995) found that the majority of such rims around low-FeO PO chondrules are pyroxene-rich. Flash melting of the dust in a manner analogous to coarse-grained-rim formation (Rubin and Krot, this volume) could have produced the pyroxene microchondrules and the irregularly shaped pyroxene fragments from dislodged chondrule surface materials. The modestly higher mean FeO contents and lower CaO contents of the microchondrules relative to pyroxene at the surface of the host chondrules are probably due to the melting of small amounts of FeO-rich, CaO-poor matrix-like material in the rim along with major amounts of pyroxene from the chondrule surface. The relatively rare high-FeO olivine microchondrules were probably formed from melting of the fine-grained FeO-rich dust itself.

The formation of microchondrule-bearing rims requires the presence of fine-grained matrix material in the region of chondrule formation, suggesting that matrix-like material was one of the chondrule precursor components. The presence of high-FeO olivine microchondrules coexisting with low-FeO pyroxene microchondrules in the chondrule rims indicates that the fine-grained material had already been oxidized by the time the microchondrules formed. Hence, mixing of precursors of different degrees of oxidation occurred during chondrule formation.

Although the fine-grained FeO-rich material is less refractory than low-FeO pyroxene, the predominance of low-FeO microchondrules indicates that the heating mechanism more efficiently melted coarser, denser, less-porous material (i.e., pyroxene at the chondrule surface) than fine-grained, porous matrix-like rim material. This conclusion is consistent with the chondrule-formation mechanism suggested by Eisenhour *et al.* (1994); heating by predominantly visible light causes larger, thicker, more opaque objects to reach higher temperatures because they absorb light efficiently compared to their ability to radiate energy away.

ACKNOWLEDGEMENTS

We thank G.J. MacPherson (Smithsonian Institution) and the Antarctic Meteorite Working Group for thin sections. We are grateful to J.T. Wasson, E.R.D. Scott, K. Keil, G.J. Taylor and S.G. Love for informative discussions and to I.S. Sanders and D.A. Kring for reviews. We are grateful to D.A. Kring for his unpublished data. This work was supported mainly by NASA grants NAGW-3535 (J.T. Wasson) and NAGW-3281 (K. Keil). This is Hawai'i Institute of Geophysics and Planetology Publication number 864 and School of Ocean and Earth Science and Technology, University of Hawai'i, Publication number 4019.

REFERENCES

Bischoff A., Palme H., Schultz L., Weber D., Weber H. W. and Spettel B. (1993) Acfer 182 and paired samples, an iron-rich carbonaceous chondrite: Similarities with ALH85085 and relationship to CR chondrites. *Geochim. Cosmochim. Acta* **57**, 2631–2648.

Christophe Michel-Lévy M. (1988) A new component of the Mezö-Madaras breccia: A microchondrule- and carbon-bearing L-related chondrite. *Meteoritics* **23**, 45–49.

Eisenhour D. D., Daulton T.L. and Buseck P. R. (1994) Electromagnetic heating in the early solar nebula and the formation of chondrules. *Science* **265**, 1067–1070.

Fodor R. V., Keil K. and Gomes C. B. (1977) Studies of Brazilian meteorites. IV. Origin of a dark-colored unequilibrated lithic fragment in the Rio-Negro chondrite. *Revista Bras. Geociencias* **7**, 45–57.

Grossman J. N., Rubin A. E., Nagahara H. and King E. A. (1988a) Properties of chondrules. In *Meteorites and the Early Solar System* (eds. J. F. Kerridge and M. S. Matthews), pp. 619–659. Univ. of Arizona Press.

Grossman J. N., Rubin A. E. and MacPherson G. J. (1988b) ALH85085: a unique volatile-poor carbonaceous chondrite with possible implications for nebular fractionation processes. *Earth Planet. Sci. Lett.* **91**, 33–54.

Jones R. H. and Scott E. R. D. (1989) Petrology and thermal history of type IA chondrules in the Semarkona (LL3.0) chondrite. *Proc. Lunar Planet. Sci. Conf.* **19th**, 523–536.

Krot A. N. and Wasson J. T. (1995) Igneous rims on low-FeO and high-FeO chondrules in ordinary chondrites. *Geochim. Cosmochim. Acta*, **59**, 4951–4966.

Nagahara H. (1984) Matrices of type 3 ordinary chondrites-primitive nebular records. *Geochim. Cosmochim. Acta* **48**, 2581–2595.

Prinz M., Weisberg M. K. and Nehru C. E. (1988) Gunlock, a new type 3 ordinary chondrite with a golfball-sized chondrule (abstract). *Meteoritics* **23**, 297.

Rubin A. E. (1989) An olivine-microchondrule-bearing clast in the Krymka meteorite. *Meteoritics* **24**, 191–192.

Rubin A. E., Scott E. R. D. and Keil K. (1982) Microchondrule-bearing clast in the Piancaldoli LL3 meteorite: A new kind of type 3 chondrite and its relevance to the history of chondrules. *Geochim. Cosmochim. Acta* **46**, 1763–1776.

Scott E. R. D. (1988) A new kind of primitive chondrite, Allan Hills 85085. *Earth Planet. Sci. Lett.* **91**, 1–18.

Wasson J. T., Krot A. N., Lee M. S. and Rubin A. E. (1995) Compound chondrules. *Geochim. Cosmochim. Acta* **59**, 1847–1869.

Weisberg M. K., Prinz M. and Nehru C. E. (1988a) Macrochondrules in ordinary chondrites: Constraints on chondrule forming processes (abstract). *Meteoritics* **23**, 309–310.

Weisberg M. K., Prinz M. and Nehru C. E. (1988b) Petrology of ALH85085: a chondrite with unique characteristics. *Earth Planet. Sci. Lett.* **91**, 19–32.

IV. Heating, Cooling and Volatiles

20: A Dynamic Crystallization Model for Chondrule Melts

GARY E. LOFGREN

SN-4, NASA Johnson Space Center, Houston, TX 77058, U.S.A.

ABSTRACT

A model for the crystallization of chondrule melts based on dynamic crystallization experiments shows how the complex interaction of heterogeneous nucleation and cooling rate result in the extensive array of chondrule textures. Nucleation on relict crystals that result from incomplete melting of precursor crystalline material produces porphyritic textures. Barred and radial textures develop when melts are heated to superliquidus temperatures eliminating all viable nuclei. Nucleation ultimately occurs on sub-critically sized, crystalline embryos at a degree of supercooling that is smallest for barred textures and larger for radial. Based on this model, the range of melting temperatures is 1200 to 1750°C for minutes to seconds and up to 1900°C on rare occasions for a few seconds. Cooling rates define the time available for crystallization of the melts, and the shapes and zoning profiles in olivine phenocrysts in chondrules compared to experimental equivalents suggest cooling rates in the range 10 to 1000°C/hr. Thus minimum cooling times of 9 minutes to 1.5 hours would be common for many porphyritic chondrules with cooling times up to 30 hours possible. The melting and cooling histories, as determined by experiment, require rapid heating and sustained cooling events that are capable of producing the high melting temperatures and cooling times much slower than simple radiative cooling.

INTRODUCTION

Chondrules are some of the most primitive particles available for study in our solar system. They are the principle component in most chondrites and thus, indirectly a major building block for planets. There is a consensus that most chondrules are crystallized melt droplets. Here the consensus ends. There is no unanimity of opinion about how and where they were formed. The source of heat to melt the material that forms the droplets is as highly contentious as the precise location and the mechanism to concentrate material in the nebula (Wood, this volume). A step in the path to a workable model is to establish the crystallization characteristics of melt droplets and to determine the details of the process such that the required quantity of heat and duration of the event can be determined. While we understand little of the nebular environment in which chondrules may form, we understand very well the nucleation and crystallization of a melt droplet. We can describe with considerable certainty what happens to a melt droplet once it forms until it finishes crystallizing. Experimentation on chondrule melts

conducted for more than a decade has provided an extensive understanding of the chondrule crystallization process and allows us to set firm requirements on the temperatures needed to melt and the subsequent time necessary to crystallize the droplets.

In this paper, a model will be described for the crystallization of melt droplet chondrules. Based on this model, it will be possible to set firm limits on the heating and cooling regimes. No attempt will be made to review all the experiments that have been completed (see Hewins, 1988 review), but their results will be used to build the total model. The model will explain one of the more surprising results of the experiments, that quite different textures, e.g. porphyritic and radial, can be produced at the same, relatively slow cooling rate. An understanding of the melting and crystallization processes in the model will reveal how the different observed textures develop and explain the apparent contradictions in the effect of cooling rate on texture. In the early studies of chondrule crystallization, there was difficulty crystallizing porphyritic textures (Blander *et al.*, 1976; Tsuchiyama *et al.*, 1980; Hewins *et al.*, 1981; Planner and Keil, 1982). The difficulty of experimentally producing porphyritic chondrule textures disappeared when attention was paid to the heterogeneous nucleation potential of the melt (Lofgren 1982, 1989; Lofgren and Russell, 1986; Lofgren and Lanier, 1990; Radomsky and Hewins, 1990; Connolly and Hewins, 1991b).

While the crystallization model is well defined, it unfortunately presents only limited insight into the nature of the heating event itself because the nucleation and growth processes are independent of the nature of the forming event. All that is required is heat from the event to melt precursor material and form droplets. Once formed, the droplets crystallize in response to the number of nuclei produced during melting and the cooling rate. Both of these parameters are a function of the forming event and set limits on what the event must produce in terms of heat and duration. How the heat is generated is not integral to chondrule melt crystallization, only the amount of heat, the rapidity with which it is generated, and in turn, the rapidity with which it is dissipated.

Those aspects of heterogeneous nucleation and crystal growth that are important to understanding the development of chondrule textures will be reviewed first. A general model for chondrule crystallization will be presented based on these processes and extensive experimental work on the duplication of chondrule textures. This model will be used to explain the formation of the most common chondrule textures. The most important part of this discussion will be determining the constraints on the chondrule forming process that can be derived from the model and the experiments.

NUCLEATION AND GROWTH

Understanding nucleation and growth processes in silicate melts is fundamental to understanding how textures develop in chondrules. I will not review the complete subject matter (see Kirkpatrick, 1981 for review), but will explain some important processes upon which I will build a chondrule crystallization model.

Nucleation

Nucleation can be either homogeneous or heterogeneous. Nucleation in silicate melts is a sluggish process and in nearly all cases requires heterogeneous nucleation (Lofgren, 1983, Uhlmann and Chalmers, 1965). In chondrules, porphyritic textures form with difficulty unless heterogeneous nuclei are present. Heterogeneous nucleation requires a substrate of crystalline material suitable for the growth of the pertinent phase that is of critical size so as to accommodate stable growth. Recognition that the formation of chondrule melt droplets most likely involves incomplete melting of crystalline precursor material provides a logical mechanism to provide nuclei. The identity of the nuclei varies for the phase under consideration. Of the important silicate phases, olivine will nucleate most readily on the broadest variety of substrates. The ability to nucleate appears to be related to the complexity of the crystal structure. The observed sequence for ease of nucleation begins with olivine, then pyroxene, then the feldspars which appear to nucleate only on feldspar nuclei. The importance of heterogeneous nucleation became obvious in the study of basalts (Lofgren, 1983) which would crystallize feldspar, and, in most cases, pyroxene with typi-

cal basaltic textures only when appropriate nuclei were present in the melt when cooling began.

One facet of heterogeneous nucleation important to understanding the origin of the dendritic and radial (including excentroradial) or spherulitic as opposed to porphyritic or granular textures, is the concept of the embryonic nucleus (embryo) versus the super-critical nucleus (Turnbull, 1950a,b; 1956, 1969; and as applied to silicate melts, Lofgren, 1983). The embryo is crystalline material that is smaller than the critical size for nucleation and will not act as a nucleation site. All melts have an embryo population in which individual embryos are constantly changing in size with a few becoming nuclei as they spontaneously and randomly increase in size. The actual size of the critical nucleus has not been determined. In experiments completed to date, critical nuclei are smaller than can be resolved in an optical microscope. This would mean they are less than approximately 0.1 micrometers. Embryos would be considerable smaller, probably no more than a few crystal unit cell dimensions, i.e., 50–100 angstroms. Crystals that greatly exceed the critical size and are readily visible in the microscope, do not usually function as nuclei that dictate the texture. They are usually fewer in number, and while they increase in size during cooling, they appear to be incidental to the final texture.

In more slowly cooled melts with a sparse population of nuclei, the random graduation of embryos to nuclei can be an important source of nuclei. By contrast, in rapidly cooled melts completely free of nuclei, embryos may be the only source of nuclei for the ultimate crystallization of the melt. The precise size of a critical nucleus is a function of the kinetic driving forces present (Turnbull, 1956, 1969). As the degree of supercooling (ΔT) increases, the critical size of a nucleus decreases so that embryos can become nuclei simply as the ΔT increases during the rapid cooling of a chondrule melt droplet. When one of the embryos becomes supercritical in size, crystallization of the droplet will begin. The ΔT at which growth begins then dictates the texture (details of ΔT versus crystal shape will be discussed below). If the embryo population is severely diminished or the critical size for nuclei required to nucleate the stable phases in the melt is large, nucleation may never take place.

Crystal growth

Crystals grow in a silicate melt in response to an externally imposed driving force, usually the cooling rate, but the shapes, sizes and distribution of the crystals are controlled by the ΔT during growth and the spatial density or number of nuclei. The shapes of the crystals are determined by the relative rates of growth and diffusion of the crystal components in the melt. As ΔT increases the rate of growth increases relative to the rate of diffusion and the interface of a crystal with the melt becomes unstable and changes from planar to skeletal to dendritic (barred) to radial or spherulitic (Lofgren, 1974; 1980; Donaldson, 1976). The size of the crystals is controlled by the number of crystals present and the mutual interference during growth, and thus the spatial density of nuclei.

It is important to recognize that the cooling rate only indirectly controls the shapes of the crystals, and thus, the texture. The temperature at which nucleation takes place during cooling determines the ΔT of the melt when crystal growth begins which in turn controls the shape of the crystals. The insensitivity of textures that have been grown experimentally in silicate melts to the cooling rate imposed on that melt, has been demonstrated many times in the already referenced experimental studies. Textures can only be related to cooling rate in the most general way with fast cooling rates producing dendritic or spherulitic textures and slower rates producing textures with more equilibrium-like features. The interaction of cooling rate with the nucleation characteristics of the melt is a distinctive function of the near-liquidus phases as determined by the composition of the melts. It is this relationship that will form the basis for the chondrule crystallization model described in the next section.

While the cooling rate may only indirectly be able to control the texture of a chondrule, it does control the length of the chondrule-forming event. It is important then to determine the cooling rates that produce such crystals and textures based on the relationship between the growth rate and the shapes of the crystals in the chondrules. Because, as will be shown below (also see reference to experimental studies mentioned above), quite different textures such as porphyritic vs. radial can be produced at the same cooling rate depending primarily on the melting history of the droplet and thus the nucleation characteristics of the melt. Therefore, it is essential that we fully understand the nucleation and growth behavior of chondrule melts before we can develop a chondrule growth model. Only then can we determine actual cooling rates for the formation of chondrules and thus the duration of the chondrule-forming event.

THE MELT DROPLET CHONDRULE GROWTH MODEL

The chondrule-crystallization model presented below is based on general concepts of nucleation and growth described above and nearly two decades of experimental studies by several groups. The model can be divided into two basic processes that interact intimately: [1] The incomplete melting of the precursor material to form the melt droplet containing the unmelted nuclei or embryos that result from that process followed by [2] the crystallization of the droplet during cooling and the development of a chondrule texture. **It should be emphasized that the melting process is as important as cooling in determining the resulting texture of a chondrule.**

While nucleation and growth in the silicate melts described in this model are relatively independent of the environment in which they occur, it is assumed that these processes are taking place in the solar nebula and that there are pre-existing solids. The solid material initially must have been condensates, but recycled chondrule material must also play a large or even dominant role (Jones, this volume). As will be demonstrated below, there must be a heating event sufficient to bring the solid material to the near liquidus and even superliquidus temperatures of the chondrules. Subsequently it is necessary to have an environment that allows cooling of the resulting melts over periods up to a few hours. The model is summarized in the flow diagrams in Fig. 1 which can be used in con-

junction with the text description of the model to follow the sequence of events described in the model.

Chondrule melting

Melting and the production of nuclei

The exact nature of the melting event is not known, and for purposes of this discussion, we do not need to know the exact elapsed time or temperature of melting. We can say with confidence that the melting event is of relatively short duration, seconds to minutes (Connolly *et al.* 1993a,b), and the temperatures of melting must nearly reach or even exceed the liquidus temperatures of chondrule melts (1200–1900°C). This range of temperatures is large because the compositional range of chondrules is large (Hewins and Radomsky, 1990; Connolly and Hewins, 1991b). These conditions will produce a large variety of melting behaviors ranging from melts with numerous nuclei and relict crystals, to melts with fewer nuclei and without relicts, to melts with no stable, supercritical nuclei, only embryos. The most important distinction is between melts that have stable nuclei and those that do not have stable, supercritical nuclei, only embryos.

Effect of precursor grain size

The number, distribution, and kinds of nuclei, which all play a role in determining texture, are themselves determined by the grain size distribution of the crystalline precursor material and its melting history. The nature and grain size of the precursor material is not well constrained. However, based on observations in chondrites and their chondrules, it must be dominantly olivine and pyroxene with smaller amounts of metal, sulfides, other minor minerals such as feldspar, and glass. The grain size must be variable. Presumably there is a large portion of fine-grained material, i.e. < 0.1 μm. Based on the size of some relict olivine and pyroxene, the maximum size of crystals in this material must be on the order of 10's to 100's of microns. Individual aggregates of precursor materials must have differing size distributions, some uniformly fine-grained and some with a more normal size distribution (Connolly and Hewins, 1991a; this volume). The effect of precursor grain size on texture will be discussed below.

Other sources for nuclei

There can be external sources of nuclei in the form of dust grains that impact the surface of the melt droplets. Such dust grains could initiate nucleation and crystallization of a melt droplet. The excentroradial texture is a logical consequence of such an event; barred textures could also be produced in this manner depending on the ΔT of nucleation (see Connolly and Hewins, 1995). We understand little of the dynamics of dust-melt droplet interaction, but it is considered unlikely that dust would penetrate uniformly into a droplet in such manner so as to produce the relatively uniform distribution of nuclei that appears to precede crystallization of the typical porphyritic texture. Connolly and Hewins (1995), however, have produced porphyritic textures in their surface seeding experiments suggesting that dust seeds may not have to penetrate the entire sphere to produce porphyritic textures. Dust is the most likely

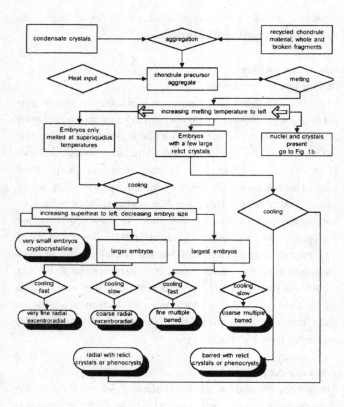

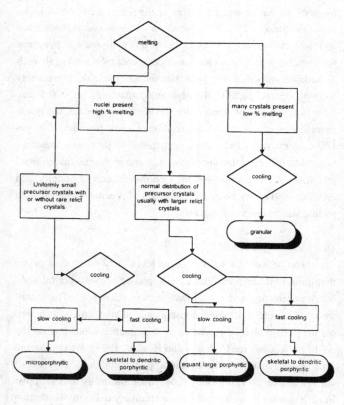

Fig. 1. A flow diagram of the model for chondrule crystallization showing the essential steps in the model. **a**. The aggregation and crystallization of melts free of nuclei. **b**. Crystallization of melts containing nuclei at the beginning of cooling.

source of nuclei for excentroradial and some barred chondrules, but this represents less than 15 percent of chondrule textures.

Chondrule cooling and texture

Once the melting event has occurred and the droplet has formed, it is the complex relationship between the nuclei and/or embryos (and particularly embryo size) in the melt and the cooling of the melt that produces the array of chondrule textures. There are two basically different melt conditions. Melts that start cooling with nuclei present which tend to produce porphyritic textures (Fig. 2); and melts that begin cooling without nuclei, only embryos present, which tend to produce barred or radial textures (Fig. 3). If nuclei are present when cooling is initiated, the nuclei begin to grow immediately. If the cooling rate results in a modest ΔT, there will be a modest growth rate and the resulting crystals will grow with planar interfaces and produce crystals with shapes generally referred to as equilibrium or near equilibrium (textural, not chemical) phenocrysts (Figs. 2a and b). The size of the crystals will be a function of the number of nuclei per unit volume, i.e. the spatial density of nuclei. With ever increasing cooling rates and, if nuclei are present, the ΔT will increase during growth with the concomitant increase in growth rate. As the growth rate increases, the crystal interface with the melt becomes less planar and projections or areas of retarded growth will occur and the crystals will assume a skeletal shape (Figs. 2d and f). With ever increasing cooling rates, the crystals will become increasingly skeletal and ultimately, dendritic (Fig. 3). It is characteristic of the textures in rapidly cooled melts that have many nuclei, that there will be many highly skeletal or dendritic crystals. Classic barred, radial or spherulitic chondrule textures are not generally produced from melts containing many nuclei when cooling begins no matter how rapid the cooling rate.

If nuclei are not present when cooling begins, no crystals will begin to grow when the liquidus is reached, and the ΔT of the melt will increase as cooling continues below the liquidus until embryos in the melt become supercritical, stable nuclei. The size distribution of embryos in the melt at the initiation of cooling will depend on the intensity of the melting and its effect on destruction of the crystalline material. The smaller the embryos, the more the ΔT will need to increase before nucleation takes place. The textures that develop in these kinds of melting situations are coarsely dendritic or barred (dendrites with only a few bars) (Fig. 3d) when ΔT is relatively small at the instant of nucleation. Finer, more complexly barred dendrites (Fig. 3c), and ultimately, ever finer radial and spherulitic textures (Fig. 3a) form as the ΔT at the instant of nucleation increases. For the coarsest dendrites, the ΔT could be a few 10's of degrees. For the finest radial and spherulitic textures, a few 100's of degrees of ΔT are most likely. Connolly and Hewins (1995) have recently demonstrated these relationships in a series of seeding experiments.

In general, the cooling rate of the chondrule has less influence on the texture than nucleation. Nucleation density determines the number and spacing of crystals. The timing of nucleation usually determines the initial ΔT which controls the growth rate and ultimately determines the crystal shape. For melts with olivine on the

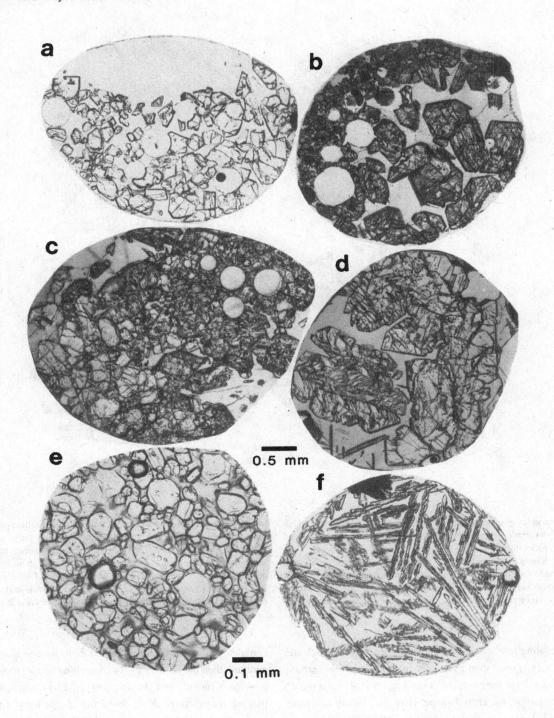

Fig. 2. Photomicrographs of porphyritic textures produced in dynamic crystallization experiments. All of these textures were produced by cooling melts that contained nuclei when cooling began. The 0.5 mm scale bar refers to a-d and f, the 0.1 mm scale bar refers to e. **a**. Olivine microporphyritic textures crystallized from an olivine rich chondrule melt, the matrix is glass. **b**. Pyroxene porphyritic texture with typical variation in grain size reflecting variations in nucleus size and distribution. **c**. Olivine-pyroxene porphyritic texture with significant variations in grain size reflecting distribution of nuclei. **d**. Pyroxene porphyritic texture, phenocrysts are skeletal and few in number suggesting that very few nuclei were present and were very close to critical size, if not slightly sub-critical, so that some degree of supercooling developed before significant growth began producing the skeletal shape. **e**. Starting material of ground minerals was melted at subliquidus temperature (1430°C) producing a modest amount of melting and a "granular" texture; the matrix is glass. If this charge were cooled slowly instead of quenched, the glass would crystallize to a true granular texture. **f**. Olivine porphyritic texture with elongate, skeletal olivine phenocrysts in a matrix of glass. Many, evenly distributed nuclei together with a rapid cooling rate produced skeletal crystals. Contrast with the situation where slower cooling produced equant crystals. More rapid cooling would produce more acicular crystals, but not barred dendrites because there are too many nuclei.

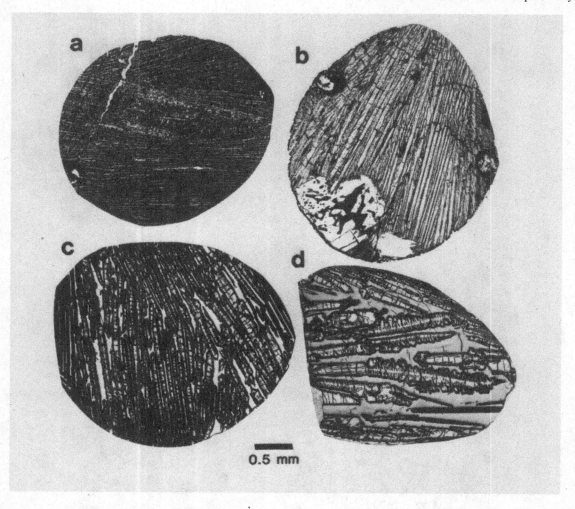

Fig. 3. Photomicrographs of barred and radial textures produced in dynamic crystallization experiments. All of these textures were produced by cooling melts that contained no nuclei when cooling began, only embryos. **a**. Radial pyroxene that grew at a high ΔT (> 100°C) relative to any of the other textures in b-d. Melting was complete and the embryo size distribution was very small. **b**. Relict pyroxene in a coarsely radial texture. The large pyroxene grew during the subliquidus melting phase of the exper-

iment, but could represent a relict crystal from a crystalline precursor. The melt surrounding the large crystal did not contain any nuclei upon cooling and the radial texture developed when embryos in the melt became stable nuclei as the ΔT increased. **c**. Barred olivine texture with multiple dendrites or groups of bars. Nucleation occurred at a modest ΔT which was less that in a or b and greater than in d. **d**. Coarse barred olivine texture with dendrites that contain only a few bars that would begin to grow at a modest ΔT.

liquidus, cooling rate is more important because olivine nucleates so much more readily than pyroxene. In the porphyritic textures where nuclei exist at the initiation of cooling, cooling rate is important in determining the crystal shape. Here the cooling rate influences the ΔT during growth. Hewins *et al.* (1981) observed a direct relationship between dendrite width and composition with cooling rate and thus ΔT. The most obvious case of nucleation being the determining factor is the case where radial textures develop at very slow cooling rates, e.g. 5°C/hr. The reluctance of pyroxene to nucleate when the embryos present are small is the determining factor in producing the large ΔT and radial growth in spite of the slow cooling rate. Such textures are often indistinguishable from radial textures formed in rapidly cooled melts.

Chondrule composition and texture

The composition of the chondrule melt determines what minerals will crystallize, their order of appearance, and the temperature a

complete melt forms. If all chondrule melts formed at a single temperature, then the primary factor controlling texture would be composition (Connolly and Hewins, 1991b). Those compositions with melting temperatures at or above this single temperature would tend to be incompletely melted and have nuclei and thus produce primarily porphyritic textures. Those compositions with melting temperatures significantly below this single temperature would all tend to be completely melted and have no nuclei and thus produce primarily dendritic to radial textures. The fact that the major chondrule chemical types display most of the observed chondrule textural types suggests there is a range of melting temperatures. The olivine-rich chondrules have the highest melt temperatures and, with increasing amounts of pyroxene in the melt, the melt temperatures are lower. As a consequence, the olivine-rich chondrules tend to display fewer radial textures while the pyroxene-rich chondrules tend to have a higher percentage of radial and excentroradial textures. This difference reflects the greater likelihood that pyroxene-rich chondrules will be completely melted.

The effect of composition during the crystal growth process is seen in the variation in the viscosity of the melt and its effect on diffusion rates of the various components. For a given ΔT and growth rate, the shapes assumed by the crystals will assume increasingly nonequilibrium shapes as the viscosity of the melt increases. Thus higher viscosity melts will tend to develop radial textures at a lower ΔT.

Growth of specific textures

There is almost an infinite number of textural variations that can be explained by subtle differences in how the crystallization parameters change and interact, but such explanations are beyond the scope of this paper. It is important to understand the differences between a few of the basic textural types. Among the porphyritic varieties, it is important to understand why microporphyritic textures develop versus the coarsely porphyritic types and to understand the gradation to granular textures. Among the rapidly crystallized textures, it is important to appreciate the difference between the single and multiple barred dendrites, the diameters of the fibers in the radial textures, and the dendritic or radial textures that contain large phenocrysts or relict crystals versus those that do not.

Porphyritic textures

The nuclei in melts that crystallize porphyritic textures are derived from precursor relict crystals. There appears to be a basic distinction between the precursor material necessary to produce two basic textural types, microporphyritic versus porphyritic. For the microporphyritic textures similar to the experimentally produced chondrule in Fig. 2a, the relict crystals must be uniformly small and evenly distributed except for the occasional large relict (Connolly and Hewins, 1991a). More-coarsely porphyritic chondrules, similar to the experimentally produced chondrules in Figs. 2b-d, have fewer, more-irregularly distributed nuclei from a precursor material with a much more heterogeneous grain size that may include some quite large crystals, on the order of 100's of microns. For the largest relicts, their former characteristics are recognizable, while the more typical nuclei or relicts are reduced to sizes less than a micron during melting and are no longer recognizable as such. Both textures are common, so that there needs to be a mechanism to produce both kinds of precursor grain-size distributions. Perhaps the uniformly fine-grained aggregates mainly represent condensate crystals and the more variable-sized material contains a larger percentage of recycled chondrule material.

Granular-porphyritic, or textures that show a limited degree of melting, are a logical part of the sequence of melting and cooling events that produce porphyritic textures which can be explained by the model. For most melt droplets that are identified with igneous textures, the assumption is that melting was nearly complete. If the process is a continuous one, there should be chondrules with lesser degrees of melting. The olivine and pyroxene would be rounded, for the most part, with possible euhedral overgrowths. These chondrules most likely experienced 50–90 % melting, enough to consolidate the precursor material, but leaving a high percent of relict crystals (Fig. 2e). When these subsequently cool, most of the melted material would crystallize either as overgrowths producing a "classic" granular texture, if cooling is slow enough in the near-solidus range, or as fine-grained, rapidly crystallized material, if cooling is rapid.

Dendritic (barred) and spherulitic textures

The chondrule textures that result from the cooling of complete melts, namely those with barred and radial textures, present a different problem. All the supercritical-sized nuclei are eliminated during melting, and the important textural differences are a product of the embryo size distribution. Barred textures develop in melts with larger embryos that become nuclei at smaller ΔT's, and the stable crystal shape is dendritic, with the dimensions of the bars decreasing and the number of bars increasing with increasing ΔT during growth (compare Fig. 3c, high ΔT, and 3d, low ΔT during growth). The number of larger embryos will determine the number of barred dendrites. The classic barred olivine chondrule that is a single dendrite and usually has a rim that is optically continuous and part of the single dendrite indicates a rare single nucleation event. This texture has to date escaped experimental duplication, although it has been closely approached by Lofgren and Lanier (1990, Fig 5b) and Radomsky and Hewins (1990, Fig. 15b). To produce a melt with a single embryo suitable to become a nucleus at a ΔT that will produce a single barred dendrite is a rare event and may represent an instance of an externally derived nucleus. In this case the ΔT at which the external nucleus impacts the droplet would have be appropriate for the growth of a shell of olivine enclosing the droplet with subsequent growth of the barred dendrite throughout the droplet. Small variations in the ΔT of nucleation would control the width of the bars in the olivine dendrite that comprises the barred chondrule. Upon close examination, it is clear that barred chondrules with multiple barred dendrites are more common that those with a single barred dendrite (Weisberg, 1987; Lofgren and Lanier, 1990). The transition from barred textures to radial is caused by more complete melting, such that the embryo size distribution is even smaller. The smaller embryos become stable nuclei at even larger ΔT's resulting in radial not dendritic growth. The radial fibers in the chondrules become finer as the ΔT of growth becomes larger (compare Fig. 3a, high ΔT, and 3b, low ΔT during growth). The ultimate end member of this sequence would be the ultra-fine-grained spherulitic texture grading to a cryptocrystalline texture. Excentroradial textures would develop if the nucleation resulted from a dust particle on the surface of the droplet as opposed to embryos inside the droplet. The size of the fibers in the excentroradial texture would be controlled by the temperature, which determines the ΔT, of the droplet when it was impacted by the dust particle. An important and complicating experimental result is observed here. If the embryo size distribution in the melt has been severely reduced, embryos may not become stable nuclei until large ΔT's are reached, even at very slow cooling rates. Thus it is possible to get fine radial textures even at cooling rates as slow as 5°C/hr, rates which are considered slow even for the production of porphyritic textures.

Barred and radial textured chondrules with large phenocrysts or relict crystals, however, would not tend to form at slow cooling rates. If the large crystals are relict, the initial grain size of the precursor material must be highly variable containing some large crystals. These crystals may not melt while stable nuclei are eliminated from the rest of the melt. Upon cooling, the relicts may develop some overgrowths, but otherwise would remain a bystander to the nucleation and growth of the enclosing melt. The dendritic or radial textures enclosing such relicts tend to be coarser (Fig. 3b), because the degree of melting required to produce smaller embryos consistent with finer-grained textures would most likely melt even large relict grains. Similar arguments can be advanced to limit the melting times, temperature, and cooling rates of chondrules which have phenocrysts instead of large relict crystals.

Effect of growth parameters on compositional zoning

Compositional zoning in minerals is sensitive to the crystal growth rate, which, as discussed above, is an indirect function of the cooling rate. In rapidly grown olivines, the zoning profile is linear, while in more slowly grown crystals, the profile assumes a Rayleigh distribution profile. Because of the indirect relationship between cooling rate and growth rate, the cooling rates for a given melt composition must be determined experimentally. Experiments on a porphyritic olivine composition (Lofgren, 1989) show Rayleigh distribution profiles that resemble closely the zoning profiles in olivines in chondrules of similar composition in Semarkona (Jones and Lofgren, 1993). Based on zoning profiles, the cooling rates for the olivines in Semarkona are 10–100°C/hr. Cooling rate estimates based on crystal shapes give similar values. In a study on a similar olivine-rich composition, however, Radomsky and Hewins (1990) found no zoning in olivine grown in experiments cooled at 10°C/hr, but did observe zoning at 1000°C/hr. Yu and Hewins (1995) were able to produce Rayleigh profiles using nonlinear cooling histories. Their experiments were cooled initially at 5000°C/hr and the rate slowed to near 500°C/hr at lower temperatures. Their results suggest that zoning can be produced in a number of ways and may not be as diagnostic of cooling rate as previously thought, but in the case of Semarkona where the zoning profiles and the crystal shape give consistent results, I feel there can be confidence in the result.

CONSTRAINTS ON CHONDRULE FORMING PROCESS

Melting temperatures and times

The predominance of porphyritic-textured chondrules indicates that melting temperatures cannot have greatly exceeded the liquidus temperatures of the chondrule melts for a significant period of time. If they had, barred and radial chondrules would dominate. Thus, there is an upper limit on temperature and time to preserve the nuclei necessary to produce porphyritic chondrules. It is difficult, however, to give precise temperature limits on the melting process because the range of chondrule compositions results in a range of liquidus temperatures of almost 700 degrees (1200–1900°C). The nominal upper temperature limit is set by the most refractory chondrules, those with essentially pure forsterite (approximately 1900°C) and

that temperature cannot be exceeded for more than a few seconds to a few minutes. Even higher temperatures can probably occur, but for no more than a few seconds because Greenwood and Hess (this volume) have found the kinetics of melting to be very rapid.

Chondrules with barred and radial textures were completely melted, i.e. no viable nuclei remain, and suggest even higher melting temperatures than can be determined based on porphyritic chondrules. Radial textures suggest higher degrees of superheat than barred textures, based the on the previous discussion of embryo size at nucleation. Chondrules with radial textures, however, tend to be pyroxene-rich chondrules with lower liquidus temperatures (Lofgren and Russell, 1986; Hewins and Radomsky, 1990). The more olivine-rich, barred chondrules tend to be barred because they have higher liquidus temperatures and thus are not as readily completely melted, and the embryos are larger and become nuclei at lower ΔT upon cooling. Radial olivine textures would suggest even higher melting temperatures than barred textures, but their scarcity (Connolly and Hewins, 1995) among chondrule textures sets another constraint on the upper temperature limit of melting. The olivine-rich compositions are not melted sufficiently above their liquidus temperatures in order to suppress nucleation until large enough ΔT's develop to stabilize the radial olivine texture. Thus, the presence of barred olivine textures suggests higher melting temperatures than their porphyritic counterparts.

The temperature/time melting constraint is best stated in reference to a particular melt composition and its liquidus temperature. The temperature cannot be exceeded for more than a few 10's of degrees for minutes or several 10's of degrees for seconds, if porphyritic textures are to be the result of cooling. The times and temperatures will vary with the grain size, as it is obvious that larger crystals will take longer to melt (Connolly and Hewins, 1995). Without detailed experiments on a given composition, whose starting material has a specific grain-size distribution, the numbers cannot be more precise. Based on experiments on the most refractory compositions, an upper temperature limit for chondrule melting can be placed in the range 1650 to 1750°C for less than 15 minutes and up to 1890–1900°C (c.f. Hewins and Connolly, this volume; Greenwood and Hess, this volume) for a few seconds.

"Flash melting" is a popular concept for the melting event, but it is not clear just what constitutes flash melting or how it is generated. Presumably the melting occurs in a matter of seconds. For most compositions there is no intrinsic limit on melting time imposed by the experiments for the production of a particular texture. Melting for longer periods of time, up to days at temperatures near the liquidus, can produce the same result as melting for short periods at higher temperatures. Other considerations, such as the loss of volatiles, suggest the shorter melt times (Hewins, 1991; Yu et al., this volume) although there are other ways to keep Na from being lost such as a Na-rich vapor in the chondrule-forming environment (Lewis et al. 1993).

Cooling rates and duration of the event

The cooling of chondrules takes longer than melting and plays a larger role in determining the length of the overall chondrule-forming event. The range of cooling rates determined by the experiments, 10–3000°C/hr, result in estimates of cooling times of 10's of hours to a few minutes. The important question is how long do the porphyritic textures require to crystallize since they make up the bulk of the chondrule textures? Two factors go into determining the precise range of cooling rates, the shapes of the crystals and the chemical-zoning profiles. The sizes and shapes and zoning profiles of olivine phenocrysts in one study suggest cooling rates on the order of 10–100°C/hr (Lofgren, 1989, Jones and Lofgren, 1993). Other studies suggest cooling rates for phenocrysts up to 1000°C/hr (Radomsky and Hewins, 1990). Linear cooling rates > 1000°C/hr will not produce appropriate phenocryst characteristics. The crystallization interval is 150 to 300°C, so the maximum cooling time based on a 10°C/hr cooling rate would be 30 hours. A more realistic cooling time would be considerably less, since the porphyritic textures are established and mostly crystallized in the first 150°C of the crystallization interval. Most phenocrysts probably grew at 100 to 1000°C/hr. Thus, a firmly established minimum cooling time to produce porphyritic textures is at least 9 minutes at the faster cooling rate and 1.5 hours at the slower rate.

Studies of the crystallographic properties of the pyroxenes in chondrules suggests even slower rates with 10°C/hr as the upper limit (Weinbruch and Muller, 1994). These rates, however, refer to events in the temperature regime near the solidus, not the liquidus. Cooling rates in this range in the temperature regime around the solidus are compatible with the slightly faster cooling rates in the liquidus range. The combination of the dynamic-crystallization-determined cooling rates with the crystallographically-determined cooling rates places strong constraints on the environment in which cooling is taking place. It must be slower than commonly used in current nebular models.

CONCLUSIONS

Heating mechanisms and cooling environments responsible for the formation of chondrules need to be consistent with experimental evidence. The heating mechanisms must produce temperatures in the liquidus to superheated range for chondrules (1200 to 1750°C) for seconds to minutes and perhaps as high as 1900°C for a few seconds. The cooling environment must allow cooling to take place, over a period of up to a few hours, to near-solidus temperatures of 1150 to 1250°C. We can define a minimum and maximum time for production of porphyritic textures of 9 minutes to 30 hours. The cooling environment must be more dense than the usual nebula so that cooling can be slowed.

ACKNOWLEDGEMENTS

I wish to thank Oscar Mullins, Jr., Dennis Smith, and Al Lanier who helped with the experiments and the operation of the laboratory over the last 2 decades. I also wish to thank the reviewers Larry Taylor and J.P. Greenwood for helpful reviews.

REFERENCES

Blander M., Planner H.N., Keil K., Nelson L.S. and Richardson N.L. (1976) The origin of chondrules: Experimental investigation of metastable liquids in the system Mg_2SiO_4-SiO_2. *Geochim. Cosmochim. Acta* **40**, 889–896.

Connolly H.C.Jr. and Hewins R.H. (1991a) The effect of precursor grain size on chondrule textures. *Meteoritics* **26**, 329.

Connolly H.C.Jr. and Hewins R.H. (1991b) The influence of bulk composition and dynamic melting conditions on olivine chondrule textures. *Geochim. Cosmochim. Acta* **55**, 2943–2950.

Connolly H.C.Jr. and Hewins R.H. (1995) Chondrules as products of dust collisions with totally molten droplets within dust-rich nebular environment: An experimental investigation. *Geochim. Cosmochim. Acta* **59**, 3231–3246.

Connolly H.C.Jr., Hewins R.H. and Lofgren G.E. (1993a) Flash melting of chondrule precursors in excess of 1600C. Series I: Type II (B1) chondrule composition experiments. *Lunar and Planetary Science XXIV*, Lunar and Planetary Institute, Houston, 329–330.

Connolly H.C.Jr., Hewins R.H. and Lofgren G.E. (1993b) Possible clues to the physical nature of chondrule precursors: An experimental study using flash melting conditions. *Meteoritics* **28**, 338–339.

Donaldson C.H. (1976) An experimental investigation of olivine morphology. *Contr. Mineral. Petrol.* **57**, 187–213.

Hewins R.H. (1988) Experimental studies of chondrules. In *Meteorites and the Early Solar System* (eds. J.F. Kerridge and M.S. Mathews), pp. 660–679. Univ. Arizona Press.

Hewins R.H. (1991) Retention of sodium during chondrule melting. *Geochim. Cosmochim. Acta* **55**, 935–942.

Hewins R.H. and Radomsky P.M. (1990) Temperature conditions for chondrule formation. *Meteoritics* **25**, 309–318.

Hewins R.H., Klein L.C. and Fasano B.V. (1981) Conditions of formation of pyroxene excentroradial chondrules. *Proc. Lunar Planet. Sci. Conf. 12B*. 1123–1133.

Jones R.H. and Lofgren G.E. (1993) A comparison of FeO-rich, porphyritic olivine chondrules in unequilibrated chondrites and experimental analogues. *Meteoritics* **28**, 213–221.

Kirkpatrick R.J. (1981) Kinetics of crystallization of igneous rocks. In *Reviews in Mineralogy volume 8* (eds. A.C. Lasaga and R.J. Kirkpatrick), pp. 321–398. Mineralogical Society of America.

Lewis R.D., Lofgren G.E., Franzen H.F. and Windom K.E. (1993) The effect of Na vapor on the Na content of chondrules. *Meteoritics* **28**, 622–628.

Lofgren, G.E (1974) An experimental study of plagioclase crystal morphology: Isothermal crystallization. *Amer. J. Sci.*, **274**, 243–273.

Lofgren, G.E. (1980) Experimental studies on the dynamic crystallization of silicate melts. In *Physics of Magmatic Processes* (ed. R.B. Hargraves), pp. 487–552. Princeton Univ. Press.

Lofgren, G.E. (1982) The importance of heterogeneous nucleation for the formation of microporphyritic chondrules. *Conference on Chondrules and their Origins*. Lunar and Planetary Institute, Houston, p. 41.

Lofgren, G.E. (1983) Effect of heterogeneous nucleation on basaltic textures: A dynamic crystallization study. *Jour. Petrology*, **24**, 229–255

Lofgren, G.E. (1989) Dynamic crystallization of chondrule melts of porphyritic olivine composition: Textures experimental and natural. *Geochim. Cosmochim. Acta* **53**, 461–470.

Lofgren G.E. and Russell W. J. (1986) Dynamic crystallization of chondrule melts of porphyritic and radial pyroxene composition. *Geochim. Cosmochim. Acta* **50**, 1715–1726.

Lofgren G.E. and Lanier A.B. (1990) Dynamic crystallization study of barred olivine chondrules. *Geochim. Cosmochim. Acta* **54**, 3537–3551.

Planner H.N. and Keil K. (1982) Evidence for the three-stage cooling history of olivine-porphyritic fluid droplet chondrules. *Geochim. Cosmochim. Acta* **46**, 317–330.

Radomsky P.M. and Hewins R.H. (1990) Formation conditions of pyroxene-olivine and magnesian olivine chondrules. *Geochim. Cosmochim. Acta* **54**, 3475–3490.

Tsuchiyama A., Nagahara H. and Kushiro I. (1980) Experimental reproduction of textures of chondrules. *Earth Planet. Sci. Lett.* **48**, 155–165.

Turnbull D. (1950a) Kinetics of heterogeneous nucleation. *J. Chem. Phys.* **18**, 198–203.

Turnbull D. (1950b) Formation of crystal nuclei in liquid metals. *J. Appl. Phy.* **21**, 1022–28.

Turnbull D. (1956) Phase changes. *Solid State Physics.* **3**, 226–306.

Turnbull D. (1969) Under what conditions can glass be formed? *Comtemp. Phys.* **10**, 473–88.

Uhlmann D.R. and Chalmers B. (1965) Nucleation. *Ind. Eng. Chem.* **57**, 19–31.

Weinbruch S. and Muller W.F. (1994) Cooling rates of chondrules derived from the microstructure of clinopyroxene and plagioclase. *Meteoritics* **29**, 548–549.

Weisberg M.K. (1987) Barred olivine chondrules in ordinary chondrites. *Proc. Lunar Planet. Sci. Conf. 17th.*, *Jour. Geophys. Res.* **92**, E663–E678.

Yu Y. and Hewins R.H. (1995) Is non-linear, rapid cooling plausible for chondrule formation? Evidence from olivine zoning profiles. *Lunar and Planetary Science XXVI*, Lunar and Planetary Institute, Houston, 1545–1546.

21: Peak Temperatures of Flash-melted Chondrules

ROGER H. HEWINS and HAROLD C. CONNOLLY, JR
Department of Geological Sciences, Rutgers University, Piscataway, NJ 08855, U.S.A.

ABSTRACT

The temperatures to which chondrules were heated may be estimated from crystallization experiments, provided that the effects of very short heating times, of precursor grain size, and of cooling rates are understood. Reducing the isothermal heating time from 60 minutes to 5 minutes raises the temperature necessary to achieve total melting by over 40°C for fine-grained starting materials (20–45 μm) and over 120°C for coarser starting material (125–250 μm). Although a charge is heated to run temperature in less than 1 minute, dissolution of relict olivine grains is slow. Melting therefore takes up to hours at near liquidus temperatures, so that under flash melting conditions chondrules required higher initial temperatures, unless the precursors were very fine-grained. Even if temperatures are over 100°C higher than liquidus values, time for total melting is several minutes for coarse-grained starting material, explaining the common occurrence of large relict grains in chondrules. Melting/dissolution continues during initial cooling, so that melting is less complete for rapid cooling. Crystalline chondrules can be formed after incomplete melting with melt temperatures hundreds of degrees above the liquidus for coarse grained starting materials. Seeding of completely melted spherules with dust grains (external nucleation) can produce a wide variety of chondrule textures, explaining the rarity of totally glassy chondrules even for those with low liquidus temperatures. The generally finer grained porphyritic magnesian chondrules probably had finer grained precursors than the more coarse grained ferroan chondrules, and flash heating therefore required temperatures elevated less with respect to their liquidus temperatures. Maximum temperatures for the most refractory chondrules (liquidus over 1750°C, all incompletely melted) were about 100°C higher than thought previously, or 1850°C, assuming peak temperatures lasted seconds, and cooling rates allowed several minutes at near-liquidus temperatures, as in our experiments. Complete melting at this temperature would require 2,100 J/g for heating from 200°C and melting. The least refractory chondrules (liquidus under 1400°C) were probably strongly superheated (minimum temperature 1500°C, minimum energy 1,750 J/g) and then dust-seeded. A range of peak temperatures 1500–1850°C, with corresponding energy input of 1,750–2,100 J/g, was experienced by chondrules of all compositions. A heating time measured in seconds is plausible, though not directly constrained.

INTRODUCTION

One of the major goals of the experimental simulation of chondrules is to define the temperature of melting. Chondrule textures are understood to depend on conditions of nucleation (Lofgren, 1982; this volume), which are related to extent of melting, which in turn reflects peak temperature, but also heating time and the composition, mass and grain size of the precursor aggregate. Past experimental simulations of chondrule textures were interpreted to indicate that chondrules of all compositions were heated relatively close to their respective liquidus temperatures (Lofgren and Russell, 1986, Hewins, 1988, Lofgren, 1989, Radomsky and

Hewins, 1990). Liquidus temperatures show an enormous range, ~1200 – ~1900°C, but there are composition-texture correlations (Hewins and Radomsky, 1990). Chondrules with liquidus temperatures in the range 1200–1400°C have textures indicating complete melting. Those with liquidus temperatures in the range 1750–1900°C indicate partial melting.

One possible problem with published chondrule temperature estimates is that they were based on experiments with a long duration melting step (30 min – 2 hr), whereas the chondrule formation process very probably involved a flash-heating mechanism (Boss, this volume). There is no rigorous definition of the term flash melting, but durations of about a minute or much less are implied. The chondrule temperatures may exceed the liquidus considerably, and thermodynamic equilibrium may not be achieved before the onset of cooling. Our understanding of melting comes primarily from phase equilibrium experiments and thermodynamic theory, yet kinetic melting effects have already been observed in experiments like those used in estimating chondrule temperatures. Textures in runs with the same heating temperature and cooling rate have been shown to vary depending on precursor mass (Radomsky and Hewins, 1990), precursor grain size (Radomsky and Hewins, 1990) and heating time in minutes (Connolly and Hewins, 1991). Melting, dissolution and destruction of nuclei take significant amounts of time, yet in many chondrule formation models currently under consideration the duration of the heat pulse is only seconds to minutes (e.g. Hood and Horanyi, 1993, Eisenhour and Buseck, 1995, Horanyi and Robertson, this volume). Chondrule peak temperatures were estimated from the degree of melting suggested by the textures, based on long-duration melting experiments, and might be higher than those given by Hewins and Radomsky (1990).

A second difficulty in estimating melting temperatures is that many of the less refractory chondrules may have totally melted, and collided with dust grains which acted as nuclei generating porphyritic and other textures (Hewins and Radomsky, 1990). This mechanism has been successfully demonstrated to occur in experiments (Connolly and Hewins, 1995). As this raises the melting temperatures for which formation of porphyritic chondrules is possible on cooling, it must be considered in estimating peak temperatures for chondrules.

The purpose of the present paper is to re-evaluate the temperature estimates given by Hewins and Radomsky (1990) in the light of probable nebular melting conditions. We therefore review relevant new experiments with very short heating times (Jones, 1991; Lerro, 1993) as well as experiments in which the nuclei responsible for developing the textures were introduced by seeding total melts (Connolly and Hewins, 1995).

MELTING AND DISSOLUTION

As it is not generally realized that melting is a complex process and solids can be superheated, we summarize below some recent observations. For a more complete discussion, see Greenwood and Hess (this volume). Even metals can be superheated a few degrees (Dages *et al.*, 1987). During heating, liquid is considered to be heterogeneously nucleated on crystal surfaces (Ainslie *et al.*, 1961). Calorimetry experiments show that albite crystals are not completely melted when heated for ten minutes 150°C above the melting point (Navrotsky *et al.*, 1989).

If a relict grain in a chondrule reaches a temperature above its own liquidus temperature, it will begin to melt and its survival time can be estimated (Greenwood and Hess, this volume). At lower temperatures, it will react with and dissolve in the bulk melt. Dissolution rates for olivine are largely controlled by the rate of diffusion of cations away from the interface into the liquid and vary from about 1 µm/hr to about 1 mm/hr for a wide variety of compositions and conditions (Donaldson, 1985; Brearley and Scarfe, 1986; Zhang *et al.*, 1989). Olivine crystals < 40 µm in size can persist for hours at 1 atm. in a basaltic melt 5–10°C above the liquidus temperature (Donaldson, 1979). Individual 1 mm olivines are clearly recognizable in melts of chondrule composition heated up to 60°C above the liquidus temperature for 30 minutes (Radomsky and Hewins, 1990). The number of nuclei remaining in chondrule melts will be controlled by the rate of dissolution of relict grains into bulk melts.

MELTING EXPERIMENTS

To interpret chondrule textures in terms of peak temperatures and heating times, we need to know how long it takes for a mineral aggregate to melt at different temperatures above the liquidus. Most chondrules have high liquidus temperatures which make such studies difficult. A very FeO-rich analog of type IIA chondrule composition (moderate FeO content) with a liquidus temperature estimated at 1211°C from experiments of Jones (1991) was therefore prepared. This material can be heated in a conventional DelTech furnace well in excess of its low liquidus temperature and hence melted in very short times. It was prepared from 70% fayalite slag, 10% orthopyroxene, 10% diopside and 10% labradorite, with a bulk composition (weight %) of SiO_2 35.5, TiO_2 0.2, Al_2O_3 3.7, FeO 52.9, MnO 0.2, MgO 2.8, CaO 2.6, Na_2O 0.4, K_2O 0.2, total 98.5 (Jones, 1991). The liquidus olivine phase for this composition, estimated from dynamic crystallization experiments, is approximately Fa_{84}. Melting and crystallization were examined for four grain sizes, of which the coarsest fraction was 125–250 µm and the finest 20–45 µm, using 40 mg pellets suspended on Pt wires.

This fayalite-rich material was heated isothermally for various times between 5 and 60 minutes before quenching in water to determine the temperatures at which all crystals disappeared. This would be the liquidus temperature for long durations, in practice a few hours, depending on grain size. Curves for complete melting as a function of heating time and temperature given in Fig. 1, after Lerro (1993), show a strong grain size effect. The temperature required to produce a completely glassy charge for the 23–45 µm fraction is about 1227°C for 60 minute heating rising to about 1263°C for 5 minute heating (Fig. 1). For the coarsest fraction studied (125–250 µm), complete melting temperatures rise to about 1256°C (60 minute heating) and about 1343°C (5 minute heating).

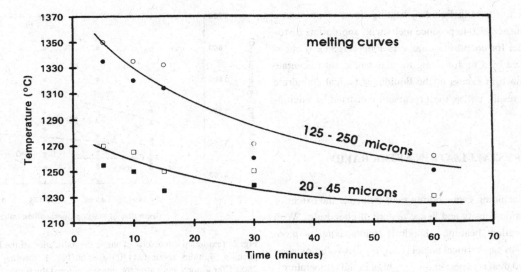

Fig. 1. Total melting curves, i.e. olivine disappearance curves, for FeO-rich chondrule analog composition as a function of temperature and heating time. Liquidus temperature is about 1211°C. Lower curve is for 23–45 μm starting material, upper for 125–250 μm. Only experiments bracketing the melting curve are shown; open symbols represent the lowest temperatures for entirely glassy charges and filled symbols the highest temperature for retention of any crystals. After Lerro (1993).

Despite studies of olivine dissolution (Zhang *et al.*, 1989), we did not anticipate that coarse grained olivine would persist metastably for several minutes 100°C above the bulk liquidus temperature. We therefore considered several possible explanations of this phenomenon, including whether the samples rapidly reached run temperature.

The surface of a massive sample heated radiatively in a solar furnace reaches peak temperature in about 1 second (Ferrière *et al.*, 1994). In our experiments, the heat is transferred to the small sample by radiation from a ceramic muffle tube. A metal wire inserted into the furnace for a temperature calibration melts "instantly" if the ambient temperature is above its melting point. Our thermocouples, which calibrations show to be accurate within one degree, take at least a minute to give a stable reading, though they heat up faster, because a sample can be melted before melting temperature is registered (S.V. Maharaj, pers. comm., 1995). A thermocouple embedded within a sample, with wires completely covered by insulating material, approaches furnace hotspot temperature closely in about one minute (S.V. Maharaj, Y. Yu, pers. comm., 1995), and the signal stabilizes shortly afterwards. In other furnaces, stable run temperature is reached in about 30 seconds (Zhang *et al.*, 1989). Radiation can pass through small silicate particles, which heat uniformly throughout, but the interiors of large particles, which have higher absorption cross sections, are heated by conduction as well as by absorption of radiation (Eisenhour, pers. comm., 1995). Heating times required for conduction can be obtained from a simple calculation (Ainslie *et al.*, 1961). For conditions relevant to chondrule formation, e.g. a 1 mm sphere of olivine raised 1600°C, the conductive heating time is about 0.02 seconds. (Heating times of milliseconds have been proposed for some models of chondrule formation, e.g. Fujii and Miyamoto, 1983; Wasson and Rasmussen, 1994). It appears certain that our grains reached run temperature in very much less than the minimum run time (five minutes).

The long melting times observed at temperatures above the liquidus are related to the kinetics of dissolution of the last solids in the bulk melt. Dissolution experiments (e.g. Donaldson, 1985; Brearley and Scarfe, 1986; Zhang *et al.*, 1989) indicate formation of a liquid boundary layer in contact with the crystals and diffusion of cations across this layer is important in controlling the rate at which crystals disappear. In most of these experiments, convection plays an important role, but it was suppressed in the runs of Zhang *et al.* (1989), as in natural chondrule melts in a microgravity setting. Brearley and Scarfe (1986) show olivine dissolution rates up to 1 mm/hr, depending on temperature, pressure and composition effects. Such rates are broadly consistent with the time to produce total melting in the present experiments, with the survival of nuclei in complementary crystallization experiments on the same composition described below, and with the disappearance rate of Mg-rich olivine in a chondrule composition liquid (type IIAB) at about 60° above the liquidus (Radomsky and Hewins, 1990). Dissolution rates of this order explain the big difference in time to produce a total melt from starting powders of different grain size.

With ambient temperatures about 100°C above the liquidus, heating times required to produce total melts are short but still in the range of several minutes for coarse-grained (> 100 μm) starting materials. If chondrules were generated by flash heating, relict grains are almost inevitable if the precursors were coarse-grained, e.g. earlier chondrules, unless peak internal temperatures were several hundreds of degrees above the liquidus. Experiments could be designed to determine whether the survival of relicts in chondrule melts requires a short duration heating event. A theoretical approach to estimating maximum heating times at elevated temperatures is given in the chapter by Greenwood and Hess. With a temperature nearer to the bulk liquidus, heating times to produce a total melt, and hence also to produce melts with appropriate distributions of nuclei for chondrules, are very long (Fig. 1). If chondrules are formed by a flash heating mechanism, it must generate temperatures much in excess of the liquidus if typical chondrule textures are to result, unless the precursors consisted of micron-sized grains.

DYNAMIC CRYSTALLIZATION AFTER RAPID MELTING

Many factors, including even cooling rate, influence the extent of melting, nucleation density and hence texture of chondrules. With very short duration heating, chondrule melts contain more unmelted relicts or supercritical nuclei (Lofgren, this volume) when cooling is started than is expected for very high initial temperatures. The dissolution of such relicts continues during initial cooling after a short duration heating pulse, during the time it takes to reach the liquidus, and possibly even longer if very little melting has occurred. A partially melted chondrule could take minutes to cool down to the liquidus temperature at the rates favored by experimentalists (Yu *et al.*, this volume). With the same peak temperature, melting is less complete for rapid cooling than for slow cooling, as demonstrated by Osborn (1993): the initial heating temperatures (peak temperatures) required for total melts (glasses) and for melts with few surviving nuclei (barred olivine textures) decrease with slower cooling. However, peak temperature, heating time and in particular precursor grain size have a greater influence on degree of melting, as shown by crystallization experiments performed by Jones (1991) and described briefly by Connolly *et al.* (1991).

For the very FeO-rich composition discussed above, the liquidus temperature was determined at 1211°C by Jones (1991) with temperatures for total melting higher depending on precursor grain size and heating time (Fig. 1). For crystallization experiments (Jones, 1991), samples were placed in the furnace at temperatures up to 400°C above the liquidus. They were immediately cooled at the maximum rate possible, about 2800°C/hr, down to the vicinity of the liquidus, spending from about 1 minute up to about 9 minutes above the olivine disappearance temperatures determined in melting experiments like those shown in Fig. 1. Melting and dissolution of relicts took place during this step. Subsequently the charges were cooled at a constant rate of 500°C/hr to permit crystallization and quenched at close to 1000°C. The total cooling interval was chosen to give a short duration melting step followed by a crystal growth step compatible with olivine Fe-Mg zonation in chondrules. (A discussion of chondrule cooling rates is given in the chapter by Yu *et al.*, who favor higher initial cooling rates ~5000°C/hr decreasing to about 500°C/hr near the solidus). Fig. 2 shows the resulting textures as a function of peak temperature and precursor grain size.

Textures previously believed to reflect peak temperatures below the liquidus were produced in these experiments with supra-liquidus initial temperatures. Microporphyritic textures were produced with initial temperatures up to about 40 degrees above the

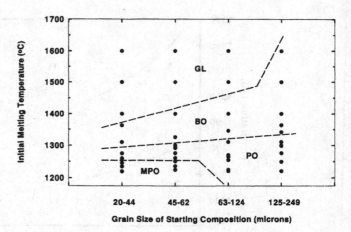

Fig. 2. Textures produced by dynamic crystallization of the FeO-rich composition (liquidus about 1211°C) used in Fig. 1. Cooling was begun at 2800°C/hr immediately after the charges entered the furnace at the temperature indicated, down to the olivine disappearance temperature, after which the cooling rate was 500°C/hr. MPO -microporphyritic olivine, PO - porphyritic olivine, BO - barred olivine, GL - glass. Textures grade from PO to BO, with transitional types including elongated hopper olivine texture. Texture is sensitive to grain size, which is given as longest dimension of starting grains.

liquidus, but for fine-grained precursors only, and porphyritic up to about 80 degrees above the liquidus (Fig. 2). In the runs with the lowest initial temperatures, the size and number of the crystals correlate directly with the properties of the starting powder. A very large fraction of the starting grains appears to have survived as nuclei and perhaps the cores of some zoned crystals were discrete relicts. A similar effect of precursor grain size on porphyry grain size was observed with an MgO-rich composition more typical of chondrules (Connolly *et al.*, 1993). Connolly and Hewins (this volume) conclude that fine-grained chondrules with granular and dark-zoned textures require incomplete melting of fine-grained precursors.

Higher initial temperatures yield the textural sequence of porphyritic, barred, and glassy (Fig. 2). There is a gradation between these textures, and transitional textures (hoppers, elongated hoppers) are plotted as porphyritic or barred in this figure. Crystals are grown for initial temperatures up to 100°C (20–45 μm fraction) or up to almost 400°C (125–250 μm fraction) above the liquidus. This reflects the observation above that melting is incomplete with short duration heating and coarser precursor grains take more time to dissolve into the melt. Olivine dissolution seems to have been complete for initial temperatures within the BO field. On crystallization, the grains are more skeletal for higher initial temperature, reflecting more undercooling before smaller embryos graduate to viable nuclei (cf. Lofgren, this volume). Nucleation in these experiments is essentially heterogeneous, in the sense that it is dependent on the degree of melting and the survival of solids even though these are embryos smaller than nuclei capable of immediate growth. With homogeneous nucleation, the growth of nuclei is due to random attachments of clusters in the liquid. What is important for the origin of chondrules is that, with short duration heating, chondrule textures (especially barred olivine) need not be formed for initial temperatures close to the liquidus, but can reflect survival of embryos and nuclei during brief heating to up to hundreds of degrees above the liquidus, especially if the precursors were coarse-grained, e.g. remelted chondrules.

DUST-SEEDING OF TOTAL MELTS

At the high temperatures required to melt Mg-rich chondrules extensively, less refractory compositions should be melted totally, yet totally glassy chondrules are rare. Nucleation as a result of collision between melt droplets and refractory dust grains is a possible explanation (Hewins, 1988, 1989). Experiments by Connolly and Hewins (1995) have shown that olivine, pyroxene, corundum, SiC, and diamond powders puffed into the furnace are effective for crystallization of complete melts, either by producing epitaxial overgrowths or by promoting nucleation. When a dust grain collides with a melt droplet supercooled about 150–450°C below its liquidus, crystallization proceeds very rapidly away from the seed producing the excentroradial texture well known in pyroxene-rich chondrules (Fig. 3). Radial olivine textures can also be produced in Si-poor compositions. The greater abundance of excentroradial pyroxene chondrules is consistent with the lower liquidus temperatures of Si-rich chondrules, making them easier to melt totally and hence to crystallize by seeding.

When dust seeding occurs at temperatures close to the liquidus, seeds can be injected at least 100 μm below the surface, and porphyritic textures are produced with olivine or pyroxene (depending on the bulk composition) phenocrysts throughout the charge, provided that enough seeds are introduced. In these experiments, barred olivine textures were produced under the same conditions as porphyritic olivine textures, except that the droplet encountered fewer dust grains. Porphyritic and barred textures then can be formed either after incomplete melting or after seeding of total

melts. However, it was not possible to introduce enough dust seeds to generate the granular textures common in type I chondrules, implying that these formed by incomplete melting (Connolly and Hewins, 1995, and this volume). Some less common chondrule textures, e.g. barred olivine porphyritic pyroxene, were reproduced by dust seeding but not by conventional dynamic-melting/crystallization experiments, strengthening the argument that dust seeding was an important process. In some heating models, dust grains reach lower temperatures than larger aggregates (Eisenhour and Buseck, 1995) and at least the most refractory ones may have survived evaporation to act as seeds. These experiments demonstrate that if nucleation is the result of melt-dust interaction, one can infer only that the peak temperature was above the liquidus value but we estimate peak temperatures for such chondrules below from energy considerations.

DISCUSSION

We would like to be able to specify the temperatures, heating times, spatial dimensions and cooling rates associated with chondrule formation, variables that are inter-related and which influence chondrule properties in ways which are difficult to disentangle. Our experiments with short duration heating indicate that chondrule peak temperatures may have been much higher than average liquidus temperatures. Consideration of melting rates at superliquidus temperatures and volatile loss rates (Greenwood and Hess, Yu *et al.*, this volume) leads to shorter heating times and higher initial cooling rates (several thousands of degrees per hour) than formerly quoted (Hewins 1988; Lofgren, 1989). Such chondrule cooling rates are obtained from calculations for radiative cooling of chondrule formation regions with dimensions on the order of 100 km (Sahagian and Hewins, 1992). Heating within very much smaller regions, implying virtual quenching of chondrules, has also been considered (Wasson, this volume). A higher cooling rate can mean less complete melting (Osborn, 1993) and tends to raise slightly the heating temperature leading to development of a particular texture.

Chondrules melted at temperatures relatively close to their liquidus temperatures. It would be very unlikely that any heating mechanism would operate so as to raise each precursor aggregate to its respective liquidus temperature in the range 1200–1900°C, but there are several effects which could operate so as to damp the degree of heating to a restricted temperature range or produce textures apparently consistent with near-liquidus temperatures. These include the buffering effects of latent heat of fusion and of differences in grain size of precursors, and the possibility of total melting followed by dust-seeding. Evaporating dust and heating gas also consume energy that would otherwise be available for heating chondrules (Scott *et al.*, this volume; Wasson, this volume).

Grossman (1988), Scott *et al.* (this volume) and Wasson (this volume) have shown that the latent heat of fusion is relatively large compared to the energy required simply to raise temperature, and should tend to act as heat sink during melting, i.e. as a thermostat. The averages of their estimates are about 1J/g to heat chondrules

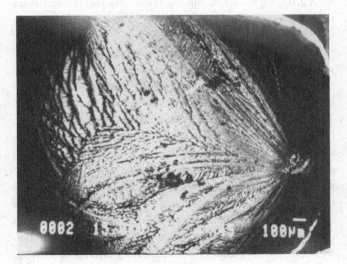

Fig. 3. An excentroradial olivine pyroxene texture produced by collision between a pyroxene grain and a totally molten droplet supercooled below its liquidus by about 200°C. From Connolly and Hewins (1995).

one degree, and about 450 J/g required as latent heat of fusion. The maximum energy input to chondrules can be estimated from the limiting composition for chondrules interpreted as being completely melted. For heating to 1750°C, the liquidus temperature (left line in Fig.3) beyond which all chondrules have textures indicating incomplete melting (Hewins and Radomsky, 1990), 2,000 J/g are required for complete melting, assuming a precursor temperature of 200°C. Similarly, a minimum energy input to chondrules, estimated for heating to 1400°C, the liquidus temperature beyond which all chondrules have textures indicating complete melting, is 1,650 J/g. A precursor with a 1750°C liquidus which received the minimum energy input would have received about 20% of the heat of fusion and could reasonably be expected to generate a fine-grained granular chondrule. One with a 1400°C liquidus which received the maximum heating energy would have been superheated 350 degrees (to the same maximum temperature as the more refractory chondrules) and could subsequently have been dust-seeded. On average 1,825±175 J/g would have been required for chondrule melting, though for flash heating and complete melting at temperatures 100°C above the liquidus, an extra 100 J/g should be allowed, see below, and additional energy may have been needed for other processes, e.g. for partial evaporation, and for heating and dissociating hydrogen (see also Scott *et al.*, Wasson, this volume). This is assuming that chondrules buffer their own cooling rates after a large scale heating event. If on the other hand chondrules can radiate to space freely, a long and gradually fading heat source, i.e. more than the minimum energy quoted above, would be required to obtain the same cooling history (Eisenhour, pers. comm., 1995). The analysis above helps in explaining chondrule temperature-composition-texture relationships, though rather simplified. However, melting would occur across a large temperature interval (~400°C between liquidus and solidus) and the latent heat would not buffer temperature very efficiently. If chondrules formed with a restricted peak temperature range, the relatively limited range of textures may be explained by the effect of short heating times, precursor grain size, olivine dissolution rates, and dust seeding.

One observation consistent with a small temperature range common to all chondrules is that in general type I chondrules (FeO-poor) are finer grained than (FeO-rich) type II chondrules (McSween, 1977; Scott and Taylor, 1983), though there are exceptions such as the fine-grained type II (FeO-rich) chondrules described by Weisberg and Prinz (this volume). The granular and microporphyritic textures typical of type I chondrules suggest that they were less melted than type IIs (Hewins and Radomsky, 1990), as expected for their more refractory compositions, but we also infer (Fig. 2) based on the results of Jones (1991) that their precursors were finer-grained. The coarse porphyritic textures common in type II chondrules could alternatively be the result of more extreme melting of fine-grained precursors due to less refractory composition, but this requires melting temperatures restricted to a very narrow range (Fig. 2). We therefore consider it more likely that they had coarser-grained precursors. Given a variety of evidence for recycling of chondrules (Jones, Rubin and Krot, Alexander, this

volume) coarse-grained precursors for many chondrules are plausible. Because of the longer olivine survival times (Fig. 1), and the corresponding evidence of porphyritic textures after melting at higher temperatures for coarser grained precursors (Fig. 2), we conclude that most FeO-rich (type II) chondrules formed at higher temperatures relative to their own liquidus temperatures than did type I.

Hewins and Radomsky (1990) derived a range of 1400–1750°C for chondrule formation temperatures, assuming that nucleation density and therefore texture were related to peak temperature in the same way as in dynamic crystallization experiments with a long duration melting step. They suggested that maximum temperatures would have been somewhat higher if dust seeding of totally melted droplets had occurred, but did not consider flash heating. Assuming short duration heating for the most refractory chondrules (liquidus greater than 1750°C) as in our crystallization experiments, the upper temperature limit (onset of complete melting) goes up by at least 80°C, depending on grain size (Fig. 2). With relatively fine-grained precursors for refractory chondrules, a 100°C increase is reasonable. This gives a maximum temperature of 1850°C and a maximum energy input of 2,100 J/g.

Connolly and Hewins (1995) showed that seeding of totally melted droplets is the probable origin for the least refractory chondrules (excentroradial and barred olivine porphyritic pyroxene textures). Seeding experiments failed to inject or generate enough nuclei to reproduce the granular textures of the most fine-grained magnesian chondrules, indicating that most type I chondrules were formed by incomplete melting, and therefore specific temperatures may be estimated for them. Such estimates are impossible for chondrules which were totally melted and then dust-seeded, but these are among those rich in moderately volatile elements (Grossman, 1988), and there is no reason to suppose they experienced higher temperatures than the FeO-poor (type I) chondrules. Some idea of the maximum temperature possible can be obtained from the range of energy inputs derived above. Applying the maximum 2,100 J/g to 1400°C liquidus chondrules puts the maximum temperature at 1850°C, the same as adopted for incompletely melted chondrules.

Because microporphyritic textures are best explained by fine-grained precursors (Fig. 2), we assume that type I chondrules (MgO-rich) had the finest grained starting material (< 45 μm) and that type II chondrules had somewhat coarser precursors (about 100 μm). We also assume that the results of Connolly and Hewins (1995) apply to the least refractory chondrules, which were all totally melted. Though these assumptions may be a great oversimplification, they allow us to redraw the final figure of Hewins and Radomsky (1990) in some detail, considering the effect of flash heating (Fig. 4). This cartoon is a pseudobinary phase diagram with the liquidus surface shown as a narrow band in T–$(SiO_2+Al_2O_3-MgO)$ space. It is divided into three composition segments, for chondrule textures indicating incomplete melting, both complete and incomplete melting, and exclusively complete melting, after Hewins and Radomsky (1990). For the most refractory samples (liquidus > 1750°C), which are all type I chondrules,

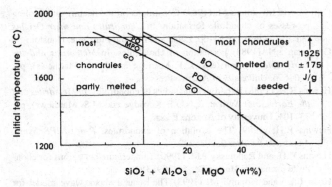

Fig. 4. Revision of a diagram from Hewins and Radomsky (1990) showing chondrule textures for peak temperature vs. wt.% ($SiO_2+Al_2O_3-MgO$). Temperatures on the equilibrium liquidus curve divide diagram into three segments: chondrules have textures indicating total melting (righthand segment), incomplete melting (lefthand segment) and both complete and incomplete melting (central segment). Textural fields extend above the liquidus, assuming the results for fine grained starting materials (Fig. 2) apply to the most magnesian chondrules, and results using coarser material apply to intermediate (type II) chondrules. GO indicates granular olivine, MPO microporphiritic olivine, PO porphyritic olivine, and BO barred olivine textures. The barred olivine upper limit is drawn in accordance with the abundance histogram for BO chondrules, but temperatures could have exceeded this limit for BO textures formed by dust-seeding. Totally glassy chondrules, which would otherwise form for initial temperatures above the barred olivine field, are rare suggesting a role for dust-seeding, the process responsible for crystallization of the least refractory chondrules (righthand segment). As discussed in the text, the upper and lower limits to the initial temperature range are 1500–1850°C: corresponding energy inputs required for heating and melting are also shown.

we plot microporphyritic and porphyritic olivine textures extending 40 and 80 °C above the liquidus line, as in Fig. 2. 1850°C is a reasonable upper limit for such chondrules because it makes porphyritic subordinate to microporphyritic, and barred olivine chondrules essentially non-existent. For the middle segment (liquidus 1400–1750°C), chondrules are mostly type II (FeO-rich), and therefore the porphyritic and barred olivine fields are extended up to 110 and 250°C above the liquidus, as in Fig. 2. However, because barred olivine chondrules become increasingly scarce as liquidus temperatures approach 1750°C (Hewins and Radomsky, 1990), the upper limit to the barred olivine field is dropped in steps proportional to steps in their abundance-liquidus histogram (though perhaps a smoothed curve would be more realistic). This results in limits for barred olivine chondrules close to 1850°C, except for the least refractory chondrules. For liquidus temperatures below 1500°C, barred olivine chondrules become abruptly less abundant because of the increased SiO_2 in the melt, and for liquidus temperatures below 1400°C (third segment, increased SiO_2), excentroradial pyroxene is the dominant chondrule textural type. For the latter two sets of chondrule compositions, total melting and seeding is possible and taken to be the rule, respectively. The barred olivine field shown is for internal nucleation, and similar textures could be generated with initial temperatures above this field in the case of dust seeding. Assumption of a heating energy of 2,100 J/g provides an upper temperature limit of 1850°C for the melted-and-seeded 1400°C liquidus material, in agreement with

results for compositions for which incomplete melting is possible.

The upper limit for the initial temperature range is derived assuming complete melting would occur at 100°C above the boundary where chondrules are last observed to show complete melting, i.e. the 1750°C point on the liquidus curve. Chondrules of all compositions then could have experienced a temperature of no more than about 1850°C, assuming they were heated for a few seconds and cooled so as to spend a few minutes at near-liquidus temperatures, as in our experiments. The lower limit is much harder to define, and in principle one might expect numerous heated but only metamorphosed particles. However, "chondrules" have igneous textures and and the least refractory chondrules have only textures indicating total melting. For our grain size assumptions, the disappearance of porphyritic textures for liquidus values below 1400°C means a minimum temperature of about 1510°C. We adopt 1500°C as the lower limit, which gives us a minimum energy input of 1750 J/g. For the more refractory compositions, this energy input is in accord with the scarcity of textures indicating extensive melting and the abundance of fine-grained little-melted chondrules. The 1500°C lower limit is consistent with data for moderately volatile elements in chondrules, e.g. the initial temperature for which S can be retained in type I chondrules (Yu *et al.*, this volume). The recommended temperature range of 1500–1850°C for chondrule melting temperatures is slightly wider than the 1600–1800°C range finally suggested by Hewins and Radomsky (1990).

CONCLUSIONS

1. Because of relatively low olivine dissolution rates, coarse-grained FeO-rich material requires longer heating time or higher temperature for total melting than fine-grained. With short duration heating, chondrule textures can be generated only with peak temperatures much above the liquidus, unless the precursors are very fine-grained; coarse-grained chondrule precursors survive as relict grains unless the initial temperature is much above the liquidus.

2. Crystallization of FeO-rich liquids after short duration heating yields porphyritic olivine textures with peak melting temperatures up to 80–120°C and barred olivine up to 150–400°C above the liquidus temperature, rather than just below and just above the liquidus as with long heating times. Microporphyritic textures are produced with fine-grained precursors (T up to 40°C).

3. Excentroradial pyroxene and barred olivine porphyritic pyroxene textures can be produced by dust seeding of total melts, and totally glassy chondrules are rare, suggesting that chondrules totally melted well above their liquidus temperatures were dust-seeded.

4. For the short duration heating of our experiments and fine grained precursors, a given degree of melting is achieved about 100°C higher than thought previously. The most refractory chondrules (liquidus over 1750°C), which were all incompletely melted, therefore experienced a maximum initial temperature of 1850°C, corresponding to an energy input

of about 2,100 J/g. The least refractory were totally melted and dust-seeded, and experienced minimum temperatures of 1500°C and energy inputs of 1,750 J/g. The range of peak temperatures 1500–1850°C applies to chondrules of all compositions.

ACKNOWLEDGEMENTS

We are indebted to B.D. Jones and C. M. Lerro for performing experiments; to C.M.O'D. Alexander, J. Greenwood, A.N. Krot and J.T. Wasson for critical reviews which materially improved the paper; to D.D. Eisenhour and J. Greenwood for helping us understand chondrule heating; to S.V. Maharaj, T.G. Osborn, Y. Yu and B. Zanda for discussions and support; and to NASA for financial support, as well as to NASA and the LPI for assistance in attending the conference on Chondrules and the Protoplanetary Disk.

REFERENCES

Ainslie N.G., Mackenzie J.D. and Turnbull D. (1961) Melting kinetics of quartz and cristobalite. *J. Phys. Chem.* **65**, 1718–1724.

Brearley M. and Scarfe C.M. (1986) Dissolution rates of upper mantle minerals in an alkali basalt melt at high pressure: an experimental study and implications for ultramafic xenolith survival. *J. Petrol.* **27**, 1157–1182.

Connolly H.C. Jr. and Hewins R.H. (1991) The influence of bulk composition and dynamic melting conditions on olivine chondrule textures. *Geochim. Cosmochim. Acta* **55**, 2943–2950.

Connolly H.C. Jr., and Hewins R.H. (1995) Chondrules as products of dust collisions with totally molten droplets within a dust-rich nebular environment: An experimental investigation. *Geochim. Cosmochim. Acta* **59**, 3231–3246.

Connolly H.C Jr., Jones B.D and Hewins R.H (1991) The effect of precursor grain size on chondrule texture (abstract) *Meteoritics* **26**, 329.

Connolly H.C. Jr., Hewins R.H. and Lofgren G.E. (1993) Possible clues to the physical nature of chondrule precursors: an experimental study using flash melting conditions (abstract). *Meteoritics* **28**, 338–339.

Dages J., Gleiter H. and Perepezko J.H. (1987) Superheating of metallic crystals. *Mat. Res. Soc. Symp. Proc.* **57**, 67–78.

Donaldson C.H. (1979) An experimental investigation of the delay in nucleation of olivine in mafic magmas. *Contrib. Mineral. Petrol.* **69**, 21–32.

Donaldson C.H. (1985) The rates of dissolution of olivine, plagioclase, and quartz in a basalt melt. *Miner. Mag.* **49**, 683–693.

Eisenhour, D.D. and Buseck P.R. (1995) Chondrule formation by radiative heating: a numerical model. *Icarus*, **108**, in press.

Ferrière A., Lestrade L. and Rouanet A. (1994) Surface glazing of plasma-sprayed thermal barrier coatings with a solar furnace. In *Advances in Inorganic films and Coatings; Proc. 8th CIMTEC* (ed. P. Vincenzi), pp. 319–326. Florence, Italy.

Fujii N. and Miyamoto M. (1983) Constraints on the heating and cooling processes of chondrule formation. In *Chondrules and their Origins* (ed. E.A. King), pp. 53–60, Lunar and Planetary Institute, Houston.

Grossman J.N. (1988) Formation of chondrules. In *Meteorites and the Early Solar System* (eds. J. F. Kerridge and M. S. Matthews), pp. 680–696. University of Arizona Press.

Hewins R.H. (1988) Experimental studies of chondrules. *In Meteorites and the Early Solar System*, (eds. J.F. Kerridge and M.S. Matthews), pp. 73–101. University of Arizona Press.

Hewins R.H. (1989) The evolution of chondrules. *Proc. NIPR Symp. Antarct. Meteorites* **2**, 202–222.

Hewins R.H. and Radomsky P.M. (1990) Temperature conditions for chondrule formation. *Meteoritics* **25**, 309–318.

Hood L.L., and Horanyi M. (1993) The nebular shock wave model for chondrule formation: one-dimensional calculations. *Icarus*, **106**, 179–189.

Jones B.D. (1991) *Effect of grain size and very short heating times on chondrule formation*. B.A. thesis, Rutgers University, 22 pp.

Lerro C.M. (1993) *The role of heterogeneous nucleation in reproducing chondrule texture: Melting vs. crystallization experiments*. B.A. thesis, Rutgers University, 30 pp.

Lofgren G. (1982) The importance of heterogeneous nucleation for the formation of chondrule textures (abstract). In *Papers presented to the Conference on Chondrules and their Origins,* LPI Contribution 493, Lunar and Planetary Institute, Houston, 67 pp.

Lofgren G. (1989) Dynamic crystallization of chondrule melts of porphyritic olivine composition: Textures experimental and natural. *Geochim. Cosmochim. Acta* **53**, 461–470.

Lofgren G.E., and Russell W.J. (1986) Dynamic crystallization of chondrule melts of porphyritic and radial pyroxene composition. *Geochim. Cosmochim. Acta* **50**, 1715–1726.

McSween H.Y. Jr. (1977) Chemical and petrographic constraints on the origin of chondrules and inclusions in carbonaceous chondrites. *Geochim. Cosmochim. Acta* **41**, 1843–1860.

Navrotsky A., Ziegler D., Oestrike R. and Maniar P. (1989) Calorimetry of silicate melts at 1773 K: Measurements of enthalpies of fusion and of mixing in the systems diopside-anorthite-albite and anorthite-forsterite. *Contrib. Mineral. Petrol.* **101**, 122–130.

Osborn T.G. (1993) *Non-linear cooling profiles and the effect on type II synthetic chondrules*. B.A. thesis. Rutgers University, 104 pp.

Radomsky P.M. and Hewins R.H. (1990) Formation conditions of pyroxene-olivine and magnesian olivine chondrules. *Geochim. Cosmochim. Acta* **54**, 3475–3490.

Sahagian D.L. and Hewins R.H. (1992) The size of chondrule-forming events (abstract). *Lunar Planet. Sci.* **XXIII**, 1197–1198.

Scott E.R.D. and Taylor G.J. (1983) Chondrules and other components in C, O and E chondrites: Similarites in their properties and origins. *Proc. Lunar Planet. Sci. Conf.* **14th**, B275–B286.

Wasson J.T. and Rasmussen K.L. (1994) The fine nebula dust component: a key to chondrule formation by lightning. In *Papers presented to Chondrules and the Protoplanetary Disk*, LPI Contribution No. 844, Lunar and Planetary Institute, Houston, 50 pp.

Zhang Y., Walker D. and Lesher C.E. (1989) Diffusive crystal dissolution. *Contrib. Mineral. Petrol.* **102**, 492–513.

22: Congruent Melting Kinetics: Constraints on Chondrule Formation

JAMES P. GREENWOOD and *PAUL C. HESS*

Department of Geological Sciences, Brown University, Providence, RI 02912, U.S.A.

ABSTRACT

The processes and mechanisms of melting and their applications to chondrule formation are discussed. A model for the kinetics of congruent melting is developed and used to place constraints on the duration and maximum temperature experienced by the interiors of relict-bearing chondrules. Specifically, chondrules containing relict forsteritic olivine or enstatitic pyroxene cannot have been heated in excess of 1901°C or 1577°C, respectively, for more than a few seconds.

INTRODUCTION

Since the discovery of relict grains by Nagahara (1981) and Rambaldi (1981), the emerging models of chondrule formation have called for an origin by the melting of pre-existing solids (e.g. Grossman, 1988). The realization that the majority of FeO-rich chondrules have lost little Na (Grossman, 1988; Hewins, 1991a; Grossman, this volume), coupled with experimental work on Na loss from chondritic melts (Tsuchiyama *et al.*, 1981) has led to the idea that chondrules were melted in flash heating events of an unknown dynamical nature (Grossman, 1988; Boss, this volume, Concise Guide).

If chondrules were formed by the melting and dissolution of minerals, then a better understanding of these two processes is needed, in order to derive constraints on the nature of the flash heating event(s). Congruent melting is the process by which a solid transforms above its melting point to a liquid of the same composition, and will be the focus of this paper. Incongruent melting occurs when a mineral solid-solution forms a liquid of different composition when heated above the solidus but below the liquidus. In a study of the incongruent melting of plagioclase (Tsuchiyama and Takahashi, 1983) it was found that the kinetics of the reaction were rate-limited by solid-state diffusion, as the solid also needs to change composition, in order to maintain equilibrium. The sluggish kinetics of this type of process suggest that it will be unimportant during flash heating. Above the liquidus, a mineral solid solution such as plagioclase or olivine will melt congruently (Greenwood and Hess, 1995), as neither the liquid nor the solid must change composition during the transition. Dissolution is the process by which a mineral dissolves into a liquid of different com-

position, and this occurs below the melting point of the solid. Dissolution should be an important process during chondrule formation (Greenwood and Hess, 1995), but will not be discussed here.

In this paper, the mechanisms and kinetics of congruent melting are examined. It is shown that the kinetics of congruently melting minerals are best described by an interface-controlled model (Wilson, 1900; Frenkel, 1932). This model is used to calculate the melting rates of possible precursor minerals (Hewins, 1991b), which leads to constraints on the durations and peak temperatures of chondrule formation. The implications of congruent melting during chondrule formation are considered below.

CONGRUENT MELTING: THEORY

Melting as a Continuous Transition

Studies of the melting transition in congruently melting materials have historically focused almost exclusively on the instability of the solid while neglecting the parallel reaction, the growth of the liquid (Boyer, 1985). At the melting point, homogeneous melting models envision the solid catastrophically transforming to liquid at all points in the crystalline lattice. Nucleation of the liquid is not necessary. Various theories have been developed to explain this bulk mechanical instability. Some examples of these are the Lindemann criterion which links the melting point to a critical amplitude of atomic vibrations (Lindemann, 1910), the vanishing of the shear modulus (Born, 1939), and the generation of dislocations (Poirier, 1986).

Continuous melting models are necessary, but insufficient components to our understanding of the melting transition. Phenomena such as superheating (exposing the solid to temperatures above T_m, the melting point) (Di Tolla *et al.*, 1995) and surface melting (melting preferentially at surfaces of the crystal) (Frenken and van Pinxteren, 1994) are not predicted by these theories. Though it has long been known that some silicates can sustain large amounts of superheat for considerable lengths of time (Day and Allen, 1905), the proponents of continuous melting models were driven by the early experimental observations that metals were melted almost instantaneously at fractions of a degree of superheat (Ainslie *et al.*, 1961). In fact, metals can be superheated by several degrees (Dages *et al.*, 1987; Di Tolla *et al.*, 1995), but due to the high rate of melting it is generally difficult to observe. Also, homogeneous melting models are at odds with the observation that melting is invariably initiated at external surfaces (Tamman, 1925; Teraoka, 1993) and internal cracks and cleavage planes (Uhlmann, 1980). The importance of the surface in initiating melting during molecular dynamics simulations of $MgSiO_3$-perovskite has also been discussed by Belonoshko (1994).

Melting as a Discontinuous Transition

The contrasting view of melting is that the solid transforms to a liquid discontinuously via a nucleation and growth mechanism, similar to crystallization (Tamman, 1925). A heterogeneous model would predict that melt will form where the barrier to nucleation of the melt phase is lowest, such as surfaces and lattice defects (Ainslie *et al.*, 1961). This agrees well with experimental observations.

If melting is considered analogous to crystallization, the principal difference being that liquids nucleate far more easily than crystals (Ainslie *et al.*, 1961), then growth of the melt phase can be modelled with existing theories of crystal growth. This approach has been utilized previously in the melting of silicates and oxides (e.g. Wagstaff, 1969; Uhlmann, 1971), and is described below.

CONGRUENT MELTING

Wilson-Frenkel Model

When a solid melts to a liquid of the same composition, it is found experimentally that the growth of the melt is inversely proportional to the viscosity of the liquid phase (Ainslie *et al.*, 1961). In melts with low viscosity, such as metals and semiconductors, growth of the melt is very fast and is generally rate limited by how fast heat can be added to the interface (Spaepen and Turnbull, 1982). In melts with high viscosity, such as silicates, growth is relatively slow and is usually rate limited by the kinetics of the solid-liquid transition. Several studies of the melting kinetics of silicates and oxides have been completed to date (e.g. quartz, Scherer *et al.*, 1970; sodium disilicate, Fang and Uhlmann, 1984; diopside, Kuo and Kirkpatrick, 1985; albite, Greenwood and Hess, 1994a). Each used a Wilson-Frenkel model for normal growth to model their data. A normal growth model is used when the interface is rough on the atomic scale (atoms can be added or removed from any site on the interface). The growth of the melt is envisioned as the propagation of the solid-liquid interface from the surface into the crys-

tal. Molecular dynamics simulations of the melting of forsterite have found that melting takes place layer-by-layer as the solid-liquid interface migrates through the crystal (Kubicki and Lasaga, 1992), in accordance with the tenets of an interface-controlled growth model. The normal growth model has been found to reproduce experimentally determined melting rates generally within an order of magnitude (Greenwood and Hess, 1994b; Table 22.1). The Wilson-Frenkel model is (Uhlmann, 1971):

$$u = (fD'/a_o)[1-\exp(-\Delta G/kT)] \qquad (1)$$

where u is the growth rate, f is the fraction of sites available (f=1 for melting; i.e. atoms leaving the crystal are not limited to rigid, fixed sites in the melt), D′ is the kinetic factor for transport at the interface, a_o is the jump distance (in this model, it is usually taken as twice the length of an important bond in the crystal structure; e.g. $2 \times$ Si-O for quartz, Ainslie *et al.*, (1961)), T is the absolute temperature, ΔG is the free energy change per atom of the transition at the temperature T, and k is Boltzmann's constant. If D′ is related to self-diffusion in the liquid, D, and the Stokes-Einstein relation for diffusion is assumed (see discussion below), then:

$$D' \approx D = kT / (3\pi a_o \eta) \qquad (2)$$

where η is the viscosity of the melt, and substituting in (1),

$$u = [fkT / (3\pi a_o^2 \eta)][1-\exp(-\Delta G/kT)] \qquad (3).$$

If $\Delta G << kT$, a condition satisfied for small superheats, we expand (3) (Fine, 1964) to

$$u = [fkT / (3\pi a_o^2 \eta)](-\Delta G/kT) \qquad (4).$$

Also, at small departures from equilibrium, $\Delta G \approx \Delta H_f \Delta T/T_m$ (Kingery *et al.*, 1976), and substituting into (4):

$$u = f\Delta H_f \Delta T / (T_m 3\pi a_o^2 \eta N_A) \qquad (5)$$

where ΔH_f is the molar heat of fusion, N_A is Avogadro's number, T_m is the melting point, and $\Delta T=T-T_m$, the amount of superheat. Notice that we are now describing the melting per mole rather than per atom. The functional form of the equation is also written as:

$$u\eta = K\Delta T \qquad (6)$$

where $K= f\Delta H_f/(T_m 3\pi a_o^2 N_A)$, and is considered a constant for small superheatings. The parameter $u\eta$ is termed the "normalized melting rate", and a plot of $u\eta$ versus ΔT should be linear with a slope equal to K for normal growth (Fang and Uhlmann, 1984). Any deviations from a linear plot would suggest a significant temperature dependence of f, ΔH_f, or a_o (Wagstaff, 1969). An experimentally determined plot of $u\eta$ versus ΔT for albite is shown in Fig. 1 (Greenwood and Hess, 1994a; Greenwood and Hess, unpublished data) and demonstrates the validity of using a normal growth model to describe the melting of albite. The normal growth model has also been shown to be appropriate for quartz (Scherer *et al.*, 1970), cristobalite (Wagstaff, 1969), sodium disilicate (Fang and Uhlmann, 1984), germanium dioxide (Vergano and Uhlmann, 1970), phosphorus pentoxide (Cormia *et al.*, 1963a), and diopside (Kuo and Kirkpatrick, 1985).

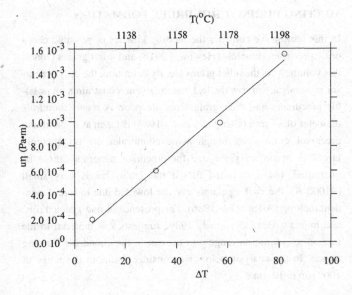

Fig. 1. Shown here is the experimentally determined normalized melting rate, uη (melting rate × viscosity) vs. ΔT (the amount of superheat) for albite. Line is a best fit to the data (open circles).

Comparison of experimental and calculated rates

A comparison of rates calculated from equation (5) with experimentally determined rates is shown in Table 1. A jump distance of 3Å was used for the calculations, except for albite and germanium dioxide, where a value of 3.5Å was used. As mentioned above, the jump distance is approximated as twice the length of an important bond in the structure.

While there is some obvious disagreement, the calculated rates are generally the same order of magnitude as the rates determined by experiment. The largest discrepancies between the calculated and experimental rates are for quartz and diopside, and illustrate some of the experimental difficulties associated with measuring melting kinetics. The melting kinetics of quartz, as well as the viscosity of liquid SiO_2, are highly sensitive to atmospheric impurities and water contamination, and large differences in melting rates have been found by different researchers for quartz (Ainslie et al., 1961; Scherer et al., 1970). The reason for the large discrepancy for diopside is not known, but may be related to the difficulty in measuring high melting rates in a small range of superheating. A similar problem was found in the determination of the melting rates of sodium disilicate, where a second study by the same research group found very different results (Meiling and Uhlmann, 1967; Fang and Uhlmann, 1984). At these high rates of melting, a heating stage may be necessary for accurate determination of the kinetics (Fang and Uhlmann, 1984).

The Wilson-Frenkel model reproduces the experimental data in five of the seven studies shown in Table 1 within an order of magnitude. While experimental difficulties may explain some of the

Table 1. Comparison of experimental melting rates with rates calculated from eqn (5) at various superheats

Mineral	ΔHf(kJ/mol)	Tm(°C)	T(°C)	η(Pa•s)	ΔT	Expt./Calc.
Quartz[†,1]	9.40[2]	1427[2]	1500	9.71 × 10[7][3]	73	20.5
			1600	1.50 × 10[7]	173	24.0
			1650	6.26 × 10[6]	223	10.4
Cristobalite[4]	8.92[2]	1726[2]	1743	1.39 × 10[6][3]	17	6.3
			1746	1.31 × 10[6]	20	7.9
			1755	1.15 × 10[6]	29	9.7
Na₂Si₂O₅[5]	37.7[2]	874[2]	875	881[5]	1	2.1
			880	805	6	2.1
			884	662	10	1.8
GeO₂[6]	15[6]	1114[6]	1119	2.95 × 10[4][7]	5	3.6
			1125.5	2.69 × 10[4][7]	11.5	2.5
			1130	5 × 10[4][8]	22	5.1
Albite[9]	64.5[2]	1118[9]	1125	4.3 × 10[6][10]	7	0.4
			1175	1.1 × 10[6]	57	0.4
			1200	5.9 × 10[5]	82	0.3
P₂O₅[11]	21.8[11]	580[11]	589	4.38 × 10[5][12]	9	0.5
			593	3.61 × 10[5]	16	0.9
			609	2.53 × 10[5]	29	0.8
Diopside[13]	137.7[2]	1391[13]	1393	0.93[14]	2	0.013
			1399	0.91	8	0.013
			1412	0.87	21	0.012

[†]Metastable melting, [1]Scherer et al. (1970), [2]Richet and Bottinga (1986), [3]Urbain et al. (1982), [4]Wagstaff (1969), [5]Meiling and Uhlmann (1967), [6]Vergano and Uhlmann (1970), [7]Fontana and Plummer (1966), [8]Sharma et al. (1979), [9]Greenwood and Hess (1994a), [10]Stein and Spera (1993), [11]Cormia et al. (1963a), [12]Cormia et al. (1963b), [13]Kuo and Kirkpatrick (1985), [14]calculated from Bottinga and Weill (1972).

discrepancy, the equation may be fundamentally flawed, and this is considered in the next section.

Discussion of Wilson-Frenkel model

The assumptions used to derive eqn. (2) are somewhat controversial. The first assumption is that the diffusion in the interfacial region can be equated to self-diffusion in the liquid. Cahn *et al.* (1964) suggest that diffusivity in the interfacial region may be as much as two orders of magnitude lower than diffusion in the bulk liquid. In their theory, lower diffusivities arise from the quasi-crystalline nature of the liquid immediately adjacent to the interface. In contrast, molecular dynamics simulations of argon suggest that the diffusivity at the interface may be higher than in the bulk liquid (Broughton *et al.*, 1982). The nature of the solid-liquid interface is not fully understood and is the subject of ongoing work (e.g. Moss and Harrowell, 1994).

The second assumption used in eqn. (2) is the application of the Stokes-Einstein equation to relate self-diffusion in the liquid and the viscosity of the liquid. There has been some success in using the Stokes-Einstein equation to model the diffusion of oxygen (Shimuzu and Kushiro, 1984) and silica (Watson and Baker, 1991) in silicate melts, but in general the quantitative agreement is poor (Kress and Ghiorso, 1995). Although the assumptions used in eqn. (2) may be the source of divergence from the experimental rates, the good agreement for the calculated rates of albite and sodium disilicate suggests that the Stokes-Einstein equation may be appropriate for the melting of silicates. The viscosities of molten albite and sodium disilicate at their melting points are 10^6 and 10^2 Pa•s, respectively (Table 1). They also have different melt structures, a fully polymerized albite melt and a somewhat depolymerized disilicate melt. Considering the large differences in viscosity, melt structure, and melting rates between these two materials, the agreement between the experimental and calculated rates harbors hope for the possibility of using the Wilson-Frenkel model as a predictive vehicle. We feel that this model can be used to predict melting rates within an order of magnitude, and as will be shown below, even two orders of magnitude difference will still lead to useful constraints on chondrule formation.

MELTING DURING CHONDRULE FORMATION

In this section we consider the melting kinetics of possible chondrule precursor minerals (Hewins, 1991b) and relict grains (Jones, this volume). As the relict grains are, by definition, the only precursor minerals to survive the last heating event, constraints on possible precursors and their grain sizes are poor. A mean chondrule diameter of ~1 mm (Grossman *et al.*, 1988) is taken as a maximum precursor grain size, though some chondrules are undoubtedly larger. A minimum grain size for precursor minerals cannot be quantified, but it is noted that if the grain size is very small (~1000 Å), the melting points may be lowered due to size-dependent melting (Allen *et al.*, 1986). The presence of fine-grained rims and matrix (Alexander *et al.*, 1989) suggests that material in the micron to submicron range was present in the chondrule forming region. In the analysis below we consider grains in the range of 1000 µm to 10 µm.

Precursor minerals

The melting rates of possible chondrule precursor minerals calculated from eqn. (5) are listed in Table 2 for superheats of 5 and 100 degrees. The jump distance is taken as 3Å (except albite; 3.5Å). Values for the viscosities of fayalite, diopside, åkermanite, and enstatite melts were determined from Bottinga and Weill (1972). For forsterite melts, viscosities were first determined by calculating values in the 1600–1800°C range by the method of Bottinga and Weill (1972), and then extrapolating to the desired temperature. The viscosities for albite and anorthite melts were determined experimentally (Stein and Spera, 1993; Cranmer and Uhlmann, 1981).

Shown in Fig. 2 are the melting rates of possible precursor minerals at 1600°C. If the melting rates for these minerals are the right order of magnitude, it becomes apparent that if any of these possible chondrule precursors are exposed to temperatures in excess of 1600°C (with the exception of forsterite, T_m = 1901°C) they will probably melt in seconds or less. In fact, for a grain radius range of 10–100 µm, even two orders of magnitude error for the calculated melting rates still leads to complete melting of chondrule precursor minerals in seconds or less. At higher temperatures, the melting

Table 2. *Melting rates calculated from eqn. (5) at superheat of 5 and 100 degrees*

Mineral	T_m(°C)	Hf (KJ/mol)	ΔT	T(°C)	Calc. rate (µm/s)	ΔT	T(°C)	Calc. rate (µm/s)
Albite*	1118[1]	64.5[2]	5	1123	4.3×10^{-5}	100	1218	4.3×10^{-3}
Fayalite	1217[2]	89.3[2]	5	1222	5×10^3	100	1317	1.7×10^5
Diopside*	1391[3]	137.7[2]	5	1396	11	100	1492	3.8×10^2
Åkermt.	1458[4]	123.6[4]	5	1463	1.6×10^3	100	1558	1.3×10^5
Anorthite	1557[2]	133.0[2]	5	1562	1.9×10^2	100	1657	8.2×10^3
Enstatite[†]	1557[5]	73.2[2]	5	1562	n.a.[†]	100	1657	2.7×10^4
Forsterite	1901[6]	142[6]	5	1906	6.3×10^3	100	2001	1.3×10^5

*Experimentally determined (see text), [†]See text for discussion of enstatite melting, [1]Greenwood and Hess (1994a), [2]Richet and Bottinga (1986), [3]Kuo and Kirkpatrick (1985), [4]Hemingway *et al.* (1986), [5]Bowen and Anderson (1914), [6]Richet *et al.* (1993).

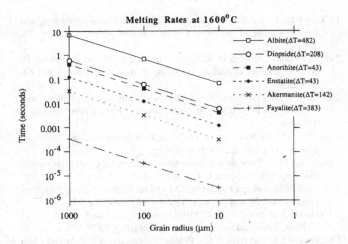

Fig. 2. Shown here is a time vs. grain radius plot for melting kinetics at 1600°C. For example, at 1600°C an enstatite grain with a radius of 1 mm will be completely melted in 0.12 seconds. (ΔT = superheat).

rates will necessarily increase rapidly (eqn. 5), suggesting that if peak temperatures were below 1901°C, the chondrule may consist of melt and relict forsterite as a result of high temperature heating. Heating above 1901°C will lead to complete melting of chondrules in seconds or less, provided that the minerals listed in Table 2 are a major proportion of the precursor assemblage.

Albite has been proposed as a chondrule precursor (Hewins, 1991a) to explain the 1:1 correspondence between Na and Al found in FeO-rich chondrules, yet relict albite is not found in chondrules. At 1600°C (Fig. 2), a 1 mm radius albite grain would be completely melted in less than 10 seconds. For grain radii in the 10 – 100 μm range, albite would be completely melted in less than 2 seconds at 1600°C. Using the experimentally determined uη vs. ΔT relationship shown in Fig. 1 for albite, the melting rates are 11μm/s and 144 μm/s at 1500°C and 1600°C, respectively. Clearly, if albite is fine-grained (1 – 10 μm) it could melt completely in a flash heating event with a peak temperature as low as 1500°C. If the peak temperature were higher (Hewins and Connolly, this volume) then it is not surprising that we do not find relict albite in chondrules.

Relict Grains

The presence of relict grains in chondrules (grains that survived the last heating event) can help constrain the maximum temperatures experienced by the host chondrules during heating. Enstatitic pyroxene and forsteritic olivine have been identified as relict grains in FeO-rich porphyritic chondrules (Jones, this volume). Dusty olivines have been identified as relict grains in FeO-poor porphyritic chondrules (Nagahara, 1981; Jones, this volume). The common existence of relict olivine in chondrules constrains the maximum temperature that the interiors of these chondrules experienced to the melting point of forsterite, 1901°C (Richet *et al.*, 1993). Chondrules containing relict enstatite could not have been heated in excess of 1577°C, the liquidus temperature of enstatite, for more than a few seconds. Enstatite melts incongruently, undergoing a peritectic reaction at 1557°C, forming a product of 95%

liquid and 5% forsterite, by weight. At 1577°C enstatite melts congruently to a liquid of pure enstatite (Bowen and Anderson, 1914). We have studied the congruent melting of an incongruently melting compound above the liquidus in the plagioclase system (Ab₈₉), and find that the congruent melting model embodied in eqn. (5) is appropriate for this type of reaction (Greenwood and Hess, 1995). We therefore feel that the estimates for the melting rate of enstatite above 1577°C are valid (Table 2, Fig. 2). As mentioned earlier, incongruent melting is generally a slow process, rate limited by solid-state diffusion. The interested reader is referred to Tsuchiyama and Takahashi (1983) for a discussion.

DISCUSSION

The constraints developed in this paper are for minerals transforming to liquids via melting. The total destruction of a relict grain and/or precursor mineral involves the incorporation of this nominally pure mineral melt into the bulk chondrule melt. This is synonomous with liquid interdiffusion, and will necessarily take extra time. For example, enstatite melts at a rate of ~3700 μm/s, at 1600°C. This enstatitic liquid would then need to interdiffuse with the chondrule melt before cooling, otherwise it will recrystallize with an anomalous core composition.

The melting rates calculated above assume there is enough energy to melt chondrule precursor minerals, and that melting is not limited by the transfer of heat to or away from the interface. The observation that some chondrules were undoubtedly completely molten (e.g. Grossman, 1988) would seem to support this assumption. Heat-flow limited melting would occur if the interface could not maintain temperature, due to the loss of heat to the interior of the mineral or its surroundings. The low thermal conductivities of silicates (Kingery *et al.*, 1976) should prevent this from happening. The interested reader is referred to Spaepen and Turnbull (1982) for a review of heat-flow limited melting.

The loss of heat from chondrule mineral surfaces to the surroundings has recently been considered by Horanyi *et al.* (1995), wherein they modelled chondrule formation in lightning discharges. They conclude that there may not be enough energy available to completely melt silicates. In their paper, they model the loss of heat from mineral surfaces to the surroundings by radiative cooling, an assumption that is in direct conflict with constraints obtained from dynamic crystallization experiments (Hewins, 1988). Also, the loss of heat to chondrule surroundings would probably only apply to minerals near the exteriors of precursor aggregates, as minerals in the interiors would be thermally insulated by the newly formed molten chondrule. These factors suggest that melting of chondrule precursor mineral grains was not limited by heat flow.

CONCLUSIONS

1. Consideration of congruent melting kinetics explains why most chondrule precursors did not survive flash heating events. A congruent melting model can be used to calculate melting rates.

2. The presence of relict forsteritic olivine and enstatitic pyroxene in chondrules provides the source of constraints on the maximum temperature that the interiors of these chondrules experienced. Chondrules containing relict grains of forsterite and enstatite probably did not exceed temperatures of 1901°C and 1577°C, respectively, for more than a few seconds.

3. Albite can be a precursor and not a relict grain, even though it has sluggish melting kinetics. Specifically, the melting rates of albite at 1500°C and 1600°C are 11 μm/s and 144 μm/s, respectively, suggesting that fine grained albite (< 10 μm) would completely melt in seconds or less for a peak temperature as low as 1500°C.

4. Chondrule formation models that attempt to explain the melting event(s) need to take into account the kinetics of melting in order to constrain their formation characteristics.

ACKNOWLEDGEMENTS

This work benefited from critical reviews of an earlier manuscript by J. Beckett, J. Longhi, and R.H. Hewins. Support was provided by a NASA Graduate Student Researchers Program Fellowship to J. Greenwood and NASA grant NAGW-3613 to P. Hess. J. Greenwood would also like to thank the Lunar and Planetary Institute for assistance to attend the Conference on Chondrules and the Protoplanetary Disk.

REFERENCES

Ainslie N. G., Mackenzie J. D. and Turnbull D. (1961) Melting kinetics of quartz and cristobalite. *J. Phys. Chem.* **65**, 1718–1724.

Alexander C.M.O., Hutchison R. and Barber D.J. (1989) Origin of chondrule rims and interchondrule matrices in unequilibrated ordinary chondrites. *Earth Planet. Sci. Lett.* **95**, 187–207.

Allen G.L., Bayles R.A., Gile W.W. and Jesser W.A. (1986) Small particle melting of pure metals. *Thin Solid Films* **144**, 297–308.

Belonoshko A.B. (1994) Molecular dynamics of $MgSiO_3$ perovskite at high pressures: Equation of state, structure, and melting transition. *Geochim. Cosmochim. Acta* **58**, 4039–4047.

Born M. (1939) Thermodynamics of crystals and melting. *J. Chem. Phys.* **7**, 591.

Bottinga Y. and Weill D. F. (1972) The viscosity of magmatic silicate liquids: A model for calculation. *Am. J. Sci.* **272**, 438–475.

Boyer L.L. (1985) Theory of melting based on lattice instability. *Phase Transitions* **5**, 1–48.

Bowen N. L. and Andersen O. (1914) The binary system $MgO-SiO_2$. *Am. J. Sci.* **37**, 487–500.

Broughton J.Q., Gilmer G.H. and Jackson K.A. (1982) Crystallization rates of a Leonard-Jones liquid. *Phys. Rev. Lett.* **49**, 1496–1500.

Cahn J.W., Hillig W.B. and Sears G.W. (1964) The molecular mechanism of solidification. *Acta Metall.* **12**, 1421–1439.

Cormia R.L., Mackenzie J.D. and Turnbull D. (1963a) Kinetics of melting and crystallization of phosphorus pentoxide. *J. Appl. Phys.* **34**, 2239–2244.

Cormia R.L., Mackenzie J.D. and Turnbull D. (1963b) Viscous flow and melt allotropy of phosphorus pentoxide. *J. Appl. Phys.* **34**, 2245–2248.

Cranmer D. and Uhlmann D.R. (1981) Viscosities in the system albite-anorthite. *J. Geophys. Res.* **86**, 7951–7956.

Dages J., Gleiter H. and Perepezko J.H. (1987) Superheating of metallic crystals. *Mat. Res. Soc. Symp. Proc.* **57**, 67–78.

Day A.L. and Allen E.T. (1905) The isomorphism and thermal properties of the feldspars. *Am. J. Sci.* **19**, 93–142.

Di Tolla F.D., Ercolessi F. and Tosatti E. (1995) Maximum overheating and partial wetting of nonmelting solid surfaces. *Phys. Rev. Lett.* **74**, 3201–3204.

Fang C.-Y. and Uhlmann D. R. (1984) The process of crystal melting. II. Melting kinetics of sodium disilicate. *J. Non-Crystalline Solids* **64**, 225–228.

Fontana E.H. and Plummer W.A. (1966) A study of viscosity-temperature relationships in the GeO_2 and SiO_2 systems. *Phys. Chem. Glasses* **7**, 139–146.

Fine M.E. (1964) *Introduction to Phase Transformations in Condensed Systems.* Macmillan, New York. 133 pp.

Frenkel J. (1932) Note on a relation between speed of crystallization and viscosity. *Z. Sovjetunion* **1**, 498–500.

Frenken J.W.M. and van Pinxteren H.M. (1994) Surface melting: Dry, slippery, wet and faceted surfaces. *Surf. Sci.* **307–309**, 728–734.

Greenwood J. P. and Hess P. C. (1994a) Superheating and the kinetics of melting: Albite and its melting point (abstract). *EOS* **75**, 370.

Greenwood J. P. and Hess P. C. (1994b) Constraints on flash heating fom melting kinetics (abstract). *Lunar Planet. Sci.* **25**, 471–472.

Greenwood J.P. and Hess P.C. (1995) Kinetics of melting and dissolution: Applications to chondrules (abstract). *Lunar Planet. Sci.* **26**, 505–506.

Grossman J.N. (1988) Formation of chondrules. In *Meteorites and the Early Solar System* (eds. J.F. Kerridge and M.S. Matthews), pp. 680–696. University of Arizona Press, Tucson.

Grossman J.N., Rubin A.E., Nagahara H. and King E.A. (1988) Properties of chondrules. In *Meteorites and the Early Solar System* (eds. J.F. Kerridge and M.S. Matthews), pp. 619–659. University of Arizona Press, Tucson.

Hemingway B.S., Evans, Jr. H.T., Nord, Jr. G.L., Haselton, Jr. H.T., Robie, R.A. and McGee J.J. (1986) Åkermanite: Phase transitions in heat capacity and thermal expansion, and revised thermodynamic data. *Can. Mineral.* **24**, 425–434.

Hewins R.H. (1988) Experimental studies of chondrules. In *Meteorites and the Early Solar System* (eds. J.F. Kerridge and M.S. Matthews), pp. 660–679. University of Arizona Press, Tucson.

Hewins R. H. (1991a) Retention of sodium during chondrule melting. *Geochim. Cosmochim. Acta* **55**, 935–942.

Hewins R. H. (1991b) Condensation and the mineral assemblages of chondrule precursors (abstract). *Lunar Planet. Sci.* **22**, 567–568.

Horanyi M., Morfill G., Goertz C.K. and Levy E.H. (1995) Chondrule formation in lightning discharges. *Icarus* **114**, 174–185.

Kingery W.D., Bowen H.K. and Uhlmann D.R. (1976) *Introduction to Ceramics.* Wiley, New York. 1032 pp.

Kress V.C. and Ghiorso M.S. (1995) Multicomponent diffusion in basaltic melts. *Geochim. Cosmochim. Acta* **59**, 313–324.

Kubicki J.D. and Lasaga A.C. (1992) Ab initio molecular dynamics simulations of melting in forsterite and $MgSiO_3$ perovskite. *Am. J. Sci.* **292**, 153–183.

Kuo L. -C. and Kirkpatrick R. J. (1985) Kinetics of crystal dissolution in the system diopside-forsterite-silica. *Am. J. Sci.* **285**, 51–90.

Lindemann F.A. (1910) Uber die berechnung molekularer eigenfrequenzen. *Phys. Z.* **11**, 609–612.

Meiling G.S. and Uhlmann D.R. (1967) Crystallization and melting kinetics of sodium disilicate. *Phys. Chem. Glasses* **8**, 62–68.

Moss R. and Harrowell P. (1994) Dynamic Monte Carlo simulations of freezing and melting at the 100 and 111 surfaces of the simple cubic phase in the face-centered-cubic lattice gas. *J. Chem. Phys.* **100**, 7630–7639.

Nagahara H. (1981) Evidence for secondary origin of chondrules. *Nature* **292**, 135–136.

Poirier J.P. (1986) Dislocation-mediated melting of iron and the temperature of the Earth's core. *Geophys. J. R. Astr. Soc.* **85**, 315–328.

Rambaldi E. R. (1981) Relict grains in chondrules. *Nature* **293**, 558–561.

Richet P. and Bottinga Y. (1986) Thermochemical properties of silicate glasses and liquids: A review. *Rev. Geophys.* **24**, 1–25.

Richet P., Leclerc F. and Benoist L. (1993) Melting of forsterite and spinel, with implications for the glass transition of Mg_2SiO_4 liquid. *Geophys. Res. Lett.* **20**, 1675–1678.

Scherer G., Vergano. P.J. and Uhlmann D.R. (1970) A study of quartz melting. *Phys. Chem. Glasses* **8**, 53–58.

Sharma S.K., Virgo D. and Kushiro I. (1979) Relationship between density, viscosity, and structure of GeO$_2$ melts at low and high pressures. *J. Non-crystalline Solids* **33**, 235–248.

Shimuzu N. and Kushiro I. (1984) Diffusivity of oxygen in jadeite and diopside melts at high pressures. *Geochim. Cosmochim. Acta* **48**, 1295–1303.

Spaepen F. and Turnbull D. (1982) Crystallization processes. In *Laser Annealing of Semiconductors* (eds. J.M. Poate and J.W. Mayer), pp. 15–42. Academic Press, New York.

Stein D. J. and Spera F. J. (1993) Experimental rheometry of melts and supercooled liquids in the system NaAlSiO$_4$-SiO$_2$: Implications for structure and dynamics. *Am Mineral.* **78**, 710–723.

Tamman G. (1925) *The States of Aggregation.* D. van Nostrand Company, New York. 297 pp.

Teraoka Y. (1993) Surface melting and superheating. *Surf. Sci.* **294**, 273–283.

Tsuchiyama A. and Takahashi E. (1983) Melting kinetics of plagioclase feldspar. *Contr. Min. Petr.* **84**, 345–354.

Tsuchiyama A., Nagahara H. and Kushiro I. (1981) Volatilization of sodium from silicate melt spheres and its application to the formation of chondrules. *Geochim. Cosmochim. Acta* **45**, 1357–1367.

Uhlmann D.R. (1980) On the internal nucleation of melting. *J. Non-Crystalline Solids* **41**, 347–357.

Uhlmann D.R. (1971) Crystallization and melting in glass-forming systems. In *Advances in Nucleation and Crystallization in Glasses* (eds. L.L. Hench and S.W. Freiman), pp. 172–197. Am. Ceram. Soc. Spec. Pub.

Urbain G., Bottinga Y. and Richet P. (1982) Viscosity of liquid silica, silicates, and aluminosilicates. *Geochim. Cosmochim. Acta* **46**, 1061–1072.

Vergano P. J. and Uhlmann D. R. (1970) Melting kinetics of germanium dioxide. *Phys. Chem. Glass.* **11**, 39–45.

Wagstaff F. E. (1969) Crystallization and melting kinetics of cristobalite. *J. Am. Ceram. Soc.* **52**, 650–654.

Watson E. B. and Baker D. R (1991) Chemical diffusion in magmas: An overview of experimental results and geochemical applications. In *Physical Chemistry of Magmas, Advances in Geochemistry Volume 9* (eds. L. L. Perchuck and I. Kushiro), pp. 120–151.

Wilson H.A. (1900) On the velocity of solidification and viscosity of supercooled liquids. *Phil. Mag.* **50**, 238–250.

23: Sodium and Sulfur in Chondrules: Heating Time and Cooling Curves

YANG YU[1], ROGER H. HEWINS[1] and BRIGITTE ZANDA[2]

[1]Department of Geological Sciences, Rutgers University, Piscataway, NJ 08855, U.S.A. [2]Muséum National d'Histoire Naturelle, 61 rue Buffon, 75005 Paris, and Institut d'Astrophysique Spatiale, 91405 Orsay, France.

ABSTRACT

New chondrule simulations have been performed using short duration heating and curved cooling paths. The heating time used is about one minute and the most rapid cooling path began at about 5000°C/hr and declined to about 500°C/hr. The resulting charges had normal chondrule textures, including porphyritic with olivine zoned from Fo_{84-90} to Fo_{60-75} very like zoned olivine in Semarkona type II chondrules. Na loss rates are higher for higher peak temperatures, lower cooling rates and lower oxygen fugacities. About 95% of the Na is retained in a relatively FeO-rich composition (type IIAB) for initial cooling rates about 5000°C/hr and oxygen fugacity near the Fe-FeO buffer. Such conditions could explain the high Na contents of type II chondrules. The more magnesian and less viscous type IA composition loses much more Na than type IIAB under the same conditions, despite being less completely melted. Some sulfide can be preserved provided that the initial cooling rate is at least 5000°C/hr or provided that the initial temperature is lower than the liquidus of the chondrule materials and the oxygen fugacity is near Fe-FeO. The restriction of primary FeS in type I chondrules in Renazzo and Semarkona to the less melted ones, with granular and dark-zoned textures, suggests that only these experienced peak temperatures under 1550°C. Considering textures, olivine zoning, and Na and S abundances, it appears that many chondrules could have experienced heating for about a minute or even less, at temperatures of 1550°C or over, with initial cooling rate about 5000°C/hr and oxygen fugacities (whether internally or externally buffered) at Fe-FeO or a few log units below.

INTRODUCTION

Experimental simulations of chondrules have been used in attempts to define both chondrule heating and cooling conditions and hence throw light on the heating mechanism and formation environment. Dynamic crystallization experiments performed in the last 10 to 20 years (e.g., Tsuchiyama and Nagahara, 1981; Planner and Keil, 1982; Lofgren, 1989; and Radomsky and Hewins, 1990) have been quite successful in reproducing the textures of natural chondrules, which can form under a wide range of conditions. The maximum heating temperature could range from approximately 1600°C to 1800°C (Hewins and Radomsky, 1990). The heating times used in these experiments range from seconds to hours. The cooling rates used in most of these experiments were linear, except in some performed by Tsuchiyama *et al.* (1980), and they varied from <100°C/hr to >1000°C/hr (Hewins, 1988). In nat-

ural cases, however, it is unclear whether chondrules were actually formed over such a large range of conditions, and the cooling process most likely followed a non-linear path.

Except for a few studies such as Lewis *et al.* (1993), Connolly *et al.* (1994), and Zanda *et al.* (1995), little emphasis was given in earlier experimental studies to the nature of chondrule precursors and their modification during melting, e.g. conditions permitting reduction of iron or loss of moderately volatile elements. Moderately volatile elements (for definition, see Palme *et al.*, 1988), such as Na, are less depleted in chondrules than expected (Grossman and Wasson, 1983, Grossman, 1988; Hewins, 1991; Jones 1994), considering the high temperature of chondrule-forming events, and Grossman (1988) therefore speculated that the linear cooling rates determined in experiments might be too low by 1 or 2 orders of magnitude. This may mean that the heating process was very brief, and that the subsequent cooling had a high initial cooling rate to minimize volatile loss, tapering off to lower values during crystallization of olivine rims. A schematic Stefan-Boltzman cooling curve of this kind was presented for chondrules by Tsuchiyama *et al.* (1981), and a similar one is shown for the case of gas drag heating by Scott *et al.* (this volume). If experiments with such thermal histories were to reproduce chondrule properties, it could provide a firm basis for the model that chondrules were formed from fractionated (Grossman, 1988; Hewins, 1991), as opposed to unfractionated precursors (Huang *et al.*, 1994).

This paper attempts to address these issues by presenting the Na and S loss data obtained from high temperature dynamic crystallization experiments. The purpose of these experiments is to put additional constraints on the range of initial temperatures and non-linear cooling profiles by reproducing satisfactory abundances of Na and S, as well as textures and olivine compositional zoning which resemble those of natural chondrules. The results suggest the chondrule heating-cooling event was somewhat shorter than inferred from previous experimental studies.

DYNAMIC CRYSTALLIZATION EXPERIMENTS

Except for some isothermal heating/quenching experiments performed between 1300°C and 1530°C, most of the dynamic crystallization experiments were designed to simulate flash heating/cooling models: the sample was inserted into the hot furnace and immediately cooled as soon as the desired temperature was reached. A thermocouple was placed 2 mm above the sample, and the total heating time needed for the thermocouple to register the desired temperature was approximately one minute. Since independent testing shows that it takes essentially the same time for both a bare thermocouple and a thermocouple enclosed in the charge to register the run temperature, the time needed for a sample interior to be heated up should be less than one minute.

The cooling of the charge was non-linear, meaning the initial cooling rate was high, decaying near the solidus to values consistent with those of crystallization experiments employing linear cooling rates. These were derived (D. Sahagian, personal communication) to be consistent with the experimentally determined linear cooling rates at near-solidus temperatures where the last of the zoned olivine grew, but to be more realistic for the complete natural cooling process. Fig. 1 shows some of the typical heating and cooling profiles employed in our experiments. The maximum heating temperatures range from 1470°C to 1750°C. The maximum cooling rate was always at the initial stage of cooling, and it ranged from > 5000°C/hr to about 500°C/hr for different cooling paths. The charges were quenched in water at about 1000°C. When approaching the quenching temperature, the cooling rates of different cooling profiles were between 250°C/hr and 500°C/hr.

To observe the textures produced under extremely fast cooling rates, some of the charges were not cooled in the furnace but taken out of the furnace immediately after they reached the desired temperature and cooled in the air all the way to room temperature (air quenching). Such a heating/quenching cycle was repeated for some

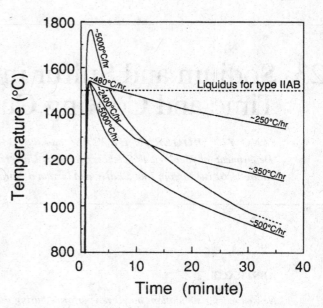

Fig. 1. Typical sample thermal histories employed in the experiments. The cooling of the charges followed different non-linear profiles. The initial and ending cooling rates are shown on each cooling curve.

of the charges 10 to 100 times at temperatures about 100°C lower than the sample liquidus.

The samples include relatively FeO-rich type IIAB, and MgO-rich type I and IA chondrule analog compositions (Radomsky and Hewins, 1990; Jones, 1994), and some of them were doped with 5% of sulfide minerals (pyrrhotite or troilite) in order to observe the S-loss behavior. The starting material consisted of powders with a grain size range of approximately 40 μm to 100 μm. The effects due to variations in precursor grain size were not investigated. The type IIAB experiments were repeated using glassy and crystalline materials. After the heating, the charges are spherical in shape with a diameter of about 3 mm. The fO_2 of the furnace gas was controlled at 0.5 to 4 log units below Fe-FeO buffer. The other detailed experimental conditions and procedures can be found in Yu and Hewins (1994) and Yu *et al.* (1994).

TEXTURES AND OLIVINE ZONING

Depending on the composition of the starting materials and the initial temperature, the charges from our experiments exhibit four different kinds of textures: total glass, barred olivine (BO), porphyritic olivine (PO), and relict olivine grains with overgrowths (only in type I composition charges). In general, with a maximum temperature above the sample liquidus, the charge will develop BO or glass textures, depending on the initial cooling rate. With a faster cooling rate, and less time for melting which continues during initial cooling, the charge is more likely to develop BO texture, as observed by Osborn (1993). PO and relict olivine grain textures can only form with a maximum temperature lower than the sample liquidus. The dependence of texture on initial temperature agrees well with previous linear cooling experiments (Tsuchiyama and Nagahara, 1981; Lofgren, 1989; and Radomsky and Hewins,

1990), though with very rapid melting and cooling, PO textures may be formed for initial temperatures a little above the liquidus (Connolly *et al.*, 1991; Hewins and Connolly, this volume).

For the type IIAB chondrule analog composition, olivine grains in the charges with PO texture are euhedral, and all of them develop noticeable compositional zoning. The olivine compositions are always Mg-rich in the core (Fo_{84-90}) and relatively Fe-rich in the rim (Fo_{60-75}). Normal zoning profiles are also observed for Cr, Mn, and Ca in olivines. Fig. 2 shows some of the FeO zoning profiles observed in olivine phenocrysts with different initial cooling rates. There is a small amount of crystal settling in the charge and some variation from grain to grain perhaps due to late nucleation. However, in general, the zoning profile is steeper for a higher initial cooling rate. The olivine composition and grain size of our charges are similar to those of Semarkona chondrules, as reported by Jones (1990) and Jones and Lofgren (1993). Our olivine zoning profiles are also strikingly similar to those in Semarkona chondrules and in their experimental charges, even though our initial cooling rates are far higher (see Yu and Hewins, 1995).

The flash heating/air quenching experiments achieved extremely fast cooling rates. An initial cooling rate of approximately 100°C/sec was recorded from the thermocouple reading, and the temperature dropped from 1470°C to 1000°C in about 10 seconds. With such high cooling rates, the final charges are composed of glass and numerous small olivine crystals with grain sizes seldom exceeding 20 μm. Repeated heating/quenching cycles at lower temperature coarsen the olivine crystals, but the coarsening process is relatively inefficient: 100 heating/quenching cycles can only double the size of olivine. The final olivine crystals produced through the multiple heating/quenching cycles are somewhat rounded and have curved embayments, and exhibit very limited compositional zonings. The difference in FeO content between the core and the rim of the olivine grain rarely exceeds 5%.

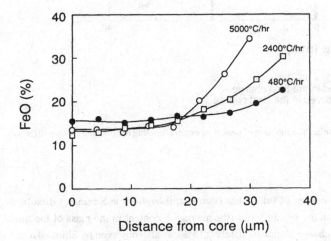

Fig. 2. Representative FeO zoning profiles observed in the olivine phenocrysts in charges with porphyritic olivine texture. The starting material is FeO-rich type IIAB chondrule analog composition. Initial cooling rate is shown for each sample. See Fig. 1 for complete cooling profiles. Higher cooling rate enhances the development of the olivine zoning.

NA LOSS PROFILES

Fig. 3 summarizes the major results of the amount of Na loss from the charges under different experimental conditions.

As in isothermal heating experiments (Tsuchiyama *et al.*, 1981), the Na loss rate of a sample is greatly affected by the peak temperature, heating time, and fO_2. However, since the charges flash heated are immediately cooled, they lose dramatically less Na than the charges isothermally heated, if other conditions are similar. For example, given the same fO_2 of 0.5 log unit below Fe-FeO ($\sim 10^{-10}$ atm at 1500°C) and peak temperature of 1530°C, a type IIAB composition isothermally heated for 10 minutes then quenched lost ~30% of its original Na content, whereas the same composition flash heated to the same peak temperature (total furnace time: 35 minutes) lost 5–25% of its original Na content, depending on the cooling rate.

Given the same peak temperatures, high cooling rate and relatively high fO_2 can dramatically reduce the Na loss from the sample. In an extreme case, with an fO_2 of $\sim 10^{-10}$ atm, a type IIAB chondrule heated to its liquidus temperature and immediately cooled at 2400–5000°C/hr initially (non-linear cooling) can preserve as much as 90–95% of its original Na content. Less Na, e.g., 50–80%, will be retained with lower fO_2 (about 10^{-11}–10^{-14} atm) with an initial cooling rate of 2400°C/hr.

Different starting compositions show similar Na-loss profiles as a function of cooling rate and fO_2, but exhibit a noticeable difference in Na-loss rate. Under similar heating and cooling conditions, type I and especially type IA composition lose Na more easily than type IIAB composition, even though they are less completely melted. Apparently, the presence of abundant relict olivine does not inhibit the escape of Na. Given that all other conditions are similar, type I chondrule composition needs higher cooling rate to preserve a similar fraction of its Na content than type II composition.

No difference in Na loss rate was seen for glassy and crystalline type IIAB starting material, despite the fact that the former was easier to melt.

S LOSS PROFILES

Figs. 4 and 5 summarize the major results of S-loss experiments, where the charges shown were either isothermally or flash heated. In isothermal heating experiments, the samples were kept at the peak temperature (slightly above or below the sample liquidus) for 5 to 30 minutes, and examination of the final charges showed that no sulfide minerals have survived such heating. Under flash heating conditions, on the other hand, some sulfide minerals did survive in charges that were either cooled extremely fast, or were heated to a relatively low maximum temperature.

For type IIAB composition (Fig. 4), with an initial cooling rate of about 5000°C/hr, all charges partially retain sulfide, even with a maximum heating temperature of 250°C above the sample liquidus. When the initial cooling rate drops to < 2500°C/hr, the surviving sulfide minerals can only be found in charges heated to temperatures below the sample liquidus. A similar phenomenon

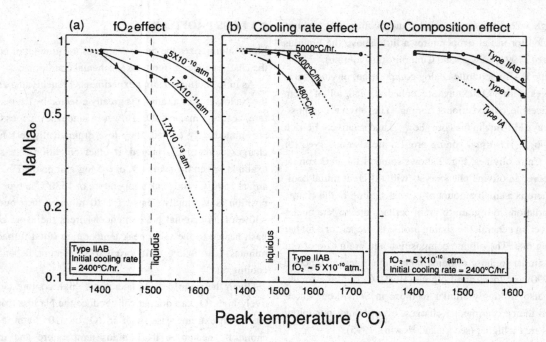

Fig. 3. (a) The Na loss profiles as a function of peak temperature and fO$_2$ for type IIAB chondrule analog composition. Charges were rapidly heated and non-linearly cooled, with an initial cooling rate of 2400°/hr. Na/Na$_o$ is the ratio between the Na content in the charge and that of the starting material. The fO$_2$ for a temperature of 1500°C is shown on each curve. (b) As (a) but Na loss is shown as a function of initial cooling rate. (c) Na loss for three different compositions. Under similar thermal history, the Na loss rate follows the order of type IA > type I > type IIAB.

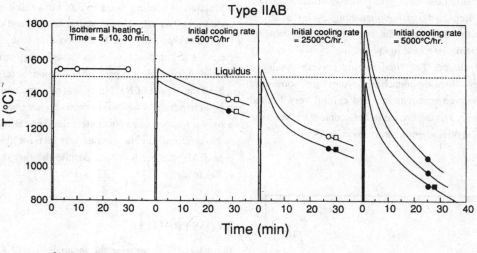

Circle: fO$_2$ = IW-0.5.   Square: fO$_2$ = IW-2.
Filled symbol: Sulfide minerals observed in the final charge.
Open symbol: No sulfide minerals observed in the final charge.

Fig. 4. The results of S-loss experiments for type IIAB chondrule analog composition as a function of peak temperature, cooling history, and oxygen fugacity.

was observed for the type I composition used in the experiment (Fig. 5), though it has a liquidus temperature almost 100°C higher than that for type IIAB composition. fO$_2$ has a similar effect on the preservation of sulfide as that observed for Na, i.e., higher fO$_2$ favors S retention.

The amount of sulfide preserved is always close to, or less than 10% of the original sulfide content, indicating that a significant amount of sulfide has been lost, though some S remains dissolved in the silicate glass (the average S content in the glass of the final charge is approximately 0.15%), as also observed by Shimaoka and Nakamura (1989). The sulfide remnants are evenly distributed inside the charge, and occur as interstitial blebs within the glass between the olivine crystals. Abundant Fe metal is also present, due to the reduction caused by the loss of sulfur.

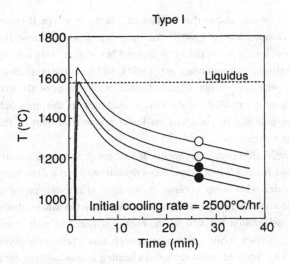

fO₂ = IW-0.5
Filled symbol: Sulfide minerals observed in the final charge.
Open symbol: No sulfide minerals observed in the final charge.

Fig. 5. The results of S-loss experiments for type I chondrule analog composition as a function of peak temperature.

DISCUSSION

Moderately volatile elements in chondrules have been a subject of tremendous interest, partly because of their seemingly paradoxical behavior. Na has a relatively low condensation temperature of about 700°C, or somewhat higher depending on assumptions about the nebular environment (Wood and Hashimoto, 1993). Some previous isothermal experiments performed in the chondrule forming temperature range show extensive Na loss from chondritic materials (e.g., Tsuchiyama *et al.*, 1981; Shimaoka and Nakamura, 1989; Radomsky and Hewins, 1990). However, chondrules are not universally depleted in Na and some (type II chondrules) even have Na contents close to or higher than CI chondrite values (Grossman, 1988; Hewins, 1991), suggesting that in certain types of chondrules, most of the Na might have survived the high temperature chondrule forming events.

Sulfur presents another example. It normally has a lower condensation temperature than Na, ~400°C, though this too depends on assumptions about the composition of the gas phase (Wood and Hashimoto, 1993). Troilite is absent from some chondrules, such as the majority of type I chondrules with textures indicating a relatively high degree of melting in Renazzo and Semarkona. But many chondrules do contain troilite. Some troilite in chondrules may have formed by recondensation onto chondrules in the nebula and by asteroidal metamorphism at temperatures much lower than temperatures of chondrule formation (Zanda *et al.*, 1995). However, in the most highly unequilibrated chondrites, Renazzo and Semarkona, there is clear evidence of troilite and metal with melt textures. Such textures are particularly common inside either type II chondrules with compositions indicating higher oxygen fugacity, or type I with textures indicative of very low degrees of melting (Zanda *et al.*, 1995). These are most probably primary sulfides which survived high temperature heating.

Unique heating processes forming chondrules could explain the retention of moderately volatile elements at very high melting temperatures. Even at 3000°C, it takes about 3 minutes to lose all of the Na content from a melt of albite composition (R. Taylor, personal communication, 1993). This means that evaporating volatiles from chondrules not only needs high temperatures, but also time at such elevated temperatures. Many people believe that the compositional differences among different types of chondrules, including volatile elements, are due to the heterogeneity of the chondrule precursor materials, not the heating events that later formed chondrules (e.g. Grossman and Wasson, 1983, Grossman, 1988; Hewins, 1991; Jones 1994). An essential premise of this model is that the heating event was brief, and a brief heating process can preserve volatile contents. The results of our dynamic crystallization experiments indicate that if the heating was intensive, but brief, and followed by a fast cooling, at least in the initial stage of the cooling process, a significant amount of Na and some sulfide minerals can indeed be preserved in the charge. Therefore, such a premise for the fractionated precursor model is justified.

High initial cooling rates favor Na and sulfide retention. Both Na and S results indicate that cooling curves beginning at 2500 – 5000°C/hr are necessary, unless the ambient gas is very enriched in these elements, to reproduce the very high Na contents (Grossman, 1988; Hewins, 1991) and primary sulfide (Zanda *et al.*, 1995) found in some natural chondrules. These cooling rates are quite high compared to the normally quoted cooling rates for chondrules, which range from 1000 to 100°C/hr or even lower (Lofgren, 1989; Radomsky and Hewins, 1990; Jones and Lofgren, 1993). The basis for previous cooling rate estimates was experimental reproduction of the textures and olivine compositional zoning profiles found in porphyritic chondrules. However, even our charges which experienced the highest cooling rates still reproduce the chondrule texture and olivine zoning profiles. These criteria are difficult to interpret in terms of unique cooling histories, and certainly do not present a strong argument against cooling rates initially as high as several thousand degrees an hour (see also Yu and Hewins, 1995).

Higher fO₂ favors Na retention. Grossman (1988) calculated the fO₂ needed to retain Na in chondrules based on Tsuchiyama's isothermal experimental data, obtained a result that is about five orders of magnitude more oxidizing than the canonical fO₂ value (10^{-16} atm at ~1500°C) and concluded that the rate at which volatiles are lost from chondrules is imperfectly understood. Compared to Grossman's results, our results for fO₂ shown in Fig. 4 are actually about three orders of magnitude lower, which is closer to, but still higher than the canonical nebula value. One might be tempted to argue that type I chondrules have suffered much greater Na loss than in our experiments (e.g. from chondritic initial Na abundance), which would require even lower fO₂. However, at the same time one has to plead for extremely high fO₂ to maintain the high levels of Na in type II chondrules. A higher than canonical nebula fO₂ due to partial evaporation of concentrated dust has been suggested for the environment forming FeO-bearing chondrules (Wood, 1985, Wood and Hashimoto, 1993). Zanda *et al.* (1994) calculated especially high fO₂ for type II chondrules, but Connolly *et al.* (1994) showed that at least under some circumstances the fO_2 can be internally buffered by the precursor minerals. Thus large dust enhancements in the chondrule-forming regions may or may not be necessary, and the sensitivity of Na loss to fO_2 could be informative. Higher fO_2 also appears to keep sulfide minerals in charges, but the details of the gas chemistry are not well understood. It is far from clear that Na and S loss data can be applied to determining nebular fO_2. Even though high fO_2 in the nebula as the result of dust evaporation is possible, we still need further experiments to determine whether volatile loss can be influenced by internally buffered fO_2.

The composition of the starting material also seems to affect Na and sulfide loss rate. A composition effect on Na vaporization has been found in several studies (e.g., Tsuchiyama *et al.*, 1981), with higher loss rates for more SiO_2-poor melts. Since Si-O tends to form tetrahedral networks in the melt (see Mysen, 1988 for review), it has been speculated that the Na vaporization rate has something to do with the degree of polymerization of the melt (Tsuchiyama *et al.*, 1981). We have observed a slightly higher Na loss rate in type I chondrule composition than that in type IIAB composition, and there seems to be a negative correlation of the degree of polymerization with the Na loss rate. But whether this is the sole reason for the composition effect is unclear. The amount of sulfide preserved in the charge, on the other hand, seems to be related to the sample liquidus, i.e., type I chondrule composition (higher liquidus) preserves its sulfide component to a higher temperature than type II composition (with lower liquidus). This differs from what we observed for Na. The vaporization of Na and sulfide from the charge seems to follow somewhat different mechanisms which are not very well understood at present. Nevertheless, although we have yet to perform experiments in which Na loss from type IA composition is trivial, our results show that in natural cases, Na contents of type IA chondrules must have been somewhat higher before melting, but might not be as high as those of type II chondrules. Our sulfide loss data also imply that a type I chondrule now lacking FeS might have been heated to a temperature above 1550°C, or have experienced cooling rates below at least 2500°C/hr (Fig. 5).

An important difference between our experiments and many previous experimental studies is that we used a non-linear cooling process for our samples, because it is more realistic compared to natural cases (see calculations by Sahagian and Hewins, 1992). The Na and S loss data can be used to constrain especially the initial cooling rate near the liquidus, because that is where Na and S are lost most easily. The studies on exsolution lamellae in clinopyroxene and plagioclase by Weinbruch and Müller (1995) suggest a cooling rate of ~10°C/hr at the temperature range of 1200 – 1300°C, where our cooling curves are in the range of several hundred °C/hr. It is unclear why the two approaches differ so much. One possibility is that the cooling rates are in fact measured on different kinds of objects: glass-free, granular type I chondrules in Allende (Weinbruch and Müller, 1995) and type IIA chondrules for olivine zonation studies. It is also unclear how such a low cooling rate relates to the higher temperature history of the objects. One of

our experiments shows that isothermal heating of a type II chondrule composition at 1300°C for three hours would cause the charge to lose at least 20% of its original Na content. This implies that continuous slow cooling near 1300°C still faces the problem of losing volatile elements. Clearly further comparisons of the two techniques are needed, on the same chondrules, as this approach has the potential to constrain the lower temperature part of the cooling curves.

Though additional experiments at low gas pressures are desirable, our Na and S loss experiments demonstrate that a flash heating model with steep cooling curves can effectively preserve volatile contents in type II chondrules. Based on the same volatile loss consideration but also on the energy required to melt chondrules, Wasson (this volume) proposes that chondrules were formed by repeated small scale flash heating events, and for each event, the cooling rate was on the order of several hundred degrees per second. With such a cooling rate at least two orders of magnitude higher than those employed in our controlled cooling experiments, it can be speculated that the volatile loss from chondrules should be even less. However, such a high cooling rate will inevitably inhibit the production of relatively large crystals observed in natural chondrules, as demonstrated by our flash heating/air quenching experiments. If one claims that repeated heating/extremely fast cooling cycles can coarsen the olivine grains, then hundreds of heating/quenching cycles are needed to grow olivine to the comparable sizes in natural chondrules, and the heating temperature has always to be lower than the sample liquidus. It is difficult to imagine any heating mechanism in the nebula that heated chondrules hundreds of times with peak temperatures always constrained within a narrow window. Furthermore, the repeated melting tends to produce more rounded olivine crystals than those found in chondrules, the olivines exhibit very limited compositional zonings, and the extent of volatile loss due to the multiple heating events is unclear. Therefore, the initial cooling rates for chondrules were probably in the range of several thousand degrees per hour, so as to both retain volatile elements and reproduce the chondrule textures.

Of course flash heating is not the only mechanism to retain volatile elements. Some previous experiments have shown that under certain conditions, Na can be preserved even if the sample experienced prolonged isothermal heating. Lewis *et al.* (1993) show that Na loss from chondrules can be prevented by high Na partial pressure in the nebula. They suggest that the Na vapor came from the evaporation of dust within a clump during the early stage of chondrule formation. It certainly cannot be ruled out that locally, due to the heterogeneity of the nebula, such a mechanism might play a role. However, the Na vapor generated in such a way could only be self buffering with the chondrules that formed from the same dust clump, and it would be hard to generate as high a Na partial pressure as used in their experiments. Greenwood and Hess (this volume) calculated that the relict grains observed in chondrules would have dissolved with a prolonged heating event. Therefore, the simplest explanation for volatile retention is flash heating with a less important role played by the composition of the gas phase.

CONCLUSIONS

The results of our volatile loss experiments have shown that short heating time plus high initial cooling rates (2400–5000°C/hr) and high oxygen fugacity can retain abundant Na in type II chondrules. Some sulfide can also be preserved under similar heating conditions, or if the peak heating temperature is lower than the liquidus of the chondrule materials. Therefore, flash heating (duration 1 minute or less) and rapid initial cooling is a plausible heating mechanism. The initial cooling rates are probably on the order of several thousand degrees per hour in order to reproduce chondrule volatile contents, textures and olivine zoning. Type I chondrules lacking FeS either were heated above 1550°C or had initial cooling rates lower than 2500°C/hr.

ACKNOWLEDGEMENTS

We thank L. C. Patino, M. J. Carr, and J. S. Delaney for their help in DCP-AES and electron microprobe analyses, and H. C. Connolly, Jr. for performing some of the furnace work at JSC-experimental petrology lab. We are indebted to D. Sahagian and M. Bourot-Denise for insights into chondrule cooling processes and the occurrence of troilite in chondrules, respectively. The manuscript was critically reviewed by R. H. Jones, H. Nagahara, and D. W. G. Sears. This study was supported by NASA (OSS NAGW-2263 and PMG NAGW-3391).

REFERENCES

Connolly H. C., Jr., Jones B. D. and Hewins R. H. (1991) The effect of precursor grain size on chondrule texture. *Meteoritics* **26**, 329.

Connolly H. C., Jr., Hewins R. H., Ash R.D., Zanda B., Lofgren G. E., and Bourot-Denise M. (1994) Carbon and the formation of reduced chondrules: an experimental investigation. *Nature*, **371**, 136–139.

Grossman J. N. (1988) Formation of chondrules. In *Meteorites and the Early Solar System* (eds. J. F. Kerridge and M. S. Matthews), pp. 680–696. University of Arizona Press.

Grossman J. N. and Wasson J. T. (1983) The compositions of chondrules in unequilibrated chondrites: An evaluation of models for the formation of chondrules and their precursor materials. In *Chondrules and Their Origins* (ed. E. A. King), pp. 88–121. Lunar and Planet. Inst., Houston.

Hewins R. H. (1988) Experimental studies of chondrules. In *Meteorites and the Early Solar System* (eds. J.F. Kerridge and M.S. Matthews), pp. 73–101. University of Arizona Press.

Hewins R. H. (1991) Retention of sodium during chondrule melting. *Geochim. Cosmochim. Acta* **55**, 935–942.

Hewins R. H. and Radomsky P. M. (1990) Temperature conditions for chondrule formation. *Meteoritics* **25**, 309–318.

Huang S., Benoit P. H., and Sears D. W. G. (1994) Group A5 chondrules in ordinary chondrites: their formation and metamorphism. *Lunar Planet. Sci.*, **25**, 573–574.

Jones R. H. (1990) Petrology and mineralogy of Type II, FeO-rich chondrules in Semarkona (LL3.0): Origin by closed-system fractional crystallization, with evidence for supercooling. *Geochim. Cosmochim. Acta* **54**, 1785–1802.

Jones R. H. (1994) Petrology of FeO-poor, porphyritic pyroxene chondrules in the Semarkona chondrite. *Geochim. Cosmochim. Acta* **58**, 5325–5340.

Jones R. H. and Lofgren G. E. (1993) A comparison of FeO-rich, porphyritic olivine chondrules in unequilibrated chondrites and experimental analogues. *Meteoritics* **28**, 213–221.

Lewis R. D., Lofgren G. E., Franzen H. F., and Windom K. E. (1993) The effect of Na vapor on the Na content of chondrules. *Meteoritics* **28**, 622–628.

Lofgren G. (1989) Dynamic crystallization of chondrule melts of porphyritic olivine composition: Textures experimental and natural. *Geochim. Cosmochim. Acta* **53**, 461–470.

Mysen B. O. (1988) *Structure and Properties of Silicate Melts*. Elsevier Science Publishing Co.

Osborn T. G. (1993) Non-linear cooling profiles and the effect on type II synthetic chondrules. B.A. thesis. Rutgers University.

Palme H., Larimer J. W., and Lipschutz M. E. (1988) Moderately volatile elements. In *Meteorites and the Early Solar System* (eds. J. F. Kerridge and M. S. Matthews), pp. 436–461. University of Arizona Press.

Planner H. N. and Keil K. (1982) Evidence for the three-stage cooling history of olivine-porphyritic fluid droplet chondrules. *Geochim. Cosmochim. Acta* **46**, 317–330.

Radomsky P. M. and Hewins R. H. (1990) Formation conditions of pyroxene-olivine and magnesian olivine chondrules. *Geochim. Cosmochim. Acta* **54**, 3475–3490.

Sahagian D. L. and Hewins R. H. (1992) The size of chondrule - forming events. *Lunar Planet. Sci.* **23**, 1197–1198.

Shimaoka T. and Nakamura N. (1989) Vaporization of sodium from a partially molten chondritic material. *Proc. NIPR Symp. Antact. Meteorites* **2**, 252–267.

Tsuchiyama A. and Nagahara H. (1981) Effects of precooling thermal history and cooling rate on the texture of chondrules: A preliminary report. *Mem. Natl. Inst. Polar Res.*, Spec. Issue, **20**, 175–192.

Tsuchiyama A, Nagahara H., and Kushiro I. (1980) Experimental reproduction of textures of chondrules. *Earth Planet. Sci. Lett.* **48**, 155–165.

Tsuchiyama A, Nagahara H., and Kushiro I. (1981) Volatilization of sodium from silicate melt spheres and its application to the formation of chondrules. *Geochim. Cosmochim. Acta* **45**, 1357–1367.

Weinbruch S. and Müller W. F. (1995) Constraints on the cooling rates of chondrules from the microstructure of clinopyroxene and plagioclase. *Geochim. Cosmochim. Acta* **59**, 3321–3230.

Wood J. A. (1985) Meteoritic constraints on processes in the solar nebula. In *Protostars and Planets II* (eds. D. C. Black and M. S. Matthews), pp. 687–702. Univ. Arizona Press.

Wood J. A. and Hashimoto A. (1993) Mineral equilibrium in fractionated nebular systems. *Geochim. Cosmochim. Acta* **57**, 2377–2388.

Yu, Y. and Hewins, R. H. (1994) Retention of sodium under transient heating conditions - experiments and their implications for the chondrule forming environment. *Lunar Planet. Sci.* **XXV**, 1535–1536.

Yu, Y. and Hewins, R. H. (1995) Is non-linear, rapid cooling plausible for chondrule formation? Evidence from olivine zoning profiles. *Lunar Planet. Sci.* **XXVI**, 1545–1546.

Yu, Y., Hewins, R. H., Zanda, B., and Connolly, H. C. (1994) Can sulfide minerals survive the chondrule-forming transient heating event? *Lunar Planet. Sci.* **XXV**, 1537–1538.

Zanda B., Bourot-Denise M., Perron C., and Hewins R. H. (1994) Origin and metamorphic redistribution of silicon, chromium and phosphorus in the metal of chondrules. *Science*, **26577**, 1846–1849.

Zanda B., Yu Y., Bourot-Denise M., Hewins R. H., and Connolly H. C., Jr. (1995) Sulfur behavior in chondrule formation and metamorphism. *Geochim. Cosmochim. Acta* (submitted).

24: Open-System Behaviour During Chondrule Formation

DEREK W. G. SEARS, SHAOXIONG HUANG and PAUL H. BENOIT

Cosmochemistry Group, Department of Chemistry and Biochemistry, University of Arkansas, Fayetteville, AR 72701, U.S.A.

ABSTRACT

The question of whether chondrules acted as open or closed systems during formation is important in understanding chondrule history and the differences in bulk composition of the chondrite classes. "Open-system" behaviour assumes that relatively volatile elements (like Fe, Na and K) were lost and the chondrules underwent chemical reactions with species in the environment during formation, while "closed-system" behaviour assumes that the various properties of the chondrules were inherited entirely from the precursors. More than 90% of the chondrules in unmetamorphosed chondrites are either groups A1 and A2, which are low-FeO and refractory chondrules, or group B1 with lithophile element ratios resembling CI proportions and high-FeO silicates. The refractory composition, low-FeO silicates, relatively high metal abundance and low-Ni content of the metal of group A1,2 chondrules are consistent with open-system behaviour. A closed-system scenario would require that chondrule precursors were the products of earlier volatility-oxidation processes. The trend in the olivine to pyroxene ratio (which decreases from group B1 to A2 and then increases again in group A1), the smaller mean size of group A1 and A2 chondrules compared to group B1 chondrules, the relationships between oxygen isotope composition and chondrule size and peak temperature, diffusion of Na into chondrules, and the greater abundance of thick fine-grained rims around group A chondrules relative to group B chondrules are consistent with major evaporative loss, first of FeO and later SiO_2, accompanying the formation of groups A1 and A2. These properties are difficult or impossible to understand in terms of closed-system behaviour. It is concluded that while group A1 and A2 chondrules formed by reduction of FeO and major evaporative loss from precursors originally resembling those of CI chondrites, evaporative loss from group B1 chondrules was restricted to only the most highly volatile trace elements like Ga, Sb, Se and Zn. Open-system behaviour was clearly very important during chondrule formation.

INTRODUCTION

Whether or not chondrules formed as open systems has been strongly debated for many decades (*e.g.* Wai and Wasson, 1977; Anders, 1977). The question is of fundamental significance in understanding chondrule formation processes and conditions. The question is also relevant to the bulk compositions and the origin of the chondrite classes, although this topic will not be dealt with here. We will restrict our discussion largely to chondrules from ordinary chondrites, noting only that chondrules in various chondrite classes are qualitatively similar, although the relative abun-

dance of the various chondrule groups may vary (Scott and Taylor, 1983; Dodd, 1981; Sears *et al.*, 1992; 1993), being predominantly FeO-poor olivine in carbonaceous, almost entirely FeO-poor pyroxene in enstatite chondrites and predominantly FeO-rich in ordinary chondrites.

THE DIVERSITY OF CHONDRULE COMPOSITIONS

Before discussing whether or not the system was open during chondrule formation, it is necessary to review the considerable diversity of chondrule properties. Our thesis is that by explaining some of the differences among chondrule groups, we can understand important aspects of how each of the chondrule groups formed.

Chondrules show a very wide range in composition, mineralogy and texture which has been summarized in the form of a number of classification schemes (Kieffer, 1975; Dodd, 1981; Gooding and Keil, 1981; McSween, 1977; Scott and Taylor, 1983; Jones, 1994). The scheme we prefer is based on the changes in the composition of the two major structural components in the chondrules; the silicate minerals (usually olivine, sometimes pyroxene) and the glassy mesostasis enclosing them (Sears *et al.*, 1992; DeHart *et al.*, 1992). These properties are determined either by electron-microprobe analysis or cathodoluminescence petrography or preferably a combination of these techniques. Unlike virtually all previous classification schemes, texture is not included but treated as an independent variable. (For a discussion of these points see Sears *et al.*, 1995). The main (and somewhat approximate) technical details of the chondrule groups and some notes on how these groups compare with previously published schemes appear in Table 1.

Some of the chondrule groups thus defined are "primary" in the sense that they only occur in essentially unmetamorphosed meteorites (A1, A2, B1 and a few A5 chondrules), while others are the result of metamorphism (A3, A4, B2, B3 and most of the A5 chondrules). Semarkona (LL3.0) is the only well-known essentially unmetamorphosed (*i.e.* type 3.0) ordinary chondrite observed fall. It is possible that the Yamato 74660 (LL) and Lewis Cliff 86134 (L) chondrites are also type 3.0 (Sears and Hasan, 1987; Sears *et al.*, 1991), but these are weathered finds and data are meagre. Unpublished TEM data suggest that Yamato 74660 resembles the Krymka LL3.1 chondrite (C. Alexander, per. comm, 1995). The percentages (by number) of each chondrule group in Semarkona are given in Table 1. In comparison, chondrules in the primitive CM carbonaceous chondrites are largely (*i.e.* ~70% by number) group A1, while virtually all the chondrules in enstatite chondrites are group A2.

CHONDRULE FORMATION: CLOSED OR OPEN-SYSTEM?

Fig. 1 is an attempt to summarize the two situations for chondrule formation that are discussed here. By "closed-system" we mean that volatiles were not lost during chondrule formation, neither did the chondrule components chemically interact with the ambient gases. This would mean that the chondrule groups are the result of mixing of precursor solids in various combinations. The advocates of chondrule formation as closed (or essentially closed) systems include, Wasson and Chou (1974), Gooding *et al.* (1983), Grossman and Wasson (1983a,b), Ikeda (1983), Rubin and Wasson (1986, 1988), Grossman (1988), Misawa and Nakamura (1988), Jones and Scott (1989), Hewins (1991a), Jones (1994). The term "open-system" refers to cases where volatiles were lost and chemical reactions occurred between chondrule components and the ambient gases. In this case, the chondrule groups are the result of processes occurring during the chondrule-forming events.

Table 1. *The chondrule groups in terms of cathodoluminescence, mineral compositions and frequency of occurrence in the Semarkona ordinary chondrite**

	Mesostases		Olivine			Freq[‡]
	CL	Composition[†]	CL	%FeO	%CaO	
A1[§]	yellow	Pl(An> 50%)	red	< 2	> 0.17	10.5
A2[#]	yellow	Pl(An> 50%)	none/dull red	2-4	0.1–0.2	25.0
A3**	blue	Pl(An> 50%)	red	< 4	> 0.2	0.0
A4**	blue	Pl(An> 50%)	none/dull red	> 4	0.16–0.3	0.0
A5[@]	blue	Pl(An< 50%)	none	> 4	< 0.25	5.0
B1[+]	none	> 30% Qtz	none	7-25	0.08–0.3	56.9
B2**	none	30–50% Qtz	none/dull red	10–25	0.08–0.3	0.0
B3**	purple	15–30% Qtz	none	15–20	< 0.08	2.6

* Phase compositions given as a rough guide since the group fields are not rectangular (see Sears *et al.*, 1995, for details). 'CL' refers to cathodoluminescence color. Electron microprobe data has the advantage of being quantitative and detecting differences missed by CL, while CL data has the advantage of being more rapid and sometimes easier to apply when textures are fine-grained and intergrown.
† Normative composition (wt%) of the mesostasis: Pl, plagioclase; An, anorthite; Qtz, quartz.
‡ Frequency (as percent by number) for 76 chondrules (DeHart *et al.*, 1992).
§ Includes some of the droplet chondrules of Kieffer (1975), some of the non-porphyritic pyroxene chondrules of Gooding and Keil (1981), the type I chondrules of McSween (1977), metal-rich microporhyritic chondrules of Dodd (1978), and the type IA chondrules of Scott and Taylor (1983).
# Includes the poikilitic pyroxene and type IB chondrules of Scott and Taylor (1983) and many of the type IAB chondrules of Jones (1994).
@ There appear to be no previous observations of this chondrule group in unmetamorphosed meteorites.
+ Dodd's (1981) "lithic" or "clastic" chondrules and Dodd's (1978) metal-poor microporphyritic chondrules are included in this group, as are the type II chondrules of McSween (1977), Scott and Taylor (1983) and Jones (1990).
** Group A3, A4, B2 and B3 chondrules are only observed in significant amounts in metamorphosed chondrites and are assumed to be the result of metamorphic alteration of the primitive group A1, A2, A5 and B1 chondrules. Common paths of evolution are A1 → A3 → A4 → A5; A2 → A5; A5 → A5; B1 → B2 → B3 → A5.

Advocates of chondrule formation as open-systems in which evaporation occurred include Larimer and Anders (1967), Walter and Dodd (1972), Osborn *et al.* (1974), Kurat *et al.*, (1983), Clayton *et al.* (1985, 1991), Nagahara *et al.* (1989; 1994), Matsuda *et al.* (1990), Lu *et al.* (1990), Huang *et al.* (1993a), Matsunami *et al.* (1993), Nakamura (1993), Scott (1994) and Alexander (this volume). Others have argued for open-system processes involving condensation (*e.g.* McSween, 1977), while Dodd has argued for closed-system behaviour for some chondrules and open-system behaviour for others (Dodd and Walter, 1972; Dodd, 1978).

Throughout the remainder of this paper we will summarize the properties of primitive chondrules, and some of the relevant laboratory experiments that have been performed, in an effort to choose between these options.

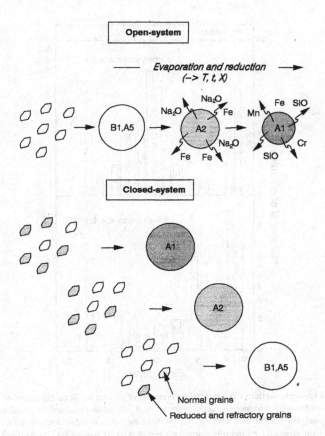

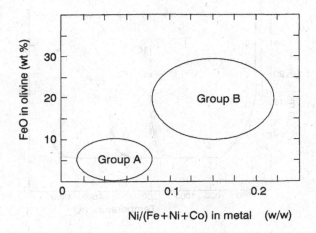

Fig. 2. A schematic summary of the FeO content of the olivine and the Ni content of the metal of group A (A1 and A2) chondrules and group B (essentially B1) chondrules. Original data from Snellenburg (1978), Jones and Scott (1989) and Lu (1992).

Fig. 1. Schematic representation of the open- and closed-system alternatives for chondrule formation. Open-system theories assume reasonably unfractionated precursor solids that are heated to produce group B1 and A5 chondrules with relatively little mass loss, while group A chondrules lost considerable mass during their formation. During the formation of group B and A5 chondrules only the most highly volatile elements (In, Tl, Bi etc) were lost or suffered modest redistribution. Reduction of FeO followed by the loss of Fe and Na_2O produced group A2 chondrules with about 40 wt % mass-loss. Further loss, this time involving elements slightly less volatile than Fe (like Mn and Si) produced the group A1 chondrules. Closed-system theories assume that all chondrules were the result of closed-system melting but that various properties of the chondrules were the result of mixing various precursors in different proportions. The extent of volatile element loss and FeO reduction would have depended on temperature, heating time and composition of the chondrule and surrounding gases during chondrule formation.

CHONDRULE PROPERTIES

1. Chondrules display a very wide range of $P(O_2)$

Group A chondrules contain silicates which are low in FeO (McSween *et al.* 1983; Dodd, 1978; Snellenburg, 1978; Scott and Taylor, 1983; Jones and Scott, 1989; Jones, 1994; DeHart *et al.*, 1992), and metal grains which are low in Ni (Snellenburg, 1978; Jones and Scott, 1989; Lu, 1992), relative to group B. Fig. 2 summarizes the ranges observed. (Unfortunately, there are few chondrules for which both metal and silicate compositions have been determined). The interiors of group A chondrules usually contain more metal than group B chondrules (Huang *et al.*, 1993b; Zanda *et al.*, 1995). These data indicate that the group A chondrules are reduced relative to group B chondrules.

The reduced state of group A chondrules compared with group B chondrules is consistent with both open and closed system

behaviour. In the case of an open-system one would argue that group A chondrules were reduced during their formation, while in the case of a closed-system one would argue that the precursors of group A chondrules were more reduced. It is difficult to discuss meaningfully the likelihood of reduced precursors, since very little is known about chondrule precursors. If we use equilibrium thermodynamics as a guide, it seems unlikely that metal as Ni-poor as that observed in the group A chondrules could be the product of nebular processes. Fig. 3, taken from the calculations of Sears (1978), which resemble those of Kelly and Larimer (1977) in their main features, shows while metal is <u>enriched</u> in Ni under all pressure and temperature conditions (≥6%, which is the value calculated from cosmic abundances), the Ni-poor metal requires the presence of a phase like schreibersite ($(Fe,Ni)_3P$) which might preferentially remove Ni. This process requires considerably more reducing conditions than cosmic (Sears, 1980; Wood, 1985; Wood and Hashimoto, 1993), whereas chondrules formed in an environment more oxidizing than cosmic (see below). On the other hand, Ni-free metal would inevitably result from the production of metallic Fe by the reduction of silicates which would form low-Ni solid solutions with existing metal.

2. Chondrules display $P(O_2)$-related patterns in their mineralogy

There are some significant, and surprisingly little-discussed, relationships between the mineralogy of chondrules and the composition of their silicate grains. The olivine to pyroxene ratio in group B1 chondrules is high, but is low in group A2 chondrules and intermediate in group A1 chondrules. Fig. 4 shows two histograms of the fayalite contents (Fe-rich olivine, Fe_2SiO_4) for chondrules from Semarkona, one for chondrules with an olivine to pyroxene volume ratio >9 and one where this ratio is <9. The chondrules whose olivine to pyroxene volume ratio is >9 display a bimodal pattern with a broad peak at around a mole fayalite content of 10% (Fa_{10}) and a narrow peak at Fa < 2 mol%, while chondrules with an

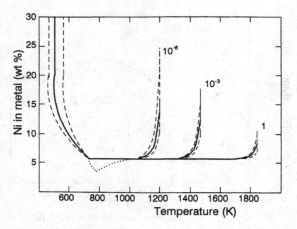

Fig. 3. The weight percent Ni in the metal phase as a function of temperature for a system with cosmic composition for total pressures of 10^{-6}, 10^{-3} and 1 atmosphere. The broken lines indicate the effects of uncertainties in the thermodynamic data. The dotted line shows the trajectory of the path in the presence of schreibersite which can accommodate some of the Ni. Curve from Sears (1978).

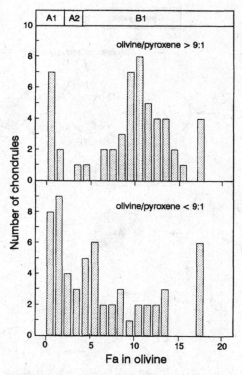

Fig. 4. Histograms of measured Fa contents of olivine for Semarkona chondrules with olivine/pyroxene ratio >9 and <9, as determined by modal analysis. Chondrules with olivine >90% show a bimodal distribution in Fa, with group B1 chondrules forming a broad peak at around Fa_{10} and group A1 chondrules forming a much narrower peak at $Fa_{<2}$. Chondrules with <90% olivine have a very different Fa histogram, with values of Fa between 2 and 4 now being occupied by the group A2 chondrules. The broad Fa_{10} peak of group B1 chondrules is missing. The trends between ol:px and Fa are expected if reduction of the FeO in the olivines during chondrule formation was accompanied first by evaporation of Fe (which lowers the olivine/pyroxene ratio) and then evaporation of Si (which increases the olivine/pyroxene ratio). Data are from Snellenberg (1978).

olivine to pyroxene ratio of <9 show a somewhat broader peak at Fa <2 (mainly due to chondrules with Fa 1–2 mol%) and perhaps a peak at about 5 mol%. The data shown in Fig. 4 are from Snellenburg (1978) but the same features are observed in data from other studies (*e.g.* DeHart *et al.*, 1992; Huang *et al.*, 1995; Sears *et al.*, 1995b). (We have no explanation for the Fa_{17} peaks in Snellenburg's data, except to note that such high values are observed in other data bases only when analysis is made close to the edge of a zoned olivine grain.)

In a closed-system scenario, one would have to propose some *ad hoc* explanation for the data in Fig. 4. For example, one would have to propose that chondrules we see in the meteorites are an incomplete sampling of the true chondrule populations and that olivine-rich chondrules with ~Fa_4 and pyroxene-rich chondrules with Fa_{10} had just been inadequately sampled. In view of the large number of chondrules examined, it would require further *ad hoc* assumptions to explain this fractionation of chondrules. Hewins (1991b) explained the mineralogical variations in terms of complex patterns of condensation, but this does not explain the size differences (that are discussed below), nor the abundant metal and sulfide in group A chondrules. Open-system chondrule formation provides a perfectly straight forward way to explain the trend between mineralogy, Fa content and size. Fig. 5 is a summary of the vaporization experiments of Hashimoto (1983). Many other authors have reported similar data (Gooding and Muenow, 1977; King, 1982; 1983; Mysen and Kushiro, 1988; Nagahara *et al.*, 1989; 1994). Thus, if a mixture of oxides approximately chondritic in composition is heated in the laboratory, loss of Fe first decreases the (Fe+Mg)/Si ratio (and lowers the olivine/pyroxene ratio) and then loss of Si begins to restore the value. A mixture originally fairly rich in normative-olivine (i.e. calculated olivine, (Fe+Mg)/Si ~ 2.0) becomes pyroxene-normative ((Fe+Mg)/Si ~ 1.0) after about 40 wt % of the charge is evaporated in the Hashimoto (1983) experiments (Fig. 5). The relevant reaction is probably of the form:

$$2Fe_2SiO_{4(l)} + Mg_2SiO_{4(l)} + 5H_{2(g)}$$
$$\rightarrow 2MgSiO_{3(l)} + 4Fe_{(g)} + SiO_{(g)} + 5H_2O_{(g)}.$$

Continued heating causes further loss of Si by evaporation, with relatively little loss in the less volatile Mg, and as evaporation exceeds about 40 wt% olivine abundances begin to increase again. This reaction is of the form:

$$2MgSiO_{3(l)} + H_{2(g)} \rightarrow Mg_2SiO_{4(l)} + SiO_{(g)} + H_2O_{(g)}.$$

The data of Hashimoto (1983) were obtained in a non-cosmic $P(O_2)$ but it seems clear from thermodynamic calculations (Larimer, 1967; Grossman, 1972; Sears, 1980; Wood and Hashimoto, 1993) that qualitatively similar results would be observed in cosmic and a wide variety of non-cosmic conditions. One would therefore expect the olivine to pyroxene ratio to vary as a function of Fa content if chondrules formed by a process involving various degrees of reduction and evaporation of material of approximately similar starting compositions. We have assumed that the reducing agent might be nebular H_2, but carbon is another

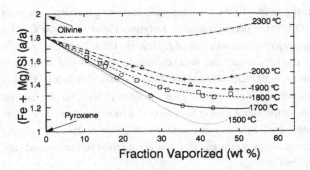

Fig. 5. The normative olivine/pyroxene ratio of solids of initially chondritic composition as a function of mass evaporated during laboratory heating experiments. Loss of Fe and partial loss of Si as the total mass evaporated approaches 30–40 wt % causes the residues to assume pyroxene compositions. Thereafter, loss of Si would cause the residues to become more olivine-like in composition. Data from Hashimoto (1983). If the starting compositions were more similar to group B chondrules than the chondritic compositions used by Hashimoto, then the starting value of (Fe + Mg)/Si would be lower and the initial straight portion of the curve would be shorter. Hashimoto (1983) obtained the curves at 1500 and 2200°C by extrapolation.

possibility, as recently discussed by Connolly et al. (1994).

The composition of the mesostasis surrounding the silicate grains also shows significant variations, some related to evaporative loss (DeHart et al., 1992). The group A chondrules, being depleted in volatile elements, contain mesostases which are refractory in nature being low in K and Na and high in Ca. Group B chondrules contain mesostasis which is not only enriched in Na and K compared to group A, but higher in SiO_2. However, the high SiO_2 reflects the history of the chondrule immediately following its formation because olivine, the first silicate to crystallize from the melt, was somehow prevented from reacting with the SiO_2 in the glass to produce pyroxene, as required by equilibrium. Apparently, group B chondrules cooled more rapidly than group A chondrules and underwent considerable supercooling (DeHart, 1989; Jones, 1990).

3. Chondrules display a range of CI-normalized abundance trends

The bulk compositions of chondrules are notoriously difficult to determine. Defocussed beam electron microprobe methods are intrinsically difficult because of the complex textures of chondrules and the two-dimensional nature of the method. Whole-chondrule bulk analyses are fairly straight forward (using neutron activation, isotope dilution or other methods) but freeing the chondrules from the meteorite is not a simple procedure. Disaggregating a sample and physically removing the fragile chondrules from their tough host material produces biases towards large strong chondrules, while chiselling the chondrules from polished surfaces avoids this bias but may result in sampling large amounts of rim material. There is no ideal way of physically separating chondrules. Nevertheless, despite differences (mainly in the relative abundance of the various chondrule populations represented and the tendency for electron-microprobe analysis towards high Na

and K values) there is reasonable agreement in the results of the various methods. Fig. 6 shows the results that were obtained by INAA. Group B1 (and A5) chondrules have flat lithophile elemental abundance patterns when normalized to CI chondrites, "lithophile" elements being those normally associated with silicates. On the other hand, groups A1 and A2 show significant depletions in volatile and moderately volatile lithophile elements like Si (not shown in Fig. 6 but accurately determined by electron microprobe), Mg, Cr, Mn, Na and K. The siderophiles and chalcophiles (elements normally associated with the metallic and sulfide phases, respectively) behave similarly in both groups A and B, showing major volatility-related depletions.

Since chondrules are mixtures of discrete phases, each with characteristic properties (volatile-rich or -poor, reduced or oxidized etc), it is possible to mathematically reproduce these compositional trends by mixing components of the requisite properties. Components of the requisite properties, to a good approximation, are also the products of the theoretical condensation sequence thermodynamically predicted for a gas of cosmic composition. However, such exercises are not particularly instructive because, as pointed out by several authors, condensation and evaporation are thermodynamically equivalent processes.

Fig. 7 summarizes the Tsuchiyama et al. (1981) data for evaporative loss of Na from chondritic material. Tsuchiyama et al. (1981) and Hashimoto (1983; Fig. 5) used different conditions for their experiments so direct comparison should be made with caution, but the results should be qualitatively applicable to a wide variety of conditions. Sodium- and Fe-loss is fairly rapid, followed by Si-loss. Sodium-loss is also dependent on oxygen fugacity, with

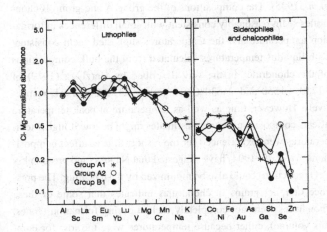

Fig. 6. The composition of chondrules from the Semarkona ordinary chondrite as determined by instrumental neutron activation analysis plotted as abundances ratioed to CI and Mg. The elements are divided into two groups on the basis of their chemical properties and are plotted in order of increasing volatility. Huang et al. (1995) data are shown, but the Grossman and Wasson (1983b; 1985) and Swindle et al. (1991) INAA data yield very similar results, as do defocussed beam electron microprobe analyses (Scott and Taylor, 1983; Jones and Scott, 1989; Jones, 1990; 1994a). Group A1 and A2 chondrules show volatility-controlled elemental abundance patterns for lithophile and siderophile/chalcophile elements, while group B1 chondrules show flat elemental abundance patterns for lithophile elements. The siderophile and chalcophile elements are depleted in both groups A and B also display volatility-related depletion patterns.

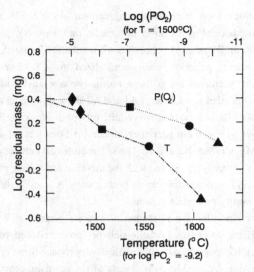

Fig. 7. The loss of Na from charges of initially chondritic composition (about 2.5 wt %) as a function of temperature and oxygen fugacity for 100-minute heating experiments. The vertical axis refers to the logarithm of the mass of Na (as wt % oxide) remaining in the charge after heating. The Na sometimes shows more than an order of magnitude loss, depending on temperature and oxygen fugacity. Data from Tsuchiyama *et al.* (1981) from which further details can be obtained.

higher $P(O_2)$ values favoring retention of Na.

Concurrent with evaporation, these experiments and thermodynamic calculations also predict reduction of Fe during chondrule formation in an open system. Fig. 8 compares the Na-loss predicted by the Tsuchiyama *et al.* data with the Fa content of the olivine predicted by thermodynamic equilibrium (Lu, 1992; Huang *et al.*, 1995). The compositions of the group A and group B chondrules seem reasonably consistent with small differences in formation temperature and the temperatures suggested seem consistent with liquidus temperatures calculated from the bulk compositions of the chondrules in the way described by Herzberg (1979) of ~1710, ~1550 and ~1630°C for groups A1, A2 and B1, respectively. However, time as well as temperature at peak temperature affects composition, so these estimates might be low if higher temperatures were experienced for too a short time to affect compositions. Yu *et al.* (1994) have suggested that Na loss (and presumably FeO reduction) could also be minimized by flash heating. The presence of relict grains in chondrules indicates that some of these chondrules were not completely melted during formation (Jones, this volume), either because temperatures were too low for complete fusion or because times spent at peak temperature were very short. In view of the facility with which Na is lost from experimental melts under a wide variety of conditions, we think it most likely that group A and B chondrules were formed in environments in which oxygen fugacities were several orders of magnitude higher than predicted by cosmic compositions (Wood, 1985).

Cooling rates differ considerably with chondrule class. Group A chondrules cooled relatively slowly and maintained a degree of equilibrium between melt and phenocrysts, while group B chondrules cooled rapidly and underwent considerable supercooling

(Huang *et al.*, 1995). Cooling rates seem to generally vary from ~100°C/h for porphyritic (e.g. Lofgren, 1989) to ~1000°C/h for non-porphyritic chondrules (e.g. Hewins, 1988; Lofgren, 1989). It has also been argued that some chondrules experienced multi-stage cooling, ~300°C/h above 1300°C and ~3000°C/h below 1300°C (Planner and Keil, 1982). That the chondrules with the highest peak temperatures during formation also have the slowest post-formation cooling rates is consistent with group A chondrules being surrounded by warm gas and dust during their formation and subsequent cooling.

There is no *a priori* reason to suspect that volatile alkali elements should directly correlate with Fa since the thermodynamics and kinetics of the two processes are very different. However, there is evidence that the barred-olivine chondrules (which arguably suffered highest temperatures during chondrule formation) do show such a correlation. Matsuda *et al.* (1990) found that Allende BO chondrules showed a positive correlation between FeO and potassium which they interpreted as being due to simultaneous loss of volatiles and reduction of FeO.

4. The chondrule groups show statistically significant differences in size.

In Fig. 9 we show histograms of chondrule sizes for the Semarkona and Krymka chondrites. Group A3 and B2 chondrules from Krymka are also included since the small level of metamorphism suffered by Krymka should not have significantly affected these data. For both chondrites, the group A chondrules are significantly

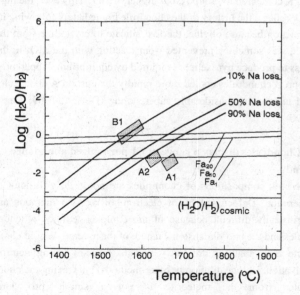

Fig. 8. Evaporation temperatures of Na-bearing phases and Fe reduction/oxidation lines for olivine (for fayalite values of 1, 10 and 20 mol%) as a function of oxygen fugacity (expressed as H_2O/H_2 ratio). The cosmic H_2O/H_2 ratio is shown for reference. Under cosmic conditions there would be considerable loss of Na and reduction of FeO. The boxes show the fields in which conditions are suitable for the formation of group A1, A2 and B1 chondrules. Na data are from Tsuchiyama *et al.* (1981), thermodynamic data for the fayalite calculation are from Robie *et al.* (1979) and the JANAF tables, and the cosmic abundance data are from Anders and Grevesse (1989).

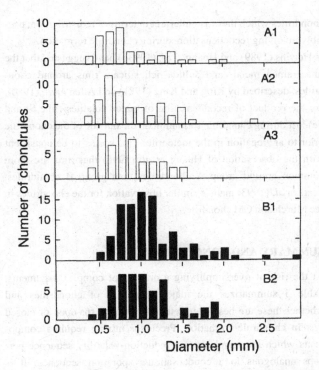

Fig. 9. Histogram of chondrule sizes as a function of chondrule group for Semarkona and Krymka. The small amount of metamorphism suffered by Krymka is not expected to affect these data. Chondrule diameters (averages of two perpendicular measurements) have been corrected for thin sectioning effects which tend to systematically underestimate diameters (Hughes, 1978). The mean diameter for group A chondrules is less than that of group B chondrules by 15–20% for both Semarkona and Krymka. Data from Huang *et al.* (1995).

Table 2. *Reduction and evaporation of major elements (wt%) to give the observed size difference between group A and B chondrules**

	B1¶ Obs.	Step 1† Calc.	Step 2‡ Calc.	A1§ Calc.	A1¶ Obs.
SiO$_2$	50	50	25	49	45
MgO	34	34	29	47	43
FeO	12	0.6	0.6	1.0	2.1
Al$_2$O$_3$	1.5	1.5	1.5	2.4	5.3
CaO	1.1	1.1	1.1	1.8	3.3
Fe (metal)	0.0	0.44	0.44	0.71	3.4
Diameter (mm)	1.05	—	—	0.89	0.81
Mass loss (wt%)	—	11	27	38	—

* Relative losses are estimated from the experimental data of Hashimoto (1983) assuming 1700°C heating and 54% total mass loss, which is sufficient to partially restore the olivine/pyroxene ratio. A 54% mass loss in Hashimoto's (1983) experiments corresponds to about a 40% in chondritic material because of the lower bulk FeO in the latter.
† Step 1: 95% of FeO is reduced, then 95% of the reduced Fe is evaporated.
‡ Step 2: 40% of SiO$_2$, and 15% of MgO, is evaporated.
§ Previous column recalculated to 100% being the calculated group A1 composition.
¶ Observed average composition.

smaller than group B chondrules by 15–20 percent. This translates to a difference in mass of about ~50 wt%. The mean sizes (in millimeters) we calculate for the Semarkona chondrules in the figure are group A1, 0.81±0.40; A2, 0.84±0.40; A3, 0.76±0.30; B1, 1.05±0.42; B2, 0.97±0.37. (Because the distributions are not normal, these standard deviations are misleadingly large).

Open-system behaviour predicts that the group A chondrules would be smaller than group B chondrules, while it is difficult to understand how closed-system behaviour would have produced such a size difference. Why would chondrule formation involving reduced and volatile-poor precursors preferentially produce smaller chondrules? In fact, making reasonable assumptions about the evaporation of major elements from the chondrules during formation the size difference is also in quantitative agreement with the compositional differences (Table 2). This is despite considerable departures from equilibrium and the possibility of recondensation of lost volatiles. We also think the calculation may be conservative; the low Al$_2$O$_3$ and CaO, high SiO$_2$ and MgO, and large size relative to observed group A chondrules indicates that we have underestimated the amount of evaporative loss.

5. Chondrules did behave as open-systems with respect to O isotope exchange, but the details are unclear.

One of the potentially most informative properties of chondrules is

their oxygen isotopic composition (Fig. 10). Unfortunately, these properties are very poorly understood. Among the Allende chondrules, those whose texture suggests complete melting (the non-porphyritic chondrules) contain isotopically heavier oxygen than those which may not have been completely melted (the porphyritic chondrules). This seems to suggest that the surrounding environment contained heavier oxygen than the chondrules. Conversely, among chondrules from ordinary chondrites there is a correlation between the abundance of isotopically light oxygen and chondrule size. Since smaller chondrules would exchange more completely with environmental gases during formation than large chondrules, it would seem that the environment contained lighter oxygen than the chondrules. (We suspect that the clustering of data for the Chainpur chondrules reflects metamorphic equilibration within this type 3.4 chondrite.) Thus it would appear that the ordinary and carbonaceous chondrule precursors were originally on opposite sides of the terrestrial line but that during chondrule formation they exchanged oxygen with a common gas plotting close to the terrestrial fractionation line. This interpretation was first offered by Clayton *et al.* (1983). More recently, a glass fragment in the ordinary chondrite Mezö-Madaras has been located which contains very heavy oxygen, leading to the suggestion that both carbonaceous chondrites and ordinary chondrites exchanged oxygen with two different reservoirs of isotopically heavier oxygen (Mayeda *et al.*, 1989; Clayton *et al.*, 1991). Silica-rich chondrules and clasts also contain isotopically heavy oxygen (Olsen *et al.*, 1981; Bridges *et al.*, 1993; 1994). Rates of exchange with silicate melts have been measured at 1 atmosphere by Yu *et al.* (1994). Thiemens (1994) has suggested that these trends are the result of chemical fractionation processes accompanying chondrule formation, rather than exchange with environmental gases.

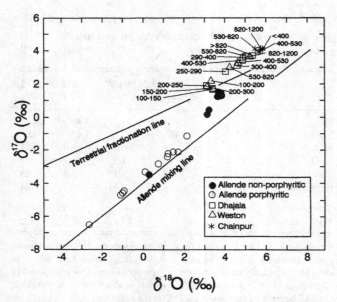

Fig. 10. Three-isotope plot for Dhajala, Weston, Chainpur and Allende chondrules. The values alongside the chondrule data refer to diameters in micrometers. Smaller chondrules from ordinary chondrites (which includes the non-porphyritic chondrules) show a tendency towards lighter oxygen while non-porphyritic chondrules from Allende show a tendency towards heavier oxygen. The two trends seem to be converging near the terrestrial line. Data from Clayton *et al.* (1983, 1991).

While the details are disputed, it seems clear from oxygen isotope properties that the chondrules were open systems during their formation.

6. Recondensation

Several chondrules in Semarkona (LL3.0) show evidence for the recondensation of volatiles prior to aggregation in the meteorite, probably during or immediately following chondrule formation. Compositional zoning in the mesostasis of certain group A chondrules can be measured directly with the electron-microprobe or observed indirectly in cathodoluminescence and thermoluminescence images. Such chondrule mesostasis zoning has been observed in ordinary chondrites (Matsunami *et al.*, 1993) and carbonaceous chondrites (Ikeda and Kimura, 1985). The possible causes of this zoning were discussed at some length by Matsunami *et al.* (1993) who concluded that it was the recondensation of volatiles lost from the chondrule during formation. Certainly, the zoning implies that the hot chondrule was in an environment enriched in the volatile elements like Na. Lewis *et al.* (1993) have demonstrated experimentally that Na can re-enter the heated chondrules very rapidly under conditions where P(Na) is considerably enhanced over cosmic.

Grossman and Wasson (1985), Huang *et al.* (1993b) and Zanda *et al.* (1995) have pointed out that sulfides are less abundant in the interiors of group A chondrules than group B chondrules. Sulfides often occur near the outer edges of group A chondrules or in their rims, suggesting major redistribution of sulfur during chondrule formation rather than complete loss. Sears and Lipschutz (1994) have discussed trace element data for highly labile elements in

chondrules which they also interpret in terms of redistribution, possibly involving recondensation, during chondrule formation.

Hewins (1989) and Huang *et al.* (1993b) have suggested that the fine-grained metal- and sulfide-rich silicate rims around chondrules, described by King and King (1981) and Allen *et al.* (1980), are the product of recondensation of major volatiles (like Si and Fe) lost during chondrule formation on the outside of the chondrule prior to aggregation in the meteorite. This seems to be consistent with the observation of Huang *et al.* (1995) that rims are more common around group A chondrules than group B chondrules. Sears *et al.* (1993) made a similar observation for the chondrules in the Murchison CM chondrite.

SUMMARY AND CONCLUSIONS

At the risk of oversimplifying a number of complex arguments, Table 3 summarizes the major properties of chondrules and whether these are best interpreted as evidence for open or closed system chondrule formation. Precursor mixing requires components which are members of a reduction-volatility sequence perhaps analogous to a condensation/evaporation sequence. It is possible, but not proven, that nebular condensation could have produced such precursors, although not the low-Ni metal observed in group A chondrules. A major problem for the precursor mixing theory is that it does not explain the size and mineralogical differences between group A and group B chondrules.

Open-system behaviour during chondrule formation explains why reduced and volatile-poor chondrules are smaller and why the mineralogy varies as a function of olivine composition. It involves a process known to have occurred (chondrule formation), and associated processes which are to be expected on the basis of laboratory and theoretical data. Open-system behaviour is also consistent with the O isotope data, the compositional zoning sometimes observed in the mesostases and the preferential occurrence of rims on group A chondrules.

Chondrule formation was a very heterogeneous process. While evaporative loss of major elements and reduction of FeO during chondrule formation were the major factors in determining the diversity in chondrule properties, the processes went to varying degrees of completion and there was considerable random heterogeneity in the precursors. It seems clear from the presence of relict grains in chondrules (Rambaldi, 1981; Nagahara, 1981; Jones,

Table 3. *Chondrule properties and open- and closed-system behavior*

Major chondrule properties	Open	Closed
1. Range of P(O$_2$)	Yes	Yes
2. P(O$_2$)-related patterns in mineralogy	Yes	No
3. Elemental abundance trends	Yes	Possibly
4. Differences in size	Yes	Doubtful
5. O isotope trends	Yes	No
6. Recondensation of volatiles	Yes	No

1995) and the restricted range of oxygen isotope compositions (Clayton *et al.*, 1991), that recycling of chondrule fragments also occurred during chondrule formation (Alexander, 1995). While significant progress has been made in identifying the many factors driving chondrule properties, and it is possible to sort some of these factors in order of importance, some of the details will be difficult or impossible to determine because of the complexity of the process. Indeed, because of the stochastic and heterogeneous nature of the processes it may not be possible to even meaningfully address some details. This is the second major aspect of chondrule formation; the process was stochastic, and the precursors heterogeneous, and this surely eliminates some of the many mechanisms for chondrule formation that have been proposed.

ACKNOWLEDGEMENTS

We appreciate the invitation of the organizers of the Chondrules and the Protoplanetary Disk for their invitation to write this paper and Rhian Jones for her considerable efforts in organizing the highly successful meeting. Extremely helpful reviews were provided by Conel Alexander and Roger Hewins, for which we are most grateful. This work was supported by grant NAGW-3519 from NASA.

REFERENCES

Allen J. S., Nozette S. and Wilkening L. L. (1980) A study of chondrule rim and chondrule irradiation records in unequilibrated ordinary chondrites. *Geochim. Cosmochim. Acta* **44**, 1161–1175.

Anders E. (1977) Critique of "Nebular condensation of moderately volatile elements and their abundances in ordinary chondrites" by C. M. Wai and J. T. Wasson. *Earth Planet. Sci. Lett.* **36**, 14–20.

Anders E. and Grevesse N. (1989) Abundance of the elements: Meteoritic and solar. *Geochim. Cosmochim. Acta* **53**, 197–214.

Bridges J. C., Franchi I. A., Hutchison R. and Pillinger C. T. (1993) A new oxygen reservoir? Cristobalite-bearing clasts in Parnallee (abstract). *Meteoritics* **28**, 329–330.

Bridges J. C., Franchi I. A., Hutchison R., Alexander C. M. D. O'D. and Pillinger C. T. (1993) An achondrite clast in Parnallee with possible links to ureilites (abstract). *Meteoritics* **29**, 449–450.

Clayton R. N., Onuma N., Ikeda Y., Mayeda T. K., Hutcheon I. D., Olsen E. J. and Molini-Velsko C. (1983) Oxygen isotopic compositions of chondrules in Allende and ordinary chondrites. In *Chondrules and Their Origins* (ed. E. A. King), pp. 37–43. Lunar and Planetary Inst., Houston.

Clayton R. N., Mayeda T. K. and Molini-Velsko C. A. (1985) Isotopic variations in solar system material: evaporation and condensation of silicates. In *Protostars & Planets II* (eds. D. C. Black and M. S. Mattews), pp. 755–771, Univ. of Arizona Press, Tucson.

Clayton R. N., Mayeda T. K., Goswami J. N. and Olsen E. J. (1991) Oxygen isotope studies of ordinary chondrites. *Geochim. Cosmochim. Acta* **55**, 2317–2337.

Connolly H. S., Hewins R. H., Ash R. D., Zanda B., Lofgren G. E. and Bourot-Denise M. (1994) Carbon and the formation of reduced chondrules. *Nature* **371**, 136–139.

DeHart J. M. (1989) Cathodoluminescence and microprobe studies of the unequilibrated ordinary chondrites. *Ph.D. Thesis*, University of Arkansas, Fayetteville.

DeHart J. M., Lofgren G. E., Lu J., Benoit P. H. and Sears D. W. G. (1992) Chemical and physical studies of chondrites X: Cathodoluminescence studies of metamorphism and nebular processes in type 3 ordinary chondrites. *Geochim. Cosmochim. Acta* **56**, 3791–3807

Dodd R. T. (1978) Compositions of droplet chondrules in the Manych (L3)

chondrite and the origin of chondrules. *Earth Planet. Sci. Lett.*, **40**, 71–82.

Dodd R. T. (1981) *Meteorites: a Petrologic-Chemical Synthesis*. Cambridge Univ. Press, 368pp.

Dodd R. T. and Walter L. S. (1972) Chemical constrains on the origin of chondrules in ordinary chondrites. In *L'Origine du System Solaire* (ed. H. Reeves), pp. 293–300. Centre National de Recherche Scientifique, Paris.

Gooding J. L. and Keil K. (1981) Relative abundances of chondrule primary textural types in ordinary chondrites and their bearing on conditions of chondrule formation. *Meteoritics* **16**, 17–43.

Gooding J. L. and Muenow D. W. (1977) Experimental vaporization of the Holbrook chondrite. *Meteoritics* **12**, 401–408.

Gooding J. L., Mayeda T. K., Clayton R. N. and Fukuoka T. (1983) Oxygen isotopic heterogeneities, their petrological correlations and implications for melt origins of chondrules in unequilibrated ordinary chondrites. *Earth Planet. Sci. Lett.* **65**, 209–224.

Grossman L. (1972) Condensation in the primitive solar nebula. *Geochim. Cosmochim. Acta* **36**, 597–619.

Grossman J. N. (1988) Formation of chondrules. In *Meteorites and the Early Solar System* (eds. J. F. Kerridge and M. S. Matthews) pp. 680–696. University of Arizona Press, Tucson.

Grossman J. N. and Wasson J. T. (1983a) The compositions of chondrules in unequilibrated chondrites: An evaluation of theories for the formation of chondrules and their precursor materials. In *Chondrules and Their Origins* (ed. E. A. King), pp. 88–121. Lunar and Planetary Inst., Houston.

Grossman J. N. and Wasson J. T. (1983b) Refractory precursor components in Semarkona chondrules and the fractionation of refractory elements among chondrites. *Geochim. Cosmochim. Acta* **47**, 759–771.

Grossman J. N. and Wasson J. T. (1985) Origin and history of the metal and sulfide components of chondrules. *Geochim. Cosmochim. Acta* **49**, 925–939.

Hashimoto A. (1983) Evaporation metamorphism in the early solar nebular-evaporation experiments on the melt $FeO-MgO-SiO_2-CaO-Al_2O_3$ and chemical fractions of primitive material. *Geochem. J.* **17**, 111–145.

Herzberg C. T. (1979) The solubility of olivine in basaltic liquids: an ionic model. *Geochim. Cosmochim. Acta* **43**, 1241–1251.

Hewins R. H. (1988) Experimental studies of chondrules. In *Meteorites and the Early Solar System* (eds. J. F. Kerridge and M. S. Matthews), pp. 660–679, Univ. of Arizona Press, Tucson.

Hewins R. H. (1989) The evolution of chondrules. *Proc. NIPR Symp. Antarctic Meteorites* **2**, pp. 200–220. National Institute of Polar Research, Tokyo.

Hewins R. H. (1991a) Retention of sodium during chondrule formation. *Geochim. Cosmochim. Acta* **55**, 935–942.

Hewins R. H. (1991b) Condensation and the mineral assemblages of chondrule presursors (abstract). *Lunar Planet. Sci.* **22**, 567–568.

Huang S, Benoit P. H. and Sears D. W. G. (1993a) The group A3 chondrules of Krymka: Further evidence for major evaporative loss during the formation of chondrules. *Lunar and Planetary Sci.* **24**, 1269–1270.

Huang S, Benoit P. H. and Sears D. W. G. (1993b) Metal and sulfide in Semarkona chondrules and rims: Evidence for reduction, evaporation and recondensation during chondrule formation. *Meteoritics* **28**, 367–368.

Huang S, Lu J., Prinz M., Weisberg M. K., Benoit P. H. and Sears D. W. G. (1995) Chondrules: Their diversity and the role of open-system processes during their formation. *Icarus* (in press).

Hughes D. W. (1978) A disaggregation and thin section analysis of the size and mass distribution of the chondrules in the Bjurbole and Chainpur meteorites. *Earth Planet. Sci. Lett.* **38**, 391–400.

Ikeda Y. (1983) Major element compositions and chemical types of chondrules in unequilibrated E, O, and C chondrites from Antarctica. *Mem. NIPR Spec. Iss.* **30**, 122–145.

Ikeda Y. and Kimura M. (1985) Na-Ca zoning of chondrules in Allende and ALHA-77003 carbonaceous chondrites. *Meteoritics* **20**, 670–671.

Jones R. H. (1990) Petrology and mineralogy of type II, FeO-rich chondrules in Semarkona (LL3.0): Origin by closed system fractional crystallization, with evidence for supercooling. *Geochim. Cosmochim. Acta* **58**, 5325–5340.

Jones R. H. (1994) Petrology of FeO-poor, porphyritic pyroxene chondrules in the Semarkona chondrite. *Geochim. Cosmochim. Acta* **58**, 5325–5340.

Jones R. H. and Scott E. R. D. (1989) Petrology and thermal history of type IA chondrules in the Semarkona (LL3.0) chondrite. *Proc. 19th Lunar Planet. Sci. Conf.*, 523–536.

Kelly W. R. and Larimer J. W. (1977) Chemical fractionations in meteorites - VIII. Iron meteorites and the cosmochemical history of the metal phase. *Geochim. Cosmochim. Acta* **41**, 93–111.

Kieffer S. W. (1975) Droplet chondrules. *Science* **189**, 333–340.

King E. A. (1982) Refractory residues, condensates and chondrules from solar furnace experiments. *Proc. Lunar Planet. Sci. Conf.* **13**, *J. Geophys. Res.* **87**: A429–A434.

King E. A. (1983) Reduction, partial evaporation, and spattering: Possible chemical and physical processes in fluid drop chondrule formation. In *Chondrules and Their Origins* (ed. E. A. King), pp. 180–187. Lunar and Planetary Science Institute, Houston.

King T. V. V. and King E. A. (1981) Accretionary dark rims on unequilibrated chondrites. *Icarus* **48**, 460–472.

Kurat G., Pernicka E., Herrwerth I. and El Goresy A. (1983) Prechondritic fractionation of Chainpur constituents: evidence for strongly reducing conditions in the early solar system. *Meteoritics* **18**, 330–331.

Larimer J. W. (1967) Chemical fractionation in meteorites-I. Condensation of the elements. *Geochim. Cosmochim. Acta* **31**, 1215–1238.

Larimer J. W. and Anders E. (1967) Chemical fractionation in meteorites-II. Abundance patterns and their interpretation. *Geochim. Cosmochim. Acta* **31**, 1239–1270.

Lewis R. D., Lofgren G. E., Franzen H. F. and Windom K. E. (1993) The effect of Na vapor on the Na content of chondrules. *Meteoritics* **28**, 622–628.

Lofgren G. E. (1989) Dynamic crystallization of chondrule melts of porphyritic olivine composition: Textures experimental and natural. *Geochim. Cosmochim. Acta* **53**, 461–470.

Lu J. (1992) Physical and chemical studies of chondrules from type 3 ordinary chondrites. *Ph.D. Thesis*, Univ. Arkansas, Fayetteville.

Lu J., Sears D. W. G., Keck B., Prinz M., Grossman J. N. and Clayton R. N. (1990) Semarkona type I chondrules compared with similar chondrules in other classes. *Lunar Planet. Sci.* **13**, 720–721.

Matsuda H., Nakamura N. and Noda S. (1990) Alkali (Rb/K) abundances in Allende barred-olivine chondrules: Implications for the melting conditions of chondrules. *Meteoritics* **25**, 137.

Matsunami S., Ninagawa K., Nishimura S., Kubono N., Yamamoto I., Kohata M., Wada T., Yamashita Y., Lu J., Sears D. W. G. and Nishimura H. (1993) Thermoluminescence and compositional zoning in the mesostasis of a Semarkona group A1 chondrule and new insights into the chondrule-forming process. *Geochim. Cosmochim. Acta* **57**, 2101–2110.

Mayeda T. K., Clayton R. N. and Sodonis A. (1989) Internal oxygen isotope variations in two unequilibrated chondrites (abstract). *Meteoritics* **23**, 288.

McSween H. Y., Jr. (1977) Chemical and petrographic constraints on the origin of chondrules and inclusions from carbonaceous chondrites. *Geochim. Cosmochim. Acta* **41**, 1843–1860.

McSween H. Y., Jr. and Labotka T. C. (1993) Oxidation during metamorphism of the ordinary chondrites. *Geochim. Cosmochim. Acta* **57**, 1105–1114.

Misawa K. and Nakamura N. (1988) Demonstration of REE fractionation among individual chondrules from Allende (CV3) chondrite. *Geochim. Cosmochim. Acta* **52**, 1699–1710.

Mysen B. O. and Kushiro I. (1988) Condensation, evaporation, melting, and crystallization in the primitive solar nebula: Experimental data in the system $MgO-SiO_2-H_2$ to 1.0×10^{-9} bar and 1870 °C with variable oxygen fugacity. *Amer. Mineral.* **73**, 1–19.

Nagahara H. (1981) Evidence for secondary origin of chondrules. *Nature* **292**, 135–136.

Nagahara H., Kushiro I., Mysen B. O. and Mori H. (1989) Experimental vaporization and condensation of olivine solid solution. *Nature* **331**, 516–518.

Nagahara H., Kushiro I. and Mysen B. O. (1994) Evaporation of olivine: Low pressure phase relations of the olivine system and its implications

for the origin of chondritic components in the solar nebula. *Geochim. Cosmochim. Acta* **58**, 1951–1963.

Nakamura N. (1993) Trace element fractionation during the formation of chondrules. In *Primitive Solar Nebula and the Origin of the Planets*. (ed. H. Oya), pp. 409–425. Terra Scientific Publishing Co., Tokyo.

Olsen E. J., Mayeda T. K. and Clayton R. N. (1981) Cristabolite-pyroxene in an L6 chondrite: implications for metamorphism. *Earth Planet. Sci. Lett.* **56**, 82–22.

Osborn T. W., Warren R. G., Smith R. H., Wakita H., Zellmer D. L. and Schmitt R. A. (1974) Elemental composition of individual chondrules from carbonaceous chondrites, including Allende. *Geochim. Cosmochim. Acta* **38**, 1359–1378.

Planner H. N. and Keil K. (1982) Evidence for the three-stage cooling history of olivine-porphyritic fluid droplet chondrules. *Geochim. Cosmochim. Acta* **46**, 317–330.

Rambaldi E. R. (1981) Relict grains in chondrules. *Nature* **293**, 558–561.

Robie R. A., Hemingway B. S. and Fisher J. R. (1979) Thermodynamic properties of minerals and related substances at 298.15 K and 1 Bar (10^5 Pascals) pressure and at high temperatures. *U.S. Geological Survey Bulletin* **1452**, pp. 456.

Rubin A. E. and Wasson J. T. (1986) Chondrules in the Murray CM2 meteorite and compositional differences between CM-CO and ordinary chondrite chondrules. *Geochim. Cosmochim. Acta* **58**, 1951–1963.

Rubin A. E. and Wasson J. T. (1988) Chondrules and matrix in the Ornans CO3 meteorite: Possible precursor components. *Geochim. Cosmochim. Acta* **52**, 425–432.

Scott E. R. D. (1994) Evaporation and recondensation of volatiles during chondrule formation. *Lunar Planet. Sci.* **25**, 1227–1228.

Scott E. R. D. and Taylor G. J. (1983) Chondrules and other components in C, O, and E chondrites: Similarities in their properties and origins. *Proc. 14th Lunar Planet. Sci. Conf.*; *J. Geophs. Res.* **88**, B275–B286.

Sears D. W. (1978) Condensation and the composition of iron meteorites. *Earth Planet. Sci. Lett.* **41**, 128–138.

Sears D. W. (1980) Formation of E-chondrites and aubrites-A thermodynamic model. *Icarus* **43**, 184–202.

Sears D. W. G. and Hasan F. A. (1987) Type 3 ordinary chondrites: A review. *Surv. in Geophys.* **9**, 43–97.

Sears D. W. G. and Lipschutz M. E. (1994) Evaporation and recondensation during chondrule formation. *Lunar Planet. Sci.* **25**, 1229–1230.

Sears D. W. G., Hasan F. A., Batchelor J. D. and Lu J. (1991) Chemical and physical studies of type 3 chondrites - XI: Metamorphism, pairing, brecciation of ordinary chondrites. *Proc. Lunar Planet. Sci. Conf.* **21**, 493–512.

Sears D. W. G., Lu J., Benoit P. H., DeHart J. M. and Lofgren G. E. (1992) A compositional classification scheme for meteoritic chondrules. *Nature* **357**, 207–211.

Sears D. W. G., Benoit P. H. and Lu J. (1993) Two chondrule groups each with distinctive rims in Murchison recognized by cathodoluminescence. *Meteoritics* **28**, 669–675.

Sears D. W. G., Huang S. and Benoit P. H. (1995a) Chondrule formation, metamorphism, brecciation, an important new primitive chondrule group and the classification of chondrules. *Earth Planet. Sci. Lett.* **131**, 27–39.

Sears D. W. G., Morse A. D., Hutchison R., Guimon R. K., Lu J., Alexander C. M. O'D., Benoit P. H., Wright I., Pillinger C. T., Xie T. and Lipschutz M. E. (1995b) Metamorphism and aqueous alteration in low petrological type ordinary chondrites. *Meteoritics* **30**, 169–181.

Snellenburg J. (1978) A chemical and petrographic study of the chondrules in the unequilibrated ordinary chondrites, Semarkona and Krymka. *Ph.D. thesis*, State Univ. New York.

Swindle T. D., Caffee M. W., Hohenberg C. M., Lindstrom M. M. and Taylor G. J. (1991) Iodine-xenon and other studies of individual Chainpur chondrules. *Geochim. Cosmochim. Acta* **55**, 861–880.

Thiemens M. H. (1994) Chemical production of chondrule oxygen isotopic composition (abstract). *Papers Presented to Chondrules and the Protoplanetary Disk*, LPI Contrib. 844, Lunar and Planetary Institute, Houston, pp. 39–41.

Tsuchiyama A., Nagahara H. and Kushiro I. (1981) Volatilization of sodium from silicate melt spheres and its application to the formation of chondrules. *Geochim. Cosmochim. Acta* **45**, 1357–1367.

Wai C. M. and Wasson J. T. (1977) Nebular condensation of moderately volatile elements and their abundances in ordinary chondrites. *Earth Planet. Sci. Lett.* **36**, 1–13.

Walter L. S. and Dodd R. T. (1972) Evidence for vapor fractionation in the origin of chondrules. *Meteoritics* **7**, 341–352.

Wasson J. T. and Chou C. L. (1974) Fractionation of moderately volatile elements in ordinary chondrites. *Meteoritics* **9**, 69–84.

Wood J. A. (1985) Meteoritic constraints on processes in the solar nebula. In *Protostars and Planets II* (eds. D. C. Black and M. S. Matthews), pp. 682–702. Univ. of Arizona Press, Tucson.

Wood J. A. and Hashimoto A. (1993) Mineral equilibrium in fractionated nebular systems. *Geochim. Cosmochim. Acta* **57**, 2377–2388.

Yu Y., Hewins R. H. and Connolly H. C. (1994) Flash heating is required to minimize sodium losses from chondrules (abstract). *Papers Presented to Chondrules and the Protoplanetary Disk,* LPI Contrib. 844, Lunar and Planetary Institute, Houston, pp. 46–47.

Zanda B., Yu Y., Bourot-Denise M., Hewins R. H. and Connolly H. C. (1995) Sulfur behaviour in chondrule formation and metamorphism. *Geochim. Cosmochim. Acta* (submitted).

25: Recycling and Volatile Loss in Chondrule Formation

CONEL M. O'D. ALEXANDER

Dept. of Terrestrial Magnetism, Carnegie Institution of Washington, 5241 Broad Branch Rd. N.W., Washington, DC 20015, U.S.A.

ABSTRACT

Here a simple model of volatile loss and chondrule recycling is used to demonstrate that a combination of these two processes can, assuming the reasonable conditions summarized below, reproduce the range of refractory and moderately volatile element compositions observed in chondrules. Assuming chondrules were flash heated and subsequently cooled at 1000°C/hr, peak temperatures of 1800–2100 K can explain the observed FeO abundances, the best understood of the major volatile components in chondrules. Although, other processes must also have played a role in establishing FeO contents. Under these conditions some SiO_2 and MgO will also be lost and most chondrule compositions can be reproduced using their vacuum volatilities, suggesting an ambient $pH_2 \leq 10^{-6}$ atm. However, some 20% of chondrules are Al-rich and have higher FeO contents than predicted. To explain them, SiO_2 and MgO would have to have evaporated up to 5 times faster, relative to FeO, than in the vacuum experiments, assuming all chondrules had similar initial compositions. Such enhancements could be achieved if these chondrules either experienced higher peak temperatures and faster cooling rates than assumed or they formed at $pH_2 \geq 10^{-5}$ atm. At present, there is no geochemical reason to prefer either nebula condensate precursors or recycling-volatility, but petrologic evidence for recycled material and volatile enrichments in chondrule rims favor the latter.

INTRODUCTION

Statistical analysis of the compositions of chondrules from ordinary chondrites (OCs) has revealed that the lithophile elements behave as three groups that appear to reflect their predicted volatility, based on thermodynamic equilibrium calculations, during nebular condensation (Grossman and Wasson, 1982, 1983). These groups are: the refractory lithophiles (such as Ca, Al, Sc, the REEs and some V and Mg), the common lithophiles (e.g. Cr, Mn and most Mg and Si) and, lastly, the volatile elements like Na and K. Most researchers have concluded that thermodynamic equilibrium was almost certainly not maintained and, since it is not possible to calculate condensation paths in all but the simplest non-equilibrium systems, the agreement between predictions of lithophile behavior during condensation and the lithophile groups observed in chondrules is, at present, only qualitative.

However, there is scant mineralogical evidence in OCs that classical nebula condensation occurred, and the presence of presolar grains in all chondrite groups (e.g. Huss, 1990) shows that at least some material escaped a nebula-wide high temperature event altogether. Recently, alternatives to condensate precursors have been proposed involving recycling of chondrule fragments (Alexander, 1994) and/or volatile loss during chondrule formation (e.g. Jones, 1990; Huang *et al.*, 1993; Alexander, 1994; Scott, 1994; Sears *et al.*, this volume).

There is good petrologic evidence, in the form of relict grains, for recycling of chondrule material (Jones, 1990, this volume). Relict grains, apparently remnants of an earlier generation of chondrules, are found in at least 15% of OC chondrules (Jones, this volume). Since their survival rate during chondrule formation is unlikely to have been 100% and some relicts are difficult to recognize, this must be a lower limit for the number of chondrules that contain recycled material. The evidence for volatile loss is more equivocal. Nevertheless, the chondrules are depleted in volatiles, such as Na, K and S, relative to the rims and matrix, which is at least consistent with volatile loss from chondrules and recondensation on to rims and matrix. Rims are also enriched in FeO (e.g. Huss *et al.*, 1981; Scott *et al.*, 1984) and SiO_2 (Alexander, 1995) compared to chondrules. FeO and SiO_2 are likely to have been two of the most volatile major constituents during chondrule formation (Hashimoto, 1983).

Simple modeling of recycling, involving random sampling of chondrule phases, can simulate many of the chemical trends seen in the chondrules (Alexander, 1994). This result is not as trivial as it might at first seem and, in fact, recycling alone cannot easily reproduce the observed correlation between Mg and Al (or any other refractory lithophile). In chondrules most of the Mg is in the refractory-poor olivine and low-Ca pyroxene, and most of the Al is in the refractory-rich glass. Random sampling fails to reproduce the Mg-Al correlation because the abundances of these three phases are independent in the sampling simulations, so Mg and Al will also be independent. To overcome this, Alexander (1994) suggested SiO_2 may be preferentially lost by volatilization during chondrule formation, thereby producing a correlation by variably increasing the abundances of the more refractory Mg and Al. Several other authors have also suggested that Si, amongst other elements, was lost during chondrule formation (e.g. Jones, 1990; Huang *et al.*, 1993; Scott, 1994; Sears *et al.*, this volume).

Here a numerical model is developed to determine whether, under reasonable conditions, volatile loss and/or recycling can reproduce the range of chondrule compositions. First, a simple model of volatile loss is derived using the experimental data of Hashimoto (1983). Then, assuming a uniform chondritic precursor composition, the volatile loss model is used to put limits on the conditions chondrules would have to have experienced to explain the chemical trends observed. As will be seen, volatility alone cannot explain the entire distribution of compositions but requires significant chemical diversity in the precursors, which the recycling of heterogeneous chondrule fragments can produce.

For use in developing this model, the bulk chondrule INAA data from six independent studies of chondrules separated from four OCs (Semarkona-LL3.0, Chainpur-LL3.4, Sharps-H3.4 and Tieschitz-H3.6) have been compiled (Gooding, 1979; Grossman and Wasson, 1982, 1983; Rubin and Pernicka, 1989; Swindle *et al.*, 1991a,b). In general, the mean compositions of chondrules from the same meteorite but different studies agree well, but there seems to be a systematic 20% difference in the mean Mg, Al, Ca and V abundances for Semarkona and Chainpur chondrules between the studies of, on the one hand, Grossman and Wasson (1982; 1983) and on the other those of Gooding (1979) and Swindle *et al.* (1991a,b). To allow comparison between these data sets, the Grossman and Wasson Mg-V data were reduced by 20%.

A MODEL FOR VOLATILE LOSS

Hashimoto (1983) measured the evaporation rates of FeO, MgO and SiO_2 from chondritic melts. The experimental charges were always compositionally uniform, implying rapid diffusion in the melts and, therefore, that the evaporation rates were determined by surface reactions. Based on these rates, for a species, i, the mass loss rate (moles/s) in vacuum is given by

$$\frac{\partial m_i}{\partial t} = -4\pi r^2 \cdot f_i \cdot \rho \cdot C_i \cdot e^{-A_i/RT} \tag{1}$$

where r is the chondrule radius, f_i the mole fraction of species i, ρ the density of the melt (moles/unit volume), C_i and A_i are the 'vacuum volatility' constants for species i (Table 1), R is the gas constant and T the temperature (K). It is important to note that eqn. 1 assumes that the fugacity of a given species in the vapor never approaches its equilibrium value. If the fugacity did approach the equilibrium value, evaporation rates would be substantially reduced.

In natural chondrules, the molten metal is immiscible in the silicate melt requiring that they be treated separately. Eqn. 1 must be modified to account for the fraction of the chondrule surface area occupied by the two melts. In the simplest case, where the chondrule is not spinning and the metal is dispersed evenly as small spherules, the fraction of the surface area occupied by each melt is proportional to their volume fractions. The volume fraction of the silicate melt is given by

$$V_{Si} = \frac{m_{Si}/\rho_{Si}}{m_{Si}/\rho_{Si} + m_{Fe}/\rho_{Fe}}, \tag{2}$$

where m and ρ are the masses and the densities of the silicate (Si) and metal (Fe) melts. The volume of the chondrule is simply the sum of the volumes of the two melts, which on rewriting gives

$$r^2 = [\frac{3}{4\pi}(m_{Si}/\rho_{Si} + m_{Fe}/\rho_{Fe})]^{2/3}. \tag{3}$$

Multiplying eqn. 1 by eqn. 2 and substituting in eqn. 3 gives eqn. 4, the new mass loss rate for species i in the silicate melt.

$$\frac{\partial m_{Si,i}}{\partial t} = -4.836 \, m_{Si}(m_{Si}/\rho_{Si} + m_{Fe}/\rho_{Fe})^{-1/3} \tag{4}$$
$$\cdot f_{Si,i} \cdot C_i \cdot e^{-A_i/RT}$$

Chondrules did not form under isothermal conditions. Chondrules are thought to have cooled at rates of between 100°C/hr and 1000°C/hr (e.g. Lofgren, 1989; Jones, 1990; Radomsky and Hewins, 1990; Alexander, 1994). Little is known of the thermal history of chondrules during melting but it is generally thought that they were 'flash heated'. In the absence of well defined thermal histories for chondrules, here it is assumed that

Table 1. *The coefficients for the vacuum volatilities of MgO, SiO_2 and FeO (Hashimoto, 1983)*

Species	C (cm/s)	A (J/mole)
MgO	4.16×10^7	5.17×10^5
SiO_2	6.40×10^8	5.50×10^5
FeO	1.59×10^6	4.06×10^5

they were heated to some peak temperature, To, instantaneously and then cooled linearly with time, t, at a rate dT. Incorporation of this thermal evolution into eqn. 4 gives

$$\frac{\partial m_{Si,i}}{\partial t} = -4.836\, m_{Si}\, (m_{Si} / \rho_{Si} + m_{Fe} / \rho_{Fe})^{-1/3}$$
$$. f_{Si,i}.C_i.e^{-A_i/R(To-t.dT)}. \qquad (5)$$

Eqn. 5 assumes an entirely molten chondrule but the presence of relict grains shows some precursors survived melting. Also, chondrules crystallized during cooling. Most relict grains are Mg-rich, as are the first olivines or pyroxenes that crystallize during cooling. The residual melt will, therefore, be enriched in FeO and SiO$_2$, which will tend to enhance their volatilities. However, since most of the volatile loss occurs at the highest temperatures when most material is molten, the above model should provide a good approximation to the actual chondrule behavior. The vacuum volatility of Fe-metal in a chondritic melt is unknown but it should be about 12 times that of FeO (e.g. Wang *et al.*, 1994). This value is adopted here. Sulfur in FeS is assumed to have been lost very rapidly (Zanda *et al.*, 1995), leaving only metal. In deriving Eqn. 5, it was assumed that diffusion in the melt is rapid compared to evaporation. For Eqn. 5 to apply to the metal, the spherules must be able to communicate with the surface via rapid diffusion through the silicate melt. If they cannot, Fe-metal evaporation will be inhibited.

The final composition of a chondrule was found iteratively by simultaneously solving for each of the four major volatile species (MgO, SiO$_2$, FeO and Fe-metal). The density of the silicate melt was calculated at the beginning of each iteration using the mole fractions and molar densities of each species in the melt. The molar densities were assumed to be comparable to their densities at room temperature. This model is able to reproduce the Na - and Fe-metal-free isothermal experiments of Hashimoto (1983) up to losses of 50–60wt%, and the addition of terms to account for changes in MgO and SiO$_2$ activities at higher losses when all FeO has evaporated (Hashimoto, 1983) allows for good agreement up to about 95wt% loss. The chondrules are rarely FeO-free and the maximum weight losses estimated below are about 50wt%, so the latter activity terms have not been included in eqn. 5.

DELINEATING THE FORMATION CONDITIONS

By comparing the chondrule INAA data with the predictions of the volatile loss model, the conditions individual chondrules would have to have experienced can be estimated. To try and isolate the contribution volatility may have made to chondrule compositions, it will be assumed that all chondrules initially had the composition given in Table 2 and that they cooled at 1000°C/hr. Only final chondrule masses of more than 0.4 mg, the smallest mass measured in the INAA studies, were considered. This leaves only the range of initial temperatures to be determined. Of the two most volatile major species in a chondritic melt, FeO and Fe-metal, the kinetic evaporation behavior of FeO is the better understood and its abundance in chondrules can be used to put limits on the range of

Table 2. *The mean initial chondrule composition (left two columns) used in the simulations is the L-chondrite composition of Wasson and Kallemeyn (1988). It is a reasonable average of the bulk compositions of the four meteorites studied, except for their higher FeO abundances. For this reason, the FeO abundance has been increased but the total Fe content kept constant. Fe$_{met}$ includes all Fe in metal and sulfide. The right three columns give the mean weight fractions of the various components, estimated by least squares fitting of their means to the bulk composition. The sum of weight fractions does not equal unity because components carrying elements like S, C etc. have not been included. The weight fractions were used to develop a simple random sampling model which produced chondrules with component abundances that were fairly uniformly distributed between the ranges shown*

Element	Conc.	Mineral	Mean wtfr.	Range
Na (mg/g)	7.00	Olivine	0.285	0.13–0.46
Mg (mg/g)	149	Low-Ca px.	0.303	0.14–0.47
Al (mg/g)	12.2	High-Ca px.	0.030	0.02–0.04
Si (mg/g)	185	Glass	0.139	0.11–0.17
K (mg/g)	0.83	FeO	0.103	0.07–0.14
Ca (mg/g)	13.1	Fe$_{met}$+Ni	0.102	0.05–0.15
Sc (ppm)	8.60			
Ti (mg/g)	0.63			
V (ppm)	77			
Cr (mg/g)	3.88			
Mn (mg/g)	2.57			
FeO (mg/g)	125			
Fe$_{met}$ (mg/g)	90			
Ni (mg/g)	12			

peak temperatures. Fe-metal spherules may be spun out of the chondrule rather than evaporate or may not be able to efficiently maintain contact with the surface via diffusion.

INAA can only measure total Fe abundances, but the FeO and Fe-metal contents of the chondrules have been estimated, where possible, using total Fe and Mg abundances and the FeO/(FeO+MgO) ratios in the silicate minerals measured by electron microprobe. All Fe not present as FeO is denoted as Fe$_{met}$ and is probably mostly Fe-metal but in many chondrules it will also include some FeS. The FeO/(FeO+MgO) ratios were not analyzed in all chondrules, but almost all had their Ni abundances measured. Ni exhibits a strong preference for Fe-metal and there is a fairly constant Ni/Fe$_{met}$ ratio of about 0.067 in the chondrules. The ratio is apparently independent of whether the chondrules came from LL- or H-chondrites. Assuming a Ni/Fe$_{met}$ ratio of 0.067, the FeO and Fe$_{met}$ contents can be estimated in those chondrules for which Ni was determined but not the FeO/(FeO+MgO) ratio.

Fig. 1 compares the FeO contents of chondrules as a function of mass with predictions for chondrules with initial bulk compositions like that in Table 2, and initial and final masses in the range 1000–0.4 mg. Fig. 1 suggests that, if volatile loss is to explain the range of FeO contents in chondrules, most chondrules would have to have experienced peak temperatures of between 1800 K and

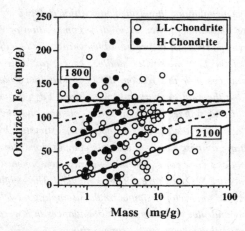

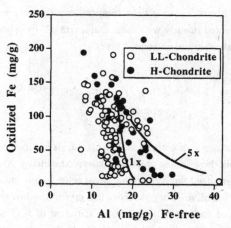

Fig. 1. A comparison of the estimated FeO abundances vs. mass in chondrules analyzed by INAA with the range expected if chondrules experienced peak temperatures of 1800–2100 K and subsequently cooled at 1000°C/hr. The initial composition used to calculate the curves was assumed to be like that in Table 2 and initial masses ranged from 1000–0.4 mg. The comparison suggests that, if volatility played a significant role in the formation of chondrules, 1800–2100 K was approximately the range of peak temperatures they experienced.

Fig. 2. Comparison of estimated FeO abundances vs. Al contents in chondrules analyzed by INAA with two curves calculated assuming chondrules experienced peak temperatures of 1800–2100 K, had initial compositions like that in Table 2 and initial masses in the range 1000–0.4 mg. The curves only show results for chondrules with final masses > 0.4 mg. The two curves used MgO and SiO_2 volatilities that were one (1×) and five (5×) times the measured vacuum volatilities of Hashimoto (1983). The Al abundances are calculated on an Fe-free basis to remove variations due solely to different Fe contents. Many chondrules have somewhat lower Al abundances than predicted due to the assumed initial composition of the simulated chondrules. Nevertheless, most chondrules parallel the 1× curve but those chondrules with higher Al- and Fe-contents apparently experienced enhanced MgO and SiO_2 volatilities.

2100 K. This is almost exactly the range of peak temperatures independently estimated from calculated chondrule liquidus temperatures (Hewins and Radomsky, 1990). Given that the cooling rates used and the peak temperatures required are within the accepted range of chondrule formation conditions, volatile loss of Fe_{met} and FeO from chondrules seems inevitable and, if recondensation occurred, is a reasonable explanation for the Fe enrichment of many chondrule rims.

However, the range of FeO and Fe_{met} abundances in chondrules cannot be explained by volatile loss alone, even if the Fe_{met}/FeO volatility ratio varied by a factor of ten. Some other process, such as precursor heterogeneity or reduction (Connolly *et al.*, 1994; Sears *et al.*, this volume), must have also influenced the FeO and Fe_{met} abundances. While there is no obvious inverse correlation between FeO and Fe_{met} abundances, as a general rule the most Fe_{met}-rich chondrules are also the most FeO-poor, and vice-versa, suggesting reduction has played a role. However, the apparently fairly constant Ni/Fe_{met} ratio argues against reduction since the more FeO that is reduced, the lower the Ni content of the Fe_{met}. Whatever the cause, the abundance of FeO in chondrules cannot be accounted for by volatile loss alone and, therefore, the range of peak temperatures obtained above should only be considered as approximate.

Nevertheless, at peak temperatures of 1800–2100 K Si and Mg will also be lost, causing an increase in refractory element concentrations. Fig. 2 plots the estimated chondrule FeO contents against the Al contents, recalculated on an Fe-free basis (the Fe abundance subtracted and the remainder retotalled to 100%), as well as the predicted curve for a uniform initial composition using the vacuum volatilities (1×). Note that most chondrules have Al contents of 13 mg/g to 18 mg/g (L-chondrite=15.6 mg) and only one has Al contents above 30 mg/g. Consequently, the great majority of chon-

drules cannot have lost more than about 50% of their Fe-free mass.

There is considerable scatter in the chondrule Al contents, much of which is probably due to precursor heterogeneity, but most chondrules more-or-less parallel the 1× curve in Fig. 2. However, those chondrules with the more extreme Al contents also tend to have higher Mg contents (Fig. 3), which recycling cannot easily reproduce. On the other hand as the curves in Fig. 3 show, volatile loss, principally of SiO_2, can produce the correlation but the range of compositions requires that SiO_2 (and to a lesser extent MgO) evaporated from these chondrules at rates that were up to five times faster, relative to FeO, than in the vacuum experiments (Figs. 2 and 3).

In Fig. 4, there appear to be two chondrule populations separated at about Al=19 mg/g. From the calculated curves, most of the Al-rich chondrules, which also have high Mg contents and require the enhanced volatilities, appear to need a rather limited range of peak temperatures (Fig. 4). Although, the lower FeO contents of the H-group chondrules in this population suggest they either had lower initial FeO contents or experienced slightly lower volatility enhancements and higher peak temperatures than the LL-chondrules.

The volatility enhancements of SiO_2 and MgO, relative to FeO, could be achieved in one of two ways. The difference between FeO and SiO_2 volatilities decreases as the temperature increases (Hashimoto, 1983). If peak temperatures occasionally reached 2500 K or higher and cooling rates to the liquidus were very fast (e.g. Yu *et al.*, this volume) it would be possible to produce the Al-rich chondrule compositions. Alternatively, the volatilities of SiO_2 and MgO, whose vaccum volatilities are kinetically inhibited,

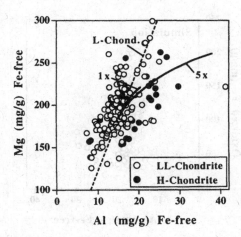

Fig. 3. A plot of Al vs. Mg abundances in chondrules analyzed by INAA. Also shown are the curves chondrules would be expected to follow if they experienced volatile loss during formation. The curves were calculated using the measured vacuum volatility rates for SiO_2 and MgO (1×) and rates that were enhanced by a factor of five (5×). The curves show results for initial masses in the range 1000–0.4 mg and final masses > 0.4 mg. Both curves follow similar paths but the enhanced rates are necessary to produce the more Mg- and Al-rich compositions.

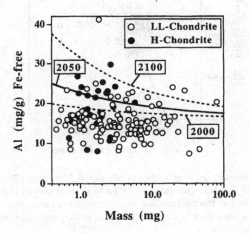

Fig. 4. The variation of Al abundance with mass in chondrules analyzed by INAA. The chondrules appear to have a bimodal distribution of Al contents. The Al-rich chondrules (> 19 mg/g) also tend to have high MgO contents (Fig. 3) and FeO contents (Fig. 2) that suggest enhanced SiO_2 and MgO volatilities of about 5× their measured vacuum rates. From the three curves, calculated assuming volatities that were enhanced by a factor of 5 and initial masses in the range 1000–0.4 mg (final masses > 0.4), the peak temperature range these chondrules experienced appears to have been quite restricted.

increases by almost two orders of magnitude as the pH_2 increases from 10^{-6} atm to 10^{-3} atm (Nagahara, 1993; Nagahara and Ozawa, 1994). The H_2 apparently enables faster surface reaction mechanisms to begin operating. Below $pH_2=10^{-6}$ atm, the volatilities are those of the vacuum experiments. Unlike SiO_2 and MgO, the vacuum volatility of FeO in a chondritic melt is close to the fastest possible evaporation reaction for FeO (e.g. Wang *et al.*, 1994) and, therefore, its evaporation rate should be less sensitive to the pH_2. If this is the case, a five fold increase in the evaporation rates of SiO_2

and MgO would require a pH_2 of about 10^{-5} atm (Nagahara, 1993; Nagahara and Ozawa, 1994) during the formation of the Al-rich chondrules compared to about 10^{-6} atm or less for most chondrules.

COMBINING RANDOM SAMPLING WITH VOLATILE LOSS

In the previous section we have delineated the range of conditions chondrules must have experienced if volatile loss is to at least partially explain the chemical trends in chondrules. That an additional process is needed can be seen in Fig. 3. The slope of the trend in Fig. 3 is similar to the ones predicted by volatile loss, but the scatter in the chondrule compositions is much larger than could be produced by evaporation and/or analytical error. The scatter almost certainly reflects precursor heterogeneity. Below, volatile loss is combined with a simple model of random sampling of chondrule fragments to provide the heterogeneity and the results are compared to the natural chondrules.

INITIAL CHONDRULE COMPOSITIONS

The relative proportions and compositions of the components being sampled are essential to any random sampling model. OCs are primarily composed of chondrules, rims and matrix and all three would presumably have been sampled by chondrules if any recycling occurred. Unfortunately, the typical proportions of these three components are not well determined and attempts to estimate them by least squares fitting of their published mean compositions to bulk OC compositions failed. Chondrules make up approximately 70–80% of an OC and the mean compositions of their silicate components are fairly well known (e.g. Alexander, 1994). In an alternative approach, it was assumed that chondrules are the only components in OCs. Using their mean compositions, the proportions of the four chondrule silicate phases (olivine, low- and high-Ca pyroxene, and glass) were estimated by performing a least squares fit to the bulk Mg, Al, Si and Ca abundances given in Table 2. The estimated proportions are given in Table 2 but the fit underestimates the abundances of Na, FeO and metal/sulfide, because they are enriched in the 10–20% of rim/matrix, and it was necessary to add them separately (Table 2).

The initial proportions of each component in the model chondrules were established by random selection from a set of ranges, and the resulting mean and range for each component are given in Table 2. In most sampling processes the compositional heterogeneity of the samples tend to increase as their mass/size decreases, i.e. the larger the chondrule the more precursor sources it can have. Predicting how the heterogeneity will change with mass requires that the precursor size distributions be known, which they are not. Instead, to try and simulate, in a simple way, a variable number of sources and fragment sizes, the compositions of each of the chondrule components were divided into two: one being the average composition and the other having the composition of an individual analysis chosen at random from the relevant component database.

As before, the mean compositions are the electron microprobe averages given in Table 1 of Alexander (1994). The individual olivine and low-Ca pyroxene compositions were taken from unpublished electron microprobe databases, while the high-Ca pyroxene and glass compositions were taken from the ion probe data of Alexander (1994). The fractional contribution an individual analysis makes to each component being added to a chondrule was selected at random from the range 0–0.5, a low value being akin to there being a large number of sources and a high value equivalent to a single large grain dominating the component.

The individual olivine and low-Ca pyroxene analyses were determined by electron microprobe but Sc and V are important tests of the model and their abundances are too low to be measured by electron microprobe. Fortunately, the few ion microprobe measurements available (Alexander, 1994) show that Sc and V in these minerals correlate with elements that can be analyzed by electron microprobe, allowing accurate estimates of their abundances. The functions used to estimate the Sc and V abundances in olivine are

$$Sc(ppm) = 3.71*Ca(mg/g)-0.309 \text{ and } V(ppm) = 66.7*Ca(mg/g),$$

and for low-Ca pyroxene

$$Sc(ppm) = 2.42*Al(mg/g)+0.492 \text{ and } V(ppm) = 24.1*Al(mg/g)+40.3.$$

The abundance of Ca in chondrule olivines are sometimes below electron probe detection limits, but Sc and V concentrations were never allowed to fall below the minimum abundances measured in the ion probe (1.5ppm and 34ppm, respectively).

Finally, a correction was applied to each chondrule to reproduce the observed range of FeO/Fe_{total} ratios in chondrules (0.15–0.95). Each chondrule was also assigned at random: (1) an initial mass, in milligrams, from the logarithmic range of –0.301 to 1.481, and (2) an initial temperature. Most chondrules were given initial temperatures from the range 1800–2100 K, but 20% were given temperatures with a mean and standard deviation of 2070±20 K and enhanced MgO and SiO_2 volatilities of 5 times the vacuum rates (Fig. 4). The model mass range reproduces the LL chondrite chondrule range fairly well but produces sizes that are generally larger than the H-chondrite chondrules.

SIMULATION RESULTS

Having assigned each model chondrule an initial composition, size and peak temperature, their final compositions were calculated as described above and the results are shown in Figs. 5–8. The correspondence between the simulated and actual chondrules is not exact but that was not the purpose of this effort. Rather, the intention was to determine whether a recycling-volatility model could, given reasonable assumptions, reproduce the general trends seen in chondrules. In this sense, the modeling can be deemed a success. The simulated distribution of FeO and Al abundances (Fig. 5) are in reasonable agreement with the natural chondrules (Fig. 2). However, the Fe_{met} is, apparently, lost too quickly in the simulations. Perhaps either (1) metal evaporation was inhibited by poor

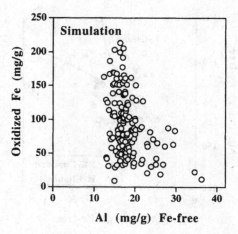

Fig. 5. The simulated distributions of FeO vs. Al. The range of simulated abundances for these elements are similar to those in natural chondrules (Fig. 2).

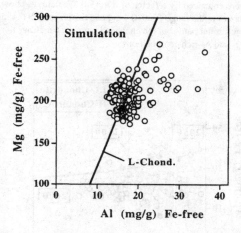

Fig. 6. The simulated distributions of Mg vs. Al. The range of simulated abundances for these elements are similar to those in natural chondrules (Fig. 3). The fact that the recycling-volatility model can reproduce the correct ranges suggests it is a viable alternative to the condensate precursor hypothesis.

diffusive contact between metal spheres trapped in the silicate melt and the evaporating surface or (2) the evaporation rates of the oxides, relative to Fe-metal, were enhanced, perhaps by a higher pH_2 than estimated earlier.

One of the prime motivations for combining volatility with recycling was to explain the correlation between Al and Mg in the natural chondrules, which the model is able to do (compare Figs. 3 and 6) but only if Mg and Si volatilities where enhanced relative to that of FeO in 20% of chondrules. It was mentioned earlier that Sc and V may be good tests of the model. They tend to partition, to varying degrees, into refractory-lithophile-poor olivine and pyroxene. Yet in the bulk chondrule INAA data, Sc correlates with refractory lithophiles, which in the context of the recycling model suggests it should be in the refractory-rich glass, and V is intermediate in behavior between refractory and common lithophiles. As can be seen in Fig. 7, Sc does correlate with refractory lithophiles like Al in the model. This is because the glass is still the single largest Sc carrier (46%) and the correlation is greatly accentuated by the volatile loss of MgO and SiO_2. The correspondence between V-Mg abundances in natural and simulated chondrules (Fig. 8) is not as good as for Sc, but the average V abundance is slightly underestimated in the model, suggesting its distribution between components is not completely understood.

DISCUSSION

Under the assumed formation conditions, which are within the range of previous estimates, the combination of volatile loss and recycling is capable of reproducing the observed range of chondrule compositions. Sears *et al.* (this volume) also argue for the importance of volatile loss. If recycling and volatile loss are responsible for the chemical trends in OC chondrules, it does not mean that nebular condensation did not occur, after all one must still explain the bulk compositions of the chondrite classes, just that the chemical signatures of condensation presumably inherited by the first chondrules have been lost as the result of reworking through several generations. Nor is it suggested that chondrules only formed from reworked chondrules, just that the dominant cause of the chemical heterogeneity in chondrule precursors were chondrule fragments from a previous generation. Rims and matrix, which contain some unprocessed nebular material as well as recondensed chondrule volatiles, must also have contributed to chondrules. Indeed, they would have been the prime means of returning oxidized and metallic Fe, the alkali metals and other volatiles to the chondrule forming system.

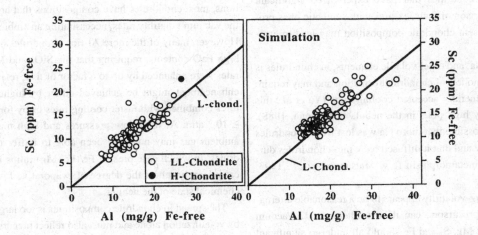

Fig. 7. Comparison of the distribution of Sc vs. Al in (a) natural and (b) simulated chondrules. Scandium is a good test of the recycling-volatility model since Sc partitions into all chondrule phases and Al predominantly into the glass (Alexander, 1994). There was, therefore, some question as to whether Sc and Al would correlate as strongly in the recycling-volatility model as they do in natural chondrules. However, as can be seen the results of the simulation are very successful at reproducing natural chondrule behavior.

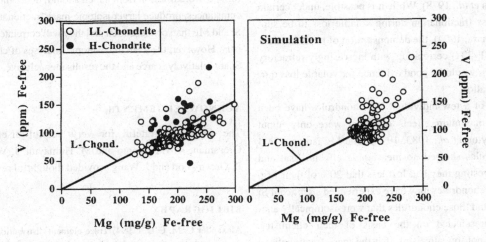

Fig. 8. Comparison of the distribution of V vs. Mg in (a) natural and (b) simulated chondrules. In chondrules, V partitions primarily into low-Ca pyroxene and, to a somewhat lesser extent, into olivine (Alexander, 1994). This behavior is similar to Mg, although Mg prefers olivine to low-Ca pyroxene. Thus, in the recycling volatility model V and Mg should have similar, though not identical, behavior but the distribution of V and Mg in the simulated chondrules is not as close to those of the natural chondrules as seen for other elemental pairs.

Superimposed on the chemical heterogeneity inherited from the precursors are the potentially dramatic effects due to volatile loss. Ca-Al-rich chondrules are rare but do occur in OCs (e.g. Nagahara and Kushiro, 1982; Bischoff and Keil, 1984) and it is possible that some of them are extreme examples of volatile loss. The ability of volatile loss to change a chondrule's chemistry may have implications for isochemical laboratory simulations of chondrule textures, which are the basis for most estimates of cooling rates. A natural chondrule may lose much of its FeO and some SiO_2 and MgO, thereby increasing its liquidus temperature. If a FeO-rich chondrule was heated to near its liquidus and held at this temperature, the loss of FeO and SiO_2 will cause its liquidus temperature to rise until the degree of supercooling is sufficient for crystallization to begin. In this case, the degree of supercooling, and therefore the texture that develops, is not determined by the cooling rate but by the rate at which volatile loss increases the liquidus temperature. Consequently, for chondrules that have experienced significant volatile loss, comparison of their textures with synthetic ones produced using the present chondrule composition may lead to erroneous conclusions.

The presence of Na, and other volatile elements, in chondrules is problematic for all models of chondrule formation and may require either faster than generally accepted cooling rates (Yu *et al.*, this volume) and/or very high pO_2s in the nebula (Grossman, 1988). Alternatively, it is possible that more Na was lost from chondrules than is apparent now and that it diffused back into chondrules during cooling and/or metamorphism (e.g. Matsunami *et al.*, 1993; Alexander, 1995).

Although recycling-volatility appears to be a reasonable alternative to condensate precursors, can it be further tested? Vacuum experiments suggest Mg, Si and Fe should all undergo significant isotopic mass fractionation during volatilization from a melt (e.g. Davis *et al.*, 1990). For instance, after 50% mass loss from a forsterite melt Si shows mass fractionation effects of $\delta^{30}Si \approx 10\permil$ (Davis *et al.*, 1990). Isotopic mass fractionations during equilibrium nebula condensation are expected to be small (e.g. $\delta^{30}Si < 2\permil$, Clayton *et al.*, 1978). While it is possible, under certain conditions, for mass fractionation during volatile loss to be suppressed (e.g. Nagahara, 1994), the demonstration of mass fractionations (e.g. $\delta^{30}Si > 1-2\permil$) correlating with increasingly refractory chondrule chemistry would be good evidence for volatile loss during chondrule formation.

The Si isotopes of a few individual OC chondrules have been analyzed but the maximum effects observed were only about $\delta^{30}Si \approx 1.5\permil$ (Clayton *et al.*, 1983, 1991). The Mg isotopes in 4 Semarkona chondrules showed no measurable effects (Esat and Taylor, 1990), suggesting they had lost less that 20% of their Mg. Only about 20% of chondrules may have experienced significant Si and Mg mass loss and those chondrules that were isotopically analyzed were not preselected on the basis of their chemistry. Consequently, it is not too surprising that the mass fractionations found to date are small compared to effects seen in some experiments. At present, the isotopic evidence for or against volatilization during chondrule formation is inconclusive.

SUMMARY AND CONCLUSIONS

Although condensate precursors can qualitatively account for the inter-element correlations observed in chondrules, there is no mineralogical evidence for these condensates in ordinary chondrites. On the other hand, there is clear evidence for recycling of chondrule material and some evidence for volatile loss from chondrules. Here a numerical model was used to demonstrate that the combination of recycling and volatile loss can, under the reasonable conditions that are summarized below, also explain the range of chondrule compositions.

Assuming chondrules were flash heated and then cooled at 1000°C/hr, abundances of FeO are consistent with peak temperatures in the range 1800–2100 K, similar to previous estimates based on calculated liquidus temperatures. However, reduction or some other unidentified process also played an important role in determining chondrule FeO contents. For these assumed conditions, most chondrules have compositions that are consistent with the vacuum volatility rates, necessitating an ambient $pH_2 \leq 10^{-6}$ atm. However, many of the more Al-rich chondrules have anomalously high FeO contents, requiring that the SiO_2 and MgO evaporation rates were enhanced by up to a factor of five, relative to FeO. This enhancement might be achieved either by higher peak temperatures combined with faster cooling rates or by formation at pH_2 of $\geq 10^{-5}$ atm. At these low pressures and high mass loss rates the ambient gas may not have been able to buffer the pO_2 in chondrules, in which case present Fe/(Fe+Mg) ratios in chondrule silicates largely reflect the degree of evaporative loss rather than the chemical state of the nebula.

The spread in chondrule compositions is too large to be explained by volatilization alone and must also reflect precursor heterogeneity. Here it is demonstrated that chondrule fragments could have provided this heterogeneity. Although volatile loss and recycling can explain chondrule chemistry and despite the petrologic observations of recycled material in chondrules, it has not been demonstrated that this, rather than melting of condensates, was how most chondrules formed. Volatilization from a melt should, under most but not all circumstances, produce larger isotopic mass fractionation effects in Fe, Si and Mg than condensation and they will correlate with bulk chemistry. However, the available isotopic analyses of chondrule Mg and Si are relatively scarce and the results inconclusive.

ACKNOWLEDGEMENTS

The author is grateful for very helpful discussions with J. Grossman, A. Hashimoto and M. Humayun. J. Wasson, S. Love, R. Greenwood and J. Wang provided thoughtful reviews.

BIBLIOGRAPHY

Alexander C. M. O'D. (1994) Trace element distributions within ordinary chondrite chondrules: Implications for chondrule formation conditions and precursors. *Geochim. Cosmochim. Acta* **58**, 3451–3467.

Alexander C. M. O'D (1995) Trace element contents of chondrule rims and interchondrule matrix in ordinary chondrites. *Geochim. Cosmochim. Acta* **59**, 3247–3266.

Bischoff A. and Keil K. (1984) Al-rich objects in ordinary chondrites: Related origin of carbonaceous and ordinary chondrites and their consitituents. *Geochim. Cosmochim. Acta* **48**, 693–709.

Clayton R. N., Mayeda T. K., and Epstein S. (1978) Isotopic fractionation of silicon in Allende inclusions. *Proc. Lunar Planet. Sci. Conf.* **9**, 1267–1278.

Clayton R. N., Mayeda T. K., Goswami J. N., and Olsen E. J. (1991) Oxygen isotope studies of ordinary chondrites. *Geochim. Cosmochim. Acta* **55**, 2317–2338.

Clayton R. N., Onuma N., Ikeda Y., Mayeda T. K., Hutcheon I. D., Olsen E. J., and Molini-Velsko C. (1983) Oxygen isotopic compositions of chondrules in Allende and ordinary chondrites. In *Chondrules and their origins* (ed. E. A. King), Lunar Planet. Inst., Houston, 37–43.

Connolly H. C. Jr., Hewins R. H., Ash R. D., Zanda B., Lofgren G. E., and Bourot-Denise M. (1994) Carbon and the formation of reduced chondrules. *Nature* **371**, 136–139.

Davis A. M., Hashimoto A., Clayton R. N., and Mayeda T. K. (1990) Isotope mass fractionation during evaporation of Mg_2SiO_4. *Nature* **347**, 655–658.

Esat T. M. and Taylor S. R. (1990) Mg isotopic composition of chondrules from the unequilibrated ordinary chondrite Semarkona. *Lunar Planet. Sci.* **XXI**, 333–334.

Gooding J. L. (1979) Petrogenetic properties of chondrules in unequilibrated H-, L-, and LL-group chondritic meteorites. Ph.D. thesis, Univ. New Mexico.

Grossman J. N. (1988) Formation of chondrules. In *Meteorites and the Early Solar System* (eds. J. F. Kerridge and M. S. Matthews), University of Arizona Press, Tucson, 680–696.

Grossman J. N. and Wasson J. T. (1982) Evidence for primitive nebular components in chondrules from the Chainpur chondrite. *Geochim. Cosmochim. Acta* **46**, 1081–1099.

Grossman J. N. and Wasson J. T. (1983) Refractory precursor components of Semarkona chondrules and the fractionation of refractory elements among chondrites. *Geochim. Cosmochim. Acta* **47**, 759–771.

Hashimoto A. (1983) Evaporation metamorphism in the early solar nebula - evaporation experiments on the melt $FeO-MgO-SiO_2-CaO-Al_2O_3$ and chemical fractionations of primitive materials. *Geochem. Journ.* **17**, 111–145.

Hewins R. H. and Radomsky P. M. (1990) Temperature conditions for chondrule formation. *Meteoritics* **25**, 309–318.

Huang S., Benoit P. H., and Sears D. W. G. (1993) The group A3 chondrules of Krymka: Further evidence for major evaporative loss during the formation of chondrules. *Lunar Planet. Sci. Conf.* **XXIV**, 681–682.

Huss G. R. (1990) Ubiquitous interstellar diamond and SiC in primitive chondrites: abundances reflect metamorphism. *Nature* **347**, 159–162.

Huss G. R., Keil K., and Taylor G. J. (1981) The matrices of unequilibrated ordinary chondrites: implications for the origin and history of chondrites. *Geochim. Cosmochim. Acta* **45**, 33–51.

Jones R. H. (1990) Petrology and mineralogy of type II, FeO-rich chondrules in Semarkona (LL3.0): Origin by closed-system fractional crystallization, with evidence for supercooling. *Geochim. Cosmochim. Acta* **54**, 1785–1802.

Lofgren G. E. (1989) Dynamic crystallization of chondrule melts of porphyritic olivine compositions: Textures experimental and natural. *Geochim. Cosmochim. Acta* **53**, 461–470.

Matsunami S., Ninagawa K., Nishimura S., Kubono N., Yamamoto I., Kohata M., Wada T., Yamashita Y., Lu J., Sears D. W. G., and Nishimura H. (1993) Thermoluminescence and compositional zoning in the mesostasis of a Semarkona group A1 chondrule and new insights into the chondrule-forming process. *Geochim. Cosmochim. Acta* **57**, 2101–2110.

Nagahara H. (1993) Evaporation in equilibrium, in vacuum, and in hydrogen gas. *Lunar Planet. Sci. Conf.* **XXIV**, 1045–1046.

Nagahara H. (1994) Why chondrules do not show and some CAIs show significant isotopic fractionation. *Lunar Planet. Sci.* **XXV**, 965–966.

Nagahara H. and Kushiro I. (1982) Calcium-aluminium-rich chondrules in the unequilibrated ordinary chondrites. *Meteoritics* **17**, 55–63.

Nagahara H. and Ozawa K. (1994) Vaporization rate of forsterite in hydrogen gas. *Meteoritics* **29**, 508–509.

Radomsky P. M. and Hewins R. H. (1990) Formation conditions of pyroxene-olivine and magnesian olivine chondrules. *Geochim. Cosmochim. Acta* **54**, 3475–3490.

Rubin A. E. and Pernicka E. (1989) Chondrules in the Sharps H3 chondrite: Evidence for intergroup compositional differences among ordinary chondrite chondrules. *Geochim. Cosmochim. Acta* **53**, 187–196.

Scott E. R. D. (1994) Evaporation and recondensation of volatiles during chondrule formation. *Lunar Planet. Sci. Conf.* **XXV**, 1227–1228.

Scott E. R. D., Rubin A. E., Taylor G. J., and Keil K. (1984) Matrix material in type 3 chondrites - occurrence, heterogeneity and relationship with chondrules. *Geochim. Cosmochim. Acta* **48**, 1741–1757.

Swindle T. D., Caffee M. W., Hohenberg C. M., Lindstrom M. M., and Taylor G. J. (1991a) Iodine-xenon and other studies of individual Chainpur chondrules. *Geochim. Cosmochim. Acta* **55**, 861–880.

Swindle T. D., Grossman J. N., Olinger C. T., and Garrison D. H. (1991b) Iodine-xenon, chemical, and petrographic studies of Semarkona chondrules: Evidence for the timing of aqueous alteration. *Geochim. Cosmochim. Acta* **55**, 3723–3734.

Wang J., Davis A. M., Clayton R. N., and Mayeda T. K. (1994) Kinetic isotopic fractionation during the evaporation of the iron oxide from the liquid state. *Lunar Planet. Sci. Conf.* **XXV**, 1459–1460.

Wasson J. T. and Kallemeyn G. W. (1988) Composition of chondrites. *Phil. Trans. Roy. Soc. London* **A325**, 535–544.

Zanda B., Yu Y., Bourot-Denise M., Hewins R. H., and Connolly H. C. Jr. (1995) Sulfur behavior in chondrule formation and metamorphism. *Geochim. Cosmochim. Acta* Submitted.

26: Chemical Fractionations of Chondrites: Signatures of Events Before Chondrule Formation

JEFFREY N. GROSSMAN

United States Geological Survey, 954 National Center, Reston, VA 22092, U.S.A.

ABSTRACT

While CI chondrites have elemental abundances like that of the Sun, all of the other chondrite groups, as well as the planets, have fractionated compositions. At least six groups of elements were involved in more-or-less independent fractionations: highly refractory lithophiles, Mg and a few less refractory lithophiles, refractory siderophiles, common siderophiles, moderately volatile elements, and highly volatile elements. Losses of the various components are much more common than gains. Earlier work suggests that most of these chemical fractionations took place prior to chondrule formation.

The recognition of two major groups of chondrules, one volatile- and FeO-poor, the other relatively unfractionated, allows a new model for nebular fractionations to be constructed, with chondrules as principal components. However, this model fails to explain many compositional characteristics of chondrites, and is almost certainly invalid.

While numerically abundant, the mass fraction of volatile-poor chondrules in ordinary chondrites is small. Despite some evidence that volatiles did, in some instances, enter some chondrules after they formed, this does not appear to be an important effect. The widely cited conclusion that chondrules are too volatile-rich to be consistent with the classic 'two-component model' still appears to be correct.

Successful models for the early solar nebula must allow for a period of high-temperature processing and chemical fractionation in chondrite-formation regions before chondrules are made.

INTRODUCTION

To a meteorite petrologist, the formation of chondrules clearly represents the single greatest event in the history of chondrites. A huge fraction of the material in these, the most primitive meteorites, has been processed through a molten or partially molten state, and solidified rapidly into mm-sized objects. To a cosmochemist, the systematic fractionations of certain groups of elements from others represent the greatest event (or events) in the histories of chondrites and the planets. Somehow, processes operating on a broad scale in the early solar system were able to separate chemically different materials from one another, and keep them apart while large bodies formed. Were the great events that formed chondrules and caused chemical fractionations one and the same? Were they related? Or, was there a series of unrelated processes operat-

ing early in the solar system's history that occurred prior to the final accumulation of planets and asteroids? The answers to these questions are essential to our efforts to understand how the solar system formed: obviously, any successful model must be able to both create chondrules and to fractionate the elements. But, it is clearly more difficult to fit a large number of processes into any model than to accommodate only a few.

Studies of the bulk compositions of chondrites provide evidence for at least 6 quasi-independent fractionations of groups of rock-forming elements: (1) highly refractory lithophile elements from Si; (2) Mg (and a handful of elements of intermediate refractoriness) from Si; (3) moderately volatile elements from Si and other more refractory elements; (4) highly volatile elements from all other elements; (5) siderophile elements from lithophiles; and (6) refractory siderophile elements from nonrefractory siderophiles. There have been at least 30 years of speculation about how, where and when each of these fractionations might have taken place, and none of the answers is known with certainty. Similarly, there has been over a century of inconclusive speculation about how chondrules formed. But, the coincidence that the CI chondrites, the only group that is essentially unfractionated relative to solar composition, is also the only group that completely lacks chondrules led earlier workers to search for links between these unknown processes.

In a pair of milestone papers, Larimer and Anders (1967, 1970) made some of the first quantitative attempts to examine the relationships between nebular fractionations and chondrule formation. In the 1967 paper, they concluded that the fractionation of moderately volatile elements was caused by the co-accretion of volatile-poor chondrules and unfractionated matrix (the 'two-component model'), thus linking one fractionation to chondrule formation. In the 1970 paper, they argued that the formation of chondrules must have *post-dated* the fractionations of refractory elements and Mg from Si, as well as the metal/silicate fractionation; this conclusion was affirmed in my later studies (e.g., Grossman and Wasson, 1982, 1983a). However, other work, especially that by Wasson and co-workers (including the author), seemed to disprove the two-component model, as chondrules were found to have too high a complement of volatiles. Wasson instead favored a model in which chondrules formed after volatile fractionation as well (e.g., Wasson, 1972; Wasson and Chou, 1974). Rubin and Pernicka (1989) concluded that the fractionation of refractory siderophiles from other siderophiles was also accomplished prior to chondrule formation. Thus, with the exception of the fractionation of highly volatile elements (e.g., In, Tl, Pb, and Bi, for which there are few conclusive data relevant to chondrules), literature through the 1980's strongly indicated that interelement fractionations all predated chondrule formation.

It has recently been suggested, however, that many of the above-cited data on chondrules might need to be re-evaluated, and that chondrule formation might indeed be linked to some chemical fractionations. Papers by the Sears group (Sears, 1993, editorial; Sears *et al.*, 1992; DeHart *et al.*, 1992) highlight their rediscovery of volatile-poor, reduced chondrules (which they call 'group A'), and suggest that earlier work on chondrules was completely non-representative of the actual population. Therefore, the conclusions of those earlier studies which relate to volatile elements might be suspect. They also claim that, on the basis of this knowledge, we must now seriously consider whether the well-known intergroup chemical fractionations seen among chondrites could be caused by chondrule formation itself. Evidence that chondrites may contain size-sorted chondrules (e.g., Skinner and Leenhouts, 1993) combined with the intrinsic size-difference between volatile-poor and volatile-rich chondrules (Huang *et al.*, 1993) have also led Haack and Scott (1993) and Scott and Haack (1993) to imply that chemical fractionations might *post-date* chondrule formation in ordinary chondrites.

In this paper, I will revisit the question of the relative timing of chondrule formation and chemical fractionations in light of this new line of evidence.

A BRIEF REVIEW OF FRACTIONATIONS

There are now 13 fairly well-characterized groups of chondrites, including seven carbonaceous, two enstatite, and three ordinary groups, plus the R chondrites, which are closest in affinity to the ordinary chondrites. The CI chondrites are indistinguishable from the Sun in the abundances of most condensable elements, and have long been assumed to represent unfractionated material from the solar nebula. I also make that assumption. Chemical fractionations of other groups can be quantified using elemental ratios in the CI group as a reference (Table 1).

For the calculations that give the data in Table 1, it was assumed that the Mg/Si fractionation was due to losses or gains of pure forsterite (following the method used by Larimer and Anders, 1970). Of course, forsterite might have had nothing to do with the actual fractionation mechanism, but it seems a reasonable choice given this mineral's high abundance in chondrites, and its predicted presence as a condensate. Later in this paper, I discuss the possibility that Si-rich vapors lost from chondrules might have been the fractionated component here. For refractory lithophiles, a pure mixture of the three most abundant refractory oxides (CaO, Al_2O_3, and TiO_2), with CI-chondrite interelement ratios was chosen for the fractionated component, and the Al/Si ratio was adjusted to the CI value by adding or subtracting this component. For the metal/silicate fractionation, Fe-Ni metal with the CI-chondrite Fe/Ni and Fe/Ir ratios was used, and the Fe/Si ratio was adjusted to the CI value by adding or subtracting it. After this calculation, the refractory siderophile fractionation was calculated as the amount of pure Ir needed to adjust the Ir/Si ratio to CI (since all refractory siderophiles are trace elements, they have no effect on the mass

Table 1. *Grams of hypothetical components that must be added to 100 g of average chondrite groups to restore them to a CI-like composition, except Ir in µg*

	Forsterite[1]	Refractories[2]	Metal[3]	Iridium[4]	Sulfur[5]	Sodium[6]
CH	+0.01	+0.3	−18	−3	+7	+0.43
CR	−0.2	−0.2	+2	−4	+7	+0.38
CM	+1.3	−0.4	+2	−7	+4	+0.20
CO	+1.2	−0.4	+3	−11	+7	+0.33
CV	−0.5	−1.7	+4	−17	+6	+0.40
CK	−2.4	−1.3	+3	−17	+7	+0.39
H	+10	+1.4	+6	−7	+9	+0.24
L	+14	+1.8	+16	+5	+10	+0.29
LL	+13	+2.0	+20	+10	+10	+0.31
R	+20	+2.3	+14	−0.5	+8	+0.37
EH	+30	+3.6	+11	+16	+7	+0.38
EL	+19	+2.7	+18	+27	+9	+0.46

[1] Amount of Mg_2SiO_4 needed to adjust Mg/Si to solar ratio.
[2] Amount of $(CaO)_{0.432}(Al_2O_3)_{0.545}(TiO_2)_{0.023}$ needed to adjust Al/Si to solar ratio.
[3] Amount of $Fe_{94.4}Ni_{5.6}$ (plus 2.41 ppm Ir) needed to adjust Fe/Si to solar ratio.
[4] Amount of pure iridium needed to adjust Ir/Si to solar ratio.
[5] Amount of pure sulfur needed to adjust S/Si to solar ratio.
[6] Amount of pure Na needed to adjust Na/Si to solar ratio.

balance, so only Ir was used). Volatile element fractionations were likewise represented by adding pure Na and S elemental endmembers, as these are the only two abundant representatives of this group. No highly volatile elements, including carbon, are shown, as these are greatly affected by metamorphism in chondrites, and "primitive" compositions are not known for all groups. In these calculations, the group mean compositions of Wasson and Kallemeyn (1988) were supplemented by data from Wiik (1969), Jarosewich (1990), Kallemeyn *et al.* (1991), and Bischoff *et al.* (1993).

As can be seen by the frequency of positive numbers in Table 1, chondrite compositions shown in this way are dominated by "missing components," i.e., material needs to be added back into chondrites to restore their interelement ratios to CI levels. (Note that to actually convert chondrite compositions to the CI composition, a large gain of oxygen is also needed in the form of water and oxidized iron.) Only the odd CH group (ALH85085-like chondrites) shows any major gains (negative numbers), in this case for metal. Most of the carbonaceous chondrites show modest gains in the refractory component and, perhaps, forsterite. All chondrite groups show significant losses of volatiles, relative to CI chondrites, and all but CH show losses of metal.

The two refractory lithophile components show a correlated degree of fractionation (Fig. 1a), although as long-ago noted (Larimer and Anders, 1970; Kerridge, 1979; Larimer, 1979), the ordinary and enstatite chondrites (to which we can now add R chondrites), seem to require a different mixture of these components than the carbonaceous groups. Similarly, the amount of fractionated Ir is correlated with the amount of fractionated refractory lithophiles (Fig. 1b), with some significant scatter indicating that the two fractionations are not perfectly tied together. The two moderately volatile elements, Na and S, also show correlated fractionations (progressing from CI to CM to the cluster of other carbonaceous, EH and R chondrites, to EL; Fig. 1c), although in this case there appears to be a very different mixture of components in ordinary chondrites, which fall somewhat off the trend. The fractionations of volatiles, refractory lithophiles and metal are mutually independent (Figs. 1d,e,f).

MAJOR-ELEMENT FRACTIONATIONS: EVALUATION OF THE REVISIONIST MODEL

The lack of intercorrelation among various chemical fractionations affecting chondrite groups argues against any simple, global mechanism causing them all. However, it is always a possibility, if not a likelihood, that the compositions of the chemical components were different in each chondrite group or family. The same *process* operating on different *materials* might thus be responsible for two or more seemingly uncorrelated fractionations. With this caution in mind, I will examine chondrules as a potential source of fractionation: I will concentrate on the three ordinary chondrite groups, H, L and LL, as we have by far the most relevant data on these, but other groups will be plotted for reference and briefly discussed.

Compositions of chemical components

Chondrules, as a whole, have compositions close to that of their host chondrite (Grossman *et al.*, 1988), a property that offers little opportunity to effect large fractionations using them as a single component. The model that will be tested here is one that takes advantage of the large compositional difference between type I and type II chondrules (generally, but not exactly, corresponding to the group A and group B chondrules of Sears *et al.*, 1992). This is essentially the model proposed independently by Sears *et al.* (1992) and Haack and Scott (1993), but not tested in either paper with regard to nebular fractionations. Type I chondrules tend to be higher in refractory lithophiles, lower in FeO, higher in metal and lower in volatile elements than type II. Thus, preferential accumulation of one type of chondrule over the other is the potential fractionation mechanism, as suggested by Lu *et al.* (1990), DeHart *et al.* (1992) and Sears *et al.* (1992), perhaps as the result of size sorting, as suggested by Haack and Scott (1993) and Scott and Haack (1993).

If type-I compositions are the result of high-temperature processing of volatile-rich, ferroan material resembling type-II chondrules, as suggested by Jones (1990), Lu *et al.* (1990), Lu *et al.* (1992), and Huang *et al.* (1993), then another chemical component in the system is the "lost component," vaporized or otherwise released during the processing. This component is complementary to type-I chondrules with respect to the starting material. Fractionation could be accomplished when elements lost from type-I chondrules exited the system, or in cases where the "lost component" from type-I chondrules in one system contaminated the system represented by a different chondrite group. Only the end-member case for which none of the lost volatiles are allowed to recondense on the source chondrules needs to be considered: any volatiles that recondensed into type-I chondrules would have been included in literature analyses of this group; and, any volatiles that recondensed onto the chondrule surfaces (and thus may have been excluded from analyses), would be linked physically to the type-I chondrules in the nebula. Either way, such volatiles could not enter into fractionations caused by separating chondrules.

The data of Jones and Scott (1989) and Jones (1990) for type-IA (the olivine-rich subset of FeO-poor chondrules) and type-II chondrules from Semarkona were used as the foundation for this model. To begin, Jones' data were adjusted to yield smooth abundance patterns: for type-IA chondrules, the Ca/Mg, Ti/Mg and Al/Mg ratios were raised to the average value for refractory element/Mg ratios found by the Sears group for group-A chondrules (which I estimate to be ~1.2 × CI; e.g. Sears *et al.*, this volume), the K/Na ratio was decreased to the CI ratio by lowering K by ~20%, and Mn was raised ~20% to produce a smooth abundance pattern in the sequence Mg-Si-Cr-Mn-Na. The reasons for doing this are that I am not interested in examining second-order effects involving interfractionation of elements of similar volatilities in this model, whether due to real, minor fractionations or simply analytical errors (e.g., Ca/Al or K/Na fractionation). Also, the neutron activation data from the Sears group are less likely to suffer from sampling errors on crystals and glass than microprobe data, but are

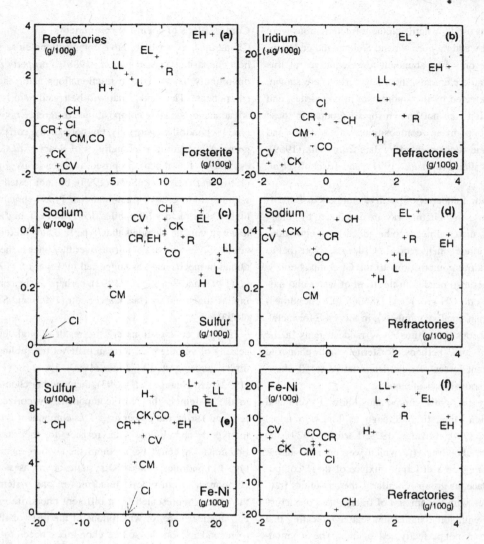

Fig. 1. Inter-comparisons of the magnitudes of different chondrite fractionations. Data are all from Table 1, and are calculated as the amounts of hypothetical components that are needed to restore chondrites to CI-like compositions. The refractory component is a mixture of $CaO+Al_2O_3+TiO_2$ with CI interelement ratios, the Fe-Ni component is a mixture of Fe+Ni+Ir with CI interelement ratios, the forsterite component is pure Mg_2SiO_4, and the other components are pure elements. The two refractory lithophile fractionations (a) are independent among carbonaceous chondrites, and correlated among other groups. Refractory siderophile fractionation, represented by Ir, correlates roughly with refractory lithophile fractionation (b). The moderately volatiles, Na and S, show crudely correlated extents of fractionation (c). All other pairs of fractionated components are uncorrelated, including Na versus refractory lithophiles (d), sulfur versus metal component (e), and metal component versus refractory lithophiles (f).

incomplete, lacking Si, Ti, S and the different forms of iron. The calculation thus produces a smoothed "compromise" between the Jones and Sears datasets for their related type-IA and group-A populations of reduced, volatile-poor chondrules. At the same time, it preserves the well-characterized chemical features of this type of chondrule common to many independent studies, and essential for understanding nebular fractionations.

Similarly, the Jones (1990) data for type-II chondrules were adjusted by minor smoothing of Ca, Al and Ti, and Na and K, to produce CI interelement ratios. Unfortunately, the resulting smoothed type-II abundance pattern (Fig. 2, data in Table 2), is not a reasonable starting material for making type-I chondrules in this model. Type-II chondrules, like the radial pyroxene chondrules (group III) measured in many chondrites (e.g., Grossman and

Wasson, 1982; 1983a), in fact have sub-CI ratios of refractories, Mg, Cr and Mn to Si, and thus are *already fractionated*. Since the whole reason for this exercise is to see if chondrule formation can create fractionations, it is pointless to start out with fractionated precursors. Instead, the starting material for type-I chondrules was taken to be material similar to type II, but with refractories/Si, Mg/Si, Cr/Si and Mn/Si raised, and Na/Si and K/Si lowered, to CI levels. This composition differs from those of CI chondrites themselves in the abundance and distribution of Fe (between oxide, sulfide and metal), and the abundances of S and Ni; all of these features are taken from the type-II chondrules. The composition of this "type-I precursor" is shown in Table 2. Type-II chondrules can still be used as a chemical component in this model, but the mechanism by which they got fractionated can not be addressed, and

Table 2. *Composition of a "lost component" that could be removed from a hypothetical precursor[1] to produce Type IA chondrules. Column C represents the difference between columns A and B, renormalized to 100%. It is assumed that no refractories (Ca, Al, Ti) or metals (Fe, Ni) are lost.*

	Jones (1990) Type II	Smoothed Type II	(A)[1] Type IA Precursor	(B) Lost Component	(C) Modeled Type IA	Adjusted & Smoothed Type IA	Jones and Scott (1989) Type IA
SiO_2	45.1	45.1	43.9	15.65	44.25	44.3	44.8
TiO_2	0.10	0.11	0.13	0.00	0.20	0.21	0.19
Al_2O_3	2.68	2.51	3.17	0.00	4.97	4.9	3.9
Cr_2O_3	0.51	0.51	0.75	0.43	0.50	0.5	0.44
FeO	14.9	14.9	14.5	13.752	1.17	1.17	1.18
MnO	0.39	0.39	0.5	0.35	0.23	0.23	0.12
MgO	31.3	31.3	31.7	6.00	40.25	40.3	40.7
CaO	1.89	1.99	2.52	0.00	3.95	3.9	3.5
Na_2O	1.65	1.65	1.29	0.97	0.51	0.51	0.52
K_2O	0.17	0.17	0.13	0.094	0.050	0.050	0.07
FeS	0.94	0.94	0.91	0.78	0.20	0.2	0.20
Fe metal	0.40	0.40	0.39	(1.87)	3.54	3.55	3.58
Ni	0.11	0.11	0.11	0.00	0.17	0.18	0.18
Total	100.1	100.1	100.0	36.15	100.00	100.1	99.38

[1] This calculated composition has CI-chondrite interelement ratios for lithophiles elements, and abundances of FeO, FeS, Fe metal and Ni based on type-II chondrules.

[2] 17.5% of the FeO listed in column B is reduced to Fe metal, and retained in column C.

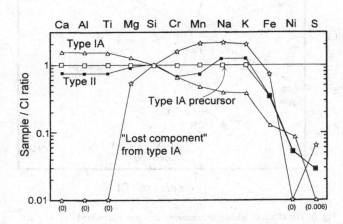

Fig. 2. Abundance patterns, normalized to CI chondrites and to Si, of all chemical components considered in the "revisionist" fractionation model. Compositions of type-IA and type-II chondrules are based on data from Jones and Scott (1989) and Jones (1990), but have been idealized by smoothing and were adjusted to be consistent with INAA data (see text). The "lost component" is calculated as the complement to the "type I precursor" component, which is material with CI-like abundances of nonvolatile lithophiles, and type-II chondrule-like abundances of Fe, Ni and S.

would remain an unexplained problem should the model be successful in explaining chondrite bulk compositions.

The composition of the "lost component" shown in Table 2 is the result of a mass-balance calculation between the idealized and smoothed type-II-like precursor of type-I chondrules (column 3) and the compromise, smoothed type-IA composition (column 5). The primary constraint of the calculation is that no refractories (CaO, Al_2O_3, TiO_2) or Ni are lost during the process (i.e., reduction and evaporation) that converts precursor material into type-I. The mechanism for Fe-loss is assumed to be evaporation of FeO rather

than loss of immiscible metal because the latter process would deplete Ni along with Fe. Measured amounts of Ni in type-I and -II chondrules force this assumption: although data are very imprecise due to inhomogeneity of metal, ambiguity over the contribution of rims, and detection limits on the microprobe, type IA compositions seem to have a small excess of Ni that is impossible to model even if none is lost from the assumed precursors. Reduction of FeO and retention of Fe metal are allowed to take place in order to give the desired Fe concentration in the modeled type-I product. A similar calculation of Sears *et al.* (this volume) differs from this one in that: 1) Sears estimates the composition of the lost component from laboratory experiments on synthetic charges, whereas here it is calculated by mass balance; 2) Sears uses average measurements for his group B1 and A1 chondrules as the starting material and desired product in his calculations, whereas here the endmembers have been idealized (to remove noise and interlaboratory discrepancies) and one end-member is hypothetical (to avoid using already-fractionated starting materials).

Testing the model: lithophile elements

Type-I chondrules, and the volatiles lost from them, each seem to have an appropriate composition to explain refractory lithophile fractionations of ordinary chondrites (Fig. 3†). Starting with CI chondrites (minus 30% volatiles), H chondrites could be explained as the loss of ~16% type-I chondrules, or the gain of ~16% of the

† Even though Fig. 3 is normalized to Si, the given amounts of components needed to produce displacements on the diagram are based on mass balance calculations. For this purpose, the very low-temperature components of CI chondrites (water plus excess oxygen associated with sulfate, carbonate and ferric iron) were removed.

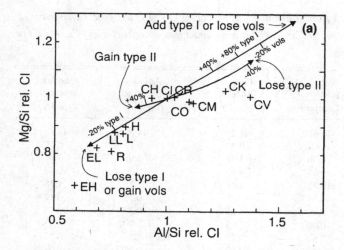

Fig. 3. The effect of adding or removing components plotted in Fig. 2 to a CI-like starting material. Concentrations of all elements in the CI material have been increased by 30% to compensate for very low-temperature components in CI (see text). Ordinary and enstatite chondrite compositions can be reproduced on this Mg/Si vs. Al/Si plot by removing type-IA chondrules or gaining the volatiles lost from these chondrules to CI. Carbonaceous chondrites cannot be explained by any of the components.

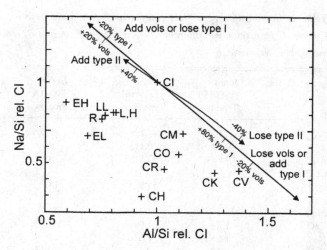

Fig. 4. The effect of adding or removing components plotted in Fig. 2 to a CI-like starting material. Concentrations of all elements in the CI material have been increased by 30% to compensate for very low-temperature components in CI (see text). No chondrite groups can be simulated on this Na/Si vs. Al/Si diagram by mixing the hypothetical components with CI.

lost component from type-I chondrules formed elsewhere. LL chondrites would have experienced ~20% loss or gain of the same components. Although gaining so much of the volatiles lost from chondrules that formed somewhere else seems ridiculous, a change of only tens-of-percent in the abundance of type-I chondrules seems acceptable, and there only needs to be a small difference between H and LL. For carbonaceous chondrites, especially CO, CM, CV and CK groups, these same components cannot explain the refractory lithophile fractionations at all: as is well-known, carbonaceous chondrites display only minimal Mg/Si fractionation, and a wide range of refractory/Si ratios (e.g., Wasson and Kallemeyn, 1988), yet McSween's (1977) data for type-I chondrules in CV chondrites show that Mg/Si and refractory/Si ratios are consistently elevated, just as in type-I chondrules from ordinary chondrites. While excesses of CAIs in these chondrites could explain much of the refractory element fractionation in this group, any significant fractionation of type-I chondrules relative to other components would greatly affect the Mg/Si ratio.

But, even the limited success of this chondrule fractionation model in explaining refractories in ordinary chondrites is negated by trends shown in other elements. Fractionation of any of the components in question, type I and type II chondrules, and the volatiles lost from type I, will produce an anticorrelation between Na and Al (Fig. 4). On this plot, only the CV chondrites have anything like a composition obtainable by fractionating CI starting material with these components. Of course, the 80% excess of type-I chondrules, or 40% loss of type IIs, that CVs would need in this case would raise their Mg/Si ratios at least 20% higher than they actually are.

The model also fails to explain the bulk iron contents of chondrites (Fig. 5). The Fe/Si ratio of ordinary chondrites is quite independent of volatiles like Na. No simple gains or losses of the

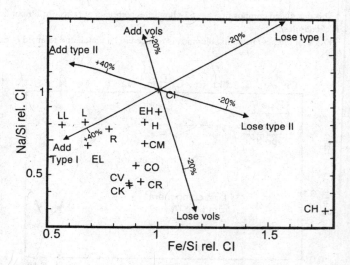

Fig. 5. The effect of adding or removing components plotted in Fig. 2 to a CI-like starting material. Concentrations of all elements in the CI material have been increased by 30% to compensate for very low-temperature components in CI (see text). This mixing model cannot adequately produce the Na/Si vs. Fe/Si relationships of chondrites shown in this diagram.

hypothetical components of the model could explain this fact.

One might also consider whether the relatively small chemical fractionations between only the three groups of ordinary chondrites might be explained by the separation of the components in Table 2. This is akin to the suggestion of Haack and Scott (1993), Scott and Haack (1993) and Scott *et al.* (this volume), that perhaps size-sorting of the same components produced differences between H, L and LL chondrites. This also seems unlikely: Fig. 6 shows that the removal of a small amount of type-I chondrules from H chondrites reproduces the LL chondrite abundance pattern for refractory lithophiles and Mg, but causes the patterns to diverge for all lithophiles more volatile than Si. One might argue that volatile-rich matrix material, which, like type-I chondrules, is finer-grained than

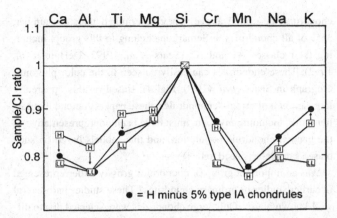

Fig. 6. Si- and CI-normalized abundance patterns of H and LL chondrites, showing that the differences between the groups cannot be explained by differences in the abundance of type-I chondrules.

other components, might have accompanied type I's during fractionation, thus compensating for the effect on volatile elements. However, there is no known difference in matrix abundance between H and LL chondrites (Huss *et al.*, 1981), and the removal of matrix (or rims), known to have low Mg/Si and low refractory/Si ratios (e.g., Huss *et al.*, 1981, Alexander, 1995), with the type-I chondrules would negate the success of the model in fractionating the refractories.

Curiously, a calculation of the composition of material that must be removed from LL chondrites to make H chondrites (or, conversely, added to H to make LL), results in something resembling type-II chondrules, but the amount of fractionation required is extreme. If LL chondrites lost 75% of their total mass in the form of material with refractory/Si ratios of 0.75–0.8CI, Mg/Si of ~0.85CI, Cr/Si, Mn/Si, K/Si and Na/Si all in the range 0.75–0.8CI, Fe^{+2}/Si around 0.4CI, and S/Si of ~0.2CI, then one can reproduce the H-group bulk composition rather well. These abundances resemble those of Jones (1990) for type-II chondrules, except Jones' data for alkalis were higher, and they also resemble the average composition for nonporphyritic chondrules reported by Grossman and Wasson (1983b). This huge loss of material would represent virtually all of the type-II chondrules in the LL group, so the H chondrites one produces in this manner would be nearly devoid of these. However, data from Sears *et al.* (1995) show that, while unequilibrated H chondrites may have fewer (by number) group-B chondrules (roughly equivalent to type II) than LL chondrites, the difference is only slight. The calculated conversion of LL into H chondrites also requires the loss of about half the metal in LL, and the addition of a significant amount of pure Fe. Such an Fe component does not correspond to any phase in LL chondrites, and would have to be the product of a chemical process like reduction; size-sorting of phases in LL chondrites could not produce this fractionation.

Testing the model: Siderophile and chalcophile elements

Type-I and type-II chondrules have significantly different amounts of metal, and different Ni/Fe ratios in their metal (Jones and Scott, 1989; Jones, 1990, Huang *et al.*, 1993): type I (and Sears' group A)

have more metal, and metal is more Ni-poor, than type II (and Sears' group B). Indeed, many low-FeO chondrules contain metal with the CI Ni/Fe ratio or lower, presumably due to reduction of FeO during heating. High-FeO chondrules almost always have Ni/Fe ratios in their metal that are higher than CI. Could these siderophile-bearing components have anything to do with the metal-silicate fractionation that affected chondrites?

From Fig. 7, a standard plot for looking at metal-silicate fractionation in chondrites, it appears unlikely that chondrules and their metal had much to do with this fractionation either. Most chondrite groups plot very close to the line of CI Ni/Fe ratio in Fig. 7, with only the metal-rich CH group being an exception. The interpretation of Larimer and Anders (1970) and Grossman and Larimer (1974), that such plots indicate the metal-silicate fractionation happened at high temperature, when most Fe was reduced (i.e., before chondrule formation), seems inescapable. First of all, given their low abundance of Ni, and high abundance of FeO, any fractionation involving type-II chondrules (or the volatiles lost from type I chondrules) would result in serious deviations in bulk Ni/Fe ratios from the CI value. Large gains of type-I chondrules or losses of metal originating in type-I chondrules would move compositions from CI toward the origin, but would produce small deviations toward excess Ni (which are not observed). But even if type-I chondrules had exactly the CI Ni/Fe ratio, LL chondrites would have to lose considerably more type-I metal than H chondrites, despite their having somewhat fewer type-I chondrules (as evidenced by their higher FeO content and larger chondrules). Finally, any metal that may have been lost from type-II chondrules, if it resembled the metal currently observed in these objects, would greatly perturb the bulk Ni/Fe ratio.

Thus, the metal in present-day chondrules appears to be more

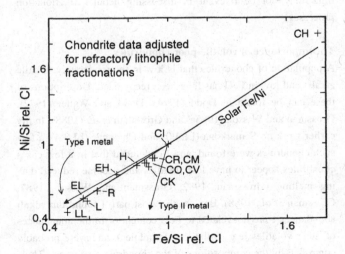

Fig. 7. Ni/Si vs. Fe/Si diagram for examining metal/silicate fractionation in chondrites, after Larimer and Anders (1970). Si data have been adjusted by adding back calculated amounts of forsterite lost (or gained) in the refractory lithophile fractionation. Chondrites show evidence for fractionation at high temperature, with compositions lying near the CI Fe/Ni ratio. The metal (or bulk) of type-I and type-II chondrules cannot be the fractionated component, as removal of these types of material would cause deviations from the solar line (arrows).

highly evolved than the more 'primitive' metal involved in metal/silicate fractionations of chondrites. There are no clear options for explaining this other than having metal/silicate fractionation occur at high temperature. Larimer and Wasson (1988) show that some moderately volatile siderophiles were condensed in metal at this time, suggesting temperatures less than 900 K. Sulfur was not fractionated along with Fe (Fig. 1e), despite the coherence of the two elements in chondritic assemblages; it was probably uncondensed at the time of fractionation, placing the temperature above 650 K. Because sulfides were among the precursor solids from which chondrules formed (Zanda *et al.*, 1995), it is the logical conclusion that chondrule formation post-dated sulfide formation, and therefore also post-dated metal/silicate fractionation.

VOLATILE-ELEMENT FRACTIONATIONS: BEFORE OR AFTER CHONDRULE FORMATION?

On the basis of the last section, it now seems highly unlikely that chondrules had anything to do with the refractory lithophile fractionations or the metal/silicate fractionation that affected bulk chondrites. There is no reason to overturn previous workers' conclusions that these fractionations predated chondrule formation. In this section, I will reassess the conclusion that volatiles were fractionated prior to chondrule formation (e.g., Grossman *et al.*, 1988). The reasons for doing this are two-fold: as stated in the Introduction, there have been suggestions in the literature that studies by myself and others of separated chondrules were biased against the relevant population of chondrules, i.e., volatile-poor ones; second, it has been suggested by several researchers that some or all of the volatile elements, including alkalis, in chondrules may be secondary in origin (Wood, this volume; Sears *et al.*, 1994, this volume; Huang *et al.*, 1994), so that chondrule compositions may not be relevant in discussing nebular fractionation processes.

The importance of volatile-poor chondrules

A population of chondrules that is low in the moderately volatile alkalis and low in FeO has long been recognized. Descriptions of these can be found in Dodd (1974), Dodd and Walter (1972), Grossman and Wasson (1983a) and Grossman *et al.* (1988). In my earlier work on Semarkona (LL3.0) and Chainpur (LL3.4), a few such chondrules were found, and I concluded that in a few cases chondrules appear to have lost volatiles and become reduced during melting (Grossman, 1982; Grossman and Wasson, 1987; Grossman *et al.*, 1988). But, for the most part, I found that alkali abundances were not related to FeO content or to the abundances of other volatiles (e.g., volatile siderophiles), and were probably controlled by the compositions of the chondrule precursors. Thus, fractionation predated chondrule formation.

In their studies of cathodoluminescence, Derek Sears and John DeHart have noted a population of chondrules in Semarkona having yellow-luminescing mesostasis and, often, reddish-luminescing olivine. These turned out to be forsteritic olivine chondrules with calcic mesostasis, i.e., the same class of reduced, alkali-poor

chondrules described above. They determined that, by number, 35% of all chondrules in Semarkona belong to this group, including their classes A1 and A2 (Sears *et al.*, 1992; DeHart *et al.*, 1992). These chondrules can easily be seen in the color photomicrograph in Sears *et al.* (1990, Fig. 4). Based on this apparently large amount of group-A chondrules, these papers contend that earlier work, including my own, must have been nonrepresentative of the general chondrule population, and that models based on such biased sample-sets need to be revised.

Was an important group of chondrules grossly underrepresented in earlier studies of ordinary chondrites? These studies had several well-known biases. Only large chondrules were selected due to difficulties in analyzing very small samples. Easily broken chondrules could also have been destroyed during disaggregation. In my work, and in Gooding's similar study (Gooding *et al.*, 1980) of separated chondrules, nonporphyritic chondrules are overrepresented by about a factor of 2 compared with abundances measured in thin sections because these chondrules tend to be large and tough (e.g., Grossman and Wasson, 1982). However, only a few volatile-poor, FeO-poor chondrules appear to have been present in these chondrule sets, indicating a severe bias (Sears *et al.*, 1992; Sears, 1993; Lu *et al.*, 1990). The reason for this bias is clearly visible in Fig. 4 of Sears *et al.* (1990) group A chondrules, with their characteristic yellow mesostasis, tend to be small, averaging about 2/3 of the radius of other chondrules (based on data plotted in Huang *et al.*, 1993). As a result of their size, these were indeed partially excluded from earlier sets of separated chondrules. However, their small size also makes group A chondrules much less important to the overall budget of volatile elements in ordinary chondrites, and to the bulk chondrule composition. Indeed, 35% percent by number becomes only ~10% by volume, given the size difference. Measurement of the area of yellow-luminescing chondrules in the Sears photomicrograph confirms this result: only 7–12% of the area of the image is covered by yellow-luminescing chondrules. So, on a weight basis, previously published chondrule sets are indeed a fair representation of all chondrules.

Late entry of volatiles into chondrules

The conclusions I and others have reached in earlier studies, that the abundances of chondrule volatiles are largely a function of the composition of the precursors, and that chondrules contain too many volatiles to be consistent with the 'two-component model' of Larimer and Anders (1967), are based on the assumption that the measured chondrules preserve their original (accretionary) compositions. Any subsequent parent-body metamorphism or metasomatism could potentially wipe out the nebular record.

Matsunami *et al.* (1993), Sears (1993) and Huang *et al.* (1994) suggest that some chondrules with zoned mesostasis provide evidence that volatiles lost during chondrule formation may have recondensed back onto the chondrules, either during chondrule formation, or afterwards (but presumably not long afterwards, and in the same nebular setting). It is conceivable that such a process might even have affected chondrules with no apparent zoning. It is important to note that if a significant fraction of the volatile elements that are lost from chondrules in the traditional 'two-component' model recondenses into the chondrules that would ultimately accrete, the potential fractionation mechanism (separation of volatile-bearing gas from depleted chondrules and pristine dust) is no longer possible. Thus, if the Matsunami *et al.* (1993) explanation is correct, these recondensed volatiles would have to be considered part of the chondrule-fraction in the 'two-component' model. Larimer and Anders (1967) clearly recognized this problem in their original model, and grappled with how one might prevent evaporated volatiles from recondensing. They pointed out that fine-grained material (dust) would be the most likely sink for volatiles due to its great surface area. Alexander (1994), accordingly, suggested that recondensation of lost volatiles onto the dust that eventually formed fine-grained rims might have occurred, and that re-entry of volatiles into the chondrules could happen during light metamorphism or reheating. This seems at least more plausible than direct recondensation onto the comparatively small surface area of chondrules, but again it makes it less appealing to use evaporative loss from chondrules as a fractionation mechanism because the rims were almost certainly acquired in the nebula, and could not be fractionated at the time of accretion.

There is other evidence that late entry of volatiles into chondrules was not a widespread phenomenon, and that the Matsunami *et al.* (1993) observations are the exception, rather than the rule. I have dealt with the question of recondensation of volatiles by examining the alkali contents of Semarkona chondrules (Grossman, 1982; Grossman and Wasson, 1983b). Alkali-rich chondrules have higher K/Na ratios than alkali-poor chondrules (the combined data from these studies and Swindle *et al.*, 1991 are plotted in Fig. 8). I initially interpreted this as evidence for recondensation of alkalis lost from some chondrules into later-formed chondrules (Grossman, 1982): it was assumed that K is more volatile than Na, and recondensation was preferred over simple partial evaporation in order to explain alkali contents that are elevated well above whole-rock (or CI) levels in alkali-rich chondrules. However, I later pointed out that the same trends could be produced by fractional condensation processes affecting only the precursors to chondrules (Grossman and Wasson, 1983b). Indeed, the lack of correlation of the K/Na ratio with chondrule size, petrographic parameters, or the content of other volatile elements are evidence against recondensation being the cause of the Na-K trends. Also, if the K/Na relationship with total alkali content (Fig. 8) was caused by recondensation, then it would require that those chondrules which lost volatiles were rarely the recipients of recondensed volatiles; otherwise, there would be a population of chondrules that lost all their alkalis and subsequently experienced partial recondensation, producing chondrules with high K/Na ratios and low total alkali contents (essentially smearing the data in Fig. 8 all around the line of solar ratio). Because only a simple trend is seen in Fig. 8, the recondensation model would have to require that chondrule melting was a one-step process, with devolatilized chondrules efficiently removed from the region of chondrule formation without any chance for recycling of materials through the chondrule-melting process. In light of voluminous evidence that recycling of material through chondrule-forming events was an important process (e.g., evidence from relict grains, some kinds of compound chondrules and coarse-grained rims), we again must conclude that the alkali contents of most chondrules were generally inherited from their precursors.

Crystal-zoning data of Jones (1994) also contradicts the idea of late entry of volatiles into chondrules, either by recondensation or metamorphism. She found that Ca-rich pyroxene in some type-I chondrules is normally zoned with respect to Ca, Al, Ti and Na, indicating that the melt had its full complement of Na during the crystallization of the major phases, which certainly happened at high temperatures. Again, the alkalis appear to be primary in the chondrules.

Finally, in a comprehensive study of the siting of troilite in chondrules from the least equilibrated chondrites Zanda *et al.* (1994, 1995) find abundant evidence for primary sulfides that survived the melting events. While they find that metamorphism can easily redistribute sulfides, this has not happened in Semarkona or Renazzo. As sulfides are certainly the primary site for sulfur, and likely the sites of some other volatile elements measured in chondrules, we again find no need to dispute published evidence against the 'two-component model.'

CONCLUSIONS

In the Introduction, I posed the questions: Were the great events that formed chondrules and that caused chemical fractionations one and the same? Were they related? Or, was there a series of unrelated processes operating early in the solar system's history that occurred prior to the final accumulation of planets and asteroids? As philosophically pleasing as the conclusion would have been, it does not appear that chondrule formation caused any of the great fractionations. Chondrules, and groups of related chondrules, simply have the wrong compositions to have served as major fractionated components.

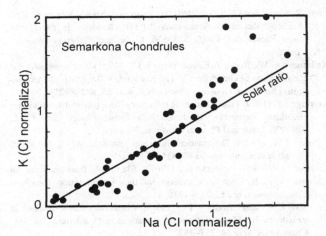

Fig. 8. CI-normalized K vs. Na plot for Semarkona chondrules, showing the systematic change in K/Na ratio with alkali content. This trend is best explained by variations in the composition of chondrule precursors, and not by evaporation during chondrule formation or re-entry of alkalis into chondrules after their formation.

The case is extremely tight that all refractory lithophile and metal-silicate fractionations were established at high temperatures, whereas chondrules formed from material rich in volatile elements. The most difficult evidence to assess is whether intergroup volatile element fractionations were already established prior to chondrule formation, or whether any volatiles lost from chondrules might have escaped the system, and caused fractionations among chondrites. But even in this case, on balance the evidence favors fractionation before chondrule formation.

Any model for the sequence of events in the solar nebula leading up to the accretion of chondrites must leave room for a high-temperature phase, however brief or localized, during which fractionations can occur prior to the widespread processing of material at low temperatures to form chondrules. Perhaps this is not unreasonable, given the evidence of Hutcheon *et al.* (1994) and Hutcheon and Jones (1995) that chondrules formed after CAIs, and that many CAIs require high-temperature processing, including fractional condensation, to explain their trace-element abundance patterns. Successful models must not only contain a mechanism to achieve separation of different chemical components, but a mechanism to keep them apart until chondrule formation and accretion are complete, as well as an explanation of what happened to all of the lost components.

ACKNOWLEDGEMENTS

I thank Rhian Jones, Horton Newsom, and Alan Rubin for their reviews, which helped produce a much more comprehensible paper, and Derek Sears for his comments, which helped to clarify my thinking. This work was supported by NASA Order No. W-18,788 to J.N. Grossman.

REFERENCES

Alexander C.M.O'D. (1994) Trace element distributions within ordinary chondrite chondrules: Implications for chondrule formation conditions and precursors. *Geochim. Cosmochim. Acta* **58**, 3451–3467.

Alexander C.M.O'D. (1995) Trace element contents of chondrule rims and interchondrule matrix on ordinary chondrites. *Geochim. Cosmochim. Acta* **59**, 3247–3266.

Bischoff A., Palme H., Schultz L., Weber D., Weber H.W. and Spettel B. (1993) Acfer 182 and paired samples, an iron-rich carbonaceous chondrite: Similarities with ALH85085 and relationship to CR chondrites. *Geochim. Cosmochim. Acta* **57**, 2631–2648.

DeHart J.M., Lofgren G.E., Lu J., Benoit P.H. and Sears D.W.G. (1992) Chemical and physical studies of chondrites: X. Cathodoluminescence and phase composition studies of metamorphism and nebular processes in chondrules of type 3 ordinary chondrites. *Geochim. Cosmochim. Acta* **56**, 3791–3807.

Dodd R.T. (1974) The petrology of chondrules in the Hallingeberg meteorite. *Contrib. Mineral. Petrol.* **47**, 97–112.

Dodd R.T. and Walter L.S. (1972) Chemical constraints on the origin of chondrules in ordinary chondrites. In *On the Origin of the Solar System* (Ed. H. Reeves), CNRS, Paris, 293–300.

Gooding J.L., Keil K., Fukuoka T. and Schmitt R.A. (1980) Elemental abundances in chondrules from unequilibrated ordinary chondrites: Evidence for chondrule origin by melting of pre-existing materials. *Earth Planet. Sci. Letts.* **50**, 171–180.

Grossman J.N. (1982) The abundance and distribution of moderately volatile elements in Semarkona chondrules (abstract). *Lunar Planet. Sci.* **XIII**, 289–290.

Grossman J.N. and Wasson J.T. (1982) Evidence for primitive nebular components in chondrules from the Chainpur chondrite. *Geochim. Cosmochim. Acta* **46**, 1081–1099.

Grossman J. N. and Wasson, J. T. (1983a) Refractory precursor components of Semarkona chondrules and the fractionation of refractory elements among chondrites. *Geochim. Cosmochim. Acta* **47**, 759–771.

Grossman J. N. and Wasson J. T. (1983b) The compositions of chondrules in unequilibrated chondrites: An evaluation of models for the formation of chondrules and their precursor materials. In *Chondrules and their Origins*, (Ed. E.A. King), 88–121, Lunar and Planetary Institute, Houston.

Grossman J.N. and Wasson J.T. (1987) Compositional evidence regarding the origins of rims on Semarkona chondrules. *Geochim. Cosmochim. Acta* **51**, 3003–3011.

Grossman J.N., Rubin A.E., Nagahara H. and King E.A. (1988) Properties of chondrules. In *Meteorites and the Early Solar System*, (Eds. J.F. Kerridge and M.S. Matthews), 619–659, Univ. Arizona Press.

Grossman L. and Larimer J.W. (1974) Early chemical history of the solar system. *Rev. Geophys. Space Phys.* **12**, 71–101.

Haack H. and Scott E.R.D. (1993) Nebula formation of the H, L, and LL parent bodies from a single batch of chondritic materials (abstract). *Meteoritics* **28**, 358–359.

Huang S., Benoit P.H. and Sears D.W.G. (1993) The group A3 chondrules of Krymka: Further evidence for major evaporative loss during the formation of chondrules (abstract). *Lunar Planet. Sci.* **XXIV**, 681–682.

Huang S., Benoit P.H. and Sears D.W.G. (1994) The group A5 chondrules in ordinary chondrites: Their formation and metamorphism (abstract). *Lunar Planet. Sci.* **XXV**, 573–574.

Huss G.R., Keil K. and Taylor G.J. (1981) The matrices of unequilibrated ordinary chondrites: Implications for the origin and history of chondrites. *Geochim. Cosmochim. Acta* **45**, 33–51.

Hutcheon I.D., Huss G.R. and Wasserburg G.J. (1994) A search for ^{26}Al in chondrites: chondrule formation time scales (abstract). *Lunar Planet. Sci.* **XXV**, 587–588.

Hutcheon I.D. and Jones R.H. (1995) The ^{26}Al–^{26}Mg record of chondrules: Clues to nebular chronology (abstract). *Lunar Planet. Sci.* **XXVI**, 647–648.

Jarosewich E. (1990) Chemical analyses of meteorites: A compilation of stony and iron meteorite analyses. *Meteoritics* **25**, 323–337.

Jones R.H. (1990) Petrology and mineralogy of type II, FeO-rich chondrules in Semarkona (LL3.0): Origin by closed-system fractional crystallization, with evidence for supercooling. *Geochim. Cosmochim. Acta* **54**, 1785–1802.

Jones R.H. (1994) Petrology of FeO-poor, porphyritic pyroxene chondrules in the Semarkona chondrite. *Geochim. Cosmochim. Acta* **58**, 5325–5340.

Jones R.H. and Scott E.R.D. (1989) Petrology and thermal history of type IA chondrules in the Semarkona (LL3.0) chondrite. In *Proc. 19th Lunar Planet. Sci. Conf.*, 523–536, Lunar and Planetary Institute, Houston,.

Kallemeyn G.W., Rubin A.E. and Wasson J.T. (1991) The compositional classification of chondrites: V. The Karoonda (CK) group of carbonaceous chondrites. *Geochim. Cosmochim. Acta* **55**, 881–892.

Kerridge J.F. (1979) Fractionation of refractory lithophile elements among chondritic meteorites. In *Proc. 10th Lunar Planet. Sci. Conf.*, 989–996, Lunar and Planetary Institute, Houston,.

Larimer J.W. (1979) The condensation and fractionation of refractory lithophile elements. *Icarus* **40**, 446–454.

Larimer J.W. and Anders E. (1967) Chemical fractionations in meteorites–II. Abundance patterns and their interpretation. *Geochim. Cosmochim. Acta* **31**, 1215–1238.

Larimer J.W. and Anders E. (1970) Chemical fractionations in meteorites–III. Major element fractionations in chondrites. *Geochim. Cosmochim. Acta* **34**, 367–387.

Larimer J.W. and Wasson J.T. (1988) Siderophile element fractionation. In *Meteorites and the Early Solar System*, (Eds. J.F. Kerridge and M.S. Matthews), 416–435, Univ. Arizona Press.

Lu J., Sears D.W.G., Keck B.D., Prinz M., Grossman J.N. and Clayton R.N. (1990) Semarkona type I chondrules compared with similar chondrules in other classes (abstract). *Lunar Planet. Sci.* **XXI**, 720–721.

Lu J., Sears D.W.G., Benoit P.H., Prinz M. and Weisberg M.K. (1992) The four primitive chondrule groups and the formation of chondrules (abstract). *Lunar Planet. Sci.* **XXIII**, 813–814.

Matsunami S., Ninagawa K., Nishimura S., Kubono N., Yamamoto I., Kohata M., Tomonori W., Yamashita Y., Lu J., Sears D.W.G. and Nishimura H. (1993) Thermoluminescence and compositional zoning in the mesostasis of a Semarkona group A1 chondrule and new insights into the chondrule-forming process. *Geochim. Cosmochim. Acta* **57**, 2102–2110.

McSween H.Y. (1977) Petrographic variations among carbonaceous chondrites of the Vigarano type. *Geochim. Cosmochim. Acta* **41**, 1777–1790.

Rubin A.E. and Pernicka E. (1989) Chondrules in the Sharps H3 chondrite: Evidence for intergroup compositional differences among ordinary chondrite chondrules. *Geochim. Cosmochim. Acta* **53**, 187–195.

Scott E.R.D. and Haack H. (1993) Chemical fractionation in chondrites by aerodynamic sorting of chondritic materials (abstract). *Meteoritics* **28**, 434.

Sears D.W.G. (1993) The case of the not-so-missing sodium (editorial). *Meteoritics* **28**, 607–608.

Sears D.W.G., DeHart J.M., Hasan F.A. and Lofgren G.E. (1990) Induced thermoluminescence and cathodoluminescence studies of meteorites: Relevance to structure and active sites in feldspar. In *Spectroscopic Characterization of Minerals and their Surfaces* (Eds. L.M. Coyne, S.W.S. McKeever and D.F. Blake), 190–222, Amer. Chem. Soc.

Sears D.W.G., Lu J., Benoit P.H., DeHart J.M. and Lofgren G.E. (1992) A compositional classification scheme for meteoritic chondrules. *Nature* **357**, 207–210.

Sears D.W.G., Shaoxiong H. and Benoit P.H. (1994) Open-system behavior during chondrule formation (abstract). In *Papers Presented to Chondrules and the Protoplanetary Disk*, LPI Contrib. 844, 37–38, Lunar and Planetary Institute, Houston.

Sears D.W.G., Shaoxiong H. and Benoit P.H. (1995) Chondrule formation, metamorphism, brecciation, an important new primary chondrule group, and the classification of chondrules. *Earth Planet. Sci. Letts.* **131**, 27–39.

Skinner W.R. and Leenhouts J.M. (1993) Sorting of chondrules by size and density–Evidence for radial transport in the solar nebula (abstract). *Meteoritics* **28**, 439.

Swindle T.D., Grossman J.N., Olinger C.T. and Garrison D.H. (1991) Iodine-xenon, chemical and petrographic studies of Semarkona chondrules: Evidence for the timing of aqueous alteration. *Geochim. Cosmochim. Acta* **55**, 3723–3734.

Wasson J.T. (1972) Formation of ordinary chondrites. *Rev. Geophys. Space Phys.* **10**, 711–759.

Wasson J.T. and Chou C.-L. (1974) Fractionation of moderately volatile elements in ordinary chondrites. *Meteoritics* **9**, 69–84.

Wasson J.T. and Kallemeyn G.W. (1988) Compositions of chondrites. *Phil. Trans. R. Soc. Lond. A* **325**, 535–544.

Wiik H.B. (1969) On regular discontinuities in the composition of meteorites. *Commun. Phys.-Math. (Helsinki)* **34**, 135–145.

Zanda B., Yu Y., Bourot-Denise M., Hewins R.H. and Connolly H.C. Jr. (1994) Chondrule precursors and cooling paths: The sulfur evidence (abstract). In *Papers Presented to Chondrules and the Protoplanetary Disk*, LPI Contrib. 844, 47–48, Lunar and Planetary Institute, Houston.

Zanda B., Yu Y., Bourot-Denise M., Hewins R.H. and Connolly H.C. Jr. (1995) Sulfur behavior in chondrule formation and metamorphism. *Geochim. Cosmochim. Acta*, submitted.

V. Models of Chondrule Formation

27: A Concise Guide to Chondrule Formation Models

ALAN P. BOSS
DTM, Carnegie Institution of Washington, 5241 Broad Branch Road, N.W., Washington, DC 20015–1305, U.S.A.

ABSTRACT

At the risk of oversimplifying a complex problem, the main arguments in support of and against the wide variety of mechanisms proposed for the formation of chondrules are briefly described. The focus is on mechanisms presented at the 1994 conference, though a few other ideas are presented as well for completeness. The goal is to assemble in one place all of the most decisive arguments, in order that a preliminary determination can be made of the most likely chondrule-forming mechanism. The constraints discussed throughout this book make it clear that a "flash heating" mechanism is required. Based on the arguments marshalled here, the opinion of at least one astrophysicist is that nebular shock waves are the most likely mechanism of chondrule formation, provided that an energetic source of nebular shock waves can be shown to have existed during the chondrule-forming epoch.

INTRODUCTION

The spheroidal shape of chondrules and their well-defined edges led Sorby (1877) to conclude that chondrules must have formed as molten droplets in free fall, "like drops of fiery rain". Over a century later, we are still trying to figure out the best way to melt a chondrule precursor. Sorby (1877) had two suggestions for where the melting occurred. The first suggestion was near the surface of the sun, with the newly formed meteorites being cast outward by solar flares. The second suggestion was melting in an environment that we would now refer to as the solar nebula – a remarkably prescient suggestion, far superior to the first.

A perusal of the abstract volume (King, 1982) from the 1982 "Conference on Chondrules and Their Origins" reveals that impact melts and a hot inner nebula were the leading mechanisms, as judged by the number of favourable abstracts. Meteor ablation, nebula lightning, and magnetic flares were also advanced as explanations. The abstract volume (Hewins, 1994) for the 1994 conference showed a clear evolution in popularity: the hot inner nebula and meteor ablation models had no advocates at all, whereas a number of new mechanisms were promoted – flash heating, FU Orionis outbursts, bipolar flows, accretion shocks, and nebular shock waves. Of the newcomers, by far the most popular (and elusive) is 'flash heating'; flash heating satisfies the cosmochemical constraints of reproducing chondrule textures and minimizing

volatile losses, but sidesteps the issue of the actual heating mechanism.

Chondrule heating mechanisms were reviewed by Wood (1988), Grossman (1988), and Levy (1988), and more recent work is described in this book. Many of the other papers in this book detail the variety of cosmochemical constraints on chondrule formation; other particularly pertinent reviews were given by Grossman *et al.* (1988), Hewins (1988), and Wasson (1993). In this chapter we shall consider each of the specific mechanisms discussed at the 1982 and 1994 conferences, in hopes of singling out the most likely chondrule heating mechanism, given the current state of our knowledge.

IMPACT MELTS

The terrestrial planets and the cores of the giant planets are believed to have formed through the collisional accumulation of progressively larger bodies. By its very nature this process involves large impacts that are energetic enough to melt much of the impactor and host rock, and thus impacts potentially could form chondrules (Urey, 1956).

Arguments for:

*Impacts were widespread and sufficiently energetic to thermally process all terrestrial planet material.

*A molten spray in free fall could produce droplets resembling chondrules, much as originally envisioned by Sorby (1877). Jetting has been invoked as a means of producing a molten spray (Kieffer, 1975). It is unclear though whether a molten spray would result in chondrules with the observed size distributions.

*The volatile inventory in chondrules could be enhanced by cometary impactors.

Arguments against:

*If impacts with planetary-size bodies produce chondrules, then lunar breccias should contain chondrules in abundance, yet they do not (Taylor *et al.*, 1983).

*A number of other potent arguments against an impact origin for chondrules are enumerated by Taylor *et al.* (1983), such as the fact that no chondrules have been found with ages significantly less than 4.4 billion years old.

*Impacts break up a larger volume of material than they melt. Collisions between previously molten bodies would produce molten sprays, but molten bodies would produce homogenized chondrules that no longer contained oxygen isotope anomalies (Taylor *et al.*, 1983; Clayton, 1993).

*The lunar regolith is dominated by impact products, such as agglutinates, that are absent from chondrites (Kerridge and Kieffer, 1977).

*Multiple cycles of concentric heating do not appear to be possible. Assume that an impact melt droplet cools into a chondrule, which is then incorporated into either the parent body, or is ejected and later accumulated into another planetesimal. Subsequent impact processing may then reheat the chondrule, but this reheating will occur while the chondrule is embedded in a host rock, and may even remelt the entire chondrule-containing rock. That is, multiple heating cycles operating on isolated chondrules would not be expected. At any rate, dusty rims would not be produced on embedded chondrules; the rims would have to form during the trajectory of the initial impact melt.

*In order for impacts to be energetic enough to result in substantial melting, relative velocities between the colliding bodies must be on the order of 5 km/sec. Such high eccentricities are only likely to develop late in the planetary accumulation phase, after most of the gas and dust is accumulated or dissipated. It seems unlikely that chondrule formation occurred after the final accumulation of Jupiter. Furthermore, without a shroud of nebular dust grains to help retain the impact energy, it is unclear if the droplets would cool too rapidly in anything other than a giant impact.

*No target rock suitable for forming chondrites through impact melts has been identified in meteorite collections (Taylor *et al.* 1983). We do not necessarily have samples representative of all inner solar system bodies, however.

*If impacts melts are similar to igenous melts, the rare earth elements (REE) would be fractionated, which is not observed. Partial melting leads to preferential segregation of REE (e.g., lanthanum, europium, and neodymium) into either the solid or the melt, which would lead to REE compositions in the melts seldom seen in chondrules (e.g., Misawa and Nakamura, 1988).

*If chondrules formed from impacts between relatively large planetesimals and planetary embryos, then chondritic meteorites would not be the truly primitive objects that we believe them to be on other grounds, such as bulk composition and oxygen isotopic variations. We would need to postulate the existence of primitive objects possibly unrepresented in our meteorite collections. This is a classic "chicken and the egg" problem: which came first, chondrules or protoplanets?

METEOR ABLATION

Meteor ablation consists of the formation of molten droplets as a large (roughly m-size or larger) solid body moves supersonically through an atmosphere and is heated by the friction associated with gas drag.

Arguments for:

*Protoplanetary atmospheres could have the high gas pressures, variable partial pressures of oxygen, and dust grains necessary to form chondrules and their rims.

*Production of chondrules inside the gravitational potential well of a sizeable protoplanet would preserve the chondrules against nebular homogenization and against disappearance into the protosun.

*Textures of terrestrial ablation melts are similar to those of chondrule rims (Podolak *et al.*, 1993).

Arguments against:

*Atmospheres are hard to retain on protoplanets unless the protoplanets are quite large, larger than any body in the present day asteroid belt.

*For melting to occur and produce stable droplets, a velocity of the large solid body relative to the protoplanet in the range of a few to 25 km/sec is necessary (Liffman 1992). Relative velocities of this size require highly eccentric orbits that develop only after the nebula gas is dissipated. However, maintaining a protoplanetary atmosphere becomes even harder in the absence of nebular gas. If chondrules could only form after the dissipation of nebular gas around 2 to 3 AU, then they would be younger than Jupiter.

*Making an opaque rim on a sub-mm-sized chondrule through ablation requires the presence of a pre-existing rim of low melting temperature matter (Podolak *et al.*, 1993), making the mechanism appear contrived, though rims have lower-melting mineralogies than cores. In addition, a very narrow range of relative velocities is necessary in order to achieve melting and avoid vaporization (Podolak *et al.*, 1993).

*An objection raised to the impact melt mechanism applies here as well – chondrites could not be primitive bodies if their production requires the presence of protoplanets.

HOT INNER NEBULA

The solar nebula may have been hot enough to account for the high temperature processing of chondrules moving (with respect to the gas) in one of the three possible directions: vertically (Cameron and Fegley, 1982), radially (Morfill, 1983), or azimuthally (Boss, 1988).

Arguments for:

*A minimum mass (0.02 $M_\odot$ inside 10 AU) solar nebula undergoing mass accretion at astronomically-inferred rates could maintain midplane temperatures close to 1000 K at 2.5 AU (Boss, 1993; Cassen, 1994), or around 1500 K in a more massive (0.13 $M_\odot$) nebula.

*A hot region located at 2.5 AU removes the need to diffuse chondrules upstream (against the inward accretion flow) from hot regions close to the sun out to the asteroid belt.

*A relatively hot inner nebula may be necessary to explain the volatile depletions in bulk compositions of the inner planets (e.g., Boss, 1993).

Arguments against:

*Nebula models cannot produce steady state temperatures at 2.5 AU high enough to match the chondrule-forming requirement of maximum temperatures of 2000 K to 2200 K.

*The cooling times for globally hot regions (many years) are much longer than the times (hours to days) over which chondrules cooled to subsolidus temperatures.

*The alternative for achieving rapid heating and cooling in a globally hot nebula is to require rapid relative motion between the chondrule precursors and the hot gas. However, the relative velocities required are too close to the speed of light for this to be considered at all possible. In fact, for mm-sized aggregates, relative velocities should be very low (Weidenschilling, 1977). Depending on the geometry of a high temperature region, it might even be necessary to alter the direction of these super high speed chondrules, in order to achieve rapid cooling after rapid heating, which seems even more implausible.

*The presence of volatiles like FeS in chondrule interiors implies formation in regions with temperatures of no more than about 650 K prior to the event that melted the precursor aggregates (Grossman, 1988; Wasson, 1993).

FU ORIONIS

About a dozen young stars have been observed to undergo a decades to centuries-long phase when their luminosity remains elevated by several orders of magnitude before decreasing again, implying the transient release of considerable energy. It is statistically possible that nearly all low mass stars like the sun undergo a similar phase a small number of times during their early evolution. Episodic dumping of mass from a disk onto the central star is a leading explanation for FU Orionis outbursts (Hartmann and Kenyon, 1985).

Arguments for:

*FU Orionis outbursts have been observed in a number of young stars and thus are well-known to be a common occurrence in low mass stars like the sun. Hypothesizing an FU Orionis phase for the sun is thus a relatively low risk proposition, compared to more speculative physical processes sometimes invoked for chondrule formation.

*The inferred mass accretion rate onto the central star of about 10^{-4} $M_\odot$/yr during an outburst is sufficient to produce intense heating of the inner protostellar disk.

*Multiple FU Orionis outbursts (about ten per star) could lead to multiple heating events.

Arguments against:

*The time scale for an FU Orionis outburst to occur is several months, followed by the several decades required for the system to return to normal luminosity. These time scales are inconsistent with the rapid heating and cooling requirements, assuming that no other processes are invoked.

*In detailed theoretical models of FU Orionis outbursts, designed to reproduce the observational light curves, the transient heating occurs very close to the star, with very little happening out at 2 to 3 AU (Clarke *et al.* 1990; Bell and Lin, 1994). Huss (1988) has proposed that an FU Orionis event melts precursors at the inner edge of the disk, which accrete into a parent body after the outburst is over. Given that the FU Orionis heating is limited to the inner few 0.1 AU of the disk, however, it does not seem possible to later transport these bodies to 2 to 3 AU.

*The FU Orionis models of Clarke *et al.* (1990) and Bell and Lin (1994) show that the inner disk (inside about 0.3 AU) regions that are heated during an outburst phase start from a temperature of about 3000 K, and reach maximum temperatures of about 100,000 K. This range of temperatures is much too high to be consistent with the presence of relict grains and volatiles in chondrules.

BIPOLAR FLOWS

A ubiquitous feature of young stars is the presence of energetic outflows with bipolar (biconical) shapes. On the largest scales (parsecs) these outflows are seen as outflowing lobes of molecular gas, while on the smallest scales (10 to 100 AU), narrow jets of collimated hot gas are optically visible. The jets and outflows occur in a direction perpendicular to that of the midplane of the protostellar disk. Some of this ejected stellar matter might return to the nebula in the form of chondrules (Skinner, 1990). It also has been proposed that the ablation of a parent body in a bipolar outflow could have produced molten droplets that became chondrules (Liffman, 1992).

Arguments for:

*As in the case of FU Orionis outbursts, bipolar flows have the great advantage of abundant observational proof of their existence.

*Enough energy is involved in powering bipolar flows (a significant fraction of the stellar luminosity) to ensure that thermal processing of grain aggregates could occur.

*Ablation of a parent body in a high velocity gas flow has some of the advantages of the meteor ablation mechanism, namely the production of molten droplets around mm-size.

Arguments against:

*Frictional drag between the ions and neutrals of an energetic bipolar flow heats the outflowing gas to about 10,000 K at a radius about 10 times that where the wind was launched (Safier, 1993), which is likely to be close to the stellar surface. Cooling below this value would only occur at great distance from the star, or for material that prematurely exits the bipolar outflow (however, entrainment of surrounding gases may dominate losses). Rapid cooling is unlikely to occur.

*Even refractory grains would be destroyed at temperatures of 10,000 K, and the resulting plasma would be chemically and isotopically homogenized. No relict grains would survive.

*Mm-sized droplets in the midst of such an outflow would hover at a fixed height (Safier, 1993); only much smaller droplets would be carried along with the outflow.

*Because the geometry of bipolar flows requires a vertical (with respect to the nebula midplane) collimation mechanism, such an outflow could not ablate matter from a parent body orbiting in the midplane of the nebula.

*In order for ablation to be efficient, a high gas density in the outflow is required, characteristic of radii very close (0.1 AU) to the star. It is unlikely that suitable parent bodies would form at the temperatures expected at 0.1 AU (at least 1800 K).

NEBULA LIGHTNING

Electric charge may accumulate in the nebula through grain-grain collisions, and may then be released catastrophically through lightning bolts, much as occurs in the atmospheres of several planets.

Arguments for:

*Lightning discharges occur quickly enough to meet the requirement of rapid heating of chondrules (Horányi *et al.* 1995), while cooling at too great a rate could be prevented by the presence of dust grains in the ambient nebula.

*Eisenhour *et al.* (1994) and Eisenhour and Buseck (1995) have shown that electromagnetic radiation can selectively heat chondrule precursors, possibly explaining the lower limit in chondrule size distributions: small grains cannot absorb the radiation efficiently and so do not melt.

*Lightning discharges are localized events that would thermally process certain grain aggregates, while leaving untouched nearby aggregates. Solids formed from both of these populations would then be thermally heterogeneous.

*Multiple cycles of thermal processing would be expected for those chondrules unlucky enough to be hit more than once by lightning, possibly accounting for the formation of multiple rims and compound chondrules.

*Lighting is well-suited to producing the rapid heating and cooling that may be required to retain volatiles such as FeS.

Arguments against:

*The electrostatic energy transferred may be insufficient to melt mm-sized grain aggregates (Love *et al.*, 1994), because the discharge probably resembles a diffuse aurora more than a terrestrial lightning bolt.

*Building up large-scale charge separations in the face of the tendency toward recombination requires dust grains to act as charge reservoirs, and processes such as dust grain collisions are needed to separate the charges (Morfill *et al.*, 1993). Based on terrestrial lightning, ice grains are well suited for the purpose of separating charges, yet ice grains could exist only high in the nebula at 2.5 AU, or outside about 5 AU if formation occurred closer to the midplane. The ambient nebula temperature in the chondrule formation region is thought to have been at least of order 500 K (Grossman, 1988), much too hot for ice to be solid. It is unclear if significant charge separation can occur for silicate and metal grains (Morfill *et al.*, 1993; Rasmussen and Wasson, 1995).

*If heating is too rapid, the surface layers will be vaporized before the interior can be melted. Rasmussen and Wasson (1995) estimate that nearby gas is heated above the liquidus for only about 1 second, insufficient time to completely melt the interior of a mm-sized precursor by conduction of heat. Rasmussen and Wasson (1995) thus propose that the precursors were loose aggregates of small grains that collapsed due to surface tension following the lightning strike and then formed droplets; energy is delivered to the precursors by energetic particles or by radiation.

*While the nebular fractional ionization (caused by ^{40}K decay) is quite low in an absolute sense, the ionization may still be high enough to short out any electrostatic charge concentrations before they become intense enough to produce a catastrophic discharge (Gibbard and Levy, 1994). Rasmussen and Wasson (1995) have proposed that a pervasive fine dust component could soak up any free electrons produced by radioactive decay, allowing significant charge buildup to occur, though what the breakdown potential would be in this case has not been calculated.

*Radiant (electromagnetic) energy is best transferred to aggregates with high proportions of metal (Eisenhour *et al.*, 1994), yet the most metal-rich chondrules often have textures indicative of less melting than the metal-poor chondrules (H. Connolly, 1994, private communication). Rasmussen and Wasson (1995) suggest instead that collisions of dust grain aggregates with discharge-accelerated ions and electrons may be the main mechanism for transferring energy.

MAGNETIC FLARES

Studies of remanent magnetism in chondritic meteorites imply the existence of magnetic fields (about 1 gauss in strength) during the cooling of these rocks below the Curie temperature (Sugiura and Strangway, 1988). Magnetic fields periodically erupt from the sun's surface and produce energetic solar flares; magnetic fields can convert their energy into thermal energy through magnetic reconnection events (Levy and Araki, 1989). A bow shock between the nebula and the sun's stellar wind has been proposed as a likely location for nebula flares capable of thermally processing chondrule precursors (Cameron, 1995). The frequency of nebular flares, should they occur, has not been estimated.

Arguments for:

*Rapid heating and cooling are possible, with a dusty nebula helping to prolong the cooling phase. Solar flares last on the order of hours, so long-lived nebula flares could also permit extended cooling periods.

*By analogy with the sun, magnetic flares probably existed on the surface of the nebula.

*Localized heating is possible, leading later to the assembly of both heated and un-heated components in the same chondrite.

*Multiple cycles of heating and cooling are likely so long as the magnetic events are not rare.

Arguments against:

*Magnetic field reconnection events should produce 1 Mev protons and electrons, which are available to heat precursor aggregates. However, these energetic particles cannot penetrate much matter before being absorbed, so the precursors must be heated well above the nebula midplane, in the nebular "corona" (Levy and Araki, 1989) where the flares occur themselves. The magnetic field strength required to melt chondrules in the corona is then at least 5 gauss (Levy and Araki, 1989), comparable to or higher than the inferred maximum midplane value of about 1 gauss at 2.5 AU (Sugiura and Strangway, 1988). The magnetic field strength is likely to fall off strongly with vertical height, making the minimum coronal field strength of 5 gauss hard to achieve (Morfill *et al.*, 1993).

*Magnetic field generation by the dynamo mechanism is hindered by the generally low degree of ionization, and may

be prevented altogether at 2 to 3 AU by the existence of a neutral gap between the warm inner regions heated by thermal ionization, and the outer regions which are penetrated by cosmic rays (Stepinski and Reyes-Ruiz, 1993). The recent detection of a magnetic moment per unit mass for the asteroid Gaspra (Kivelson *et al.*, 1993) in the same range as that of chondritic meteorites (Sugiura and Strangway, 1988) may mean that chondrites obtained their remnant magnetism not as isolated mm-sized spherules in the solar nebula, but as part of much larger parent bodies, which may have themselves been immersed in a nebula-wide magnetic field, or may have obtained their remnant field through some process involving the parent body. In the latter case, we would no longer have any evidence for a significant nebular magnetic field.

*Given that heating must occur in the nebula corona, precursor aggregates must be transported upwards and out of the nebula to the corona for thermal processing. Turbulent convective motions have been invoked as a possible lofting mechanism (Levy and Araki, 1989), but it is unlikely that this mechanism is efficient enough to loft significant numbers of precursor aggregates out of the nebula and into the nebular corona (Morfill *et al.* 1993).

ACCRETION SHOCKS

Matter infalling from the parent molecular cloud core onto the protoplanetary disk must pass through an accretion shock, where the kinetic energy of infall (sufficient to heat the gas and dust into a plasma) is converted into heat as the infalling gas is brought to rest. Pre-existing dust grains continue through the shocked gas until they are slowed by gas drag. Frictional heating from the gas drag could heat precursor grains to melting (Wood, 1984).

Arguments for:

*Adiabatic shocks do not produce enough of a post-shock density increase to be effective at heating grains (Wood, 1988). However, radiative cooling behind the accretion shock increases the post-shock density to higher values that increase the amount of gas drag heating of precursor grain aggregates (Ruzmaikina and Ip, 1995).

*Unlike other shock fronts postulated in the nebula, the accretion shock is certain to have existed, given that protostellar collapse occurs.

Arguments against:

*Precursor aggregates with sizes on the order of a mm (or much larger if the aggregates are fluffy) must be formed somewhere upstream of the accretion shock in order to provide the raw material for chondrules. Dust grains originate in stellar atmospheres and may undergo subsequent growth. One analysis of grain growth in the interstellar medium (where mean dust grain sizes are less than 0.1 micron) and during protostellar collapse found that growth could occur to sizes of at most 10 microns (Cassen and Boss, 1988). A more detailed analysis found that growth in a dense cloud core and during collapse leads to grains of 1 to 100 micron-size (Weidenschilling and Ruzmaikina, 1994); essentially no grains formed with masses comparable to chondrules, unless extreme parameter choices are made in the model. The standard astronomical wisdom is that the maximum size of dust grains consistent with cosmic abundances and observed interstellar reddening of starlight is about 1 micron, but it must be admitted that these observations are insensitive to the presence of much larger grains.

*In order to heat precursors to melting at 2 to 3 AU, disk mass accretion rates of at least $10^{-4} M_\odot$/yr are required (Ruzmaikina and Ip, 1995), a rate which is a factor of 10 to 100 times larger than that usually inferred for the formation of low mass stars. Rates this high might only be achieved during transient bursts of accretion.

*Only a single cycle of thermal processing is possible, because matter passes through the accretion shock only once. Making compound chondrules or rims is thus difficult if not impossible.

NEBULAR SHOCKS

A shock wave propagating through the interior of the nebula would overtake precursor aggregates and could heat them to the melting point, through heating processes such as gas drag and gas thermal energy lost by gas-grain collisions, and to a lesser extent through thermal radiation from the shock front and from the other dust grains (Hood and Horányi, 1991; 1993).

Arguments for:

*Shock wave processing inside the nebula ensures the presence of precursor aggregates grown within the high density nebula. Shock waves need not propagate all the way to the disk midplane in order to process grains, because even a small amount of turbulence is sufficient to keep most small (< mm-size) grains lofted above the nebula midplane (Weidenschilling 1988; Cuzzi *et al.* 1993; Dubrulle *et al.* 1995).

*Shock waves propagating with speeds of 4 to 6 km/sec can result in the rapid heating required for chondrules (Hood and Horányi, 1991; 1993), and the presence of an enhanced dust density could account for the observed cooling rates. Hood and Horányi (1993) find that melting temperatures (nominally 1600 K) can be reached within about 100 sec, and that grain temperatures can drop back down below the solidus within a similiar or shorter time period.

*Because heating of dust grains by shock wave processing occurs over periods of order 100 sec (Hood and Horányi, 1993), heat conducted inward from hot grain surfaces is able to melt the centers of even solid mm-sized precursors.

*Given an episodic source of nebula shock waves, multiple cycles of heating would occur.

*Depending on how the shock waves are generated and dissipated, varying degrees of thermal processing would occur, depending on whether the precursor was located close to where the shock wave was generated, or close to where it finally dissipated.

*On the scale of the nebula, shock waves are quite localized, and would not be expected to homogenize their products. Chondrules processed by shock fronts at different locations would be expected to have a range of properties.

Arguments against:

*Shock waves require supersonic motions, which dissipate quickly in the absence of regeneration. A plausible source of shock waves internal to the nebula must be specified. FU Orionis outbursts might launch shock waves at the inner edge of the disk, but it is uncertain if they could propagate out to 2 or 3 AU. Growth of nonaxisymmetry (such as bars or spiral density waves) could lead to shock waves in the azimuthal direction, but such growth probably requires a nebula that is cooler and more massive than is normally considered likely (Boss, this volume). There is astronomical evidence for the episodic accretion of clumps of gas and dust onto the nebula (Boss and Graham, 1993), but it is unclear if these clumps really exist and are massive enough to drive shock waves into the nebula (Hood and King, this volume).

CONCLUSIONS

Astrophysicists may not yet be able to reach a consensus about the precise chondrule heating mechanism, but cosmochemists seem to be quite insistent that rapid "flash" heating was required. Still, one must ask the uniqueness question – is flash heating really the *only* way to explain chondrule textures and volatile inventories? If the answer is a resounding "yes", then many models of chondrule heating can be summarily tossed out. In this astrophysicist's somewhat biased opinion, the most promising means of heating chondrules is with shock waves propagating within the nebula, though it is conceivable that future work may ameliorate some of the objections raised against some of the other contenders. The challenge for those who take shock wave processing seriously is clear: find evidence and arguments for an energetic, episodic source of nebula shock waves.

ACKNOWLEDGEMENTS

I thank John Wasson and an anonymous referee for their many important improvements to the manuscript. This work was partially supported by the NASA Planetary Geology and Geophysics Program under grant NAGW-1410.

REFERENCES

Bell K. R., and Lin D. N. C. (1994) Using FU Orionis outbursts to constrain self-regulated protostellar disks. *Astrophys. J.*, **427**, 987–1004.

Boss A. P. (1988) High temperatures in the early solar nebula. *Science* **241**, 565–567.

Boss A. P. (193) Evolution of the solar nebula. II. Thermal structure during nebula formation. *Astrophys. J.* **417**, 351–367.

Boss A. P. and Graham J. A. (1993) Clumpy disk accretion and chondrule formation. *Icarus* **106**, 168–178.

Cameron A. G. W. (1995) The first ten million years in the solar nebula. *Meteoritics* **30**, 133–161.

Cameron A. G. W. and Fegley M. B. (1982) Nucleation and condensation in the primitive solar nebula. *Icarus* **52**, 1–13.

Cassen P. (1994) Utilitarian models of the solar nebula. *Icarus*, **112**, 405–429.

Cassen P. and Boss A. P. (1988) Protostellar collapse, dust grains, and solar system formation. In *Meteorites and the Early Solar System* (eds. J. F. Kerridge and M. S. Matthews), pp. 304–328. University of Arizona Press.

Clarke C. J., Lin D. N. C. and Pringle J. E. (1990) Pre-conditions for disc-generated FU Orionis outbursts. *Mon. Not. R. Astron. Soc.* **242**, 439–446.

Clayton R. N. (1993) Oxygen isotopes in meteorites. *Ann. Rev. Earth Planet. Sci.*, **21**, 115–149.

Cuzzi J. N., Dobrovolskis A. R. and Champney J. M. (1993) Particle-gas dynamics in the midplane of a protoplanetary nebula. *Icarus* **106**, 102–134.

Dubrulle B., Morfill G. and Sterzik M. (1995) The dust subdisk in the protoplanetary nebula. *Icarus* **114**, 237–246.

Eisenhour D. D., Daulton T. L. and Buseck P. R. (1994) Electromagnetic heating in the early solar nebula and the formation of chondrules. *Science* **265**, 1067–1070.

Eisenhour D. D. and Buseck P. R. (1995) Radiative heating and the size distribution of pre-chondrule aggregates of dust. *Lunar Planet. Sci.* **XXVI**, 365–366.

Gibbard S. G. and Levy E. H. (1994) On the possibility of precipitation-induced vertical lightning in the protoplanetary nebula. In *Papers Presented to Chondrules and the Protoplanetary Disk*, LPI Contrib. 844, p. 9. Lunar and Planetary Institute, Houston.

Grossman J. N. (1988) Formation of chondrules. In *Meteorites and the Early Solar System* (eds. J. F. Kerridge and M. S. Matthews), pp. 680–696. University of Arizona Press.

Grossman J. N., Rubin A. E., Nagahara H. and King E. A. (1988) Properties of chondrules. In *Meteorites and the Early Solar System* (eds. J. F. Kerridge and M. S. Matthews), pp. 619–659. University of Arizona Press.

Hartmann L. and Kenyon S. J. (1985) On the nature of FU Orionis objects. *Astrophys. J.* **299**, 462–478.

Hewins, R. H. (1988) Experimental studies of chondrules. In *Meteorites and the Early Solar System* (eds. J. F. Kerridge and M. S. Matthews), pp. 660–679. University of Arizona Press.

Hewins, H. (1994) *Papers presented to Chondrules and the Protoplanetary Disk*. Abstract volume. Lunar and Planetary Institute Contribution No. 844.

Hood L. L. and Horányi M. (1991) Gas dynamic heating of chondrule precursor grains in the solar nebula. *Icarus* **93**, 259–269.

Hood L. L. and Horányi M. (1993) The nebular shock wave model for chondrule formation: one-dimensional calculations. *Icarus* **106**, 179–189.

Horányi M., Morfill G., Goertz C. K. and Levy E. H. (1985) Chondrule formation in lightning discharges. *Icarus*, **114**, 174–185.

Huss G. R. (1988) The role of presolar dust in the formation of the solar system. *Earth, Moon, Planets* **40**, 165–211.

Kerridge J. F. and Kieffer S. W. (1977) A constraint on impact theories of chondrule origin. *Earth Planet. Sci. Lett.* **35**, 35–42.

Kieffer S. W. (1975) Droplet chondrules. *Science* **189**, 333–339.

King E. A. (1982) *Papers presented to the Conference on Chondrules and their Origins*. Abstract volume. Lunar and Planetary Institute Contribution No. 493.

Kivelson M. G., Bargatze L. F., Khurana K. K., Southwood D. J., Walker R. J. and Coleman P. J. (1993) *Science* **261**, 331–334.

Levy E. H. (1988) Energetics of chondrule formation. In *Meteorites and the Early Solar System* (eds. J. F. Kerridge and M. S. Matthews), pp. 697–711. University of Arizona Press.

Levy E. H. and Araki S. (1989) Magnetic reconnection flares in the protoplanetary nebula and the possible origin of meteorite chondrules. *Icarus* **81**, 74–91.

Liffman K. (1992) The formation of chondrules via ablation. *Icarus* **100**, 608–620.

Love S. G., Keil, K. and Scott, E. R. D. (1994) Formation of chondrules by electrical discharge heating. In *Papers Presented to Chondrules and the Protoplanetary Disk*, LPI Contrib. 844, p. 21. Lunar and Planetary Institute, Houston.

Misawa K. and Nakamura N. (1988) Highly fractionated rare-earth elements in ferromagnesian chondrules from the Felix (CO3) meteorite. *Nature* **334**, 47–50.

Morfill G. (1983) Some cosmochemical consequences of a turbulent protoplanetary cloud. *Icarus* **53**, 41–54.

Morfill G., Spruit, H. and Levy E. H. (1993) Physical processes and conditions associated with the formation of protoplanetary disks. In *Protostars & Planets III* (eds. E. H. Levy and J. I. Lunine), pp. 939–978. University of Arizona Press, Tucson.

Podolak M., Prialnik D., Bunch T. E., Cassen P. and Reynolds R. (1993) Secondary processing of chondrules and refractory inclusions (CAIs) by gasdynamic heating. *Icarus* **104**, 97–109.

Rasmussen K. L. and Wasson J. T. (1995) A lightning model of chondrule melting. *Earth Planet. Sci. Lett.*, submitted.

Ruzmaikina T. V. and Ip W. H. (1995) Chondrule formation in radiative shock. *Icarus*, **112**, 430–447.

Safier P. N. (1993) Centrifugally driven winds from protostellar disks. I. Wind model and thermal structure. *Astrophys. J.* **408**, 115–147.

Skinner W. R. (1990) Bipolar outflows and a new model of the early Solar System. Part II: the origins of chondrules. *Lunar Planet. Sci.* **XXI**, 1168–1169.

Sorby H. C. (1877) On the structure and origin of meteorites. *Nature* **15**, 495–498.

Stepinski T. F. and Reyes-Ruiz M. (1993) Magnetically controlled solar nebula. *Lunar Planet. Sci.* **XXIV**, 1351–1352.

Sugiura N. and Strangway D. W. (1988) Magnetic studies of meteorites. In *Meteorites and the Early Solar System* (eds. J. F. Kerridge and M. S. Matthews), pp. 595–615. University of Arizona Press.

Taylor G. J., Scott E. R. D. and Keil K. (1983) Cosmic setting for chondrule formation. In *Chondrules and Their Origins* (ed. E. A. King), pp. 262–278. Lunar and Planetary Institute, Houston.

Urey H. C. (1956) Diamonds, meteorites, and the origin of the solar system. *Astrophys. J.* **124**, 623–637.

Wasson J. T. (1993) Constraints on chondrule origins. *Meteoritics* **28**, 14–28.

Weidenschilling S. J. (1977) Aerodynamics of solid bodies in the solar nebula. *Mon. Not. Roy. Astron. Soc.* **180**, 57–70.

Weidenschilling S. J. (1988) Formation processes and time scales for meteorite parent bodies. In *Meteorites and the Early Solar System* (eds. J. F. Kerridge and M. S. Matthews), pp. 348–371. University of Arizona Press.

Weidenschilling S. J. and Ruzmaikina T. V. (1994) Coagulation of grains in static and collapsing protostellar clouds. *Astrophys. J.* **430**, 713–726.

Wood J. A. (1984) On the formation of meteoritic chondrules by aerodynamic drag heating in the solar nebula. *Earth Planet. Sci. Lett.* **70**, 11–26.

Wood J. A. (1988) Chondritic meteorites and the solar nebula. *Ann. Rev. Earth Planet. Sci.* **16**, 53–72.

28: Models for Multiple Heating Mechanisms

L. L. HOOD and D. A. KRING

Lunar and Planetary Laboratory, The University of Arizona, Tucson, AZ, 85721, U.S.A.

ABSTRACT

Two classes of multiple transient heating mechanisms that appear to be broadly consistent with meteoritic constraints on chondrule formation are electrostatic discharges (lighting) and gas dynamic shock waves (shocks). Both mechanisms are capable of providing relatively short heating times, localization of heating zones within the nebula, and multiple heating events. Theoretical models of charge separation and discharge in the tenuous solar nebula raise doubts about whether nebular lighting was widespread and energetic enough to explain the abundance of chondrules in chondrites. Astronomical observations of time-dependent phenomena near young stellar objects suggest that episodic nebular shock waves were a common occurrence in the proto-planetary nebula. However, the precise nature of the shock generation mechanism(s) and the extent to which shock heating can be quantitatively consistent with meteoritic constraints remain incompletely established. The clumpy disk accretion model of Boss and Graham implies shocks up to Mach $\simeq 8$ in strength penetrating to within a few tenths of an AU from the midplane at 2–3 AU. Precursor aggregates lofted to these altitudes and concentrated by turbulent eddies could have been thermally processed in a manner consistent with observed chondrule properties. Detailed models of shock structure, including the effects of post-shock cooling, combined with simulations of dust particle heating and cooling histories are needed to more fully test this possible mechanism for chondrule formation.

1. INTRODUCTION

As reviewed in much greater detail in this volume and elsewhere, chondrules are small (approximately 0.2 to 1 mm in diameter), usually spheroidal-shaped, silicate particles constituting a major component (0 – 75% by volume) of chondritic meteorites [Wasson, 1985; Grossman et al., 1988; Hewins, 1988]. The heating events responsible for the melting of chondrule precursor grains were evidently widespread and efficient at a time prior to and during the formation of meteorite parent bodies. The nature of these heating events is therefore a basic problem whose solution could yield insights into the manner in which the agglomeration of solid material in the early solar system first occured. An analogous, and possibly related, problem is the nature of events responsible for the melting of Type B and C calcium-aluminium-rich inclusions (CAI's) in chondrites [MacPherson et al., 1988].

Most detailed analyses have concluded that chondrules formed in localized heating events in a relatively cool, dusty nebular environment [Taylor et al., 1983; Wood, 1984; Hewins, 1988; Grossman, 1988]. For example, formation from volcanic melts is eliminated by the closely chondritic compositions of chondrules. Formation as impact melt droplets on planetesimal surfaces is disfavored by the general lack of similarity of chondrules to components of lunar and meteorite parent body regoliths. In addition, compositional evidence (e.g., the covariation of lithophile elements in chondrules) is most consistent with formation from heterogeneous nebular condensates rather than from homogeneous or processed (igneous or metamorphic) solids [Grossman and Wasson, 1983]. The ambient nebula environment must have been cool (< 650 K) to account for the presence of FeS in most chondrules. The nebula was apparently dusty (> 1 – 10 chondrule-sized particles per m³) because of the number of collisions between plastic chondrules [Gooding and Keil, 1981], the high inferred nebular oxygen fugacity [Wood, 1985; Kring, 1988] and the dusty rims present on many chondrules [Grossman *et al.* 1988; Metzler *et al.*, 1992]. Finally, textural and compositional evidence indicates that the heating of chondrule precursors occurred rapidly (minutes or less) while experimental simulations show that chondrule formation temperatures peaked in excess of 1873 K and remained above 1423 K for approximately 1 hour [Hewins, 1988; Hewins and Radomsky, 1990]. The relatively short inferred cooling times, in particular, require that the heating events were localized rather than nebula-wide. One recent estimate for the spatial scale of heating events is < 10 km, based on petrologic evidence for the accretion of hot material onto cold chondrules [Kring, 1991]. However, the inferred cooling rates are still much less than that for direct radiation to space (about 10^6 K per hour), implying the persistence of relatively hot surroundings.

Although pre-solar solids (interstellar dust) cannot be completely ruled out as precursors of chondrules, most detailed evaluations have concluded that nebular condensates are the most probable or dominant precursors [Grossman, 1988; see also Kerridge, 1993]. Compositional trends and interelement correlations in chondrules can be understood in terms of models involving condensation and reaction in the solar nebula [Grossman and Wasson, 1983]. While the compositional distributions of interstellar dust are incompletely known, it is unlikely that these precursors would have the correct composition and variability to produce the observed chemical trends and correlations. Moreover, it is uncertain whether interstellar dust, which is typically submicron in size, could have aggregated to millimeter size in sufficient numbers prior to infall into the protosolar accretion disk [Cassen and Boss, 1988; but see also Weidenschilling and Ruzmaikina, 1994].

There is increasing evidence that many chondrules have been thermally processed repeatedly and thus contain recycled fragments of previous generations of chondrules [e.g., Alexander, 1994]. Among this evidence are the textures and chemical and chemical compositions of relict grains [Kring, 1988] and the juxtaposition of high-temperature rims around chondrule cores [Kring, 1991]. Many chondrules may record at least three separate heating events: An initial one to explain some relict grains, a second one to explain the host chondrule, and a final one to explain igneous rims on the host chondrule. Similarly, refractory rims on CAI's [Wark and Lovering, 1977] most probably imply the occurrence of a secondary "flash" heating event of no more than a few seconds' duration [Palme and Boynton, 1993].

Studies of chondrules themselves have not uniquely identified the nature of the primary energy source that was involved. Electromagnetic radiation, gas-dust energy transfer, and energetic electron and ion bombardment have all been proposed as candidate heat sources. Although the textures of certain opaque inclusions in chondrules are consistent with heating by electromagnetic radiation [Eisenhour *et al.*, 1994], it is possible that a combination of other energy sources and electromagnetic radiation (from, e.g., surrounding dust and hot gas) could also produce these textures.

Finally, a subtle but potentially important clue to the circumstances of chondrule formation is that the net energy transfer to precursor particles was apparently limited in some manner. As summarized, for example, by Wasson [1994], most chondrules were not completely melted and most were not heated sufficiently to allow FeS to decompose. Also, the fact that chondrules were not homogenized by thermal processing implies that the number of heating events was limited. Thus, the chondrule heating event(s) deposited a relatively narrow range of energy per unit mass in chondrules. The amount of specific energy transferred to chondrule precursors does not seem to be dependent on mass since there is no observed correlation between chondrule mass and texture or composition [Grossman, 1988]. For a silicate specific heat of about 1 J/gm/K, the approximate minimum energy required to raise the temperature of a precursor dust assemblage to just above its liquidus temperature (~ 1800 K) is approximately 1500 J/gm. Apparently, either the heating events were always only marginally capable of melting grains or a negative feedback process acted to prevent further heating after liquidus temperatures were reached.

As summarized above, a successful chondrule transient heating mechanism must apparently be capable of operating locally and repeatedly in a cool dusty solar nebula. Among the many chondrule formation mechanisms that have been proposed [see, e.g., Grossman, 1988], only two classes of mechanisms appear to be consistent with this general requirement: Electrostatic discharges (.lightning) and gas dynamic shock waves (shocks). Consequently, only these two classes of mechanisms are currently receiving serious study by more than one investigator or group of investigators. Among the few potentially viable alternatives that have been proposed are magnetic flares [Sonett, 1979; Levy and Araki, 1989] and highly exothermic chemical reactions in presolar grains after entry into the nebula [Clayton, 1980]. Lightning has the advantage of localization and simplicity but there are questions about whether the nebula was sufficiently resistive to allow it to have been more widespread. There are also questions about whether lightning discharges in the tenuous solar nebula would have produced a sufficient energy flux to have been an efficient mechanism for heating and melting precursor aggregates. Shocks have the advantage of being capable of efficiently processing large numbers of dust grains, but there are questions about how shocks were generated in the nebula and whether the resulting heating events were sufficiently localized. Shocks have the added advantage of being potentially relatable to astrophysical observations of protostars and their associated accretion disks. In sections 2 and 3 of this paper, analytic methods are applied to examine further the extent to which nebular lightning and shocks are likely to provide adequate models for chondrule formation.

2. ELECTROSTATIC DISCHARGES

Since the initial proposal by Whipple [1966], lightning within a dusty solar nebula has been considered as a variable chondrule formation mechanism by a series of authors [Cameron, 1966; Rasmussen and Wasson, 1982; 1995; Fujii and Miyamoto, 1983; Pilipp *et al.*, 1992; Morfill *et al.*, 1993; Horanyi and Robertson, this volume]. From the standpoint of meteoritic constraints, the major advantages of lightning as energy for chondrule formation are the relatively short heating times, the localization of heated regions to the zones of the discharges, and the likelihood of multiple events. However, from a theoretical standpoint, there are questions about whether electrostatic discharges could have occurred with sufficient frequency and energy flux in the solar nebula to have melted large numbers of millimeter-sized dust particles.

The first issue that should be considered is whether processes existed in the solar nebula that would produce lightning discharges. By analogy with the mechanism responsible for charge separation in terrestrial thunderclouds, one may suppose that collisions between ice particles (or dust particles) of differing sizes in an insulating environment would result in the accumulation of a net charge of opposite sign on large and small particles [see, e.g. , Uman, 1987]. Gravitational segregation of such particles (in the assumed absence of turbulence) would then lead to charge separation and a large-scale electric field [e.g., Morfill *et al.*, 1993]. However, the degree of charge separated is limited by the small but finite electrical conductivity of the nebula itself which is believed to be determined (at least near the midplane) mainly by the decay of radioactive isotopes such as ^{40}K. Gibbard and Levy [1994] have recently investigated this problem and calculate that ionization of the nebula is sufficient to preclude any significant charge separation by this mechanism. They therefore conclude that precipitation-induced vertical lightning in the nebula is unlikely although the possibility of smaller-scale radial lightning discharges should be investigated further.

If it is assumed that dust charging processes not currently understood were successful in separating substantial amounts of charge in the nebula, the remaining issue is whether the resulting electrical discharges would have been energetic enough to have melted milimeter-sized silicate particles. As noted near the end of section 1, the minimum energy required to melt a precursor silicate aggregate is approximately 1500 J/gm. For an idealized spherical grain with typical diameter of 0.5 mm, cross-sectional area 2×10^{-3} cm^2, and mass $\simeq 0.2$ mg, the minimum required time-integrated energy flux is about 120 J/cm^2. In reality, energy loss by radiation from the particle, latent heat loss due to vaporization of small dust grains adhered to the particle, etc., would raise this value significantly. One approach toward investigating whether hypothesized nebular lightning would be capable of providing such a minimum energy per unit area is to construct a discharge model that combines current understanding of terrestrial lightning with estimates of the properties of the solar nebula. Love *et al.* [1995] have recently constructed such an approximate model and conclude that at the low gas densities characteristic of the protoplanetary nebula, electrical discharges analogous to terrestrial lightning would have energy flux densities too low to have melted millimeter-sized chondrule precursors.

In order to illustrate the problems encountered by nebular lightning models for chondrule formation, it is useful to consider the following simplified scaling calculations. A typical single-stroke terrestrial lightning discharge has a duration of ~ 100 milliseconds, a length of ~ 1 km, a width of ~ 2 mm, and transfers roughly 30 C of charge through a typical cloud electric field of ~ 0.1 MV/m [e.g.,

Uman, 1987]. The 30 C of charge (equivalent to 1.9×10^{20} electronic charges) is moved over a total potential drop of about 0.1 MV/m $\times$ 1 km = 100 MV. This yields a net energy transfer of $\simeq 1.9 \times 10^{28}$ electron volts (eV) or $\simeq 3 \times 10^9$ J. The cross-sectional area of the discharge is typically ~ 0.03 cm^2 so the time-integrated energy flux is $\simeq 10^{11}$ J cm^{-2}. A millimeter-sized particle exposed to such a discharge would obviously be vaporized almost instantaneously. However, because the protoplanetary nebula gas density was much less than that in the terrestrial troposphere, a nebular electrical discharge is expected to have significantly different characterstics as compared to those of a terrestrial lightning bolt. To first order, the discharge thickness should be proportional to the mean free path for electron-neutral collisions which is in turn inversely proportional to the gas number density [Pillipp *et al.*, 1992; Morfill *et al.*, 1993]. In the terrestrial troposphere, the number density is of order 10^{19} cm^{-3} while that in the solar nebula at 2–3 AU was in the approximate range 10^{13} to 10^{14} cm^{-3} [e.g., Wood and Morfill, 1988]. Consequently, a simple linear scaling yields an estimate for the thickness of a nebular discharge of 0.2 to 2 km. If we initially assume that the net energy transfer in a nebular discharge was comparable to that in a terrestrial lightning discharge (about 3×10^9 J), the increase in cross-sectional area (to $\simeq 10^9$–10^{11} cm^2) reduces the time integrated energy flux to $\simeq 0.03$ to 3 J cm^{-2}. Such an energy input falls short of that required to melt a precursor particle even if energy losses are neglected.

However, it is unclear whether the net energy transfer in a nebular discharge would have been as large as that in terrestrial lightning discharges. One basic parameter is the breakdown electric field which is an idealized measure of the minimum electric field above which an "avalanche" of ionization and current flow can occur. As discussed by Pilipp *et al.* [1992] and Love *et al.* [1995], the breakdown electric field magnitude E_{BD} can be roughly estimated as that which accelerates an electron to $\simeq 1$ eV of energy between molecular collisions. Hence, $eE_{BD}l_{en} \simeq 1$ eV where e is the electron charge and E_{BD} is the breakdown electric field. l_{en} is the electron-neutral mean free path given by $l_{en} \simeq (\sigma_{en} n)^{-1}$ where $\sigma_{en} \simeq 10^{-15}$ cm^2 and n is the neutral number density. In the terrestrial troposphere, $l_{en} \simeq 10^{-4}$ cm and $E_{BD} \simeq 10^6$ V/m. Although this value is comparable to laboratory-measured values for dry air [Uman, 1987], it is about 1 order of magnitude larger than the observed electric field magnitudes in terrestrial thunderclouds. Evidently, local electric filed maxima or other triggering processes act to reduce the minimum large-scale electric field magnitudes required for discharges to occur. Hence, E_{BD} probably represents an upper limit for actual mean large-scale electric fields in a given medium. In the solar nebula, $l_{en} \simeq 10$–100 cm and $E_{BD} \simeq 1$–10 V/m. Thus 10 V/m probably represents a firm upper limit for large-scale electric fields in the solar nebula.

For a maximum electric field magnitude of 10 V/m, a nebular discharge of 30 C would need to occur over a length of $\simeq 10,000$ km to deliver the same amount of energy as a terrestrial lightning bolt. Alternatively, if the length of the discharge is fixed at 1 km, the net charge transferred would need to be increased to about 300,000 C. Thus, nebular lightning must have occurred on a much larger scale in terms of either length or charge transfer to have provided a viable energy source for chondrule formation. Larger charge transfers are problematical in view of the difficulty of separating charge in a finitely conducting nebula [Gibbard and Levy, 1994]. To examine the plausibility of a larger length scale for chondrule-forming nebular discharges, it is necessary to consider the time scale over which nebular discharges would have deposited their energy.

A rough estimate for the time required to transfer charge along the discharge path in the solar nebula can be made by scaling from the terrestrial case assuming that the time is proportional to that between electron-neutral collisions and, hence, inversely proportional to the gas neutral density [e.g., Morfill *et al.*, 1993]. As mentioned above, a $\simeq 1$ km length single-stroke terrestrial lightning discharge has a typical duration of order 100 ms [e.g., Uman, 1987]. Assuming a nebular number density at 2–3 AU of 10^{13}–10^{14} cm^{-3}, the expected duration of a $\simeq 1$ km length nebular discharge is 10^4–10^5 s. (Note that this time scale differs from the 10–100 s estimate of Morfill *et al.* [1993] who adopted a millisecond time scale for terrestrial lightning.) Such a time scale is significantly longer than the "minutes or less" inferred from chondrule properties (section 1). If nebular discharges actually occurred over an increased length as hypothesized above to increase the total potential drop and energy flux, the discharge time scale would presumably be increased in proportion, reducing the net rate of energy transfer.

One possible way to increase the efficiency of nebular lightning as a chondrule heating mechanism is to suppose that the nebula has a large component of very fine (< 0.1 μm radius) dust particles in the chondrule formation region [Rasmussen and Wasson, 1995]. Such a fine dust component could increase the breakdown electric field E_{BD} by inhibiting the current flow that would occur in a tenuous gas alone. However, fairly large dust/gas mass ratios of 10 to 100 are required according to the numerical results of Rasmussen and Wasson. For dust particles with radii of 25 nm and a midplane gas density at 2.5 AU of 3.2×10^{-11} gm cm^{-3} (see section 3.1), the implied dust number density is of the order of 10^6 to 10^7 cm^{-3}. Such a large concentration of fine dust could only be achieved near the midplane in a completely quiescent nebula with absolutely no turbulence. In contrast, Cuzzi *et al.* [1993; this volume] find that particles smaller than 1 cm in size would not experience appreciable settling toward the midplane for the currently accepted range of nebula Reynolds numbers. The latter are estimated assuming turbulence generated by either thermal convection or differential rotation.

While the above scaling arguments do not rigorously exclude nebular lightning as a plausible energy source for chondrule formation, they indicate that a conceptually simple lightning model for chondrule formation is not yet at hand. Further work (both theoretical and experimental) is needed to reduce some of the uncertainties cited above. Previous experimental investigations of electrical discharge heating of chondrule-sized dust particles [Wdowiak, 1983] were not encouraging in that a much wider range of artificial chondrule properties were produced (e.g., extensive vesiculation) than

are observed in real chondrules. Following Love *et al.* [1995], it is interesting to note that a possible better analogue for nebular discharges than tropospheric lightning may be provided by recently identified high-altitude flashes or "sprites" apparently induced by electrons accelerated in electric fields between thunderstorms and the ionosphere [Lyons, 1994]. The observed luminous structures occur at altitudes as high as $\simeq 90$ km where gas number densities are comparable to those of the protoplanetary nebula. However, they are believed to be due to emission from ionized oxygen and nitrogen atoms which were relatively rare in the solar nebula. Although it is doubtful whether these optical flashes are sufficiently energetic to melt silicate dust particles, further characterization of their properties could lead to a better understanding of discharge phenomena that could have occurred in the protoplanetary nebula.

3. GAS DYNAMIC SHOCK WAVES

The possibility that gas dynamic shock waves were responsible for chondrule formation has been mentioned in the literature for decades but has received serious study only during the past 10 years [Wood, 1984; 1986; Huss, 1988; Hood and Horanyi, 1991; 1993; Boss and Graham, 1993; Ruzmaikina and Ip, 1994]. From the standpoint of meteoritic constraints, the main advantages of shocks are similar to those of lightning: Relatively short heating times, localization of heating to regions of shock compression and relative motion, and the possibility of multiple shock passages. From a theoretical standpoint, the major issues to be addressed are: (1) the nature and origin of shocks sufficiently strong to have melted dust particles in the nebula; and (2) the extent to which shock heating can be consistent with meteoritic constraints (chondrule heating rates, cooling rates, compositions, presence of secondary igneous rims, limitation of heating, etc.).

3.1 Generation of nebular shocks

The most well-known example of a shock wave expected to be produced in disk formation is the putative "accretion shock" formed when infalling gas from the rotating presolar molecular cloud core joins supersonically with pre-existing nebular gas (schematically illustrated in the upper panel of Fig. 1). Chondrule formation by viscous ("drag") heating of interstellar grain aggregates falling into a protosolar nebula accretion shock has been investigated by Wood [1984] and by Ruzmaikina and Ip [1994]. Both of these papers are valuable for initially exploring aspects of the physics of dust grain heating in gas dynamic nebular shocks. Wood [1984], in particular, provided a major stimulation for later work on nebular shock models for chondrule formation. Ruzmaikina and Ip [1994] showed that post-shock cooling may significantly enhance the ability of a shock with a given Mach number to heat mm-sized particles (see below). However, as summarized in section 1, the current assessment of meteoritic constraints is not very compatible with the overall model scenario of interstellar dust aggregates falling into an accretion shock. Compositional trends and interelement correlations in chondrules are most directly understood if precursors were nebular condensates rather than interstellar dust. In addition, it is unclear how this model scenario could account for multiple heating of chondrules, chondrules made from recycled previous generations of chondrules, etc. Aside from meteoritic constraints, numerical simulations of the formation of an accretion shock from the infall of a rotating, homogeneous molecular cloud core [Ruzmaikina *et al.*, 1993] indicate that supersonic infall velocities may not be achieved except at large radial distances (> 40 AU) and very near to the protosun (< 1 AU). Consequently, a true accretion shock may occupy only a small part of the disk surface and may not have been present in the 2–5 AU radial distance range where meteoritic chondrules were produced.

An alternate shock scenario for chondrule formation that would allow the heating of nebular condensate aggregates has been explored qualitatively by Huss [1988]. In this scenario, a supersonic protostellar wind flows isotropically outward from the sun and interacts strongly with the protoplanetary nebula, forming a bow shock at a radial distance where pressure equilibrium is attained. Condensed particles near the disk midplane are exposed to the shock-heated gas between the bow shock and the stagnation point where the gas velocity approaches zero. From an astrophysical standpoint, the main problem with this model is that bipolar mass ejections or winds from protostars are apparently powered by disk accretion. For example, bipolar outflows are observed to be the most intense when the accretion rate from the molecular cloud core onto the disk and star is the largest [Hartmann *et al.*, 1993]. In order for accretion onto the star to occur and provide the energy source for the bipolar outflows, the disk must extend down essentially to the surface of the star [Shu *et al.*, 1991]. Thus, a model in which protostellar winds are isotropic at the star and stand off the accretion disk may not be self-consistent. From a meteoritic standpoint, the main problem with the model is the long heating and cooling times of dust particles exposed to such a quasi-stationary shock. In addition, shock heating occurs only at the inner edge of the accretion disk so it is unclear how particles would be heated repetitively or how particles at larger radial distances would be heated.

On a smaller scale, shocks were undoubtedly generated in the form of bow shocks upstream of planetesimals or cometary bodies in eccentric orbits in the nebula after substantial accretion had already occurred [e.g., Hood and Horanyi, 1991]. However, since chondrule formation is thought to have preceded accretion, it is somewhat dissatisfying to propose a chondrule origin model that presupposes the existence of accreted objects. In addition, shock-heated regions upstream of planetesimals (hundreds of km in size) would have dimensions of only tens of km or less. As will be seen below, larger dimensions than this are needed to bring mm-sized grains to melting temperatures. Possibly, bow shocks of this type may have been responsible for the formation of some secondary igneous and/or sintered rims on chondrules and CAI's prior to their incorporation into meteorite parent bodies.

In general, the possibility that gas dynamic shocks may propagate within dusty protoplanetary nebulae is supported by observations of time-dependent phenomena in the near vicinities of young

stellar objects. These phenomena include bipolar mass ejections Edwards *et al.*, 1993] and relatively short-term irregular brightness variations attributed to obscuration by circumstellar cloud clumps [Boss and Graham, 1993]. The source of this variability appears to be changes in mass accretion rate perhaps resulting from inhomogeneities in the surrounding molecular cloud core. The most intense variation occurs during FU Orionis outbursts when the rate of disk accretion can increase by several orders of magnitude within a year leading to high disk temperatures persisting for decades or more [Hartmann *et al.*, 1993]. Statistics lead to the inference that most protostars probably experience multiple (~ 10) FU Orionis outbursts. The repetition time scale is of the order of $10^4 - 10^5$ years, perhaps increasing with time during the evolution of a young star.

It is interesting to note that studies of the bulk compositions of chondritic meteorites indicate that temperatures of 1200 – 1400 K existed in the inner solar system, including the asteroid belt, for time periods of at least tens of years [Palme and Boynton, 1993]. For these mid-term nebular heating events, FU Orionis episodes represent a likely candidate mechanism. Although these episodes may be too large in spatial and temporal scale to be appropriate for chondrule formation, their occurrence suggests that smaller inhomogeneities in disk accretion rate could have produced smaller-scale transient heating events.

Boss and Graham [1993] have specifically proposed that shocks generated by collisions of interstellar gas clumps with the protoplanetary nebula were responsible for chondrule formation. As already alluded to above, they cite several types of astronomical observations suggesting that optically thick clumps of gas move at high velocities within a few AU of young stellar objects. For example, rapid variations in stellar luminosity and in the surface brightness of nearby reflection nebulae are observed that are difficult to explain without such opaque clumps near to the star. Typical masses, sizes, and velocities of $\simeq 10^{22}$ gm, $\simeq 10^{12}$ cm, and $\simeq 50$ km/s are estimated by Boss and Graham. Using rough estimates for the masses, velocities, and impact rate of clumps onto a protoplanetary nebula, they argued that a substantial fraction of the nebular mass could be thermally processed in an episodic, variable manner as required to explain the properties of chondrules.

The origin of the inferred clumps, which are not gravitationally bound, is not well understood. They might originate from inhomogeneous infall of the residual molecular cloud core or possibly from the return of matter ejected during bipolar outflows. The possibility that the gas clumps inferred to occur in the near proximity of young stellar objects represent inhomogeneities in the infalling molecular cloud core is indicated schematically in the lower panel of Fig. 1. Accretion onto the disk is assumed to be inhomogeneous. Clumps of gas impact randomly and locally on the disk surface resulting in localized shocks penetrating to the interior of the nebula.

Because the hypothesized density inhomogeneities are not massive enough to be gravitationally bound, it is likely that their lifetimes are relatively short (< 10^5 years). In fact, if they are not maintained in some manner (e.g., as shocks embedded in the accre-

Classical Homogeneous Disk Accretion

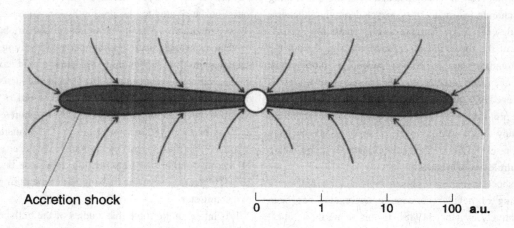

Accretion shock

Inhomogeneous ("Clumpy") Disk Accretion

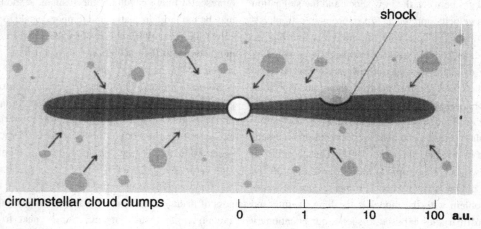

circumstellar cloud clumps

Fig. 1. The top panel is a schematic illustration of the accretion shock formed at the surface of a nebular disk by the supersonic addition of infalling gas from a homogeneous molecular cloud core. The arrows indicate approximate velocity components in the plane of the diagram. The lower panel schematically illustrates the formation of more localized shocks in the nebula by the infall of gas clumps from an inhomogeneous protosolar cloud.

tion flow), they would tend to dissipate at a rate comparable to the sound speed. For example, a 10^{12} cm clump at a temperature of 10 K would theoretically disappear in only a few years. On larger scales, as reviewed by Boss and Graham [1993], inhomogeneities in the interstellar medium are observed in both continuum and line emission from nearby molecular clouds. The detectable scale sizes are as small as 7000 AU. Such inhomogeneities may extend to smaller scales as well although a considerable extrapolation is involved. Possible origins of inhomogeneities in molecular clouds include shocks resulting from virialized motions [Elmegreen, 1990] and from the interaction of stellar winds with molecular cloud gas. The latter processes may help to explain the relatively high typical clump velocity of 50 km/s (almost twice the escape velocity at 2.5 AU).

To illustrate the production of a localized nebular shock wave by

a gas cloud clump similar to that considered by Boss and Graham [1993], we present some results of a two-dimensional numerical hydrodynamic simulation in Fig. 2 and 3. Fig. 2 shows the initial conditions and Fig. 3 shows the resulting shock disturbance 30 days later (about 24 days after impact of the cloud clump onto the surface of the nebula). The basic state nebula has a midplane number density $n(r, 0) = 5 \times 10^{13} (1 \text{ AU}/r)^2$ and a z-independent temperature distribution $T(r, z) = 400 (1 \text{ AU}/r)^{\frac{1}{2}}$ K. The nebula is in hydrostatic equilibrium in the z direction and, for simplicity, rotates at the Keplerian speed. For a disk radius of 100 AU, the disk mass is 0.054 solar masses. The midplane mass density at 2.5 AU is 3.2×10^{-11} gm cm^{-3}, the column mass density at this radial distance is $\simeq 150$ gm cm^{-2}, and the sound speed is about 1.1 km/s.

We consider the impact onto the nebula of a "standard" molecular cloud clump as defined by Boss and Graham [1993]. The clump

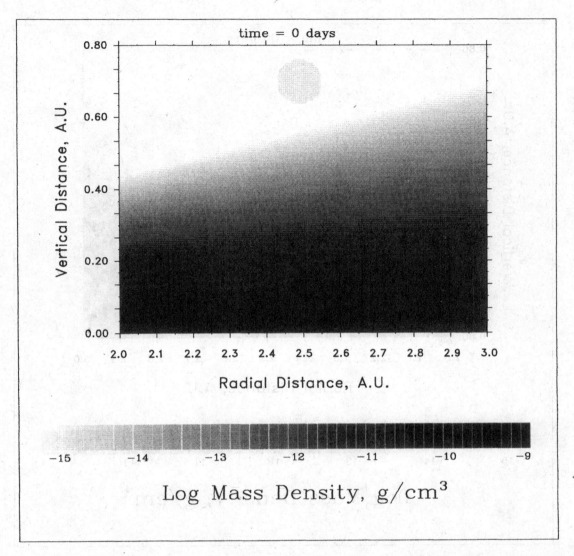

Fig. 2. Initial conditions for a two-dimensional numerical simulation of the clumpy disk accretion shock model of Boss and Graham [1993]. A "standard" clump with mass density 3×10^{-15} gm/cm³, and radius ~ 0.06 AU has an initial downward velocity of 50 km/s. The nebula is initially in hydrostatic equilibrium in the vertical direction with parameters given in the text.

has a mass density of ~ 3×10^{-15} gm cm⁻³, a radius of $\simeq 0.06$ AU, and impacts vertically onto the disk at $r = 2.5$ AU with a speed of 50 km/s. The resulting hydrodynamic disturbance is calculated using a two-dimensional piecewise parabolic method that is especially accurate for shocks [Woodward and Colella, 1984]. The numerical code is a modified version of that written by B. Fryxell [private communication, 1994; see also Malagoli *et al.*, 1990]. The grid spacing is 0.01 AU so the sharpness of the shock transition is not perfectly reproduced. As shown in Fig. 3, the shock has propagated approximately 0.13 AU in 24 days with a mean velocity of about 9 km/s corresponding to a Mach ~ 8 shock. Further integration shows that the shock weakens and penetrates no closer than about 0.3 AU from the midplane. According to Boss and Graham, more massive clumps (up to $\simeq 10^{24}$ gm) than the standard clump considered here are allowed by observational uncertainties. Additional simulations assuming larger clump mass densities (up to 3×10^{-13} gm cm⁻³) result in shocks that penetrate almost to the midplane.

3.2 Dust particle heating in nebular shocks

In general, the gas dynamic jump conditions for an ordinary adiabatic shock can be written as [e.g., Landau and Lifshitz, 1987],

$$\rho_2/\rho_1 = (\gamma + 1)M^2/[(\gamma - 1)M^2 + 2] \qquad (1)$$

$$T_2/T_1 = [2\gamma M^2 - (\gamma - 1)][(\gamma - 1)M^2 + 2]/[(\gamma + 1)^2 M^2] \qquad (2)$$

where ρ_1, T_1 are the initial gas mass density and temperature, ρ_2, T_2 are the post-shock gas parameters, γ is the (presumed constant) ratio of specific heats ($\gamma = 7/5$ is appropriate for a diatomic gas), and the sonic Mach number M is the ratio of the shock velocity v_s to the initial sound speed c_1. The sound speed is given by $c_1 = (\gamma k T_1/m)^{<<1/2>>}$, where k is Boltzmann's constant and $m \simeq 4 \times 10^{-24}$ gm in the solar nebula. For example, with $T_1 = 300$ K, $c_1 = 1.2$ km/s. for a strong shock ($M >> 1$), $\rho_2/\rho_1 = (\gamma + 1)/(\gamma - 1) = 6$ and $T_2/T_1 = 2\gamma(\gamma - 1)M^2/(\gamma + 1)^2/(\gamma + 1)^2 = 0.194 M^2$. The post-shock gas velocity v_2 is less than the shock velocity and is given by

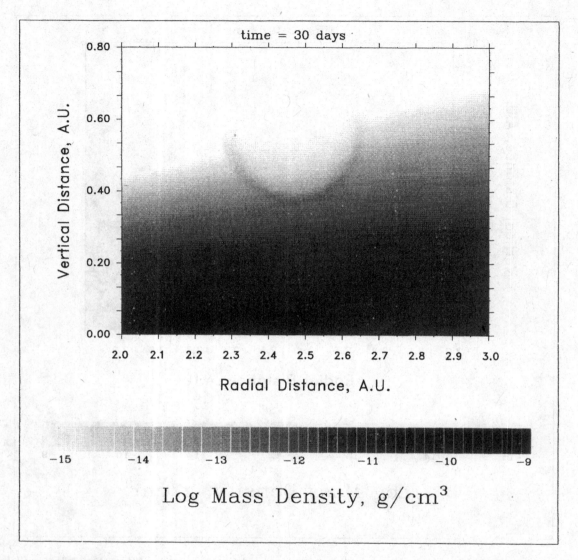

Fig. 3. Mass density distribution in the nebula 30 days after the initial conditions shown in Fig. 2. The impact of the standard clump onto the nebula surface occurred at a time of approximately 6 days. The mean speed of the shock during this time interval was about 9 km/s.

$$v_2 / v_s = \frac{2}{M^2} \frac{M^2 - 1}{\gamma + 1}. \tag{3}$$

For strong shocks, $v_2 / v_s = 2/(\gamma + 1) = 5/6$. For example, the jump conditions and post-shock velocity for a Mach 5 shock are $\rho_2/\rho_1 = 5$, $T_2 / T_1 = 5.8$, and $v_2 / v_s = 0.8$.

Although the gas density, temperature, and velocity immediately behind a shock front of Mach number M are approximately described by equations (1)–(3), post-shock radiative cooling can significantly modify these properties with distance behind the shock [Ruzmaikina and Ip, 1994; Hollenbach and McKee, 1979]. Specifically, the temperature decreases and the gas density increases rapidly with distance from the shock due to dissociation of hydrogen molecules and radiative cooling by gas molecules and small dust particles in the gas. Numerical results presented by

Ruzmaikina and Ip [1994] indicate cooling and densification by several orders of magnitude within a distance of approximately 100 km from the shock front (see their Figs. 2 and 3).

As an example, the Mach ~ 8 shock shown in Fig. 3 occurs in a gas with ambient molecular number density $\simeq 3 \times 10^{11}$ cm^{-3} and temperature $\simeq 250$ K. The immediate post-shock density and temperature are $\simeq 2 \times 10^{12}$ cm^{-3} and $\simeq 3100$ K while the post-shock gas velocity is about 7.3 km/s. The numerical results presented by Ruzmaikina and Ip [1994] include calculations for a gas with comparable density and velocity but with a somewhat higher temperature (10^4 K). If these results are assumed to apply also (to first order) to the shock shown in Fig. 3, post-shock radiative cooling could result in an increase in density to $\simeq 10^{14}$ cm^{-3} and a simultaneous decrease in temperature to $\simeq 300$ K within a few hundred km behind the shock. As will be seen below, such an increase in density, combined with the post-shock gas velocity, could be sufficient to bring mm-sized particles to melting temperatures within a few tens of seconds after passage of the shock.

In the free molecular flow approximation (mean free path much greater than the particle size), the equations of gas-dust energy and momentum transfer can be written for a spherical dust particle of radius a as

$$\frac{4}{3}\pi a^3 \rho_d C \frac{dT_d}{dt} = 4\pi a^2 q_d + 4\pi a^2 \varepsilon_{abs} J_r - 4\pi a^2 \varepsilon_{em} \sigma T_d^4 \qquad (4)$$

$$\frac{4}{3}\pi a^3 \rho_d \frac{dV_d}{dt} = \pi a^2 \frac{C_D}{2} \rho_g V_d^2 \qquad (5)$$

where T_d, C, and ρ_d are the dust temperature, specific heat, and mass density, respectively, q_d is the gas-dust heat transfer rate per unit surface area, J_r is the external radiation energy flux, σ is the Stefan-Boltzmann constant, and ε_{abs} *and* ε_{em} are the dust absorption and reemission coefficients. In (5), ρ_g *is* the gas mass density and C_D is the drag coefficient. In general, q_d and C_D include the effects of both gas-dust translational motion and thermal collisional heating [Gombosi *et al.*, 1986; Hood and Horanyi, 1991]. In general, J_r describes radiation from any source including surrounding dust particles, shocked gas, etc.

For the special case in which thermal collisional heating is neglected, $q_d \rightarrow \frac{1}{8}\rho_g V_d^3$. If external radiation heating is also neglected and V_d is assumed to remain constant, (4) implies that the particulate temperature approaches a maximum value given by

$$T_{max} = \left(\frac{\rho_g}{8\varepsilon_{em}\sigma}\right)^{\frac{1}{4}} V_d^{\frac{3}{4}} \qquad (6a)$$

$$= 1041\left(\frac{n_g}{10^{14}}\right)^{\frac{1}{4}}\left(\frac{V_d}{10\text{km/s}}\right)^{\frac{3}{4}} \text{Kelvin} \qquad (6b)$$

where n_g is the gas number density in molecules cm^{-3}, V_d is in km/s, and where we have conservatively taken $\varepsilon_{em} \simeq 0.75$. Note that this maximum temperature is independent of particle size and mass. I f the particle in question is assumed to be at the center of a spherical cloud of radius R consisting of similar particles with number density N heated to a common temperature, $J_r \rightarrow (\varepsilon_{em}/\varepsilon_{abs})[1-\exp(-\pi a^2 \varepsilon_{abs} NR)]\sigma T_d^4$ [Hood and Horanyi, 1991]. Conse-quently, (6a) is modified so that ε_{em} is replaced by $\varepsilon_{em}\exp(-\pi a^2 \varepsilon_{abs} NR)$. Choosing a moderately optically thick dust cloud such that $\pi a^2 \varepsilon_{abs} NR \simeq 2$, (6b) becomes

$$T'_{max} = 1717\left(\frac{n_g}{10^{14}}\right)^{\frac{1}{4}}\left(\frac{V_d}{10\text{km/s}}\right)^{\frac{3}{4}} \text{Kelvin} \qquad (6c)$$

Equation (6c) shows that even if heating sources other than drag heating are neglected, gas number densities of 10^{14} cm^{-3} and gas-grain relative velocities of 10 km/s (assumed constant) can bring a silicate particle to its approximate melting temperature provided that it is part of a sufficiently dense cloud of similar particles. Alternatively, an isolated particle would require a somewhat larger velocity and/or number density.

In order for equations (6) to be valid, V_d must be approximately constant during the time required for a particle to reach its maximum temperature. A rough estimate for the heating time scale can be obtained by retaining only the first term on the r.h.s. of (4),

$$\frac{dT}{dt} \simeq \frac{3}{8aC}\frac{\rho_g}{\rho_d}V_d^3 \qquad (7)$$

This equation is a reasonable approximation for $T < T_{max}$ because the second and third terms on the r.h.s. of (4) are proportional to T^4 (neglecting other sources of radiative heating) and therefore do not become comparable to the first term except when $T \simeq T_{max}$. For example, the time required to reach a temperature of $0.75\,T_{max}$ can be estimated as

$$t_{\frac{3}{4}} \simeq \frac{8aC}{3}\frac{\rho_d}{\rho_g}\frac{(\frac{3}{4}T_{max}-T_o)}{V_d^3} \qquad (8)$$

where T_o is the initial temperature of the particle.

For maximum temperatures to be reached, $t_{\frac{3}{4}}$ as given by (8) must be much less than the time required for the gas-dust relative velocity to decrease significantly. We define the stopping time t_s as that required for the gas-dust relative velocity V_d to decrease to $1/e$ of its initial value. Integrating (5) directly yields

$$t_s = \frac{4}{3}\frac{(e-1)a\rho_d}{\rho_g V_d} \qquad (9)$$

where Vd is now the initial gas-dust relative velocity. The stopping distance is given approximately by xs D tsVd D 2ard/rg. Combining (8) and (9) then yields

$$t_{\frac{3}{4}} \simeq \frac{2C}{(e-1)}\frac{(\frac{3}{4}T_{max}-T_o)}{V_d^2}t_s. \qquad (10)$$

In order for (6) to be valid, we must have $t_{\frac{3}{4}} \ll t_s$. Taking $C \simeq 10^7$ erg gm^{-1} K^{-1}, $T_{max} = 1717$ K, $T_o = 500$ K, and $V_d = 10$ km/s, (10) gives $t_{\frac{3}{4}} \simeq 0.01 t_s$ and the condition for (6) to be a reasonable approximation is satisfied. For $a = 0.05$ cm and $n_g = 10^{14}$ cm^{-3}, $t_s \simeq 716$ s and $x_s = 7160$ km. From these approximate calculations, we expect that the time to reach melting temperatures is of the order of 10 s for the given parameters and this is much less than the stopping time. For shock velocities of ~ 10 km/s, the minimum thickness of a shocked gas region required to melt a chondrule-sized particle (for $n_g = 10^{14}$ cm^{-3}) is $\simeq$ 100 km. However, the scale sizes of heated dust regions may be smaller than this due to variations of dust density (see section 4).

A minimum time scale for cooling of a particle exposed to drag heating in a nebular shock wave is given by the stopping time t_s (equation 9). For a 1 mm diameter dust particle with $\rho_d = 2.5$ gm cm^{-3}, (9) becomes

$$t_s = \frac{700}{(n_g/10^{14})(V_d/10 \text{ km s}^{-1})} \text{seconds} \qquad (11)$$

where n_g is again in molecules cm^{-3} and V_d is again in km/s. Thus for $n_g \sim 10^{14}$ cm^{-3} and $V_d \simeq 10$ Km/s, the minimum cooling time scale is 700 s. This is somewhat shorter than indicated by meteoritic constraints (~ hours). Reducing either the gas density or velocity would increase t_s but then drag heating would be insufficient to bring a particle to melting temperatures.

The particle cooling time scale could be increased in several ways. First, if the particle is part of a cloud of similarly heated particles, radiation from surrounding dust will increase the cooling time to an extent dependent on the size of the cloud and the dust density [Hood and Horanyi, 1991; Sahagian and Hewins, 1992]. Second, the cooling time would be increased if the shock compressed region has a finite thickness. The particle could then pass through to a lower gas density where it would be decelerated (and cooled) at a reduced rate. In support of this possibility, the shock-compressed zone in Fig. 3 is clearly finite in extent. Although the numerical calculation indicates a thickness of several hundredths of an AU ($\simeq 3 \times 10^6$ km), part of this large thickness is due to the coarse grid size (0.01 AU). Accurate determination of the thickness of the compressed zone would require a much finer grid resolution and a complete treatment of post-shock cooling effects as a function of distance behind the shock [Ruzmaikina and Ip, 1994]. In addition to finite thickness of the compressed zone, other heating sources besides drag heating and radiation from surrounding particles should be included in an accurate calculation of the particle cooling time.

Finally, it should be emphasized that while analytic methods can yield useful insights, detailed numerical calculations are normally required for more realistic simulations. For example, a numerical simulation has been carried out for the interaction of a one-dimensional gas dynamic shock wave with a dust cloud in the solar nebula that was assumed to be finite in dimension parallel to the direction of shock propagation [Hood and Horanyi, 1993]. For the specific case of a 200 km thick dust cloud with a density of 3 particles m^{-3} in a nebula with initial number density 10^{14} cm^{-3} and temperature 500 K, it was found that shock Mach numbers > 4 were sufficient to melt most mm-sized particles in the cloud. As shown in Fig. 4, for the stated parameters a substantial fraction of dust particles reach melting temperatures within a few tens of seconds. Particles located well inside the upstream boundary of the cloud are heated most rapidly by the sum of radiative heating from surrounding grains and drag heating by the gas. Larger dust densities tended to shield the interior of the cloud from drag heating while smaller dust densities required larger Mach numbers to melt grains. While generally useful, this simulation was deficient in that post-shock radiative cooling with distance from the shock was not considered. Accounting for increased gas densities in the radiatively cooled zone behind the shock could allow melting of dust particles for much lower nebula gas number densities [Ruzmaikina and Ip, 1994].

4. DISCUSSION

Although both the lightning and shock models for chondrule formation are still characterized by large uncertainties, the shock model has reached a point where quantitative tests of its validity are beginning to be possible. One specific shock generation mecha-

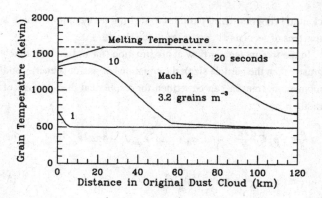

Fig. 4. Dust particle temperature at several times after a passage of a Mach 4 shock wave past the upstream edge of a 200 km thick one-dimensional dust cloud. Particles in the cloud have radii of 1 mm and number density 3.2 particles per m^3. The initial nebula number density and temperature were assumed to be 10^{14} cm^{-3} and 500 K, respectively. (from Hood and Horanyi [1993]).

nism that can currently be directly investigated is the clumpy disk accretion model of Boss and Graham [1993]. However, other mechanisms may be identified when the time-dependent evolution of protostellar accretion disks is better understood.

One problem for the clumpy disk accretion shock model is that the "standard" clump mass and velocity produce shocks that penetrate no closer than $\simeq 0.3$ AU from the mid-plane (Fig. 3). At this height, the nebular gas density is relatively low so drag heating of particles behind the shock is less effective. A more accurate calculation of the shock structure including post-shock cooling and densification [Ruzmaikina and Ip, 1994] is required to determine whether "standard" clump accretion shocks are capable of melting particles. If not, then shocks generated by more massive clumps may need to be considered. Assuming that the resulting shocks are sufficiently strong to melt dust particles, the remaining question is whether substantial quantities of dust were lofted to altitudes of several tenths of an AU above the midplane. If large-scale turbulence driven by, e.g., differential rotation and thermal convection was operative, chondrule-sized particles would probably be distributed throughout the vertical extent of the nebula [Cuzzi *et al.*, this volume]. Turbulent eddies may also have served to concentrate dust particles in local regions (i.e. dust-rich zones). Dust-rich zones would in turn be more effectively heated by passing shock waves because of reduced radiative losses [Wood, 1984; Hood and Horanyi, 1991]. As originally pointed out by Wood [1984], the high oxidation states of chondrule minerals (e.g., large concentrations of Fe^{2+}) could also indicate heating in a dust or dust/ice rich zone where partial vaporization of ice or silicates would drive up the O/H ratio in the adjacent nebular gas.

The extent to which the nebular shock model can be consistent with other meteoritic constraints on chondrule formation must also be addressed. As discussed in section 3.2, the time-scale for heating a mm-sized particle to its melting temperature by drag heating in a shock wave is sufficiently short to be consistent with the "minutes or less" requirement derived from textural and compositional evidence. The time-scale for cooling of a melted particle in a shock wave is comparable to the stopping time (equation 9) and is significantly longer than the radiative cooling time in the absence of heating. However, detailed modelling of dust particle passage through a realistic shock structure (including post-shock cooling) is necessary to determine whether the cooling time is as long (~ hours) as has been inferred from experimental studies of chondrules. The presence of some secondary high-temperature rims on chondrules (and CAI's) requires relatively brief heating events that may not be explicable in terms of the passage of large-scale shocks such as that shown in Fig. 3. As noted in section 3.1, relatively thin bow shocks upstream of planetesimals in eccentric orbits represent one candidate heat source for these secondary rims.

The limitation of heating to that marginally required to melt chondrule precursors (section 1) can potentially be understood as due to interaction of a gas-dynamic shock with a dust-rich zone [Wood, 1984; see also Hood and Horanyi, 1991]. As melting temperatures were marginally reached by mm-sized particles, partial vaporization of minor grain components and/or smaller particles could have acted to limit further heating in several ways. First, the removal of small particles would reduce the opacity of the dust cloud allowing more rapid cooling by thermal radiation. Second, if drag heating was dominant, then the release of dense silicate vapor combined with momentum conservation could significantly reduce the gas-dust relative velocity, thereby reducing the drag heating rate.

Although the minimum thickness of shocked gas required to melt a chondrule-sized particle appears to be > 100 km (section 3.2), the minimum scale-size of a chondrule formation zone may be less than this because of the dependence of particle heating rate on dust density. Specifically, local concentrations of dust (perhaps produced by turbulent eddies as suggested by Cuzzi *et al.* [this volume]) would be heated more easily to melting temperatures than nearby isolated particles. Thus, inferred separations of as little as 10 km between melted and unmelted, colder particles [Kring, 1991] do not necessarily contradict the shock model.

Finally, the observed distribution of chondrule diameters (centered on 0.2 to 1 mm) may be a secondary effect of aerodynamic sorting rather than a direct consequence of the formation event itself. As first proposed in detail by Dodd [1976], aerodynamic sorting of formed chondrules during the accretion of ordinary chondrite parent bodies is one means for producing the observed narrow size distribution (see also Rubin and Keil [1984]). In addition, Cuzzi *et al.* [this volume] have recently calculated that mm-sized particles are preferentially concentrated in turbulent eddies for the currently estimated range of nebula Reynolds numbers. On the other hand, the lower limit of ~ 0.2 mm may be due in part to reduced ability of grains smaller than this size to efficiently absorb infrared radiation [Eisenhour *et al.*, 1994]. More detailed numerical simulations of shock wave heating for a realistic precursor dust size distribution are needed to investigate this issue.

ACKNOWLEDGEMENTS

Supported by the Origins of Solar Systems research program under NASA grant NAGW-2315. Constructive critical reviews by H. McSween, P. Cassen, and an anonymous reviewer are greatly appreciated.

REFERENCES

Alexander C. M. O'D. (1994) Chondrules from chondrules? An ion probe trace element study, in *Lunar Planet. Sci. Conf.* XXV, pp. 7–8, Lunar and Planetary Institute, Houston.

Boss A. P. and Graham J. A. (1993) Clumpy disk accretion and chondrule formation. *Icarus* **106**, 168–178.

Cassen P. and A. Boss (1988) Protostellar collapse, dust grains and Solar System formation, in *Meteorites and the Early Solar System*, (eds. J. F. Kerridge and M. S. Matthews), pp. 304–328. University of Arizona Press.

Clayton D. D. (1980) Chemical energy in cold-cloud aggregates: The origin of meteoritic chondrules, *Astrophys. J.* **239**, L37–L41.

Cuzzi J. N., Dobrovolskis, and Champney J. M. (1993) Particle-gas dynamics in the mid-plane of a protoplanetary nebula, *Icarus* **106**, 102–134.

Dodd R. T. (1976) Accretion of the ordinary chondrites, *Earth Planet. Sci. Lett.* **30**, pp. 281–291.

Edwards S., Ray T. and Mundt R. (1993), Energetic mass outflows from young stars, in *Protostars and Planets III* (eds. E. H. Levy and J. Lunine), p. 567–602. University of Arizona Press.

Eisenhour D. D., Daulton T. L. and Buseck P. R. (1994) Electromagnetic heating in the early solar nebula and the formation of chondrules, *Science* **265**, 1067–1070.

Elmegreen B. G. (1990) A wavelike origin for clumpy structure and broad linewings in molecular clouds, *Astrophys. J.* **361**, L77–L80.

Fujii N. and Miyamoto M. (1993) Constraints on the heating and cooling processes of chondrule formation, in *Chondrules and their Origins* (ed. E. A. King), pp. 53–60. Lunar and Planetary Institute, Houston.

Gibbard S. G. and Levy E. H. (1994) On the possibility of precipitation-induced vertical lightning in the protoplanetary nebula, in *Papers Presented to Chondrules and the Protoplanetary Disk*, LPI Contrib. 844, pp. 9–10, Lunar and Planetary Institute, Houston.

Gombosi T. I., Nagy A. F. and Cravens T. E. (1986) Dust and neutral gas modeling of the inner atmospheres of comets. *Rev. Geophys.* **24**, 667–700.

Grossman J. N. (1988) in *Meteorites and the Early Solar System* (eds. J. F. Kerridge and M. S. Matthews), pp. 680–696. University of Arizona Press.

Grossman J. N., Rubin A. E., Nagahara H. and King E. A. (1988) in *Meteorites and the Early Solar System* (eds. J. F. Kerridge and M. S. Matthews), pp. 680–696, University of Arizona Press.

Grossman J. N. and Wasson J. (1983) in *Chondrules and Their Origins* (ed. E. A. King), pp. 88–121. Lunar and Planetary Institute, Houston.

Hartmann L., Kenyon S., and Hartigan P. (1993) Young stars, episodic phenomena, activity and variability, in *Protostars and Planets III* (eds. E. H. Levy and J. Lunine), pp. 497–520. University of Arizona Press.

Hewins R. H. (1988) in *Meteorites and the Early Solar System* (eds. J. F. Kerridge and M. S. Matthews), pp. 660–679. University of Arizona Press.

Hewins R. H. and Radomsky P. M. (1990) Temperature conditions for chondrule formation, *Meteoritics*, **26**, 309–318.

Hollenbach D. and McKee C. F. (1979) Molecule formation and infrared emission in fast interstellar shocks. I. Physical processes, *Astrophys. J. Suppl.*, **41**, 555–592.

Hood L. L. and Horanyi M. (1991) Gas dynamic heating of chondrule precursor grains in the solar nebula. *Icarus* **93**, 259–269.

Hood L. L. and Horanyi M. (1993) The nebular shock wave model for chondrule formation: One-dimensional calculations, *Icarus* **106**, 179–189.

Huss G. R. (1988) The role of presolar dust in the formation of the solar system, *Earth, Moon, and Planets* **40**, 165–211.

Kerridge J. F. (1993) What can meteorites tell us about nebular conditions and processes during planetesimal accretion? *Icarus* **106**, 135–150.

Kring D. A. (1988) *The petrology of meteoritic chondrules: Evidence for fluctuating conditions in the solar nebula*, Ph. D. Thesis, Harvard Univ., Cambridge, Mass.

Kring D. A. (1991) High temperature rims around chondrules in primitive chondrites: Evidence for fluctuating conditions in the solar nebula, *Earth Planet. Sci. Lett.* **105**, 65–80.

Landau, L. D., and E. M. Lifshitz (1987) *Fluid Mechanics*, Pergamon, New York. 539 pp.

Levy E. H. and Araki S. (1989) Magnetic reconnection flares in the protoplanetary nebula and the possible origin of meteorite chondrules, *Icarus* **81**, 74–91.

Love S. G., Keil K. and Scott E. R. D. (1995) Electrical discharge heating of chondrules in the solar nebula, *Icarus*, in press.

Lyons W. A. (1994) Characteristics of luminous structures in the stratosphere above thunderstorms as imaged by low-light video, *Geophys. Res. Lett.* **21**, 875–878.

MacPherson G. J., Wark D. A. and Armstrong J. T. (1988) Primitive material surviving in chondrites: Refractory inclusions, in *Meteorites and the Early Solar System* (eds. J. F. Kerridge and M. Matthews), 746–807.

Malagoli A., Rosner R. and Fryxell B. (1990) Numerical simulations of thermal instabilities in stratified gases, *Mon. Not. R. Astr. Soc.* **247**, 367–376.

Metzler K., Bischoff A. and Stöffler D. (1992) Accretionary dust mantles in CM chondrites: Evidence for solar nebula processes, *Geochim. Cosmochim. Acta*, **56**, 2873–2897.

Morfill G., Spruit H. and Levy E. H. (1993) Physical processes and conditions associated with the formation of protoplanetary disks. In *Protostars and Planets III* (eds. E. H. Levy and J. Lunine), pp. 939–978. University of Arizona Press.

Palme H. and Boynton W. V. (1993) Meteoritic constraints on conditions in the solar nebula. In *Protostars and Planets III* (eds. E. H. Levy and J. I. Lunine), pp. 979–1004. University of Arizona Press.

Pilipp W., Hartquist T. W. and Morfill G. E. (1992) Large electric fields in acoustic waves and the stimulation of lightning discharges, *Astrophys. J.*, **387**, 364–371.

Rasmussen K. L. and Wasson J. T. (1982) in *Papers Presented to the Conference on Chondrules and Their Origins*, LPI Contrib. 844, p. 53, Lunar and Planetary Institute, Houston.

Rasmussen, K. L. and Wasson J. T. (1995) A lightning model of chondrule melting, *Earth Planet. Sci. Lett.*, submitted.

Rubin A. E. and Keil K. (1984) Size distributions of chondrule types in the Inman and Allan Hills A77011 L3 chondrites, *Meteoritics* **19**, 135–143.

Ruzmaikina T. and Ip W. (1994) Chondrule formation in radiative shock *Icarus* **112**, 430–447.

Ruzmaikina T., Khatuncev I. V. and Konkina T. V. (1993) Formation of the low-mass solar nebula, in *Lunar Planet. Sci. Confl. XXIV*, pp. 1225–1226. Lunar and Planetary Institute, Houston.

Sahagian D. L. and Hewins R. H. (1992) The size of chondrule-forming events, in *Lunar Planet. Sci. Conf. XXIII*, pp. 1197–1198. Lunar and Planetary Institute, Houston.

Scott E. R. D., Taylor G. J., Rubin A. E., Keil K. and Kracher A. (1982) In *Papers Presented to the Conference on Chondrules and Their Origins*, pp. 54–55, Lunar and Planetary Institute, Houston.

Shu F., Ruden S. P., Lada, C. J. and Lizano S. (1991) Star formation and the nature of bipolar outflows, *Astrophys. J.*, **370**, L31–L34.

Sonett C. P. (1979) On the origin of chondrules, *Geophys. Res. Lett.* **6**, 677–680.

Taylor G. J., Scott E. R. D. and Keil K. (1983) in *Chondrules and Their Origins* (ed. E. A. King), pp. 262–278. Lunar and Planetary Institute, Houston.

Uman M. A. (1987) *The Lightning Discharge*, Academic Press, Orlando. 377 pp.

Wark D. A. and Lovering J. F. (1977) Marker events in the early solar system: Evidence from rims on Ca-Al-rich inclusions in carbonaceous chondrites. In *Proc. Lunar Sci. Conf. 8th*, pp. 95–112. Lunar and Planetary Institute, Houston.

Wasson J. T. (1985) *Meteorites: Their Record of Early Solar System History*, Freeman, New York. 277 pp.

Wasson J. T. (1994) Chondrule origins: Constraints from chondrule properties and cosmo-chemistry. In *Papers Presented to Chondrules and the Protoplanetary Disk*, LPI Contrib. 844, pp. 42–43, Lunar and Planetary Institute, Houston.

Weidenschilling S. J. and Ruzmaikina T. V. (1994) Coagulation of grains in static and collapsing protostellar clouds, *Astrophys. J.*, **430**, 713–726.

Whipple F. L. (1966) Chondrules: Suggestion concerning their origin, *Science* **153**, 54–56.

Wood J. A. (1984) On the formation of meteoritic chondrules by aerodynamic drag heating in the solar nebula, *Earth Planet. Sci. Lett.* **70**, 11–26.

Wood J. A. (1985) Meteoritic constraints on processes in the solar nebula: An overview, in *Protostars and Planets II*, (eds. D. C. Black and M. S. Matthews), pp. 687–702, University of Arizona Press.

Wood J. A. (1986) High temperatures and chondrule formation in turbulent shear zone beneath the nebula surface. In *Lunar Planet. Sci. XVII*, pp. 956–957, Lunar and Planetary Institute, Houston.

Wood, J. A. and G. E. Morfill 1988. A review of solar nebula models. In *Meteorites and the Early Solar System* (J. F. Kerridge and M. S. Matthews, Eds.), pp. 329–347. Univ. of Arizona Press, Tucson.

Woodward P. R. and Colella P. (1984) The Piecewise Parabolic Method for two-dimensional hydrodynamics, *J. Comp. Phys.* **54**, 115–128.

29: Chondrule Formation in the Accretional Shock

T. V. RUZMAIKINA[1] and W. H. IP[2]

[1] *Lunar and Planetary Laboratory, University of Arizona, Tucson, AZ 85 721, U.S.A.* [2] *Max Planck-Institute für Aeronomie, D-3411 Katlenburg-Lindau, Germany.*

ABSTRACT

We discuss the possibility of chondrule formation in the accretional shock which inevitably is produced by the infalling gas when it hits the forming solar nebular. This mechanism of chondrule formation requires the presence of relatively large dust aggregates – chondrule precursors – in the infalling gas, and their melting by gas drag in the postshock region. This mechanism was first suggested by Wood in 1984, but since that time it seems not to have been considered efficient enough at the radial distances of the asteroid belt to be interesting. Recently we found that gas cooling behind the shock front significantly intensifies the heating of chondrule precursors. The cooling results in a sharp increase of gas density in the postshock region. Millimeter-size grains cross the region of cooling before being decelerated, and are heated by gas drag in the cooler and denser gas. It makes possible chondrule formation in the accretional shock at the distance of the asteroid belt.

We speculate about the formation of chondrule precursors in the presolar cloud, and suggest that they could be formed first as fluffy aggregates. Then they are compressed by mutual collisions as masses and relative velocities of the aggregates increase, and finally become dense enough to avoid (because of decreased cross-section) further destructive collisions with aggregates of comparable size before entering the shock front.

Collisions between chondrules and between chondrules and smaller grains occur in the postshock regions. These could be responsible for the formation of compound chondrules and chondrule rims, fragmentation of some chondrules, and also the evaporation and recondensation (predominantly on small particles and fluffy aggregates having large surface area) of some fraction of the solid materials.

INTRODUCTION

Chondrules – sub-millimeter to millimeter-sized spherical-shaped grains, composed of silicates – are the most abundant constitutents of most groups of chondrite meteorites. The physical, mineralogical, and isotopic properties of chondrules strongly indicate that they were formed by the rapid melting and resolidification of pre-existing solids followed by cooling at a rate 10^2 to 10^3 K/hr (Grossman 1988).

The great abundance of chondrules in almost all types of chondrites demonstrates the high efficiency of some energy source for conversion of precursor material into liquid droplets.

Many mechanisms have been suggested for the chondrule formation. Among them: lightning discharges (Whipple 1966, Cameron 1966, and the most recent development of the idea by Horanyi *et al,*

1995) flares in the solar nebula (Sonett, 1979; Levy and Araki, 1989), and heating of the chondrule precursors by friction with gas as they were decelerated in the accretional shock (Wood 1984, Ruzmaikina 1990) or in a shock within the solar nebula (Hood and Horanyi 1991). In spite of this great diversity of ideas, a mechanism capable of converting a significant fraction of the dispersed solids into chondrules has not been identified. Flares and lightning could produce some chondrules. Flares occur in the solar nebula at high altitudes (the smaller the intensity of the magnetic field, the higher the altitude at which magnetic pressure dominates the thermal one and flare activity can occur), and only a small fraction of solar nebula matter is situated in regions of possible flare activity (Levy 1988). Chondrule precursors, which are relatively massive grain aggregates, could be uplifted to the flare site only in the presence of rather intense turbulence, and would settle quickly to the central plain if the turbulence decayed. It is unclear therefore if a significant fraction of solids could be converted into chondrules by this mechanism. Lightning discharges in principle could occur throughout the solar nebula, and definitely can melt precursors within discharge channels and close to them (Horanyi *et al.*, 1995). However Gibbard and Levy (1994) have demonstrated recently that the mechanism of charge separation by the pull of gravity is not efficient enough to make lightning possible. These problems make it difficult to argue that flares and lightning are dominant mechanisms of chondrule formation, even if these processes may have played some role.

In this regard, large-scale shocks in the solar nebula have a significant advantage. They can affect a large area and reprocess a significant fraction of the solid material that passes through the shock. The accretional shock associated with the formation of the solar nebula is in this class of shocks. The shock is produced by infalling gas of the presolar cloud when it hits the forming solar nebula. It is inevitable that such a shock developed during the formation of the solar nebula, and a significant fraction of the infalling presolar gas and dust came through this shock. The ratio of total velocity of infalling gas to sound speed in the disk is $Ma = v / c_s \simeq 20$. Hence, a rather strong shock occurs in regions where infalling gas encounters the disk at nearly normal angles (Fig. 1).

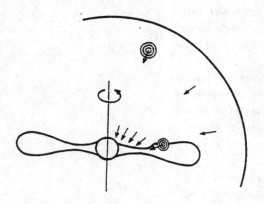

Fig. 1. Schematic structure of the solar nebula during the infall stage of its formation. Vertical line represents the rotational axis of the forming Sun and solar nebula. The fraction of the circle shows a part of the presolal cloud. The solid curve shows the surface of the solar nebula, which could be determined as the position of the shock front, produced by the infalling gas. The surface could be smooth only if the infalling gas is perfectly homogeneous. Inhomogeneities in the infalling gas would make it rippled. Arrowed circles represent denser clumps of infalling gas. Hitting the solar nebula, these clumps penetrate into the denser layers of the solar nebula, producing a shock which is especially strong in the direction of clump motion where $v_\perp \approx v$.

Arrows (which are shown only in the first quadrant for simplicity of the figure) denote velocities of the infalling matter. They are directed radially at large heliocentric distances where the centrifugal force, associated with the rotation of the envelope, is negligible compared with the gravitational pull of the protosun. At smaller distances, the centrifugal force is comparable with the solar gravity. Therefore the horizontal component of the velocity decreases, and the direction of the infalling gas tends to become vertical. The radial extension of this region (R_{ce}) depends on the value (J_{SN}) and the distribution of the angular momentum in the presolar cloud, and on the fraction of the mass which is still in the envelope. The radius of the solar nebula would not exceed R_{ce} if the tangential tensions were absent. However it can grow to much larger size if the angular momentum is efficiently redistributed in the solar nebula, e.g. by the turbulent viscosity or by the magnetic torque. The figure shows such a case.

Wood (1984) was the first to considered the possibility of chondrule formation in the accretional shock by heating of infalling large dust aggregates (chondrule precursors) by gas drag. The drag occurs when a chondrule precursor continues to move by inertia (initially with the same velocity) through gas decelerated in the shock front. He concluded, however, that the density in the infalling gas is much lower than needed to melt silicates at the distance of the asteroid belt if the matter had the cosmic ratio of dust to gas. Melting of chondrule precursors is difficult because of their effective cooling by the thermal radiation. Suppression of the radiative cooling of individual grains in dust swarms, which are opaque to the thermal emission, was considered to be the only possible means of chondrule formation in solar nebula shocks. This pessimistic conclusion decreased interest in this mechanism, and it was forgotten for years.

More recently Ruzmaikina and Ip (1994) have shown that effective cooling of the shocked gas can significantly intensify the heating of chondrule precursors. Fast cooling of gas in the postshock region (by molecular hydrogen, dipole molecules, and dust emission) results in a rapid increase in the gas density by more than an order of magnitude. Millimeter-size grain aggregates cross the region of cooling before being decelerated, and are heated by gas drag more strongly than they would be heated without gas cooling. Still, to melt silicates, at the distance of the asteroid belt, v needs to be about the free-fall velocity, and the density of the infalling gas must be a few times larger than the average gas density, providing a rate of accretion of ~ 10^{-5} $M_\odot$ yr^{-1}. Also, this mechanism requires that dust aggregates massive enough to be chondrule precursors are embedded in the infalling gas. While the principle of the possibility of formation of large dust aggregates before or during collapse was demonstrated by Cameron (1975), and more recently by Weidenschilling and Ruzmaikina (1994), this problem remains to be solved. We discuss it in Section 3.

Boss and Graham (1993) and Hood and Kring (1995) have considered another aspect of the accretional shock, produced by clumpy accretion. Penetrating into the solar nebula, clumps generate shocks which can melt dust aggregates that have been agglomerated and uplifted in the nebula. However the role of this process in the production of chondrules depends on the density of the infalling clumps. If clumps are just a few times denser than the infalling gas on average, then the penetration is shallow, and it is hard to expect that a significant fraction of the dust aggregates in the solar nebular could be processed. (These are only those aggregates which have been uplifted to high nebula altitudes. Vertical mixing and uplifting of a new set of dust aggregates may be unable to increase the fraction of chondrules to that found in chondrites, because the infalling gas introduces new dust.) Still, this mechanism can work in combination with the mechanism which melts chondrule precursors agglomerated in the presolar cloud. These newly formed chondrules and unmelted aggregates float for some time in the upper layers of the solar nebula, and could be heated (or reheated) by the shock generated by another infalling clump.

In this paper we summarize earlier results and open questions related to chondrule formation in the accretional shock. After the introduction we briefly discuss the possible formation of chondrule precursors in protostellar clouds before and during collapse (Sec. 3). Next we describe a model for the infall stage of solar nebula formation, and the structure of the shock front (Sec. 4). Then (Sec. 5) we consider the efficiency of heating, and study the conditions for melting of millimeter-size grain aggregates (both dense and fluffy). Finally (Sec. 6) we speculate about the formation of compound chondrules and other aggregation processes in the postshock region.

FORMATION OF CHONDRULE PRECURSORS

Models of chondrule formation in the accretional shock pose the question: can precursor grains as big as ~ 1 mm in size (~ 10^{-3} g in mass) form before entering the solar nebula?

Numerical treatment of the coagulation of interstellar dust particles in protostellar clouds was undertaken by Weidenschilling and Ruzmaikina (1994). These simulations and analytical estimations have shown that the number of collisions between grains, and hence the rate of their growth, strongly depends on how the cross-section of the aggregate changes with time. We showed that dense aggregates would not experience enough collisions to reach the size/mass of chondrule precursors during 10^7 yrs of the precollapse history of the presolar cloud, or during the collapse. However, reasonably strong aggregates with fractal dimension $D = 2.11$ can reach the mass of chondrules faster, and can grow even more massive. Here we estimate the efficiency of coagulation of these and denser dust aggregates.

These calculations have shown also that the size-spectrum of growing aggregates is relatively flat; most of the mass is concentrated in larger grains. Therefore we assume further in this section that all aggregates have the same radius r_{gr} and mass m_{gr}, and that this assumption provides qualitatively correct results of grain coagulation.

Considering grains to be homogeneous spheres with density ρ_{gr}, and taking $n_{gr} m_{gr} = q \rho$, where q is the mass fraction of the grains, and n_{gr} is their number density, the coagulation equation can be written in the form

$$\frac{d\rho_{gr} r_{gr}}{dt} = \frac{3}{2} q V_{rel} \rho, \tag{1}$$

where V_{rel} is an average relative velocity of the random motion of grains with respect to one another. In a perfectly quiet gas V_{rel} is the velocity of Brownian motion of grains. In the turbulent gas this is the velocity of vortices whose period is equal to the time of the change of a grain's velocity t_e in response to the changing velocity of the gas. For turbulence with the Kolmogorov spectrum, $V_{rel} \simeq (t_e/t_t)^{1/2} V_t$ in a significant range of grain sizes, where V_t and t_t are the turbulent velocity and turnover time in the main length scale. Then integration of Eq. (1) yields

$$\rho_{gr} r_{gr} = \left(\rho_{gr0}^{1/2} r_{gr0}^{r/2} + \frac{3}{32} \left(\frac{\pi}{8} \right)^{1/2} q \left(\frac{\rho}{c_s t_t} \right)^{1/2} V_t t \right)^2, \tag{2}$$

where the subscript $gr0$ refers to initial values of the parameters.

The radius and mass to which aggregates grow in a given time depends on the relationship between aggregate density and grain size. For aggregates of fractal dimension D, the ratio between the density and its radius is $\rho_{gr} \propto r_{gr}^{D-3}$ and the mass of the aggregates can be expressed as

$$m_{gr} = m_{gr0} \left(1 + \frac{3}{32} \left(\frac{\pi}{8} \frac{\rho}{c_s t_t \rho_{gr0} r_{gr0}} \right)^{1/2} q V_t t \right)^{\frac{2D}{D-2}} \tag{3}$$

The numerical values of the parameters for a typical molecular cloud are $c_s = 2 \times 10^4$ cm s^{-1}, $\rho_{gr0} r_{gr0} \approx 10^{-4}$ g cm^{-2}, and $\rho = 5 \times 10^{-19}$ g cm^{-3}. We take also $V_t \approx c_s$, $L_t \sim 10^{-1} r \sim 10^{16}$ cm for the main length-scale of the turbulence, and $q = 4 \times 10^{-3}$, which is about a factor 5 less than the solar abundance of heavy elements, taking into account only the nonvolatile elements that comprise chondrule material.

If the aggregate has a constant density, then $r_{gr} \propto t^2$ for large t. For fractal aggregates with dimensions $D = 2.75,. 2.5, 2.25$, and 2.11, we have density $\rho_{gr} \propto r_{gr}^{D-3} \sim r_{gr}^{-0.25}$, $r_{gr}^{-0.5}$, $r_{gr}^{-0.75}$, and $r_{gr}^{-0.89}$. For $t > 10^6$ yrs in our case, the mass of aggregates increases with time $\propto t^{D/D-2} \propto t^{7.3}$, t^{10}, t^{18} and t^{38}, for $D = 2.75, 2.5, 2.25$, and 2.11, respectively. Perfectly sticking aggregates with $D = 2.75$ reach in ~ 10^7 yrs a typical mass of 3×10^{-5} g, which is the mass of a chondrule ~ 0.3 mm in diameter. The aggregates with $D = 2.5$ grow to 10^{-3} g, *i.e.*, to the mass of 1 mm-size chondrules, in 8×10^6 yrs. Fractals with smaller D grow faster.

As follows from Eq. (2), the average relative velocities between two aggregates of equal size can be expressed in terms of their masses as

$$V_{rel} \simeq \left(\frac{\pi}{8} \right)^{1/4} \left(\frac{\rho_{gr0} r_{gr0}}{\rho c_s} \right)^{1/2} \frac{V_t}{t_0^{1/2}} \left(\frac{m_{gr}}{m_{gr0}} \right)^{\frac{D-2}{2D}}. \tag{4}$$

This velocity is ~ 10^4 cm s^{-1} for fractals with $m_{gr} \sim 10^{-4}$ and $D = 2.75$, or $m_{gr} \sim 10^{-3}$ g and $D = 2.5$; it is also $\approx 5 \times 10^3$ and 3×10^3 cm s^{-1}, for fractals with $m_{gr} \sim 10^{-3}$ g, and $D -= 2.25$ and 2.1 respectively. The corresponding kinetic energies, between 10^7 to a few times 10^8 erg g^{-1}, are the same order of magnitude as the specific energy of the fragmentation threshold for a 1 cm target composed of rocky materials (Fujiwara *et al.* 1989). (The experiments were undertaken only for cm-size and larger targets; in general the fragmentation threshold tends to grow with the decreasing of the target size.) Therefore if the effective strength of fluffy aggregates is of the same order of magnitude, then these fractal aggregates, or at least those with $D < 2.5$, can reach chondrule masses before their disruption due to mutual collisions becomes important.

The danger of collisional destruction of the aggregates may be less than just estimated for the following reason. Before a catastrophic disruption occurs, previous collisions will have caused deformation and compaction of the aggregates. (The possibility of the compaction of small aggregates under energetic collisions has been demonstrated by the recent numerical simulations by Dominik and Tielens 1995). It may happen that an aggregate is compressed by the collisions and decreased in radius enough for it to avoid further collisions with aggregates of comparable size, and thereby avoid destruction. This possibility suggests the following scenario for the formation of chondrule precursors: The aggregates could grow initially as fractals of low density and reach the masses of chondrule precursors before they begin to disrupt each other. When the kinetic energy of impacts approaches the strength of the material but is still less than it, collisions might cause compaction of the aggregates instead of destruction. The more compact aggregates experience a smaller number of collisions, and tend to avoid both further increases of mass and catastrophic destruction. (It follows from Weidenschilling and Ruzmaikina (1994) that solid grains with radius $> 10^{-3}$ cm experienced few, if any, collision for 10^7 yrs.) Such a scenario for accumulation seems capable of producing a population of dust aggregates with a relatively narrow size range, determined by the strength of the aggregates.

Dust aggregates that have formed in molecular clouds can acquire significant amounts of volatiles, such as water ice. In the infalling envelope, dust grains and aggregates loose the volatiles as they move inward and are heated progressively stronger by the diffuse radiation of the hot vicinity of the star. Volatiles could be also evaporated by collisions between the aggregates.

The formation and survival of chondrule precursors in the presolar cloud and during collapse is not unrealistic, and the questions deserve further investigation.

THE SOLAR NEBULAR DURING INFALL

Scenario of solar nebula formation

To consider chondrule formation in the accretional shock it is necessary to specify a model of the solar nebula, during the infall stage of its formation. Reconstruction of the solar nebula by augmentation of the planets with H and He in the amount needed to restore the solar composition results in a mass of the solar nebular $M_{SN} \simeq 0.01$ to 0.1 M$_\odot$, and a total angular momentum $J_{SN} \simeq 3 \cdot 10^{51}$ to $2 \cdot 10^{52}$ g cm^2 s^{-1} (Hoyle, 1960; Weidenschilling, 1977). Including the angular momentum of the early Oort cloud increases the angular momentum of the solar nebula to 1 to 3×10^{52} g cm^2 s^{-1} (Weissman 1991). These are reasonable minimal angular momenta for the presolar cloud. Similar parameters have been estimated from observational data for disks surrounding many T Tauri type stars (Sargent and Beckwith, 1987).

In this paper we assume that the angular momentum of the presolar cloud (*PC*) is close to or a few times larger than the minimal angular momentum of the solar nebula. Such a model results in higher densities of the infalling gas and, hence, a stronger shock at the distance of the asteroidal belt than models with much larger angular momenta, provided that the rate of infall is the same (Ruzmaikina and Ip, 1994). Therefore we consider models with relatively low angular momentum as being most favorable for chondruable formation.

The dynamical pressure of the infalling gas in these models contributes remarkably to the compression and heating of the inner solar nebula (Boss, 1993). Redistribution of the angular momentum causes an increasing of the radius of the solar nebula within the infalling envelope (Ruzmaikina, 1981; Cassen and Moosman, 1981; Ruzmaikina and Maeva, 1986; Ruzmaikina *et al.*, 1993; Cassen, 1994). Even weak subsonic turbulence with $\alpha \sim 10^{-2}$ can trigger an increase in the disk radius to the current size of the solar system over the time scale of infall, which is usually taken to be equal to 10^5 years in accordance with Shu's (1977) estimate of the accretion rate of a molecular cloud with a temperature of 10 K. Fig.

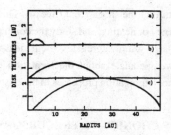

Fig. 2. Evolution of the solar nebula during accretion when the mass of the protosun is 0.1 M$_\odot$ (a), 0.5 M$_\odot$ (b) and about 1 M$_\odot$ (c). The angular momentum of the presolar cloud is 2×10^{52} g cm^2 s^{-1}. The disk is turbulent with the viscosity $\alpha \approx 10^{-2}$. The rate of infall $\dot{M} = 10^{-5}$ M$_\odot$yr^{-1}.

2 shows the height versus radius of the forming solar nebula at three successive times, when the mass of the central core is 0.1, 0.5, and ≈ 1M$_\odot$ and the rest of mass is in the infalling gas, according to Ruzmaikina *et al.* (1993).

Hitting the solar nebula, infalling gas generates shock whose intensity depends on the distance from the Sun and the angle between the lines of flow of the gas and levels of constant pressure in the solar nebula. A smooth "surface" on the solar nebula is obtained only if the gas density in the infalling envelope is homogeneous. Inhomogeneities in the infalling gas would make it rippled, and this produces sites with $v_\perp \approx v$. An infalling denser clump penetrates into the solar nebula and generates a shock with $v_\perp \approx v$ ahead of itself (in the direction of its motion), independent of the angle of infall. Therefore we assume that the shock always is produced by gas with v. (In the context of chondrule formation we consider a relatively late stage of infall, when the most of the Sun has formed and the velocity of the infalling gas $v \sim \sqrt{GM_\odot / R}$).

Structure of the shock front

When gas crosses the shock, its temperature, density and pressure increase in a thin region of adiabatic compression, which is followed by a broader postshock region of cooling (Zeldovich and Raizer, 1967). For the ratio of specific heats $\gamma = \frac{7}{5}$ (to take into account the rotational excitation of hydrogen molecules), the jump conditions for the strong shock are $n_2 = 6n_1$, $T_2 = \frac{7}{36} \frac{\mu m_H}{k_T} u_1^2$, and $u_2 = \frac{1}{6}u_1$; n_1 and n_2 are number densities of the gas before and after the shock front, and $u_1 \equiv v$. Subscripts 1 and 2 mark values before and after the shock.

Behind the shock the temperature decreases rapidly. At high temperatures ($T > 5 \times 10^3$ K) the dissociation of hydrogen molecules dominates radiative cooling. Cooling by vibrational transitions of the hydrogen molecules, rotational transitions of the dipole molecules CO, CH, OH, H$_2$O, and HCl, and thermal radiation of small dust particles determine the rate of gas cooling at lower temperatures (the cooling rates by these mechanisms were taken from Hollenbach and McKee 1979). For more details see Ruzmaikina and Ip (1994).

Fig. 3 shows the stationary distribution of gas temperature and density, in the post-shock region of cooling. (Cooling associated with molecular dissociation has not been taken into account.)

The thermal radiative temperature in the postshock region is approximately equal to the temperature of small dust particles that

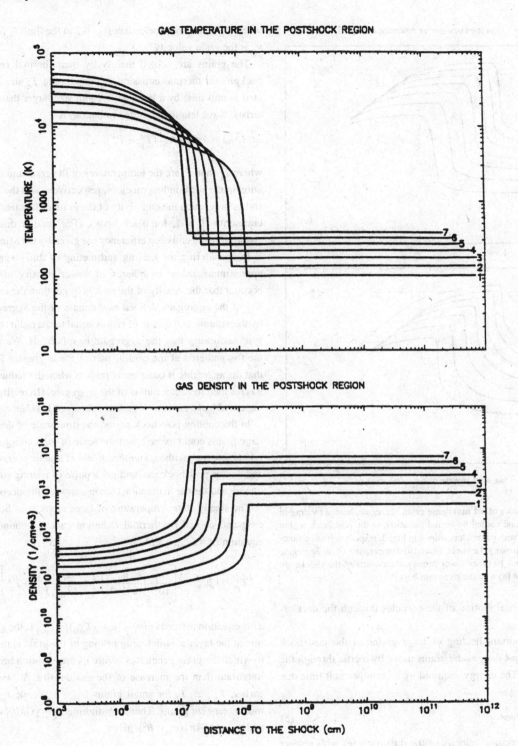

Fig. 3. (a) Gas temperature distribution behind the cooling accretion shocks for velocities ranging from 15 km s^{-1} (line 1) to 33 km s^{-1} (line 7), in 3 km s^{-1} increments. These velocities correspond to escape velocities at distances of 8.73 (line 1), 6.07 (2), 4.46 (3), 3.41 (4), 2.70 (5), 2.18 (6), and 1.80 (7) AU from the Sun, respectively. Densities of the infalling preshocked gas are taken to be 10 times larger than for a stationary rate of infall 10^{-5}M$_\odot$ yr^{-1}. (b) Gas density behind the same shocks for which the temperature is shown in Fig. 3a.

dominate the opacity (horizontal portion of solid lines in Fig. 4). This is much lower than the maximum temperatures to which large grains are heated. Therefore the radiative cooling of large dust aggregates in the postshock region is similar to cooling in a cold environment.

MAXIMUM TEMPERATURE OF DUST AGGREGATES IN THE POSTSHOCK REGION

Three main processes contribute to the heating of solid particles in the vicinity of the shock front. These are absorption of ultraviolet radiation and the thermal IR radiation emitted by other particles, thermal collisions with gas atoms and molecules, and drag associ-

GRAIN TEMPERATURE IN THE POSTSHOCK REGION

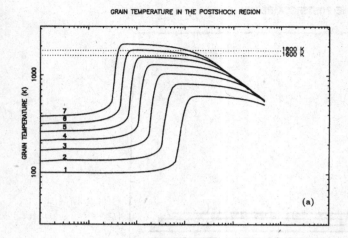

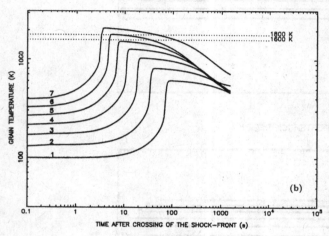

Fig. 4. (a) Temperature of a 1 mm dense grain aggregate, heated by drag in the postshock gas and cooled by termal radiation, in the postshock region for cooling shocks with parameters shown in Fig. 3, respectively, as a function of time after crossing the shock front. (b) Temperature of an aggregate of the same mass but 10 times lower density. Parameters of the shocks are the same as in Fig. 4 (a) with the same numbers.

ated with the inertial motion of the particles through the decelerated gas.

The most important heating of large grains in the postshock region is due to gas drag as the grains move by inertia through the decelerated gas. The energy acquired by a grain per unit time due to the drag heating is

$$L^{+} = 4\pi r_{gr}^{2} \frac{\xi}{8} \rho v_{gg}^{3} \tag{5}$$

where v_{gg} is the relative velocity of the dust particles with respect to the gas, and the coefficient $\xi \approx 1$ in the cooled and compressed postshock region, provided that $v_{gg} \gg c_s$; in general, ξ could be expressed in terms of the adiabatic "recovery" temperature T_{rec} and the Stanton number C_H given, e.g., by Hood and Horanyi (1991).

The value for the particle-to-gas relative velocity just after crossing the density discontinuity is $v_{gg2} = u_1 - u_2 \simeq \frac{5}{6} u_1$. The evolution of the velocity of a spherical particle with radius r_{gr} and density ρ_{gr} is described by the equation

$$\frac{dv_{gr}}{dt} = -\frac{3}{4} \frac{c_D}{2} \frac{\rho v_{gg}^2}{\rho_{gr} r_{gr}}, \tag{6}$$

where c_D is the drag coefficient ($c_D \approx 2$ in the limit $v_{gg}/c_s \gg 1$), and v_{gr} is the grain velocity.

The grains are cooled mainly by their thermal emission. For background thermal emission at temperature T_r, the energy radiated in unit time by a large grain (with size larger than the characteristic wave length of emitted radiation) is

$$L^{-}_{rad} = 4\pi r_{gr}^{2} \zeta \sigma_{ST} (T_{gr}^{4} - T_{r}^{4}), \tag{7}$$

where T_{gr} and T_r are the temperatures of the grain and thermal radiation in the surrounding medium, respectively, and the coefficient ζ shows how much the emissivity of the grain differs from blackbody emissivity ($\zeta = 1$, for black body). (For smaller dust grains the absorption and emission efficiency are given by the Mie theory).

In considering the heating and meting of fluffy aggregates, we approximate them as spheres of lower density and take into account that the density of the melt is larger than the effective density of the aggregate, defined as the mass of the aggregate divided by the volume of a sphere of radius equal to the radius of the aggregate (assuming that the aggregate is spherical). We assume that melt is gathered at the central part of the aggregate (Fig. 5), and that the aggregate is completely molten when the radius of the melt has reached the outer radius of the aggregate. (In reality the gathering could leave cavities, which are observed in some chondrites.)

In the cooling postshock region the time scale of deceleration of large grains could exceed the time scale of gas cooling. In this case grains penetrate without significant loss of kinetic energy into cooler and denser postshock gas, and get a pulse of heating which is more intense and shorter in duration than in a shock without cooling.

The steady-state temperature of large aggregates heated by gas drag and cooled by thermal radiation can be obtained from the equation $L^{+} = L^{-}_{rad}$, yielding

$$T_{gr} \approx 120 \zeta^{-1/4} \left[\frac{n_1}{10^{10}} \left(\frac{u_1}{2 \cdot 10^6} \right)^3 \left(\frac{u_1}{c_s(T_{min})} \right)^2 + T_r^4 \right]^{1/4}. \tag{8}$$

This equation assumes $n/n_1 \approx [u_1/c_s(T_{min})]^2$; T_{min} is the gas temperature at the layer at which drag heating of the grain is maximal, i.e., to which the grain penetrates before its deceleration becomes more important than the increase of the gas density. As was discussed earlier, $T_{min} \simeq T_2$ for small grains ($r_{gr} \sim 10^{-5}$cm); $T_{min} \simeq T_r$ for millimeter-size grains. Then, substituting $u_1 = \sqrt{GM/R}$ into Eq. (8), and $n_1 = \dot{M}/(4\pi \mu m_H v_r R^2)$, gives

$$T_{gr} = 2.8 \cdot 10^3 \chi \left(\frac{\dot{M}}{10^{-5} M_\odot yr^{-1}} \right)^{1/4} \cdot \left(\frac{M_\odot}{M} \right)^{-1/4} \left(\frac{R}{1AU} \right)^{-1} K, \tag{9}$$

where it is assumed that $v_r/c_s = 20$ at T_{min}. ξ is a nondimensional coefficient equal to the ratio of the density inside a clump to the average density in the infalling envelope.

For $\xi = 1$, Eq. (9) yields $T_{gr} \geqslant T_{melt} \simeq 1600$ K at heliocentric distances

$$R \lesssim 1.75 \chi \left(\frac{\dot{M}}{10^{-5} yr^{-1}} \right)^{1/4} \left(\frac{M_\odot}{M} \right)^{-1/4} \left(\frac{T_{gr}}{1600K} \right)^{-1} AU. \tag{10}$$

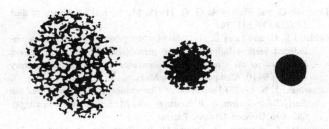

Fig. 5. Schematic diagram, showing compaction of a fluffy aggregate during melting.

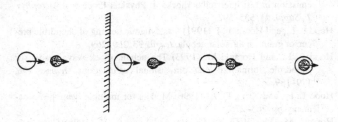

Fig. 6. Schematic diagram, illustrating the process of formation of a composite chondrule by coalescense of chondrules in the postshock region. A smaller chondrule precursor is melted by gas drag heating, then decelerates and solidifies. A larger grain which crosses the shock front (vertical line) later is still in the molten state when it reaches and catches the smaller one. This happens because the time-scale of deceleration and hence cooling is longer for larger grains. The lenghts of vectors reflect grain velocities.

It follows from this equation that the region of melting of silicate grain extends to 2 AU, and to 4 AU if the density of infalling gas is, respectively, 1.7 and 27 times larger than the average density of infalling gas for the rate of accretion $10^{-5}\,M_\odot\,\mathrm{yr^{-1}}$ and $M \simeq M_\odot$.

This estimate is in a reasonable agreement with the numerical results of Ruzmaikina and Ip (1994), and more recent work. Figure 4 (a-b) show the results of numerical simulations of evolution of the temperature for a solid grain (a) with radius 1mm and density 2.5 g cm^{-3}, and for a fluffy aggregate (b) of the same mass and 10 times smaller density. When the grain is melted, the density of the melt is equal to the density of the rock (Fig. 5). One can see from the comparison of Figures 4a and 4b that fluffy aggregates experience larger drag and, hence, are decelerated faster than dense aggregates. This difference decreases, however, if the aggregates begin to melt, because of the decreasing size and increasing density of the aggregates.

COAGULATION IN THE POSTSHOCK REGION

The general sphericity of chondrules indicates that most of them were solid at the time of agglomeration. However, properties of chondrites and chondrules themselves show signs of fast aggregation. Some chondrules have craters that appear to be dents caused by low velocity collisions between partially molten chondrules. Also, the difference in types of chondrules that make up different chondrite types shows the lack of efficient mixing of chondrules before agglomeration of larger bodies (Grossman 1988). Herein we qualitatively analyze some aspects of coagulation in the postshock region.

After crossing the shock front, grains continue to move by inertia through the decelerated gas until they are braked by gas drag. The time-scale of braking depends on the size and density of the grain or grain aggregate ($\propto (\rho_{gr}\,r_{gr})^{-1}$). Larger or more dense grain aggregates (or larger drops) have a greater velocity at the same distance from the shock. Therefore they can overtake slower moving smaller grains (or drops). The overtaking can result in the formation of compound chondrules (Fig. 6). To illustrate this, imagine that a smaller chondrule precursor is melted by gas drag heating, then decelerated, cooled, and solidified. A larger grain, which has crossed the shock front later, can still be in the molten state when it reaches and catches the smaller one. (This can happen because the time-scale of deceleration, and therefore of cooling of the larger grain, is larger). The velocity of the compound chondrule drops after the coalescence (because of conservation of momentum), and this decrease of velocity could be sufficient to drop the temperature of the compound chondrule below the melting point.

The probability (number per unit time) of collision of a given chondrule (denoted by subscript 1) with other chondrules, smaller grains, or grain aggregates (subscript 2) can be estimated from the equation

$$\frac{dN}{dt} = \sigma_{gr2} n_{gr2} \Delta v_{1,2} = \frac{3}{2} \frac{q\rho}{\rho_{gr2} r_{gr2}} \Delta v_{1,2} \qquad (11)$$

where $\Delta_{1,2} = v_{gr1} - v_{gr2}$ is the relative velocity of chondrule 1 with respect to other objects 2; we suppose that the latter have uniform radius and density. It follows from Eq. (6) that the dependence of grain velocity on time in a postshock region of constant density is

$$v_{gr} = \left(v_{gr0}^{-1} + \frac{3}{2} \frac{c_D}{2} \frac{\rho}{\rho_{gr} r_{gr}} t \right)^{-1}, \qquad (12)$$

where v_{gr0} is the "initial" velocity of the grain, at $t = t_0$. This formula can be used to estimate the evolution of the grain velocity when gas cooling is finished and the gas density has reached a stationary value. In this case v_{gr0} is the velocity of the grain at the end of the gas cooling phase. For conditions in the accretional shock front, v_{gr0} is between $5/6u_1$ and u_1 for mm-size grains, and $v_{gr0} \ll u_1$ for µm-size particles.

Substituting these formulas into Eq. (1) and integrating yields

$$N = \frac{2q(r_{gr2})}{c_D} \left[Y_{gr} \log\left(1 + \frac{3}{2} \frac{c_D}{2} \frac{\rho v_{gr10}}{\rho_{gr1} r_{gr1}} (t - t_{10}) \right) \right.$$
$$\left. - \log\left(1 + \frac{3}{2} \frac{c_D}{2} \frac{\rho v_{gr20}}{\rho_{gr2} r_{gr2}} (t - t_{20}) \right) \right] \qquad (13)$$

where $Y_{gr} = (\rho_{gr1} r_{gr1})/(\rho_{gr2} r_{gr2})$, and $q(r_{gr2})$ is the weight fraction of grains with radius r_{gr2} with respect to gas. Assuming that $t_{10} \approx t_{20}$, v_{gr20}, $v_{gr10} = v_{gr20} = 3 \times 10^6$ cm s^{-1}, $\rho_{gr1} r_{gr1} = 10^{-1}$ g cm^{-2}, $q(r_{gr2}) = 5 \times 10^{-3}$, and $Y_{gr} = 10/9$ and 3/2, we estimate that the probability of a grain colliding with smaller grains during 3×10^2 to 10^3 s ranges between 1.6% and 4.4%. The relative velocities of the grains by these times are 0.75 and 0.45 km s^{-1}, respectively. The probabilities of collision are equal to 8.5% and about 22% when $Y_{gr} = 3/2$ (ρ_{gr2} $r_{gr2} \approx 0.0667$ g cm^{-2}),. and other parameters are the same. The relative velocities of the grains are 2.1 and 1.6 km s^{-1}, respectively. One can see that the number of collisions between chondrules over a time-scale $\leq 3 \times 10^2$ s (when large chondrules are molten or in a plastic state) is sufficient to explain the fraction of composite chondrules. The larger the difference in sizes (more precisely in $\rho_{gr} r_{gr}$), the larger the relative velocities of grains, and hence the probability of collisions when the number of grains is the same.

The velocities of impacts between chondrules and very different sizes, and between chondrules and smaller grains, might become large enough to result in catastrophic consequences. The possible outcomes of such collisions include catering or destruction, and partial evaporation of chondrule or projectile materials. The maximal amount of the chondrule material which could be evaporated is equal to $m_{ev} = m_{gr2} v^2_{gr2}/(2 \lambda_{evap})$, provided that all the kinetic energy of the impacting particle is converted into latent heat. An impact with a 1 μm-size ($m_{gr2} \sim 10^{-12}$ g) particle can cause the evaporation of a tiny fraction of the chondrule mass, $\sim 10^{-9}$ to 10^{-11} g, if the relative velocity of grains ranges between 3 and 30 km s^{-1}. High velocity collisions with larger ($\geq 10^{-2}$ cm) grains might be more catastrophic, causing partial or complete evaporation of the chondrule. Evaporated material will be recondensed again predominantly on fine matrix particles, having a much larger total surface area. This possibility requires further quantitative exploration.

After the grain motion is converted from directed to chaotic, the process of coagulation of grains becomes similar to that in the turbulent solar nebula, but with a heterogeneous initial distribution of grain sizes which differs from the usually assumed homogeneous distribution with initially submicron dust particles (Weidenschilling and Cuzzi 1993). Now the distribution includes chondrules and dust aggregates together with smaller dust particles. The relative velocities of chondrules and smaller grains in this relatively dense gas are low enough for them to coagulate during collisions and cause rapid growth of larger bodies. This could plausibly explain the differences between chondrules in different types of meteorites.

We conclude that the investigation of the efficiency of coagulation of small particles to form chondrule precursors, and efficiency of coagulation of dispersed chondrules and matrix material, are now among the "hottest" problems of chondrule formation.

REFERENCES

Boss A. P. (1993) Evolution of the solar nebula. II. Thermal structure during nebula formation. *Astrophys. J.*, **417**, 351–367.

Boss A. P., and Graham J. A. (1993) Clumpy accretion and chondrule formation. *Icarus*, 153–154.

Cameron A. G. W. (1966) The accumulation of chondritic material. *Earth Planet Sci. Lett.* **1**, 93–96.

Cameron A. G. W. (1975) Clumping of interstellar grains during formation of primitive solar nebula. *Icarus* **24**, 128–133.

Cassen P. (1994) Utilitarian models of the solar nebula. *Icarus* **112**, 405–430.

Cassen, P. M., and Moosman, A. (1981) On the formation of the solar nebula. *Icarus* **81**, 353–376.

Dominik C. and Tielens A. G. G. (1995) Mechanical properties of dust *LPSC XXVI* 341–342.

Gibbard S. G. and Levy E. H. (1994) On the possibility of precipitation-induced vertical lightning in the protoplanetary nebula, in *Papers Presented to the Conference: Chondrules and the Protoplanetary Disk*, pp. 9–10, Albuquerque, New Mexico.

Grossman, J. N. (1988) Formation of Chondrules. In *Meteorites and the Early Solar System*, (j. F. Kerridge, and M. S. Matthews, Eds.), pp. 680–696. Univ. of Arizona, Tucson.

Fujiwara, A., Cerroni, P., Davis, D. R., Ryan, E., Di Martino, M., Holsapple, K., and Housen, K. (1989) In *Asteroids II*, (R. P. Binzel, T. Gehrels, and M. S. Matthews, Eds.), pp. 240–268. Univ. of Arizona, Tucson.

Hollenbach, D., and McKee, C. F. (1979) Molecule formation and infrared emission in fast interstellar shocks. I. Physical Processes. *Astrophys. J., Suppl.* **41**, 555–592.

Hood L. L., and Horanyi M. (1991) Gas dynamic heating of chondrule precursor grains in the solar nebula. *Icarus* **93**, 259–269.

Hood L. L., and Horanyi M. (1993) The nebula shock wave model for chondrule formation: one dimensional calculations. *Icarus* **106**, 179–189.

Hood L. L., and Kring D. A. (1996) Models for multiple heating mechanisms. pp. 265–276.

Horanyi M., Morfill G. K., Goertz, and Levy E. H. (1995) Chondrule Formation in Lightning Discharges. *Icarus* **114**, 174–185.

Hoyle, F. (1953) On the origin of the solar system. *Quart. J. Roy. Astron. Soc.* **1**, 28–55.

Landau, L. D., and Lifshitz, E. M. (1987) *Fluid Mechanics*, Pergamon, New York.

Levy E. H. (1988) Energetics of Chondrule Formation. In *Meteorites and the Early Solar System*, (J. F. Kerridge, and M. S. Matthews, Eds.), pp. 697–714. Univ of Arizona, Tucson.

Levy E. H. and Araki S. (1989) Magnetic reconnection flares in the protoplanetary nebula and the possible origin of meteorite chondrules. *Icarus* **81**, 74–91.

Ruzmaikina, T. V. (1981) On the role of magnetic field and turbulence in the evolution of the presolar nebula. *Adv. Space Res.* **1**, 49–53.

Ruzmaikina, T. V. (1990) Chondrule formation in the accretional shock. *Lunar Planet. Sci. XXI*, 1053–1954.

Ruzmaikina T. V., and Ip W. H. (1994) Chondrule formation in the accretional shock. *Icarus* **112**, 430–447.

Ruzmaikina, T. V. and Maeva S. V. (1986) Process of formation of the solar nebula. *Astronomicheskii Vestnik* **20**, 212–226.

Ruzmaikina, T. V., Khatuncev I. N. and Konkina T. V. (1993) Formation of the low-mass solar nebula. *Lunar Planet. Sci. XXIV*, 1225–1226.

Sargent, A. I., and Beckwith S. V. (1987) Kinematics of the circumstellar gas of HL Tauri and R Monocerotis. *Astrophys. J.* **323**, 294–305.

Shu, F. H. (1977) Self-similar collapse of isothermal spheres and star formation. *Astrophys. J.* **214**, 488–497.

Sonett C. P. (1979 On the origin of chondrules. *Geophys. Res. Lett.* **6**, 677–680.

Weidenschilling S. J. (1977) The distribution of mass in the planetary system and solar nebula. *Mon. Not. Roy. Astron. Soc.* **51**, 57–70.

Weidenschilling S. J., and Cuzzi J. N. (1993) Formation of planetesimals in the solar nebula. In *Protostars and Planets*, (E. H. Levy and J. I. Lunine Eds.), pp. 1031–1060, The University of Arizona Press.

Weidenschilling S. J., and Ruzmaikina T. V. (1994) Coagulation of grains in static and collapsing protostellar clouds. *Astrophys. J.*, 713–726.

Weissman P. R. (1991) The angular momentum of the Oort cloud. *Icarus* **89**, 190–193.

Wood J. A. (1984) On the formation of meteoritic chondrules by aerodynamic drag heating in the solar nebula. *Earth Planet. Sci. Lett.* **70**, 11–26.

Zeldovich Ya. B., and Raizer Yu. P. (1967) *Physics of Shock Waves and High-Temperature Hydrodynamic Phenomena* (W. D. Hayes, and R. F. Probstein. Eds.) Academic Press, NY.

30: The Protostellar Jet Model of Chondrule Formation

KURT LIFFMAN[1†, 2] *and MICHAEL J. I. BROWN*[2]

[1]*CSIRO/DBCE, P.O. Box 56, Highett, Victoria 3190, Australia.* [2]*Astrophysics Group, School of Physics, University of Melbourne, Parkville, Victoria 3052, Australia.*
†*To whom all correspondence should be addressed*

ABSTRACT

A chondrule formation theory is presented where the chondrule formation zone is located within 0.1 AU of the protosun. This hot, optically thick, inner zone of the solar accretion disk is coincident with the formation region of the protosolar jet. It is suggested that chondrules are ablation droplets produced by the interaction of the jet wind with macroscopic bodies that stray into the jet formation region.

Provided these droplets are small enough, they will be swept up by the jet wind and subsequently cool at approximately the same rate as the expanding gas in the jet. There is a critical gas density ($\sim 10^{-11}$ g cm^{-3}) below which chondrules will undergo large, damped oscillations in altitude and thereby suffer reheating.

We claim that it is in the cooler, high altitude regions above the midplane of the inner accretion disk that compound chondrules are formed, and the collisional fragmentation of chondrules takes place. Since these processes take place in the jet flow, one can make a prediction for the expected structure of triple compound-chondrules. For such chondrules, it is suggested that two "relatively large" secondary chondrules will avoid each other. This prediction is valid only if the gas-flow is sufficiently laminar or if the "spin-down time" for a double compound chondrule is less than the inter-chondrule collision time.

The model assumes that particles, ranging in diameter from 1 μm to 1 cm, can be ejected from the inner-accretion disk by the jet flow, and that the angular momentum of this material is sufficient to eject it from the jet flow. Given these assumptions, any material so ejected, will fly across the face of the accretion disk at speeds greater than the escape velocity of the system. This material can only be recaptured through the action of gas drag. Such a capture process naturally produces aerodynamic size sorting of chondrules and chondrule fragments, while the ejection of refractory dust provides a possible explanation for the observed complementarity between matrix and chondrules.

This transfer of material will result in the loss of angular momentum from the upper atmosphere of the outer accretion disk and thereby facilitate the accretion of matter onto the protosun.

PREFACE

In this paper we discuss a relatively new chondrule formation model, which uses a hypothetical protosolar jet as the main chondrule formation mechanism. As this model is in the earliest stages of development, this particular monograph should not be treated as the forever, enduring, last word on the subject. Instead, we hope that our discussion will demonstrate the explanative and predictive power of this model, and possibly prompt our colleagues to view chondrule formation from a different point of view as compared to other theories.

Some readers would argue that fewer, not more, chondrule formation theories are required. Certainly with nearly twenty different published theories, how can one determine which theory, if any, is the correct one? The answer, of course, is to embrace the theory that can not only explain many of the observed data, but also predict new facts. So in this study, we will attempt to explain:

(1) the chondrule formation process;

(2) the chondrule cooling rate;

(3) the collisional fragmentation of chondrules and the formation of adhering compound chondrules, where the secondary (and smaller) chondrule was plastic at the time of collision;

(4) chondrule reheating;

(5) the chondrule upper size limit;

(6) chondrule size sorting,
    and

(7) the complementary chemical composition of matrix and chondrules.

We will also discuss a possible link between protostellar jets and disk accretion, where a protostellar jet may inject material with low angular momentum into the upper atmosphere of an accretion disk, thereby enhancing stratified accretional flow, i.e., the upper layers of the disk accrete onto the protostar more readily than do the layers adjacent to the mid plane of the accretion disk.

As for predictions, we will describe the expected structure of triple compound chondrules, i.e., where two small secondaries are adhering to a large primary. We suggest that, for such a compound chondrule, the two secondary chondrules will tend to "avoid" each other. We give the mathematical expression for the minimum avoidance angle between the two secondaries and discuss under what circumstances this prediction will break down. We also show that the experimental data collected so far (Unfortunately, for only four such chondrules) have structures that are consistent with the predicted structure. It is our fond hope, that we will inspire some of our colleagues to collect much more data, so that the validity or otherwise of this prediction can be confidently determined.

1. JET ABLATION

In the late 1970's, it was found that protostellar systems formed bipolar outflows of material, i.e., the gas flows travelling in two opposing directions, approximately perpendicular to the plane of the accretion disk (for a review, see Beckwith and Sargent 1993). These flows were detected via the CO rotation lines and consisted mostly of molecular material. The flow speeds of the bipolar outflows were found to be around 20 km s^{-1}, and by dividing their length by the observed flow speed, one could deduce a "dynamic" lifetime for these flows of around 10^5 years. Later, unbiased surveys of protostellar systems suggests that *all* protostellar systems undergo some form of bipolar outflow stage (Fukui *et al.* 1993).

Often, but not always, one can find within a bipolar outflow a faster, more collimated flow known as a protostellar jet. Protostellar jets are usually detected in the SII, OI and Hα lines and are consequently referred to as optical jets. They have observed wind speeds in the range of 100 – 400 km s^{-1}, they eject a large amount of gas (total mass 10^{-3} – 10^{-1} M$_\odot$), are quite energetic

(total kinetic energy 10^{44} – 10^7 erg), long-lived phenomena (10^6 – 10^7 years) that exist at the very earliest stages of star formation (Cabrit *et al.* 1990, Edwards, Ray and Mundt 1993). They occur not only in the massively active FU-Ori stages of star formation, but they also are to be found in the more quiescent, Classical T-Tauri Star (CTTS) stages. They appear to be generated by the interaction between protostars and accretion disks, within 10 stellar radii (R_*) of the protostar (Hartmann 1992).

What is the connection between bipolar outflows and protostellar jets? There is a growing consensus that bipolar outflows are a byproduct of protostellar jets (Snell *et al.* 1980, Shu *et al.* 1993), where the protostellar jet sweeps up ambient molecular cloud material into two thin shells, which manifest themselves as the observed bipolar lobes of CO emission. Once the molecular cloud material has been swept away (on a timescale of 10^5 years), the bipolar outflow disappears, leaving the protostellar jet to erratically fire away for a further 10^6 – 10^7 years.

Our interest in protostellar jets is sparked by the following mass processing argument. Suppose the solar nebula formed a protosolar jet. Such a jet would have ejected 10^{-5} – 10^{-3} M$_\odot$ of "rocky" material (i.e. all elements excluding H and He) from the solar nebula. If only 10% of this material were to fall back to the solar nebula then we would have 10^{-6} – 10^{-4} M$_\odot$ of rocky (possibly refractory) material being contributed to the solar nebula over a 10^6 – 10^7 year period. Given that the "rock" mass of the planets is of order 10^{-4} M$_\odot$, a protosolar jet may have made a significant contribution to the chemical structure of the solar nebula. Indeed, it is possible that ejecta from the jet flow may have been incorporated into the best preserved samples of the solar nebula: the chondritic meteorites

Of course, such an argument does not prove that protostellar jets formed chondrules, but only provides a plausible basis for constructing a theory of chondrule formation. To create such a theory, our first task is to investigate the thermal environment of the jet formation region. This is a function of the formation distance of a protosolar jet from the protosun. A distance which is uncertain, but model fits to the Li 6707 Å, Fe 1 4957 Å, Fe II 5018 Å lines in Fu Ori, plus the lack of extinction and infrared excess in the wind, suggests a formation distance of ≤ 10R_* (≈ 0.1 AU for the solar nebula), where R_* is the radius of the protostar (Hartmann 1992). The temperature of the disk surface (T_s) at such distances, is given (approximately) by the formula (Frank, King and Raine 1992)

$$T_s(r) = 850\,\text{K} \left\{ \frac{(M/M_\odot)(\dot{M}/10^{-7}\,M_\odot\,\text{yr}^{-1})}{(r/0.1\,\text{AU})^3} \right\}^{1/4}$$
$$\left[1 - (R_*/r)^{1/2} \right]^{1/4}, \tag{1.1}$$

where M is the mass of the protostar, $\dot{M}$ is the mass accretion rate, and r is the radial distance from the protostar's centre. The maximum temperature of the midplane of the accretion disk (T_m) can be approximately determined from the formula

$$T_m \approx T_s(\eta\tau)^{1/4}, \tag{1.2}$$

where η is some number of order 1, and τ is the optical depth (*ibid.*, and Cassen 1993). Since τ may have values as high as 10^4, it is probable that at distances of order 0.05 AU from the protostar, the temperature at the midplane maybe high enough to melt or vaporise any rocky bodies that happen to be in that section of the accretion disk. This raises the possibility that the protostellar jet winds may ablate droplets from the surface of rather warm "rock" bodies and if the droplets are small enough, they may be ejected from the inner accretion disk by the drag force of the jet flow.

It may be asked, however, why embrace the idea that chondrules are ablation droplets? We are encouraged to advance such a hypothesis, because chondrules appear to have been extruded or drawn from one or more extended magma bodies (Dodd and Teleky 1967). Also, an ablative process readily forms chondrule-like spheres from meteors (Brownlee *et al.* 1983). These "ablation-spheres" are chondrule-like in size and shape. They also share some physical similarities, e.g., they often contain relict grains. Despite this supportive evidence, there are, however, some significant problems to overcome before one can believe the ablation hypothesis.

First, we require the appropriately sized rocky-bodies to enter into the jet formation region. Protostellar systems generally have accretion disks and large bodies could be caught up in the inward accretional flow. However, the existence of chondrules in single ordinary chondrites that have (admittedly uncertain) ^{129}Xe*/^{127}Xe ages ranging over 10^7 years (Swindle *et al.* 1991, Swindle and Podosek 1988), suggests that the accretional flow in the solar nebula was small enough to allow the survival of the chondrite precursors. Indeed, it is possible that such accretional inflow may have been dependent on height within the accretion disk, where only the gas at higher altitudes took part in the accretional flow (see §5).

Another mechanism for radial migration is that of gas drag due to the velocity difference between nebula gas and the near Keplerian motion of large bodies around the protosun. The infall velocity for material subject to this gas drag is (Whipple 1973, Weidenschilling 1977)

$$\frac{dr}{dt} \approx -\frac{2t\Delta V}{V_{Kep}\, t_D}, \qquad (1.3)$$

where K_{Kep} is the Keplerian velocity, ΔV is ($\approx 10^{-3} V_{Kep}$) the difference between the gas angular velocity and V_{Kep}, while t_D is the time scale for gas drag to influence the motion of the body. In the Stokes drag regime,

$$t_D = \frac{2 r_{fod}^2 \rho_b}{9 \eta_g}, \qquad (1.4)$$

where ρ_b is the mass density of the body and η_g is the viscosity of the gas, and r_{fod} is the radius of the "chondrule-fodder" body. Using the appropriate formula for the gas viscosity (e.g. Eq. (2.3.5) of Liffman 1992), one can obtain, from Eq. (1.3), the timescale for orbital decay (τ_{decay})

$$\tau_{decay} \approx 10^5 \left\{ \frac{(r_{fod}/10^3\,\text{cm})(\rho_b/1\,\text{g cm}^{-3})}{((T_g/100)^{1/2}\,\text{K})} \right\}\text{yrs}, \qquad (1.5)$$

with T_g being the temperature of the gas.

So, objects with a radius of around 10 m would fall into the Sun on a timescale of about 10^5 years. This analysis, however, ignores the change in the size of the body due to accretion of dust from the solar nebula. If the dust to gas mass ratio is of ~ 1, accretion of material will increase the size of the body and eventually stop its infall into the Sun (Weidenschilling 1988). On the other hand fragmentation due to inter-body collisions would have been a source of smaller material which may have eventually reached the boundary layer between the protosun and the solar nebula.

There may have been other mechanisms that brought macroscopic material into the inner regions of the solar nebula. We will simply note that it is a plausible assumption and one we require for our model.

Once we have these large bodies in the jet-formation region, we are faced with the following scenario: the high speed jet-flow will probably occur at $z \geq$ the scale height of the accretion disk, H, since the outflow jet must be governed by the conservation of mass equation, and so high velocities will only occur when $\rho_g(z) \ll \rho_g(0)$. On the other hand, our "chondrule-fodder" bodies will be located on or near the midplane of the nebula ($z = 0$). This separation between the critical points in the wind and the chondrule-fodder bodies would appear to be the death-knell for our ablation hypothesis – for how else can one produce ablation droplets if the hypothetical windflow is nowhere near our chondrule-producing planetesimals?

To answer this, we examine protostellar jet theory.

Most protostellar outflow models assume that magnetic fields provide the main coupling mechanism between the accretion disk and the outflow (for recent reviews, see Bicknell 1992, Königl and Ruden 1993 plus Shu *et al.* 1993). In the most recent models, the magnetic driving mechanism involves the interaction between the dipole field of the protosun and the inner accretion disk (e.g. see Lovelace *et al.* 1991, and Shu *et al.* 1994). Studies of this protosun-nebula interaction can be traced back to the work of Freeman (1977), who assumed that the dipole field of the protosun was able to thread the inner (partially ionized) disk as shown in Fig. 1. In effect, the magnetic field of the protosun is "tied" to its surrounding accretion disk.

In Freeman's model, the dipole field of the protosun rotates (approximately) with a rigid body velocity, so there exists a point away from the protosun ~ 0.04 AU), where the speed of the field sweeping over the disk equals the (approximately) Keplerian velocity of the disk. In Fig. 1, this position is denoted as the "synchronous orbit". For distances r greater than this synchronous distance, the magnetic field of the protosun has a greater velocity than the accretion disk and so the magnetic field becomes "wrapped up", i.e. the purely poloidal magnetic field of the protosun is converted into a toroidal field in the disk.

In Fig. 1, we show a side view of the resulting field structure. The toroidal field becomes the dominant field in the disk, and its direction reverses as one traverses the central plane. for such a configuration, the central plane of the nebula can become a "current sheet" where magnetic fields reconnect and the magnetic energy so released is converted into particle energy.

As discussed in Priest (1994) the particle flow velocity (v_f) obtained from the merging of magnetic fields (B) is close to the

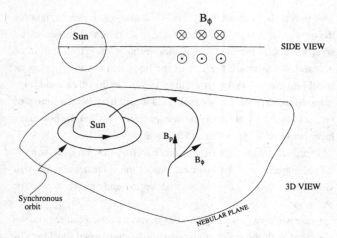

Fig. 1. The dipole field of the protosun interacts with the accretion disk, such that for distances greater than a critical distance, (~ 0.01 AU) from the protosun, the poloidal field of the Sun is "wrapped-up" by the accretion disk. This process may cause the central plane of the inner accretion disk to become a "current sheet" produced by the merging magnetic field lines. This merging will be due to magnetic diffusion and the switch in the sign of the toroidal field as one traverses the central plane.

characteristic speed of a magnetic medium (*i.e.*, the Alfvén speed C_A), which in mathematical notation has the form

$$v_f \approx C_A = \frac{B}{\sqrt{\mu_0 \rho_g}} = 3\left(\frac{B}{100\,\text{G}}\right)\left(\frac{10^{-8}\,\text{g cm}^{-3}}{\rho_g}\right)^{1/2} \text{km s}^{-1} \qquad (1.6)$$

where ρ_g is the mass density of the gas and μ_0 is the permeability of free space. In the accretion disk, the direction of the reconnection flow would be roughly parallel to the central plane, with a small component in the z direction.

To obtain ablative behaviour from gas flows with gas densities in the range of 10^{-6} to 10^{-8} g cm^{-3}, the required gas speeds range from 1 to 20 km s^{-1} (Liffman 1992). Comparing this to Eq. (1.6) suggests that these "reconnection" flows may have the required densities and speeds to ablate the planetesimals. Some of the ablation droplets so produced would then, presumably, be caught up in the flow field that eventually becomes the main protostellar jet flow.

It may turn out, however, that outflows are not powered by "wrapped-up" toroidal fields. Our purpose in describing this particular model is to demonstrate that, in the outflow region, the central plane of the accretion disk may be a highly active region, where conditions are conducive to chondrule formation by ablation.

2. CHONDRULE COOLING

A long standing problem in chondrule formation is how to explain the slow rate of chondrule cooling. Experimental simulation of chondrule formation suggests that chondrules cooled at a rate of 5 to 2000 °C/hour (Hewins 1988). Such cooling rates are 3 to 5 orders of magnitude smaller than those expected for an isolated black body radiating heat directly into space. It would appear that chondrules were formed in a hot optically-thick medium or were

produced in close proximity to other chondrules so that mutual radiation could damp the cooling rate.

In our model, chondrules are produced in the hot, optically-thick midplane of the inner accretion disk around a protostar. They are produced by the ablative interaction between the initial stage of a protostellar jet wind and rocky chondrule "fodder" bodies that happen to stray into the jet formation region. If the subsequent droplets are small enough, they will be swept up by the wind and begin to move with the jet flow.

As the jet flow moves away from the midplane of the accretion disk, it is likely that the gas in the flow will cool, since work is being done to expand the gas. Consequently, if the molten droplets are close to thermodynamic equilibrium with the surrounding gas flow, they must also cool at just about the same rate as the expanding gas flow.

Naturally, to model this process we require a model of a protostellar jet. Unfortunately, a comprehensive theory of protostellar jets is unavailable at this time. To partially circumvent this difficulty, we adopt a simple parameterised model of an outflow. First, we describe our system with cylindrical coordinates, the plane $z = 0$ being the midplane of the accretion disk with the protostar residing at the origin (see Fig. 11). With this coordinate system, the steady state form of the continuity or mass conservation equation ($\nabla \cdot (\rho_g \mathbf{v}_g) = 0$) is easily solved, for an axisymmetric flow, to give the equation

$$2\bar{\rho}_g(z)\bar{v}_{gz}(z)\pi r(z)^2 = \text{constant} = \dot{M}_o, \qquad (2.1)$$

where $\bar{\rho}_g(z)$ is the "r-averaged" gas mass-density as a function of z, $r(z)$ is the cylindrical radius of the outflow, $\bar{v}_{gz}(z)$ is the z component of the "r-averaged" gas velocity, and $\dot{M}_o$ is the mass-loss rate of the outflow. The factor of two in Eq. (2.1) arises, because the protostellar jet is produced from both sides of the accretion disk.

To model the gas density and gas velocity profile within the disk, we assume that the disk constrains the radial size of the protostellar jet. if we now let

$$\bar{\rho}_g(z) \approx \bar{\rho}_g(z_1)\left(\frac{z_1}{z}\right)^m, \quad z \lesssim z_1, \quad m > 0. \qquad (2.2)$$

where z_1 is the distance above the disk midplane where we can best define the flow variables. We can substitute Eq. (2.2) into Eq. (2.1) and obtain

$$\bar{v}_{gz}(z) \approx \bar{v}_{gz}(z_1)\left(\frac{z}{z_1}\right)^m, \quad z \lesssim z_1, \quad m > 0. \qquad (2.3)$$

These solutions break down as $z \to 0$, this being the price we pay for not solving the momentum and energy equations. Equations (2.2) and (2.3) are only a first order fit to the mass conservation equations. However, they do, approximately, satisfy the density-velocity relationship observed in thermally driven winds, i.e., as ρ_g decreases, v_{gz} increases.

If the gas pressure (P_g) can be globally modelled as a polytropic gas, then

$$P_g = \kappa \rho_g^\gamma, \tag{2.4}$$

where κ and γ are constants (γ is the ratio of the specific heats if the jet is adiabatic). Using the ideal gas law with Eqns (2.2) and (2.4), we can obtain an expression for the gas (and chondrule) temperature (T_g) as a function of z.

$$T_g(z) \approx T_g(z_1) \left(\frac{z_1}{z} \right)^{m(\gamma-1)}. \tag{2.5}$$

The gas flow and entrained chondrules decrease in temperature as they travel away from the central plane of the accretion disk.

If we assume that the particles are entrained with the flow, and they start from $z = 0$, then we can use Eqns (2.3) and (2.5) to give the gas and chondrule temperature as a function of time (t). Solving for t in Eq. 2.3 gives

$$\frac{z}{z_1} = \left[\frac{t}{\tau_1} \right]^{1/(1-m)} \quad 0 < m < 1, \tag{2.6}$$

where $\tau_1 = z_1/((1-m)\bar{v}_{gz}(z_1))$. Substituting the above equation into Eq. (2.5) gives,

$$T_g(t) = T_g(z_1) \left[\frac{\tau_1}{t} \right]^{\frac{m(\gamma-1)}{(1-m)}} \quad 0 < m < 1. \tag{2.7}$$

The gas and chondrule cooling timescale is determined by the value of τ_1 and $T_g(z_1)$. As we shall show in the next section, for $T_g(z_1) \sim 1200$ K, then $z_1 \sim 10^4 - 10^5$ km and $v_{gz}(z_1) \sim 0.1 - 1$ km s^{-1}. As a consequence, a minimum parameterization for τ_1 is

$$\tau_1 = 10^4 \, \text{s} \, \frac{(z_1/10^4 \, \text{km})}{(1-m)(\bar{v}_{gz}(z_1)/1 \, \text{km s}^{-1})}. \tag{2.8}$$

The general decrease in gas temperature, and chondrule temperature, as a function of z within or near the accretion disk has important consequences for the collisional interactions between chondrules. Indeed, a simple analysis of this phenomenon leads to a model which can possibly explain the observed structure of compound chondrules and leads to a prediction for the physical structure of triple compound chondrules.

3. CHONDRULE COLLISIONS

3.1 Hover particles

Let us consider a chondrule that has just been ablated from its parent body. We assume, that the chondrule will be accelerated by the gas jet so that its motion is, initially, in the z direction. The equation of motion for the particle, parallel to the z axis, is given by

$$m_p \ddot{z} = \frac{C_D}{2} \rho_g (v_{gz} - \dot{z})^2 \, \pi a_p^2 - \frac{GM m_p z}{(z^2 + r^2)^{3/2}}, \tag{3.1.1}$$

where C_D is the coefficient of gas drag, while m_p and a_p are the mass and radius of the particle, respectively.

Suppose the particle reaches a state where the gas drag is balanced by gravity, so that $\ddot{z} = 0$, and $\dot{z} = 0$. Our equation of motion becomes (for $z \ll r$)

$$0 \approx \frac{C_D}{2} \rho_g v_{gz}^2 \, \pi a_p^2 - \frac{GM m z_h}{r^3}, \tag{3.1.2}$$

where z_h is the value of z for the hovering particle. For z close to the midplane, we should expect that v_{gz} will be less than the sound speed, so if the mean free path of the gas (l) satisfies the relation

$$l / 2a_p \gtrsim 10 \tag{3.1.3}$$

then C_D has the Epstein (1924) form

$$C_D \approx \frac{8}{3 v_{gz}} \sqrt{\frac{8kT_g}{\pi m_g}}, \tag{3.1.4}$$

where k is the Boltzmann constant and m_g is the mass of a gas particle (in this case we assume monatomic hydrogen). Using Eqns (3.1.2), (3.1.4) and (2.1) we find

$$z_h \approx \left(\frac{8kT_g}{\pi m_g} \right)^{1/2} \frac{\dot{M} r}{2\pi GM \rho_p a_p} \tag{3.1.5}$$

or

$$z_h \approx 5 \times 10^4 \, \frac{(\dot{M}/10^{-8} \, \text{M}_\odot / \text{yr})(r/0.1 \, \text{AU})(T_g/10^3 \, \text{K})^{1/2}}{(M/\text{M}_\odot)(\rho_p/1 \, \text{gm cm}^{-3})(a_p/0.1 \, \text{cm})} \, \text{km}. \tag{3.1.6}$$

Thus, $z_h \sim 10^{-4}$ AU $\ll 0.1$ AU $\sim r$ as is required from the assumption leading to Eq. (3.1.2). Now, z_h is inversely proportional to the radius of the particle (a_p), so smaller particles will have higher hover heights than larger particles of the same density. To reach their respective hover altitudes, the particles will move relative to each other and may undergo collision interaction. if the collisional velocities are small enough and the temperature high enough, we may have chondrules fusing together to form compound chondrules.

3.2 Compound chondrule formation

Studies of compound chondrules have classified them into two general types: *enveloping*, where one chondrule envelopes the other and *adhering*, where one chondrule forms a "bump" on the other (Wasson 1993). In this paper, we restrict our attention to adhering compound chondrules.

Most adhering compound chondrules consist of a smaller chondrule stuck to a larger particle, where the smaller chondrule was plastic at the time of collision (Wasson *et al.* 1995). This fact immediately poses a major problem for most chondrule formation theories which use simple "flash" heating scenarios. Since, if the chondrules were all formed in the one flash heating event, then we should expect the larger chondrules to be plastic at the time of collision.

In the Jet model, chondrules are continuously formed by an ablative process as chondrule fodder bodies stray into the jet formation region. As discussed in the derivation of Eq. (3.1.6), it is reasonable to assume that 0.1 cm particles will be supported by the jet flow at a z distance of 10^{-4} to 10^{-3} AU above the midplane of the accretion disk. Let us suppose that at such distances, the temperature of the gas in the jet stream is around 1200 K – the approximate solidus temperature of a chondrule.

Assuming a fairly steady jet flow, a hover particle in such a position should have sufficient time to equilibrate with the gas temperature and be relatively non-plastic. The hover particle will only see smaller particles flying past it in the jet flow, since larger particles of approximately the same mass density will have a lower hover altitude.

As these smaller particles fly past the hover particle, the thermal inertia of these particles will give them higher temperatures, and therefore greater plasticity than the hover particle. If the hover particle collides and fuses with these particles, then the observed structure of adhering compound chondrules will be obtained (for a schematic depiction of this process, see Fig. 2).

To give this idea a quantitative context, we note that the equation for the rate of temperature change of a spherical chondrule has the form

$$\frac{4}{3}\pi a_p^3 \rho_p C_V \frac{dT_S}{dt} = \Lambda 4\pi a_p^2 q + 4\pi a_p^2 \varepsilon_a \sigma T_e^4 - 4\pi a_p^2 \varepsilon_e \sigma T_S^4,$$

(3.2.1)

(Liffman 1992 and references therein).

The left hand side of the equation describes the time rate of change of the heat energy of the body, where a_p is the radius of the particle, ρ_p is the density of the particle, C_V is the specific heat per mass of the body ($\sim 10^7$ erg g^{-1} K^{-1} for chondrules), T_S is the sur-

face temperature of the body, and t the time. The first term on the right hand side describes the energy added to the body by the gas/body interaction, where Λ is the heat transfer coefficient and q is the gas/body heat transfer rate per unit surface area. The last two terms are radiation terms, where T_e is the radiation temperature of the surrounding environment, $\varepsilon_{a/e}$ is the absorptivity/emissivity of the surface (absorptivity and emissivity are assumed to be the same in this case), and σ is the Stefan-Boltzmann constant. In the following discussion, we will assume local thermodynamic equilibrium applies and that $T_e = T_g$.

For chondrule formation, the size of the chondrules (< 1 cm) and the expected low gas density (< 10^{-6} g/cc) indicate that the mean-free path of the gas is large relative to the size of the chondrules. For this "free molecular" flow regime $\Lambda \approx 1$ and q has the form

$$q = \rho_g |v_g - v_p|(T_{rec} - T_s)C_H,$$

(3.2.2)

(Hayes and Probstein 1959, Probstein 1968) where ρ_g is the mass density of the gas, v_g is the gas velocity, v_p is the particle velocity, T_{rec} is the recovery temperature, and C_H is the heat transfer function for free molecular flow. T_{rec} and C_H have the forms

$$T_{rec} = \frac{T_g}{\gamma+1}\left[2\gamma + 2(\gamma-1)s_{gp}^2 - \frac{\gamma-1}{0.5 + s_{gp}^{-2} + s_{gp}\pi^{-0.5}\exp(-s_{gp}^2)/\mathrm{erf}(s_{gp})}\right],$$

(3.2.3)

and

$$C_H = \frac{\gamma+1}{\gamma-1}\frac{k}{8m_g s_{gp}^2}[\pi^{-0.5}s_{gp}\exp(-s_{gp}^2) + (0.5 + s_{gp}^2) \times \mathrm{erf}(s_{gp})],$$

(3.2.4)

(*ibid.*) where T_g is the gas temperature, m_g the mean gas particle mass ($\sim 1.66 \times 10^{-24}$ g), γ is the ratio of specific heats, k is Boltzmann's constant, erf(s) is the error function and s_{gp} is the ratio of the relative streaming gas speed and the most probable Maxwellian gas speed,

$$s_{gp} = \frac{|v_g - v_p|}{\sqrt{2kT_g/m_g}}.$$

(3.2.5)

We are considering flows at the very base of the protostellar jet, so we should expect that $s_{gp} \ll 1$. Thus, q becomes

$$q \approx \rho_g(T_g - T_s)\left(\frac{\gamma+1}{\gamma-1}\right)\sqrt{\frac{k^3 T_g}{8\pi m_g^3}}.$$

(3.2.6)

or

$$q \approx 5\times10^7\left(\frac{\rho_g}{10^{-8}\ \mathrm{gcm^{-3}}}\right)\left(\left(\frac{T_g}{10^3\ \mathrm{K}}\right) - \left(\frac{T_s}{10^3\ \mathrm{K}}\right)\right) \times \left(\frac{\gamma+1}{\gamma-1}\right)\sqrt{\frac{T_g}{10^3\ \mathrm{K}}}\ \mathrm{ergs^{-1}\ cm^{-2}\ s^{-1}}.$$

(3.2.7)

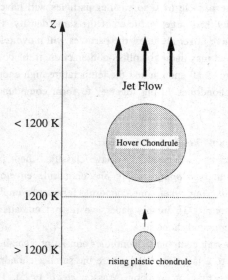

Fig. 2. Schematic depiction of compound chondrule formation in a jet flow. The gas flow is depicted within the accretion disk, where the gas temperature decreases as z (the distance from the midplane of the disk) increases. A "large" hovering chondrule, with a temperature near or at the solidus temperature of 1200 K, collides with smaller chondrules that are entrained in the gas flow. Due to thermal inertia, the smaller chondrules are warmer and more plastic than the hover chondrule. In this way, one can obtain the observed phenomena that smaller chondrules were plastic at the time the compound chondrule was formed.

When $\rho_g \gtrsim 10^{-7}$ g cm^{-3}, $\gamma = 5/3$ (monatomic gas), and $T_g \approx 1200$ K, then the q term in Eq. (3.2.1) dominates the radiative terms, since

$$\sigma T_g^4 \approx 5.7 \times 10^7 \left(\frac{T_g}{10^3 \, K} \right)^4 \text{ergs}^{-1} \text{cm}^{-2} \text{s}^{-1}. \qquad (3.2.8)$$

So Eq. (3.2.1) becomes a first-order linear differential equation in T_s, and has the solution

$$T_s(t) \approx T_g + (T_s(0) - T_g) \exp(-t / \tau), \qquad (3.2.9)$$

where τ has the form

$$\tau = \frac{a_p \rho_p C_V}{3 \Lambda \rho_g} \left(\frac{\gamma - 1}{\gamma + 1} \right) \sqrt{\frac{8 \pi m_g^3}{k^3 T_g}}. \qquad (3.2.10)$$

or

$$\tau \approx 5 \frac{(a_p / 0.1 \text{cm})(\rho_p / 3 \text{g cm}^{-3})(C_V / 10^7 \, \text{erg g}^{-1} \, \text{K}^{-1})}{(\rho_g / 10^{-8} \, \text{g cm}^{-3})}$$

$$\times \left(\frac{T_g}{10^3 \, \text{K}} \right) s. \qquad (3.2.11)$$

So, if we have $\rho_g \gtrsim 10^{-7}$ g cm^{-3} then particles with a radius between 0.01 cm and 0.1 cm will have a minimum temperature equilibration timescale of 0.5 to 5 s. Similar timescales are obtained for the low density case ($\rho_g \ll 10^{-7}$ g cm^{-3}), where the radiative terms in Eq. (3.2.1) dominate the "q" convective term. In this case, Eq. (3.2.1) can only be solved numerically. The results

from this calculation are shown in Fig. 3. As can be seen, The e-folding timescales in the radiative case turn out to be about the same as those for the pure conduction case. One should note that these are minimum timescales, because we have neglected the latent heat of fusion and the diffusion time for heat to travel from the surface of the particle to its centre.

Despite these caveats, it is probable that the timescale for thermal equilibrium are not much greater than as stated in Eq. (3.2.11). In such circumstances, we can only expect the smaller chondrule to be plastic at the time of collision if, in the collision zone, there is a sharp decline in gas temperature. We suggest that a temperature drop of 10 to 100 K s^{-1} is required for compound chondrules to form. We also expect that the collision zone for compound chondrule formation must be quite thin, since the maximum time that the smaller secondary can remain plastic is ~ 20 s and the maximum collision velocity is ~ 0.1 km s^{-1} (a higher collision velocity would fragment the chondrules, see Vedder and Gault (1974)). These numbers give a maximum length scale for he thickness of the compound chondrule formation region of around 1 km.

Now, suppose we have a chondrule hover-particle in a region of the flow where the ambient temperature of the gas flow is much less than the solidus temperature of the particle. If this solid hover-particle collides with smaller particles which are entrained in the jet flow, and if the relative speed of the two particles is greater than 0.1 km s^{-1} then it is likely that one or both of the particles will undergo some damage from the collision. Depending on the relative speed of collision, this damage can range from slight chipping to complete fragmentation.

It is well known that many chondrules have undergone chipping

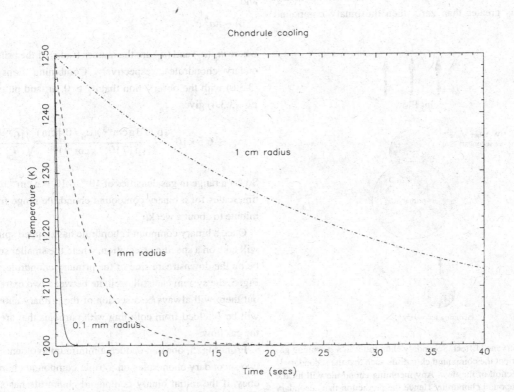

Chondrule cooling

Fig. 3. The radiative cooling of 3 particles of different radii from 1250 K to 1200 K.

and fragmentation. In the Jet model, one can provide a cause for this damage: collisions between chondrules. One can also give a site for where this damage will take place: in the jet flow at higher (and hence colder) altitudes than the compound chondrule formation region (see Fig. 11). In this way, the jet model can provide a unified scheme linking both compound chondrule formation and chondrule fragmentation.

Of course, like all theoretical mutterings in this field, the above results and ideas should be (and will be) treated with a reserved caution. A physical theory is of little use, unless it can be used to predict as well as to explain. So far we have attempted the latter, in the next section we will try the former. Let us throw caution to the winds (no pun intended) and predict the physical structure of triple compound chondrules.

3.3 THE WEATHER VANE EFFECT

If compound chondrules were formed in protostellar jets, then gas drag would orient the compound chondrules immersed in the jet flow. This is simply because the spherical symmetry of the simple chondrule is destroyed. A secondary chondrule will act like the tail on a weather vane, orienting a binary compound chondrule such that the smaller secondary will be on the "downstream" side of the primary. A subsequent collision and fusion with another chondrule will produce a triple compound chondrule where the secondary chondrules are separated by a minimum avoidance angle. This situation is shown schematically in Fig. 4.

This simple scenario does have some initial complications, the major being that if the impact parameter (the offset distance between the centres of the particles, see Fig. 4) between the colliding chondrules is greater than zero, then the binary compound

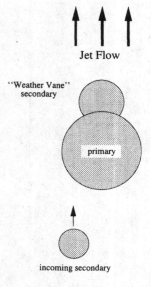

Jet Flow

"Weather Vane" secondary

primary

incoming secondary

Fig. 4. The "Weathervane" Effect. Secondary chondrules in a streaming gas flow, will tend to orient the compound chondrule such that the secondary is pointing in the direction of the flow. Any incoming chondrule will hit the rear end of the compound chondrule. Hence the prediction that secondary chondrules in a triple compound chondrule will tend to avoid each other.

chondrule will be set spinning. This rotational motion will, however, be damped, because the secondary chondrule will be travelling half the time with the gas jet, and half the time against the jet. If the secondary chondrule has a cross-sectional area A, then the rate of change of the rotation rate, ω, is given by

$$I\dot\omega \approx \frac{C_D}{2}\rho_g a_{ps}(v_{gz}-a_{ps}\omega)^2\frac{A}{2}-\frac{C_D}{2}\rho_g a_{ps}$$
$$\times(v_{gz}+a_{ps}\omega)^2\frac{A}{2}, \qquad (3.3.1)$$

where I is the moment of inertia of the system, and a_{ps} is the distance between the primary and secondary chondrules. If we substitute the Epstein solution for C_D (see Eq. 3.1.4), we obtain

$$\dot\omega \approx \frac{-8}{3}\sqrt{\frac{8kT_g}{\pi m_g}}\left(\frac{\rho_g a_{ps}^2}{I}A\right)\omega, \qquad (3.3.2)$$

which has the solution

$$\omega \approx \omega_0\exp(-t/\tau_{sd}), \qquad (3.3.3)$$

where ω_0 is the initial rate of rotation, and τ_{sd} – the e-folding "spin-down time" – has the form

$$\tau_{sd} = \frac{3I}{8\rho_g A a_{ps}^2}\sqrt{\frac{\pi m_g}{8kT_g}}. \qquad (3.3.4)$$

To obtain an estimate for τ_{sd}, we note that

$$I \sim (m_p+m_s)a_{ps}^2 \lesssim 2m_p a_{ps}^2, \qquad (3.3.5)$$

and

$$A \sim \pi a_s^2, \qquad (3.3.6)$$

where m_p (a_p) and (a_s) are the masses (radii) of the primary and secondary chondrules, respectively. Combining Eqns (3.3.5) and (3.3.6) with the observation that $a_s \gtrsim 0.1a_p$ and putting it all into Eq. (3.3.4) gives

$$\tau_{sd} \lesssim 6.5\times10^3\frac{(\rho_p/3\,\mathrm{g\,cm^{-3}})(a_p/0.1\,\mathrm{cm})}{(\rho_g/10^{-8}\,\mathrm{g\,cm^{-3}})}\sqrt{\frac{10^3\,\mathrm{K}}{T_g}}\,s. \quad (3.3.7)$$

So for a range in gas densities of 10^{-6} to 10^{-10} g cm^{-3} the spin-down timescales for a binary compound chondrule range from around a minute to about a week.

Once a binary compound chondrule has stopped spinning, then it will take on a specific orientation, where the smaller secondary will be on the downstream side of the primary chondrule. As shown in Fig. 5, the system can still oscillate between two extreme positions but there will always be a section of the primary chondrule which will be shielded from colliding with particles that are entrained in the gas flow.

From Fig. 5, one can deduce a minimum avoidance angle θ that two secondary chondrules on a triple compound chondrule should obey, if the initial binary compound chondrule has stopped spinning. Given that the distance between the centres of the primary

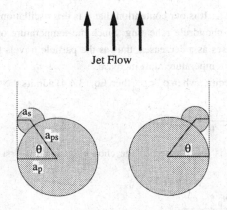

Fig. 5. The extreme positions of the secondary chondrule while oscillating from side to side.

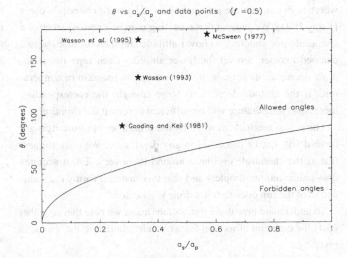

Fig. 6. Comparison of theory vs observations for triple compound chondrules. We plot the angle between the centres of secondary chondrules vs the radius ratio of the largest secondary to the primary. The largest secondaries were used for a_s values, as they would show 'disallowed' chondrules, if any were present. As can be seen, all the observed angles lie in the 'allowed' zone. The line separating the 'forbidden' and 'allowed' zones is given by Eq. (3.3.8).

and secondary chondrules is $a_{ps} = a_p + fa_s$, where f is the deformation factor of the secondary chondrule, then

$$\cos(\theta) = \frac{a_p - a_s}{a_p + fa_s} = \frac{1 - a_s / a_p}{1 + fa_s / a_p}. \qquad (3.3.8)$$

A plot of a solution to this equation is shown in Fig. 6. If the weather vane effect did not occur then we would expect a uniform distribution of angles between secondaries for each value of a_s/a_p. If the weather vane effect does occur, then the distribution should be skewed towards the 'allowed angles' shown in Fig. 6. This is what we see in the few data we have been able to scour from the literature.

Again, we should make clear that the weather vane effect will only be observed if the timescale for a binary compound chondrule to stop spinning is less than the chondrule-chondrule collisional timescales. It will also only be observed in triple compound chondrules (i.e. one primary and two adhering secondaries). Compound chondrules that have two or more adhering secondaries will have unpredictable orientations and so an additional secondary may land anywhere on such a compound chondrule.

Finally, we should point out that we have ignored gas flow fluctuations that are parallel to the central plane of the accretion disk and that the derivation of Eq. (3.3.2) assumes v_{gz} remains constant during one revolution of the spinning compound chondrule. This latter assumption is probably true just after the formation of the compound chondrule, but may not be true as the compound chondrule stops spinning.

To see this, we note that a compound chondrule will stop spinning when it receives enough torque from the gas jet to remove all the angular momentum and reverse the chondrule's initial spinning direction. The length of time the secondary will be moving against the gas jet is π/ω. The torque is approximately $C_D \rho_g v_{gz}^2 A a_p/2$. So the minimum rotation rate ω_{min}, is given by

$$I\omega_{min} \simeq \frac{C_D}{2} \rho_g v_{gz}^2 A a_p \frac{\pi}{\omega_{min}}. \qquad (3.3.9)$$

Thus

$$\omega_{min} \simeq \left(\frac{\pi C_D \rho_g v_{gz}^2 A a_p}{2I} \right)^{1/2}. \qquad (3.3.10)$$

Using the Epstein drag law (Eq. 3.1.4) plus the approximate values for I and A (Eqs (3.3.5) and (3.3.6), respectively). One can show that

$$\omega_{min} \approx 0.015 \frac{(a_s / 0.1\,\text{cm})}{(a_p / 0.1\,\text{cm})^2}$$

$$\left(\frac{(v_{gz} / 0.1\,\text{km s}^{-1})(\rho_g / 10^{-11}\,\text{g cm}^{-3})}{(\rho_p / 3\,\text{g cm}^{-3})} \right)^{1/2}, \text{rad Hz}$$

$$(3.3.11)$$

where we have assumed that $T_g = 1000$ K and that $m_g = m_H$. So the period of rotation $(2\pi/\omega)$ of this slowly spinning compound chondrule is approximately 400 seconds. In a 0.1 km s^{-1} gas flow, this gives a scale length (L) of order 10 to 100 km.

So the relevant Reynolds number (Re) of the flow (see Liffman 1992) is given by

$$Re \approx 8.7 \times 10^3 \frac{(L / 10^3\,\text{km})(v_{gz} / 0.1\,\text{km s}^{-1})(\rho_g / 10^{-11}\,\text{g cm}^{-3})}{\sqrt{(T_g / 10^3\,\text{K})(m_H / m_g)}}$$

$$(3.3.12)$$

thus, with Re ranging from 100 to 1,000, v_{gz} may undergo fluctuations which may modify the spin-down time of Eq. (3.3.4).

3.4 Chondrule reheating

If we combine the theory that we have acquired from §2 & §3, we can present a scenario for the phenomenon of chondrule reheating, where high temperature rims are observed around chondrule cores (Kring 1991). We suggest that reheating may be a consequence of a chondrule overshooting its hover altitude and then oscillating, in a damped manner, around the hover altitude. Each time the chondrule decreases its altitude, it will undergo an increase in temperature. If the altitude decrease is large enough, the corresponding increase in temperature will be sufficient to remelt the chondrule.

The actual mechanism whereby these high temperature rims are formed will not be discussed in any depth here. We only suggest that as the chondrule oscillates around its hover point, it accretes dust-grains/molten droplets and that this material forms the foundation of the rim once the chondrule is reheated.

To understand how these oscillations arise, we note that (see also §3.1) the equation of motion for a particle, parallel to the z axis, is given by

$$m_p\ddot{z} = \frac{C_D}{2}\rho_g(v_{gz}-\dot{z})^2\,\pi a_p^2 - \frac{GMm_pz}{(z^2+r^2)^{3/2}}. \tag{3.4.1}$$

As noted in §3.1, the drag coefficient (C_D) is likely to have the Epstein form

$$C_D \approx \frac{8}{3v_{gz}}\sqrt{\frac{8kT_g}{\pi m_g}}. \tag{3.4.2}$$

Combining Eqs (3.4.1) and (3.4.2) and assuming $z \ll r$, we have:

$$m_p\ddot{z} + \frac{4}{3}\left(\frac{8\pi kT_g}{m_g}\right)^{1/2}\rho_g\pi a_p^2\dot{z} + \frac{GMm_pz}{r^3} \approx \frac{GMm_pz_h}{r^3}, \tag{3.4.3}$$

where z_h is the hover height given by Eq. (3.1.5).

Let $x = z - z_h$, then Eq. (3.4.3) becomes

$$m_p\ddot{x} + c\ddot{x} + kx = 0, \tag{3.4.4}$$

where

$$c = \frac{4}{3}\pi a_p^2\rho_g\left(\frac{8\pi kT_g}{m_g}\right)^{1/2}, \tag{3.4.5}$$

and

$$k = \frac{GMm_p}{r^3}. \tag{3.4.6}$$

If we assume that c and k are approximately constant then $c^2 < 4km_p$ gives solutions of Eq. (3.4.6) that oscillate, with decreasing amplitude, as a function of time (t). If $c^2 = 4km_p$ or $c^2 > 4km_p$ then the solutions of Eq. (3.4.6) are critically or strongly damped and the particle approaches its hover height without oscillation.

The case $c^2 = 4km_p$ translates into an equation which gives the critical density for particle oscillation (ρ_{gc}):

$$\rho_{gc} = 2.7\times10^{-12}\frac{(a_p/0.1\mathrm{cm})(\rho_p/1\mathrm{gm\,cm^{-3}})(M/M_\odot)^{1/2}}{(r/0.1\mathrm{AU})^{3/2}(T/10^3\mathrm{K})^{1/2}}\ \mathrm{gcm^{-3}}. \tag{3.4.7}$$

If $\rho_g < \rho_{gc}$ then $c^2 < 4km_p$ and the particle will oscillate around the

hover point z_h. It is our contention that it is this oscillation process that causes chondrule reheating. Since the temperature of the jet flow increases as z decreases, thus as the particle travels to lower altitudes its temperature must increase.

To be specific, when $\rho_g < \rho_{gc}$ then Eq. (3.4.4) admits the solution

$$z = z_h + \frac{\dot{z}(z_h)}{\mu}e^{-bt}\sin\mu t, \tag{3.4.8}$$

where $\dot{z}(z_h)$ is the z speed of the chondrule when it first reaches $z = z_h$,

$$\mu^2 = \frac{4m_pk - c^2}{4m_p^2}, \tag{3.4.9}$$

and

$$b = \frac{c}{2m_p}. \tag{3.4.10}$$

Thus the period of the oscillation is

$$\frac{2\pi}{\mu} \gtrsim 11.6\frac{(r/0.1\mathrm{AU})^{3/2}}{(M_\odot)^{1/2}},\ \mathrm{days} \tag{3.4.11}$$

while the damping timescale is given by

$$\frac{1}{b} = 5\frac{(a_p/0.1\mathrm{cm})(\rho_p/1\mathrm{gm\,cm^{-3}})(m_g/m_H)^{3/2}}{(\rho_g/10^{-12}\,\mathrm{gm\,cm^{-3}})(T_g/10^3\,\mathrm{K})^{1/2}},\ \mathrm{days} \tag{4.3.12}$$

and the amplitude of the oscillation has the form

$$\frac{\dot{z}(z_h)}{\mu} = 1.6\times10^4\frac{(\dot{z}(z_h)/0.1\mathrm{kms^{-1}})(r/0.1\mathrm{AU})^{1/2}}{(M_\odot)^{1/2}}.\ \mathrm{km} \tag{4.3.13}$$

Comparing Eqs (3.4.13) and (3.1.6) shows that if $\dot{z}(z_h)$ is comparable to the gas flow speed, then the amplitude of the oscillations can be a significant fraction of the hover height. The corresponding temperature variations may also be significant.

Of course, from Eq. (3.4.5), c is a function of ρ_g and T_g and cannot be considered a constant, so the above analysis should only be treated as an informative approximation to the complete system. A slightly more realistic analysis can be obtained by constructing a computer simulation of the flow. We were able to accomplish this by assuming a z velocity for the outflow wind of the form

$$\bar{v}_g(z) \approx 0.1\left(\frac{z}{z_h}\right)^m\mathrm{kms^{-1}},\quad z \lesssim z_h, \tag{3.4.14}$$

with $m = 0.1$, with the orbital (angular) velocity of the gas assumed to be Keplerian. The density structure of the gas flow could then be easily deduced from Eqs (2.1) and (3.4.14), and had the form

$$\bar{\rho}_g(z) \approx 4.5\times10^{-12}\frac{(\dot{M}/10^{-8}\,\mathrm{M_\odot/yr})}{(\bar{v}_g(z)/0.1\mathrm{kms^{-1}})(r/0.1\mathrm{AU})^2}\mathrm{gcm^{-3}}, \tag{3.4.15}$$

while the temperature of the gas was given by

$$T_g(z) \approx 1200 \left(\frac{z_h}{z} \right)^{m(\gamma - 1)} \text{K}, \qquad (3.4.16)$$

where we set $\gamma = 5/3$.

As can be seen from the above formula, $T_g(z)$ will be greater than 1900 K when $z \lesssim 0.001 z_h$. In such cases, we simply set $T_g(z)$ = 1900 K. Finally, to compute the temperature of the particle, we did a full time integration of Eq. (3.2.1).

In Fig. 7a we show the motion of a chondrule-like particle (a_p = 0.1 cm, ρ_p = 3.8 g cm^{-3}) released from very close to the midplane ($z = 100$ cm) into a jet flow with a total mass loss rate of 10^{-9} M$_\odot$ yr^{-1}. The particle was released at a distance (r) of 0.05 AU from the centre of a solar mass protostar, where 0.05 AU is also the assumed radius of the protostellar jet.

The particle underwent a series of damped oscillations the period of which was slightly less than 6 days, while the damping timescale was approximately 30 days. The corresponding temperature of the particle is shown in Fig. 7b. The particle temperature has a damped "saw-tooth" pattern, the maximum temperature of the resulting temperature peaks being approximately 1420 K. Of course higher temperature values can be obtained if the temperature gradient is steeper than what we have assumed (*e.g.*, if $m > 0.1$).

While this oscillating motion is of interest, the particles can also undergo a critically damped trajectory. In particular, if $\rho_g > \rho_{gc}$ then Eq. (3.4.4) gives the solution

$$z = z_h + \frac{\dot{z}(z_h)}{2\sqrt{-\mu^2}} (e^{(-b+\sqrt{-\mu^2})t} - e^{(-b-\sqrt{-\mu^2})t}), \qquad (3.4.17)$$

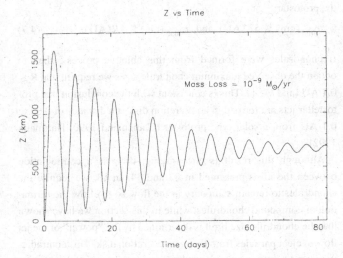

Z vs Time

Fig. 7a. A chondrule-like particle is released into a protostellar jet flow at a distance of 0.05 AU from a protostar. The mass loss rate of the flow is set at 10^{-9} M$_\odot$ yr^{-1}. Such a low mass loss rate ensures the gas density is lower than the critical density given in Eq. (3.4.7.). In such a circumstance the particle will oscillate, in a damped manner, around the hover point (see Eq. (3.4.8)).

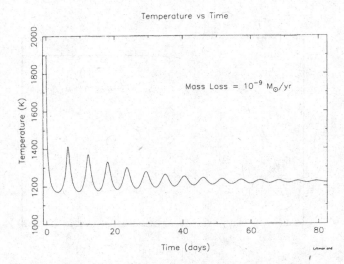

Temperature vs Time

Fig. 7b. The temperature of a particle as it undergoes the trajectory shown in Fig. 7a.

The trajectory and temperature of such a particle are shown in Figs 8a & b, where now the mass loss rate of jet is assumed to be 2 × 10^{-8} M$_\odot$ yr^{-1}.

From this simple model, we would suggest that chondrule reheating occurs when the jet flow is declining in mass flux, *i.e.*, during the later stages of a CTTS's evolution.

4. THE CHONDRULE SIZE LIMIT

Why do chondrules have an upper size limit? Possible solutions such as size-limited precursor dust balls to aerodynamic sorting have been suggested. In the Jet model, one has to eject particles from the inner accretion disk and ensure that they arrive in the outer parts of the accretion disk. Clearly, there must be some size limit to this process and this size limit must be dependent on the "strength" of the jet flow, which in turn is dependent on the gravitational potential of the inner accretion disk. To derive a quantitative relationship between chondrule and accretion disk, we must first describe the behaviour of a droplet in a flow.

Suppose we have a droplet of molten material that is subject to a streaming gas flow. The surface drag of the flow will tend to "rip" the droplet apart, while the surface tension of the melt will try to minimise the exposed surface area of the droplet and keep the droplet together. The balance between these two conflicting forces produces a stable droplet of maximum radius a_p, the formula for which is

$$a_p \approx \frac{\gamma We_0}{C_D \rho_g v_{gp}^2}, \qquad (4.1)$$

where γ is the surface tension of the molten material, and We_0 is a dimensionless factor called the "critical Weber number". We_0 accounts for the non-uniformity of the gas drag pressure over the surface of the droplet. Typically, $We_0 \approx 10$ (Bronshten 1983).

The streaming gas flow will not only determine the stable size of the droplet, but it will also subject the particle to a drag force given

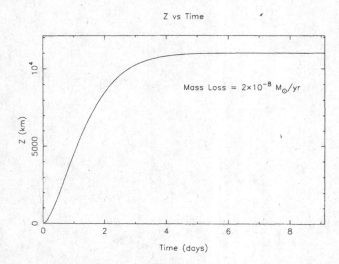

Fig. 8a. A chondrule-like particle is released into a protostellar jet flow at a distance of 0.05 AU from a protostar. The mass loss rate of the flow is set at 2×10^{-8} M. Such a low mass loss rate ensures the gas density is higher than the critical density given in Eq. (3.4.7). In such a circumstance the particle will approach the hover point in a critically damped manner (see Eq. (3.4.17).

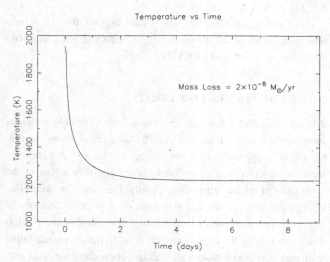

Fig. 8b. The temperature of a chondrule-like particle as it undergoes the trajectory shown in Fig. 8a.

by $C_D / 2 \, \rho_g \, v_{gp}^2 A_p$, where A_p is the cross sectional area of the particle that is facing the gas flow.

One can show (Liffman and Brown 1995) that the work done, W, by the wind-flow in ejecting a particle is

$$W = \frac{C_D^*}{2} A_p < \rho_g \, v_{gp}^2 > L, \tag{4.2}$$

where C_D^* is a "representative" value of the drag coefficient during the propulsion phase, $< \rho_g \, v_{gp}^2 >$ is the mean value of $\rho_g \, v_{gp}^2$, and L represents the actual length of the propulsion stage.

For a particle to escape the disk, we require that

$$< \rho_g \, v_{gp}^2 > = \frac{kGMm_p}{RC_D^* A_p L}, \tag{4.3}$$

where k is ≥ 1, G is the gravitational constant, m_p is the mass of the particle, and R is the initial radial (or semi-major axis) distance of the particle from the protostar. Substituting Eq. (4.3) into Eq. (4.1) gives

$$a_p \approx \frac{We_o C_D^* \gamma R A_p L}{kC_D(0)GMm_p}, \tag{4.4}$$

where $C_D(0)$ is the value of the gas drag coefficient at the position where the particle is formed. If we assume that our particle is spherical (as are most unfragmented chondrules) and that our propulsion distance L is proportional to the height of the disk at a distance R away from the protostar, i.e. $L = \lambda R^\beta$ then

$$a_p \approx \left[\frac{3\lambda We_o C_D^* \gamma R^{1+\beta}}{4C_D(0)kGM\rho_p} \right]^{1/2}. \tag{4.5}$$

Now C_D, generally, decreases with increasing gas-flow speed. So, we should expect that $C_D^*/C_D(0) \lesssim 1$, since C_D^* is an 'average' gas drag coefficient over the entire propulsion stage, while $C_D(0)$ samples the gas flow at the beginning of the propulsion stage, where the gas flow is probably at its slowest. Setting $\lambda \approx 0.01$, $\beta \approx 1$, (typical, approximate values for the disk height), $k \approx 1$, $M \approx M_\odot$, and $We_o \approx 10$, we obtain

$$a_p (\text{cm}) \lesssim 0.4 \left[\frac{\gamma}{\rho_p} \right]^{1/2} R(\text{AU}). \tag{4.6}$$

The values for the surface tension, γ, and the density, ρ_p, are material dependent. We are interested in Fe-Ni and silicate chondrules, so we shall use the surface tensions of meteoric iron and stone, which are $\gamma_{iron} = 1,200$ and $\gamma_{stone} = 360$ gs^{-2} (Allen et al. 1965) with corresponding mass densities of $\rho_{iron} = 7.8$ and $\rho_{stone} = 3.4$ gcm^{-3}. Substituting these values into Eq. (4.6) gives us the approximate radius of iron and silicate droplets as a function of distance from the protostar:

$$r_{iron} (\text{cm}) \lesssim 5R(\text{AU}) \quad \text{and} \quad r_{stone} (\text{cm}) \lesssim 4R(\text{AU}). \tag{4.7}$$

If chondrules were formed from this ablative process, then to obtain the observed maximum chondrule sizes we require that $R \lesssim 0.1$ AU (and $k \geq 1$). This is consistent with the conclusion that protostellar jets are formed in an accretion disk at a distance of 0.05 to 0.1 AU from a solar-type protostar (Camenzind 1990, Hartmann 1992).

Although this result is encouraging, there is a contradiction between the ideas presented in §3 and §4. In §3, we require the chondrules to remain stationary in the flow so to allow the formation of compound chondrules, while in this section we have shown that the chondrule size limit is determined by the "power" of the jet flow to eject particles from the inner accretion disk. This contradiction would appear to severely limit, perhaps destroy, our theory. After all, how can one require particles to be stationary relative to the accretion disk and also expect them to be ejected at speeds $\geq$ the escape speed of the protostellar system?

The only way out of this contradiction is for the jet flow to be highly variable in both density and velocity. Compound chondrules would presumably form when the flow is relatively quiescent, while chondrule ejection would perhaps occur when there is a major increase in the density and/or velocity of the flow. Observations do suggest that jet flows are highly variable in their behaviour (Edwards *et al.* 1993). For example, Mundt (1984) obtained observational evidence of variations in young stellar winds that vary on time scales of months. Despite this observational support, the resolution of this problem must await a coherent theory of protostellar jet formation, which in turn must await high resolution observations of protostellar jets.

Besides determining a maximum size limit for chondrules, the ejection of chondrules by a protostellar jet also, indirectly, causes chondrules and chondrule fragments to be size sorted.

5. SIZE SORTING AND DISK ACCRETION

It has been known for many years (Dodd 1976) that chondrules, both silicate and metal, are size-sorted in meteorites, i.e., particles in a particular chondrite satisfy the relation (Skinner and Lennhouts 1993)

$$\rho_g \, a_p \approx \text{constant}. \tag{5.1}$$

It has recently become apparent that size-sorting also applies to fragments of chondrules (Skinner and Leenhouts 1991). This shows that chondrites act as "size-bins". Chondrules and their fragments were formed, mixed and later sorted by size within the chondrite forming regions of the solar nebula.

The usual explanation for this phenomenon (e.g. Dodd 1976 and references therein), is some form of aerodynamic drag effect, where larger and/or denser particles can travel further into a resistive medium than can smaller and/or less dense particles. We too shall use this idea, for it arises naturally from the Jet model.

To see this, suppose that a particle is ejected from the inner accretion disk, at speeds close to or exceeding the escape velocity. Suppose further that the angular momentum of particle is high enough to eject the particle from the gas flow and allow it to move across the face of the accretion disk. If the particle is not subject to gas drag, from the upper-atmosphere of the accretion disk, it will simply move out of the system and into interstellar space. If the particle is subject to gas drag then, given sufficient drag, the particle will be recaptured by the protostellar system and may fall into the outer parts of the accretion disk.

A stream of such particles will be size-sorted. Since the smaller, less dense particles will fall close to the protostar, while the larger, more dense particles will fall further away from the protostar. Thus, we should expect that meteorites from the outer parts of the solar system will contain larger particles than meteorites from the inner solar system, and indeed, chondrules from the carbonaceous chondrites are (usually) larger than chondrules from ordinary chondrites.

There is an additional consequence of this model which has to do with angular momentum transfer. A particle in Keplerian orbit around a star has an angular momentum that is proportional to the square root of the distance between the particle and the star. Thus, a particle that is ejected from the inner accretion disk, and subsequently stopped by gas drag, will simply fall back into its original orbit unless angular momentum is transferred from the disk to the particle. If gas drag allows the disk to transfer angular momentum to the particle, then the particle can fall to the outer parts of the disk.

Clearly, if the disk gas gives up angular momentum, then it must move in towards the protostar. Since protostellar jets are fuelled by disk accretion, one obtains the schematic picture of a jet flinging out material into the disk such that the disk will accrete onto the protostar and refuel the jet. Heuristically, one can think of chondrules and other associated Jet ejecta as delayed protostellar Jet fuel.

To turn this qualitative speculation into quantitative speculation, we have to model the particle ejection process. Unfortunately, this is a difficult thing to do, since there is no consensus on how protostellar jets work. So, we make the following assumptions:

(1) We suppose that our chondrule particles are, initially, in a circular Keplerian orbit of radius $R(\leq 0.1 \text{ AU})$ from the protostar.

(2) We assume that the protostellar jet gives the particles an initial "boost" velocity $z(0)$ that is comparable to the protostar's escape velocity at that point. Our tentative justification for this assumption is that protostellar jets are observed to have speeds comparable to the maximum escape velocity of a protostellar system. So, we presume that particles initially entrained in such a flow may also obtain similar speeds.

(3) Finally we assume that the particles have, initially, no radial velocity, and that the self-gravity of the disk is negligible compared to the gravity of the protostar. The former of these two assumptions comes from the observation that jet flows tend to be perpendicular to their respective accretion disks. Protostellar jets do have a nonzero radial velocity, but we have, for simplicity, ignored this component. The validity or otherwise of these ideas is discussed at some length in Liffman and Brown (1995).

Given these assumptions, the equations of motion for a particle become

$$\ddot{r} = \frac{h^2}{r^3} - \frac{GMr}{[r^2 + z^2]^{3/2}}, \tag{5.2}$$

$$v_\theta = r\dot{\theta} = \frac{h}{r}, \tag{5.3}$$

and

$$\ddot{z} = -\frac{GMz}{[r^2 + z^2]^{3/2}}, \tag{5.4}$$

where h is the specific angular momentum of the particle and has the value

$$h = \sqrt{GMR} \tag{5.5}$$

The value of h is a constant, since it is assumed that there are no external torques, parallel to the z axis, acting on the system.

These equations are not difficult to model numerically and can be solved by standard techniques. Although a brief analysis of the above equations shows that if the particles are travelling at speed greater than the escape velocity of the system, then the initially vertical path of the projectile will quickly turn into a "horizontal" path across the face of the accretion disk. A computer simulation of this phenomenon is given in Fig. 9 (see also Liffman and Brown (1995)).

In Fig. 9, particles are ejected at $r = 0.04$ AU with different vertical velocities ranging from 149 to 174 km s^{-1}. The particles are assumed to move out of the gas outflow at $r = 0.1$ AU, whereupon they encounter the gas halo of the accretion disk and their subsequent motion is governed by gas drag plus the gravitational force from the protostar.

As the particles move through the halo gas of the accretion disk, they will acquire, by gas drag, angular momentum from the disk. If we assume that the centrifugal force of the halo gas balances the radial component of the protostar's gravity, then the angular speed of the halo gas, $v_{\theta, gas}$, will be given by

$$v_{\theta, gas} = \frac{r\sqrt{GM}}{(r^2 + z^2)^{3/4}}.$$

(5.6)

Once the particle has come to rest, relative to the halo gas, it will have the angular velocity given by Eq. (5.6) and a specific angular momentum, h, given by $h = rv_{\theta, gas}$. As can be deduced from Eq. (5.2), such a specific angular momentum implies that $r = 0$, and the only force acting on the particle will be the z component of the gravitational force, which will point towards the accretion disk. As the particle moves towards the accretion disk, the gas density and angular velocity of the gas will increase, thereby keeping $r \approx 0$. As can be seen from Fig. 9, the subsequent path of the particle is roughly parallel to the z axis.

The paths of the captured particles, shown in Fig. 9, can be approximated to that shown in Fig. 10, where the ascending path length, l, is given by $l = K\chi$, with K being a number in the range of 4 ± 1, (Liffman and Brown 1995) and χ is the "stopping distance" as defined by the equation,

$$\chi = \frac{4a_p\rho_p}{3\rho_g}.$$

(5.7)

As a justification for the above equation, we note that a macroscopic particle will come to rest when it has encountered a total gas mass approximately equal to its own mass. The "stopping distance", so defined by this prescription, is easily shown to be the χ length scale as given in the above equation. So, for a constant gas density, the larger or more dense a particle is, the further it will be able to travel before it comes to rest.

The range of a recaptured projectile is simply given by the formula

$$r \approx l\cos(\theta),$$

(5.8)

where θ is the angle between the ascending path of the particle and the midplane of the accretion disk.

The simplicity of Eq. (5.8) suggests that aerodynamic size sorting may occur to particles that are recaptured by the protostellar system. To see this, we suppose that we have two particles with different mass densities: ρ_1 and ρ_2. Suppose further that these two particles fall back to the accretion disk at the same distance from the protostar. Then we can write the radius ratio of particle 1 to particle 2 as

$$\frac{a_1}{a_2} \approx \frac{\rho_2}{\rho_1}Q(r),$$

(5.9)

where

$$Q(r) = \frac{K_2}{K_1}\frac{\rho_{g1}}{\rho_{g2}}\frac{\cos(\theta_2)}{\cos(\theta_1)}.$$

(5.10)

The factor $Q(r)$ is dependent on the initial ejection speeds of the particles (through K and θ), the initial distance of the particles from the protostar (again K and θ) and the scale height of the accretion

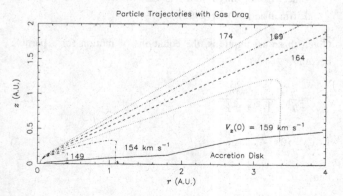

Fig. 9. Simulation data showing chondrule-like particles travelling above the disk and falling back into it. A solar-mass protostar is located at $r = 0$, $z = 0$. Surrounding the protostar is an accretion disk, the scale height of which is shown in profile. Chondrule-like particles (radius = 0.1 cm, density = 3.5 g cm^{-3}) are given a velocity boost in the z direction, the magnitudes of which (in km s^{-1}) are shown next to the trajectories. Particles that are subject to sufficiently high gas drag are later recaptured.

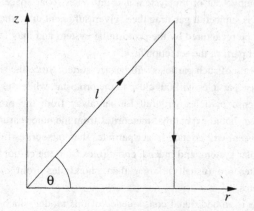

Fig. 10. An approximation of chondrule trajectories as a triangular path.

disk (through ρ_g). If the particles are created at about the same distance from the protostar then we would expect that K, θ and ρ_g would be similar for both particles, since the average flight path of both particles would be about the same. This would imply that $Q(r) \approx 1$. So, from the collection of particles that fall to the accretion disk at r, the denser particles should have smaller radii.

As has been discussed, size sorting is density dependent in that dense Fe-Ni chondrules are always smaller than less dense silicate chondrules. The mass densities of the two types of chondrules are (Skinner and Leenhouts 1993) $\rho_{Si} \approx 3.8$ g cm^{-3}, and $\rho_{Fe} \approx 7.8$ g cm^{-3}, which implies that $\rho_{Si}/\rho_{Fe} \approx 0.5$, and so if Eq. (5.2) is applicable to chondrules, we should expect that

$$\frac{a_{Fe}}{a_{Si}} \approx 0.5Q(r). \tag{5.11}$$

Using published data for the meteorite types H, L, LL, (Dodd, 1976) and CR (Skinner and Leenhouts 1993), we can compute the average Fe to Si size ratio for all these different types of meteorites (excluding Bjurböle)

$$\left\langle \frac{<a_{Fe}>}{<a_{Si}>} \right\rangle_{H,L,LL,CR} = 0.52 \pm 0.16. \tag{5.12}$$

The corresponding approximate mean Q values is

$$<Q_{H,L,LL,CR}> = 1.04 \pm 0.32. \tag{5.13}$$

Of course, the agreement between the above Q value and our theoretical model should be treated with caution, since we have only presented the bare beginnings of a quantitative model. Nonetheless, it does illustrate the potential of the "Jet" model to explain the phenomenon of size-sorting.

Finally, we return to our discussion of angular momentum transfer and disk accretion. Let us consider a ring of material in an accretion disk at a distance r from a protostar. The angular momentum of this material is

$$L(r) = m(r)\sqrt{GMr}, \tag{5.14}$$

where $m(r)$ is the mass of the ring of material at r. Now suppose that material falls onto the accretion disk, and this infalling material has essentially zero angular momentum. In such a case, the angular momentum of the ring is conserved, and the mass of the ring becomes a function of time, i.e., $m \equiv m(r, t)$. For such a case, we can differentiate Eq. (5.14) to obtain

$$\frac{dr}{dt} = -\left(\frac{2r}{m}\right)\frac{dm}{dt}. \tag{5.15}$$

or

$$r(t) = r(0)\left(\frac{m(0)}{m(t)}\right)^2. \tag{5.16}$$

Thus, if infalling material doubles the mass of the ring, it will move from its initial position $r(0)$ to $r(0)/4$. It is via this mechanism of mass and angular momentum transfer that disk accretion may be,

in part, mediated. Indeed, because it is the halo gas of the accretion that will be transferring most of the angular momentum to the "Jet projectiles", it is possible that the upper layers of the accretion disk are the ones that undergo most of the accretion, leaving the midplane relatively untouched.

This type of process may explain an implicit contradiction between observations and meteoritics. Observations suggest that protostars keep on accreting material from their disk for periods of up to 10^7 years (Cabrit *et al.* 1990). Radiometric data from the decay of ^{129}I and ^{26}Mg suggest that meteorites accreted material for periods of order 10^6–10^7 years. How could the meteoritic material have been preserved if a major portion of the solar nebula was accreted onto the protosun? The answer, we suggest, is that disk accretion was, in part, altitude dependent. Material flung from the protosolar jet mediated the angular momentum transfer and one component of this mass transfer was the chondrule.

6 CHONDRULE-MATRIX COMPLEMENTARITY

In this paper, we claim that chondrules have the physical characteristics expected of ablation droplets that have been formed and ejected by a protosolar jet, and then recaptured by the solar nebula through the action of the gas drag.

Such a model, however, is immediately confronted with the complementary composition of matrix and non-matrix material in meteorites. For example, Wood (1985) discusses the case of Murchison, where the matrix has an Fe/Si ratio of 1.23, while the non-matrix material (chondrules, CAIs, isolated crystals) is ~ 0.2. These two dissimilar components combine to give an Fe/Si ratio of 0.81, which is close to the solar value of 0.9. This is unlikely to be accidental, and is clear evidence for the local formation of chondrules, and the refactory component of chondritic material.

Before we throw out our wind-transport model, however, one should note that chondrules, at least in this model, form in a relatively small region of the solar nebula, i.e., in or near the boundary layer of the protostellar system. If the chondrules have a low Fe/Si value then the surrounding material will, by mass balance, have a high Fe/Si ratio. If this latter material comes in a non-gaseous form (e.g. 10 μm dust, CAI &c) then it too will be ejected with the chondrules. One will obtain the desired mass balance if all this material lands back into the solar nebula, in a uniform manner, and over a long period of time.

For such a model to work, we require a fairly large component of matrix material to be made from dust that has been recycled through the protostellar jet. There are at least two consequences if this idea is correct. First, this recycled dust would have to be more refractory than CI material, since the jet formation region of the inner solar nebula would have been far warmer than the regions where most chondrites were formed. Second, if matrix material were formed from dust that had been lofted into the upper atmosphere of the solar nebula then this may be an observable phenomenon.

To understand this latter point, we need an estimate for the amount of dust that should be resident in the upper atmosphere of the solar nebula at any particular time.

Protostellar jets have average mass loss rates of order 10^{-8} M$_\odot$ yr^{-1} and since the mass of dust to gas in the Interstellar Medium (ISM) is 1/100, this implies that $\lesssim 10^{-10}$ M$_\odot$ yr^{-1} of dust is blown out by the protostellar jet. So, an upper limit for the amount of dust that is lobbed into the upper atmosphere of an accretion disk in a Classical T Tauri Star (CTTS) per year is 10^{-10} M$_\odot$ yr^{-1}.

The dust settling timescale (τ_{settle}) can be deduced from Eq. (3.1.1) with $v_{gz} = 0$, *i.e.*,

$$m_p \ddot{z} = -\frac{C_D}{2} \rho_g \dot{z}^2 A_p - \frac{GMm_p z}{[r^2 + z^2]^{3/2}}. \tag{6.1}$$

The motion of the dust is subsonic, so the drag coefficient takes the Epstein form for C_D (Eq. 3.1.4). To compute the velocity of the particle, we note that $\ddot{z} \approx 0$, which implies

$$\dot{z} \approx \frac{-\omega_K^2(r, z)\rho_p a_p}{\rho_g \bar{v}} z, \tag{6.2}$$

where $\omega_K(r, z) = \sqrt{GM}/[r^2 + z^2]^{3/4}$ is the Keplerian angular velocity at the point (r, z), and $v = \sqrt{8kT_g / \pi m_g}$ is the mean Maxwellian speed of the gas particles.

If $z \ll r$ then $\omega_K(r, z) \approx \omega_K(r) = \sqrt{GM/r^3}$ and we can compute the settling time:

$$\tau_{settle} \approx 8,000 \frac{(\rho_g / 10^{-11}\,\text{g cm}^{-3})(\bar{v} / 1\,\text{km s}^{-1})(r / 1\,\text{AU})^3}{(a_p / 1\,\mu\text{m})(\rho_p / 1\,\text{g cm}^{-3})}\,yr. \tag{6.3}$$

Thus, we have characteristic dust-settling timescales in the range $10^3 - 10^4$ years, which means that up to $10^{-7} - 10^{-6}$ M$_\odot$ of dust will be in the upper atmosphere of a CTTS accretion disk at any one time.

As discussed in Natta (1993), 10^{-7} M$_\odot$ of high-altitude dust may produce the observed "flat-temperature distributions" in CTTSs. These temperature distributions are a surprisingly common phenomenon in CTTSs. They arise when the temperature of the disk does not decrease as rapidly with distance from the protostar as one would predict from standard accretion disk theory. Natta suggested that a spherical halo of dust around a CTTS would reflect light from the protostar and into the accretion disk, thereby increasing the temperature of the outer disk.

Of course, the idea that protostellar jets can loft dust into the outer parts of the surrounding accretion disk requires a quantitative investigation to determine whether dust grains can be ejected from the jet flow. There will be a size limit where particles smaller than a certain size will simply be entrained in the jet flow and ejected from the system. We simply note, that we require dust ejection similar to that shown in §5 to account for chondrule-matrix complementarity.

CONCLUSIONS

Chondritic meteorites are typically an agglomeration of igneous rocks, i.e., chondrules and refractory inclusions (formation temperatures 1500–2000 K), surrounded by sedimentary material that, in some cases, has never experienced temperatures greater than 500 K. This unusual structure has prompted theorists to develop heating mechanisms (e.g. lightning) that can provide brief, intense impulses of energy in the otherwise cold outer regions of the solar nebula. These energy impulses are presumed to have melted small dust aggregates into chondrules, which were then incorporated into larger dust aggregates that eventually formed meteorites.

Such energy impulse theories are not required in the 'Jet' model of chondrule formation. Chondrules are formed in the hot inner regions of the accretion disk adjacent to the protostar. There is no difficulty in obtaining the required temperatures, because in or near this region the accretion disk dumps around half of its gravitational energy. The protostellar jet ejects the chondrules from the hot inner disk and gas drag brings these particles back to the cooler outer regions, where they could be incorporated into growing aggregates of cool nebular material. The "plum pudding" structure of chondritic meteorites is a natural consequence of this model.

One of the many problems in chondrule formation is the deduced low cooling rate (~ 1–1000 K/hour) for these particles. This cooling rate is many orders of magnitude smaller than that expected for a particular radiating directly into space. In our model, chondrules are formed in the optically thick regions of the inner accretion disk. They will, therefore, be in thermodynamic equilibrium with the gas and their temperature variations will be damped.

Of course if the chondrules were to remain in this environment, their temperatures would not decrease. Chondrules, however, are produced by the ablative interaction between a streaming gas flow (perhaps produced from the merging of magnetic field lines) and molten material. Particles, that are small enough, will be swept up with the flow. As the gas flow moves away from the midplane of the inner accretion disk, it will expand and therefore probably cool. Particles that are entrained in this gas flow will also cool at the same rate.

Particles that are moving with the gas flow may be ejected from the accretion disk, but it is possible that a particle may simply hover at some distance away from the midplane of the disk. If the gas density in the flow is below a critical gas density (~ 10^{-11} g cm^{-3}) then the particles will undergo damped oscillations around their hover points. It is due to these oscillations, we suggest, that chondrules can undergo reheating. Smaller particles of the same mass density will still move past the hovering particles, and so may collide with these particles.

Such a scenario allows for the formation of adhering compound chondrules. These chondrules pose a major problem for chondrule formation theories as it is nearly always the small chondrules that were plastic at the time of collision. Smaller particles will lose heat more readily than larger particles, since they have a larger surface to area ratio. If chondrules were formed in a single flash heating event, we should expect the smaller particles to become solid before the larger particles. That we actually see the opposite behaviour (i.e., the large particles were solid, while the small particles were semi-molten), strongly suggests that single flash-heating models require some modification.

In the Jet model, adhering compound chondrules are formed when small chondrules collide with larger chondrules that are stationary in the flow. The smaller particles that are moving with the flow will be warmer than the larger hovering particles, because the latter particles have had time to equilibrate with the local temperature of the gas. If the temperature gradient in the flow is sufficiently steep (~ 10 to 100 K/s) at this collision point, compound chondrules with the observed structure will be formed.

Once a compound chondrule has been formed, and has stopped spinning, the smaller secondary chondrule will orient the entire compound chondrule, just like a weathervane, such that the secondary chondrule is pointing in the direction of the flow. Any incoming secondary chondrule will hit the rear end of the primary chondrule. The resulting triple compound chondrule, will have to secondaries chondrules that will tend to "avoid" each other. Of course, this prediction implicitly assumes a fairly steady gas flow, an assumption that may be incorrect. However, if this prediction is found to be valid, it would be major piece of evidence in favour of the Jet model, as it is difficult to produce such an effect with other chondrule formation theories.

The fragmentation of chondrules is a simple extension of compound chondrule formation, where instead of one particle being plastic at the time of collision, we now have two solid particles colliding at higher velocities. We expect the fragmentation zone to be 'above' the compound-chondrule formation zone, since the gas flow will be cooler and the flow speed higher as one moves further away from the midplane of the disk.

The chondrule size range is yet another aspect of chondrule formation which has not been satisfactorily explained. In the Jet model, the sizes of the ejected droplets are determined by the balance between gas drag and the surface tension. A jet flow with high energy density will tend to make molten droplets smaller, while molten materials with higher surface tensions will tend to form larger droplets. Observations suggest that protostellar jets are produced within 0.1 AU of solar-mass protostars. The minimum energy density of a wind than can eject Si and Fe droplets from such close proximity to the protostar is such that the radii of these droplets is ≤ 1 cm. Protostellar winds that form larger droplets, will have a lower energy density and will not be able to eject them from the jet formation region and we will not see them in meteorites.

This ejection mechanism may also explain the complementary chemical structure of chondrules and their surrounding matrix. For this to occur, we require that refractory dust as well as chondrules are ejected by the jet flow so that mass balance will be obtained once the dust settles back to the solar nebula. This leads to a simple calculation, which suggests that up to 10^{-7}–10^{-6} M. of dust will be in the upper atmosphere of a CTTS accretion disk at any one time. These figures appear to be consistent with those deduced from observation (Natta 1993).

The transfer of dust and chondrules from the inner to outer portions of the accretion disk, necessarily requires the removal of angular momentum from the upper atmosphere of the accretion disk. This will increase the viscosity of the disk and hasten the stratified infall of disk material, *i.e.*, the upper atmosphere of the

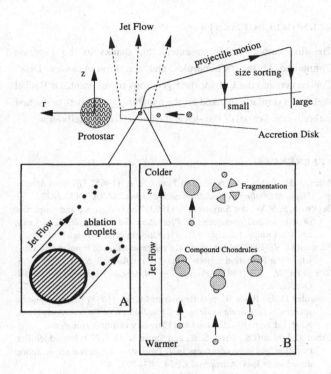

Fig. 11. A pictorial summary of the ideas presented in this paper. An accretion disk surrounds a protostar. At a distance within 0.1 AU of the protostar, the disk produces a protostellar jet (only one side of which is shown). Infalling m-sized bodies move into this hot "jet zone" and are subsequently ablated by the jet wind. The resulting droplets, if small enough, move with the flow and away from the ablating body. It is our claim that these ablation melt-droplets eventually become chondrules.

At higher altitudes, within the accretion disk, some of the larger particles may hover in the gas flow. Smaller particles will still move with the flow, however, and may collide with the hover particles. These chondrule-chondrule collisions produce compound chondrules and chondrule fragments. The fragmentation of chondrules occurs at higher altitudes relative to the compound-chondrule formation zone, because the flow cools as it increases in altitude and the two colliding chondrules will be solid at the time of collision.

If the protostellar jet is sufficiently powerful, it will eject particles from the inner accretion disk, and if the orbital angular momentum of the ejected particles is large enough, the particles will move out of the jet flow and travel across the face of the accretion disk. With sufficient gas drag from the upper atmosphere of the accretion disk, these (now size-sorted) particles will be brought down to the outer accretion disk, where they will be incorporated into planetesimals.

disk will accrete more readily than the central plane of the disk. Finally, aerodynamic size sorting of the ejected particles is a natural consequence of the Jet model. Ejected particles that are captured by the accretion disk must suffer aerodynamic size sorting, since the particles will be subject to gas drag. Particles that are not subject to gas drag will simply leave the protostellar system, since their initial velocities were higher than the escape velocity of the system.

It is for all these reasons that we consider chondrules to be ablation droplets formed by a protosolar jet in the first 10^6–10^7 years of the solar system.

ACKNOWLEDGEMENTS

The first author wishes to extend his thanks to the Program Committee of the "Chondrules and the Protoplanetary Disk" Conference and the LPI for their generous travel grant. Dr Rachel Webster is similarly thanked for the matching grant given to the first author by the School of Physics at the University of Melbourne.

REFERENCES

Allen, J. H., Baldwin B. S. Jr., and James N. A. (1965) *Effect on Meteor Flight of Cooling by Radiation and Ablation.* NASA TN D-2872.

Beckwith, S. V. W., and Sargent A. I. (1993) The occurrence and properties of disks around young stars. In *Protostars and Planets III*, (E. H. Levy, and J. I. Lunine, Eds), pp. 521–541. University of Arizona Press.

Bicknell G. V. (1992) Mechanisms for the production of bipolar flows and jets in star formation regions. *Aust. J. Phys.* **45**, 513–529.

Bronshten V. A. (1983) *Physics of Meteoric Phenomena.* Reidel, Dordrecht.

Brownlee D. E., Bates B., and Beauchamp R. H. (1983) Meteor ablation spheres as chondrule analogs. In *Chondrules and Their Origins* (E. A. King, Ed.), pp. 10–25. Lunar and Planetary Institute, Houston.

Cabrit S., Edwards S., Strom S. E., and Strom K. M. (1990) Forbidden-line emission and infrared excess in T Tauri stars: Evidence for accretion driven mass loss? *Astrophys. J.* **354**, 687–700.

Camenzind M. (1990) Magnetized disk-winds and the origin of bipolar out-flows. In *Accretion and Winds*, (G. Klare, Ed.), Rev. Modern Astron. **3**, 234–265. Springer-Verlag.

Cassen P. (1993) Why convective heat transport in the solar nebula was inefficient. *Lunar Planet. Sci. Conf.* **24**, 261–262.

Dodd, R. T. (1976) Accretion of the ordinary chondrites. *Earth Planet. Sci. Lett.* **30**, 281–291.

Dodd, R. T., and L. S. Teleky (1967) Preferred orientation of olivine crystals in porphyritic chondrules. *Icarus* **6**, 407–416.

Edwards S., Ray T., and Mundt R. (1993) Energetic mass outflows from young stars. In *Protostars and Planets III*, (E. H. Levy, and J. I. Lunine, Eds), pp. 567–602. University of Arizona Press.

Epstein P. S. (1924) On the resistance experiment by spheres in their motion through gases. *Phys. Rev.* **2**, 710–733.

Frank J., King A., and Raine D. (1992) *Accretion Power in Astrophysics.* Cambridge University Press, Cambridge. 294 pp.

Freeman J. (1977) The magnetic field in the Solar nebula *Proc. Lunar Sci. Conf. 8th* 751–755.

Fukui Y., Iwata T., Mizuno A., Bally J. and Lane A. P. (1993) Molecular Outflows. In *Protostars and Planets III*, (E. H. Levy, and J. I. Lunine, Eds), pp. 603–639. University of Arizona Press.

Gooding, J., and K. Keil (1981) Relative abundances of chondrule primary textual types in ordinary chondrites and their bearing on conditions of chondrule formation. *Meteoritics* **16**, 17–43.

Hartmann L. (1992) Winds from protostellar accretion disks. In *Nonisotropic and Variable Outflows from Stars*, (L. Drissen, C. Leitherer, and A. Nota, Eds), A.S.P. Conference Series, **22**, pp. 27–36. Astronomical Society of the Pacific.

Hewins R. H. (1988) Experimental studies of chondrules. In *Meteorites and the Early Solar System* (eds J. F. Kerridge and M. S. Matthews), pp. 660–679. University of Arizona Press.

Königl, A., and Ruden S. P. 1993. Origin of outflows and winds. In *Protostars and Planets III*, (E. H. Levy, and J. I. Lunine, Eds), pp. 641–687. Univ of Arizona Press, Tucson.

Kring D. A. (1991) High temperature rims around chondrules in primitive chondrites: evidence for fluctuating conditions in the solar nebula. *Earth and Planetary Science Letters* **105**, 65–80.

Liffman, K. (1992) The formation of chondrules by ablation. *Icarus* **100**, 608–619.

Liffman, K. and Brown M. (1995) The motion and size sorting of particles ejected form a protostellar accretion disk. *Icarus* **116**, 275–290.

Lovelace, R. V. E., H. L. Berk, and J. Contopoulos (1991) Magnetically driven jets. *Astrophys. J.* **379**, 696–705.

McSween H. (1977) Chemical and petrographic constraints on the origin of chondrules and inclusions in carbonaceous chondrites. *Geochim. Cosmochim. Acta.* **41**, 1843–1860.

Mundt R. (1984) Mass loss in T Tauri stars: Observational studies of the cool parts of their stellar winds and expanding shells. *Ap. J.* **280**, 749–770.

Natta A. (1993) The Temperature Profile of T Tauri Disks. *Ap. J.* **412**, 761–770.

Priest E. R. (1994) Magnetohydrodynamics. In *Plasma Astrophysics*. (A. O. Benz and T. J.-L. Courvoisier, Eds), pp. 1–109. Springer-Verlag.

Shu F., Najita J., Galli D., Ostriker E., and Lizano S. (1993) The collapse of clouds and the formation and evolution of stars and disks. In *Protostars and Planets III*, (E. H. Levy, and J. I. Lunine, Eds), pp. 3–45. University of Arizona Press.

Shu F., Najita J., Ostriker E., Wilken F., Ruden S., and Lizano S. (1994) Magnetocentrifugally driven flows from young stars and disks. I. A generalized model. *Astrophys. J.* **429**, 781–796.

Skinner, W. R., and J. M. Leenhouts (1991) Implications of chondrule sorting and low matrix contents of type 3 ordinary chondrites. *Meteoritics* **26**, 396.

Skinner, W. R., and J. M. Leenhouts (1993) Size distributions and aerodynamic equivalence of metal chondrules and silicate chondrules in Acfer 059. *Lunar Planet. Sci. Conf.* XXIV, 1315–1316.

Snell R. L., Loren R. B. and Plambeck R. L. (1980) Observations of CO in L1551: evidence for stellar wind driven shocks. *Astrophys. J. Lett.* **239**, 17–22.

Swindle, T. D., Caffee M. W., Hohenberg C. M., Lindstrom M. M. and Taylor G. J. (1991) Iodine-xenon studies of petrographically and chemically characterized Chainpur chondrules. *Ceochim. Cosmochim. Acta* **55**, 861–880.

Swindle T. D., and Podosek F. A. (1988) Iodine-Xenon dating. In *Meteorites and the Early Solar System* (eds J. F. Kerridge and M. S. Matthews), pp. 1127–1146. University of Arizona Press.

Vedder J. F., and Gault D. E. (1974) A chondrule: evidence of energetic impact unlikely. *Science* **185**, 278–379.

Wasson J. T. (1993) Constraints on chondrule origins. *Meteoritics* **28**, 14–28.

Wasson J. T., Krot A. T., Lee M. S., and Rubin A. E. (1995) Compound Chondrules. *Geochim. Cosmochim. Acta* **59**, 1847–1869.

Weidenschilling S. J. (1977) Aerodynamics of solid bodies in the solar nebula. *Mon. Not. Roy. Astron. Soc.* **180**, 57–70.

Weidenschilling S. J. (1988) Formation processes and timescales for meteorite parent bodies. In *Meteorites and the Early Solar System* (eds J. F. Kerridge and M. S. Matthews), pp. 348–371. University of Arizona Press.

Whipple F. L. (1973) Radial pressure in the solar nebular as affecting the motions of planetesimals. NASA-SP-319, 355–361.

Wood, J. A. (1985) Meteoritic constraints on processes in the solar nebula. In *Protostars and Planets II*, (D. C. Black, and M. S. Matthews, Eds), pp. 687–702. Univ of Arizona Press, Tucson.

31: Chondrule Formation in Lightning Discharges: Status of Theory and Experiments

MIHÁLY HORÁNYI[1] and SCOTT ROBERTSON[2]

[1]*Laboratory for Atmospheric and Space Physics, University of Colorado, Boulder, CO 80309, U.S.A.* [2]*Department of Astrophysical, Planetary and Atmospheric Sciences, University of Colorado, Boulder, CO 80309, U.S.A.*

ABSTRACT

In this paper we summarize our recent theoretical progress on chondrule formation in lightning discharges in the early solar system and also report on preliminary results from our laboratory experiments. We describe our theoretical model that follows the dynamics of an expanding and cooling plasma discharge channel, and give estimates of the energy flux that could heat embedded solid particles. Solving the heat diffusion equation we show that theses fluxes can melt chondrule precursor particles. We also discuss results from our laboratory experiments where chondrule precursor grains are exposed to plasma conditions similar to lightning discharges. Based on our simple theoretical estimates and on our ongoing experiments we conclude that lightning is a viable mechanism to produce chondrules in the early solar system.

INTRODUCTION

Chondrules represent a significant mass fraction of primitive meteorites. These millimeter sized glassy droplets appear to be the products of intensive transient heating events. Their size distribution, chemical and mineral composition, texture and isotope composition suggest that chondrules were produced as a result of short duration melting followed by rapid cooling of solid precursor particles. Gas dynamic heating (Wood 1984; Hood and Horányi 1991, 1993, Ruzmaikina and Ip 1994), magnetic reconnection (Sonett 1979; Levy and Araki 1989) and electrostatic discharges (Cameron 1966; Whipple 1966) are thought to be the leading candidates to explain chondrule formation. In this paper we explore electrostatic discharges both theoretically and experimentally.

Differential settling of various sized grains towards the midplane of the nebula and also size sorting of dust particles due to turbulence are suspected to build large scale charge separations that episodically relax via the electric breakdown of the nebular gas. The electrostatic discharge is analogous to lightning in the Earth's atmosphere. In this paper we do not discuss the complex physical processes that may lead to electric field generation. Instead, assuming that this may occur due to a number of inductive and non-inductive processes, we use the expected initial conditions in such a discharge. We then follow the expansion of the initially energetic plasma column as it expands, cools and recombines. We calculate

the energy flux reaching the surface of an embedded dust grain and also its subsequent heating. We show that, within the range of expected initial plasma conditions, lightning is a viable mechanism for chondrule formation.

We have also modified an existing laboratory research facility to allow us to introduce chondrule precursor solid grains into a plasma discharge. Here we report on the status of these experiments and also discuss our future experimental plans.

STATUS OF THEORETICAL STUDIES

Present models of the early solar system describe a turbulent, differentially rotating nebula where the random motion and superimposed organized transport of dust grains could have produced large scale electrostatic charge separations (for a comprehensive review see Morfill *et al.* 1993). The build-up of large electrostatic potentials is limited by the gas conductivity (due to cosmic ray ionization and radioactive decays), by the available gravitational energy from settling towards the midplane (vertical lightning) or accretion onto the central body (radial lightning) and finally by the breakdown potential of the nebular gas. At the onset of a discharge, electrons are accelerated beyond energies required for ionization before they strike a neutral. The resulting electron avalanche relieves the large scale charge separation. The processes that result in lightning were recently reviewed by Morfill *et al.* (1993). The existing models provide only a qualitative picture for lightning generation in the early solar system and they are often questioned by comparing the low pressures in the nebula to conditions in fluorescent light bulbs and TV tubes. Though we do not have detailed models of the processes for the generation of lightning, these analogies are misleading. In the case of terrestrial lightning, for example, the initial diameter of the discharge channel is $10^3 - 10^4$ times larger than the electron neutral mean free path, a condition which cannot be met in these devices. However, the expected low breakdown potential of the nebula might impair sufficient charge separations (Gibbard and Levy 1994; Love *et al.* 1994). It is perhaps reassuring to note that the generation of lightning even in the Earth's atmosphere, though frequently observed, is still lacking a full theoretical description (Uman 1987).

Independently of the details of their generation, lightning discharges, if they occur at all, must produce an ionized plasma column. The initial plasma parameters (density, electron and ion temperature and the size of the discharge channel) fully determine the subsequent evolution. In a recent paper (Horányi *et al.* 1995), we quantitatively described the history of the discharge channel simultaneously with the thermodynamics of the embedded dust grains. Below we briefly summarize the results.

The expansion of the discharge channel

The build up of expected large scale charge separations is limited by the breakdown of the nebular gas. In the presence of sufficiently large electric fields a few free electrons can initiate a cascade of electrons. If electrons are accelerated during one electron-neutral collisional mean free path to energies such that they can ionize an atom (or H_2 molecule), the ensuing chain reaction builds a cascade that shorts the charge separation. Upon completion of the discharge, an expanding plasma column of hot electrons and cold ions is left behind. The expansion of the discharge channel into the cold ambient gas generates a shock wave as it displaces the surrounding cold neutral gas. Assuming no mixing through the shock we can follow the expansion of a thin massive shell, filled with the ambient neutral hydrogen gas. Considering the relevant time scales involved we anticipate that in the expanding channel the electrons and protons will first come to thermal equilibrium and then cool together. At some later stage, when the temperature becomes low enough, neutrals will appear as product of radiative recombination.

The conservation of mass of an axially symmetric discharge channel - since we ignore mixing - has a simple form

$$\frac{d}{dt}\left(\left(n_p + n_n\right)r^2\right) = 0 \tag{1}$$

where r is the radius of the channel, and n_p and n_n are the number density of the plasma (i.e., the electron or ion number density) and neutral particles, respectively.

Electrons and protons can recombine to form neutrals that can be ionized again by electron impacts

$$\frac{dn_p}{dt} = -v_r n_p^2 + v_{e,n} n_e n_n - 2\frac{n_p}{r}\frac{dr}{dt} \tag{2a}$$

$$\frac{dn_n}{dt} = +v_r n_p^2 - v_{e,n} n_e n_n - 2\frac{n_n}{r}\frac{dr}{dt} \tag{2b}$$

where $v_r = 2.4 \times 10^{-10}\, T_e^{-0.7}$ cm^{3}s^{-1} is the radiative recombination rate, T_e is the electron temperature in K units (Banks and Kockarts 1973) and $v_{e,n}$ is the electron impact ionization rate, also a function of T_e. For $v_{e,n}$ we interpolated the tabulated values of Cravens *et al.* (1987). The last terms on the right hand sides of eqns. (2a) and (2b) describe the number density changes due to the expansion of the discharge channel. Using similar equations for the conservation of momentum and energy, we can derive a set of differential equations that describe the evolution of the expansion velocity, the electron, ion and neutral densities and temperatures, and the UV photon production. Naturally, we also need to define the initial conditions.

The parameters for presolar nebulae are uncertain; here we assume a main stream working model where hydrostatic equilibrium is assumed perpendicular to the mid-plane and the temperature is calculated from the energy balance between local viscous heating and radiative losses (Wood and Morfill 1988). Typical mid-plane values for distances between 3–10 AU from the central body are $n_o = 10^{14}$cm^{-3} and $T_o = 100$ K (these parameters correspond to a pressure of $P = n_o k T_o = 1.38 \times 10^{-6}$ bar). Inside of this distance range the nebula is probably too hot, so that thermal ionization eventually results in too large a conductivity for efficient charge separation. Outside of this range gravity or turbulence may not be able to support sufficiently large electrostatic fields (Morfill *et al.* 1993). The initial conditions inside the discharge channel are also uncertain. We assume that initially the gas inside is fully ionized, $n_p(t = 0) = n_o$ and $n_n(t = 0) = 0$. The electron temperature must be just high enough to create an ionization cascade. Here we assume $15 \le T_e(t = 0) \le 60$ eV and $T_i(t = 0) = T_o$. The most important process determining the initial size of the channel is electron impact ionization. On the Earth the radius of a typical lightning bolt is about 10^3 times the ionization mean free path, $\lambda_{e,n}$. Clearly, the processes may be quite different in the Earth's atmosphere but as a guide we assume approximate scaling $10^3\lambda_{e,n} \le r_o \le 10^4 \lambda_{e,n}$, so that $1 \le r_o \le 10$ km.

In Fig. 1 the evolution of the expansion velocity and the plasma parameters are plotted for $kT_e(t = 0) = 30$ eV with $r_o = 10^3 \lambda_{e,n}$. One can recognize the characteristic time scales of the various processes involved. As anticipated, the electron–ion thermalization is the fastest process, clearly separated from the expansion. Electrons and ions have the same temperature in less than 10^{-4} sec, independent of the initial size of the discharge channel. The slowest process is the radiative recombination; the build up of the neutral density is delayed until the electrons become cold enough so that their ionization rate of neutrals is significantly lowered.

Thermodynamics

A dust grain embedded inside the expanding discharge channel is exposed to intense UV radiation, and bombarded by charged and neutral particles.

Assuming a low albedo for the dust grains the UV energy flux

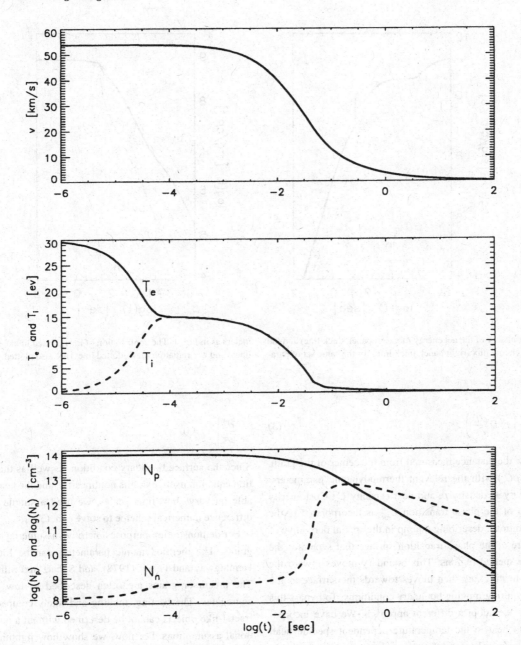

Fig. 1. The evolution of the expansion velocity (top), electron and ion temperatures (middle) and the plasma and neutral densities (bottom) assuming an initial electron temperature $kT_e = 30$ eV and discharge channel radius $r_o = 10^3$ electron neutral impact ionization mean free path.

simply becomes $q_{uv} = F_{uv}\Delta E$, where $\Delta E \approx 10$ eV is the energy of a Lyman-α photon. In our model these photons are produced via radiative recombination and electron impact excitation collisions.

Grains inside the discharge channel initially will be heavily bombarded by electrons. They will rapidly become negatively charged and the flux of incoming electrons will eventually slow down due to the electrostatic repulsion between the free electrons and the ones already attached to the grain's surface. On the other hand, the originally small flux of protons will be enhanced as the negative charge on the grain builds up. The time history of the plasma density and temperature determines the electrostatic charging history of the embedded dust particles. UV photons, ions and neutrals bombarding a grain deliver energy to its surface generat-

ing a heat flux (the electron contribution is negligible). The basic issue of our discussion is to point out that the energy flux reaching the surfaces of grains inside a discharge channel can be sufficient for partial or even full melting, providing a viable scenario for chondrule formation.

Fig. 2 shows the time history of the energy flux delivered by UV photons, ions and neutrals $F = q_{uv} + q_i + q_n$, and the integrated energy flux $E = \int_o^t F dt$, onto a grain with $a = 4$ mm. Initially ion bombardment represents the bulk of the energy flux. As the channel expands and cools the density of the neutral particles increases resulting in an enhanced UV radiation.

The evolution of the temperature distribution inside a grain is described by the heat conduction equation

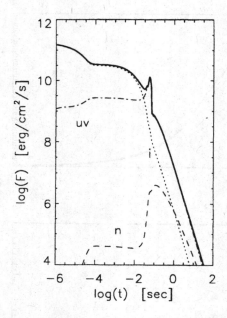
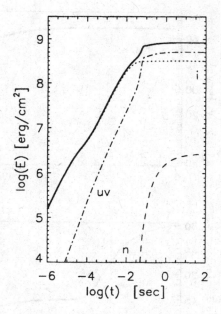

Fig. 2. The time history of the net energy flux (left panel, thick line) and the total integrated energy flux (right panel, thick line) for the same set of parameters as in Fig. 1. The contribution of ions (dotted lines), neutrals (dashed lines) and UV radiation (dash-dotted lines) are also plotted.

$$\frac{\partial T}{\partial t} = \frac{\partial}{r^2 \partial r}\left(r^2 D \frac{\partial T}{\partial r}\right). \tag{3}$$

where r is now the distance measured from the center of the grain, and $D = \kappa / (\rho C_p)$ with the relevant thermodynamic parameters: heat conductivity κ, density ρ, and heat capacity C_p. Customarily the latent heat of the phase transition L_{melt} is incorporated in the boundary conditions, describing a jump in the spatial derivative of the temperature at the phase transition surface that separates the solid from the melted regions. This boundary moves toward the center during heating and then moves towards the surface as the grain cools, requiring moving boundary conditions. To bypass this difficulty, here we adopt a different approach. We have incorporated the latent heat in the temperature dependent specific heat. This effective specific heat is defined as

$$C_p^{eff}(T) = C_p(T) + L_{melt}\delta(T - T_{melt}) \qquad erg/g/K, \tag{4}$$

where δ is the Dirac delta function. For numerical purposes we approximated δ with a square pulse, carefully selecting the minimal width (ΔT) suitable in our finite difference scheme. Naturally, the height of the pulse, l, was selected to ensure that $l\Delta T = L_{melt}$.

On the surface of the grain with radius a, the energy balance equation defines the boundary condition, the incoming energy flux is in part radiated away and in part is conducted into the interior

$$q = \sigma T_s^4 - \kappa \frac{\partial T}{\partial r}\bigg|_{r=a}, \tag{5}$$

where σ ($= 5.67 \times 10^{-5}$ erg/cm²/s/deg⁴) is the Stefan-Boltzmann constant, and T_s is the surface temperature. In the very center we kept

$$\frac{\partial T}{\partial r}\bigg|_{r=0} = 0. \tag{6}$$

Since the surface boundary condition, as well as the heat conduction equation itself, exhibits nonlinearity, and the heat input is variable on very short time scales, we used a simple explicit finite difference numerical scheme to solve eqn. (3).

For demonstration purposes, here we assume fayalite (Fe_2SiO_4) grains. The thermodynamic parameters can be found in Robie, Hemingway and Fisher (1978), and Schatz and Simmons (1972). The precursor grains are often described as low density fluffy aggregates. The heating, melting and likely compaction of these fractal-like objects cannot be described without a number of additional assumptions. For now, we show how lightning might produce chondrules assuming compact spherical fayalite precursors and plan a more complete investigation of other nebular solids and precursor structures in the future.

Fig. 3 shows the temperature history of grains with radii $a = 0.3$ and 0.4 mm assuming the heat flux of a discharge event with $r_o = 10^3\lambda_{e,n}$ and $T_e(t = 0) = 30$ eV (Figs. 1 and 2). Melting is strongly sensitive to the surface-to-volume ratio of the grains. The 0.3 mm radius grain becomes fully melted, but the melt/solid interface penetrated only ≈ 0.1 mm into the grain with $a = 0.4$ mm. The grains reach their maximum surface temperature in less than 0.1 sec. The subsequent cooling of the fully melted grain reflects the effect of the phase transition, which now serves as a heat source slowing the cooling rate of the particle, producing the shoulder-like features in the middle left panel of Fig. 3.

These calculations indicate that lightning might melt precursor dust particles. These ideas can also be tested in laboratory experiments.

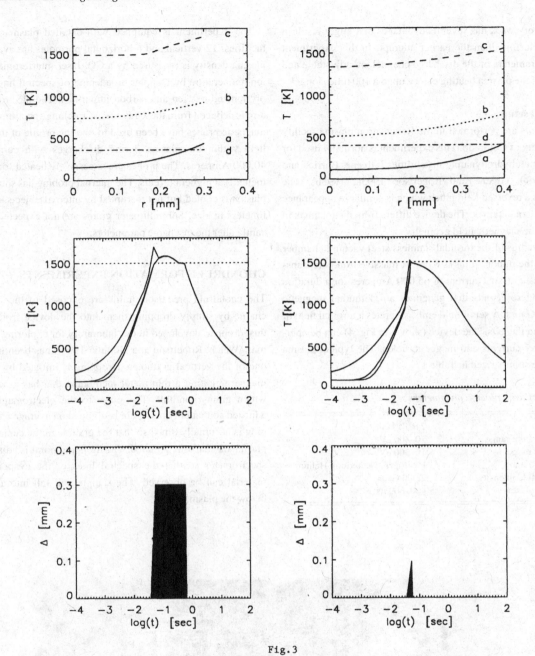

Fig.3

Fig. 3. The thermal histories of a 0.3 mm (left panels) and a 0.4 mm (right panels) size grains. The top panels show the snapshots of the temperature distribution inside the grains at $t = 0.001$ (a); 0.01 (b); 0.05 (c); and 60 (d) sec. The middle panels display the history of the temperatures monitored at $r = a$ (surface); $a/2$; and 0 (center). The bottom panels show the melt-layer thickness. The parameters of the discharge channel were the same as in Fig. 1 (from Horányi et al. 1995).

STATUS OF EXPERIMENTAL STUDIES

In our laboratory experiments we expose chondrule precursor material to plasma conditions similar to lightning discharges. These experiments are similar, in spirit, to one previous attempt by Wdowiak (1983). His pioneering work came to a negative conclusion: the artificial chondrules produced were vesiculated (contained void volumes). He argued that the very strong temperature dependence of the viscosity ($\eta \sim T^{-(10-30)}$) coupled with rapid radiative cooling ($dT/dt \sim -T^4$) was responsible for the bubble entrap-

ment. The finding of vesiculation in the samples led to the conclusion of this experiment, suggesting that short duration impulse heating and radiative cooling could not produce them. However, as discussed before, there is a growing body of evidence that suggests transient heating events are responsible for chondrule formation. We have also learned that the cooling rates could have been slower than calculated for radiation into free space. We suspect that the bubbles might have been a result of overheating the samples ('boiling') and the very short duration of the heating event (we estimate the discharge lifetime on the order of ≈ 5 μsec).

Our laboratory work has several advantages that might result in different conclusions than the earlier attempt. In this experiment the plasma parameters of the discharge channel are adjustable and the duration of the plasma heating can be up to a 100 times longer.

Experimental setup

The experiments are performed in the Reversatron plasma facility at the University of Colorado. This experiment is normally used for studying magnetohydrodynamic instabilities (Greene, Barrick and Robertson 1990; Greene and Robertson 1993a, 1993b). The Reversatron is a reversed field pinch which is similar in appearance and operation to a tokamak. The device differs from the tokamak in the details of the magnetic field geometry.

The major radius of the toroidal stainless steel vacuum chamber is 50 cm, and the minor radius is 8 cm. A massive iron-core transformer induces a plasma current of 65,000 Amperes for a duration of 400–600 microseconds that generates an azimuthal magnetic field of 1,600 Gauss. A set of field coils generates a toroidal field of 1,000 Gauss on the axis. The device (shown in Fig. 4) can be operated with lower currents and magnetic fields. The typical plasma parameters are summarized in Table 1.

Table 1. *Reversatron plasma parameters*

Plasma density	2×10^{14} cm^{-3}
Electron temperature range	50–300 eV
Ion temperature range	10–300 eV
Working gas options	Hydrogen, Deuterium, Helium
Peak magnetic field intensity	2,000 Gauss
Duration of the plasma	400–600 μsec

The experiment is equipped with detailed plasma diagnostics facilities: 12 vertical and 6 horizontal portholes are available. The plasma density is measured by a CO_2 laser interferometer and the ion temperature by Doppler broadening of spectral lines of multiply ionized oxygen and carbon impurities. The electron temperature is deduced from the x-ray bremsstrahlung spectrum. Insertable magnetic probes have been used to find the profile of the magnetic field within the plasma during discharges with currents up to 40,000 Amperes. The probes are excessively heated (on occasions melted!) at higher currents. The internal probing has shown that the plasma is cooled but not disrupted by internal objects several millimeters in size. Submillimeter grains are not expected to significantly alter the discharge parameters.

CHONDRULE FORMATION EXPERIMENTS

The chondrule precursor grains are exposed to the plasma discharge by simply dropping them into the device from above. A dust dropper, developed in our laboratory for exploring dusty plasmas (Walch, Robertson and Horányi 1994), has been mounted on one of the vertical portholes. Dropping is initiated by an electromagnet agitating a thin metal membrane that has a hole through which grains can fall. The pulse to the electromagnet can be adjusted so that only one grain is dropped on average and the dropping is accurately timed so that the grain is in the center when the plasma discharge is initiated. The dropping point is adjacent to the spectrometer so that the spectral lines of the evaporated grain material can be observed. The sample will fall into a receptacle below the plasma.

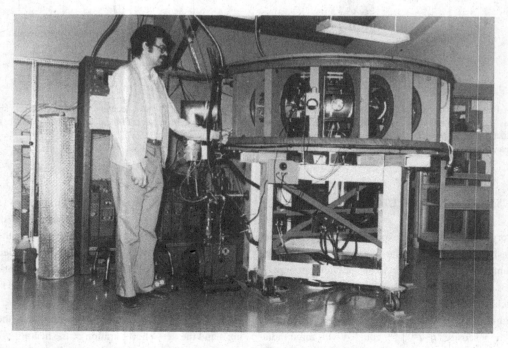

Fig. 4. The Reversatron plasma facility at the University of Colorado with one of the authors (S. Robertson).

Our first experiments were designed only to test our dust dropper, the timing device and ultimately the presence of dust in the plasma. In Fig. 5 the left panel shows the history of the plasma currents in two very similar discharge events. The right panel shows the history of the observed UV light (100–900 Angstroms) collected by a filtered photodiode with and without a dust particle (50 micron radius glass sphere). The light is produced from the evaporated and sputtered atoms and molecules and also other impurities in the plasma. The presence of dust in the plasma is clearly verified by the greatly enhanced light emission. We are now replacing the broad-band UV detector with an existing spectrometer in the near UV to monitor specific lines.

The parameters of our discharge plasma (composition, density and temperature, see Table 1). can be adjusted to match the range of expected conditions in the presolar nebula . However, the size of our device limits the duration of our discharge. In our numerical example, total melting of a 0.3 mm radius particle required about 0.1 sec, that is much longer than the duration of our experimental plasma discharge ($\leq 0.6 \times 10^{-3}$ sec). To overcome this shortcoming of the experiments we can adjust the initial electron temperature to much higher values and/or use smaller dust particles. Setting $n_p = 10^{14}$ cm^{-3} and $kT_e \approx kT_i \approx 300\ eV$ we could melt/vaporize grains up to a centimeter in radius. We plan to have a series of experiments to examine the scaling with both the size of the precursor particles and also the initial plasma parameters. This will test our theoretical models for a wide range of parameters and build confidence in the necessary extrapolations for nebular conditions. Similarly, the magnetic field in our experiment is much stronger than it is expected in the early solar system. Clearly, we will not be able to reproduce the remnant magnetization observed in chondrules. However, we will be able to estimate the efficiency with which a cooling chondrule can record the magnetic field intensity of its environment.

At the moment we are using commercially available uniformly sized grain samples, but we also plan to use chondrule precursor mineral assemblages. These samples will be provided by Dr. Hewins (Rutgers University), one of our collaborators. The samples will be recovered from the experiment and returned to him for chemical and mineralogical analysis.

SUMMARY

There are a number of effects not considered here. Lightning discharges are likely to occur in the high dust density regions of the solar nebula where dust grains might become the major charge carriers reducing the conductivity and hence increasing the breakdown potential. Based on our calculations, we expect that the part of the particle population that is smaller than the characteristic chondrule size will fully vaporize and produce plasma and neutral gas, mass loading the original plasma. In addition, dust particles can get entrained and take up part of the momentum of the expansion. The number density of the dust grains is also intimately related to the optical depth. Clearly, higher opacities – as opposed to the rapid cooling due to radiation into free space as discussed here – can slow the cooling rates to the values suggested by dynamical crystallization experiments (Hewins 1988). Motivated by the results of our "test particle model" presented here, we will develop these ideas further in the future. In continuing collaboration with Drs. G. Morfill and E.H. Levy we plan to model the expansion of a dusty-plasma discharge. The energy deposition, heating and mass loss from the grains will be coupled with continuity equations for the mass, momentum and energy of the expanding plasma.

The experiments and our theoretical models will iteratively arrive at the plasma parameters and precursor materials that are the most likely candidates representing chondrule formation regions

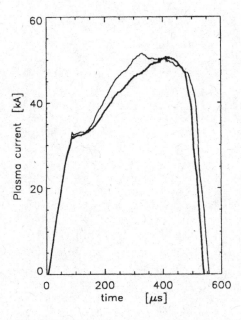
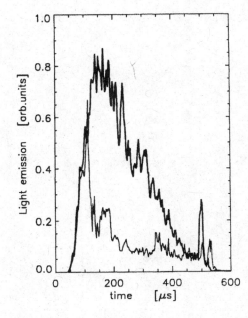

Fig. 5. The history of the plasma currents (left panel) and the observed light emissions (right panel) with (thick lines) and without (thin lines) an embedded dust particle.

and materials in the early solar system. Our simple theoretical estimates and preliminary experimental results presented here suggest that lightning remains a leading candidate to explain chondrule formation worthy of more complex theoretical and also laboratory investigations.

ACKNOWLEDGEMENT

The theoretical calculations were done in collaboration with Drs. G. Morfill and E.H. Levy and our late colleague C.K. Goertz. D. Alexander provided crucial help in the laboratory. Comments by Drs. S. Araki, P. Cassen and S. Love were very helpful. This work was supported by the Origins of Solar Systems Program of NASA.

REFERENCES

Banks P.M. and G. Kockarts (1973) *Aeronomy* , Academic, Orlando, Fla.

Cameron A.G.W. (1966) The accumulation of chondritic material, *Earth Planet. Sci. Lett.* **1** , 93.

Cravens T.E., J.U. Kozyra, A.F. Nagy, T.I. Gombosi and M. Kurtz (1987) Electron impact ionization in the vicinity of comets, *JGR* **92** , 7341.

Gibbard S.G. and E.H. Levy (1994) On the possibility of precipitation-induced vertical lightning in the protoplanetary nebula, in: *Papers presented to the Conference: Chondrules and the Protoplanetary Disk*, LPI Contrib. 844, p. 9, Lunar and Planetary Institute, Houston.

Greene P., G. Barrick and S. Robertson (1990) *Phys. Fluids* **B 2** , 3059.

Greene P., and S. Robertson (1993a) *Phys. Fluids* **B 5** , 550.

Greene P., and S. Robertson (1993b) *Phys. Fluids* **B 5** , 556.

Hewins R.H. (1988) Experimental studies of chondrules, in: *Meteorites and the Early Solar System* , The Univ. of Arizona Press, 680.

Hood L.L. and M. Horányi (1991) Gas dynamic heating of chondrule precursor grains in the solar nebula, *Icarus* **93** , 259.

Hood L.L. and M. Horányi (1993) The nebular shock wave model for chondrule formation: One-dimensional calculations, *Icarus* **106** , 179.

Horányi M., G. Morfill, C.K. Goertz and E.H. Levy (1995) Chondrule formation in lightning discharges, *Icarus* **114**, 174.

Levy E.H., S. Araki (1989) Magnetic reconnection flares in the protoplanetary nebula and the possible origin of meteorite chondrules, *Icarus* **81**, 74.

Love S.G., K. Keil and E.R.D. Scott (1994) Formation of chondrules by electrical discharge heating, in: *Papers presented to the Conference: Chondrules and the Protoplanetary Disk*, LPI Contrib. 844, p. 21, Lunar and Planetary Institute, Houston.

Morfill G., H. Spruit and E.H. Levy (1993) Physical processes and conditions associated with the formation of protoplanetary disks, in: *Protostars & Planets III*, The Univ. of Arizona Press.

Robie R.A., B.S. Hemingway and J.R. Fisher (1978) Thermodynamic Properties of Minerals and Related Substances, *Geological Survey Bulletin* **No: 1452**.

Ruzmaikina T.V., W.H. Ip (1994) Chondrule formation in the accretion shock, *Icarus* **112**, 430.

Schatz J.F., G. Simmons (1972) Thermal conductivity of Earth materials at high temperatures *J. Geophys. Res.* **77**, 6966.

Sonett C.P., On the origin of chondrules (1979) *J. Geophys. Res.* **6**, 677.

Uman M.A. (1987) The Lightning Discharge, *International Geophysics Series* **39**, Academic Press, Inc.

Walch B., M. Horányi and S. Robertson (1994) Measurements of charging of individual dust grains in a plasma, IEEE Trans. Plas. Sci. **22** , 97.

Wdowiak T.J. (1983) Experimental investigation of electrical discharge formation of chondrules, in: *Chondrules and their Origins*, Lunar and Planetary Institute, Houston.

Whipple F.L. (1966) Chondrules: Suggestion concerning their origin, *Science* **153**, 54.

Wood J.A. (1984) On the formation of meteoritic chondrules by aerodynamic drag heating in the solar nebula, *Earth Planet. Sci. Lett.* **70** , 11.

Wood J.A and G.E. Morfill (1988) A review of solar nebula models, in: *Meteorites and the Early Solar System*, The Univ. of Arizona Press.

32: Chondrules and Their Associates in Ordinary Chondrites: A Planetary Connection?

ROBERT HUTCHISON

Mineralogy Department, The Natural History Museum, London SW7 5BD, U.K.

ABSTRACT

The ordinary chondrites (OCs) are aggregates of chondrules and other particles of stone, metal and sulfide. Gas-rich OCs formed after a protoplanetary disk had gone. In this paper, chondrules are defined as objects that were wholly or partly molten before or during accretion, with no limit on size, composition or shape. Droplet chondrules, clast chondrules and igneous objects are distinguished. Some igneous objects accreted together with hot, plastic chondrules.

Silicate droplet chondrules have textures produced by rapid cooling. Analogous textures occur in metal-sulfide chondrules in Tieschitz. Some silicate chondrules have unremelted, relict grains. Droplet chondrules cooled faster than clast chondrules, which may have pre-accretionary fabrics and may be chemically fractionated, like igneous objects. Most chondrules have unfractionated refractory lithophile elements, but some examples of most chondrule types have non-chondritic Ca/Al ratios, so volatility was not a major compositional control. Chondrules range up to 5 cm in size. The parent melts of some clast chondrules and igneous objects were more voluminous than ~ 10m³. Chondrules and other objects in unequilibrated OCs may have fine-grained opaque rims. Rims and interchondrule matrix are the product of chondrule fragmentation, with a minor nebular fraction. The limited age data indicate that chondrules and igneous objects formed together, when 100 kilometer-sized bodies existed. From these observations I argue that the OCs are from secondary bodies that formed by the re-accretion of the debris from the break-up of partly molten differentiated primary bodies within the remnant of a protoplanetary disk.

INTRODUCTION

The ordinary chondrites (OCs) are rocks which are aggregates of chondrules and other particles of stone, metal and sulfide or mixtures thereof. Some OCs contain solar-wind implanted gases that testify to their sojourn on exposed parent body surfaces after any shielding protoplanetary disk had gone (Bunch and Rajan, 1988). These gas-rich meteorites are excluded from this discussion. There is further selectivity in that emphasis is given to the observations deemed by the author to be critical to our understanding of chondrule origins, while others, deemed more equivocal, are neglected. For example, among the topics not considered are the bulk chemistry of OCs and volatile redistribution between chondrules and matrices.

Silicate chondrules, the dominant objects in OCs, are accompanied by various types of particles such as Fe-Ni metal, metal-sulfide, chondrule fragments, mineral fragments and rare pieces of igneous rock. If, as is often assumed, the OCs accreted directly from a protoplanetary disk and remained essentially unaltered since then, logic dictates that their diverse constituents co-existed in the disk. It then follows that chondrules, metal, mineral and igneous rock fragments formed within the disk, so the process or processes that produced them operated within the disk.

This paper advocates a minority view that chondrules formed by the break-up of partly molten differentiated small planets. Sanders (this volume) favors a similar scenario, but he investigates the constituents of chondrites on the premise that they formed by planetary collision. In contrast, I argue that the presence of igneous objects and fractionated chondrules in OCs implies that they had differentiated planetary parents.

Definitions

Chondrule

The definition of 'chondrule' is an artificial but important constraint on theories of chondrule origin(s). It is recognised that sub-spherical, millimeter-sized, mainly silicate objects, 'chondrules', constitute up to seventy-five volume per cent of the OCs (Grossman *et al.*, 1988). All definitions seem to include the view that a chondrule passed through a completely or partially molten stage before accretion (Dodd, 1982; Grossman and Wasson, 1983; Taylor *et al.*, 1983), but thereafter the definitions diverge. Grossman *et al.* (1988, p. 622) state that a chondrule 'formed as an isolated droplet' but exclude 'nonchondritic droplets' formed by impact on the Moon or on achondritic bodies. Also excluded are refractory chondrules abnormally rich in Al and/or Ca, 'metallic chondrules' (Wood, 1967) and 'metal-troilite chondrules' (Bevan and Axon, 1980). In searching for a single process in the formation of chondrules, Grossman *et al.* (1988, p. 623) suggest that 'related chondrules should show a continuum' of properties such as size, mineralogy and texture. Although the extremes of chondrule types appear to have been eliminated as 'constraints on chondrule formation', Grossman *et al.* (1988) state that 'any complete model must account for all of the objects found in a single place'. This last view was espoused by Hutchison and Bevan (1983) at the first chondrules conference. Chondrites were regarded 'as rocks, coexisting assemblages of silicate, metal and sulfide minerals that, at least in part, shared the same history' (p. 162). This comprehensive approach is the essence of the present paper. A theory for the origin of chondrules should be compatible with the contemporaneous formation of chondrules and their associated metal and igneous particles.

Hutchison and Bevan's (1983, p. 163) definition of 'chondrule' is retained, although slightly modified: Chondrules are objects that were wholly or partly 'liquid before or during the accretion period(s) that led to the formation of chondrites'. There is a continuum from silicate, through silicate plus metal and/or sulfide to metal/sulfide types, which appears to satisfy one criterion of Grossman *et al.* (1988). It is recognised that most chondrules are, or were, sub-spherical, but irregularly shaped objects are included if it can be demonstrated that they had formed from total or partial melts (see below). Pieces of sub-spherical objects are clearly chondrule fragments. The definition imposes no constraint on size, shape, mineralogy, composition or texture, but the texture must be consistent with a wholly or partly molten stage in a chondrule's history.

Droplet chondrules and clast chondrules

A distinction is made between droplet chondrules and clast chondrules. Following Dodd (1981), a droplet chondrule has a near-spherical shape and a texture that indicates that it cooled rapidly from a melt. Its present size is essentially that of the molten droplet. Clast chondrules 'have irregular to more or less rounded shapes and coarser internal textures, which suggest that they are abraded fragments of larger rocks' (Dodd, 1981, p. 34). By this definition, a clast chondrule could be a slightly abraded droplet chondrule. At the other extreme, a clast chondrule may be only a fraction of the size of its precursor rock, so an abraded, sub-spherical fragment of igneous rock is a clast chondrule. In the present paper, angular fragments of igneous rocks are excluded from the definition, although they may have escaped abrasion by chance and so avoided becoming clast chondrules.

Igneous objects

The term 'igneous' is reserved for objects inferred from their textural or chemical properties to have been derived from melts that were considerably more voluminous than droplet chondrules, possibly > 10 m³. For example, a microgabbro fragment in Parnallee (LL3) is texturally and chemically similar to lunar mare basalt or mid-ocean ridge basalt (Kennedy *et al.*, 1992), so it is described as igneous. The term is also applied to a pyroxene-free, metal-poor olivine-rich pebble in the L6 chondrite, Barwell (Hutchison *et al.*, 1988a). In Bovedy, L3, a 2 mm fragment of Ca-feldspar glass is 'igneous' because it is light rare earth element (LREE) enriched and has a large positive Eu anomaly like lunar highlands anorthosite (Graham *et al.*, 1976). No distinction is made between impact and other causes of melting in the genesis of objects deemed igneous.

Co-accretion of chondrules and igneous objects

Isotopic and textural data indicate that at least some exotic objects were not added to unshocked and unbrecciated OCs tens of millions of years after the main period of accretion. For example, the olivine-rich pebble in Barwell, L6, underwent some metamorphic equilibration with its host. Host and pebble have indistinguishable I-Xe ages, 3.5 Ma older than Bjurböle chondrules, although the pebble has some seven times more iodine than the host (Hutchison *et al.*, 1988a). Lithification and metamorphism probably were contemporaneous and occurred when ^{129}I was very much 'alive'.

Chondrule deformation

There is observational evidence that the various types of physically identifiable components of unbrecciated OCs co-existed as separate entities just before, or during, accretion. In Tieschitz, H3, for example, all types of object, including droplet chondrules, clast chondrules and metal-sulfide chondrules are enveloped in fine-grained opaque rims. The rims were added to the enclosed objects before or during accretion. This is true of an angular igneous fragment (Hutchison and Bevan, 1983, Fig. 5), whose Ca-feldspar is LREE enriched and which has a large, positive Eu anomaly (I.D. Hutcheon, personal communication, 1989). Fine-grained opaque rims also coated objects when they were plastic and in contact with neighboring, near-spherical chondrules. Such objects and their rims deformed together and in some, crystallisation of high-temperature, anhydrous minerals occurred after the plastic deformation (see for example, Hutchison and Bevan, 1983, Figs. 1 and 2). These deformed, once molten objects are considered to be deformed chondrules. The outlines of chondrules, therefore, were not determined by pressure-induced diffusion at low temperature (Skinner, 1989a,b); chondrules and matrices of OCs did not react like carbonate spherules in brine, under hydrostatic pressure.

As outlined above, the physically identifiable components of unshocked and unbrecciated OCs co-existed as separate entities before or during accretion. Each component, therefore, was formed in an environment before or during accretion, so may hold clues to the process(es) that operated within it. One or more of these processes made chondrules. The properties of chondrules and associated objects in OCs are now outlined, to try to identify or eliminate the processes that may have led to their formation.

CONTENTS OF THE ORDINARY CHONDRITES

Chondrules

Chondrule textures

Many silicate ferromagnesian droplet chondrules have porphyritic, barred or cryptocrystalline textures that testify to rapid cooling. In Tieschitz, the textures of metal-sulfide chondrules were interpreted by Bevan and Axon (1980) as analogues of those in rapidly cooled silicate chondrules. Intergrowths of zoned crystallites of taenite with troilite, and metallic chondrules composed dominantly of polycrystalline taenite, were thought to have been produced by quenching melts. The evidence indicates that silicate chondrules, silicate plus metal and sulfide chondrules, and metal-sulfide chondrules had similar thermal histories. The conditions required to produce the textures of ferromagnesian droplet chondrules are known from experiments on chondrule analogues (Hewins, 1988; Connolly and Hewins, 1990). The maximum temperature attained and the survival or absence of crystal nuclei proved to be as important as cooling-rate in governing the final texture. In general, however, maximum temperatures of 1400–1600°C appear to have been followed by cooling at 100–2000°C/hr.

Some chondrules contain phenocrysts - relict grains - that were not in equilibrium with their liquid hosts, as attested by more magnesian rims on dusty, FeO-bearing cores (Wasson, 1993, Fig. 3a,b). In most porphyritic chondrules phenocrysts are 'normally' zoned from magnesian cores to more FeO-rich rims. These relicts may be 'seed' crystals that did not melt during the heating that produced the chondrule. Porphyritic clast chondrules < 4 cm in size have been found (Binns, 1967) and ten proved to have a preferred orientation of the olivine phenocrysts (Dodd, 1969). In poikilitic clast chondrules, olivine globules are enclosed in often interlocking, tabular Ca-poor pyroxene crystals with or without Ca-pyroxene and Ca-feldspar (Dodd, 1981). This texture is indicative of slower cooling than porphyritic chondrules, which is confirmed by experiment. Lofgren and Russell (1986, Fig. 2) produced a poikilitic texture in a charge cooled at 5°C/hr from 1500 to 1200°C, from which it was quenched. A similar texture resulted from cooling at 10°C/hr from 1470–1250°C, followed by quenching (Radomsky and Hewins, 1990, compare Fig. 15[k] with Fig. 15[l]). Such cooling rates are within the range estimated for Apollo 15 quartz-normative basalts, albeit from lower starting temperatures (Lofgren et al., 1975); the basalts may have come from a flow 2–3 m thick.

Some clast chondrules have chemical signatures consistent with their being fragments of igneous rocks (Hutcheon et al., 1989; Hutchison, 1992). Chondrule CC-1, in Semarkona (LL3), is about 1 mm across and has olivine dendrites (Fo$_{76-72}$) poikilitically enclosed in pyroxene in the pigeonite to augite range. Anorthite (CaAl$_2$Si$_2$O$_8$), present in the interstices, is heavy REE (HREE) depleted, with a large positive Eu anomaly (Hutcheon et al., 1989, Fig. 2). The pyroxene has a complementary REE pattern, consistent with crystallisation from a melt with ~8xCI REE. The partition of the REE between pyroxene and anorthite is typical of an igneous rock. CC-1 proved to have cooled to Mg-Al closure temperature when ^{26}Al (t$_{1/2}$ = 0.72 Ma) was 'live' (Hutcheon and Hutchison, 1989), which attests to its antiquity.

Textural relationships between chondrules

The chondrites are classified on both chemical and textural criteria (see Sears and Dodd, 1988). OCs with strongly intergrown crystals and few apparent chondrules belong to petrographic (textural) type 6. Types 5 and 4 are intermediate between the textural extremes of 6 and 3. Chondrites of type 3 have numerous well defined chondrules and little intergrowth between their mineral constituents. In addition, minerals such as olivine have variable composition and these meteorites are known as the unequilibrated ordinary chondrites (UOCs). The UOCs are further subdivided into types 3.0–3.9 (see Sears and Dodd, 1988), depending on their degree of equilibration, which is lowest in 3.0. Transmission electron microscopy (TEM) by Alexander et al. (1989) revealed that Bishunpur (LL3.1) has interchondrule matrix with amorphous feldspathic 'glue'. It is little recrystallised and Bishunpur may be the most pristine UOC. Alexander et al. (1989) also found that even in Tieschitz, type 3.6, mineral fragments in the opaque rims of chondrules have secondary overgrowths only in the micrometer range, so primary textures survive.

In Tieschitz (Hutchison et al., 1979; Holmen and Wood, 1986), Krymka, LL3.1, Parnallee, LL3.6 (Fig. 1) and Udaipur, H3 (Fig. 2), many chondrule pairs, triplets and quadruplets are mutually indented, so they aggregated when at least one member was hot and plastic. In fact, the exactness of fit between the various sub-spherical, angular and deformed or indented objects in many UOCs indicates that large portions of them accreted at an average temperature of about 800°C (Hutchison et al., 1979; Hutchison and Bevan, 1983). The temperature is limited by the Fe-FeS eutectic at 988°C, above which melting in metal-sulfide would have been widespread and observable. A lower limit is required to maintain plasticity in deforming chondrules. It is possible that groups of chondrules and fragments clumped together in space before accretion, which progressed by the settling of clumps on parent body surfaces. In this case, too, there must have been enough heat to prevent quenching of the plastic member(s). It is noted also that many objects were brittle on accretion, as shown by their angular outlines. Brittle fragmentation was common and preceded or accompanied the accretion of the OCs.

Chemical composition of chondrules

Most silicate chondrules are ferromagnesian and strongly olivine-normative or pyroxene-normative (Dodd, 1981), with some normative sodic feldspar. (A 'norm' is a theoretical mineral assem-

Fig. 1. Photomicrograph, plane polarised transmitted light, thin-section of the margin of the Parnallee macrochondrule. Width of field 2.5 mm. The black arrow indicates dark, devitrified glassy material adjacent to the margin of the macrochondrule (Binns, 1967), which runs from top left to lower center. Fifteen white triangles in the host chondrite indicate locations where chondrules mutually indent.

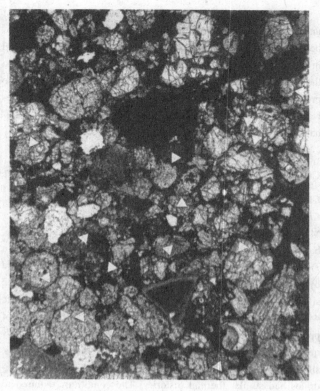

Fig. 2. Photomicrograph, plane polarised transmitted light, thin-section of the Udaipur, H3, chondrite. Width of field 2.0 mm. Fifteen white triangles indicate locations where chondrules mutually indent.

blage calculated from the bulk chemical composition of a rock.) There is, however, a continuum from ferromagnesian chondrules to extremes which may be highly quartz-normative (some have silica minerals) or nepheline-normative (Hutchison *et al.*, 1979; 1988b). Quartz-normative chondrules often have normative Ca-feldspar and corundum (Al_2O_3).

Most chondrules have Ca/Al near the cosmic ratio, but pyroxene-rich chondrules tend to have high Ca/Al and olivine-rich chondrules tend to have low Ca/Al, especially if Na-rich and nepheline-normative (Hutchison *et al.*, 1988b, Table 1 and Fig. 1). The Ca/Al (atomic) ratio of bulk chondrules has a minimum range of 0.1–5.0 (Hutchison *et al.*, 1988b). Aluminium and Ca are refractory and should have been almost completely condensed from a gas of solar composition when Mg began to condense (Larimer, 1988). This indicates that condensation or volatilisation in a nebular environment could not have produced the observed range in Ca/Al ratios in Mg-rich chondrule precursors.

Palme (1994) emphasises that there is 'large compositional variability of chondrules from a single meteorite, reflected in major variations of Mg/Si ratios, of Al and other refractory element abundances'. A theory for the origin of chondrules must account for the wondrous situation whereby each naturally aggregated gram-sized mass of chemically diverse chondrules, metal and matrix, in each chondrite class, has uniform, near solar, inter-element ratios.

Chondrule sizes – macrochondrules

Some authors, for example Wasson (1993), state that chondrules have a 'limited range of sizes', which could constrain the size of chondrule precursors. Weisberg *et al.* (1988) give the range as 0.1–3.8 mm, but describe 'macrochondrules' as big as 5 cm.

The 4 cm, porphyritic clast chondrule (see above) in Parnallee, LL3 has 'no chilled zone ... near its border, although in places it is surrounded by a narrow glassy selvedge, indented by adjacent chondrules' (Binns, 1967, p. 321). Some olivine phenocrysts are truncated at the margin. The textural relationships indicate that this macrochondrule was solid when it made contact with still plastic melt which was indented by chondrules before it chilled to become the glassy selvedge (Fig. 1). The preferred orientation of the phenocrysts (Dodd, 1969) indicates that this object, and nine others in Hallingeberg and Krymka, is an abraded porphyry rock of unknown initial size. Dodd (1969) interpreted the fabric as the result of flow in partly crystalline magma. The composition and mineralogy of the 4 cm clast chondrule resemble those of 'normal', millimeter-sized porphyritic olivine chondrules. The macrochondrule has the oxygen isotopic ratios of an L-group chondrite (I.A. Franchi, personal communication, 1993) and unfractionated REE (N. Nakamura, personal communication, 1993). It has no Ca-poor pyroxene and little metal or sulfide (Binns, 1967), so it is fractionated relative to bulk L- or LL-group chondrite. This 4 cm object with a pre-accretionary fabric sets a lower limit to the size of rocks that existed when Parnallee accreted.

Chondrule sizes - microchondrules

Microchondrules, with diameters less than 40 μm, (Krot and Rubin, 1994), occur in a number of UOCs (and gas-rich breccias). Most microchondrules seem to be pyroxene-rich types (Rubin et al., 1982), but olivine-rich types are known, for example in a clast in Krymka, LL3 (Rubin, 1989). The fine-grained opaque rims of chondrules and the interchondrule matrix (see below) are the usual setting for microchondrules, but they may be abundant in rare fine-grained clasts.

Chondrule rims and interchondrule matrix

In addition to chondrules and rare fragments of igneous rock, UOCs contain opaque, fine-grained material. This most commonly occurs as rims some tens of micrometers thick that envelop whole, deformed or fragmented silicate chondrules, metal-sulfide chondrules and igneous inclusions (see for example, Hutchison and Bevan, 1983). In addition, fine-grained material occurs as a filling between chondrules and clasts, when it is known as interchondrule matrix. Matrix rarely occurs as lumps that appear to be unrelated to chondrules or other objects (Scott et al., 1984).

At resolutions from the millimeter to the sub-micrometer range, optical microscopy, scanning electron microscopy and TEM have shown that opaque rims and matrix are dominated by fragments of chondrules and of grains of olivine and twinned Ca-poor pyroxene, commonly set in feldspathic material (Alexander et al., 1989). It was argued that pyroxene-rich chondrules selectively fragmented, induced by shrinkage along the c-axis during rapid inversion of protopyroxene to Ca-poor monoclinic pyroxene. Reworking of pyroxene fragments and adhering mesostasis might have produced microchondrules.

Although a nebular origin has been suggested (Nagahara, 1984), the direct observations, by TEM, of Alexander et al. (1989) showed that even the finest-grained opaque rims and interchondrule matrix are mainly composed of the products of chondrule break-up. Most material in rims and matrices originally formed from melts above about 1200°C. No primary low-temperature component was identified. Clay and other minerals were produced by hydrous alteration, in Semarkona for example (Hutchison et al., 1987). The abundance of a low-temperature, nebular component is limited to 3 volume per cent of bulk Bishunpur, which has 15 volume per cent matrix (Huss et al., 1981). A high D/H ratio in Semarkona and other isotopic evidence, however, indicate that a tiny pre-solar component is present, often as grains, in many UOCs (Alexander et al., 1990; Huss, 1990).

Fragments of igneous rock

Rare silicate fragments with the properties of igneous rocks occur in OCs (see Hutchison, 1992, for a review). The best example, in Parnallee, is a 2 mm microgabbro which has extreme zoning in pyroxenes and a REE distribution like that in lunar mare basalt (Kennedy et al., 1992). A 1 mm olivine norite clast chondrule, CC-1 in Semarkona (LL3) has already been discussed. Three L6 chondrites contain centimeter-sized, olivine-rich inclusions with H-group chondrite oxygen isotopic signatures (Hutchison et al.,

1988a; Nakamura et al., 1994). Each inclusion is metal-poor and lacks Ca-poor pyroxene. One, the olivine-rich 'pebble' in Barwell, is 1.4 cm across, depleted in siderophile elements and HREE and has a positive Eu anomaly (Hutchison et al., 1988a). It is partly 'digested' in that olivine has homogenised with its host but Cr-spinel and Ca-feldspar have compositional gradients over the outer 0.4 and 0.2 cm, respectively, of the pebble. The Ar-Ar and I-Xe systematics of inclusion and host are indistinguishable. This differentiated rock of H-group chondrite parentage existed before the end of accretion and metamorphism of an L-group chondrite body.

Ages of chondrites, chondrules and igneous inclusions

Phosphates in H4–5 chondrites closed to U-Pb exchange about 10 Ma before phosphates in L5–6 chondrites (Göpel et al., 1994). This is consistent with the formation, melting and break-up of an H-group body before the end of accretion and metamorphism of an L6 parent body. The relative time-scale is demanded by the presence of the partly digested H-group pebble in the metamorphosed fabric of the unshocked chondrite, Barwell (Hutchison et al., 1988a).

There are few other unambiguous quantitative data for the relative ages of chondrites and their chondrules and igneous inclusions. The 4 cm clast chondrule in Parnallee has no radiogenic ^{129}Xe (Gilmour et al., 1995), which indicates that it had a complex history. The simplest explanation of the textural relationships, described above, is that clast chondrule, surrounding melt and chondrules accreted together. Loss of radiogenic ^{129}Xe may have been the result of shock, which Swindle et al. (1991a) took to be the cause of disturbed I-Xe systematics among Chainpur, LL3, chondrules. Swindle et al. (1991b) found the I-Xe systematics of Semarkona, LL3, chondrules to be inconclusive, but suggested that chondrule formation may have occurred about 4 Ma before Bjurböle, L4, chondrules, which are used as an arbitrary standard in I-Xe work. The 'age' of formation inferred for Semarkona chondrules is indistinguishable from that of the oldest Chainpur chondrule (Swindle et al., 1991a) and from the age of host and igneous pebble in Barwell, 3.5 Ma before Bjurböle chondrules (Hutchison et al., 1988a).

The Al-Mg systematics of the noritic clast chondrule, CC-1, in Semarkona testify to its antiquity. This millimeter-sized rock cooled to the closure temperature of the Al-Mg system when the ^{26}Al/^{27}Al ratio was 7.7×10^{-6} (Hutcheon and Hutchison, 1989). Because of the short half-life of ^{26}Al (0.72 Ma) compared with ^{129}I (16 Ma), the noritic inclusion must be at least as old as the oldest chondrules and chondrites. CC-1 certainly seems to be older than Sainte Marguerite, H4, feldspars which have an initial ^{26}Al/^{27}Al ratio of 2.0×10^{-7} (Zinner and Göpel, 1992). This meteorite has phosphate with a Pb-Pb age of 4.563 Ma (Göpel et al., 1994), the oldest yet determined in an OC.

DISCUSSION

Chondrules probably formed by complete or partial melting of pre-existing solids. If chondrules formed in a protoplanetary disk, their precursors also formed within it. These were composed of silicate, of metal and/or sulfide, and of mixtures of all three. The silicate had a range of compositions, reflected in Ca/Al ratios, that was not greatly controlled by vaporisation or condensation from a gas of solar composition. The following discussion outlines arguments that the compositional range in ferromagnesian chondrules requires igneous differentiation by crystal/liquid fractionation on a parent body.

Differentiation is required to produce the precursor rocks of clast chondrules or igneous fragments with fractionated REE abundances. Moreover, many chondrules are depleted in metal and sulfide, as exemplified by the macrochondrule in Parnallee. Relative to an H-group parent, the olivine-rich pebble in Barwell (Hutchison *et al.*, 1988a) lost siderophiles and Ca-poor pyroxene. Its REE abundances demand that it formed by a complex process involving fractional melting and crystallisation. Other clast chondrules or inclusions in Bovedy, L3, Tieschitz, H3, Semarkona, LL3 (Hutcheon *et al.*, 1989), and Parnallee (Kennedy *et al.*, 1992) have fractionated REE. The parent liquids of these objects had REE abundances like those that gave rise to basaltic (eucritic) meteorites, which formed by partial melting, presumably on an asteroid. Thus the igneous objects and clast chondrules in OCs testify to partial melting and fractional crystallisation on planetary bodies. The simplest scenario, based on textural and age and isotopic evidence, is that the variety of silicate, metal and metal-sulfide objects and matrix that constitute the OCs, formed contemporaneously and accreted together. Differentiated bodies existed when chondrules formed.

It is possible that essentially unfractionated droplet chondrules are unrelated to the differentiated planetary bodies. Chondrules may not be from a single source or from a single melting process (Dodd, 1981). Some chondrules may be nebular and others planetary (Hutchison *et al.*, 1988b). I believe, however, that an origin by planetary collision or gravitational disruption probably accounts for all chondrules, because only one process is required. Simplicity is one of the strengths of this hypothesis, as emphasised by Sanders (this volume).

The question follows - was the heat for melting internal, perhaps from the decay of ^{26}Al, or by an external process such as impact? Could the Parnallee macrochondrule and igneous inclusions in OCs be impact melts? The answers are largely irrelevant in their implications for the minimum size of body inferred to have been present in a protoplanetary disk. Igneous differentiation requires the existence of bodies massive enough to have sufficient gravity to separate liquids from solid(s) of different density. Although there are uncertainties, for example in temperature, viscosity and density of the materials involved, a rough indication of the minimum size of body may be obtained and melting by internal heating or by impact yields similar results.

For an internally heated parent body, segregation of 5 per cent eucritic melt through solid residue with 25 mm crystal-size indicates that the parent body was at least 20 km in diameter. If the crystals were smaller, 2.5 mm, the minimum diameter required is 200 km (Walker *et al.*, 1978).

Impact is inefficient at producing melt. Velocities in the km/sec range are required and to avoid total disruption the projectile must be much less massive than the target (Stöffler *et al.*, 1988). On Earth, melt sheets voluminous enough for igneous differentiation seem to be present beneath impact craters over 15 km in diameter (Grieve and Cintala, 1992). The minimum diameter for a body to sustain such impacts approaches 100 km. Thus, igneous differentiation requires the existence of asteroid sized bodies, regardless of how they were melted. Impact melting differs from melting by internal heating in being local, not widespread.

Hughes (1991) suggests that between Mars and Jupiter there may have been eight differentiated Mars-sized bodies and 640 asteroids larger than Ceres, but they were 'cleaned out' by Jupiter's gravity. For planets in the first few million years of the Solar System, differentiation and impact probably were the rule. Chondrules, opaque rims and interchondrule matrices might therefore be the products of repeated reworking (Alexander, 1994).

Impact or gravitational break-up of primary bodies (Urey, 1959), if partly molten (Zook, 1980; Wänke *et al.*, 1981; Hutchison and Bevan, 1983), could have produced the chemical, mineralogical and textural range observed among chondrules. If the bulk of each primary body was formed of partly molten, olivine-rich mantle, fragmentation would have yielded an abundance of the common types of chondrule, such as partially melted porphyritic or wholly melted barred olivine, with unfractionated lithophile elements. Taylor *et al.* (1983) rejected impacts on cold bodies, such as the Moon, for chondrule formation, but the scenario favored here involves partly molten bodies.

Metallic chondrules and metal-sulfide chondrules, from the core, and clast chondrules and rarer fragments of near-surface igneous rocks and minerals would also have resulted. Re-accretion of the products of a disrupted primary body or bodies could reproduce the original bulk composition, but incomplete disruption or re-accretion of metal from the cores of primary bodies could have fractionated metal from silicate. Secondary OC bodies might otherwise have inherited the near solar inter-element ratios of their predecessors.

The products of comminution of wholly or partly crystalline chondrules, plus rare pre-existing solids such as SiC and diamond from the remnant of a protoplanetary disk, formed rims on chondrules and associated objects and accreted as interchondrule matrix.

Since the first chondrules conference, evidence for very early melting and core formation in asteroids has strengthened. Members of four chemical groups of magmatic iron meteorites cooled to Pd-Ag closure temperature within about 10 Ma of the synthesis of ^{107}Pd (Chen and Wasserburg, 1990). Thus, the timespan for the formation of magmatic irons encompasses that for chondrule formation based on the I-Xe system (Swindle *et al.*, 1991a). Cooling of the magmatic irons 'much faster than 150K/Ma' implies that they formed in parent bodies less than about 60 km in diameter (Chen and Wasserburg, 1990). Alternatively, these meteorites may have formed in larger bodies that were disrupted to produce fragments that cooled rapidly through perhaps 1000°C, more slowly through 500°C, to produce the Widmanstätten pattern.

CONCLUSIONS

Chondrules formed and the OCs accreted when primary, differentiated bodies existed. All chondrules are probably planetary. Less likely, most chondrules formed by melting clumps of grains in a protoplanetary disk, but extreme types such as Al-rich, Ca-poor, formed by planetary processes. There was a planetary connection between chondrules and the protoplanetary disk, which probably still existed after primary bodies had differentiated, because presolar grains in the OCs indicate that dust accreted on to secondary bodies. Alternatively, the grains may have been contributed by fragments of primitive (cometary?) material. The hypothesis outlined above seems to be self consistent and preferable to the wide demands that must be made for chondrule formation in a nebula.

ACKNOWLEDGEMENTS

I thank John Bridges for helpful suggestions on this paper, and colleagues past and present, such as Stuart Agrell, Conel Alexander, John Ashworth, Howard Axon, David Barber, Alex Bevan and John Bridges for discussion, support, criticism and comment over the years. The comments and criticism of Marty Prinz and Timothy McCoy led to improved clarity of the final version.

REFERENCES

Alexander C.M.O'D. (1994) Trace element distributions within ordinary chondrite chondrules: implications for chondrule formation conditions and precursors. *Geochim. Cosmochim. Acta* **58**, 3451–3467.

Alexander C.M.O'D., Hutchison R. and Barber D.J. (1989) Origin of chondrule rims and interchondrule matrices in unequilibrated ordinary chondrites. *Earth Planet.Sci. Lett.* **95**, 187–207.

Alexander C.M.O'D., Arden J.W., Ash R.D. and Pillinger C.T. (1990) Presolar components in the ordinary chondrites. *Earth Planet. Sci. Lett.* **99**, 220–229.

Bevan A.W.R. and Axon H.J. (1980) Metallography and thermal history of the Tieschitz unequilibrated ordinary chondrite- metallic chondrules and the origin of polycrystalline taenite. *Earth Planet. Sci. Lett.* **47**, 353–360.

Binns R.A. (1967) An exceptionally large chondrule in the Parnallee meteorite. *Mineralog. Mag.* **37**, 319–324.

Bunch T.E. and Rajan R.S. (1988) Meteorite regolith breccias. In *Meteorites and the Early Solar System* (eds. J.F. Kerridge and M.S. Matthews), pp. 144–164. University of Arizona Press.

Chen J.H. and Wasserburg G.J. (1990) The isotopic composition of Ag in meteorites and the presence of [107]Pd in protoplanets. *Geochim. Cosmochim. Acta* **54**, 1729–1743.

Connolly H.C., Jr. and Hewins R.H. (1990) The production of chondrule textures by introducing refractory dust to superheated melts. *Meteoritics* **25**, 354–355 (abs.).

Dodd R.T. (1969) Petrofabric analysis of a large microporphyritic chondrule in the Parnallee meteorite. *Mineralog. Mag.* **37**, 230–237.

Dodd R.T. (1981) *Meteorites: A chemical-petrologic synthesis.* Cambridge University Press, New York. 368 pp.

Dodd R.T. (1982) Objects we call chondrules. In *Papers presented to the Conference on Chondrules and their Origins.* Lunar and Planetary Institute Contribution 493, p. 15.

Gilmour J.D., Ash R.D., Hutchison R., Bridges J.C., Lyon I.C. and Turner G. (1995) Iodine-xenon studies of Bjurböle and Parnallee using RELAX. *Meteoritics* **30**, 405–411.

Göpel C., Manhes G. and Allegre C.J. (1994) U-Pb systematics of phosphates from equilibrated ordinary chondrites. *Earth Planet. Sci. Lett.* **121**, 153–171.

Graham A.L., Easton A.J., Hutchison R. and Jérome D.Y. (1976) The Bovedy meteorite; mineral chemistry and origin of its Ca-rich glass inclusions. *Geochim. Cosmochim. Acta* **40**, 529–535.

Grieve R.A.F. and Cintala M.J. (1992) An analysis of differential melt-crater scaling and implications for the terrestrial impact record. *Meteoritics* **27**, 526–538.

Grossman J.N. and Wasson J.T. (1983) The compositions of chondrules in unequilibrated chondrites: An evaluation of models for the formation of chondrules and their precursor materials. In *Chondrules and Their Origins* (ed. E.A. King), pp. 88–121. Lunar and Planetary Institute.

Grossman J.N., Rubin A.E., Nagahara H. and King E.A. (1988) Properties of chondrules. In *Meteorites and the Early Solar System* (eds. J.F. Kerridge and M.S. Matthews), pp. 619–659. University of Arizona Press.

Hewins R.H. (1988). Experimental studies of chondrules. In *Meteorites and the Early Solar System* (eds J.F. Kerridge and M.S. Matthews), pp. 660–679. University of Arizona Press.

Holmen B.A. and Wood J.A. (1986) Chondrules that indent one another: Evidence for hot accretion? *Meteoritics* **21**, 399 (abs.).

Hughes D.W. (1991) The largest asteroids ever. *Q. Jl. R. astr. Soc.* **32**, 133–145.

Huss G.R. (1990) Ubiquitous interstellar diamond and SiC in primitive chondrites: abundances reflect metamorphism. *Nature* **347**, 159–162.

Huss G.R., Keil K. and Taylor G.J. (1981) The matrix of unequilibrated ordinary chondrites: implications for the origin and history of chondrites. *Geochim. Cosmochim. Acta* **45**, 33–51.

Hutcheon I.D. and Hutchison R. (1989) Evidence from the Semarkona ordinary chondrite for [26]Al heating of small planets. *Nature* **337**, 238–241.

Hutcheon I.D., Hutchison R. and Kennedy A.K. (1989) Mg isotopes and rare earth abundances in plagioclase from ordinary chondrites: A search for [26]Al. *Lunar Planet. Sci.* **XX**, 434–435 (abs.).

Hutchison R. (1992) Earliest planetary melting - the view from meteorites. *J. Volcanol. Geotherm. Res.* **50**, 7–16.

Hutchison R. and Bevan A.W.R. (1983) Conditions and time of chondrule accretion. In *Chondrules and their Origins* (ed. E.A. King), pp. 162–179, Lunar and Planetary Institute.

Hutchison R., Bevan A.W.R., Agrell S.O. and Ashworth J.R. (1979) Accretion temperature of the Tieschitz, H3, chondritic meteorite. *Nature* **280**, 116–119.

Hutchison R., Alexander C.M. O'D. and Barber D.J. (1987) The Semarkona meteorite: first occurrence of smectite in an ordinary chondrite, and its implications. *Geochim. Cosmochim. Acta* **51**, 1875–1882.

Hutchison R., Williams C.T., Din V.K., Clayton R.N., Kirschbaum C., Paul R.L. and Lipschutz, M.E. (1988a) A planetary, H-group pebble in the Barwell, L6, unshocked ordinary chondrite. *Earth Planet. Sci. Lett.* **90**, 105–118.

Hutchison R., Alexander C.M.O'D. and Barber D.J. (1988b) Chondrules: Chemical, mineralogical and isotopic constraints on theories of their origin. *Phil. Trans. R. Soc. Lond.* **A325**, 445–458.

Kennedy A.K., Hutchison R., Hutcheon I.D. and Agrell S.O. (1992) A unique high Mn/Fe microgabbro in the Parnallee (LL3) ordinary chondrite: nebular mixture or planetary differentiate from a previously unrecognized planetary body? *Earth Planet. Sci. Lett.* **113**, 191–205.

Krot A.N. and Rubin A.E. (1994) Microchondrules in ordinary chondrites: Implications for chondrule formation. In *Papers presented to Chondrules and the Protoplanetary Disk*, pp. 17–18. Lunar and Planetary Institute Contribution 844.

Larimer J.W. (1988) The cosmochemical classification of the elements. In *Meteorites and the Early Solar System* (eds. J.F. Kerridge and M.S. Matthews), pp. 375–389. University of Arizona Press.

Lofgren G. and Russell W.J. (1986) Dynamic crystallisation of chondrule melts of porphyritic and radial pyroxene composition. *Geochim. Cosmochim. Acta* **50**, 1715–1726.

Lofgren G., Donaldson C.H. and Usselman T.M. (1975) Geology, petrology, and crystallization of Apollo 15 quartz-normative basalts. *Proc. Lunar Sci. Conf. 6th*, 79–99.

Nagahara H. (1984) Matrices of type 3 ordinary chondrites - Primitive nebular records. *Geochim. Cosmochim. Acta* **48**, 2581–2595.

Nakamura N., Morikawa N., Hutchison R., Clayton R.N., Mayeda T.K., Nagao K., Misawa K., Okano O., Yamamoto K., Yanai K. and Matsumoto Y. (1994) Trace element and isotopic characteristics of inclusions in the Yamato ordinary chondrites Y-75097, Y-793241 and Y-794046. *Proc. NIPR Symp. Antarctic Meteorites, No. 7*, 125–143. National Institute of Polar Research, Tokyo.

Palme H. (1994) Formation of chondrules and CAIs by nebular processes. In *Papers presented to Chondrules and the Protoplanetary Disk*, p. 30. Lunar and Planetary Institute Contribution 844.

Radomsky P.M. and Hewins R.H. (1990) Formation conditions of pyroxene-olivine and magnesian olivine chondrules. *Geochim. Cosmochim. Acta* **54**, 3475–3490.

Rubin A.E. (1989) An olivine-microchondrule-bearing clast in the Krymka meteorite. *Meteoritics* **24**, 191–192.

Rubin A.E., Scott E.R.D. and Keil, K. (1982) Microchondrule-bearing clast in the Piancaldoli LL3 meteorite: A new kind of type 3 chondrite and its relevance to the history of chondrites. *Geochim. Cosmochim. Acta* **46**, 1763–1776.

Scott E.R.D., Rubin A.E., Taylor G.J. and Keil K. (1984) Matrix material in type 3 chondrites - occurrence, heterogeneity and relationship with chondrules. *Geochim. Cosmochim. Acta* **48**, 1741–1757.

Sears D.W.G. and Dodd R.T. (1988) Overview and meteorite classification. In *Meteorites and the Early Solar System* (eds. J.F. Kerridge and M.S. Matthews), pp. 3–31. University of Arizona Press.

Skinner W.R. (1989a) Cold vs hot accretion of Tieschitz and other chondrites. *Lunar Planet. Sci.* **XX**, 1018–1019 (abs.).

Skinner W.R. (1989b) Compaction and lithification of chondrites. *Lunar Planet. Sci.* **XX**, 1020–1021 (abs.).

Stöffler D., Bischoff A., Buchwald V.F. and Rubin A.E. (1988) Shock effects in meteorites. In *Meteorites and the Early Solar System* (eds. J.F. Kerridge and M.S. Matthews), pp. 165–202. University of Arizona Press.

Swindle T.D., Caffee M.W., Hohenberg C.M., Lindstrom M.M. and Taylor G.J. (1991a) Iodine-xenon studies of petrographically and chemically characterized Chainpur chondrules. *Geochim. Cosmochim. Acta* **55**, 861–880.

Swindle T.D., Grossman J.N., Olinger C.T. and Garrison D.H. (1991b) Iodine-xenon, chemical, and petrographic studies of Semarkona chondrules: Evidence for the timing of aqueous alteration. *Geochim. Cosmochim. Acta* **55**, 3723–3734.

Taylor G.J., Scott E.R.D. and Keil K. (1983) Cosmic setting for chondrule formation. In *Chondrules and their Origins* (ed. E.A. King), pp. 262–278. Lunar and Planetary Institute.

Urey H.C. (1959) Primary and secondary objects. *J. Geophys. Res.* **64**, 1721–1737.

Walker D., Stolper E.M. and Hayes J.F. (1978) A numerical treatment of melt/solid segregation: Size of the eucrite parent body and stability of the terrestrial low-velocity zone. *J. Geophys. Res.* **83**, 6005–6013.

Wänke H., Dreibus G., Jagoutz E., Palme H. and Rammensee W. (1981) Chemistry of the Earth and the significance of primary and secondary objects for the formation of planets and meteorite parent bodies. *Lunar Planet. Sci.* **12**, 1139–1141 (abs.).

Wasson J.T. (1993) Constraints on chondrule origins. *Meteoritics* **28**, 14–28.

Weisberg M.K., Prinz M. and Nehru C.E. (1988) Macrochondrules in ordinary chondrites: Constraints on chondrule forming processes. *Meteoritics* **23**, 309–310 (abs.).

Wood J.A. (1967) Chondrites: Their metallic minerals, thermal histories, and parent planets. *Icarus* **6**, 1–49.

Zinner E. and Göpel C. (1992) Evidence for ^{26}Al in feldspars from the H4 chondrite Ste. Marguerite. *Meteoritics* **27**, 311–312 (abs.).

Zook H.A. (1980) A new impact model for the generation of ordinary chondrites. *Meteoritics* **15**, 390–391 (abs.).

33: Collision of Icy and Slightly Differentiated Bodies as an Origin for Unequilibrated Ordinary Chondrites

MASAO KITAMURA[1] *and AKIRA TSUCHIYAMA*[2]

[1]*Department of Geology and Mineralogy, Graduate School of Science, Kyoto University, Sakyo, Kyoto 606–01, Japan.*
[2]*Department of Earth and Space Science, Faculty of Science, Osaka University, Toyonaka 560, Japan.*

ABSTRACT

Microtextures of relict minerals in chondrules of unequilibrated ordinary chondrites are consistent with shock-melting being a primary mechanism in chondrule formation. Recent shock recovery experiments conducted on powdered chondrite to attempt to reproduce the texture, abundance and bulk composition of chondrules and fine-grained aggregates suggest that unequilibrated ordinary chondrites may have been formed by shock compression of a porous body. The porous body consisting of refractory minerals was interpreted not to be porous initially but to have included ices of highly volatile elements in the pores. The redox conditions of unequilibrated ordinary chondrites in chondrule formation can be explained by interaction with H_2O ice in the body. A two-component model involving a collision between an icy and a slightly differentiated body is suggested as a possible origin of unequilibrated ordinary chondrites.

INTRODUCTION

Ordinary chondrites, consisting of three chemical groups (H, L and LL), are the most abundant types of chondrites. Unequilibrated ordinary chondrites have not been chemically homogenized during and/or after their accretion to a parent body and therefore preserve more information about the formation processes than the equilibrated ordinary chondrites. Unequilibrated ordinary chondrites consist of chondrules, lithic fragments, mineral fragments, and matrix. These constituents have been extensively studied to try to understand the origin of chondrites.

Chondrules, which are rapidly cooled melts or molten droplets, are the main constituent of unequilibrated ordinary chondrites, and many models have been proposed for their origin (e.g., Grossman, 1988). The finding of relict minerals in chondrules, which survived melting during chondrule formation (Nagahara, 1981; Rambaldi, 1981; Jones, this volume) has given new insight into our understanding of the origin of chondrules, and has helped place constraints on the nature of the precursor materials to the chondrules. The existence of relict grains, up to one hundred μm in size, indicates that at least some chondrules were not derived from a gaseous phase, but were formed by melting of precursor minerals that were originally larger than the relicts.

Three possible sources of energy have been proposed to account for the melting of the precursor materials (e.g., Levy, 1988): (a) gravitational infall of the nebula (infall model); (b) the dissipative evolution of the nebula disk itself, such as lightning or magnetic flares (dissipative evolution model); and (c) solid-body impacts (impact model). The former two processes are nebular, and the third process is planetary. In spite of much research, the heat source responsible for chondrule formation is still in debate.

Planetary processes are attractive in that they can help account for the textures and chemical variations of chondrules and other constituents (Hutchison, this volume). Shock impact models of the planetary processes have been proposed by several workers. For instance, Hutchison and Bevan (1983) have suggested that the precursors can be treated as a two-component system and that a collision took place between differentiated and gaseous bodies. However, a number of inconsistencies with other constraints on chondrule formation have been pointed out (e.g., Taylor *et al.*, 1983; Kluger *et al.*, 1983; Grossman, 1988; Levy, 1988). These include the dissimilarity of chondrules to the lunar regolith brec-

cias and the difficulty of producing sufficient heat by such a collision.

In the following we describe textural observations of relict minerals and the results of shock experiments on chondritic materials,
which provide new information on the origin of chondrules. On the
basis of the experimental results, we propose a new shock impact
model for the origin of chondrules in unequilibrated ordinary chondrites. A scenario for the chondrite formation process will also be
presented to explain the origins of chondrules, matrix and other
constituents in unequilibrated ordinary chondrites.

BASIS OF A SHOCK IMPACT MODEL

Textural observations of relict minerals in chondrules

Microtextures of the relict minerals in chondrules may provide a
key to understanding the events during and before chondrule formation, and were studied using a transmission electron microscope
(Watanabe *et al.*, 1984; Ruzicka, 1990). Relict olivine in unequilibrated ordinary chondrites appears as dusty grains with inclusions
surrounded by clear overgrowths grown during chondrule formation. Watanabe *et al.* (1984) first found that the relict grains have a
higher dislocation density (10^9 cm^{-2}) than the overgrown part
(10^8 cm^{-2}) (Fig. 1). They interpreted the high dislocation density of
the relict as resulting from shock deformation before the formation
of the overgrowth, and suggest a shock melting origin for chondrules. Similar observations were also made by Ruzicka (1990).
He interpreted the deformation texture as reflecting high deviatoric

stress caused either by cooling of chondrules or by shock deformation. Relict pyroxene grains have also been shown to have many
stacking faults and high dislocation densities (Watanabe and
Kitamura, 1989; Ruzicka, 1990).

The textural observations of relict minerals in chondrules in
unequilibrated ordinary chondrites are therefore consistent with the
shock-melting origin of chondrules. The fact that some of the relict
minerals have no apparent shock texture does not necessarily contradict a shock-melting origin, because dislocation migration and
partial annealing may have occurred before the chondrules cooled
down. It is also possible that shock compression was inhomogeneous and only a certain proportion of grains experienced high
deviatoric stresses.

Shock-melting experiments of porous chondritic materials

Many shock experiments have been conducted on minerals and
rocks (e.g., Reimold and Stöffler, 1978; Ahrens, 1980); however,
relatively few experiments have been carried out on chondrites or
related materials (Fredriksson *et al.*, 1963; Sears *et al.*, 1984;
Arakawa *et al.*, 1987; Kitamura *et al.*, 1992). Shock recovery
experiments on chips and powders of a chondrite (the Cedar chondrite; H6) (Kitamura *et al.*, 1992) were conducted to test the idea
that shock-melting could be the origin for chondrites, as suggested
from the micro-textures in relict minerals (Watanabe *et al.*, 1984;
Ruzicka, 1990). In the experiments with shock pressures of 25 and
45 GPa, where the speeds of the projectile were 1.20 and
1.94 km/sec respectively, shock veins and a small amount of

Fig. 1. Mosaic of dark-field transmission electron micrographs of an olivine
grain in a chondrule of the ALH-77015 chondrite (L3). Right hand side of
the figure is the core part of the grain, which is relict olivine. This part
shows the reverse zoning in Fe/Mg and higher dislocation density in the
order of 10^9 cm^{-2}. The left hand side is the rim part grown in chondrule formation which has the normal zoning and lower dislocation density in the
order of 10^8 cm^{-2}. The significant difference of dislocation density between
the relict and overgrown parts suggests an shock-melting origin for the
chondrule. (after Watanabe *et al.*, 1984).

shock-induced melt were observed in a chip of the starting chondrite. In contrast, at the same shock pressures, powders of the chondrite with high porosities of 28 and 43 % showed no shock veins but appreciable amounts of melt pockets and aggregates of fine-grained materials (Fig. 2). Shock melts and aggregates of fine-grained materials in the shock experiments on porous materials are heterogeneous on small scales but average compositions are similar to that of the starting chondrite material. Chemical compositions of natural chondrules also vary from one to the next, but scatter in a range not far from the bulk chondrite composition. The close resemblance between the bulk compositions of the shock melted materials and the natural chondrules is consistent with a shock impact model for chondrule formation.

In similar shock experiments at pressure of 65 GPa (the speed of the projectile was about 3 km/sec), powder (a porosity of 45 %) of mineral mixture with a chondritic composition completely melted (Arakawa *et al.*, 1987), while a rock chip of a chondrite (the Ställdalen chondrite) formed a small amount of shock melt (Fredriksson *et al.*, 1963). These experimental results as well as our experiments indicate that a porous starting material favors the generation of large amounts of shock melt, as also shown by similar experiments on granulated lunar basalt (Schaal *et al.*, 1979). If chondrules form through shock melting, it is suggested that porous materials are the most suitable precursor material for chondrules, especially in cases where high shock pressures are not expected.

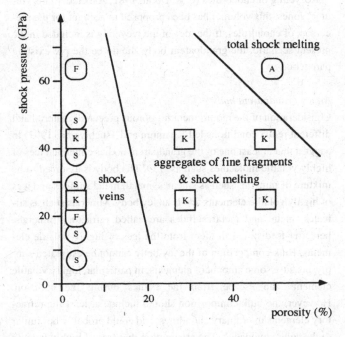

Fig. 2. Schematic diagram of shock deformation textures in relation to shock pressure on sample containers and porosity of starting chondritic materials. Experimental conditions were plotted. F: Fredriksson *et al.* (1963), S: Sears *et al.* (1984), A: Arakawa *et al.* (1987), K: Kitamura *et al.* (1992). Regions characterized by shock veins, shock melting and aggregates of fine fragments, and total melting in the recovered specimens are also shown.

Fine-grained aggregates and their precursor

Unequilibrated ordinary chondrites commonly contain fine-grained aggregates as separate clusters which have an irregular shape and are composed of minerals with grain sizes intermediate between those in chondrules and fine-grained matrices (Ikeda, 1980). While the aggregates have quite different external and internal textures than chondrules, they consist mainly of olivine, pyroxene and plagioclase and have bulk compositions similar to those of chondrules. A transmission electron microscopic study of minerals in fine-grained aggregates (Watanabe *et al.*, 1987) showed that there is glass between the minerals of the aggregates, and that pyroxene has chemical zoning and spinodal decomposition textures similar to those in chondrules (Kitamura *et al.*, 1983). The similarity of the zoning and microtextures indicates that the fine-grained aggregates have experienced the heating and cooling processes similar to those of chondrules. The aggregates are interpreted by Watanabe *et al.* (1987) to represent chondrules with a low degree of partial melting. The lower degree of partial melting resulted in an irregular shape, while a greater degree or indeed complete melting would result in a spherical shape of the chondrules due to surface tension acting on the abundant melt. Since the bulk compositions of the aggregates are within the range of those of chondrules, the precursor materials are probably common to both types of chondrules. Because precursors of fine-grained aggregates seem to be finer particles than those of relict minerals in chondrules, Watanabe *et al.* (1987) suggest that there may have been both coarse and fine grained precursors. Aggregates with textures and chemical compositions similar to those of natural fine-grained aggregates were reproduced by the shock experiments of porous powders of a chondrite (Kitamura *et al.*, 1992). Since the powders were not such fine-grained minerals, the experiments indicate that fine-grained aggregates can be produced from coarse precursors through fragmentation and a low degree of shock-induced partial melting. Therefore, the interpretation that natural fine-grained aggregates are chondrules with low degrees of partial melting is also consistent with a shock-melting origin for chondrules, and also with the idea that the precursors for chondrules and the fine-grained aggregates are the same.

Lithic and mineral fragments in matrix

Lithic fragments with an igneous texture in unequilibrated ordinary chondrites are generally larger than chondrules. The lithic fragments are generally considered to have been derived from igneous rocks which existed before chondrule formation (e.g., Kennedy *et al.*, 1992; Hutchison, 1992). Some lithic fragments show shock deformation textures which developed before accretion (e.g., Ashworth, 1981). In a shock impact model, these lithic fragments can be explained simply as igneous rocks fragmented by the shock event and incorporated into the chondrite during the subsequent accretion.

The matrix of unequilibrated ordinary chondrites is composed mainly of sub-micron size minerals which are opaque in thin (about 30 μm) sections. The minerals in the matrix have been interpreted to be either condensates from the solar gas (e.g., Nagahara,

1984), or fragments of chondrule minerals (Ashworth, 1977; Alexander *et al.*, 1989). The latter interpretation is based on the size distribution which obeys Rosin's law (Ashworth, 1977) or the power law (Alexander *et al.*, 1989), which is expected in a fragmentation process. Since the textures of matrices in unequilibrated ordinary chondrites differ from chondrite to chondrite, both cases are possible. The condensation origin of matrix minerals will be discussed in relation to a shock impact model in a later section.

Some of mineral fragments in the matrices of unequilibrated ordinary chondrites are large enough for optical microscopy and electron probe microanalysis. Such fragments have been considered to be clasts of chondrule minerals (e.g., Alexander *et al.*, 1989). Some of the olivine and pyroxene fragments in three unequilibrated ordinary chondrites were shown to have compositions closer to those of relict minerals in chondrules than minerals crystallized in the chondrule formation (Fujita, 1993; Fujita and Kitamura, 1994). This result indicates that at least some of the mineral fragments originated from the brecciation of precursor materials common with those of the chondrules, and are thus consistent with a shock impact origin. Compositional variations of the pyroxene fragments form a chemical trend from Ca-poor pyroxene to pigeonite and augite, which is similar to the variation of pyroxenes crystallized in more slowly cooled melts than the chondrules of the petrologic type 3, such as chondrules of type 4–5, lunar basalts and achondrites. Therefore, at least some of the pyroxene crystals may represent precursors to the chondrules and mineral fragments that were formed by crystallization not from a gas but from a liquid phase. This result also suggests a liquid origin of the precursors.

CONSTRUCTION OF A COLLISION MODEL

Grandparent bodies

As described above, three models have been proposed to account for the energy required to melt the chondrule precursors; (a) infall model, (b) dissipative evolution model, and (c) impact model. The results of textural observations and chemical analyses of precursor materials summarized above suggest that some of the precursors for chondrules, and lithic and mineral fragments are not direct condensates from a solar nebula gas, but crystallized from a liquid prior to chondrule formation. We also suggest that some of these precursors were melted or brecciated by a shock event during chondrule formation. These observations and inferences are most readily explained by an impact model within the framework of the planetary model for the origin of chondrules (Hutchison, this volume).

In a shock impact model, a body must clearly have existed before the shock event. In the following we refer to this pre-existing body as the grandparent body. This is a useful concept to help distinguish two states of a body before and after the shock event. For instance, in the case of a collision between two bodies, both bodies are grandparent bodies and the united body after the collision is the parent body for chondrite. The nature of these grandparent bodies is discussed below.

Any model involving a shock impact origin for chondrules has to account for the following observations: (1) porous material is required to form significant amounts of shock melt; (2) redox conditions of chondrule/matrix formation processes different from those expected in the solar nebula; and (3) some of the relict minerals were crystallized from a liquid phase. In order to satisfy these three constraints, we accepted a two-component model, including slightly differentiated and primitive materials, for the precursors, as in Hutchison and Bevan (1983). These authors state that a two-component model is required to account for the oxygen isotope data of Clayton (1981).

A slightly differentiated grandparent body

A liquid origin for some of the precursors of chondrules, lithic and fine-grained mineral fragments, mentioned as constraint (3), requires the existence of a differentiated grandparent body. Since ordinary chondrites have a bulk composition similar to the solar abundance except for highly volatile elements, any igneous process in the grandparent body occurred only on a local scale. Igneous processes on a large scale, such as large scale fractional crystallization results in the formation of differentiated bodies such as the parent bodies of achondrites. Such an igneous process can occur on a small scale by either local shock heating in the grandparent body (e.g., Misawa *et al.*, 1992), and/or local heating by radio-active elements such as ^{26}Al (Sanders, this volume). It is not important for our argument whether the differentiated body was partially molten at the time of impact or not. In this paper, 'differentiated' refers only to a local phenomena and does not imply chondrite formation involving large scale differentiation from the solar abundance.

Recycling of chondrules (e. g., Dodd, 1981; Alexander, this volume; Jones, this volume) has been proposed to account for the precursors of chondrules. If the idea of the recycling is included in the present scenario, the grandparent body should be the pre-existing parent body.

An icy grandparent body

Consideration of the requirement for porous precursor material and different redox conditions led Kitamura and Tsuchiyama (1991) to suggest that at least one of the grandparent bodies contained ices of highly volatile molecules such as H_2O. The body is assumed to be mixture of minerals such as silicates, metal, metal sulfides and ices of highly volatile elements and is an icy body. Minerals such as silicates, metal and metal sulfides are called refractory minerals, hereafter to distinguish them from the ices of highly volatile elements. Bulk composition of the icy body cannot be estimated simply, because some amount of elements, in particular, highly volatile elements can escape from the bodies during the collision. However, the bulk composition should include at least the refractory elements in ordinary chondrites, and could probably be similar to the solar abundance. The assumption that ices of highly volatile molecules such as H_2O coexisted with refractory minerals in grandparent bodies in the asteroid belt at the early solar system is suggested by the observation that some carbonaceous chondrites have actually suffered hydrous alteration in the parent body (e.g., Zolensky and McSween, 1988).

The icy body which we propose differs in the state of highly volatile elements from the gaseous protoplanet which is assumed as one of the collided bodies in previous shock impact models (Hutchison and Bevan, 1983). The two main reasons that we envisage an icy body rather than a gaseous body are as follows:

(1) **Porosity of the body :** When a solid rock is shocked and brecciated, the brecciated fragments should have characteristics similar to those of regolith breccias. Moreover, the constituents of chondrites, such as chondrules, do not show such characteristics. This dissimilarity between chondrite and lunar and other regoliths has been used as an argument against previously proposed shock models (Kerridge and Kieffer, 1977; Taylor et al., 1983). Our shock experimental results show that a porous material can provide much more abundant melt than a material with low porosity even when shock pressures are the same (Kitamura et al., 1992). The generation of abundant melt even under low shock pressure is suitable for chondrule formation, because a high-velocity collision is unlikely in the primitive solar nebula (e.g., Levy, 1988).

The size of a grandparent body is considered to be on the order of planetesimals, as mentioned below, and not very different from the parent body of chondrites. Even when two bodies with similar sizes collide with each other and unite into one parent body, the radius of the parent body increases by less than 26 %. A large change in the porosity from a grandparent to parent bodies due to contraction under the influence of gravity is therefore not expected. The porosity of the parent body, i.e., ordinary chondrites is very low. The grandparent body must therefore also had low porosity, because there are no other reasons for the porosity to significantly decrease after a shock event. In order to satisfy this further constraint, the gaseous grandparent body in the Hutchison and Bevan's model should have a core with a low porosity consisting of refractory minerals and a mantle of gas, and is not suitable for production of abundant melt by a shock impact.

On the other hand, the icy body in our model contains ices of highly volatile elements in the interstices of refractory minerals and results in having low porosity as a whole. During shock compression of the mixtures of refractory minerals and ices, ice crystals can be compressed more easily than the refractory minerals and would change to vapor due to heating by compression even at high shock pressure. The high compressibility can allow for the ice to act as pores for the refractory minerals in the shock process. The large amounts of vapor formed from ice during a collision will escape from the collided bodies. The following accretion of brecciated and recondensed refractory minerals should result in a solid body with low porosity as observed in chondrites. Thus, an icy body can satisfy the two apparently contradictory conditions that the grandparent body should have low porosity but be porous during the shock compression. Chondrule formation should then have a different mechanism from the formation of the lunar and achondrite regolith breccias.

(2) **Redox conditions in chondrule formation :** Redox states of chondrites are different from each other, and the order of oxidation state for the chemical groups of chondrites is C > LL > L> H> E (Rubin et al., 1988). Ordinary chondrites were formed under more oxidized conditions than those expected in the solar abundance. Rubin et al. (1988) stated that formation of fayalitic olivines coexisting with metallic Fe in the matrix of primitive chondrites may have occurred at low temperatures in a gas of solar composition or at higher temperatures in a more oxidizing gas. They further suggested that the former alternative may not be kinetically feasible, whereas the latter requires an unspecified mechanism for producing a highly oxidizing region. Condensation of fayalitic olivines have been discussed from thermodynamic calculations under oxidizing conditions (Palme and Fegley, 1990; Wood and Hashimoto, 1993). The change of redox conditions to an oxidation state for formation of fayalitic olivines and other minerals at high temperature can be explained by extraction of H_2-rich gas from the solar gas or an addition of ice such as H_2O, according to phase relations (Tsuchiyama and Kitamura, 1995). In the case of a gaseous grandparent body, extraction of H_2 gas from the system may be possible by a shock event. It is unlikely that the body was able to maintain an H_2-rich atmosphere before the shock event, since the body must have been at high temperature (at least higher than the vaporous of volatile elements). On the other hand, an icy body consisting of mixtures of refractory minerals and H_2O ice can be easily expected when the body cooled down below the vaporous temperature of the ice. During heating by a shock event, some amounts of refractory minerals would be expected to vaporize (Yamamoto et al., 1989), as described below, as well as vaporization of H_2O ice. The H_2O vapor can easily react with residual refractory minerals, shock melt and the vapor of refractory elements, and provide the oxidized conditions required for the formation of fine-grained matrix fayalitic olivines as recondensates from an oxidizing vapor.

Collision of icy and slightly differentiated bodies
Two separate bodies existed one slightly differentiated and the other icy. For the purposes of this discussion we shall take the simplest scenario of a collision between an icy and a second slightly differentiated planetesimal.

In this model, the size of grandparent bodies is limited by the need to explain the subsequent accretion to the parent body and cooling rates of chondrules. Yamamoto et al. (1989) carried out a calculation of chondrule formation process based on a model where chondrules were formed through condensation of vapor resulting from collisions of planetesimals. In their calculation, the energetics of the cloud with chondritic composition produced by the collision is determined by cooling due to thermal emission and heating due to absorption of incident solar radiation. During the condensation period, latent heat released by grain growth heats the cloud. The cooling due to work done by the expansion is negligible, since the gaseous pressure of the cloud is much higher than the ambient pressure of the solar nebula. The condensation process is treated on the basis of the theory of homogeneous nucleation and growth (Yamamoto and Hasegawa, 1977; Kozasa and Hasegawa, 1987), which is a kinetic theory and allows us to calculate the radius of condensed grains. Some of conclusions by these workers are: (1) A cooling rate of less than 1 $K \cdot sec^{-1}$ for chondrules is obtained for collisions producing a vapor mass larger than 10^{21} g.

The grains condensed under this condition will not escape from a sphere of influence of the planetesimal gravity, when the planetesimal mass is larger than 10^{24} g. (2) Grains pass through a liquid phase during condensation. Latent heat accompanying condensation plays a key a role in the formation of a liquid phase. (3) Grains in sizes similar to chondrules are produced for a vapor mass of 10^{22} g or less. Heterogeneous condensation on dust particles ejected during the collisions may have to be taken into account in determining the grain radius. Taking the results of these calculation, the collisions in the present model are required to be between substantial bodies with the sizes required in the above calculation, i.e., planetesimal-planetesimal collisions.

DISCUSSION

Inconsistency pointed out for shock models previously proposed

Shock impact models previously proposed have been criticized for being inconsistent with some constraints on chondrule formation (e.g., Grossman, 1988; Levy, 1988). Three of the main inconsistencies are discussed below, in the content of the present collision model.

1. One of the inconsistencies is that the low production rate of melt spherules coupled with the high production of comminuted material and agglutinates in the lunar regoliths and achondrite regolith breccias is inconsistent with the apparent high efficiency of the chondrule forming process (Kerridge and Kieffer, 1977; Taylor *et al.*, 1983). This problem arises from the assumption that grandparent bodies are similar to the moon and achondrite parent bodies which have low porosity and contain no ice. In the present model, one of the grandparent bodies is assumed to be icy with the ability to produce large quantities of melt and vapor by collision. Even in the case of a lunar basalt, if it is granulated before being shock compressed, a large population of melt can be produced experimentally (Schaal *et al.*, 1979). Therefore, by introducing a new model of an icy grandparent body, the dissimilarity between chondrules and lunar regolith breccias is not considered to be contradictory to the shock model.

2. The second criticism is based on the uniformly old ages of chondrules (Taylor *et al.*, 1983). Some chondrites show shock deformation textures but shock-melt pockets are rare (e.g., Stöffler *et al.*, 1988), indicating the lack of shock events energetic enough to produce large quantities of melt after the accretion as a function of porosity. The probability of collisions between small bodies should not change significantly before and after the formation of chondrites. Actually, Taylor *et al.* (1983) point out that the uniformly old ages of chondrules are inconsistent with impact melting for chondrule formation, because this process should have continued at least as long as bombardment in the lunar highlands (until 3.9 Gyr ago). If chondrules were formed by a shock event, the event must have occurred only once under specific conditions that were uniquely characteristic of the grandparent body. Since an icy body is assumed in the present model, a large amount of vapor of ices should be produced by the collision. If the vapor escapes from the collided bodies, the parent body can end in the state of chondrites with low porosities. Even if the accretion of vapor onto the parent body happened after the accretion of refractory minerals, the parent body could then consist of a core of refractory minerals only and a mantle of ices. In both cases, the resulting parent body is not a mixture of refractory minerals and ices. If these bodies were impacted again by a solid body, the shock impact would not produce a large amount of melt. If, on the other hand, such bodies were impacted again by an other icy body, recycling of chondrules (e.g., Dodd, 1981; Alexander, this volume; Jones, this volume) can be expected. Thus, we consider that the body should remain as a parent body of chondrites after the last collision with an icy body.

3. Levy (1988) pointed out that in order for a collision model to be viable, the velocity dispersion must exceed 3 km/sec, and that collisions must have occurred after the dissipation of the nebula gas. He further suggested that the required collision velocities could have been produced by the gravitational disturbance of Jupiter, provided chondrule formation occurred after giant-planet formation. In the present model, the formation of such an icy body should take place during the last stage of the cooling of the solar nebula. It can therefore be reasonably expected that chondrule formation occurred after the dissipation of the nebula gas and giant-planet formation. Since the velocity of about 3 km/sec can produce abundant melt as suggested by the shock experiments mentioned above, an icy grandparent body is a more suitable candidate for this situation than a gaseous body.

Cooling and accretion processes

As mentioned by some workers (e.g., Levy, 1988), it remains to be shown that a workable scenario for the subsequent accumulation of chondritic meteorites after the collision can emerge from the confines of the various constraints. The most distinct characteristic in the present model from the previous shock model is that the cloud produced by a collision contained vapor from ices of highly volatile molecules such as H_2O ice. Thus, the cloud consisting of vapor of highly volatile, moderately and refractory elements, melt droplets and fragments could be denser than that expected in the previous shock model. The effect of such a vapor on the cooling and accretion processes of ordinary chondrites can not be estimated quantitatively in the present study, because the cooling and accretion processes are complex and cannot be simply formulated. However, a few possible effects on these processes will be described below.

A shock model is suitable for providing a short duration of heating for chondrule formation, which is strongly suggested by the texture and chemical compositions of chondrules such as the slight loss of moderately volatile elements such as Na (e.g., Tsuchiyama *et al.*, 1981). Since the loss of Na is prevented under oxidized conditions (Yu *et al.*, 1994), the cloud containing H_2O produced by a shock event is suitable for explaining the slight loss of Na in chondrules. The undepleted nature of moderately volatile elements in chondrules can also be preserved if the evaporation of these elements occurred in a nearly closed system of a dense cloud.

Cooling rates of chondrules in unequilibrated ordinary chondrites have been estimated to be $10^3 - 10°C/h$ at liquidus temperatures (1100–1200°C) (e.g., Tsuchiyama *et al.*, 1980; Lofgren, this volume) and of the order of 10°C/h at around the temperature (about 1000°C) for the spinodal decomposition of pyroxene (Kitamura *et al.*, 1983), indicating that the rates decrease as temperature decreases. The cooling rates of chondrules indicate that chondrules cooled not in a vacuum but in a medium with high opacity (blanket effect) (Tsuchiyama *et al.*, 1980). The dense cloud can be interpreted as such a blanket in the present case. Kluger *et al.* (1983) showed by model calculation that the cooling rate of condensed matter from $10^4 - 10^{15}$ g generated by impact is shorter than 60 seconds, and concluded that chondrule formation by impact seems to be an unlikely event. However, as mentioned above, Yamamoto *et al.* (1989) carried out the similar calculation and showed that the cooling rates (less than 1 K/sec) of chondrules in unequilibrated chondrites can be explained by such a cooling of the cloud, when the vapor mass of 10^{21} g was presented. Therefore, the present shock impact model is consistent with the model calculation for unequilibrated ordinary chondrites.

In the present model, all the constituents of an unequilibrated ordinary chondrite should accrete onto a parent body simultaneously. Since fine-grained minerals consisting of matrix are explained in the present model as having formed by recondensation from a cloud and by fragmentation of precursor materials, they must interact mechanically with chondrules during accretion. The mechanical interaction between chondrules and matrix materials has been observed as abrasion and adhesion of chondrules by matrix materials (e.g., Kitamura and Watanabe, 1986). These observations suggest that the accretion processes have occurred successively after chondrule formation as expected in the present model.

CONCLUDING REMARKS

1. A shock impact model is consistent with both textural observations and shock experiments of rock chips and powders of a chondrite. Chemical trends of some of fine-grained pyroxene fragments derived from common precursors with chondrules indicate that some of the precursors must have a liquid origin. The slightly differentiated materials therefore must be one of the components for the precursors. The results of shock experiments showing that porosity increases melt production indicate that one of the grandparent bodies must be porous. We suggest the porous body is icy, to explain the lack of a significant change in the porosity before and after collision and the redox conditions for formation of minerals such as fayalitic olivines. The simplest scenario incorporating these conclusions is a collision between a slightly differentiated body and an icy body. We refer to these as grandparent bodies.

2. The present collision model is basically applicable to the origin of enstatite chondrites and carbonaceous chondrites, because these chondrites contain relict minerals in chondrules and lithic and mineral fragments as in ordinary chondrites. However, some modifications are required when the model is applied to enstatite and carbonaceous chondrites. In the case of enstatite chondrites which show highly reduced conditions during chondrule formation, an icy body must include abundant CH_4 ice (Tsuchiyama and Kitamura, 1995). In the case of carbonaceous chondrites, the grandparent body must contain H_2O ice to explain the highly oxidized conditions for the formation of matrix minerals, and a few constituents of chondrites such as Ca- and Al-rich inclusions must also be included in the grandparent body. It is possible that H_2O accreted after the collision could have reacted with refractory minerals as aqueous alteration of a parent body of carbonaceous chondrite.

ACKNOWLEDGEMENTS

We are grateful to Drs. S. Watanabe, T. Fujita, T. Yamamoto and T. Kozasa for their discussions, and to Dr. S.R. Wallis for critical reading of the manuscript. We benefited greatly from constructive reviews by Drs. E.R.D. Scott, M.K. Weisberg, and an anonymous reviewer.

REFERENCES

Ahrens T.J. (1980) Dynamic compression of earth materials. *Science* **207**, 1035–1041.

Alexander C.M.O'D., Hutchison R. and Barber D.J. (1989) Origin of chondrule rims and interchondrule matrices in unequilibrated ordinary chondrites. *Earth Planet. Sci. Lett.* **95**, 187–207.

Arakawa M., Sawamoto S., Urakawa S., Kato M., Mizutani H. and Kumazawa M. (1987) Experiment of shock-compression for the chondritic materials. *Koatsu-toronankai* pp. 286–287 (Abstract in Japanese).

Ashworth J.R. (1977) Matrix textures in unequilibrated ordinary chondrites. *Earth Planet. Sci. Lett.* **35**, 25–34.

Ashworth J.R. (1981) Fine structure in H-group chondrites. *Proc. Roy. Soc. London* **A374**, 179–194.

Clayton R.N. (1981) Isotope variations in primitive meteorites. *Phil. Trans. Roy. Soc. London* **A303**, 339–349.

Dodd R.T. (1981) *Meteorites: A Petrologic-Chemical Synthesis.* Cambridge University Press, New York. 368 pp.

Fredriksson K., De Carl P.S. and Aarämae A. (1963) Shock-induced veins in chondrites. *Space Research* **3**, 974–983.

Fujita T. (1993) Origin of fine pyroxene fragments in matrices of unequilibrated ordinary chondrites. *Doctoral thesis, Kyoto University.*

Fujita T. and Kitamura M. (1994) Crystallization trends of precursor pyroxene in ordinary chondrites - implications for igneous origin of precursor. In *Chondrules and the Protoplanetary Disk.* LPI Contribution No. 844, Lunar and Planetary Institute, Houston. pp. 8–9.

Grossman J.N. (1988) Formation of chondrules. In *Meteorite and the Early Solar System.* eds. Kerridge J.F. and Matthews M.S. Univ. Arizona Press, Tucson, Arizona. pp. 680–696.

Hutchison R. and Bevan A.W.R. (1983) Conditions and time of chondrule accretion. In *Chondrules and Their Origins*, ed. King E.A., Lunar Planetary Institute, Houston, pp. 162–179.

Hutchison R. (1992) Earliest planetary melting - the view from meteorites. *J. Volcano. Geotherm. Res., 50*, 7–16.

Ikeda Y. (1980) Petrology of Allan Hills-764 chondrite (LL3). *Mem. Natl. Inst. Polar Res., Spec. Issue* **17**, 50–82.

Kennedy A.K., Hutchison R., Hutcheon I.D. and Agrell S.O. (1992) A unique high Mn/Fe microgabbro in the Parnallee (LL3) ordinary chondrite: nebular mixture or planetary differentiate from a previously unrecognized planetary body ? *Earth Planet. Sci. Lett.*, **56**, 82–88.

Kerridge J.F. and Kieffer S.W. (1977) A constraint on impact theories of chondrule formation. *Earth Planet. Sci. Lett.* **35**, 35–42.

Kitamura M., Yasuda M., Watanabe S. and Morimoto N. (1983) Cooling history of pyroxene chondrules in the Yamato-74191 chondrite (L3): An electron microscopic study. *Earth Planet. Sci. Lett.* **63**, 189–201.

Kitamura M. and Watanabe S. (1986) Adhesive growth and abrasion of chondrules during the accretion process. *Mem. Natl. Inst. Polar Res., Spec. Issue* **41**, 222–234.

Kitamura M. and Tsuchiyama A. (1991) Icy grandparent bodies for shock origin of ordinary chondrites. *Lunar Planet. Sci. Conf.* **XXII**, 723–724.

Kitamura M., Tsuchiyama A., Watanabe S., Syono Y. and Fukuoka K. (1992) Shock recovery experiments on chondritic materials. In *High-Pressure Research: Application to Earth and Planetary Sciences*. eds. Syono Y. & Manghnani M.H., 333–340, Terra Scientific Publishing Company, Tokyo, pp. 333–340.

Kozasa T. and Hasegawa H. (1987) Grain formation through nucleation process in astrophysical environments II. *Prog. Theor. Phys.* **77**, 1402–1410.

Kluger K., Weinke H.H. and Kiesl W. (1983) Chondrule formation by impact ? The cooling rate. In *Chondrules and Their Origins*, ed. King E.A., Lunar Planetary Institute, Houston, pp. 188–194.

Levy E.H. (1988) Energetics of chondrule formation. In *Meteorites and the Early Solar System*. eds. Kerridge J.F. and Matthews M.S. Univ. Arizona Press, Tucson, Arizona. pp. 697–711.

Misawa K., Watanabe S., Kitamura M., Nakamura N., Yamamoto K. and Masuda A. (1992) A noritic clast from Hedjaz chondritic breccia: implications for melting events in the early solar system. *Geochem. J.* **26**, 435–446.

Nagahara H. (1981) Evidence for secondary origin of chondrules. *Nature* **292**, 135–136.

Nagahara H. (1984) Matrices of type 3 ordinary chondrites - primitive nebular records. *Geochim. Cosmochim. Acta* **48**, 2581–2595.

Palme H. and Fegley B.Jr. (1990) High-temperature condensation of iron-rich olivine in the solar nebula. *Earth Planet. Sci. Lett.* **101**, 180–195.

Rambaldi E.R. (1981) Relict grains in chondrules. *Nature* **293**, 558–561.

Reimold W.U. and Stöffler D. (1978) Experimental shock metamorphism of dunite. *Proc. Lunar Planet. Sci. Conf.* **9th** 2805–2824.

Rubin A.E., Fegley B. and Brett R. (1988) Oxidation state in chondrites. In *Meteorites and the Early Solar System*. eds. Kerridge J.F. and Matthews M.S. Univ. Arizona Press, Tucson, Arizona. pp. 488–511.

Ruzicka, A. (1990) Deformation and thermal histories of chondrules in the Chainpur (LL3.4) chondrite. *Meteoritics* **25**, 101–113.

Schaal R.B., Hörz F., Thompson T.D. and Bauer J.F. (1979) Shock metamorphism of granulated lunar basalt. *Proc. Lunar Planet. Sci. Conf.* **10th**, 2547–2571.

Sears D.W., Ashworth J.R., Broadbent C.P. and Bevan A.W.R. (1984) Studies of artificially shock-loaded H group chondrites. *Geochim. Cosmochim. Acta* **48**, 343–360.

Stöffler D., Bischoff A., Buchwald V. and Rubin A.E. (1988) Shock effects in meteorites. In *Meteorites and the Early Solar System*. eds. Kerridge J.F. and Matthews M.S. Univ. Arizona Press, Tucson, Arizona. pp. 165–202.

Taylor G.J., Scott E.R.D. and Keil, K. (1983) Cosmic setting for chondrule formation. In *Chondrules and Their Origins*, ed. King E.A., Lunar Planetary Institute, Houston, pp. 262–268.

Tsuchiyama A., Nagahara H. and Kushiro I. (1980) Experimental reproduction of textures of chondrules. *Earth Planet. Sci. Lett.* **48**, 155–165.

Tsuchiyama A., Nagahara H. and Kushiro I. (1981) Volatilization of sodium from silicate melt spheres and its application to the formation of chondrules. *Geochim. Cosmochim. Acta* **45**, 1357–1367.

Tsuchiyama A. and Kitamura M. (1995) Phase diagrams describing solid-gas equilibria in the system Fe-Mg-Si-O-C-H, and its bearing on redox states of chondrites. *Meteoritics*, **30**, 423–429.

Watanabe S., Kitamura M. and Morimoto N. (1984) Analytical electron microscopy of a chondrule with relict olivine in the ALH77015 chondrite (L3). *Mem. Natl. Inst. Polar Res. Spec. Issue* **35**, 200–209.

Watanabe S., Kitamura M. and Morimoto N. (1987) Fine-grained aggregates in L3 chondrites. *Earth Planet. Sci. Lett.* **86**, 205–213.

Watanabe S. and Kitamura M. (1989) High dislocation densities of relict minerals in chondrules. *Abstract 14th Symp. Antarctic Meteorites* pp. 73–74.

Wood J.A. and Hashimoto A. (1993) Mineral equilibrium in fractionated nebular systems. *Geochim. Cosmochim. Acta* **57**, 2377–2388.

Yamamoto T. and Hasegawa H. (1977) Grain formation through nucleation process in astrophysical environment. *Prog. Theor. Phys.* **58**, 816–828.

Yamamoto T., Kozasa T., Honda R. and Mizutani H. (1989) Chondrule formation by planetesimal - planetesimal collisions. *Proc. 22nd ISAS Lunar Planet. Symp.* 135–140.

Yu Y., Hewins R.H. and Connolly H.C.Jr. (1994) Flash heating is required to minimize sodium losses from chondrules. In *Chondrules and the Protoplanetary Disk*. LPI Contribution No. 844, Lunar and Planetary Institute, Houston. pp. 46–47.

Zolensky M. and McSween H.Y.Jr. (1988) Aqueous alteration. In *Meteorites and the Early Solar System*. eds. Kerridge J.F. and Matthews M.S. Univ. Arizona Press, Tucson, Arizona. pp. 114–143.

34: A Chondrule-forming Scenario Involving Molten Planetesimals

IAN S. SANDERS

Department of Geology, Trinity College, Dublin 2, Ireland

ABSTRACT

This contribution aims to revive a dormant hypothesis (Zook 1980, 1981) in which chondrules are frozen droplets released by the collision and splashing of molten planetesimals. It is envisaged that planetesimals between about 30 and 100 km in radius had accreted by 1 Ma and had been melted internally by the decay of ^{26}Al before 2 Ma. Collisions involving these planetesimals yielded enormous clouds of incandescent spray which cooled, in minutes to hours, to form chondrules. The hot spray acted as a powerful transitory heat source which sintered and melted intermixed fragments and dust ejected from the cool, unmelted, near-surface regions of the planetesimals. Impact velocities were similar to escape velocities so, sooner or later, the cold chondrules and other ejecta re-accreted onto planetesimals (including the residual target planetesimal) at roughly the same heliocentric radius as the collision. Chondrule production and re-accretion were repeated, perhaps many times. Old chondrules became re-worked and broken. The re-accreted chondrule-rich debris blanketed the hot, molten interiors of the growing planetesimals, and the debris therefore underwent thermal metamorphism.

The above scenario is shown to be consistent, at least in a qualitative way, with established constraints on chondrule formation. These constraints include energy needs and rates of energy transfer, the age of chondrules, isotopic and chemical features, objects associated with 'normal' chondrules, and the metamorphism and melting of meteorite parent bodies.

INTRODUCTION

Despite widespread agreement that chondrules were formed by the rapid melting of dust clumps in the solar nebula before the existence of planetesimals, this view is not held unanimously. A small number of dissenters, most notably Hutchison (e.g. this volume), have consistently favoured Urey's (1967) concept that chondrules are secondary objects which formed during collisions between earlier generations of primary, 'grandparent' planetary bodies.

The invocation of planetary collisions for the production of chondrules became unfashionable around the time of the first chondrules conference when arguments against such mechanisms seemed compelling. Almost by default, therefore, a nebular setting for chondrule formation became, and still remains, generally accepted (Taylor *et al.* 1983, Grossman 1988, Grossman *et al.* 1988).

However, one particular planetary hypothesis (the one advanced by Zook (1980, 1981)) makes so many predictions which accord with observation, that it should not, perhaps, have been tarred by the brush which fairly disposed of other planetary hypotheses. Zook postulated that chondrules are 'splash ejecta' formed by the collision of planetesimals that were already molten (as opposed to being formed by the energetic impact-melting of solid planetesimals). The melting was attributed to rapid heating by the decay of radioactive ^{26}Al.

Bearing in mind that more than a decade of intensive effort has failed to produce an acceptable nebular mechanism for chondrule production (Wood, this volume), Zook's planetary hypothesis justifiably merits re-appraisal. This paper examines and develops Zook's hypothesis in three stages. The first stage presents a complete scenario of events envisaged for the region of the protoplanetary disk where the ordinary chondrite parent bodies were accreting. The second stage briefly justifies the choice of physical parameters used in the scenario. The third stage assesses the compatibility of the scenario with a wide range of established chondrule-forming constraints. The scenario is shown to be reconcilable, at least in broad terms, with all the constraints considered. Arguments against the scenario are found to be weak. It is concluded that Zook's concept of colliding molten planetesimals provides a basis for a plausible answer to the vexing question of the origin of chondrules.

THE MOLTEN PLANETESIMAL SCENARIO FOR THE PROTOPLANETARY DISK

The scenario of events envisaged is presented in the following paragraphs. No attempt is made at this stage to justify the various assumptions and proposals that are made. The main objective here is to present a clear picture of what is imagined. Justification will follow in subsequent sections of the paper.

The first planetesimals accreted soon after the beginning of the solar system (i.e. after $^{26}Al/^{27}Al$ was 5×10^{-5}) from largely unprocessed interstellar dust. Small bodies collided gently with each other and coalesced into larger ones. By about one million years, many planetesimals had already grown to between about 30 and 100 km in radius.

At the same time as the bodies were growing, they were being heated strongly by the decay of short-lived isotopes, particularly ^{26}Al. The outer, uncompacted layers of dust acted as an efficient thermal insulator, so the internal temperature rose steadily and melting began. By 2 Ma, possibly earlier, many planetesimals had totally molten interiors. On collision these planetesimals 'splashed' and the ejecta was released to space as a huge incandescent cloud of molten droplets. As the expanding cloud radiated heat to space, the droplets cooled over a period of minutes to hours and became chondrules.

Collisions disrupted the cool dusty outer layers of the bodies as well as the molten interiors. The dispersed dusty material became heated intensely and sintered by radiation from the abundance of hot droplets with which it mixed. The sintered dust formed agglomeratic chondrules and chondrule rims. The hot, turbulent environment of the chondrule cloud also permitted the formation of compound chondrules. Volatile elements like Na, Mn and even Fe underwent partial vaporization and later condensation onto cool chondrules and dust

By about two million years the rate of heat generation had declined to about one eighth of its initial value and was now too weak to melt planetesimals from cold. However, chondrule production by collision and splashing continued for several million years more while the planetesimal interiors remained molten and insulated.

Collision velocities were not generally in excess of the planetesimals' escape velocities (in the order of 10 to 100 m s^{-1} for bodies with radii between about 10 and 100 km respectively), so most of the chondrules and other ejecta underwent gravitational re-assembly, or at least remained at about the same heliocentric radius and

eventually re-accreted to other planetesimals in similar orbits. The cumulative result of successive impacts was an increasing proportion of recycled and broken chondrules in the re-accreted chondritic debris.

The re-accreted layers of cold chondrules, fragments and dust became heated substantially by thermal conduction from the planetesimal's molten interior, and acquired metamorphic features. The more deeply buried layers of debris equilibrated internally, whereas material in the near-surface (youngest) layers remained fairly cool and did not equilibrate.

As time advanced, orbital perturbations became stronger and relative velocities between planetesimals began to exceed escape velocities significantly. Many bodies were disrupted and dispersed at this stage. The chondrule-rich covering on the surviving planetesimals experienced impact shock and brecciation after metamorphism. Beneath the layers of chondrule debris, the molten interiors of the bodies had already segregated into metal cores and silicate mantles. During the subsequent 50 Ma or so the planetesimals, with their chemically differentiated interiors and their metamorphically zoned and somewhat brecciated carapaces of chondritic rock, cooled down to become the meteorite parent bodies.

JUSTIFICATION OF THE CHOICE OF PHYSICAL PARAMETERS

1. Distribution of ^{26}Al in primitive nebular dust

MacPherson *et al.* (1992) showed that calculated initial ratios of $^{26}Al/^{27}Al$ in non-FUN CAIs have a bimodal cluster with values close to zero and close to an upper limit of 5×10^{-5}. They inferred from these data that ^{26}Al was probably uniformly distributed in the solar nebula, at least in the region where the CAIs' host meteorite parent bodies accreted. Their interpretation is adopted here, and the upper limit of 5×10^{-5} is taken as the ratio of $^{26}Al/^{27}Al$ at the start of the solar system.

2. Timescale for accretion of planetesimals

The scenario contends that planetesimals with radii > 30 km had accreted by the time the solar system was 1 Ma old. This assumption is consistent with quoted timescales for the formation and growth of planetesimals, though these timescales are themselves poorly constrained. Weidenschilling *et al.* (1989) suggest that the formation of the first, small planetesimals (< 10 km) from dispersed dust took more than 10^4 years but less than 1 Ma. Starting from a swarm of planetesimals with radii up to about 10 km, Wetherill (1989) showed that, with perturbations preventing runaways, bodies with radii of 100 km would appear within about 2×10^5 years. On this basis the suggestion that planetesimals of > 30 km radius existed by 1 Ma is conservative.

3. Thermal evolution of planetesimals

Using the assumptions made in the preceding two paragraphs, a simple one-dimensional thermal model has been developed. The model uses a Schmidt plot using a Fourier number of 0.5 (Holman 1992). Temperature profiles from core to surface have been computed at time intervals of 10^5 years from the time of accretion (Fig. 1). Two different accretion times (0.5 and 1 Ma) and three different radii (10, 30 and 100 km) were selected.

The thermal model assumes an ambient nebular temperature of 300 K, constant thermal diffusivity (7×10^{-7} m^2 s^{-1}), a specific heat capacity of 800 J kg^{-1} K^{-1}, and instantaneous planetesimal growth. The decay energy and half-life of ^{26}Al are taken as 4 MeV and 0.74 Ma respectively (Lederer *et al.* 1967), and Al is assumed to comprise 0.9% by weight of the material accreted. An outer coating, several metres thick, of unconsolidated dust is assumed. Uncompacted dust in a near-vacuum environment is an excellent insulator below 800K, but above this temperature sintering becomes important (Wood 1979, p.883) and the thermal conductivity increases by two or three orders of magnitude (Yomogida and Matsui 1984). Thus it is assumed that all radiogenic heat was retained within the planetesimals as they were heated to 800K, and thereafter 800K remained the *effective* ambient temperature, with a

very steep temperature gradient through the top few metres of dust to the planetesimal surface. Latent heat of fusion is not considered, and therefore calculated temperatures above the beginning of melting (1400K, Fig. 1) will exceed the actual temperature. A calculated temperature of about 2300K is presumed to correspond to total melting.

Real planetesimals were obviously more complex than those modelled here. They presumably continued to grow in size while they were heating up. Collisions undoubtedly accelerated heat loss. Melt may have migrated to the surface, extracting the heat-generating aluminium from the interior. Nevertheless, the thermal profiles in Fig. 1 demonstrate that the present scenario is at least energetically plausible, particularly where accretion was early, and the radius large. Evidently a radius of 10 km would have permitted little or no melting, but a body of radius 30 km accreted by 0.5 Ma could have become substantially molten. Bodies 100 km in radius would probably have retained most of their heat and become com-

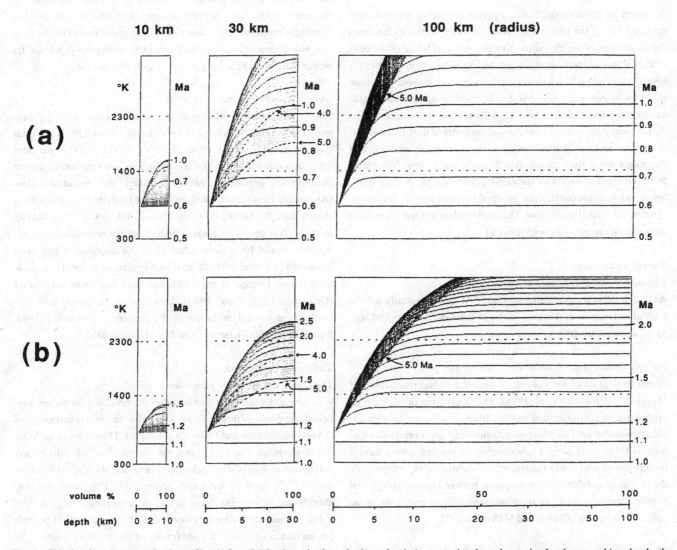

Fig. 1. Calculated temperature-depth profiles before 5 Ma through six hypothetical planetesimals during heating (continuous lines) and cooling (dotted lines grading into shaded area). Profiles are at 0.1 Ma intervals following instantaneous accretion at (a) 0.5 Ma and (b) 1.0 Ma after the formation of CAIs for planetesimals of radius 10, 30 and 100 km. Note that the horizontal axis is proportional to planetesimal volume, making the depth scale non-linear. Temperatures of 1400 K and 2300 K (picked out) approximate the onset and completion of melting respectively. See text for further details.

pletely molten, apart from the outer few kilometers, in less than 1 Ma of accretion, and they could have remained in this state for several million years more.

4. Impact velocities of planetesimals

The progressive increase in collision velocity with time envisaged in the scenario is adopted from the models proposed by Wetherill (1978, 1989) for the accretion of the terrestrial planets and the asteroids. Initial collision velocities must have been similar to, or less than, local escape velocities otherwise accretion could not have occurred. Later on, as the planetesimals grew in size, orbital perturbations became progressively more extreme and collisions led to the dispersal (rather than the accretion) of the material in the asteroid belt.

RECONCILIATION OF THE SCENARIO WITH CONSTRAINTS ON CHONDRULE FORMATION

A wealth of observational and experimental data, painstakingly acquired during the past twenty years or more, serves to constrain the parameters of the chondrule-forming process. Some of the constraints relate to the energy budget and to energy transfer mechanisms, some relate to the timing of chondrule formation, and some relate to the isotopic ratios and the chemical compositions of individual chondrules and whole chondrites. Yet other constraints derive from petrographic features such as sintered dust-balls, inclusions, caps and fine-grained rims, and the presence in meteorites of fragments other than chondrules. Finally some constraints arise from the inferred nature of meteorite parent bodies. A large number of independent constraints, grouped together into six broad categories, are considered below. The molten planetesimal scenario is shown to be reconcilable with them all.

Energy constraints

The energy required to melt chondrules

As discussed in the preceding section, ^{26}Al was potentially able to melt planetesimals and hence, in the context of the proposed scenario, to melt chondrules.

The proportion of dust converted to chondrules

About two-thirds of the volume of ordinary chondrites consist of chondrules (Grossman *et al.* 1988). The conversion of such a high proportion of primitive dust to chondrules is compatible with the collision model provided that planetesimals of appropriate size had formed sufficiently early. Carbonaceous chondrites, which have a high proportion of dusty matrix, and few chondrules, are possibly the re-assembled debris from collisions between planetesimals that mostly accreted too late, or remained too small, to have undergone radioactive heating (Grimm and McSween 1993).

The limited range of energy per gram in chondrules

A fundamental constraint which is seldom addressed in nebular models is the uniform amount of energy per gram in chondrules (Grossman 1988). Most chondrules appear to have reached the same subliquidus temperature of within about 100 K of 1800 K; few if any of them seem to have experienced higher temperatures. What kind of heating process could have given rise to such an even distribution of energy in chondrules? It is as though the rise in temperature stalled shortly after melting began (Levy 1988). The scenario proposed here can explain this observation. The internal temperature of molten planetesimals is unlikely to have exceeded the liquidus. As liquidus temperatures were approached, the outer, insulating carapace of solid material would have thinned down, thereby accelerating the loss of heat through the planetesimal surface and buffering the internal temperature. It is also possible that convection set in or that melt flowed to the surface, again serving to keep the internal temperature below the liquidus.

The rate of heating of chondrules

The retention of volatile elements such as Na in chondrules suggests that chondrules remained at their maximum temperature for only a brief period, perhaps only a few seconds according to Wasson (1993). In a nebular setting, this observation implies extremely rapid heating. However, in the present model the heating rate is irrelevant since volatiles would have remained within the parent body during the long period of radioactive heating.

Chondrule cooling rates

The experimental simulation of chondrule textures strongly suggests that cooling rates were in the order of 100 to 2000 K per hour (Radomsky and Hewins 1990). Rates of 1000 to 2000 K per hour are independently suggested by disequilibrium rare earth element fractionation between phenocryst phases and chondrule glass (Alexander 1994). Such rates are three or four orders of magnitude slower than the cooling of isolated hot chondrules by free radiation in space. It is generally acknowledged that the retardation of cooling rate would have been achieved if the chondrules had been immersed in a 'thermal bath' in which enormous numbers of chondrules were formed at the same time and irradiated each other (Grossman 1988, Wood 1984). This kind of environment is precisely what one might expect in the proposed 'chondrule cloud' following a collision between molten planetesimals.

Timescale constraints

The timing of chondrule formation

Available evidence hints that the onset of chondrule formation happened some two or three million years after the initial formation of CAIs (MacPherson and Davis, this volume). The evidence includes the observation that CAIs have the lowest ^{87}Sr/^{86}Sr ratio of any solar system objects (Podosek *et al.* 1991), and the highest levels of excess ^{26}Mg (derived from radioactive ^{26}Al). Plagioclase-bearing chondrules, in contrast, have lower, or no, excess ^{26}Mg. A time delay of the order of two million years in the onset of chondrule production is, of course, a basic feature of the present scenario. The delay is needed to give the planetesimals time to have grown, to have heated up and melted, and finally to have collided.

The level of Al-correlated excess ^{26}Mg detected in plagioclase-bearing chondrules of the H4 chondrite, Ste Margeurite (Zinner and Göpel 1992) suggests, incidentally, that these chondrules had formed, accreted, been metamorphosed, and had begun cooling before about 5 or 6 Ma had elapsed. This short period of time is not only consistent with the collision scenario but, as Zinner and Göpel pointed out, it also corresponds (within error) with the time interval independently suggested by the U-Pb age of CAIs (4566±2 Ma) and the Pb-Pb model age of phosphate in Ste Margeurite (4563±1 Ma).

Oxygen isotope constraints

The range of oxygen isotope ratios of chondrules in a single meteorite

It has been claimed that the observed spread of oxygen isotope signatures within the chondrule population of a single meteorite is incompatible with the collision of molten planetesimals (Taylor *et al.* 1983). However, as Grossman (1988) noted, this argument fails if the chondrules in a single meteorite are the products of many collisions involving planetesimals with different oxygen isotope signatures. In the present scenario it is likely that the chondrules accreting to a particular meteorite parent body were, indeed, the products of many different collisions.

Distinctive oxygen isotopic signatures of chondrite groups

Each chondrite group has its own distinctive position on a three oxygen isotope plot. This observation appears to rule out nebula-wide mixing and homogenization of precursor materials. The absence of large-scale mixing and the consequent preservation of isotopic differences between parent bodies is, perhaps, easier to contemplate if accretion took place at a very early stage, as in the scenario proposed here, than if accretion was delayed for two or three million years as is apparently required by nebular models.

Chemical constraints

The quasi-solar chemistry of chondrites

The constraint most enthusiastically cited by those who claim that chondrules were formed in a nebular setting is probably the primitive, quasi-solar chemistry of chondritic meteorites. The preservation of elements with differing chemical properties in fixed, solar proportions on a local (few millimetres) scale must, it is claimed, rule out large scale processes of chemical fractionation such as would occur with the melting of planetesimals. In defence of the present scenario, however, the envisaged disintegration of planetesimals, and the dispersal and re-accretion of the chondrules and fragments, amount to an efficient, mechanical mixing process capable of recombining separate chemical fractions on a small scale. Unless there exists a mechanism for adding fractionated material to, or removing it from, a nebular region where planetesimals are accreting, then it would seem difficult to avoid regenerating an average chondritic composition in the clastic mixture that accretes.

Metal depletion in L and LL chondrites

Following from the last point, chondrites belonging to the L and LL groups have a distinct shortfall, relative to primitive (CI) chondrites, of those elements (Fe, Co and Ni) which preferentially enter the metal phase (e.g. Newsom, 1994). The L and LL parent bodies appear, therefore, to have lost metal. An explanation for this observation, compatible with the present scenario, is that some collisions did not lead to total disruption and re-assembly of the planetesimals, but left the metal core of the larger planetesimal (the eventual meteorite parent body) intact.

Flat rare earth element (REE) patterns of individual chondrules

Many chondrules in unequilibrated chondrites have REE abundance patterns which are flat and unfractionated relative to primitive (CI) chondrites (Grossman *et al.* 1988). Such patterns are claimed to rule out crystal fractionation, and hence to rule out the derivation of chondrules from partially molten planetesimals. However, this argument fails in the present case where planetesimal interiors are completely molten. The argument also fails where partially molten planetesimals have olivine as the only liquidus phase. Olivine takes up a negligible proportion of rare-earth elements, and therefore its crystallization and removal from a chondritic melt will simply increase the overall concentration of rare-earth elements in the remaining melt without significantly changing their relative proportions. Similarly, olivine addition as a cumulus phase will uniformly reduce REE abundances. Both enrichment and depletion of flat REE patterns are observed in chondrules (Grossman *et al.* 1988).

Fractionation trends in chondrule chemistry

It has been suggested that if chondrules had formed by the collision of molten planetesimals, then a batch of individual chondrules would preserve compositional trends and inter-element ratios reflecting fractional crystallization. The absence of such compositional features (e.g. Grossman 1988) has been taken as evidence against the present scenario. However, fractionation trends may be masked if the chondrules originated from many planetesimals with varying bulk compositions.

Correlations have been observed between refractory lithophile elements in Semarkona chondrules by Grossman *et al.* (1988) who inferred three reservoirs of precursor dust that mixed in varying proportions before being melted. According to Alexander (1994) these refractory lithophile correlations are also consistent with chondrule precursors being derived from earlier generations of chondrules. Alexander's concept of chondrule recycling would seem to be broadly consistent with the present scenario.

The retention of moderately volatile elements in chondrules

A puzzling feature of chondrules is that they have undepleted levels of moderately volatile elements such as Na and Mn. Such elements might be expected to have been lost by evaporation while the chondrules were molten. An explanation for their retention, consistent with the present scenario, is that during the immediate aftermath of a collision the chondrule density may have been sufficiently high for the vapour between them to have become saturated with volatile elements. This would have inhibited further volatilization while temperatures were at their highest (Lewis *et al.*

1993). Also, re-condensation of the volatiles onto chondrules and dust may have accompanied cooling, thus explaining the observed enrichment of these elements near chondrule margins in unequilibrated chondrites (Matsunami *et al.* 1993). On the same subject, it is suggested that Fe enrichment in chondrite matrix relative to chondrules may be attributed to FeO vapourization at high temperature followed by re-condensation onto matrix grains (Nagahara *et al.* 1994, McCoy *et al.* 1991).

Constraints set by the textures, sizes, shapes and other aspects of chondrules and clasts

Chondrules which appear to be sintered dust clumps

Agglomeratic olivine (AO) chondrules are dark, very fine grained aggregates of broken crystals, mainly olivine (Weisberg and Prinz, this volume). They appear to have been sintered but not melted, and therefore, in the context of a nebular setting for chondrule formation, they provide evidence for a link between molten chondrules and their precursor dust clumps. AO chondrules are, however, also consistent with the present scenario. Their immediate precursors could be fragments ejected from the cool, surface layers of dust on the colliding bodies. Alternatively the dust may, perhaps, have been present already in a local dusty disk that was already orbiting one, or both, of the planetesimals at the time of collision. In either case, sintering would have occurred as the cold dust became caught up in the hot chondrule cloud. The chondrule cloud would have provided precisely the kind of powerful, transient heat pulse commonly invoked for flash melting in the nebula. The chondrule cloud would have been, in effect, a rather special kind of nebular environment - a short-lived, circum-planetary nebula. Thus those textural features of chondrules that suggest a nebular origin are also likely to be consistent with the 'planetary' model proposed here.

Chondrule rims, compound chondrules and matrix lumps in chondrules

Following on from the last point, fine-grained, sintered rims on chondrules may have formed when hot chondrules passed through particularly dusty zones within the chondrule cloud. Their formation process would presumably have been broadly analogous to the formation of accretionary lapilli in the ash column above terrestrial volcanic erruptions.

As well as colliding with dust, molten droplets could have collided with each other to produce compound chondrules and, more rarely perhaps, they could have collided with, and enveloped, clumps of dust to produce matrix lumps inside chondrules. Such matrix lumps, contrary to the views of Taylor *et al.* (1983), are no evidence against the present scenario.

Clasts apparently derived from differentiated planetary bodies

Rare clasts in chondrites have textures and compositions suggesting that they are fragments of igneous rock that formed by differentiation on a planetary body. An example of such a clast is the microgabbro chip in the Parnallee meteorite described by Kennedy *et al.* (1992). These clasts provide compelling evidence that accre-

tion on some planetesimals happened after other planetesimals (the sources of the clasts) had melted and differentiated. The evidence is difficult to reconcile with a nebular setting for chondrule formation, but is entirely consistent with the present scenario.

Large, isolated mineral grains

Some chondrites contain single crystals of olivine, perhaps 1 mm across, both inside chondrules and as isolated clasts. Such olivine grains may be interpreted as phenocrysts that had already formed in partially molten planetesimals before collision occurred.

Enormous chondrules

One of the arguments put forward by Taylor *et al.* (1983) against the present scenario is that large blobs of melt would be expected but have not been identified as meteorites. However, chondrules up to 4 cm in size are known (Binns 1967, Dodd 1978). Such chondrules are obvious candidates for the predicted large melt blobs.

These so-called macrochondrules represent, incidentally, a serious obstacle to the idea of flash melting of dust clumps in the nebula by mechanisms involving radiative heat transfer. Radiation predicts an inverse relationship between chondrule radius and the temperature reached, since the increase in temperature is a function of surface area and an inverse function of volume. Thus, if a millimetre-sized clump were heated through 1500 K then, in the same heating event, a centimetre-sized clump should be warmed through a mere 150 K. This inverse relationship is not seen. Macrochondrules have the same porphyritic texture, and evidently reached the same temperature, as normal sized chondrules.

The presence of metal and sulphide inside chondrules

Grossman (1988) argued that it would be impossible to produce metal-bearing chondrules from collisions between molten planetesimals, claiming that metal and sulphide would rapidly separate from molten silicate. However, such efficient segregation need not have been the case. For example, segregation of metal from silicate would possibly have been inhibited if the planetesimal interiors (or parts of their interiors) were only partially molten, with a consistency resembling slush. Collision and splashing of such planetesimals would have yielded slush droplets with silicate, metal and sulphide. Even where molten planetesimals were internally segregated, it would seem possible that the violence of impact, and the turbulent environment of the chondrule cloud, could have caused small-scale re-mixing of silicate, metal and sulphide. Also, disaggregated dust and grains from the unmelted crust of the planetesimal could, presumably, have become engulfed during micro-impacts with the molten silicate droplets.

Size sorted chondrules and clasts

The restricted size range of chondrules and clasts suggests aerodynamic sorting. In the context of the present scenario, aerodynamic sorting may have been achieved just prior to accretion. As the chondrules and clasts drifted into the gravitational field of the parent body, and spiralled down to its surface, they were possibly subjected to a steady wind of nebular gas which preferentially carried away much of the fine grained fraction. The wind would have resulted from a difference in the orbital velocity of the gas and that of the parent body around the sun (Dodd 1976).

Hot accretion textures

The existence of chondrule fragments bounded by brittle fracture surfaces, and the survival of pre-solar grains in chondrite matrix, testify to accretion at a low temperature. However, in some meteorites the chondrules are moulded tightly against one another in a manner reminiscent of pillow basalt (e.g. Holmén and Wood 1986). While such textures are usually attributed to the pummelling of chondrules into voids by post accretional impact-induced shock (Sneyd *et al.* 1988) their origin is not entirely resolved, and some workers (e.g. Hutchison and Bevan 1983) believe that the chondrules were hot and plastic at the time of accretion. The molten planetesimal scenario, while obviously compatible with the widely accepted idea of cold accretion (Haack *et al.* 1992), can also accommodate hot accretion. Hot accretion would have resulted from gravitational re-assembly of splash ejecta almost immediately after collision, before the chondrules had had time to harden.

Constraints relating to meteorite parent bodies

Thermal metamorphism

The metamorphism of chondrites is seldom discussed in the context of chondrule production. Perhaps this is because attempts to identify a mechanism for metamorphic heating have run into problems enough of their own. Accretion onto an already hot parent body is perhaps the simplest means of causing the metamorphic changes observed in chondrites (Heyse 1978), and such a process is an inevitable consequence of the molten planetesimal scenario.

In this context, the popular 'onion shell' model for the structure of the chondrite parent bodies requires modification, for the new scenario predicts the existence of a molten, and presumably segregated, interior beneath a carapace of typical chondritic material. The concept of metamorphic zoning, with type 3 material on the outside, still remains. However, the parent body in cross section would now resemble a clove of garlic, perhaps, with a thin chondritic skin wrapping a once-molten mantle and core, rather than an onion composed entirely of shells of chondritic material.

The age of differentiated meteorites

The U-Pb age of 4563±15 Ma for zircon crystals in a eucrite enclave in the Vaca Muerta mesosiderite (Ireland and Wlotzka 1992) confirms the likely existence of differentiated parent bodies, and hence of molten planetesimals, at a very early stage in the history of the solar system.

CONCLUSIONS

The scenario proposed here is the logical consequence of a single, plausible assumption, namely that the growth of sizeable planetesimals by low velocity collision and aggregation occurred so soon after the start of the solar system that ^{26}Al was still potent enough

to cause internal melting. Once molten, continued collisional accretion of the planetesimals presumably involved splashing followed by cooling and gravitational re-accretion of the resulting spray. If this spray did *not* form chondrules, then one might reasonably ask what it did form, and why evidence for it is not preserved. It has been shown here that very many of the established constraints on chondrule formation are reconcilable, at least in a broad qualitative way, with what the model seems to predict. At the very least, therefore, it would seem that the model is worth pursuing.

However, it would be dishonest to pretend that the model is without its problems. Amongst the questions which will need to be addressed in the future are the following. Were the planetesimals completely molten internally (to eliminate chemical fractionation of chondrules), or were they partially molten (suggested as one way to retain metal in chondrules)? Or indeed were they layered like garlic cloves with a partially molten transition zone separating a completely molten core from an unmelted crust? In the last case, where are all the meteorites that represent the transition zone? How frequent, and of what velocity, were the collisions, and of what size were the planetesimals as a function of time? Did the heat lost in chondrule production leave the residual planetesimal too cool to contribute to further chondrule-producing collisions? How was heat transferred from the interior to the planetesimal surface? How does the origin of CAIs fit into the model? What pattern of cooling might be expected in the postulated droplet cloud? Was the gas density in the vicinity of accretion high enough to cause aerodynamic size sorting? Answers to these questions will either make the model more robust, or relegate it to the dormant state it has hitherto endured.

Despite these outstanding problems, Zook's (1980) hypothesis appears to offer plausible answers to so many of the puzzling questions of chondrule formation that it can no longer be ignored. As a corollary, the widely held belief that chondrules were formed in a nebular setting, in the absence of planetary bodies, is beginning to look a little less secure, perhaps, than is often assumed.

ACKNOWLEDGEMENTS

I greatly appreciate the advice and encouragement given by Ed Scott, Robert Hutchison, and Jerry Delaney. I also thank Hugh Hill whose insatiable enthusiasm for meteorites proved to be infectious. Constructive comments by Rhian Jones, Hap McSween and an anonymous reviewer improved the paper significantly.

REFERENCES

Alexander C.M.O'D. (1994) Trace element distributions within ordinary chondrite chondrules: Implications for chondrule formation conditions and precursors. *Geochimica et Cosmochimica Acta* **58**, 3451–3467.

Binns R.A. (1967) An exceptionally large chondrule in the Parnallee meteorite. *Mineralogical Magazine* **37**, 319–324.

Dodd R.T. (1976) Accretion of the ordinary chondrites. *Earth and Planetary Science Letters* **30**, 281–291.

Dodd R.T. (1978) The composition and origin of large microporphyritic chondrules in the Manych (L–3) chondrite. *Earth and Planetary Science Letters* **39**, 52–66.

Grimm R.E. and McSween H.Y.Jr. (1993) Heliocentric zoning of the asteroid belt by aluminium–26 heating. *Science* **259**, 653–655.

Grossman J.N. (1988) Formation of chondrules. In *Meteorites and the Early Solar System* (eds. J.F. Kerridge and M.S. Matthews) pp 680–696. Univ. Arizona Press.

Grossman J.N., Rubin A.E., Nagahara H. and King E.A. (1988) Properties of chondrules. In *Meteorites and the Early Solar System* (eds. J.F. Kerridge and M.S. Matthews) pp 619–659. Univ. Arizona Press.

Haack H., Taylor G.J., Scott E.R.D. and Keil K. (1992) Thermal history of chondrites: hot accretion vs. metamorphic reheating. *Geophysical Research Letters* **19**, 2235–2238.

Heyse J.V. (1978) The metamorphic history of LL-group chondrites. *Earth and Planetary Science Letters* **40**, 365–381.

Holman J.P. (1992) *Heat transfer*. McGraw Hill International. 713 pp.

Holmén B.A. and Wood J.A. (1986) Chondrules that indent one another: evidence for hot accretion? *Meteoritics* **21**, 399.

Hutchison R. and Bevan A.W.R. (1983) Conditions and time of chondrule accretion. In *Chondrules and Their Origins* (ed. E.A. King) pp 162–179. Lunar and Planetary Institute.

Ireland T.R. and Wlotzka F. (1992) The oldest zircons in the solar system. *Earth and Planetary Science Letters* **109**, 1–10.

Kennedy A.K., Hutchison R., Hutcheon I.D. and Agrell S.O. (1992) A unique high Mn/Fe microgabbro in the Parnallee (LL3) ordinary chondrite: nebular mixture or planetary differentiate from a previously unrecognized planetary body? *Earth and Planetary Science Letters* **113**, 191–205.

Lederer C.M., Hollander J.M. and Perlman I. (1967) *Table of Isotopes*. New York. John Wiley and Sons.

Levy E.H. (1988) Energetics of chondrule formation. In *Meteorites and the Early Solar System* (eds. J.F. Kerridge and M.S. Matthews) pp 697–711. Univ. Arizona Press.

Lewis R.D., Lofgren G.E., Franzen H.F. and Windom K.E. (1993) The effect of Na vapor on the Na content of chondrules. *Meteoritics* **28**, 622–628.

MacPherson G.J., Davis A.M. and Zinner E.K. (1992) Distribution of ^{26}Al in the early solar system - a reappraisal. *Meteoritics* **27**, 253–254.

Matsunami S., Ninagawa K., Yamamoto I., Kohata M., Wada T., Yamashita Y., Lu J., Sears D.W.G. and Nishimura H (1993) Thermoluminescence and compositional zoning in the mesostasis of a Semarkona Group A1 chondrule and new insights into the chondrule-forming process. *Geochimica et Cosmochimica Acta* **57**, 2101–2110.

McCoy T.J., Scott E.R.D., Jones R.H., Keil K. and Taylor G.J. (1991) Composition of chondrule silicates in LL3–5 chondrites and implications for their nebular history and parent body metamorphism. *Geochimica et Cosmochimica Acta* **55**, 601–619.

Nagahara H., Kushiro I. and Mysen B.O. (1994) Evaporation of olivine: Low pressure phase relations of the olivine system and its implication

for the origin of chondritic components in the solar nebula. *Geochimica et Cosmochimica Acta* **58**, 1951–1963.

Newsom H.E. (1994) Siderophile elements and metal-silicate fractionation in the solar nebula. In *Chondrules and the Protoplanetary Disk*. LPI Contribution No. 844, 27–28. Lunar and Planetary Institute, Houston.

Podosek F.A., Zinner E.K., MacPherson G.J., Lundberg L.L., Brannon J.C. and Fahey A.J. (1991) Correlated study of initial ^{87}Sr/^{86}Sr and Al-Mg isotopic systematics and petrological properties in a suite of refractory inclusions from the Allende meteorite *Geochimica et Cosmochimica Acta* **55**, 1083–1110.

Radomsky P.M. and Hewins R.H. (1990) Formation conditions of pyroxene-olivine and magnesian olivine chondrules. *Geochim. cosmochim. Acta* **54**, 3475–3490.

Sneyd D.S., McSween H.Y., Sugiura N., Strangway D.W. and Nord G.L.Jr. (1988) Origin of petrofabrics and magnetic anisotropy in ordinary chondrites. *Meteoritics* **23**, 139–149.

Taylor G.J., Scott E.R.D and Keil K. (1983) Cosmic setting for chondrule formation. In *Chondrules and Their Origins* (ed. E.A. King) pp 262–278. Lunar and Planetary Institute, Houston.

Urey H.C. (1967) Parent bodies of the meteorites. *Icarus* **7**, 350–359.

Wasson J.T. (1993) Constraints on chondrule origins. *Meteoritics* **28**, 13–28.

Weidenschilling S.J., Donn B. and Meakin P. (1989) Physics of planetesimal formation. In *The Formation and Evolution of Planetary Systems* (eds. H. Weaver and L. Danly) pp 131–150. Cambridge University Press.

Wetherill G.W. (1978) Accumulation of the terrestrial planets. In *Protostars and Planets* (ed. T. Gehrels) pp. 565–598. Tucson. Univ. Arizona Press.

Wetherill G.W. (1989) Origin of the asteroid belt. In *Asteroids II* (eds. R.P. Binzel, T. Gehrels and M.S. Matthews) pp 661–680. Tucson. Univ. Arizona Press.

Wood J.A. (1979) Review of the metallographic cooling rates of meteorites and a new model for the planetesimals in which they formed. In *Asteroids* (ed. T. Gehrels) pp 849–891. Univ. Arizona Press.

Wood J.A. (1984) On the formation of meteoritic chondrules by aerodynamic drag heating in the solar nebula. *Earth and Planetary Science Letters* **70**, 11–26.

Yomogida K. and Matsui T. (1984) Multiple parent bodies of the ordinary chondrites. *Earth and Planetary Science Letters* **68**, 34–42.

Zinner E. and Göpel C. (1992) Evidence for ^{26}Al in the H4 chondrite Ste. Marguerite. *Meteoritics* **27**, 311–312.

Zook H.A. (1980) A new impact model for the generation of ordinary chondrites. *Meteoritics* **15**, 390–391.

Zook H.A. (1981) On a new model for the generation of chondrules. In *Lunar and Planetary Science XII*, pp. 1242–1244. Lunar and Planetary Institute, Houston.

Glossary

Definitions are derived or adapted largely from papers in this volume, *Origin of the Moon* (eds. W. K. Hartmann, R. J. Phillips and G. J. Taylor), Lunar and Planetary Institute, 1986, and glossaries compiled by Melanie Magisos in *Meteorites and the Early Solar System* (eds. J. F. Kerridge and M. S. Matthews), University of Arizona Press, 1988 and *Protostars and Planets II* (eds. D. C. Black and M. S. Matthews), University of Arizona Press, 1985.

Ab	abbreviation for albite, the Na-rich plagioclase feldspar.
ablation	erosion of material from the surface of a body; e.g., thermal erosion of a meteoroid during atmospheric entry.
accretion	the accumulation of matter to form larger objects such as planetesimals, moons, planets and stars.
accretion disk	a disk-shaped cloud of gas and solids that is accreting on to a central protostar or other object.
achondrite	a stony meteorite with a non-chondritic composition that formed from a melted or partly melted asteroid.
agglutinate	aggregate of soil particles bonded by glass that formed by micrometeorite bombardment in a regolith.
albite	$NaAlSi_3O_8$, the Na-rich plagioclase feldspar.
alkali element	the univalent elements, Li, Na, K, Rb and Cs.
alkali feldspar	feldspar rich in Na and/or K.
An	abbreviation for anorthite, a plagioclase feldspar end member.
anorthite	$CaAl_2Si_2O_8$, the Ca-rich plagioclase feldspar.
asteroid	a rocky object that is sub-kilometer to kilometers in size in heliocentric orbit; most lie between the orbits of Mars and Jupiter.
AU	astronomical unit. The mean distance of the Earth from the Sun; 1.50×10^{11} m.
basalt	a dark-colored, fine-grained igneous rock composed mainly of plagioclase and pyroxene, which crystallized on or near a planetary surface.
basaltic achondrite	stony meteorite that resembles terrestrial basalts.
bow shock	a shock wave in front of a body moving supersonically through a fluid or gas.
breccia	a consolidated fragmental rock. Lunar and meteorite breccias are formed by meteoroid impact and consist of angular broken fragments of rocks and minerals set in a fine-grained matrix, which may be smaller fragments or impact melt.
Ca-, Al-rich inclusions	typically mm-cm sized particles in chondrites that are enriched in the refractory elements such as Ca, Al and Ti.
CAI	Ca-, Al-rich inclusions.
carbonaceous chondrite	originally a C-rich chondrite; now used for chondrites that are rich in refractory lithophile elements.
CC	carbonaceous chondrite.
chalcophile element	an element that tends to be concentrated in the sulfide phase, e.g., S, Se, Cd, Zn.
chondrite	a meteorite that contains or once contained chondrules, or is chemically similar to such meteorites.
chondrules	particles composed mostly of olivine and pyroxene that are found in primitive meteorites and were once wholly or partly molten. Chondrules are typically mm-sized and may be droplet-shaped.

circumstellar disk   flattened disk of gas and solids in orbit around a star or protostar.

clast   a fragment of rock or mineral embedded in another rock.

clinopyroxene   a pyroxene mineral with monoclinic symmetry (one two-fold axis) and a composition typically close to those in the system $CaMgSi_2O_6$- $CaFeSi_2O_6$- $Fe_2Si_2O_6$-$Mg_2Si_2O_6$. The term is also used specifically for a proxene close in composition to $CaMgSi_2O_6$.

comet   a body with a highly elliptical or parabolic orbit composed of ices and dust that formed in the outer solar system and has a coma and tails when near the Sun.

compound chondrule   an object composed of two or more chondrules that were fused together.

cosmic-ray exposure age   period of time for which a meteoroid was exposed to cosmic rays; the total time spent within a few m of the surface of a body.

cryptocrystalline   a rock texture in which individual crystals are less than a few micrometers in size.

differentiation   the separation of minerals or phases in a planet into layers that differ chemically, mineralogically or physically.

density wave   a wave, usually with a spiral form, that perturbs the density of gas and solids in a disk.

drag heating   frictional heating of an object moving through a gas.

dynamic crystallization   term used by experimental petrologists for crystallization experiments in which furnace temperatures were decreased continuously during crystallization; c.f. isothermal crystallization.

ejecta   material excavated from an impact crater by a meteoroid impact.

electron microprobe   instrument in which a beam of electrons is accelerated and focussed on a sample producing X-rays. The chemical composition of the sample is determined by analyzing these X-rays with spectrometers.

enstatite   the Mg-rich pyroxene, $MgSiO_3$.

enstatite chondrite   collective name for the EH and EL chondrite groups that are largely composed of enstatite.

eucrite   a class of achondritic meteorites resembling terrestrial basalts.

euhedral   used to describe a crystal that is completely bounded by planar faces.

extinct nuclides   radioactive nuclides such as ^{26}Al and ^{129}I that have short half lives compared with the age of the solar system, were present when a meteorite or meteorite component formed and have since decayed below detection limits.

Fa   abbreviation for fayalite.

fall   a meteorite seen to fall.

fayalite Fe_2SiO_4, the Fe-rich olivine.

feldspars   a group of chemically and structurally related minerals most of which belong to the albite-orthoclase-anorthite system, $NaAlSi_3O_8$-$KAlSi_3O_8$-$CaAl_2Si_2O_8$. Feldspars are the most abundant mineral in the Earth's crust.

ferrosilite $FeSiO_3$, the Fe-rich pyroxene.

find   a meteorite that was not seen to fall.

Fo   abbreviation for forsterite.

forsterite Mg_2SiO_4, the Mg-rich olivine.

fractionation   the concentration or separation of one phase, element or isotope from an initially homogeneous system.

Fs   abbreviation for ferrosilite, the Fe-rich pyroxene.

fugacity   a measure of the chemical potential of a gaseous species; the equivalent for a non-ideal gas of the partial pressure of an ideal gas.

FUN   acronym for fractionation and unknown nuclear effects; CAIs that show large mass fractionated and non-mass fractionated isotopic effects are said to be FUN inclusions.

FU Orionis   a young, pre-main-sequence star about 500 pc away that brightened by 6 magnitudes in 1936 and has slowly faded since 1960.

FU Orionis outburst   an episodic increase in brightness shown by about a dozen young stellar objects. All low-mass stars like the Sun may have experienced these outbursts before they became main-sequence stars.

gas drag   the frictional force due to collisions with gas molecules.

half-life   the length of time for half of the atoms of a radioactive nuclide in a sample to decay.

HREE   heavy rare earth elements; Gd to Lu.

INAA   instrumental neutron activation analysis. An analytical technique in which samples are irradiated by neutrons creating unstable isotopes that decay with gamma rays having characteristic energies. Samples are not processed chemically to concentrate elements of interest (see RNAA); gamma rays from different nuclides are resolved solely by instrumental means.

ion microprobe an analytical instrument in which a finely focused beam of ions ionizes atoms in the sample and ejects them for analysis with a mass spectrometer.

isochemical without change in bulk chemical composition.

isochron a line on an isotopic ratio plot passing through data for samples that appear to have formed at the same time.

isotopic anomaly unexpected deviation in isotopic abundance from solar-system value.

kamacite Fe-Ni mineral with less than 7% Ni that is found in nearly all metal-bearing meteorites.

liquidus the line or surface on a phase diagram above which the systems are completely liquid.

lithophile element an element that tends to be concentrated in the silicate phase, e.g., Si, Mg, Ca, Al, Na.

LREE light rare earth elements; La to Sm.

M$_\odot$ Solar mass; 1.99×10^{30} kg.

M$_\oplus$ Earth mass; 5.98×10^{24} kg.

mafic used for silicate minerals with cations that are predominantly Mg and/or Fe, e.g. forsterite, enstatite.

main-sequence star star with a core in which H is being converted to He.

magmatic that associated with molten rock.

mass fractionation fractionation of isotopes or elements that is dependent on their masses.

matrix the fine-grained predominantly silicate material that is found between chondrules and inclusions in chondrites.

mesostasis the interstitial, fine-grained material between larger minerals in a chondrule or another igneous rock; generally the last material to crystallize.

meteorite a natural object of extraterrestrial origin that survives to reach the surface of the Earth or other planetary body.

meteoroid a natural object in heliocentric orbit prior to impact with the Earth or other planetary body.

microchondrule a tiny chondrule, generally taken to be 40 μm or less in diameter.

monomict breccia a breccia formed from a single rock type that has been shattered or crushed.

noble gases the gaseous elements He, Ar, Kr, Ne, Xe and Rn, which rarely undergo chemical reactions; also known as inert and rare gases.

norite a type of igneous rock containing plagioclase in which orthopyroxene predominates over clinopyroxene.

norm theoretical mineral proportions calculated from the bulk chemical composition of a rock.

OC ordinary chondrite.

olivine the mineral $(Fe, Mg)_2SiO_4$, which consists of the forsterite-fayalite series. (The term may also be used for isostructural minerals such as $CaMgSiO_4$, which are rare in meteorites.)

ordinary chondrites three closely related groups of chondrites (H, L and LL) that together constitute 80% of all observed meteorite falls.

orthoclase $KAlSi_3O_8$, the K-rich feldspar.

orthopyroxene pyroxene mineral in the $MgSiO_3$-$FeSiO_3$ series that has orthorhombic symmetry (three mutually perpendicular two-fold axes).

parent body body from which meteorites were derived by impact.

parsec 1 parsec is the distance at which 1 AU subtends an angle of 1 arcsecond; equivalent to 3.26 light years or 3.08×10^{16} m.

partial melting the process whereby rocks melt to form liquids different in composition from the original unmelted rock and residual unmelted crystals.

pc abbreviation for parsec.

petrologic type a measure of the degree of aqueous alteration (types 1–2) or metamorphism (types 3–6) experienced by a chondrite. Type 3 chondrites are further subdivided into types 3.0–3.9. (=petrographic type).

phenocryst a crystal in an igneous rock that is large relative to the groundmass crystals and crystallized from the parent melt of the rock.

phyllosilicate a silicate mineral with layers of linked SiO_4 tetrahedra that invariably contains water.

plagioclase feldspar in the albite-anorthite series, $NaAlSi_3O_8$-$CaAl_2Si_2O_8$.

planetesimal a small body (perhaps meters to kilometers in size) that formed in the solar nebula.

poikilitic
a rock texture in which many small grains of one mineral are enclosed within a larger crystal of another mineral.

polymict breccia
a breccia composed of a variety of materials, e.g., fragments of different types of bedrock, preexisting breccias and impact melts.

porphyritic
a texture common to many igneous rocks in which relatively large crystals called phenocrysts are set in a fine-grained or glassy groundmass or mesostasis.

ppb
parts per billion (usually by weight); also written ng/g.

ppm
parts per million (usually by weight); also written μg/g.

protoplanetary disk
flattened disk of solids and gas orbiting a star from which planets can form.

pyroxenes
a group of minerals with chains of linked SiO_4 tetrahedra, many of which belong to the $MgSiO_3$-$FeSiO_3$-$CaSiO_3$ system.

radiogenic
a product of radioactive decay.

radionuclide
a nuclide that decays radioactively.

REE
rare earth elements; those with atomic numbers from 57 to 71; La, Ce etc to Lu.

refractory element
an element that would be vaporized at high temperatures in the solar nebula, e.g., Al, Ca and the REE.

refractory inclusion
synonym for CAI.

regolith
a surface layer of unconsolidated, fragmental material that covers consolidated bedrock. This term was originally used for terrestrial rock debris of all kinds including volcanic ash. Regoliths on the Moon and asteroids result instead from the continuous impact of meteoroids and the bombardment of charged particles.

relict grain
a crystal in a chondrule that survived the melting event that formed the chondrule; i.e. it did not crystallize *in situ*.

RNAA
radiochemical neutron activation analysis. Like INAA except that samples are chemically treated before or after irradiation to concentrate elements of interest.

SEM
scanning electron microscope. A microscope in which a focused beam of electrons is scanned across an object so that magnified images can be obtained from backscattered or secondary electrons.

shock wave
an abrupt perturbation in the temperature, pressure and density of a solid, liquid or gas that propagates faster than sound.

siderophile element
an element that tends to be concentrated in the metallic phase, e.g., Ni, Co, Fe, Au, Ir.

sintering
formation of a coherent aggregate of grains by heating without melting.

solar nebula
the disk of dust and gas that surrounded the protosun from which the planets formed. Protoplanetary disk is used as a synonym for solar nebula.

solar wind
energetic, charged particles (mostly ionized hydrogen) that stream radially outwards from the solar corona.

solidus
the line or surface in a phase diagram below which the systems are entirely solid.

taenite
Fe-Ni mineral with >10% Ni.

TEM
transmission electron microscope. A beam of electrons is focused on a thin rock sample so that high resolution images of minerals can be obtained from transmitted diffracted beams.

troilite
the mineral FeS, which is common in nearly all chondrites.

T-Tauri star
a pre-main sequence star of approximately solar mass, commonly characterized by a broad spectral energy distribution (excess ultraviolet and infrared emission) and evidence for extensive mass loss. The prototype is T Tauri.

UOC
unequilibrated ordinary chondrite; petrologic type 3 ordinary chondrite.

volatile element
an element that would evaporate at relatively low temperatures in the solar nebula (or another system); e.g., S, Na.

wollastonite
the mineral $CaSiO_3$.

xenolith
fragment in a rock or meteorite that formed apart from the host material.

Contributors

C. M. O'D. Alexander
Department of Terrestrial Magnetism
Carnegie Institution of Washington
Washington, District of Columbia, U.S.A.

P. H. Benoit
Department of Chemistry and Biochemistry
University of Arkansas
Fayetteville, Arkansas, U.S.A.

A. Bischoff
Institut für Planetologie
Universität Münster
Münster, Germany.

A. P. Boss
Department of Terrestrial Magnetism
Carnegie Institution of Washington
Washington, District of Columbia, U.S.A.

A. J. Brearley
Institute of Meteoritics
University of New Mexico
Albuquerque, New Mexico, U.S.A.

M. J. I. Brown
Astrophysics Group,
University of Melbourne
Victoria, Australia.

P. Cassen
Space Science Division
NASA - Ames Research Center
Moffett Field, California, U.S.A.

H. C. Connolly, Jr
Department of Geological Sciences
Rutgers University
Piscataway, New Jersey, U.S.A.

J. N. Cuzzi
Space Science Division
NASA - Ames Research Center
Moffett Field, California, U.S.A.

A. M. Davis
Enrico Fermi Institute
University of Chicago
Chicago, Illinois, U.S.A.

A. R. Dobrovolskis
Lick Observatory
University of California
Santa Cruz, California, U.S.A.

J. P. Greenwood
Department of Geological Sciences
Brown University
Providence, Rhode Island, U.S.A.

J. N. Grossman
United States Geological Survey
Reston, Virginia, U.S.A.

L. Hartmann
Harvard-Smithsonian Center for Astrophysics
60 Garden Street, MS-15,
Cambridge, Massachusetts, U.S.A.

P. C. Hess
Department of Geological Sciences
Brown University
Providence, Rhode Island, U.S.A.

R. H. Hewins
Department of Geological Sciences
Rutgers University
Piscataway, New Jersey, U.S.A.

R. C. Hogan
SYMTECH, Inc.
Ames Research Center,
Moffett Field, California, U.S.A.

C. M. Hohenberg
McDonnell Center for Space Sciences
Washington University
St. Louis, Missouri, U.S.A.

L. L. Hood
Lunar and Planetary Laboratory
University of Arizona
Tucson, Arizona, U.S.A.

M. Horányi
Laboratory for Atmospheric and Space Physics
University of Colorado
Boulder, Colorado, U.S.A.

S. Huang
Department of Chemistry and Biochemistry
University of Arkansas
Fayetteville, Arkansas, U.S.A.

R. Hutchison
Mineralogy Department
The Natural History Museum
London, U.K.

W. H. Ip
Max-Planck-Institut für Aeronomie
Katlenburg-Lindau, Germany.

R. H. Jones
Institute of Meteoritics
University of New Mexico
Albuquerque, New Mexico, U.S.A.

M. Kitamura
Department of Geology and Mineralogy
Kyoto University
Kyoto, Japan.

D. A. Kring
Lunar and Planetary Laboratory
University of Arizona
Tucson, Arizona, U.S.A.

A. N. Krot
Hawai'i Institute of Geophysics and Planetology
University of Hawai'i at Manoa
Honolulu, Hawai'i, U.S.A.

K. Liffman
CSIRO/DBCE
Highett, Victoria, Australia.

G. E. Lofgren
NASA Johnson Space Center
Houston, Texas, U.S.A.

S. G. Love
Division of Geological and Planetary Sciences
California Institute of Technology
Pasadena, California, U.S.A.

G. J. MacPherson
Department of Mineral Sciences
Smithsonian Institution
Washington, District of Columbia, U.S.A.

K. Metzler
Institut für Mineralogie
Museum für Naturkunde der Humboldt-Universität
Berlin, Germany.

K. Misawa
Department of Earth and Planetary Sciences
Kobe University
Kobe, Japan.

N. Nakamura
Department of Earth and Planetary Sciences
Kobe University
Kobe, Japan.

L. E. Nyquist
NASA Johnson Space Center
Houston, Texas, U.S.A.

M. Prinz
Department of Earth and Planetary Sciences
American Museum of Natural History
New York, U.S.A.

S. Robertson
Department of Astrophysical, Planetary and Atmospheric Sciences
University of Colorado
Boulder, Colorado, U.S.A.

A. E. Rubin
Institute of Geophysics and Planetary Physics
University of California
Los Angeles, California, U.S.A.

T. V. Ruzmaikina
Lunar and Planetary Laboratory
University of Arizona
Tucson, Arizona, U.S.A.

I. S. Sanders
Department of Geology
Trinity College
Dublin, Ireland.

E. R. D. Scott
Hawai'i Institute of Geophysics and Planetology
University of Hawai'i at Manoa
Honolulu, Hawai'i, U.S.A.

D. W. G. Sears
Department of Chemistry and Biochemistry
University of Arkansas
Fayetteville, Arkansas, U.S.A.

T. D. Swindle
Lunar and Planetary Laboratory
University of Arizona
Tucson, Arizona, U.S.A.

M. H. Thiemens
Department of Chemistry
University of California, San Diego
La Jolla, California, U.S.A.

A. Tsuchiyama
Department of Earth and Space Science
Osaka University
Toyonaka, Japan.

J. T. Wasson
Institute of Geophysics and Planetary Physics
University of California
Los Angeles, California, U.S.A.

M. K. Weisberg
Department of Earth and Planetary Sciences
American Museum of Natural History
New York, U.S.A.

J. A. Wood
Harvard-Smithsonian Center for Astrophysics
Cambridge, Massachusetts, U.S.A.

Y. Yu
Department of Geological Sciences
Rutgers University
Piscataway, New Jersey, U.S.A.

B. Zanda
Muséum National d'Histoire Naturelle
Paris, France.

Index

(Bold page numbers denote first pages of chapters in which this term is a major topic.)